CRC
HANDBOOK OF
LABORATORY
SAFETY 4TH EDITION

A. KEITH FURR

CRC Press
Boca Raton   London   Tokyo

Library of Congress Cataloging-in-Publication Data

Furr, A. Keith
  CRC handbook of laboratory safety—4th ed. / editor, A. Keith Furr
      p.   cm.
  Includes bibliographical references and index.
  ISBN 0-8493-2518-8
  1. Chemical laboratories—Safety measures. I. Furr, A. Keith, 1932–   . II. Title: Handbook of
laboratory safety
QD51.C73   1995
542′.1′0289—dc20   94-47603
                                                                                                                        CIP

DEDICATION

I wish to thank my son Robin Furr for his invaluable help with the sophisticated software used to prepare the text and figures for this edition of the book. I also wish to thank Mr. John Cure and Mr. Larry Doucet for their contributions to this edition, in terms of discussions and actual material. The patience exhibited by my wife in allowing me to create and maintain chaos in a portion of our home for the many months the manuscript was in preparation must be acknowledged. I also wish to acknowledge Ms. Sharon Ray who must be given the credit for the final appearance of the book. Finally, I must thank those of you who gave the previous edition its success and encouraged me and the publisher to move forward with this edition.

At the time of preparation of the third edition of the Handbook, a number of significant evolutionary changes affecting laboratory safety had just been made and a number of others were pending but it was impossible to predict how soon these would become law. Among the latter were the OSHA Laboratory Safety Standard and changes in NRC radiation safety regulations. An attempt was made to anticipate the future requirements for these as well as others, such as the impact of AIDS on laboratories using human blood and other human tissues and fluids. The public concern involving infectious waste showing up on beaches and landfills was treated in order to assist managers to treat these materials responsibly. The manner in which chemical wastes could be disposed of was also undergoing changes as well as the requirements for using hazardous materials. The material in the third edition was reasonably successful in predicting the course of regulations affecting laboratory operations so that the third edition has continued to be a useful reference.

Since the publication of the previous edition, the anticipated realization of numerous standards has become fact and the present edition contains material on all of these. There have been numerous other actions that have impacted on laboratory safety: changes in Department of Transportation regulations on the packaging of hazardous materials, EPA actions on the release of radioactive materials into the environment, and OSHA actions covering hazardous waste operations and emergency response. OSHA has finally taken actions to use revised industrial standards instead of the original ones on which the original OSHA Act was based. On the other hand, an attempt by OSHA to update the contaminant limits for air pollutants suffered a setback in court so these are not changed, although some states have independently adopted more recent standards. The bloodborne pathogen standard is now being firmly enforced, and the possibility now exists that similar explicit action may be necessary to stem the rise of resistant strains of tuberculosis. The Centers for Disease Control has issued new versions of their guidelines for research with microbiological hazards. Although the federal government has chosen not to take national actions covering infectious waste, virtually every state has done so. As a result a substantial growth has occurred in firms developing alternative means of processing infectious waste materials.

There does not appear to be as much pending for the immediate future. The 1986 guidelines for working with recombinant DNA materials appear likely to be revised shortly, and action may be taken on indoor air quality within the next few years. The first of these areas seems to be causing less concern than originally thought so the impact on laboratory safety may decrease instead of increase as is often the case. Indoor air quality rules are not likely to impact research facilities as much as other types of construction for those facilities built to the current standards on ventilation for laboratory facilities. Coverage has been extended to millions of other workers as the Department of Energy and the military have adopted their own versions of OSHA equivalent standards.

Every attempt has been made to make the current edition as up-to-date as possible by continually reviewing current regulatory standards. There is a great amount of new material in this edition as represented by the additional pages, but every article has been revised to make sure that it reflects the newest changes. Topics may be similar but the content may have changed significantly. It is the author's and the publisher's hope that the user will find this edition to be factually accurate, current, readable, and helpful.

A. Keith Furr, Ph.D., was, until his retirement in late 1994, Head of the Department of Environmental Health and Safety at Virginia Polytechnic Institute and State University, Blacksburg, Virginia, and Professor of Nuclear Science and Engineering. He received an A.B. degree, cum laude from Catawba College in 1954, an M.S. degree from Emory University in 1955, and a Ph.D. from Duke University in 1962. From 1960 until 1971, he was in the Department of Physics where he attained the rank of Professor. In 1971 he transferred to Engineering as Professor of Nuclear Science and Engineering. In 1975 he established the environmental health and safety program at the University. In addition to other assignments, he was Director of the Nuclear Reactor Facility and Head of the Neutron Activation Analysis Laboratory. During the early 1970s, he created an undergraduate program in Radiation Safety and recently he has supported a broad program in Industrial Safety in the University's Department of Industrial and Systems Engineering. In recent years he has played a leadership role in developing a unique program in correcting indoor air quality problems in the University. He belongs to the Health Physics Society, the Campus Safety Association, the National Safety Council, and the National Fire Protection Association. He has published over 60 articles in refereed journals, many in the area of environmental studies, and three encyclopedia articles and was editor and principal contributor to the previous edition of this handbook. Dr. Furr has been active in working with public bodies to develop programs that respond to environmental emergencies; that address the disposal of hazardous materials, infectious wastes, and other solid wastes; and that are environmentally and economically sound.

Daniel R. Conlon, Instruments for Research & Industry, Cheltenham, Pennsylvania

John W. Cure, III, Health Physics Consultation, Lynchburg, Virginia

Richard F. Desjardins, M.D., Wilmington, North Carolina

Lawrence G. Doucet, P.E., Doucet & Mainka, P.C., Peekskill, New York

Caldwell N. Dugan, Division of Institutional Resources, National Science Foundation, Washington, D.C.

Frank A. Graf, Jr., Research Scientist, Westinghouse Hanford Company, Richland, Washington

Scott A. Heider, Division of Institutional Resources, National Science Foundation, Washington, D.C.

G.H. Herrmann, M.D., Medical Director, Pontchartrain Plant, E.I. du Pont de Nemours & Co., La Place, Louisiana

Harold Horowitz, Division of Institutional Resources, National Science Foundation, Washington, D.C.

Alvin B. Kaufman, Litton Systems Divison, Litton Industries, Woodland Hills, California

Edwin N. Kaufman, Senior Scientist, Douglas Aircraft Co., Woodland Hills, California

David M. Moore, D.V.M., Director, Laboratory Animal Resources, College of Veterinary Medicine, Virginia Polytechnic Institute and State University, Blacksburg, Virginia

Gail P. Smith, Ph.D., Director, Technical Staff Services, Corning Glass Co., Corning, New York

Eric W. Spencer, Brown University, Providence, Rhode Island

William L. Sprout, M.D., Medical Consultant, Haskell Laboratory, E.I. du Pont de Nemours & Co., Newark, Delaware

Norman V. Steere, Laboratory Safety and Design Consultant, Minneapolis, Minnesota

M.A. Trevino, M.D., Medical Director, Quimca Fluor, S.A. de C.V., Matamoras, Tamaulipas, Mexico

Paul Woodruff, Environmental Resources Management, Inc., West Chester, Pennsylvania

Chapter 3
Laboratory Facilities—Design and Equipment 71

CHAPTER 1

Introduction

Safety and health programs for industrial organizations began to be significant early in the 20th century, so there have been significant improvements in the industrial working environment. In the early 1970s, the Federal Occupational Health and Safety Act was passed as a national program to establish minimum standards for safety for industry. Corporations in the United States have had to create internal safety organizations to comply with federal and state regulations. The more enlightened of these firms, recognizing the value to themselves and their employees of keeping employees safe and well, strongly support these internal units. Smaller firms sometimes have made a lesser commitment proportionally to safety, although there are many exceptions. The standards incorporated in the regulations were based on previously developed industrial standards created by various organizations. Some important groups were not covered for a time, including public employees and, effectively, laboratory research workers. For the latter group, the standards, based on industrial processes, simply did not translate well to the laboratory environment. Some industrial organizations proceeded to develop their own safety programs for laboratory personnel that were quite effective. Many progressive universities have created good safety programs, but because of unique features of the academic environment, programs affecting research operations typically were less comprehensive than in industry. However, even some larger educational institutions, and a very high percentage of moderate to smaller schools and research laboratories, have had minimal safety programs, often limited to a fire safety group in maintenance departments. In consequence, safety may not have been stressed as much as it should have been in academic research institutions. Many students graduate without having received any formal safety training, or in some cases without even being exposed to good safety practices. Since attitudes, once established, are very difficult to change, the attitude that safety is of secondary importance has been carried over into professional careers for a significant fraction of current laboratory research personnel.

Over a period of years, the regulatory environment has grown to encompass more of the research environment, beginning with the users of radioactive materials in the 1940s and 1950s. If an organization used radioactive materials, it was required to have a radiation safety officer or committee, because these were mandated by the Nuclear Regulatory Commission, or its antecedents, under their licensing regulations. Approximately half the states have assumed this responsibility as a surrogate for the NRC. In recent years, additional regulations have been passed explicitly covering laboratory operations affecting the disposal of chemicals, infectious waste, exposure to human blood, tissue and other body fluids, the use of animals and human subjects in research, and work involving recombinant DNA. Concern about risks to workers and the environment from chemicals led to regulations requiring hazard information to be given to individuals exposed to chemicals, and to local communities, where the potential exposure to hazardous chemical incidents existed. These last regulations covered industrial operations as well as research personnel. Finally, in 1991, OSHA passed a standard specifically designed for the laboratory environment which required operators of research laboratories to provide protection to the employees equivalent to that enjoyed by industrial workers covered by the OSHA General Industry Standard. This last standard is a performance standard, which is a virtual necessity in view of the tremendous variation in laboratory operations. Although it provides flexibility, it is no less demanding than specific standards. Additional regulations can be expected. The 1992 Bloodborne Pathogen standard probably affects more laboratory workers than emergency care providers, the group that most persons immediately think of in the context of accidental exposure to blood, tissue, and other bodily fluids. Infectious waste rules already in force in many states affect laboratories performing basic research in the life sciences as well as medical facilities. The Americans With Disabilities Act has many implications for the laboratory environment.

Part of the difficulty in developing an effective laboratory safety program lies in the nature of laboratories. Research activities conducted within laboratories normally are extremely varied and change frequently. The processes and materials in use may present unidentified problems. Research materials may be being synthesized for the first time or may be being used in novel ways. Flammable solvents are probably the most common class of chemicals in use, and there are ample sources of ignition in most laboratories. Because of the changing needs in the laboratory and the scale of most reactions over a short period, even a well-managed facility tends to accumulate a large and varied inventory of chemicals. In one moderate, not atypical, research laboratory at the author's institution, a recent inventory found 1041 different chemicals on the shelves, only a handful of which were currently in use or anticipated to be in use in the near future. Many will never be used since research personnel no longer trust their quality. Laboratory equipment is often fabricated or modified within the laboratory in an instrument shop maintained by the facility. Devices manufactured or modified locally obviously are extremely unlikely to have been tested by any certified safety testing organization. Laboratory facilities that may have been well designed for the initial use may easily become wholly inadequate in terms of electrical services, ventilation, or special equipment such as hoods, as programs change or as new occupants move into the space. Subsequent changes are more often made solely to accommodate more activities rather than to improve safety. However, it is also quite possible that a laboratory was not designed properly initially. Relatively few architectural firms appear to really know how to design laboratories. Engineering factors are often the ones sacrificed to achieve other goals. Buildings and spaces originally built to house classrooms in a university are often converted to serve expanding research programs. These converted spaces rarely serve well as laboratory structures without substantial and difficult to obtain sums being needed to renovate them properly. Even with the best of intentions, physical solutions to safety problems that may work well in large-scale industrial processes are often extremely difficult to scale to the laboratory environment and alternatives may be expensive.

Individuals managing laboratories are usually very capable persons who tend to be strongly goal oriented, often to the unintended exclusion of other factors. Scientists designing an experiment may inadvertently neglect some peripheral factors affecting safety because they are not directly applicable to attaining the research goal. They also may forget that the technicians or students may not have the years of experience and training that research directors do. They often do not consider all of the consequences if some item of equipment were to fail or make errors. It is true, too, that familiarity all too often can lead to a casual approach to routine procedures. Also, and unfortunately, as with any other large group, the research community is not immune to the presence of a few persons who cut corners to serve their own interests. The growing competitiveness in finding funds for research, to publish, and to obtain tenure, for example, in academic environments must exacerbate this problem. The pressures can be intense. It is probably unwise to depend wholly on the professional expertise of the research scientist or his voluntary use of safety programs. There are too many other factors affecting priorities. Yet, in the new laboratory standard, much of the responsibility for laboratory safety is placed squarely on the shoulders of the laboratory research director. It is essential that means are provided to inform laboratory managers and laboratory staff of appropriate means to create an effective program and to monitor compliance with the program.

The organizational structure of many research institutions, especially at academic institutions, also contributes to the variability and general weakness in the safety programs, due to fragmentation of responsibility. Strong line organizations found in businesses and factories are often very weak in universities. Individual research programs in an academic laboratory are typically defined solely by the laboratory director, virtually independently of his department head. The typical academic department head may or may not establish the broad areas within the department that will be emphasized, but will intervene to a minimal degree in the conduct of research in individual laboratories. Intervention is construed as an infringement of their academic freedom by some scientists. Similar gaps exist between the department head and his dean and so on through successive administrative layers. Thus, carrying out a safety program is a difficult logistical problem in colleges and universities. The situation in commercial laboratories is better, since programs are less often defined solely by the laboratory director. However, even in industry, the laboratory director usually has more autonomy and discretion than a line supervisor in a factory, so that communication and carrying out of safety policies are somewhat more difficult in commercial laboratories, also.

None of the difficulties associated with laboratory safety programs cited in the previous paragraphs have disappeared in the past few decades. The complexity and cost of effective laboratory safety programs have increased too, adding to the problem. Competitive pressures have significantly increased both in academia and in commercial facilities. New mandatory regulations and standards have been enacted, so "doing what has to be done" has become a much larger task. The acceptance of the status quo has become increasingly unfashionable and unpalatable. Expectations of laboratory workers for their personal well-being have continued to increase. They have become very concerned about the effects of their working environment on their health and safety, and they expect something to be done to eliminate perceived problems. The public is less likely now to accept the premise that the scientists "know what they are doing" and may be concerned about releases of dangerous materials into the environment. Acquiring liability, followed by litigation initiated by an employee, due to a weak safety program is a possibility that management cannot ignore. Lack of resources for meeting safety issues is not an acceptable defense. A good case can be made that without these higher expectations and possible repercussions, many regulations governing health and safety in the workplace may not have been passed. Lack of resources for meeting safety issues is not an acceptable argument or defense.

Because of the new regulations and new attitudes, it has become necessary for management of research institutions to take an increasingly active role in safety programs, although the new lab standard places much of the direct responsibility on the research manager. Organizations can no longer legally choose to have an effective safety program or not. They must provide support to the laboratory directors to aid them in creating their own safety programs and to monitor that the laboratory's programs are being properly done. They have to provide training to employees and take positive steps to encourage employees at all levels to actively support and apply standard safety policies for everyone.

Resources have had to be identified to comply with more comprehensive and complex regulations. Safety departments often have had to be enlarged at a time when available funds were likely to be decreasing. However, it cannot be overlooked that individual laboratory directors have been *assigned* the primary responsibility in managing their operations safely. Therefore, they are more important than ever as key persons in carrying out the institutional safety program. This is not a role to which many of them are accustomed, and many of them are not comfortable in it. Many are concerned about the liability implicit in these new responsibilities, as they should be. Old habits are difficult to shed, and there will be a definite transition period before safety is as high a priority in the laboratory as it should be, even under the impetus of the new standards. It will be a major responsibility of management to assist the laboratory directors and, where necessary, ensure that they do comply.

A critical component of an organization's safety effort is an effective safety department. The primary objective of such a department is to provide guidance to the remainder of the organization, including upper management. Due to the complexity of modern research operations and the regulatory climate, it is the responsibility of the members of the safety and health organization to be knowledgeable of current regulations and their interpretation. They must also be aware of technological advances so that the advice and training which it is their responsibility to provide will be the best possible in terms of effectiveness and efficiency. A second major responsibility of the safety organization is to monitor the performance of the employees and management of the organization. The latter may not realize this function is required, but it is a major responsibility. Safety and health departments must be given enforcement authority by the parent organization to properly monitor compliance with safety issues. It is the enforcement function that must be managed most sensitively by safety and health personnel. It is critical that it be done well. If done heavy-handedly, the scientific personnel will be resentful, but if safety personnel are too timid, they will be ignored by both workers and management. Without active and visible support by management, it is very likely that even excellent safety and health programs will not succeed. Operational duties are also now assigned to most safety and health departments, such as disposal of hazardous waste. Finally, in the current litigious climate, the safety and health department must keep superb records to protect workers, themselves, and their organization.

Safety personnel must understand in defining their role that they are not the primary function of the organization of which they are a part. A widget manufacturer makes widgets as its first priority. A university has students to educate and scholarly research to perform. A good safety program, done well, will facilitate attainment of these primary goals, not hinder it. A safety department must act so that this is true and is perceived to be true by all of their client constituencies. Safety departments need to establish themselves with an image of a strong emphasis on service. Service as an operating premise does not mean that the safety department is to take the safety burden entirely upon itself. It cannot. The department must provide an appropriate structure for both employees and management to conduct operations safely.

Without the support of management, no safety department can succeed. Support entails many factors. Resource support is obviously very important in physically enabling the department to function. If the safety department is not provided with a reasonable level of support, protestations of support by management will seem insincere and hollow. Conversely, when management "puts its money where its mouth is," the conclusion will be drawn by the employees that management does see value in the programs affecting their safety and health. There are other ways that management can illustrate support. To whom does the head of the department report? Is it a mid-level manager, or is it a senior person such as a vice president? If confrontations do arise (and these may occur, especially in an academic research laboratory where full professors are accustomed to operating with few restraints), will the safety professional be supported when he is right? Does the safety department participate in making operational decisions or policy decisions? The status of the safety department will be enhanced if the answers to these questions are positive. If the answers are negative, the effectiveness of the organization's safety program will be seriously diminished.

There is a management tool called "strategic planning" which can be used to define the function of any organization and, if used properly, can make the entire organization function better. Management structures come and go, but strategic planning should be adaptable to any type of organization. A first step in doing strategic planning is to define the organization's clients and stakeholders. Most persons take a too limited view of whom these are. For a safety and health department, the most obvious "clients" are the employees and management. Other internal units within the organization also can be considered clients since it frequently requires collaborative efforts to achieve specific goals. Other important clients are regulatory agencies, the local community, and the public. The perceived needs of each client group should be explicitly identified.

Once the clients have been defined, the objectives and goals need to be defined. Goals are usually more generalized, such as to reduce the number of accidents within the organization, while an objective may be more specific, such as to reduce the number of back injuries by 25% within the next 12-month period. When doing this exercise, it may be found that there may be dozens of goals, and even more objectives. Not all of these will necessarily be equally important.

The strengths and weaknesses of the department should then be determined. Among the strengths would be the skills of departmental personnel, the funds and equipment available to the department, the relationships between the department and its clients, and the relationships among departmental personnel. The same things could be weaknesses as well if deficiencies exist. However, it is too easy to blame lack of success on lack of resources. Failure to make optimum use of available resources is a measure of inefficiency, which the strategic planning process can help minimize.

Once the preliminary work is done, the department can define its programs to maximize their effectiveness. The programs need to be ranked. Not all are equally productive or attainable in enhancing safety and health. Note that in each of the preceding few paragraphs, it is the department that is to do all these things, not the organization, nor the department head. Everybody in the department and, desirably, representatives of the clients should be involved. Everyone needs to "buy in" to the process and have a share in the result. The department head and the more experienced persons can and should provide guidance and leadership. However, there is obvious value in fresh ideas and approaches that may be provided by newer personnel. A younger person can be as productive in developing new ideas as persons "set in their ways," but they may need to be encouraged to contribute their ideas. No one individual should dominate the strategic planning process.

The cooperation of all the client groups is important, but that of the employees is essential. Employees must see the actual value of the safety program to themselves. Reasonable persons are not going to seriously advocate doing things unsafely, but many honestly feel that they are doing things as safely as possible and resent being told otherwise. Many conscientiously feel that a formal safety program, with the attendant rules, regulations, and all the accompanying "red tape" is not necessary and is counterproductive. An astute safety professional, with the active support of management, can overcome this attitude. This can be done by education and training, so that people know what is expected of them and why; by actively involving the employees in developing good approaches to achieve compliance with standards; by minimizing the administrative burden on individuals; by demonstrating that the safety program helps the organization's personnel solve their problems; and by making sure that the safety program clearly provides a safer environment. Training in management techniques to enhance a safety professional's "people skills" is a sound investment by management. Training for safety professionals is frequently thought of only in terms of technical skills. In scientific research, typically some employee clients will have advanced degrees, especially those ultimately in charge of a laboratory, while many safety professionals will not have comparable credentials. This can be a hurdle which with proper training of both parties can be overcome. A good, meaningful laboratory safety program is possible and, under today's regulations, mandatory. Under the new OSHA laboratory safety standard, a research organization must require each laboratory within it to have a satisfactory written "industrial hygiene plan" or, in other words, an effective safety and health program for that facility. According to the standard, the "plan" is the responsibility of the person ultimately in charge of the laboratory, with the organization required to monitor the performance of the laboratory compliance with the plan. Since so much responsibility is placed on the laboratory manager and the laboratory staff, the goal of this handbook is to help the individual research scientist define the requirements of a successful program. Recommendations usually will be based on current regulations or accepted practices. Sometimes the recommendations will go further where this appears desirable or where the opportunity exists for innovative approaches. No one should be afraid to try an innovative and more cost-effective program to achieve a desirable goal. In the latter case, the differences between the required procedures and the modified approach should be clear. The area of emphasis for much of the material will be the chemistry laboratory, but there will be separate sections for several other important areas. Where corresponding material was included in the third or earlier editions of this book, the corresponding material will have been thoroughly reviewed and brought up to date, or projected into the future, where changes can be anticipated clearly.

1.1. LABORATORY SAFETY AS A COOPERATIVE RESPONSIBILITY

The new OSHA Laboratory Safety Standard has changed the ground rules governing laboratory safety. As noted in the previous section, much of the responsibility for developing and carrying out a laboratory safety program is assigned to the individual ultimately responsible for the laboratory. This standard preempts most, but not quite all, of the general industry standards. It is a performance standard that permits substantial flexibility in how the health and safety of the laboratory employees are to be assured. Laboratories do not exist in a vacuum, although they are treated as individual units for the purposes of the standard. Other agencies and individuals within the organization must share much of the responsibility for achieving a successful program. Compliance with the requirements of the standard will be discussed in detail in later chapters, but the roles of other organizational units will be briefly discussed in this section.

The personnel, employee relations, or human services department is the first contact that most employees have with an organization. This department should have the responsibility of informing the new employee of the basic policies of the organization in an initial orientation program. It is important that they include the rights and responsibilities of the employee involving safety as well as working hours, benefits, etc. Employees should receive written statements of the general safety policies of the organization, where additional information can be obtained, and to whom the employee can turn to express safety concerns. The initial impression of the importance placed on safety is made at this time, and every effort should be made to ensure that it is positive. For laboratory employees, the role of the central organization in supporting the safety programs of the individual work-places and monitoring compliance with work place safety procedures needs to be clearly explained. The safety and health department is the logical one to provide the orientation material in this area, and may be asked to provide a presenter.

Detailed safety procedures should not be covered at the initial orientation program point where laboratory safety is concerned, since laboratory activities vary widely from laborato-ry to laboratory. Information concerning safety procedures for a given laboratory is the responsibility of the person in charge of the facility.

The personnel department also often is the department charged with handling worker's compensation claims. Thus, it has a direct concern in reduction of injuries to minimize the cost of such claims. In some organizations the equivalent department is Risk Management as far as insurance claims are concerned. The safety department may also be affiliated with the risk management department. Clearly, the personnel department plays a critical role in an effective laboratory safety program.

In many organizations, the group charged with assuring that no discrimination occurs and that everyone is assured of an equal opportunity is also in the personnel department. In others, this group is autonomous or reports through a different chain of command, to avoid any conflict of interest. There are obvious general safety implications for disabled persons, such as both audible and visual fire alarms or safe, unblocked evacuation routes. However, one of the biggest laboratory-related safety issues for the EEO group is that of exposure of younger women to chemicals in the laboratory. Fetal vulnerability is an example of a very sensitive topic where this applies. In some areas, information is suffic-iently developed to allow well-defined policies. Substantial information is available on absorption of radioactive material by the fetus or the effects of external radiation on fetal growth. As a result, the rights and responsibilities of fertile women have been delineated as an NRC Regulatory Guide for several years and have been incorporated in the new version of the Federal Title 10, Part 20. On the other hand, information about the effect on the fetus or the reproductive process of most of the thousands of chemicals in use is very limited. In such a situation, women may have legitimate concerns about their expo-sures and about possible unfair exclusion from some areas of employment. It is the responsibility of the EEO office to provide them with a means of expressing these con-cerns. EEO officers must work with the employees and other parts of the organization to provide adequate safeguards to protect the rights of women employees. The new Ameri-cans With Disabilities Act of 1990 has placed some real teeth into requirements for making appropriate provisions for disabled employees and others. The definition of disabilities is no longer necessarily limited to those with obvious physical dysfunctions. An individual who is hypersensitive to certain chemical fumes may also be considered disabled and thus eligible for accommodation under this statute, if these chemicals are part of the work environment. Further, Congress is considering possible legislation on indoor air quality that might be applicable in such a case. It will be necessary for research facilities to work closely with the personnel/EEO departments to ensure that they comply with this political-ly sensitive issue.

There are many occasions for the legal office of an organization to concern itself with laboratory safety. Many of these are straightforward, concerned with contract terms and conditions for sponsored research. Contracts for public academic institutions, for example, usually cannot contain a "hold harmless and indemnification clause" since this in effect waives the protection provided by the immunity to suit claimed for many state agencies. The legal counsel must find alternative language or an alternative procedure for satisfying this contract condition. There are many restrictions built into law or policies for various fund granting agencies, and it is the responsibility of the legal office to ensure that these terms are met. The question of personal liability often arises as well. Laboratory managers often ask about the risk of personal exposure to litigation as a result of the manner in which they enforce the organizations safety program. It is obviously impossible to monitor every action by laboratory staff to assure compliance with safety and still have time to do research. However, what are the boundaries of adequate supervision? Finally, most organization's have grievance and disciplinary procedures for actions that involve alleged violations of safety procedures. The legal counsel will necessarily become involved in these.

Besides these issues, academic institutions are increasingly vulnerable to citations by regulatory agencies, many including substantial fines. The organization's legal representative should carefully review citations to determine if there is a problem with the language or the application of specific regulations to the activity being cited. It is unlikely that they will be sufficiently knowledgeable about the technical aspects of the basis of the citation to adequately represent the organization or to prepare a response. The safety and health department, the scientist, and the legal counsel must all cooperate in preparing a response, to minimize the level of the citation and to avoid excessive costs or too restrictive abatement agreements.

There are opportunities to have informal meetings or discussions with most regulatory agencies before making a formal response. An attorney should be a participant in these meetings and may be the appropriate person to represent the organization in negotiations. The final reply should definitely be carefully reviewed by an attorney representing the organization. Inappropriate or imprecise language is the natural breeding ground for litigation. Lawsuits by outside groups, such as citizens concerned about the environment, represent similar legal challenges, again requiring a coordinated response. Many organizations now include a legal representative on their safety committees. Sensitive decisions by these committees should be reviewed by a legal advisor.

Liability insurance—the ability to obtain it or not and the cost of the insurance when obtainable—has become a major issue. The insurance manager has become a key person in processing contracts where safety is implicated. For example, contractors handling hazardous laboratory waste for an organization are typically required to carry substantial amounts of liability insurance, in addition to general liability insurance, to cover their errors in handling the waste. The contracts that waste contractors offer also often contain the phrase "indemnify and hold harmless." Most public facilities cannot sign a contract containing this clause since it, in effect, waives some of their constitutional rights. A $5,000,000 policy is a typical requirement, and many small firms cannot afford the insurance. This may limit the organization's ability to provide services at reasonable costs or may limit access to firms providing specialized critical expertise. With the support of the legal counsel, the purchasing department, and the risk manager, occasionally it is reasonable to go unprotected where the risk is judged to be sufficiently small by the technical safety advisor. The insurance manager also usually is eager to work with the safety group and personnel departments to stem the rising tide of worker's compensation claims and other safety-related claims.

In most organizations, the purchasing department processes orders for every item purchased. The cooperation of the purchasing department can facilitate the creation of safe laboratory conditions. With their assistance, it is possible to add safety specifications, and sometimes to limit eligibility of vendors of products to ensure that the items ordered will function safely. This may occasionally increase costs, but can significantly enhance safety. A good example of how additional quality specifications can be used to enhance client acceptance involves the choice of chemical splash goggles. Chemical splash goggles that will minimally meet ANSI standards can be purchased very cheaply. However, the least expensive units are typically uncomfortable, are hot, fog quickly, or have a combination of these undesirable qualities. For short-term, very limited wear, they are reasonably satisfactory, but after a short interval the wearer usually removes them or slides them up on the forehead. In either case, they would not offer the needed protection. With the assistance of the purchasing department, additional specifications for anti-fogging coatings, better air flow than that provided by the typical side-ports, hypoallergenic materials, etc., can be added to the specifications. Limiting the choice to a single brand can still allow bidding for the best price if there are a number of distributors. If the improvement in quality is sufficiently clear, then it is often possible to justify use of specifications even when there is only a single source. Major items such as refrigeration units can be required to be safe for flammable material storage. Fume hoods can be required to be reviewed by the safety department and the organization's engineering staff to ensure that blowers and other features are adequate. Chemical purchases, with new computer technology, can be tracked from the time of order to the point at which they are fully used or disposed of as surplus or waste, if chemical purchases can be ordered and delivered to a single point of receipt. This last possibility will allow compliance with many regulations involving chemicals, which would be otherwise virtually impossible.

A key issue in the new OSHA laboratory safety standard is for research facilities to be suitable for the purpose for which they are to be used. To many research personnel, this simply means sufficient space with appropriate equipment and services. They do not think of their area in the context of the whole facility or of the many safety-related codes that affect its design. The user's considerations are largely involved with their research program. Research personnel depend on architects and contractors or internal equivalents from planning, physical plant, and maintenance departments to design or renovate their spaces to comply with codes and regulations, if it is not too inconvenient or expensive. They have a point. They are the *raison d'etre* for the space being required. However, current regulations make specific demands on the design and construction of facilities. Unfortunately, other than the fire code requirements, of which most design firms are reasonably well aware, many firms do not take measures to assure that they are current on changes in other regulatory requirements involving safety. With their lack of background in the specific disciplines involved, they may not even know the right questions to ask which would be applicable to a given facility. It is critical that at least three parties are involved in facility design and construction: the users, the architects (both external and internal), contractors, and the safety department. The need to involve safety goes beyond regulations. There are many issues that are covered only by guidelines, and others that have been revealed by experience to be important where even adequate guidelines have not been developed. For example, until quite recently, the needs for ventilation were assumed to be met if sufficient fresh air were made available. The characteristics of the supplied air have been found to be much more critical than previously thought. Low levels of various contaminants, well below OSHA PELs or the ACGIH TLVs, have been found to cause problems for exposures over extended periods. Wherever it is feasible, within reason, facilities should meet more than minimum standards, especially if life cycle costs can be reduced and future, more stringent regulations can be anticipated.

The physical plant department has the responsibility to provide heat, light, utilities, and custodial services, make repairs, and do renovations and most of the other services that are needed to keep buildings functional. However, few laboratory workers really appreciate what they are asking of these persons. Many persons have had the experience of custodial workers being afraid to enter a laboratory with a radiation sign on the door and make allowances for their fears. However, the same persons expect a mechanic to work on a chemically contaminated exhaust motor to a fume hood without question. Maintenance workers are becoming increasingly concerned about their exposure to toxic materials while performing maintenance in laboratory facilities. The risk may not be great, but they do not really *know* that it is not. The new hazard communication standard that went into effect on May 25, 1986, requires the facility manager to inform maintenance personnel of potential risks. Maintenance personnel may become even more reluctant to perform laboratory maintenance, unless they can be honestly assured of the lack of risk to themselves or of being protected from exposure. In their own areas, maintenance personnel have many skills and considerable knowledge, but they are not scientists. The lack of familiarity with the potential risks in the research being conducted in the facility where they are to work may lead to perhaps unreasonable fears. It will require a combination of education, tact, and accommodation by laboratory personnel to assuage these fears so they may continue receiving needed laboratory maintenance services.

In order to comply with the spirit of the law for both the organization's own maintenance personnel and personnel of outside contractors, procedures are needed to assure that maintenance personnel are properly informed of exposure risks. One of the most difficult situations involves work on or near fume hood exhausts. In a research building that can have separate blowers for hoods from many different laboratories, at any given time it is hard to know what is currently being exhausted or what residues are present from times past. Individuals working on fume hood exhaust components have been known to have severe reactions to these residues. A method to control access to such areas must be established, so that steps can be taken to prevent unnecessary exposures to protect the worker and minimize liability to the organization.

This section would not be complete without mentioning the vital role of central management. They establish the atmosphere and provide the needed resources. A positive attitude on their part in support of an effective laboratory health and safety program is critical. However, the other units must understand that not all problems can be solved efficiently by resources alone, and they cannot measure this support solely by the level of physical support assigned to safety. In one sense, measuring quality of programs by measurement of resources made available is to measure the programs by how inefficient they are. Parkinson's Law, which involves the principle that work will expand to use all resources provided, is certainly applicable to safety programs as with any other type. However, management must provide *sufficient* resources to make it possible to develop and implement a *cost-effective* program. They must also provide the moral support for the program to be managed with full regard toward meeting its goal of striving for a safe and healthy environment for everyone.

1.2. ORGANIZATIONAL STRUCTURE

A major and valid justification for OSHA to adopt the laboratory safety standard to supersede the general industry standards and the hazard communication standard for the laboratory environment is that the laboratory environment is radically different from industry and typical offices. The organizational structure for research institutions is not nearly so structured, especially in the academic environment, and operations are different in size and character. The laboratory standard uses this difference as a criterion as to whether the laboratory standard is to be applied to a program.

A major factor that is important to consider in developing a safety and health program is the way directives are transmitted from one level to another. Laboratory supervisors in universities and in many commercial research institutions do not have the same degree of communication with, or responsibilities to, higher level management as do typical industrial supervisors. They are much more independent and tend to resent "unwarranted" interference. In a university, the term "academic freedom" is often called into use when a researcher feels imposed upon. Scientific personnel especially do not want outsiders to attempt to intervene in their program of research. Indeed, a scientist has some justification for being concerned about anyone becoming too familiar with their work, as priority of discovery is an extremely important aspect of their work. Even a nonspecialist in a field, such as a safety professional, may inadvertently provide some key information to a competitor. A major role for safety personnel, with these factors in mind, is to define the organizational constraints for safe practices for the laboratory supervisor, who has been assigned the responsibility to integrate safety concepts within a written health and safety program under the laboratory standard.

Since every laboratory operation differs in detail, safety personnel cannot be expected to prepare a written plan for each facility. For example, a research university, is likely to have a thousand or more different laboratories. Attempting to draft a separate plan for each laboratory is far too big a task for most safety departments. It is reasonable, however, to expect a safety professional to design a standardized template and *help* the laboratory manager develop his own plan and written operational procedures. Even this effort will require a major effort for both the manager and the laboratory safety staff of most health and safety departments. When a new regulation or policy needs to be introduced, generally it is not feasible to send it through a chain of command structure. As noted earlier, in a university deans or department heads have extremely limited control over the academic staff, outside of classroom teaching assignments. In industry, the chain of command is better defined, since the general mission is usually profit oriented. This factor typically defines research targets more explicitly, but in facilities engaged in basic research, there is often similarity to the academic environment.

Due to the relative independence of the research investigator, the laboratory environment leads to the need for much self-supervision, especially for the laboratory director or manager, who is typically judged primarily by the quantity and quality of his work. The laboratory staff interacts with the organization primarily through the person ultimately responsible for their employment, so it this person who legitimately is held responsible for carrying out of the safety plan for the unit. A laboratory represents a considerable investment for academic and other research institutions, so success is critical to the career of a research scientist. This difference may encourage the laboratory scientist to take abnormal risks to achieve desired results. This usually is not done maliciously but because scientists are typically strongly goal oriented. In addition, they work under a great deal of stress, so "extraneous" safety factors generally are not first among those considered.

The interface between the safety and health department and laboratories has evolved rapidly in recent years. Not too long ago the role of a safety department was primarily to advise and train employees, mainly those providing support functions. As regulations became more complex, new sophisticated services were needed, dealing with health and safety issues more germane to laboratory work. By default, safety and health departments assumed many of these functions. They were delegated the functions and powers of the federal and state agencies whose regulations were applicable to the organization. They were also charged with the responsibility of managing the resources to achieve compliance with the regulations. With the new laboratory standard, the pendulum has swung back to a degree, in that the responsibility for implementing the standard has been explicitly assigned to the individual laboratory. However, the authority to ensure that the laboratories fulfill their obligations has been assigned to the organization. The organization, in turn, has passed the responsibility on to the safety and health department.

There are any number of possible working arrangements for a safety program. The alternatives should have several common characteristics to be effective. Among these are

- Assignment of safety responsibility to a senior executive, such as a vice president, with the manager of safety and health programs reporting to this person.
- The formation of a safety and health department, staffed by professionals and with sufficient resources and personnel to perform their function. In an academic environment, there may be some advantages for the head (and possibly some senior staff) of the safety department to have academic credentials. Since most laboratory heads at a university have a doctorate and often judge themselves and others by their scholarly activity, an individual with similar credentials may find it easier to gain and hold their respect. In larger, more sophisticated commercial research organizations, the same argument may hold true. Under the new OSHA Laboratory Standard, an individual, presumably from the safety and health department, is required to be identified as the organizational chemical hygiene officer.

This individual clearly should have appropriate credentials in chemistry as well as chemically related safety practices. Most larger research organizations represent major potential environmental risks for the area in which they are located. A representative of the safety and health department should be represented on the Local Emergency Planning Committee, required under the provisions of the SARA (Superfund Amendments and Reauthorization Act) Title III EPA regulations.

- The creation of one or more safety committees at the institutional level. In an industrial situation, one safety committee may suffice, but in a research facility, especially one with a wide variety of programs, it could be desirable for these to be more specialized. Examples could include a radiation safety committee, an institutional biosafety committee, a general laboratory safety committee (possibly separated into chemical and life science areas), a human subjects review board, a fire and emergency committee, and an animal care committee. All of these are either mandated or recommended by various standards if the organization has research programs in corresponding areas. Some of these special interest committees will have more authority than others, due to the underlying strength of the applicable regulatory standard. For example, the use of radioactive materials is very strictly regulated by the Nuclear Regulatory Commission or, in agreement states, by the state surrogate agency. Failure by the institution to comply with the regulations and the terms of the institution's license can, and frequently does, result in substantial fines, as well as national publicity. A comprehensive committee, formed of the heads of the special interest committees, may be a useful vehicle to recommend or define broad safety policies to ensure consistency in the various areas of responsibilities. Each committee should have a definite, written charge, as well as definite rules of procedures. An appropriate person from the health and safety department should serve as an *ex officio* member on each of these committees. A major benefit of an effective committee is that it assumes, along with the safety professional, part of the responsibility for actions, procedures, and policies that could be perceived by the regulated group as being onerous. It partially insulates the safety professional from being held solely responsible for the constraints imposed by the regulations.
- Because of the independence of the various divisions, or schools, in larger organizations, the separate internal administrative units may find it desirable to establish their own safety committees. The unit committees would adapt the overall institutional safety policies which must be followed in the context of their own operations and areas of responsibility. It would probably be desirable to identify an individual as the division chemical hygiene officer to act as liaison with the individual represent-

ing the central authority. This position is not required, however, for a subunit of an organization under the OSHA laboratory standard.

- Each department that includes chemical laboratories as a normal function should identify a single individual, equivalent within the department to the organizational chemical hygiene officer. This person would act as liaison with the institutional safety department, the safety committees, and the laboratories within the department. In many organizations, serving in such a position is not especially beneficial to the career of the individual, so some organizations tend to shift such responsibilities onto less productive personnel. This is not appropriate. The assignment should be rotated among active and productive individuals. A laboratory hygiene officer might be appointed as the primary contact on safety issues, besides the laboratory manager, if a laboratory unit is large enough by itself. Often the person ultimately in charge of a laboratory facility spends very little time in the laboratory, leaving day-to-day operations under the direction of a lab supervisor or a senior technician. The senior scientists typically have many other responsibilities to distract them, not the least of which is obtaining funds to support the laboratory.

Selection of suitable members of the committees at the various levels is extremely critical. As just noted, less productive individuals are often given the more demanding and less glamorous committee assignments. Safety committees, if performing as they should, are very demanding of their members. There is also reluctance by many very active and productive persons to accept responsibilities not immediately germane to their own programs. The members should be drawn from active, productive, and respected laboratory personnel, for limited terms if necessary, but the persons whose programs are to be most affected should play a major role in defining policies. Policies and procedures adopted by such respected, active research scientists will be much more likely to be accepted by their peers than would otherwise be the case. Candidates for a committee position should be encouraged to look at the assignment as an opportunity to positively influence a program that is important to them. Most persons will be flattered to be asked and will accept appointment if this is the approach taken.

Some care must be taken to define "productive" personnel. The most productive individuals are not always the most visible or the most vocal. The actual achievers may be too busy to promote themselves and simply are quietly effective. This latter class should be sought out and the former passed over. A safety committee member must be able to consider issues objectively and be willing to act, even if occasionally their own immediate self-interests might be adversely affected. "Difficult" individuals should not necessarily be excluded. Often, if they are given an opportunity to participate, such persons can become a committee's strongest advocate. The background, credentials, and references of candidates for membership on a safety committee should be checked as carefully as if the candidate were a job applicant.

Many regulatory agencies now require that independent "lay" persons be added to the regulatory committees. These individuals, if they are to act as a "conscience" for the committee or at least present a different point of view, as is really their role, should truly be independent. They should have no possible conflict of interest ties to the organization. Among the affiliations that might disqualify an individual would be a personal relationship (spouse, brother or sister, life-partner, etc.) with a person employed by the organization, or a financial relationship (a retired former employee perhaps, a paid consultant, or one with a significant investment in a commercial firm). They should also be sufficiently technically qualified to understand the area of responsibility assigned to the committee. Unfortunately, these two provisions often make it difficult to obtain suitable lay members.

Nothing will destroy a committee more rapidly than for the members to feel that the work of the committee serves no useful purpose. The committee must be provided with a specific charge which must include clearly defined objectives and expectations. There must be meaningful work for the committee to do at each meeting. The results of their work must be taken seriously and not simply disappear into files. Each meeting should be structured, with a definite agenda and definite goals. Much of the responsibility for making a committee work depends upon an effective chair and good staff work. The staff must prepare the working documents and distribute them in sufficient time for review by the members to allow them to be prepared at the time of the meeting. A chair has the responsibility to conduct the meeting fairly, with everyone having enough time to contribute, but the chair must also see that the business is conducted expeditiously. Meetings should be no longer than necessary.

It would be very desirable for chairs of the various safety committees to have administrative experience. They should have experience in managing a budget and managing people so that they can provide guidance to the committee should it take actions that might demand resources that would be hard to obtain or cause personnel problems. They should, of course, be knowledgeable in the area of the committee's responsibility. There are occasions when not all these qualities can be found in a single individual, usually the specialized knowledge or recent active laboratory experience being the missing factors. In such cases, senior professional persons who have had to obtain and manage funded projects would be reasonably satisfactory alternatives to administrators. It would be less desirable for the safety professional to be committee chair because of a possible perception of a conflict of interest. There are circumstances, due to experience and training, when exceptions to this "rule" are appropriate. There also can be logistic factors when it would be helpful for the chair to take a directly active role in a safety area, which could make it acceptable for a safety professional to chair a committee.

In the preceding paragraphs, there has been considerable emphasis placed on defining a functional committee structure, but in a research institution, it is unlikely that the busy professionals making up committees would be willing to meet more than once a quarter, except in emergencies. Therefore, safety committees cannot be assigned the day-to-day task of directly managing a safety program. They are, in effect, somewhat analogous to the legislative branch (and on occasion, the judicial branch) of the safety "government," with the safety department being the administrative branch. Safety department staffs must have the authority to act as necessary between committee meetings, with the knowledge that they will be held accountable for their actions by the committees and their administrative superiors. It is up to the head of the safety department to prepare budgets, administer the programs and personnel assigned to the department, and to provide leadership in the area of safety.

Figure 1.1 is a simplified organizational chart that embodies the concepts discussed in this section. Some duties and layers of responsibility might be combined in a small firm or academic institution, while in very large organizations additional functions or layers might be needed. Note that the safety committees are appointed by a senior executive and have direct access to him.

1.3. THE SAFETY DEPARTMENT

The safety department was seen as largely advisory in older descriptions in the literature of its role in the scientific work environment. Under the OSHA laboratory safety regulations, each laboratory does have the responsibility for developing and carrying out its own chemical safety plan. However, the central administration of an organization has the responsibility to ensure that the industrial hygiene safety plans for individual laborato-

SAMPLE SAFETY PROGRAM ORGANIZATION

Figure 1.1. In an academic institution, the operations vice president would be the provost and the major divisions would be colleges or schools.

ries meet the employees' needs. This requires that the organization must provide a means to monitor the individual programs. Since it is difficult, if not wholly impractical, for an individual laboratory to independently meet all of the obligations and responsibilities required of them under the plan, access to many support services must be provided too. There are major areas involving regulations that are mandated by external agencies, such as disposal of chemical, infectious, and radioactive waste, that can only be administered efficiently by a central organization. Administration of organizational policies should be consistent for all of the employees, and the oversight function should be consistent across disciplines. Creation of a safety and health department is the mechanism adopted by most organizations. According to a survey performed by the Bureau of National Affairs, by 1990 over 93% of the organizations surveyed had one or more persons whose primary responsibilities involved occupational health and safety. The safety and health department represents the central administration to internal agencies under the corporate or institutional umbrella for safety and health issues. Thus, the safety department has assumed the character of an administrative rather than a purely advisory department to oversee the individual laboratories and to provide access to centralized services. Because of shifting emphasis toward health and the environment, both in the workplace and elsewhere, the common name for most organizational safety departments now is likely to include "Environmental Health & Safety" to reflect this change. However, because the Laboratory Safety Standard is a performance standard, the role of the department does include advisory aspects since more than one approach can be effective to achieve the same end. Perhaps the safety department in this context more closely corresponds to an environmental consulting group hired to oversee a program than any other prototype.

Maintaining an awareness of the current status of all environmental health and safety-related regulations that affect operations demands more time and specialized knowledge than a busy laboratory director or most administrators have available. It has become necessary to develop sharply defined specialists, even within the environmental health and safety discipline, because of the growing complexity of the field. This again has led to a strengthening of the role of the safety department because both management and the client groups are often forced to defer to the expertise of the department. Without comparable competence and experience, it could be foolhardy from a standpoint of liability for a laboratory manager or a senior administrator to ignore the opinion of a specialist.

The safety department is often forced to assume another unfamiliar role by directly providing support services, instead of simply monitoring others, to see that operations are done correctly. The skills, knowledge, and specialized equipment that are needed are not normally available from other support units. A comprehensive safety department needs some laboratory facilities of its own and the trained personnel to use them. However, exclusive use of in-house analyses can lead to concerns of conflicts of interest. Outside analytical services should be used where this is a possibility. A direct consequence of the provision of safety-related services is the need to fund these services. Resources to meet safety needs are typically funneled through the safety department instead of being directly allocated to the research departments. For example, monitoring to see that legal environmental PELs or action levels are not exceeded usually requires special equipment and the training to use it not normally available to individual laboratories. Environmental health and safety departments now typically provide access to medical surveillance programs, training, waste disposal, emergency planning, and a host of other programs to laboratory personnel. The following is a list of common functions of a safety department in a research organization. Not all safety departments would provide all these services, while there are some departments would provide these and more.

*Functions Relevant to Laboratories for the Safety Department**

- **Training**
  Safety orientation, organizational policies
  General work practices
  Safe work habits
  Confined spaces
  Electrical safety
  Machine guarding
  Facility-related issues
  General requirements of the laboratory safety standard
  Corporate responsibilities
  Supervisor responsibilities
  Individual responsibilities
  Training of maintenance and support personnel for work involving laboratories
  Emergency procedures
  Fire
  Precautions involving chemicals
  Use of portable extinguishers

---

*   Some of these functions, e.g., confined spaces, asbestos abatement, normally impact laboratory operations only indirectly.

Evacuation procedures
Medical
First aid
Cardio-pulmonary resuscitation
Serious injuries, physical and chemical
Universal precautions/OSHA regulations for bloodborne pathogens contacts
Chemical, radioactive materials, biological toxins spills and releases
Corporate institutional generic safety policies/procedures as applicable for
    Radiation safety
    Chemical laboratory safety
    Biological laboratory safety
    Recombinant DNA laboratory safety
Animal care facilities
Chemical, infectious, radioactive waste procedures
Personal protective equipment usage and care
Publication of safety manuals
Other OSHA required training

- **Waste Disposal**
  Chemical wastes (including permissible storage, processing, and redistribution)
  Radioactive wastes (including mixed waste)
  Infectious waste
  Contingency planning for emergencies
  Liaison with federal, state, and/or local waste officials

- **Building Safety**
  Building inspections
  Fire system inspections, testing, calibration, and maintenance
  Monitoring building evacuation programs
  Interactions with local fire and emergency response groups
  Code review of all renovations and new construction
  Consultation on building design, participates on building committees
  Interactions with on-site construction contractors and architects
  Liaison with fire marshals and building code officials

- **Environmental Protection and Industrial Hygiene**
  Air quality monitoring, testing
  Testing of fume hoods and biological safety cabinets
  Laboratory inspections
  Review laboratory design proposals
  Review research proposals for potential problems
  Chemical accountability (regulated chemicals, chemical inventories)
  Maintains controlled substance licenses
  Reviews and specifies personal protective equipment
  Review selected equipment purchases for safety features
  Maintain laboratory census for safety features
  Provides consultation services
  Investigate laboratory accidents
  Asbestos testing (bulk and air sampling) in existing facilities, equipment
  Monitoring of asbestos removal projects
  Release of chemicals to the environment (community right-to-know)
  Maintenance of records, database management

- **Medical Program**
  Sets criteria for employee participation in medical surveillance
  Initiates medical surveillance/counseling as required
  Provides staff support for occupational physicians
  Maintains all medical records
  Works with personnel department for pre-employment examinations
  Analyzes accident records, prepares reports
  Investigates serious accidents
  Acts as liaison with local medical emergency groups

- **Radiation Safety**
  Inspects and surveys radiation-using facilities
  Reviews all internal applications for use of ionizing radiation
  Performs hazard analyses for *all* experimental procedures involving usage of ioniz-
      ing radiation
  Reviews usage of non-ionizing radiation
  Processes all orders for radioactive material
  Receives and checks all radioactive material deliveries
  Maintains radioactive material inventory
  Monitors all personnel exposures
  Maintains calibration of all survey instruments
  Maintains all required bioassays
  Manages or oversees all radioactive waste disposal
  Maintains all required records
  Acts as liaison with regulatory agencies

- **Organizational Hazard Awareness Program**
  Maintains material safety data sheet files
  Tracks chemical purchases/employee participation in training programs
  Provides organization's emergency coordination
  Represents organization on SARA, Title III local emergency planning committee
  Responsible for safety guidance for outside contractors involved with laboratory en-
      vironments

- **Miscellaneous Responsibilities**
  Participates in all organizational-wide safety committees
  Provides staff support to safety committees
  Investigates employee complaints
  Anticipates potential regulatory actions
  Provides sanitation, health evaluations of facilities

Not all of the areas listed above are the sole responsibility of the safety and health department. Many involve or require collaboration with other support departments, with the participation of safety differing in degree. The list reflects a strong direct involvement of safety departments in the affairs of their parent organizations. Because of the increasingly complex knowledge, skills, and resources needed to meet current safety standards, there is really no choice except to centralize the responsibility for dealing with many of these standards. The alternative of making everyone comparably knowledgeable and able to manage their affairs sufficiently safely to avoid potential liability for themselves, as well as be technically proficient, is impractical. The goal is to preserve, in the area of safety, as much local autonomy and a sense of personal responsibility as possible among the organization's other employees. Safety and health personnel can define the programs needed and in many important areas provide required support services, but they cannot be present all the time to see that everyone acts safely. Safety ultimately must be a local responsibility.

Despite the sense of the previous few pages identifying broad areas of safety department responsibility, environmental safety departments should not be too eager to accept additional responsibilities from other service groups. Responsibility often will be shifted without a commensurate transfer of resources, so that the departmental resources, including personnel, physical, and financial resources, may be severely strained. Available resources may be difficult to obtain, especially in times of retrenchment, for most public and private organizations. Care must be taken to set achievable goals and priorities. It is better to do a few things well instead of attempting to do too much and failing to do anything properly. An example involves underground storage tanks. The EPA published regulations in 1988 requiring all underground storage tanks containing petroleum products or other substances that could be hazardous to the environment to be tested and modified, if necessary, to prevent environmental contamination. The owner would be required to clean up the site if environmental contamination were found to have occurred already. It seems logical at first that this "environmental" problem would be the concern of an "environmental" safety department. However, it is really a maintenance problem. The role of the environmental health and safety department should be limited to a monitoring of the maintenance department's program. This would ensure that the testing and remediation program would be done properly so that existing problems were corrected and additional problems not created.

1.4. DEPARTMENTAL RESPONSIBILITIES

A department is an entity within an organization with which individuals with more or less similar interests and goals identify themselves and are recognized by the organization as a unit for administrative purposes. As such they are the logical unit to establish and administer common safety-related procedures and practices suitable for the laboratory programs in the discipline identified with the department. The leadership of the department provides a natural channel for communication with laboratory personnel and provides a vehicle for enforcing organizational policies or for directing concerns by the department about these policies to higher administrative levels. In the context of the present discussion, the major virtue of a department's leadership is that they are physically present on the scene and represent a recognized source of authority. No safety department is sufficiently large to be present everywhere. Safety personnel must rely on the help of the local departments to implement safety practices and policies. The local department's leadership must accept the need for a strong safety program and see that a concern for safety is the standard expected of all departmental personnel.

At a minimum, even small departments should designate an individual to be a safety coordinator for the department, who may also act as liaison to the institutional or corporate safety committee. Larger departments preferably should establish an internal safety committee with representatives from each major division within the department. Individuals at various levels should be included on the committee, such as laboratory supervisors, laboratory chemical hygiene officers, and technical staff to ensure that all points of view are fairly represented. In an academic department, it might be desirable to include an experienced graduate student. One function of the committee would be to interpret the corporate or institutional policies in the context of the departmental operations and the operations of the individual laboratories. The committee can also advise individuals and make recommendations on improvements for safe operations in laboratories. The support of a committee often can facilitate obtaining needed resources, whereas an individual acting alone might not succeed. Finally, the committee should establish a frequent schedule of inspection of facilities of the laboratories and the support infrastructure for the laboratories. The supervisor responsible for the problem area should be informed if conditions are found which need correction, and a follow-up inspection should be done after a reasonable interval to ensure that the corrections have been made.

Departments also can and should function as a resource center. For example, the Laboratory Safety Standard, under OSHA, requires that certain information be readily available to the employees in the department, e.g., a copy of the organization's industrial hygiene plan and Material Safety Data Sheets. A central hard copy file of these, in a large research organization or university, could easily encompass 5000 to 10,000 records in a massive and difficult-to-manage file. A departmental file would normally be expected to be much smaller and more manageable, although in a large academic chemistry department, the distinction might disappear. Similarly, the department would be a logical group to provide much of the technology-specific training required for the OSHA standard for the department's laboratory employees, since as noted previously, the department is normally organized around a common discipline.

The department should also accept some responsibility for the safety of its personnel in terms of resources. There are legitimate questions about the limits of the institutional or organizational responsibility and those of the department or laboratory. It is clear, for example, that such things as utilities, custodial services, and building maintenance are the purview of the corporation or university. Specialized equipment such as an electron microscope or a highly specialized laboratory environment is equally clearly the responsibility of the local department. However, who is responsible for eyewash stations and deluge showers, fume hoods, glove boxes, chemical waste disposal, respirators, safety glasses, flammable material storage cabinets, and refrigerators, etc.? Some of these would not be needed if it were not for specific research programs or grants. Should not the department or laboratory manager incorporate these costs or at least a portion of them in their own budget? To some degree, the imposition of safety standards and regulations has inculcated, in many persons, a sense that safety is not their own responsibility. Some persons have the attitude that someone besides themselves is responsible for not only the rules, but to provide the means to achieve compliance with the rules. Awareness that safety is, to a major degree, its own reward seems to have been lost to an extent. As a minimum, it would appear that in soliciting a grant the investigator and the departmental leadership should bear the responsibility of assuring in advance that the resources are available or attainable, so that the research can be done safely and with appropriate regard for the protection of the environment.

1.5. LABORATORY RESPONSIBILITIES

Ultimately, the responsibility for being safe and working safely falls to the laboratory personnel. The laboratory supervisor, who may or may not be an active participant in the daily, routine work of the laboratory, still must set the standards of performance expected of everyone within the laboratory. The laboratory supervisor must make it clear by example or by direction that carelessness in safety is no more acceptable than careless and sloppy science. The supervisor has a responsibility to those under his direction to establish safe work practices and to ensure that the employees are given the opportunity to be informed and fully understand any risks associated with the program of work. No one should be asked to perform an act posing a substantial degree of personal risk, unless that is a normal aspect of their work, such as expected of firefighters or police. This is not to say that laboratory work must be made totally free of risk because this ostensibly desirable goal is unattainable. It does mean that all reasonable and practical steps that can be done to minimize risks have been taken. The concept of **ALARA** (**A**s **L**ow **A**s **R**easonably **A**chievable), used in the nuclear field, is a good, pragmatic guideline to follow. Equipment must be maintained in good repair and adequately designed to work properly. Written standard operating procedures are required under the OSHA Laboratory Standard, and individuals must be fully trained in the procedures that apply to their duties. Work should be carefully analyzed to foresee potential accidents or failures. The "what if" criterion should be applied to every procedure used in the facility. Even low probability bets do pay off on occasion. Contingency plans must be developed to meet at least the most likely emergencies. To fail to do these things could expose the laboratory director and in turn possibly the department and the university or corporation to charges of willful negligence.

Employees, on the other hand, have an equal share in the responsibility. They must adhere to the safety policies that have been established. They must take the initiative to ensure that they are knowledgeable about good laboratory safety practices, and they must not diverge from these practices because they are too time consuming, too much trouble, or inconvenient at the moment. If they are uncertain of the proper procedure, they must not be reluctant to admit it, but should seek clarification. Often, especially if the laboratory supervisor is not a reasonably active participant in the work in progress, a knowledgeable employee may understand the actual potential risks better than his supervisor. Any employee should offer suggestions to improve safety in the workplace. Unfortunately this ideal may be a bit unrealistic in that many subordinates are sometimes wary of "making waves" or contradicting their superiors, and not all supervisors are appreciative of suggestions from subordinates. It is this situation for which the ability to make anonymous complaints, which are certain to be investigated, was incorporated in the Federal OSHA Act and subsequently passed on to the state standards, where states have adopted their own OSHA statutes. Ideally, the team of a caring, conscientious laboratory supervisor working in cooperation with competent, intelligent, and imaginative laboratory workers should make it possible to conduct laboratory work with minimal risk.

REFERENCES

1. **Petersen, D.,** *Techniques of Safety Management,* 2nd ed., McGraw-Hill, Inc., New York, 1978.
2. Prudent Practices for Handling Hazardous Chemicals in Laboratories, National Academy Press, Washington, D.C., 1981.
3. *Safety in Academic Chemistry Laboratories,* 4th ed., American Chemical Society. Washington, D.C., 1985.
4. **Steere, N.V.,** Responsibility for laboratory safety, in *CRC Handbook of Laboratory Safety,* 2nd ed., Steere, N.V. (Ed.), CRC Press, Inc., Cleveland, 1971, 3.
5. **Becker, E.I., and Gatwood, G.T.,** Organization for safety in laboratories, in *CRC Handbook of Laboratory Safety,* 2nd ed., Steere, N.V. (Ed.), CRC Press, Inc., Cleveland, 1971, 11.
6. **Songer, J.R.,** Laboratory safety program organization, in *Laboratory Safety: Principles and Practices,* Miller, B.M., Gröschel, D.H.M., Richardson, J.H., Vesley, D., Songer, J.H., Housewright, R.D., and Barkley, W.E. (Eds.), American Society for Microbiology, Washington, D.C., 1986, 1.

7.   **Bilsom, R.E.,** Torts Among the Ivy: Some Aspects of the Civil Liability of Universities. Asst. V.P. for Administration. University of Saskatchewan, Saskatoon, Saskatchewan, 1986.

8.   **Campbell, S.L.,** Expanding environmental duties, in *Occupational Health & Safety*, **61** 11, 18, 1992.

9.   **Jefferson, E.G.,** Safety is good business, in *Chem. & Eng. News,* November 17, 1986, 3.

10. **Koshland, Jr., D.E.,** The DNA Dragon 1, in *Science,* **237** No. 4821, 1987, 1397.

11. **Walton, M.,** *The Deming Management Method,* Perigee Books, 1986, 12. *OSHA Exposure to Hazardous Chemicals in Laboratories,* 29 CFR 1910.1450, 55 FR 3327, January 31, 1990 and FR 7967, March 6, 1990.

12. *OSHA Hazard Communication Standard,* 29 CFR Parts 1910.38, 1910.120, 1910.1200, 1910.1450 and 1910.1500, as amended.

13. *DOT Hazardous Materials, Substances and Waste Regulations*, 49 CFR Parts 171-177, as amended.

14. *EPA Hazardous Waste Regulations,* 40 CFR Parts 260-265, 268, 302, 303, 311, 355, and 370, as amended.

15. *OSHA Occupational Exposure to Human Pathogens*, 29 CFR 1910.1030, 56 FR 64004, December 6, 1991.

16. *EPA Standards for the Tracking and Management of Medical Waste*, 40 CFR 259.

17. *Americans With Disabilities Act of 1990*, Titles I-V, July 26, 1992.

18. *DHEW Guidelines for Research Involving Recombinant DNA* Molecules, FR 51, 16957, May 7, 1986, as amended.

19. *DHHS Protection of Human Subjects,* 45 CFR 46.

20. *NRC Standards for Protection Against Radiation Revision*, 10 CFR Part 20.

21. *Prevalence of Safety and Health Specialists,* Bureau of National Affairs, Occupational Health and Safety Reporter.

CHAPTER 2

Emergencies

Emergencies are, by definition, not planned. However, planning for emergencies can not only be done, but is an essential component of laboratory safety. There are many regulatory standards that now *require* that organizations using chemicals in laboratories and elsewhere, or that produce chemical waste, to have formal emergency plans. These plans must cover emergency evacuation and response plans, emergency equipment to be kept on hand, security, training of personnel handling hazardous chemicals, reference materials, identification of emergency personnel, and access to external resources, including aid agreements with local emergency organizations. For example, every facility with laboratories that comes under the OSHA laboratory standard must meet this obligation under Title 29 Part 1910.1450. This includes even small organizations. In addition, OSHA Industry General Standards under 1910.38, 120, and 1200 also provide for emergency planning. The recent OSHA Bloodborne Pathogen Standard (29 CFR 1910.1030) has provisions for emergency actions in case of an accidental exposure. The RCRA Act considers all organizations that generate more than 100 kg of hazardous waste per month as large generators. Title 40 CFR Part 265.16 and Parts 311 and 355 defines the emergency requirements of RCRA. The Americans With Disabilities Act, Titles II and III, 28 CFR, impacts accessibility for disabled individuals. Each of the regulatory acts will be discussed in more detail in later chapters. However, these regulations simply provide the specifics for a legal mandate to do what every organization handling hazardous material should do anyway.

A realistic appraisal of the circumstances that can lead to emergencies in a laboratory will reveal many foreseeable and controllable problems. Some problems that can be expected to occur might include:

- Fires
- Chemical spills
- Generation of toxic fumes and vapors
- Inhalation, ingestion, or absorption of toxic materials
- Release of compressed toxic, anesthetic, explosive, asphyxiating, and corrosive gases locally or beyond the boundaries of a facility
- Release of radioactive materials
- Release of pathogens and restricted biological materials
- Power failure, involving loss of lights or ventilation
- Electrical shocks
- Explosions, or runaway reactions
- Failure of a facility exhaust system
- Physical injuries to individuals
- Consequences of natural disasters
- A combination of any of these simultaneously

This list is not intended to be complete, and some events are more likely to create immediate and pressing problems than others. It is impossible to anticipate all classes of problems that can occur. Some events are recognized as emergencies more readily, while others may not be identified for extended periods of time. Some involve the threat of personal injury, while others impact the environment with little likelihood of immediate injurious effects. Emergency personnel often mention that many emergencies in which they have been involved were not anticipated and would have been unlikely to have been considered, even by the most careful planning. There is no limit to the variations that human ingenuity and the vagaries of fate can take to modify the factors that control our lives. The inability to foresee all possible emergencies should not inhibit the development of plans to cope with those that can be anticipated, or to provide a basic emergency response infrastructure that can be used, even for unanticipated types of emergencies.

The scope of this chapter will be to examine the general principles of emergency preparedness to serve as a guide for preparation of individual, specific action plans and to provide some useful information to be used in various classes of emergencies. Planning and preparation are necessary to help in identifying and finding the resources needed to support a flexible, effective, and, if needed, rapid response to laboratory emergencies. Injuries, property, and environmental damage can be limited if emergency procedures already exist and are practiced regularly. Practice is essential to expose deficiencies in the procedures and to familiarize personnel involved with them. Plans that are developed and then filed away are worse than useless. They can provide a false sense of security. In a real emergency, it is essential to *know* immediately what to do or, often, what not to do. Time to read a manual may not be available.

Before developing the theme suggested in the preceding paragraph, there are some caveats that need to be introduced that are applicable to the contents of this entire chapter. There are advantages in not overreacting. It is easy for well-meaning and knowledgeable individuals to turn a relatively minor event into a major and expensive incident by acting too quickly, without full awareness of the total situation and without consulting other persons involved. No serious worsening of a situation might result in doing absolutely nothing until the situation has been discussed and a plan of action developed. Evacuation, containment, and exclusion of nonessential personnel are the appropriate initial actions in almost every emergency. Unless a situation is clearly deteriorating and is becoming out of control, a review of the situation and an examination of the response options by emergency and operational personnel is desirable.

One other note of caution: no one is expected in the normal course of their work to go to extreme measures, risking their own lives, to cope with an emergency when the risk is certain to be too great. The more responsible action often is to leave the scene when the situation is obviously beyond an individual's capabilities. Doing so makes it possible for emergency response groups to have a competent source of information about the situation when they arrive. It is difficult to do when lives are involved, instead of only property, but there is no point in adding to the loss when the situation is clearly hopeless. It is a judgment call that can only be made at the time by persons present.

Despite the two cautionary paragraphs immediately preceding, there are steps that individuals and local groups can and should take, when appropriate, to confine and minimize the impact of emergencies. The first few moments of an emergency are frequently the ones that are the most crucial. Actions should be based on training, knowledge, and a due regard for priorities. Protection of life and health should come before protection of property, or reputation. Unfortunately, many persons do not seek help or take inappropriate actions for fear of being blamed for a problem until too late, often allowing the situation to worsen until out of control. Trained and knowledgeable personnel are less likely to make these mistakes.

2.1. COMPONENTS OF EMERGENCY PREPAREDNESS

Emergency preparedness should not be considered the responsibility of any one individual or group. In an actual emergency, there are key persons or groups who will usually respond to the situation. There are external agencies such as fire departments, police, rescue squads, hospitals, hazardous material response teams, and local emergency planning groups. Internal groups may include comparable agencies in very large organizations. Fire brigades may have been established. Hazardous waste personnel are likely to have received special training. An emergency coordinator may have been appointed. In laboratory emergencies, every person present has an important role, from the laboratory supervisor to graduate students in academic institutions. A framework for emergency preparedness should be created in every organization of every size, even if it is little more than "evacuate and call 911 for help."

2.1.1. Initial Conditions

Basic conditions should exist to ease meeting emergency responsibilities. Some of these conditions stem from before a building is constructed. For example, in the initial planning, the building should have been designed by architects, in cooperation with the persons responsible for the programs to be housed in the building, to incorporate safety codes and regulations. Codes represent minimum requirements and may very likely need to be exceeded. Appropriate fixed and movable equipment must be installed or provided, consistent with the concept of a facility that could be operated safely. Emergency equipment must be available, but decisions must be made about what design features and equipment should be mandatory, what is desirable, and what would be a luxury. If all of the correct decisions were made so that the building, the furnishings, and the appropriate emergency equipment are all available, then personnel responsibilities should be considered next. It is necessary to define the role of each person in a facility and to specifically designate which individuals and groups should have the leadership responsibility for emergency planning and emergency response. It should not be necessary to develop an impromptu plan or seek one buried in a file cabinet.

The emergency plan for a given facility should be a subset of a plan for the entire organization. The infrastructure and planning available to the entire organization can be adapted to local needs. Decisions must be made as to who is responsible for providing equipment, and obviously with this decision the need arises to decide the source of funds. A major decision is to define the type of command structure that will be used and who will be involved. The responsibilities of the key individuals and groups must be delineated and boundaries established between local responsibility, institutional responsibility, and outside emergency response agencies. Finally, based on all of the applicable factors, each individual facility can establish a written emergency response plan for itself with specified responses to anticipated classes of emergencies for the facility. The organizational plans as well as the plans for smaller units all must be sufficiently flexible to provide responses to unanticipated emergencies.

The following material will elaborate on these points.

2.1.2. Facilities, Fixed and Movable Equipment

Most discussions of emergencies assume existing buildings with all of their characteristics as given and superimpose emergency planning on this existing system, but if the opportunity arises, there is much that can be done to reduce the severity of later emergencies when designing, building, and equipping a facility. The results of past decisions are difficult and expensive to change, but it is possible to avoid making new errors. Renovating existing structures to make them earthquake resistant, unnecessary in many areas while very important in others, is very expensive, but it is possible to do it at much less cost for new buildings. In order to facilitate the design and construction of safe buildings, fire and building codes have been established in most localities that govern new construction and renovations to existing buildings. Generally, under these codes, research laboratories come under the classification of a business use occupancy, or possibly as a hazardous use occupancy, each of which incorporates different safety requirements. OSHA also has standards in the area of fire safety, as well as ventilation, which must be met. OSHA standards, where applicable,* are consistent in every state, but building codes vary from locality to locality, often depending upon interpretations of a local code official. The requirements for access for the disabled under the ADA clearly affect emergency movements. As a result, fire alarms now require intense strobe lighting devices as well as audible signals. Braille instructions may be required for the blind in parts of a facility. Special chairs may have to be provided for the physically disabled. Places of refuge to which disabled persons can go while awaiting help must be identified. For several other types of risk, special regulations, such as the classification system for recombinant DNA research facilities, also have safety restrictions that must be included in the building design. This latter set of safety restrictions will be reserved to later chapters dealing with these special topics.

Concerns which should be addressed in the designs of laboratory buildings to enhance emergency response depend upon the classification. For example, if the building is a hazardous use occupancy, most codes will require a sprinkler or other fire suppression systems. If a sprinkler or alarm system is required by a local fire code, then OSHA 1910.37(m&n) requires maintenance and testing. Also for this classification, OSHA will require under 1910.37(f)(2) that the doors swing in the direction of exit travel, yet most building codes have restrictions on doors swinging into corridors to avoid creating obstructions to corridor traffic. In order to satisfy both requirements, doors should be recessed into alcoves inside the laboratory. To meet some code standards, even existing facilities may have to be upgraded.

The size of a building, the number of floors, and the relationship to other structures all enter code decisions affecting safety in emergencies. Addition of equipment to a laboratory, such as a hood, can have serious fire safety implications. Is there adequate makeup air? If not, where can it be obtained? Halls should not be used as a plenum or as a supply of makeup air for more than a few hundred ft^3 per minute (cfm) for each laboratory space. Even a small, 4-foot fume hood discharges about 800 cfm, so that one cannot draw the required makeup air in through louvers in the door. Usually one must go outside for a source of makeup air, but what is the relation of this new inlet air intake to the exhaust system? Toxic fumes could be drawn back into a building. A fume exhaust duct penetrating a floor could allow a fire to spread from one floor to another. Therefore, most codes require fume hood ducts to be enclosed in a fire-rated chase. Because of the expense of constructing a chase, the cost of avoiding worsening the fire separation in a building could preclude installation of the hood, which in turn could preclude using the space for the intended research. One option, to allow future flexibility, is to incorporate external chases

---

* About 25 states have adopted their own state OSHA plans which are required to be as stringent as the federal standards, however, public employees may not be covered by the OSHA standards.

as an architectural feature in the design. Energy loss considerations can impact the design of a laboratory. Auxiliary air hoods have been used to reduce the amount of tempered air being "wasted," but there are a number of reasons why this type of hood is less desirable. An alternative is to design a ventilation system for a laboratory to maintain a constant volume of air through a hood while in use, but to use a proximity sensor to reduce the volume to a lower level if no one is standing in front of the hood for a predetermined interval. Ventilation will be discussed in much more detail in Chapter 3.

The interior arrangements of a laboratory are critical in permitting safe evacuation from the laboratory. The types of accidents listed earlier could pose much more serious risks to individuals should they occur between an individual and the exit from the room. A simple solution for these potential emergencies for larger laboratories is to have two well-separated exits. This is not always possible, especially in smaller laboratories. An alternative would be to evaluate what components of a laboratory are most likely to be involved in an incident and which would increase the hazard if they became involved in an ongoing emergency. These components should be located so that an escape route from the normal work area does not pass by them. Also, portable fire extinguishers, fire blankets, respirators, and other emergency equipment should be located on this same escape route. Eyewash stations and deluge showers should be located close to where injuries are likely to occur, so an individual will not have to move substantial distances while in intense pain or blinded. Aisles should be wide (typically a minimum of 42 to 48 inches), straight, and uncluttered with excess equipment to ease movement in emergencies. A laboratory should have emergency lighting, but many do not. The considerable dangers posed to an individual stumbling around in a pitch dark laboratory should the power fail are obvious. Inexpensive, battery-powered rechargeable units are a potential solution here and are not expensive, even in retrofitting a facility.

Many regulations found in OSHA standards include features that will minimize the scope and impact of an emergency, such as a fire. For example, restrictions in 1910.106 on container sizes of flammable liquids and the amounts of these materials that are permitted to be stored outside flammable material storage cabinets are designed to limit the amount of fuel available to a fire and to extend the time before the material could become involved.

Every action should be considered in terms of what would result if the worst happened. In large projects, this is often part of a formal hazard analysis, but this concept should be extended to virtually every decision within a laboratory. For example, a common piece of equipment found in most laboratories is a refrigerator. A unit suitable for storing flammables, i.e., containing no internal sources of ignition, costs about two to three times as much as a similar unit designed for home use. It is tempting, especially if money is tight and the immediate need does not require storage of flammables, to save the difference. However, the average lifetime of a refrigeration unit is roughly 15 to 20 years. Who can say what materials research programs will entail over such a long period? If flammable vapors within an ordinary refrigerator should be ignited, a violent explosion is very likely to occur. Employees could be injured or killed and the laboratory, the building, and the product of years of research could be destroyed. Not only would there be immediate problems, but in most cases, replacing laboratory space would be very expensive, currently in the vicinity of $130 to $300 per square foot. Actual construction of replacement space for buildings as complex as most laboratories, from the time of planning to completion of construction, typically takes up to 4 years after the money is available.

Many actions are influenced by the costs involved, as in the preceding example. A continuing question involves who should be responsible for paying for safety facilities and equipment. Under the new OSHA laboratory standards, the adequacy of a facility to allow the work to be done safely is a key condition. There are some straightforward guidelines that can be used:

1. For new construction, safety should be integrated into the building design and the choice of all fixed equipment. The latter should be incorporated in the building furniture and equipment package. This would include major items such as fume hoods, since these are relatively expensive units to retrofit.
2. Certain equipment and operational items common to the entire organization (e.g., fire extinguishers, emergency lighting, deluge showers, eyewash stations, and fire alarm systems) and maintenance of these items should be just as much an institutional responsibility as provision of utilities.
3. Items which are the result of operations unique to the individual laboratory or operations should be a local responsibility. This would include equipment such as flammable material refrigeration units, flammable material storage cabinets (if these are not built in), and specialized safety equipment such as radiation monitors, gas monitors, etc. Some major items which might be included under fixed equipment in new construction might have to be provided by the individual if renovation of a space were to be involved. For example, it might be necessary to construct a shaft to enclose a fume hood duct and to provide a source of additional makeup air for the hood. Personal protective equipment, such as goggles, face masks, respirators, and gloves, should also be provided at either the laboratory or departmental level.

It is unlikely that any individual, whether it is the laboratory supervisor, safety professional, planner, or architect, will alone be sufficiently knowledgeable or have the requisite skills to make appropriate decisions for all of the factors discussed in this section. In addition, every one of these persons will have their own agenda. The inclusion of emergency preparedness features should be explicitly included as one of the charges to the building or project design committee so that these needs can be integrated with function, efficiency, esthetics, and cost.

It was not the intent at this point to elaborate on all the implications of codes as safety issues but, by a few examples, to draw attention to the idea that the root cause of an emergency and the potential for successfully dealing with it could well lie with decisions made years earlier. The point that *was* intended to be made was that laboratory safety and the capability to respond to emergencies does not start with teaching good laboratory technique and the adoption of an emergency response plan after beginning operations.

2.2. INSTITUTIONAL OR CORPORATE EMERGENCY COMMITTEE

In most organizations, there are many support groups that have been assigned specific responsibilities in dealing with emergencies which extend beyond those associated only with laboratories. Among these are safety, police or security, maintenance, communications, legal counsel, and media or public relations. Unlike the laboratory supervisor, departmental chair, or individual laboratory employee who is primarily concerned with his research or administrative duties, these groups are directly concerned with one or more aspects of emergency response. In larger organizations, fire departments, physicians or medical services, or even more specialized groups may exist. Each of these groups have their own expertise, their own dedicated resources, and their own contacts with outside agencies. Representatives from these agencies will be the ones normally called to the scene of an emergency, and will be the ones expected to cope with the situation. This group should form the nucleus of the emergency planning committee. It should have participation from the remainder of the organization, and in the current context, this participation should include comprehensive coverage of the various areas of the corporate or institutional research programs. The committee should have direct access to upper levels of management, and it should also interact closely with safety committees associated with each broad research area, e.g., chemical, radiation, biosafety, and animal care. This committee also needs to coordinate its efforts with nonorganizational support groups such as local, state, and federal police authorities, fire departments, rescue units, local emergency planning groups, environmental regulatory agencies such as EPA and local or regional water, air, and waste management agencies, and safety regulatory groups such as OSHA.

The emergency committee, once formed and its charge clearly defined, should meet periodically, at least once a year and preferably more often, to review the status of the organization's emergency preparedness, to plan for practice sessions, to review drills that have been conducted, and to investigate and review incidents that occur. Reports of these meetings, along with the findings, should be presented to management and to the individual safety committees.

2.3. EMERGENCY PLAN

The initial order of business for the emergency committee is to develop an emergency response plan (ERP). In developing the ERP, the committee should analyze the types of emergencies which could happen, their relative seriousness, and their relative probability of occurrence. The emergencies to be considered should specifically include releases of hazardous and toxic chemicals to the environment, as required under SARA, Title III.* Once the classes of emergencies have been defined, each should be analyzed as to the resources, equipment, training, and manpower which would be needed for an adequate response. An integral part of this analysis would be provisional plans for using these resources to respond to potential emergencies. The analysis should include both internal and external resources. Finally, a critical evaluation should be made of the current status of the institutional resources and a recommendation made to correct deficiencies. Based on the preliminary studies, the final plan should be drafted, circulated for review, amended if required, and implemented. The support of management is critical, or this exercise would be futile.

Plans should be developed which would be operative at differing levels. A basic plan should be short and easy to grasp and implement. The simple sign in Figure 2.1 is effective for most emergencies. The caller is expected to be guided by the person (usually a dispatcher) at the other end of the line for specific guidance for responding to a given emergency. The major caveat is that the time to make such a call may not be available prior to evacuation for emergencies representing immediate and worsening emergency situations. Occupants of a facility should be trained to recognize when this condition exists and know how to initiate an evacuation of as large an area as necessary.

2.3.1. Laboratory Emergency Plan

Workers in most laboratories are intelligent and knowledgeable and can cope with many small emergencies such as a spill of a liter of sulfuric acid or a small fire if they

---

* SARA – Superfund Amendments and Reauthorization Act of 1986.

Figure 2.1.   A sign such as this placed on or near all telephones is an effective way to enable reporting of emergencies.

have received appropriate emergency training. Such training is mandatory under the new OSHA laboratory standard. A laboratory emergency response plan is required under current standards for the risks associated with operations within the facility. The plan needs to include risk recognition, means of internal responses to small to moderate emergencies, and evacuation training. All employees in the laboratory must receive instruction on these points at the time of beginning work in the facility or when any new procedure or operation is introduced posing different risks. In order to identify risks, a detailed, thorough hazard analysis needs to have been done, based on the things that could go wrong, not just the risks associated with normal operations. Among information which must be included in the plan is where an employee can get not only the laboratory specific plan, but also the organization's overall plan. Another key ingredient of the plan is where safety and health information for the chemicals used in the laboratory, as represented by Material Safety Data Sheets (MSDSs), can be readily provided.

A written emergency plan for an individual laboratory might, in outline, resemble the following:

I.   In bold letters, the basic emergency number to call for emergencies.
II.   A defined line of authority. This should provide the names and home and work telephone numbers of several individuals authorized to make decisions for the facility. They should be persons with direct knowledge of laboratory operations and, at least at the top of the list, persons who can make financial commitments.
III. List of external persons/groups, with telephone numbers, who can provide emergency assistance relevant to the risks associated with operations. Such a list should include the following:

Emergency telephone number—911, if available in the area
University police or corporate security, if not available through 911 number
Local government police, if not available through 911 number
Fire department number, if not available through 911 number
Emergency medical care (rescue squad), if not available through 911 number
Nearest Poison Control Center
Nearest hospital

Safety department

Spill control group, if not available through 911 dispatcher or Safety Department

Maintenance department number(s)

Laboratory supervisor business and home telephone number

Secondary laboratory authorities business and home telephone numbers

Departmental or building authority numbers

IV. A list of normally required safety procedures appropriate to laboratory operations.

V. A simplified list of emergency actions to take for most likely emergencies.

VI. Evacuation instructions, including a map of at least two alternative evacuation routes. The primary route should be identified and normally should be the shortest, most direct means of egress from the facility. A gathering area should be identified to which evacuees would normally go. This is important to allow a "head count" to ensure that everyone did successfully evacuate, and to provide a location where external agencies could come in order to receive information concerning the emergency.

VII. Location of MSDS's and other safety and health reference materials.

VIII. Location of organization's emergency plan.

IX. Procedures for expanding the emergency response to additional areas of the building and organization when the emergency is a "large" one extending beyond the immediate area. The location of one or more telephones outside of the affected facility but readily accessible should be clearly identified.

Two items need to be placed on or adjacent to the laboratory door to assist emergency responders when lab personnel are not immediately available during an incident: the line of authority, listed in Item II above, and indications of the types of hazards to be found within the laboratory. Some areas have ordinances requiring the use of the NFPA Diamond for the latter purpose, but unfortunately most laboratories would have at least some material with high-risk ratings in all categories. Pictographic labels identifying classes of hazards within a facility are also used. Eventually, the best way to alert firefighters, for example, would be to have laboratory inventories on a computer database and provision made for emergency response groups to have electronic access to this information. Software is available, although not yet in wide use, which does this.

This plan incorporates some aspects of the Laboratory Industrial Hygiene Plan as required under OSHA, which could be deleted, since the written industrial hygiene plan must be maintained. However, items I, II, III, V, VI, and VII are essential.

The plan just described should be reviewed with each new employee and at least annually for all occupants of a laboratory. An annual practice drill is strongly recommended.

2.3.2. Organizational Emergency Plan

There is some overlap between planning for responses to local emergencies in individual laboratories and the response to large-scale emergencies. At the extremes, the distinction is clear. A minor spill or a trash can fire obviously is a minor emergency, while a fire that involves an entire building or a major spill where hazardous materials are released into the environment clearly is beyond the capacity of laboratory personnel. Planning needs to provide guidelines to cover the transition between the two levels to ensure that an appropriate response does occur.

A comprehensive plan is intended to provide a general infrastructure for all classes of emergencies. Detailed plans are essential for organized emergency groups, but for the use of the general public a basic emergency plan is to evacuate the area or building, and call for emergency help. Often, evacuation will be more than is actually needed, but it is usually a conservative and safe approach. The essential information to enable this can be placed on a single page for a facility. Normally, planning for large-scale emergencies will be the responsibility of the corporate or institutional Emergency Committee, working with internal groups and the Local Emergency Planning Committee (required under SARA Title III) and nearby support agencies.

A basic means of reacting to virtually any emergency for untrained persons would be to place a sign, such as is shown in Figure 2.1, on or near every telephone. In this case, it is up to the individual at the other end of the telephone line, normally a dispatcher, to give verbal directions for subsequent actions. The dispatcher needs to be well trained and provided with a list of individuals and groups whom they would notify of the incident, in an appropriate priority. These individuals, groups, and priorities are defined in the master emergency plan for the organization.

Following is a simplified table of contents for an emergency plan established for an area containing a university, major commercial activities including chemically related industries, transportation sources (highway, rail, and air), and the usual variety of emergency support groups.

1.0   Charge
1.1   Assignment of legal authority and responsibilities
        Charge
        Members of governing body
1.2   Purpose of plan, functional description
1.3   Instruction on how to use the plan
1.4   Initial conditions
        Demographics
        Geographic description
        Natural risks
        Climate
        Time factors
        Local hazard sources
        Utilities
        Local administrative units
        Local emergency units
        Local resources
1.5   Communications
        Notification procedures
        List of agencies/personnel requiring notification
        Telephone lists
            Key personnel and alternates
                Telephone tree
            Emergency assistance numbers
                Local
                Regional
                State
                National
                Commercial
            Regulatory agencies
        Alternative communication options
            Authorized radio coordination procedure

1.6   Incident recognition/response
Identification of incident
Response protocol
Emergency command structure (see Figure 2.2)
    Command center, normal
    At-scene control center
    Emergency coordinator
    On-scene commanders
1.7   Responsibilities of emergency support groups (initial response)
Fire/rescue/haz-mat teams
Law enforcement
Medical
Communications (public notification/media relations)
Logistics support
    Transportation
    Public works
    Emergency housing/refuge centers
Administrators (government/corporate/institutional)
Agencies (regional/national/regulatory)
Emergency committee
1.8   Ongoing and completion
Assessment of conditions
Containment
Termination
Recovery
Critique
1.9   Continuing processes
Training
Practice drills
Resource development
Plan review
2.0   **Appendices**
Incident forms
Mutual aid agreements
Current emergency rosters
Evacuation centers
Hospitals/medical assistance
Social agencies
Emergency equipment lists
Likely incident locations
Cleanup contractors
Experts
Testing laboratories
Maps/overlays
Radio/TV/newspaper contacts
Copies of regulations

All of the groups likely to be involved in the emergency response should possess a copy and be familiar with the organization's emergency response manual. The manual should spell out in detail, but still as simply and as flexibly as possible, the correct response to the classes of emergencies incorporated in the ERP.

It is always the intent of every organization that no emergency will ever occur and for the more unusual situations considered in the ERP, long intervals may pass between

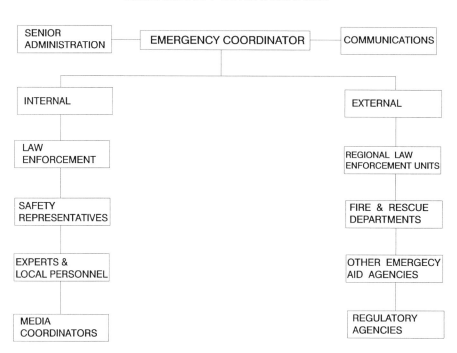

Figure 2.2. A typical military-type command structure for responding to a substantial emergency.

incidents. However, it is essential to include in every emergency plan provisions for periodic review and practice.

2.3.2.1. Emergency Plan Components

A partial list of some of the more common laboratory-related emergency situations was given in Section 2.1. A written response plan should be provided for each of these situations, identifying the likely locations where these classes of problems would be likely to occur, the characteristics of the locations, accessibility, probable means of response, local resources available, contact persons, outside agencies that would need to be notified, and possible refuge areas to which the occupants would evacuate. Important characteristics or questions which need to be addressed would include: is it a multiple story building, what type of construction (combustible or fire resistant), does an alarm and/or sprinkler system exist, are there standpipe connections or hydrants nearby, what is the typical occupancy level at various times of day, are there disabled persons in the building requiring special assistance, are there hazardous materials in the facility and the kinds and quantities of these materials and if so what is the potential impact on adjacent structures for various environmental conditions, among other factors. This type of information requires a great deal of time to compile. However, the information should be placed in a well-organized appendix to the main body of the plan, so that it would not be necessary to wade through what would necessarily be a massive amount of data for larger organizations.

The management structure is critical to controlling emergencies. This needs to be defined in advance. If the organization is sufficiently large, the plan may include managing virtually every aspect internally, unless the scope of the emergency extends beyond the area of the organization's control. In such cases outside agencies must be notified, and they *may* assume partial responsibility for management of the emergency response, but an emergency extending beyond the controlled boundaries will definitely invoke mandatory notification of outside agencies. A large organization may have its own fire brigade, police force, safety department, hazardous material response team, rescue squad, and access to experts internally. Most larger corporations and universities have some of these, but typically not all. Smaller firms and colleges might have only a security force and a small safety department, which might be combined.

Most emergency plans employ a pseudo-military organization, at least for coordinating the initial response. An individual, with alternates, is identified as the emergency coordinator. If the organization is highly structured, a command center, again with alternates, is identified to which the emergency coordinator and other key individuals will go when an emergency of sufficient scope occurs. This command center should have radio and telephone communication capability, which would be less vulnerable to loss of power and normal communications channels. Radio contact on emergency frequencies should be available to fire and rescue units, nearby hospitals, local and state police, and state emergency response agencies. In large-scale emergencies, even these channels can become overloaded, as will normal telephone lines. Cellular telephone service is an alternative becoming very widely available that does not depend upon hard-wired telephone communications. A chart is shown in Figure 2.2 which reflects this typical command center operation.

The emergency coordinator is a key individual and must be someone who will be accepted as a command figure. The individual ideally should be one to grasp information quickly, be able to integrate it, and come up with appropriate responses. Since the most often employed emergency response structure is semi-military in nature, a person often designated as the emergency coordinator will be the public safety director. In the context of laboratory emergencies, most public safety managers are likely to have had police training, not scientific training, so having knowledgeable persons present to make technically correct recommendations is very important. These may be from the safety department and individuals from the scene of the incident. In addition to the structured internal departments, major resources available at any research-oriented institution are the scientists and technicians who work there. The ones most likely to be helpful for the types of emergencies anticipated in developing the emergency plan should be identified and a master list of their office and home telephone numbers maintained. A copy of the current list should be maintained by the key internal organizations involved in the emergency response plan. A copy of the list should also be personally maintained by the key individuals in these latter organizations, both in their offices and at home. Alternates should always be designated for these key persons, so that backups are available at all times. Radios, cellular telephones, or beeper systems to allow these key persons to be reached when not at their usual locations would be highly desirable.

Organizations having the capability for a response at this level will have some type of security or police force. These individuals are very likely to be the very first "outsiders" arriving on the scene of an emergency and, as such, initiating a first response. Clearly, they need to receive sufficient training to permit them to make a "first response" evaluation of the incident and set the containment and response mechanisms in process. It is relatively rare, though, that they will have sufficient training to manage the response to technically involved emergencies. Some key personnel will ideally again have been given additional "hazardous-materials-incident" training to allow them to initiate or effect an evacuation of affected personnel and provide safety for themselves and for the evacuees, pending further response actions.

In many jurisdictions, the legal responsibility for management of incidents involving hazardous materials has been delegated to the fire department or to specialized hazardous material response teams. When these arrive on the scene, the management responsibility for an incident may shift so that the emergency coordinator, having the ultimate authority, will no longer be a representative of the organization or institution. In such a case, the internal side of the picture would shift to a supportive and/or advisory role. However, in many instances, the fire department, if that is the responsible agency to which authority is delegated, may choose to take a great amount of advice and guidance from the organization's team. Depending upon the incident, one or more regulatory agencies may need to be notified promptly. If a significant chemical release is involved which becomes airborne or involves a liquid spill such that hazardous materials escape from the controlled boundaries of a facility, the National Response Center must be notified as well as the local emergency response coordinator (often the sheriff, police chief, or civil disaster coordinator) and state agencies. Other agencies would also be called, as their areas of regulatory concern would become involved. Although these outside regulatory agencies will arrive on the scene, the responsibility for management of the incident normally remains a local responsibility, unless it truly becomes a massive problem. Written aid agreements need to have been worked out in detail between corporations and institutions with local emergency response organizations.

There are three groups identified in Figure 2.2 that have not been touched upon as yet. No major incident occurs without news media quickly arriving at the scene. Emergency personnel must not be distracted by these persons, so media contact persons or groups should be established with whom the news representatives may interact. The security or police may need to act to ensure that not only news media but other nonessential persons do not enter the area. In a mature response stage of an emergency, the role of the police will almost certainly have devolved from active management to control of the boundaries of the affected area. The emergency coordinator has to have some resources immediately at his disposal but is unlikely to have access to larger amounts. Typically, when or if these are needed, authorization will have to come from senior administrators with authority to make substantial financial commitments. Finally, communications has been touched upon in terms of contacting agencies, support groups, and the media. The communications team is also responsible to see that all occupants of an area affected by, for example, an airborne plume of a toxic gas are notified. Time may be critical, so the communications group must have procedures in place to communicate by all reasonable means using radio, TV, roving vehicles equipped with public address systems, and (if time and conditions permit) door to door searches.

A library of reference materials should be maintained for the use of the emergency responders. Following is a short summary of some of the more useful references, many of which are revised frequently. Although these are primarily printed books, today a number of other types of data information sources are becoming widely available for chemical products, primarily as a result of information needs evoked by the OSHA Hazard Communication Standard. An example of these, included in the list, are Material Safety Data Sheets, available directly from the chemical product manufacturer. These are provided when the chemical is first purchased and when significant new information becomes available. Compilations of these are sold as hard bound or looseleaf volumes or on microfiche. A large number of regulatory standards and guides are now available on CD-roms which can contain an enormous amount of information extremely compactly. Many of these provide quarterly upgrades at reasonable costs so there is little reason for not being informed.

- *ACGIH, American Conference of Industrial Hygienists—Threshold Limit Values (TLV) for Chemical and Physical Substances*
  P.O. Box 1937
  Cincinnati, OH 45201
- *Chemical Hazards Response Information* (CHRIS manuals; four books)
  U.S. Coast Guard
  400 7th St., SW
  Washington, DC 20590
- *Dangerous Properties of Industrial Materials*
  Sax, Irving
  Van Nostrand Reinhold
  New York, NY 10602
- *Department of Transportation Emergency Response Guidebook,* DOT Publication No. P5800.3
  Materials Transportation Bureau
  Research and Special Programs Administration
  U.S. Department of Transportation
  Washington, DC 20590
- *Effects of Exposure to Toxic Gases, First Aid & Medical Treatment*
  Matheson Gas Products
  P.O. Box 85
  East Rutherford, NJ 07073
- *Emergency Medical Treatment for Poisoning*
  National Poison Center Network
  125 Desoto St.
  Pittsburgh, PA 15213
- *Farm Chemicals Handbook*
  Meister Publishing Co.
  Willoughby, OH 44094
- *Fire Prevention Guide on Hazardous Materials*
  National Fire Protection Association (NFPA)
  Batterymarch Park
  Quincy, MA 02269
- *First Aid Manual for Chemical Accidents*
  Lefevre, Marc J.
  Dowden, Hutchinson & Ross
  Stroudsburg, PA 18360
- *Guide to Safe Handling of Compressed Gases*
  Matheson Gas Products
  P.O. Box 85
  East Rutherford, NJ 07073
- *Hazardous Materials*
  Department of Transportation
  Office of Secretary Transportation
  Washington, DC 20590
- *Hazardous Materials Emergency Planning Guide-NTR-1*
  Hazmat Planning Guide WH-562A
  401 M. Street, SW
  Washington, DC 20460
- Material Safety Data Sheets Master File for Chemicals in Use at the Institution.
  (Available from chemical manufacturer or generic database. Note that there are now a number of commercial providers of generic databases, either in hard copy form or in various computer formats.)

- *Merck Index*
    Merck & Co. Inc.
    Rahway, NJ 07065
- *NIOSH/OSHA Pocket Guide to Chemical Hazards,* DHHS (NIOSH) Publication No. 78-210
    U.S. Government Printing Office
    Washington, DC 20402
- *Physicians' Desk Reference*
    Medical Economics Company
    Oradell, NJ 07649
- *Prudent Practices for Handling Hazardous Chemicals in Laboratories*
    National Academy Press
    2101 Constitution Avenue, NW
    Washington, DC 20418
- *Handbook of Chemistry and Physics*
    CRC Press, Inc.
    2000 Corporate Blvd., NW
    Boca Raton, FL 33431
- *Laboratory Safety, Principles and Practice*
    American Society for Microbiology
    1913 I St., N.W.
    Washington, DC 20006

Internal resources will not always be sufficient to handle an emergency. Therefore, a list of external emergency organizations should be maintained by the organizational emergency groups as well. The following are among the most likely to be useful and available. Any others that might be useful to you and are available should be identified and added to the list. Currently available telephone numbers are given in some cases. These are subject to change and should be verified before incorporating them in a plan.

- Regional emergency group/coordinator
- Arson and/or bomb squad, if not otherwise identified
- Civil Defense coordinator, if not otherwise identified
- Commercial analytical laboratories
- Commercial environmental emergency response firms
- Law enforcement organizations, e.g., city or county Police Chief or Sheriff, state police, F.B.I.
- Centers for Disease Control, phone no. 404-633-5313
- CHEMTREC (for chemical and pesticide spills), phone no. 800-424-9300
- Compressed Gas Association, phone no. 212-354-1130
- Manufacturing Chemists Association, Inc., phone no. 202-483-6126
- National Fire Prevention Association, phone no. 617-328-9290
- National Response Center (USCG and EPA), phone no. 800-424-8802
- Nuclear Regulatory Commission, phone no. 301-492-7000
    (also state or regional federal office)
- Occupational Safety and Health Administration, phone no. 202-245-3045
    (also state or regional federal office)
- Poison Control Center, phone no. 502-432-9516

Many of these are sources of information only, and normally do not provide actual assistance for the emergency response. The ones likely to have the capability to do so are the first six. However, the two commercial groups listed represent profit-making organizations and the institution or corporation must be willing to pay for their services. Since ultimately the organization (or their insurers) will bear the bulk of the costs for the emergency response, authority must be provided to pay for these services.

2.3.3. Emergency Equipment

Another important step in preparing for an emergency is acquiring appropriate equipment, which is kept readily available for use. Some of this should be located in the laboratory area and every laboratory should be furnished with it. Other equipment, because of the cost and relatively rare occasions when it is likely to be needed, should be maintained at a central location. Even at the central location the equipment maintained needs to be realistically selected. It is neither necessary nor desirable for every organization to maintain an expensive, fully equipped hazardous-material emergency response team. Some very large organizations may find them essential but most institutions will not be able to justify the cost.

Some of the emergency equipment needs to be built in, as part of the fixed equipment in the laboratory. Included in this group are the following items:

Eyewash stations—At least one of these, meeting ANSI standard Z358.1-1990, must be placed in an easily accessible location. The travel distance to a unit should be no more than 100 feet according to the standard and travel time should not exceed 10 seconds. According to Andrew Munster, M.D., Secretary of the American Burn Association, "time is critical" and Russell Kilmer of the Polymer Products Division Of the E.I. DuPont Experiment Station in Wilmington, DE, is quoted as saying "Every laboratory in their facility is equipped with an emergency shower or eyewash station to meet their safety requirements...." It is very undesirable for an injured person, possibly blinded by a chemical, to have to find a way to units outside the immediate room. Proposed standards for disabled individuals have recently been proposed as ANSI standard 117.1-1992, establishing access clearances and other physical limits. Eyewash stations should be mounted on a plumbed water line, rather than the small squeeze bottles that are sometimes used for the purpose. The bottles do not contain enough water to be effective. OSHA inspectors are likely to cite a facility in which the bottles represent the only source of water for flushing contaminants from a person's eyes. Where plumbed water lines do not exist, such as in the field, larger self-contained units are available which do provide sufficient water flow for an extended period. Cold water itself can, be uncomfortable to the eye, so if possible the eyewash supply should have a holding tank to ensure that the water is at least near room temperature. In many of the colder areas of the country, tap water is frequently well below room temperature for several months of the year.

Deluge shower—Eyewash stations and deluge showers ideally should be installed as a unit. The standards cited in the preceding paragraph apply to emergency showers as well. Although the eyes are the probably the most critical exposed organs susceptible to damage, chemicals splashed on the face may also splash on the body. A deluge shower should be capable of delivering about a gallon per second with a water pressure of 20 to 50 psi. A common error is to plumb the unit into too small a line incapable of delivering an adequate flow. The water supply should be at least a 1-inch line. Although a floor drain is desirable, it is not essential. One can always mop up afterward.* Care must be taken to

---

* There should be a timed cutoff, however, at about 15 to 20 minutes, after which the unit would need to be reactivated. Cases have occurred where, as an act of vandalism, a deluge shower was activated and rigged so that it would continue to run. In one case, before the problem was discovered, over 30,000 gallons of water flooded the facility. The unit was in the hall outside ensure that the water from the shower cannot come into contact with electrical wiring, either directly from the shower or by coming into contact with extension cords improperly running across the floor. Again, the units should always be placed in an easily accessible location. Care is essential to maintain clear accessibility. In laboratories, many instances have been noted where limited floor space has resulted in equipment being placed immediately under the showers. The ANSI standards meeting ADA requirements for the disabled cited in the previous section must be maintained.

Fire extinguishers—OSHA requires that every flammable material storage area be equipped with a portable class B fire extinguisher. The standard does not specify the amount of a flammable material which makes a room a storage facility, so in effect most laboratories face the need to comply with the standard. The unit should be at least a 12-lb unit and it should not be necessary to travel more than 25 feet to reach it from any point in the laboratory. This specific requirement in the General Industry Standard may be preempted by the OSHA Laboratory Standard, but requirements of that standard provide for emergency response training, which is construed to include training in how to use portable fire extinguishers. If it is intended that employees may attempt to put out small fires and not simply evacuate immediately, then the employees should be trained in the proper use of an extinguisher at the time of employment and receive refresher training annually. Class B extinguishers are, of course, intended for flammable solvents. Other classes of fire extinguishers are class A, intended for combustible solid materials, such as paper or wood, class C, where electrically live equipment is involved, and class D, where reactive metals, such as sodium, are used. Combination units such as AB or ABC are available, which, although not equally effective for all types of fires, can be used where mixed fuels are involved. More information on fire extinguishers will be found in a later section.

Fire blanket—A fire blanket is a desirable unit to have permanently mounted in a laboratory. These are usually installed in a vertical orientation so that a user need only grasp the handle and roll themselves up in it in order to smother the fire. Some blankets include asbestos in their manufacture; these should not be installed, and existing units should be replaced. The concern is that they could become a source of airborne asbestos fibers, which have known carcinogenic properties. Unfortunately, the heavy woolen blankets most often used as alternatives are likely to be stolen. There are fire blankets using fiberglass or special fire-resistant synthetics instead of asbestos or wool available. If a fire situation is a distinct possibility, consideration should be given to providing a woolen blanket saturated with a water-soluble, oil-based gel. This not only protects against fires and aids in escape through an active fire, but can be useful in the emergency treatment of burn victims.

Emergency lights—Emergency lighting to enable safe evacuation must be provided by some mechanism. One alternative is to have two sources of power to the lighting circuits in a building. This can be achieved by having a second source external to the building or secondary power sources within the building, but this alternative is defeated in power outages covering a wide area. There are several alternative types of internal power sources including emergency generators; large, uninterruptible power supplies to provide power for lights for a substantial area which depends on batteries to provide power for a fairly limited interval; and individual trickle-charged battery-powered lights in individual laboratories. Generator units require frequent testing under load and thus are a maintenance problem. Uninterruptible power supplies are best suited for maintenance of power to equipment such as computers, where a controlled shutdown is almost essential. The most

---

the laboratory, another argument for placing units inside laboratories.

economical alternative especially in retrofitting an older facility is the individual trickle-charged battery-powered units that come on when the power fails.

First aid kit—One of these needs to be in every laboratory and should be kept in a pre-determined fixed location. They are intended to be used for minor injuries or basic treatment while awaiting more advanced care for major injuries. Access to appropriate emergency medical care is required under OSHA standard 1910.134. Kits should be relatively small units. Packaged units are sold that are adequate for five or six persons. There is little value in having larger units, since in the event of an emergency involving more persons help definitely will be needed from trained emergency care provider units, including rescue squads and physicians. Present in the kits should be a variety of bandages, adhesive tapes, alcohol swabs, gauze, perhaps some protective creams, and a few cold packs. Special situations could require special items to be available to provide treatment. Items such as iodine, methiolate, and tourniquets are no longer recommended for inclusion in most cases. It is essential that a maintenance program be established to ensure that the kit is always adequately supplied. It is all too easy to use up the supplies without replenishing them.

Fire alarm pull station—The location of the nearest pull station should be familiar to everyone in the laboratory.

Special safety equipment—There are many specialized research areas which require special safety items such as explosion-proof wiring, combustible gas monitors, and explosion venting for laboratories working with highly explosive gases. The possibilities are too many to dwell on at this point.

Some emergency equipment need not be built in but should be available. Among these items are the following:

Absorptive material—Probably the most common laboratory accident is a spill from a beaker or a chemical container. The volume is typically fairly small, rarely exceeding more than 4 or 5 liters and usually much less. Of course, there are spills which would require immediate evacuation of the area or even the building, but more frequently the spilled material simply must be contained and cleaned up as quickly as possible. **THIS IS NOT THE RESPONSIBILITY OF THE CUSTODIAL STAFF**. They are not trained to do it properly or safely. Spill kit packages are available commercially to neutralize acids and bases, and to absorb solvents or mercury. Although it is possible to put together similar packages, the commercial packages are convenient to obtain and store. After being used, the materials should be collected and disposed of as hazardous waste.

Personal protective equipment and janitorial supplies—Several miscellaneous items are needed to clean up an area. Among these are plastic and metal buckets, mops, brooms, dust pans, large, heavy-duty polyethylene bags, kraft paper boxes (for broken glass), plastic-coated coveralls, shoe covers, duct tape, and an assortment of gloves. If not kept in an individual laboratory, at least one set should be kept on each hall or floor of a building. Custodians have some of these materials, but they are not always available to laboratory personnel, especially outside normal working hours.

Respirators—Fumes and vapors from many irritating and dangerous materials can be protected against by the use of respirators with appropriate cartridges or filters. If operations are sufficiently standardized so that a standard respirator combination would be effective, they should be kept in an emergency kit. However, cartridge respirators are not intended for protection against materials which are immediately dangerous to life and health (IDLH). Whatever units are provided. laboratory personnel must be trained in the appropriate use of the units and the units must be maintained properly. Respirators should be assigned to specific individuals.

Supplied air escape units—Supplied air units, such as emergency squads might use, are expensive and require a significant level of training to be able to put them on quickly and use them properly. However, small air-supplied units are available at very reasonable prices which only need to be pulled over one's head and activated to provide 5 minutes of air. This is usually sufficient time in which to escape the immediate area of an accident.

Virtually any small to moderate chemical emergency can be handled with the equipment described above.

A few major items of equipment should be readily available from the safety department, fire department, security force, or perhaps the emergency medical team. Their availability is by no means certain, and the institution or corporation should maintain a set of these major items. Many of these items require special training to be used safely.

Oxygen meter—A portable meter should be available to ensure that the oxygen level is above the acceptable limit of 19.5%. It is important to be able to detect oxygen-deficient atmospheres, where the levels are significantly less than the acceptable level.

Combustible gas and toxic fume testing equipment—A number of different types of equipment are sold to test for the presence of toxic fumes. A common type, frequently combined with an oxygen meter, is a device to detect "combustible gases." Specialized units are built to detect other gases such as carbon monoxide and hydrogen sulfide. Very elaborate and, consequently, expensive units, such as portable infrared spectrometers, gas chromatographs, and atomic absorption units, can detect a much greater variety of chemicals, often to very low concentrations. A less expensive alternative is a hand pump, used to pull known quantities of air through detector tubes containing chemicals selected to undergo a color change upon exposure to a specific chemical. All of these can be used to obtain an instantaneous or "grab" reading. Where a longer duration sample is desired, powered pumps can be used to collect samples, and for some chemicals passive dosimeters can be worn which can be analyzed in a laboratory. Equipment to meet local needs should be selected. Although sophisticated testing devices are available, there are tens of thousands of possible chemical contaminants. It is essential for emergency personnel to know what to test for to ensure rapid identification. Emergency medical care may be delayed or limited to supportive treatment until positive identification is obtained. Therefore, a list of possible hazardous materials should be maintained by the laboratory and be available to emergency responders.

Supplied air breathing units—These are not to be confused with the escape units previously described and usually will be available from the fire department or, for larger institutions or corporations, from the chemical safety division of the safety department. There are two types, one of which provides air from a compressed air tank just as does SCUBA gear. The most common size tank is rated at 30 minutes, which, under conditions of heavy exertion, may last only 20 minutes or less. The second type uses pure oxygen recirculated through a chemical scrubber to extend the life of the supply to 1 hour, or longer for some units. The first of these two types have been available for a longer period and more emergency personnel have been trained to use them. The second does offer a significantly longer working interval. This could be very important. Pairs of either type should be owned so that in the event an individual entering an emergency area is overcome, it would be possible to effect a rescue.

Fire-resistant suits—Special fire-resistant suits are needed to enter burning areas. There are different grades of these which provide varying degrees of protection to fire. Some protect against steam or hot liquids as well. They normally require a self-contained supplied air system to be worn during use.

Chemical-resistant suits—Protection is frequently needed in chemistry incidents for protection against corrosive liquids and vapors. In standardized situations, these can be custom selected for maximum protection for the specific chemicals of concern. Where a variety of chemicals such as acids, bases, and frequently used solvents are involved, a butyl rubber suit is often a reasonable choice. Combination units of these and fire entry suits are available.

Clean air supply system—An alternative to self-contained air or oxygen tanks is a compressor system capable of delivering clean air through hoses from outside the area involved in the incident. Persons inside the work area would wear masks connected to the system. Personal air-powered units are available which use small, battery-powered packs to draw local air through a filter and maintain a positive air pressure within the face mask.

High-efficiency particulate and aerosol (HEPA) filtered vacuum cleaner—Ordinary vacuum cleaners, including wet shop vacuums, do not remove very small particulates from the air. They remove larger particles, but the smaller ones pass through the internal container or filter and return to the room. In several instances this can actually worsen the situation. For example, droplets from a mercury spill can be dispersed back into the air in the form of much smaller droplets and cause the mercury vapor pressure in the air to increase. (A mercury vacuum has additional special requirements.) In another actual case, in a carpeted room where large quantities of forms and computer paper were processed, vacuuming with an ordinary vacuum cleaner during normal working hours increased the number of respirable particles suspended in the air to a level such that several individuals who were allergic to dust had to be sent out of the area. HEPA filters will remove 0.9997 of all particles from the air which have a diameter of 0.3 microns or greater. They will remove a smaller fraction of particles of smaller sizes, but the smaller particles have difficulty reaching the deep respiratory system, so they are less of a problem.

Radios—Radio communication between persons entering an accident area and those outside is highly desirable. Emergency groups will have portable radios with frequencies specifically assigned to them. Cellular phones are a recent alternative which provide access through the telephone system to virtually any external resource.

Fire suppressant materials—In addition to water and the usual materials available in portable fire extinguishers, most fire departments now have available foam generators which can saturate a fire area.

Containment materials—In order to prevent the spread of large amounts of liquid chemicals, a supply of diking materials needs to be maintained. Ready access to a supply of bales of straw is a great asset. Straw is cheap, easily handled, and easy to clean up afterward. In the event of a spill reaching a stream, floating booms and skimmers are useful in containing and cleaning up the spill. Booms are not effective for materials more dense than water and not water soluble.

Radiation emergency—Many laboratories use radioisotopes. For emergencies involving these units, in addition to the other emergency equipment, radiation survey instruments should be available or maintained in an emergency kit. The radiation safety office will be able to supply additional units. These should include instruments capable of detecting both low levels of gammas and low energy betas as well as high-range instruments. Although low levels are not necessarily dangerous, normally very restrictive levels of contamination are permissible under established safety limits for most organizations, according to ALARA (**A**s **L**ow **A**s **R**easonably **A**chievable) guidelines.

Miscellaneous clothing—Items needed include a variety of coveralls, including (but not limited to) chemically resistant suits in a range of grades; disposable Tyvek™ coveralls; gloves with different chemical resistances; regular work gloves; Kevlar™, Nomex™, or Zetex™ gloves for hot use; rubber and neoprene boots and shoe covers; head covers; hardhats; chemical splash goggles; safety glasses; and masks.

Miscellaneous tools and paraphernalia—A variety of small tools could be needed as well as shovels, pickaxes, axes, rope, flares, emergency lights, sawhorses, a bullhorn, a chain saw, a metal cutting saw, a bolt cutter, and a "jaws of life" metal spreader. Special nonsparking tools may be required where sparks may ignite vapors.

Victim protection—In equipping an emergency kit, the emphasis is usually on protecting the emergency response personnel. In order to bring a victim out through a fire or chemically dangerous area, blankets, disposable coated Tyvek™ overalls, loose-fitting chemically resistant gloves, and the 5-minute escape air units should be available.

All of the equipment listed in this section must be maintained properly, and a definite maintenance schedule must be established. For example, the integrity of the chemical protective suits must be verified on a 6-month schedule. A maintenance log must be kept in order to confirm that the maintenance program has been done on schedule.

A fire hose is specifically *not* included as a desirable item of emergency equipment that should be available to the usual occupants of a building. Although standards are provided in OSHA for fire brigades, in general, if a fire is sufficiently large to require a fire hose to control it, it is usually too large for anyone except professionals. Building codes frequently require installation of 1 1/2-inch emergency hose connections; in many instances building officials encourage the owners to request a variance to permit this requirement to be deleted. Many fire departments question the value of such connections or, even if available, whether a hose of this size would be sufficiently useful. Those institutions or corporations that do choose to establish a fire brigade will need to provide training beyond the scope of this book.

2.3.4. Basic Emergency Procedures

A list of several common types of emergencies that might occur in a laboratory was given in the introductory section to this chapter. Many of these emergencies, as well as others not mentioned in the list, share common characteristics of the initial response which are important to do *first*. The following material will, for the most part, be in the context of a fire incident, but the recommendations would be the same if a substantial release of a toxic chemical were released and became airborne.

1. Make sure everyone in the immediate vicinity is made aware of the problem. In a busy, active laboratory, an accident can occur in one part of the laboratory and personnel in other areas within the same laboratory could be temporarily unaware of the event. This is especially likely if the space is subdivided or if there are no obvious effects associated with the event, such as a loud sound from an explosion.

2. Confine the emergency, if reasonably achievable. Many emergencies can be readily confined if quick action is taken. Small quantities of a spilled chemical can be contained with absorbent materials or toweling by the persons directly involved if the chemical is not immediately dangerous to life and health (IDLH). Individuals should be trained to take these actions, and appropriate containment materials for the materials in use should be conveniently available. In the event that the emergency includes a fire, laboratory personnel, if properly trained, can and should put out small fires with portable fire extinguishers, but a very serious question of judgment is involved. What, precisely, is a small fire? Unless there is a reasonable certainty that the fire can be controlled, then evacuation of the building should be strongly considered and implemented as soon as the situation appears to be deteriorating. Time is likely to be critical if the volume of solvents often available as fuel in a typical laboratory is considered. If more than one person is available, there may be

more flexibility. One or more persons may attempt to contain the fire, while others are taking initial steps to evacuate the building. Where it is necessary to evacuate an area larger than a single laboratory, the building's evacuation plan should include measures to ensure that all spaces are checked, including restrooms, janitors closets, etc.

3. Evacuate the building. Whenever the situation is obviously serious, such as a major fire, a moderate-to-large spill of an IDLH material, a rupture of a large gas cylinder, or large spills of ordinarily dangerous materials, such as strong acids, then evacuation procedures for the area or the building must be initiated as soon as possible. Any measures taken in such a case to confine the emergency situation should provide extra time for the evacuation to be carried out safely.

Evacuation is a conservative step and should be implemented whenever any doubt exists of the severity of the situation at hand. It is inconvenient and is disruptive to work activities, but the alternative is far worse if a fire cannot be controlled. The first few minutes of a fire are very important and any significant delay can make the job of the fire department much more difficult. Once a fire takes hold, it is often very difficult to bring under control. In a laboratory situation, the involvement of additional chemicals can convert a straightforward fire into one which could involve the generation of extremely toxic vapors. Most fire departments are inadequately trained to handle complex chemical fires. Their chemical incident training usually includes situations involving only a single material. Even if the fire is out before they arrive, there are things that the fire department needs to do. They need to check the area to ensure that it is really out. A fire department is usually the organization charged with the legal responsibility for managing and terminating hazardous material incidents. They also need to determine the cause of the fire in order to prepare an accurate report. The information they obtain will be needed to determine how to prevent subsequent fires due to the same cause. Where property and personal injuries losses occur, their report will normally be needed by the insurer of the property to determine the amount of cost recovery available.

Normally, primary evacuation routes from an area within a building should follow the shortest and most direct route, along corridors designed and constructed to meet standards for exitways. However, since in an emergency, any given path may be blocked, alternate secondary routes should be designated. In no instance should an evacuation plan include elevators as part of the evacuation procedure, even for a disabled person. In the case of a fire, elevators should be designed to immediately go to the ground floor and be interlocked to stay there until the danger is over. There are convenient evacuation chairs which a single individual can use to assist disabled persons. One of these should be available and one or more persons designated to provide the required aid.

In any evacuation procedure, provisions for closing down operations should be included, *if* there is sufficient time to implement them. Gas should be turned off, along with electric and other types of heaters. Valves on gas cylinders should be turned off, especially if they contain flammable or toxic materials. High voltage equipment should be turned off. Closing sashes on fume hoods may be desirable. Certainly, any flammable material storage cabinets should be closed.

Even in the worst situation, there are some simple things which can be done by individuals evacuating the building to confine and minimize the emergency. The highest priority is to protect personnel, so the first thing is to actuate the building alarm, assuming that one exists. If not, then air horns should be used or, failing that, verbal warning. Doors to the laboratory should be closed on the way out. Doors between floors should be closed behind those evacuating. Stairwells serve very well as chimneys to carry smoke and fire to upper floors if the doors are not closed. If the building has been built according to code, as briefly discussed earlier in this chapter and covered in much more detail in Chapter 3, then these last two simple steps can significantly retard the spread of a fire or spread of fumes.

If a laboratory is under negative pressure, as most should be, then the negative pressure will also tend to confine the emergency to a single room. In order to maintain a negative pressure, it may be desirable to leave the sash of a hood open or to leave the hood working, even though there might appear to be concern about the fire spreading through the hood duct. If a hood has been installed properly, the exhaust will be at a negative pressure with respect to the space surrounding the exhaust duct and, as noted earlier, is either going directly outside the building without passing through an intervening floor level or is enclosed in a fire-rated chase. Under either of these conditions, fire being drawn through a hood exhaust should not cause fire to spread to other floors. The door to the room being closed will further reduce the amount of fresh air available to support a fire. If an air exhaust is turned off, any air intake should also be turned off to avoid creating a positive pressure in the room and thus possibly causing extension of the emergency by leakage into corridors.

Evacuation should be done as quickly as possible, but in such a way as to not engender a panic situation. This can best be achieved by having it be a frequently practiced procedure, so that everyone is familiar with the routes. In a corporate situation with a stable personnel complement in the building, drills two or three times a year will quickly accomplish the purpose. In an academic environment, the problem is much more complicated. In most colleges and universities, as many as 8 to 12 classes per day may be held in the same classroom. Classrooms may be assigned by some central authority. This may result, for example, in a professor of economics being assigned a class in a chemistry building for one quarter during a year, and who may not have a class in that building for the remainder of the year. During the course of an academic year, the population in the building may change in large part every quarter or semester. Because of all these complicating factors, a single drill per academic session could prepare as little as 10% of the population in a building for an actual emergency. Under these circumstances, unusual care should be taken to clearly mark evacuation routes from buildings, and to train those individuals who form the "permanent" population in the building to take charge during an evacuation.

Complicated maps placed at intersections to show evacuation routes are often used, but are difficult to read and interpret in the press of events occurring during a serious emergency situation. A simple but very effective evacuation system is illustrated in Figures 2.3 and 2.4. A distinctive, high contrast, standardized symbol, employed only for marking primary evacuation routes, is placed directly across a corridor from every door opening onto a corridor, at appropriate intervals (30 to 50 feet) along the corridor, and at every branching point along the path of egress. A person totally unfamiliar with a building need only follow the symbols to be conducted to the nearest exit. Smoke tends to rise, so these should be placed a short distance off the floor, so that they would remain visible when signs placed above doors might be obscured. Power can fail, even in buildings equipped with separate emergency power for lights, so the directional symbol can be made with a phosphorescent paint and will remain visible for 2 or more hours, this being ample time to evacuate almost any building. Printing of the signs on a fragile substrate which cannot be removed intact will minimize the theft of the signs to be used as decorations in dormitories and residence halls. This system should be used as a supplement to a code-conforming system of exit lights rather than a substitute. The maps mentioned above are useful when time permits.

A standard part of any emergency evacuation plan should include a previously chosen point of assembly for those evacuating. This should be a location generally

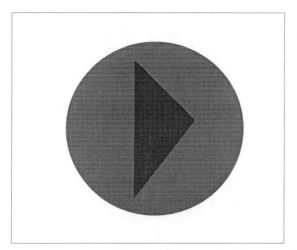

Figure 2.3. A simple sign such as this is easily recognizable as a directional guide.

Figure 2.4.   The directional sign, placed about 2 feet above the floor, at opposite doors, and at about 30-foot intervals along the evacuation route, provides a guide to an exit. If the arrow segment is phosphorescent, it would be visible even in a power failure.

upwind from the building being evacuated. Obviously, the wind does not always blow from the same direction, so alternative locations should be chosen. Those individuals most directly involved with the emergency, and presumably the most aware of the circumstances, should remain at the evacuation location and make themselves known to the emergency response groups when they arrive in order to assist these groups.

There should be a clearly defined line of communication among the persons responsible for the facility. Individuals in authority and with assigned responsibility for the space involved should also go to the assembly point, if not already on the scene, and remain available to assist the emergency personnel in managing the evacuees. Often the police will be among the first emergency groups arriving on the scene and their aid will be invaluable in crowd control. If other senior corporate or institutional officials involve themselves, they should inform the building authorities that they have arrived and they may wish to assume responsibility, although they will typically be less knowledgeable of local circumstances than building and laboratory officials. Procedures should be employed to account for the individuals who are known to have been in the building, and reports of any persons suspected to still be in the building should be made to the authority figures to pass on to the emergency responders. Any decision to allow reentrance to the building approved by the fire department or other emergency group should be disseminated by these persons.

4. Summon aid. Current building codes normally require building alarms to be connected to a central location manned 24 hours per day. Unless the building alarm system is connected to a central station, pulling the alarm will not alert any external agencies. The more sophisticated systems commercially available will pinpoint almost the exact location in a building and will even provide a map showing the best route to the facility for the emergency group. Such systems greatly facilitate the response of an emergency group. Unfortunately, current building codes apply only to new construction, so older buildings may have much less sophisticated systems or none at all. In such cases, using the almost universally available 911 system, would have to suffice.

It is imperative that requests for assistance be initiated quickly. However, in a serious, life-threatening emergency, evacuation should not be delayed to call for assistance. Calls can be made from a point outside the area affected by the emergency. Where personnel are available, one individual can be designated to make the appropriate telephone calls while others are engaged in other aspects of the emergency response.

As noted earlier, a key group in responding to most institutional or corporate emergencies is the police department. If they are not the group initially contacted, it is probable that even small departments monitor emergency radio frequencies and will arrive either at nearly the same time as the emergency group summoned or even before. They should have training in a number of key areas, such as how to use fire extinguishers most effectively and how to give first aid and CPR, and should have the capability of independently causing a building to be evacuated. The police should also have at least basic hazardous materials training (usually called "first responder haz-mat" training). This basic training essentially trains them in how to identify a hazardous material incident situation. It does not provide for managing the incident response. It would be desirable if at least a cadre of a police department could receive higher levels of training. Once the fire department, rescue squad, or other emergency group arrives and assumes the responsibility for their duties, the police are needed for crowd control and communications.

In any type of incident in which a spill of a hazardous material has occurred, the standard procedure is to establish a command center outside the periphery of the area affected by the accident and establish a controlled access point for emergency response and, later, decontamination personnel entering the area. All materials and personnel entering and leaving the area should pass through the control point. When the remediation stage is reached, unless there are overriding considerations, decontamination should begin at the periphery and the work program designed to progressively constrict the affected area.

Everything collected at the control point, including waste materials, contaminated clothing, and equipment which cannot be cleaned and reclaimed, should be immediately packaged for disposal according to standards applicable to the contaminant. Information and status reports should flow to the command point and overall direction of the response should come from the command center. Emergency response, to be effective, needs central coordination and a clearly defined chain of command.

2.3.5. Emergency Procedures for Selected Emergencies

2.3.5.1. Spills

A chemical spill is probably the most common accident in the laboratory, and in most cases can be cleaned up by laboratory personnel with minimal effort or risk. According to the requirements of the OSHA Hazard Communication Standard, laboratory personnel should be trained in the risks associated with the chemicals with which they are working and should know when it is safe to clean up a minor spill. Workers should be especially sure to be familiar with the risks and the corrective actions to be taken in an emergency for chemicals labeled on the container "DANGER" or "WARNING." If personal protective equipment is needed, personnel required to wear it must receive appropriate training in how to use and maintain the equipment.

Paper towels, assuming that the paper would not react with the spilled materials, or absorbent and/or neutralizing materials can be used to clean up minor spills, with the residue being placed in an appropriate container for later disposal. Work surfaces should be nonabsorbent and chemically inert, but in order to avoid possible decontamination of work surfaces, it is often convenient to protect them with plastic-backed absorbent paper. Relatively few materials associated with a cleanup should be placed in the trash. Most should be disposed of as hazardous waste. Individuals (including students as well as employees) handling the waste materials should have documented training in the handling of chemical wastes in order to conform with both OSHA and RCRA standards. Chemical containers should not be placed in ordinary trash for disposal. In one actual instance, two different, partially empty chemical containers placed in the ordinary trash in two separate buildings combined in the trash vehicle and resulted in a fire. The fire caused the worker on the truck to be overcome with fumes, requiring emergency medical treatment. The nature of the fumes was unknown until later, when it was possible to retrieve the containers. It should be a general policy that no chemical container should be disposed of in the general trash. Trash handlers have become very concerned when they see containers with chemical names and chemical hazards on the labels.

Even small spills can often be dangerous if the spilled chemical interacts with the body. Strong acids and bases, as is well known, can cause serious chemical burns to tissue by direct contact. Eyes are especially vulnerable. Chemicals can cause serious injury by ingestion, inhalation, or absorption through the skin. For example, phenol is readily absorbed through the skin, and in relatively small quantities is quite toxic. Vapors from some spilled materials are IDLH by inhalation, even in small quantities. Obviously, work with such materials should be done in a hood where the sash can be lowered should a spill occur. However, if an accident occurs outside a hood involving these very hazardous materials, the area should be evacuated and the door to the laboratory closed and help sought from persons trained and equipped to cope with such dangerous materials. In some instances, if a material is sufficiently volatile to give off enough vapors to be dangerous from a small spill, it is often sufficiently volatile so that it will quickly evaporate. Evacuation will allow time alone to effect a remedy by allowing the vapors to be exhausted through the facility's ventilation system.

In most cases flushing the area of the body affected by a splash of a liquid chemical with copious amounts of low pressure water for 15 to 30 minutes is the best immediate treatment. The best source would be an eyewash station or deluge shower, but in an emergency, if these are not available, any other source of running water (at low pressure) should be used. If the exposure is to the eyes, check for contact lenses and remove them if found and if possible (although contact lenses should never be worn in a chemical laboratory), and hold the eyes open while they are being flushed with water. Any clothing or jewelry in the affected area should be removed to ensure thorough cleansing. No neutralizing agents should be employed. If the original exposure was due to a dry chemical, normally the best course would be to brush off loose material and then follow the same course of action.

While washing is taking place, emergency medical help should be summoned, normally by calling 911. Chemical injuries, due to their possible complexity, probably should elicit a response from a crew capable of providing advanced life support-level care. If a severe physical injury has occurred in addition to the chemical exposure, appropriate first aid measures should be taken while waiting for aid. In order of priority, restoration of breathing and restoration of blood circulation, stopping severe bleeding, and treatment for shock should be done first. These injuries are life threatening. Training in these techniques are available from many sources, such as the Red Cross, the American Heart Association, local rescue squads, and hospitals, usually at minimal or no cost.

Persons involved in the accident or the subsequent treatment of the injured person or persons should remain at the scene until emergency medical aid arrives. It is important that those treating the victim know what chemical was involved. In addition, the persons providing assistance can provide emotional support to the victim. Generally, it is preferable that transport to a hospital be done by the emergency rescue personnel. They are not only trained and qualified to handle many types of medical emergencies, but they will also have communication capability with an emergency medical treatment center. Through this radio contact, they can advise the emergency center physician of the situation and the physician can instruct the emergency team of actions they can initiate immediately. In addition, if special preparations are needed to treat the injured person upon arrival at the emergency center, these can be started during the transport interval.

Some materials, such as mercury, do not appear to pose much of an obvious hazard upon a spill and a cursory clean up may seem to be sufficient. However, mercury can divide into extremely small droplets and get into cracks and seams in the floor and laboratory furniture. Mercury remains in metallic form for a long time after a spill, capable of creating a significant concentration of mercury vapor pressure in a confined, poorly ventilated space. Exposure to these fumes over an extended period can lead to mercury poisoning. After gross visible quantities have been cleaned up by carefully collecting visible drops, preferably with an aspirator, absorbent material specifically intended to absorb mercury should be spread on the floor and left there for several hours. Afterwards, the area of the spill should be vacuumed with a special version of a HEPA filtered vacuum cleaner *adapted for mercury cleanup.* A penknife can be used to check seams in floor tiles and cracks to check if the cleanup has been thoroughly done.

The preceding material on spills assumed that the incident only involved one chemical. Figure 2.5 shows what could have been, but miraculously was not, a major disaster which could have injured several persons and even have resulted in the loss of a major academic building. A set of wall shelves, loaded with a large variety of chemicals, collapsed while no one was working in the area. Unlike the incident referred to earlier (where two small

Figure 2.5. Accident due to poorly installed and insufficiently strong shelves.

quantities of chemicals, discarded in two different buildings, reacted and injured a non-laboratory worker) here, although several bottles broke with chemicals becoming mixed, no reaction occurred and the damage was limited to the loss of the chemicals. If a vigorous reaction had occurred between the contents of any two of the broken bottles, the resulting heat might well have caused more of the unbroken containers to have ruptured and a major disaster could have resulted. Where multiple chemicals are involved, the same techniques as those used in a simple incident should be applied, with the stipulation that unnecessary mixing of chemicals should be carefully avoided.

Spills which result in a substantial release of toxic liquids or airborne vapors such that the release extends beyond the facility boundaries invoke the requirements of the Community Right-To-Know Act. Notification of the local emergency coordinator by the dispatcher would be the first legal step to get the mechanisms moving.

While all of the corrective measures are being taken, the affected area should be secured to ensure that no one is allowed in who is not needed. "Tourists" are not welcome. If necessary, help should be obtained from security or police forces to exclude nonessential persons.

2.3.5.2. Fire

A second common laboratory emergency involves fire. Laboratory fires stem from many sources, the ubiquitous Bunsen burner, runaway chemical reactions, electrical heating units, failure of temperature controls on equipment left unattended, such as heat baths, stills, etc., overloaded electrical circuits, and other equipment. With a fire, the possibility of the immediate laboratory personnel being qualified and able to cope with the emergency depends very strongly on the size of the fire. As indicated earlier, only if it is clear that the fire can be safely put out with portable extinguishers should a real attempt be made by laboratory personnel to do so. However, trained personnel temporarily can use portable extinguishers for moderate fires which are not gaining ground rapidly to gain time to initiate evacuation procedures.

In order to use an extinguisher effectively, laboratory personnel must receive training in their use. If possible this training should include hands-on experience. They should be familiar with the different types of extinguishers and the type of fires for which they would be effective.

Class A extinguishers are intended to be used on fires involving solid fuels such as paper, wood, and plastics. Generally, a class A extinguisher contains water under pressure. Water acts to cool the fuel during the extinguishing process, which has the advantage that the fuel has to regain kindling temperature once the fire has been put out. The large amount of energy required to convert liquid water into vapor places an added burden on the energy requirement to rekindle the fire in wet fuel. An extinguisher rated 1A is intended to be able to put out a fire of 64 square feet if used properly. A typical extinguisher will throw a stream of water up to 30 to 40 feet for approximately 1 minute.

Class B extinguishers, intended for use on petroleum and solvent fires, usually contain carbon dioxide or a dry chemical, such as potassium or sodium bicarbonate. The first of these puts out the fire by removing one of the essential components of a fire, oxygen, by displacing the air in the vicinity of the fire. The second uses a chemical in direct contact with the burning material. Some chemical extinguishers contain materials such as mono-ammonium phosphate or potassium carbamate, which even in small sizes have very impressive ratings for putting out a solvent fire. Chemical extinguishers are messy and can damage electronic equipment. Typical dry chemical or carbon dioxide portable units last on the order of 15 to 30 seconds, and in the case of carbon dioxide units it is necessary to be within 10 feet of the fire to use them effectively. A third type of unit, which does not have this latter negative characteristic, contains one of a class of chlorinated fluorocarbons called Halon™. The Montreal Protocol, regulating chlorinated fluorocarbons because of the deteriorating effect of these materials on the atmospheric ozone layer, will eliminate the two major types of Halon™ within a relatively few years. For the time being existing systems will continue to be acceptable, but replenishing units will become increasingly difficult as existing stocks are depleted. These two types are Halon™ 1211 and Halon™ 1301, distinguished chiefly by the fact that the first of these operates at a lower pressure than the second and thus is more common as a portable extinguisher. Permanently installed systems have tended to be Halon™ 1301. Both types work by interrupting the chemistry of the fire; however, Halon™, being gaseous, can be dissipated easily. Once the air concentration falls below the level at which it is effective, it no longer provides any residual fire protection. One way in which the Halon™ units have been used effectively has been to install them in small storage rooms as ceiling-mounted units. Reasonably priced units were available which went off automatically at temperatures set by fusible links in the heads of the units. Alternatives to these two types are being sought and hundreds of compounds have been tested. One appears promising. CF_3I appears to be a drop-in replacement for Halon™ 1301, but testing for corrosion and toxicity as well as its possible effect on stratospheric ozone is too incomplete to be sure that it can replace the older materials.

Class C extinguishers are intended for electrical fires, which, because of the potential shock hazard, preclude the use of water. Many class B extinguishers are also rated for use on electrical fires.

Class D extinguishers are used primarily for reactive metal fires and a few other specialized applications. Due to the extra cost of these units, only those laboratories which actively use reactive metals need to be equipped with class D units.

As has already been noted in several instances, training is required to use a portable extinguisher effectively, since the available supply of fire suppression materials last less than 1 minute in most cases. To be most effective, the extinguishing material should be aimed at the base of the fire and worked from the point immediately in front of the extinguisher operator progressively toward the rear of the fire, away from the operator. If more than one person is present, additional extinguishers should be brought to the scene so that as one is used up, another can be quickly brought into use to prevent the fire from regaining vigor. More than one unit at a time can be used, of course. About half of all fires that can be put out with portable extinguishers require only one, but conversely, the other half require more than one.

To be effective an extinguisher must be full. Units can leak, and unfortunately individuals with juvenile mentalities apparently feel that extinguishers are toys, provided for their amusement. This seems to be an attitude especially prevalent on college and university campuses (most of the problems exist in resident dormitories, but not exclusively so). Therefore, extinguishers in laboratories should be checked frequently by laboratory personnel as well as by fire safety staff. If the unit has a gauge, it should be in the proper range. Empty and full weights are indicated on the extinguisher, so weighing will confirm if the unit is full or not. Breakable wire or plastic loops through the handles, which are broken when the unit is used, should be checked to see if they are intact. If a loop is found to be broken, the unit should be checked. Any units found to be discharged should be replaced immediately, preferably as a practical matter within one working day.

Since a hood is where most hazardous laboratory operations should be carried out, a substantial number of laboratory fires occur in them. In the event of a fire in a hood, a simple and often effective procedure to control the fire is to close the sash. This serves two purposes: it isolates the fire from the laboratory and reduces the amount of air available to support combustion. Since a properly installed hood exhausts either immediately to the outside or through a fire-rated chase, in many instances a fire in a hood can safely be left to burn itself out, or at least can reasonably be counted upon not to spread while an extinguisher is obtained. If the risk of a fire within a hood is substantial, automatic extinguishers are available that can be mounted within the hood.

In the event a person's clothing catches on fire, it is important not to run because this provides additional air to support the flames. Many authorities recommend that a person aflame should roll on the floor to attempt to smother the flames. In a crowded laboratory there is often a risk of involving solvents and other materials in the fire, however. A deluge shower is an effective way to put out the fire if it is in the immediate area, or, if a fire blanket is available, the fire can be smothered by the person quickly wrapping himself in it. If others are present, they can help smother the flames or they might employ a fire extinguisher to put the fire out. As with any other type of injury or burn, call for emergency medical assistance as quickly as possible. Perform whatever first aid is indicated, if qualified, while waiting for assistance.

2.3.5.3. Explosions

Among many other possibilities, an explosion may result from a runaway chemical reaction, a ruptured high-pressure vessel, reactive metals coming into contact with moisture, degraded ethers set off by friction or shock, or perhaps ignition of confined gases or fumes. Fortunately, explosions are less common in the laboratory than a fire but they still occur too frequently. Where the potential is known to be appreciable, the use of protective shields and personal protective equipment should be mandatory. Heavy gloves with gauntlets will offer protection to arms and hands. A mask *and* goggles should be used to protect the eyes, face, and throat. When an explosion does occur, in addition to the shock wave and the extreme air pressures which also may occur, flying debris, possibly secondary fires, and spilled chemicals may exacerbate the situation and feed a fire or lead to further reactions. Often there are toxic fumes released which may be the most serious hazard, not only to the persons immediately involved but to others outside the area and to emergency personnel. Initiation of procedures to handle resultant fires and chemical spills are appropriate if the situation is manageable. The most likely physical complications are personal injuries, including injuries to the eye, lacerations, contusions, broken bones, and loss of consciousness. Toxic fumes may cause respiratory injuries, possibly leading to long-lasting, permanent effects, possibly even death. In addition, chemicals may be splashed over the body even more extensively than in a spill, so it may be even more imperative to wash them off. However, it is essential to establish priorities. If breathing is impaired, artificial respiration should be administered, and if heavy bleeding occurs, pressure should be applied to the wound to stop it. These two problems are immediately life threatening. If there is time, and if it appears safe to do so, i.e., it does not appear that the spine has been injured or that other injuries will be worsened by the movement, then injured persons should be removed from the immediate vicinity of the accident. This is partially to protect the rescuer as well as the victim from the effects of chemicals, fumes, and smoke. Basically the same criteria apply as in a fire. Unless it is possible to safely handle the situation with the personnel present, then at least the immediate area should be evacuated, if necessary, the building as well, and the fire department and other professional aid summoned. Care should be exercised by the emergency responder that in their efforts to assist injured personnel, that they do not incur injuries to themselves, such as coming into contact with spilled acids.

For most fire departments, a fire or an explosion in a laboratory represents an uncommon occurrence. It would be highly desirable, in the absence of a knowledgeable person immediately on the scene, if information on the contents of the laboratory could be found posted either on the door or close by. Preferably this information should be brief, legible from a distance, and be in a format already familiar to fire personnel. Many localities have attempted to meet these needs by requiring the laboratory to be posted with the NFPA universal hazard diamond in which the degree of danger for reactivity, flammability, and health effects are indicated by a numerical rating, with the numerical rating referring to the contents of the laboratory instead of a specific chemical.

An example of an NFPA symbol is shown in Figure 2.6. There are four small diamonds, which together are assembled into a larger one. The four smaller diamonds are blue for health or toxicity, red for flammability, yellow for reactivity, and white for special warnings, such as radiation or carcinogenicity. Printed in each segment is a prominent black number showing the degree of hazard involved, ranging from 0 to 4.

The numerical ratings are

0 = according to present data, no known hazard
1 = slight hazard
2 = moderate hazard
3 = severe hazard
4 = extreme hazard

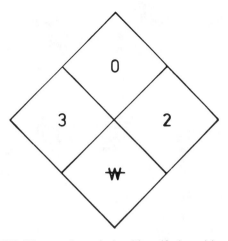

Figure 2.6. NFPA Diamond symbol, with sulfuric acid as an example.

Although this system appears simple, it is difficult to implement meaningfully in practice. In a typical laboratory, there may literally be hundreds of chemicals on the shelves. How should the rating for the laboratory be established? Should it be determined by the rating of the worst material present for each category or should the rating also depend upon the total amount of each of the chemicals present? For example, if the most flammable chemical present in a lab were ether, there would be a substantial difference in risk to firemen responding to a laboratory fire where the amount present was a single 500-mL container compared to one in which several 200-L containers were present. For a worst case criterion, both would have the same flammability rating. An alternative would be a subjective rating, combining both the worst-case type of chemical with the amount present to give a rating which in the judgment of the individual doing the rating properly takes into account both factors.

Another problem with the use of the NFPA symbol alone is that it may be too concise. Obviously, it does not inform fire personnel of what is present. Under SARA Title III, corporations and institutions are required to provide information to the fire department on the locations and quantities of their hazardous chemical holdings. However, there are some important exceptions, one of the most important of which applies to research laboratories. A hazardous chemical used in a laboratory, under the direction of a competent scientist, even in excess of a reportable quantity (which may range from 1 to 5000 pounds, depending upon the chemical), need not be reported. Reporting all of the contents of laboratories in a major research facility could overwhelm the ability of a fire department to absorb data. In a major research institution, there may be literally hundreds of laboratories, each with potentially hundreds of different chemicals, with the inventory changing daily. Although a record of the contents would be helpful, even if not completely current, it would be very clumsy to use as a first response tool. In a later chapter, a means of providing computerized data to emergency groups will be discussed.

Possible alternatives that offer the advantage of providing information in a form with which fire departments are familiar would include posting symbols such as those shown in Figure 2.7, or use the DOT placard system, with a space to fill in the approximate amounts of each class present. An advantage of this latter alternative is that every fire and rescue group will have an emergency response guide in their vehicles at all times. The recommended responses are very conservative, which is generally desirable.

Figure 2.7.   Additional symbols such as these posted on the outside of laboratory doors can inform emergency responders of possible risks and needed precautions within the facility.

2.3.5.4. Toxic Air Quality

An uncommon laboratory emergency situation that does need to be mentioned because it often leads to a fatality when it occurs is the danger of entering a space filled with a toxic gas or which is deficient in oxygen. OSHA has recently issued an updated confined space standard which offers some guidance, although most laboratories would not be anticipated to fall under the provisions of the standard. However, as a result of a fire, a spill of an IDLH substance, a leaking gas cylinder, or an improperly vented experiment releasing toxic fumes, it would be possible for a laboratory to be full of fumes and gases which would be fatal. Even a cylinder full of a nontoxic gas, such as nitrogen, can rupture and displace normal air sufficient to cause asphyxiation. The common practice of riding in an elevator with a 30-L dewar full of liquid nitrogen could prove fatal should the dewar rupture. The volume of the elevator is small, there is no rapid means of escape, and the speed of many freight elevators could mean that it could take far longer to reach the intended floor than most persons could hold their breath. The same concern could exist in riding elevators with full gas cylinders. Not all gases which may be found fairly commonly in use in the laboratory have adequate warning properties. No one should enter a space where this could conceivably be a problem without using a self-contained air breathing apparatus, nor should an individual go in such a space without others being aware of it. There should always be a backup set of self-contained breathing equipment with personnel available, trained, and able to use it to effect a rescue if necessary.

2.3.5.5. Radioactive and Contagious Biological Material Releases

Releases of radioactive material and active contagious biological materials represent two different types of emergencies which cause unusual concern because of the potential danger perceived by the public of the problem spreading beyond the immediate scene. In almost every instance, the levels of these two classes of materials used in ordinary laboratories are sufficiently small that the risk to the general public is minimal. However, in both cases, because of the public sensitivity to risks associated with these two classes of materials, unusual care must be taken in responding to the emergency.

2.3.5.5.1. Biological Accident

In recent years, the Centers for Disease Control has established a system of classification of laboratories for biological safety defining biological safety levels 1 through 4. Research with organisms posing little or moderate risk, requires only level 1 or 2 facilities, which are essentially open laboratories. Work with organisms which do pose considerable or substantial risks, requires level 3 or 4 facilities. A characteristic of both level 3 and 4 laboratory facilities is that they are essentially self-contained, with entrance through an anteroom or airlock and with access restricted to authorized personnel. This has greatly limited the possibility of an accident spreading beyond the confines of the facility. The major risks are accidents that cause direct exposures to individuals working in the laboratories. The facilities, especially those intended for higher risk use, are built to allow ease of decontamination to minimize the chances of a continuing source of infection in the event of a spill. Whenever a possibly infectious spill occurs, the immediate emergency procedure is to obtain medical care for the potentially exposed person as quickly as possible and to perform tests to determine if in fact the person involved has received the suspected exposure. A baseline medical examination (including a medical history) for each employee at the time of employment, with a serum sample taken for storage at that time, is of great value for comparison at the time of an accident. Because there may be delayed effects, records of any suspected incident need to be maintained indefinitely. As long as contaminated materials removed from the facility are autoclaved or double-bagged followed by incineration, there is little risk to the general public from laboratory research involving biological materials. Recent concerns about the disposal of infectious waste or "regulated medical waste" (as is now becoming the acceptable term) have caused a major increase in research into alternative means of rendering these types of waste harmless and *unrecognizable* by the general public. Materials made biologically safe by steam sterilization would still have to be mechanically processed to change their appearance. The concern, of course, is based on the fear that an individual coming into contact with improperly disposed of regulated medical waste could contract a serious disease, specifically AIDS or hepatitis B. Further discussion of these processes will be found in Chapter 4. In addition, the impact of this concern about bloodborne pathogens on emergency responders will be discussed later in this chapter.

Individuals not involved directly in the accident should evacuate the laboratory, and the area must be decontaminated by persons wearing proper protective clothing. Only those individuals who have received documented training as required by the OSHA Blood-Borne Pathogen standard are allowed to clean up any materials that might be contaminated by human blood, other bodily fluids, mucous, or tissue. It may be necessary to chemically decontaminate the entire exposed space. However, each incident needs to be treated on a case-by-case basis.

If it is necessary to transfer individuals to an emergency facility, all information available should be given to the emergency response personnel and also transmitted to the personnel at the hospital facility. They may wish to activate isolation procedures to protect themselves and others.

2.3.5.5.2. Radiation Incident

Radioactive spills represent another class of accident of special concern. There are circumstances that ameliorate the risk in actual accidents. Although laboratories in which radioactive materials are used are not classified as to the degree of risk as are laboratories using pathogens, they do operate under unusually tight regulations established by the Nuclear Regulatory Commission (NRC) or an equivalent state agency, which tend to minimize the amount of material involved in a single incident and to limit the number of persons involved to authorized and experienced personnel. As a result, an individual involved in a spill generally knows to restrict access to the area of the accident and to avoid spreading the material to uncontaminated areas. Unfortunately, not all researchers exercise the required care, and as a result, there are occasions when radioactive materials may be spread unnecessarily. Every institution licensed to use radioactive materials is required to have a radiation safety program and a radiation safety officer who should be notified immediately in case of an accident. In obtaining the license to use radioactivity, the institution or corporation must demonstrate to the NRC that it has the capability of managing accidents properly. In addition, there are requirements governing reports to the NRC, or to the equivalent state agency in an "Agreement" statement, spelled out in Title 10 of the Code of Federal Regulations, Part 20, when an accident occurs. Thus, the response to an emergency involving a release of radioactive material is relatively straightforward. Individuals working with many classes of radioactive materials must wear personal dosimeters (usually a badge containing a material with a known dose response relationship), so that in the event of an incident, their total external exposure can be read from these badges. Nasal swipes can be taken to check for inhaled materials. The clothes and skin of persons in the area and those allowed to leave can be checked with survey meters, which should be present in laboratories using radioactive materials or brought to the scene by radiation safety personnel. Surface contamination within the laboratory and on personnel can be cleaned up with little risk, using proper personal protective equipment to protect those doing it. The protective equipment normally would consist of a cartridge respirator and filter, coveralls of Tyvek™ or a similar material, head and foot covers (these may need to be impregnated with an appropriate plastic material), and "impermeable" gloves (unless chemical solvents are involved, the gloves most commonly used are made of polyethylene). Duct tape is an invaluable asset to seal gaps in the protective clothing around wrists, ankles, and the front opening. If the possibility exists that anyone ingested or inhaled radioactive material, then the individual should undergo further testing. This would include a bioassay for radioactive materials and, possibly, whole body counting at a facility with this capability. Whole body counters are available as mobile units which can be brought to a site should the need be justified. A major advantage of radioactive materials is that instruments exist which can detect radiation from spilled materials to levels well below any known risk.

A situation in which personal injury is accompanied by a spill of radioactive material onto that person introduces significant complications in the emergency medical response. Radioactive material may have entered the body through a wound, and there is a possibility that both the emergency transport vehicle and the emergency room at the hospital could become contaminated. Due to the small quantities used in most laboratories, the contamination is unlikely to actually be a serious problem, but could be *perceived* as one by emergency medical personnel. In order to reassure them, a radiation safety person should accompany the victim to the emergency center, if possible, and be able to provide information on the nature of the radioactive material, the radiation levels to be expected, and advice on the risks posed by the exposure to the patient and to others. The type of radiation and the chemical or material in which it is present can have a major impact on the actions of the emergency room personnel. Some materials are much worse than others if they have entered the body. As noted above, a bioassay, other specialized tests, and a whole body count of the victim may be needed in order to ascertain that no internal contamination exists.

A sheet of plastic placed between the injured person and the backboard or stretcher and brought up around the person will effectively reduce the amount of contamination of the ambulance and the equipment being used, and will serve the same purpose later at the emergency room. If it is felt to be necessary, the emergency personnel can wear particulate masks or respirators to avoid inhalation of any contaminants. Emergency equipment used in the course of the emergency response can be readily checked and, if necessary, decontaminated after the patient has been transferred to the emergency room. The patient should be separated from any other occupants of the emergency reception area to avoid any unnecessary exposures, even if they are well within safe limits, again because of the public concern regarding exposures to radioactivity at any level. In the unusual event that substantial levels of radiation might be involved, the victim should be placed in an isolated room and emergency equipment brought to the room rather than using the normal emergency room. A possible location would be the morgue. In such an incident, it is important to document exposures for everyone involved in the emergency response. Even in low-activity situations, it is good standard practice to survey the interior of the ambulance, the parts of the emergency facility which might have been contaminated, equipment that may have been used, and the emergency personnel involved and make wipe tests for loose contamination. All radiation survey data should be carefully recorded and the records maintained for future use. The records should include estimated dose levels of all personnel participating in the event, based on the proximity to the radiation and the duration of the exposure. There could easily be a need for these data in court at a later time.

2.3.5.6. Multiple Class Emergencies*

Emergency response procedures will need to incorporate sufficient flexibility to serve in many nonstandard situations. Unfortunately, one cannot depend upon an accident being of a single type or even limited to one or two complicating factors. Consider the following hypothetical situation: a laboratory worker puts a beaker containing a volatile solvent, to which a radioactive compound has been added, into an ordinary refrigerator. Due to carelessness, it is not covered tightly. During the next several hours, the concentration of vapors builds up in the confined space and at some point the refrigerator goes through a defrost cycle. The vapor ignites explosively, the refrigerator door is blown off, strikes a worker, and knocks several bottles of chemicals off a shelf. Chemicals from the broken bottles spill onto the floor and onto the injured person. The solvent in the beaker, as well as in several other containers, spills on the floor and ignites. The radioactive material in the beaker and in some of the other containers is spread throughout the laboratory and into adjacent rooms. Although this is posed as a hypothetical situation, it could happen.

In a complicated incident such as the one described, the first priority is preservation of life, even ahead of possible future complications. In the presence of a fire which, in a laboratory containing solvents, always has at least the potential of spreading uncontrollably, evacuation of the injured party should be considered as the first priority, followed by or paralleled by initiating evacuation of the rest of the building. Note that in every case of injury, the comparative risk of further injuring a person by moving them must be compared to the risk of not moving them. Treatment of the physical and chemical injuries to the victim and summoning the fire department and other emergency personnel can be done outside the building. Once everyone is safe, then efforts to reduce property losses

---

*   This section and the remainder of this chapter were coauthored by Richard F. Desjardins, M.D., and A. Keith Furr, Ph.D.

can take priority. Assuming that the fire is controllable, then steps can be taken for cleanup and decontamination of the spilled chemicals and radioactive material. Unless there appears to be a risk that the contaminated area will spread, perhaps due to runoff of water used in fighting the fire, it is not necessary for these last steps be done in any haste. However, the surrounding area must be cordoned off until measurements and surveys are completed by trained radiation safety and, perhaps, chemical safety personnel. This isolation must be maintained until a *formal* release of the area by the individual in charge, based on the information provided by the safety specialists.

After the incident is over, a review of the causes of the accident and the emergency response should be conducted by the appropriate safety committee or committees. In this case, the laboratory safety committee and the radiation safety committee would probably jointly conduct the review. Basically, there were two root causes of the incident. Solvents should not be stored in any container which cannot be tightly sealed, but this would not have caused the explosion if the refrigerator had been suitably designed for storage of flammable materials. These are commercially available.

The subsequent review should consider if anything could have made the incident worse. For example, in the hypothetical accident, the worker could have been alone, although this was not assumed to be the case. In academic research laboratories, research workers, and especially graduate students, tend to work unusual hours as they try to work around their class schedules to meet deadlines imposed by the framework of timetables, deadlines for submission of theses and dissertations, etc. If the injured person had been alone, the potential for a loss of life would have existed.

The situation described in the previous paragraph illustrates not only that in the real world emergencies can be very complicated, but also illustrates that some emergency responses can wait but others cannot. Components of the emergency that are immediately life threatening must be dealt with promptly, but others, such as cleaning up, can wait to be done carefully and properly after appropriate planning. Any incident also should be treated as a learning opportunity. There were basic operational errors leading to the postulated incident which could be repeated in other laboratories. There were aspects to the incident which would have permitted it to be worse. These should be factored into the emergency plan if they had not already been considered. If violations of policy had occurred, then the review should point these out and recommend courses of action to prevent future violations. An emergency plan should not only cover responses to classes of emergencies which have occurred, but should have the capability of reducing the possibility that emergencies will occur.

2.3.6.   Artificial Respiration, Cardiopulmonary Resuscitation (CPR), and First Aid

In several examples of responses to various emergencies, allusions were made to emergency medical procedures which should be performed. Most of these procedures require prior training. Because of the relatively high probability of accidents in laboratories, it would be desirable if at least a cadre of trained persons was available in every laboratory building.

Both first aid and CPR classes are taught by a number of organizations in almost every community. Among these are the Red Cross, American Heart Association, rescue squads, other volunteer organizations, and many hospitals. Usually, except for a small fee to cover the cost of materials, the classes are free. In addition, new labeling regulations and the OSHA Hazardous Communication Standard require that emergency information be made available on the labels of chemical containers and as part of the training programs. Since in most cases involving a chemical injury the chemical causing the injury will be known, and thus information will be available, the following material on first aid for chemical injuries will be restricted to the case of basic first aid for an injury caused by an unknown chemical. Similarly, since formal class instruction in CPR, which will also cover artificial respiration, is almost always available, the material on CPR will be very basic. CPR should be done only by properly trained individuals, with the training including practice on mannequins. Certification in CPR is easily and readily acquired. It is also important to periodically become recertified, as new concepts and procedures are frequently evolving and presented in the training programs.

In all the following sections, it is assumed that emergency medical assistance will be called for immediately. Emergency medical personnel are trained to begin appropriate treatment upon their arrival. Depending upon the level of training and the availability of telemetry, they will have radio contact with a hospital emergency facility or a trauma center and can receive further instruction from a physician while providing immediate care and during transit to the treatment center.

The following material is a composite of the information provided by a number of different sources. Where sources differed slightly, the more conservative approach was taken, i.e., that approach which appeared to offer the most protection to an injured person, with a second priority being the approach offering the least risk to the individuals providing the assistance. A third criterion was simplicity and the feasibility of performing the procedure with materials likely to be available. It was compiled explicitly in the context of injuries that are likely to occur as a result of laboratory accidents and is not intended to provide a comprehensive treatment of emergency medical care. It has been reviewed and revised by a physician.

Except where mandated by the nature of the problem, such as removal from a toxic atmosphere, no stress was placed on evacuation. Unless there are obvious fractures, there may be injuries to the spine or broken bones that may puncture vital organs which are not immediately apparent. If it is essential to move the victim, do so very carefully. Use a backboard or as close to an equivalent as possible to keep the body straight, and support the head so that it does not shift. Any inappropriate movement of a fractured neck may sever the spinal cord and result in paralysis, death, or compromising of the patient's airway.

To repeat, before performing any of the more complicated first aid procedures, formal training classes taught by certified instructors should be taken. It is possible for an inexperienced person to cause additional injuries.

2.3.6.1. Artificial Respiration

The lack of oxygen is the most serious problem that might be encountered. If the victim is not breathing or the heart is not beating, then oxygen will not be delivered to the brain. If this condition persists for more than 4 to 6 minutes, it is likely that brain damage will occur. In this first section, it will be assumed that the heart is beating but that the victim is not breathing.

2.3.6.1.1. Artificial Respiration, Manual Method

Although mouth-to-mouth or mouth-to-nose artificial respiration is much more effective, an alternative method of artificial respiration will be discussed first. There are rare occasions when it is not safe to perform direct mouth-to-mouth resuscitation, such as when poisoning by an unknown or dangerous chemical substance is involved, or when the victim has-suffered major facial injuries which make mouth-to-mouth impossible. Since the first of these conditions can be expected to occur in some laboratory accidents, it is good to know that there is an alternative procedure available. The method considered the best alternative is described below.

1. Check the victim's mouth for foreign matter. To do this, insert the middle and forefinger into the mouth, inside one cheek and then probe deeply into the mouth to the base of the tongue and the back of the throat, finally sliding your fingers out the opposite side of the mouth. Be aware that a semiconscious patient may bite down on your fingers. It would be wise to insert a folded towel or object that would not break teeth between the teeth while you are doing your examination.
2. Place the victim on his back in a face up position. Problems with aspirating vomitus can be reduced by having the head slightly lower than the trunk of the body. An open airway is essential and can be maintained by placing something, such as a rolled up jacket, under the victim's shoulders to raise them several inches. This will permit the head to drop backwards and tilt the chin up. Turn the head to the side. *Important! Do not do this if there is any suspicion of neck trauma.*
3. Kneel just behind the victim's head, take the victims wrists, and fold the victim's arms across the lower chest.
4. Lean forward, holding onto the wrists, and use the weight of your upper body to exert steady, even pressure on the victim's chest. Your arms should be approximately straight up when in the forward position. This will cause air to be forced out of the victim's chest. Perform this step in a smooth, flowing motion.
5. As soon as step 4 is completed, take your weight off the victim's chest by straightening up and simultaneously pulling the victim's arms upwards and backwards over his head as far as possible. This will cause air to flow back into the lungs, thereby completing one equivalent breathing cycle.
6. Steps 4 and 5 should be repeated 12 to 15 times per minute to pump air into and out of the victim's lungs. Stop and check frequently that no vomitus or foreign matter has been brought up into the airway. If a helper is available, let the helper do this while you continue the pumping process.
7. Continue the procedure until normal breathing is established or until emergency medical personnel arrive. Be ready to recommence the process if breathing difficulties reoccur.

This less effective method was given first since it is human nature to question the need to learn a less effective method if a better one exists, and it has already been covered. However, it is important to limit the number of injured parties as well as to treat those already injured. Since, in the case of an unknown toxic or an especially dangerous substance, the person giving emergency treatment could be exposed to the same material, this procedure will serve in such cases.

2.3.6.1.2. Artificial Respiration, Mouth-to-Mouth Method

There is a consensus that this is the most effective method of artificial respiration. To be effective, it should be begun as quickly as possible.

1. Check the victim's mouth for foreign matter. Clear out any that is found with your fingers. A cloth such as a towel or handkerchief covering the fingers helps in removal of objects, or even some chemical, which would slide off wet fingers. The towel also will serve to protect the fingers should the patient bite down on them.

2. The victim's air passage must be open. Put your hand under the person's neck and lift. Place your other hand on the victim's forehead and tilt the victim's head back as far as it will reasonably go, essentially straightening the airway. A folded jacket or coat under the shoulders will aid in keeping the head back.

3. Maintain this position, and with the fingers of the hand being used to tilt the head backward, pinch the nostrils closed.

4. Open your mouth widely, take a deep breath, place your mouth firmly around the victim's mouth to get a good seal and blow into the victim's mouth.

5. Try to get a good volume of air into the victim's lungs with each breath.

6. Watch to see the chest rise. When it has expanded, stop blowing and remove your mouth from the victim and check for exhalation by listening for air escaping from the mouth and watching to see the chest fall. This occurs naturally from the inherent elasticity of the chest wall.

7. Steps 4 through 5 constitute one breathing cycle. Assuming it has been successful, i.e., there has been no significant resistance to the flow of air during the inflow cycle and air has been exhaled properly, repeat steps 4 through 6 at a rate of 12 to 15 times per minute.

8. If you are not getting a proper exchange of air, check again to see if there is anything obstructing the air passage and make sure you are holding the head properly so as to prevent the tongue from blocking the flow of air.

9. Continue the procedure until normal breathing is established, or until emergency medical personnel arrive. Be ready to recommence the process if breathing difficulties reoccur.

2.3.6.2. Cardiopulmonary Resuscitation

Cardiopulmonary resuscitation combines artificial respiration with techniques to manually and artificially provide blood circulation, and is to be used where the heart is not beating. It requires special training and practice and should not be done except by trained, qualified individuals. As noted earlier, receiving the training is not difficult since it is readily available from several organizations. The material presented here is intended to provide information as to the general procedures, but neither it nor the previous procedures are intended as an instructional guide.

One-person CPR will be the procedure described. In this procedure, the person performing it will provide both artificial breathing and blood circulation. When two persons are available, each person can take the responsibility for one of these functions and can switch from time to time to relieve fatigue.

2.3.6.2.1. Initial Steps

1. Check for breathing. Do this by placing your hand under the victim's neck and the other on the forehead. Lift with the hand under the neck and tilt the head back. While doing this, place your ear near the victim's mouth and look toward the chest. If the victim is breathing, you should be able to feel air on your skin as it is being exhaled, you should be able to hear the victim breathe, and you should be able to see the chest rise and fall. Do this for at least 5 seconds.

2. If the victim is not breathing, while holding the victim as in step one and pinching the nostrils closed, open your mouth widely, place it over the victim's mouth, and give two quick full breaths into the victim's mouth.

3. After step 2, repeat step 1 to see if breathing has started. If not, proceed to step 4.

4. Check the victim's pulse with the hand that had been under the victim's neck. Keep the head tilted back with the other hand on the forehead. Check the pulse by sliding the tips of the fingers into the groove on the victim's neck to the side of the Adam's apple nearest you. Again, check for at least 5 seconds.

2.3.6.2.2. Formal CPR Procedures

The victim must be on a firm surface. Otherwise, when pressure is applied to the chest, the heart will not be compressed against the backbone as the backbone is pressed into a yielding soft surface. The head of the victim should not be higher than the heart, in order for blood to flow to the brain, as is needed to avoid brain damage. Although the brain averages about 2% of your body weight, it requires 20% of the oxygen you breathe, as well as at least 40 mgm% of blood dextrose. Any concentration less will result in unconsciousness and progressive brain damage. Preferably, the feet and legs should be adjusted to be higher than the heart to facilitate blood flowing back to the heart, but this has a lower priority than commencing CPR procedures.

1. Kneel beside the victim, at breast height. Rest on your knees, not your heels.
2. Locate the victim's breastbone. Place the heel of one of your hands on the breast-bone so that the lower edge of the hand is about two finger-widths up from the bottom tip of the breastbone. Put your other hand on top of the first hand. Lift your fingers or otherwise keep from pressing with them. Improper placement of the hands can cause damage during the compression cycle.
3. Place your shoulders directly over the breastbone. Keep your arms straight.
4. To initiate the compression cycle, push straight down, pivoting at the hips.
5. Push firmly and steadily down until the chest has been compressed about 4 to 5 centimeters. Then smoothly relax the pressure until the chest rebounds and is no longer compressed, then start the compression cycle again. The compression-relaxation cycle should be a smooth, continuous process.
6. Continue the chest compression procedure for 15 cycles at the rate of 80 per minute. This should take between 11 and 12 seconds. Then quickly place your mouth over the mouth of the victim and give two quick full breaths. Then return to compressing the chest for another 15 cycles. Be sure to locate your hands properly and compress the chest as in steps 4 and 5.
7. Continue step 6, alternately compressing the chest and providing artificial respiration.
8. Quickly check for a pulse after 1 minute and then every few minutes thereafter. Watch for any signs of recovery.
9. If a pulse is found, then check for breathing. If necessary give artificial respiration only, but check frequently to be sure that the heart is still beating.
10. Continue with whatever portion of the procedure is necessary until the victim is functioning on his own, emergency medical personnel arrive, or it is obvious that efforts will not succeed. A half hour is not unreasonable as a period of actively using CPR.

A person who has not been properly trained in the techniques of CPR can injure the victim, either by driving the lower end of the breastbone into the liver, or by breaking ribs if the hands are not placed correctly or the pressure is exerted incorrectly. However, remember that ribs will heal, but a nonbreathing or heart-stopped patient dies. Thus, to repeat once more, the techniques for artificial respiration and CPR are not difficult but do require training and practice. Familiarity with these techniques is likely to be of value at any time.

2.3.6.3. First Aid

2.3.6.3.1. Severe Bleeding

A person may bleed to death in a very short time from severe or heavy bleeding; whenever this problem is involved in an accident, it is extremely important to stop it as soon as possible. Arterial bleeding may be frightening, but the muscular artery wall usually contracts to diminish or stop the flow. Venous bleeding is more insidious as it flows steadily. The relative absence of muscle in the vein wall does not help to stem the flow.

1. The most effective treatment is pressure applied directly to the wound over which a sterile dressing has been placed (Figure 2.8).
2. Apply a sterile dressing if immediately available (a handkerchief or some other cloth to the wound, if not). Then place the palm of the hand directly over the wound and apply pressure. If nothing else is available use your bare hand, but try to find something to use as a dressing as soon as possible.
3. A dressing will help staunch the flow of blood by absorbing the blood and permitting it to clot. Do not remove a dressing if it becomes blood soaked, but leave it in place and apply an additional one on top of the first in order not to disturb any clotting that may have started. Keep pressure on with the hand until you have time to place a pressure bandage over the dressing to keep it in place.
4. Unless there are other injuries, such as a fracture or the possibility of a spinal injury, cases for which the victim should be disturbed as little as possible, the wound should be elevated so that the injured part of the body is higher than the heart. This will reduce the blood pressure to the area of the wound.
5. If bleeding persists and cannot be stopped by direct pressure, putting pressure on the arteries supplying the blood to the area may be needed. In this technique, pressure is applied to the arteries by compressing the artery between the wound and the heart against a bone at the points indicated in Figure 2.9. Since this stops all circulation to points beyond the point of compression, it can cause additional injuries if continued too long. For this reason, it should be discontinued as quickly as possible and the use of direct pressure and elevation to control the bleeding should be resumed, unless this is the only effective technique.
6. As a last resort, since it stops the flow of blood to the limb beyond the point of application, a tourniquet, which should be at least two inches wide, can be applied. An example where the use of a tourniquet might be indicated is to stop bleeding from a severed limb.

2.3.6.3.2. Shock

Shock may accompany almost any type of severe injury, exposure to toxic chemicals, a heart attack, loss of blood, burns, or any other severe trauma. It can be recognized from a number of characteristic symptoms: skin cold to the touch (possibly clammy and bluish or pale), weakness, a rapid weak pulse, rapid irregular breath, restlessness, and exhibition of unusual signs of thirst. As the condition worsens, the victim will become unresponsive and the eyes may become widely dilated.

The treatment for shock is

1. The victim should be lying down, although the type of injuries may determine what is the best position. If uncertain, allow the victim to lie flat on his back. Unless it

Figure 2.8. Apply pressure directly to a wound to control bleeding.

is painful or it makes it harder for the victim to breathe, it will help if the feet are
raised 20 to 30 centimeters high.
2. Use blankets to keep the victim from losing body heat, but do not try to add heat.
3. If the victim is conscious and is not vomiting, nor appears likely to do so, then about
half a glass of liquid every 15 minutes or so will be helpful. However, do not give
fluids if the victim is unconscious or nauseated.

2.3.6.3.3. Poisoning by Unknown Chemicals

The rationale in limiting this section to unknown chemicals is that first aid information
is readily found when the chemical is known, either immediately on the label of the
container or in the MSDS for the chemical (note that current law requires chemical produc-
ers to provide MSDSs to those who purchase chemicals from them). An MSDS file should
be maintained in every laboratory.

Even if the chemical causing the injury is not known and the victim is unconscious so
that no direct information is available, an examination of the circumstances of the accident
and an examination of the victim's lips, skin, the mouth, and the tongue could provide
helpful information on whether the victim swallowed a poisonous substance or inhaled a
toxic vapor, or whether the injury was due to absorption through the skin. If the abdomen
is distended and pressing on it causes pain, the likelihood is that a corrosive or caustic
substance has been ingested. Any information on the nature of the harmful material will
be helpful to the physician who will treat the victim.

- Poisoning by Inhalation

   1. When poisoning by inhalation is suspected, evacuate the victim to a safe area as
      soon as possible. If there are fumes still suspected to be present, a rescuer should
      wear a self-contained respirator. Do not take a chance which might result in a
      second victim.
   2. Check for unusual breath odors if the victim is breathing.
   3. Loosen tight clothing around the victim's neck and waist.
   4. Maintain an open airway.

Figure 2.9. Pressure points to control bleeding in the extremities.

5. If the victim is not breathing, perform artificial respiration using the manual method. It is dangerous to the person providing aid to give mouth to mouth artificial respiration if the toxic material is not known.

• Poisoning by Ingestion

1. Examine the lips and mouth to ascertain if the tissues are damaged as a possible indicator that the poison was ingested, although the absence of such signs is not conclusive.
2. Check the mouth and remove any dentures.
3. If the victim is not breathing, perform manual artificial respiration.
4. If the victim becomes conscious, try to get the victim to vomit, unless it is possible that the poisoning is due to strong acids, caustics, petroleum products, or hydrogen peroxide, in which case additional injuries would be caused to the upper throat, esophagus, and larynx. Vomiting may be induced by tickling the back of the throat. Lower the head so that the vomit will not reenter the mouth and throat. Dilute the poison in the stomach with water or milk.
5. If the victim has already vomited, collect a sample of the vomit, if possible, for analysis.
6. If convulsions occur, do not restrain the victim, but remove objects with which he might injure himself or orient the victim to avoid his striking fixed, heavy objects.
7. Watch for an obstruction in the victim's mouth. Remove if possible, but do not force fingers or a hard object in between the victim's teeth. If a soft pad can be inserted between the victim's teeth it will protect the tongue from being bitten. A badly bleeding tongue immensely complicates the patient's problems.
8. Loosen tight clothing, such as a collar, tie, belt, or waistband.

9. If the convulsions cease, turn the victim on his side or face down so that any fluids in the mouth will drain.
10. Treat for shock if the symptoms for shock are noted.

- Poisoning by Contact

1. If the chemical got into the victim's eyes, check for and remove any contact lenses. Take the victim immediately to an eyewash fountain (if one is not available, to a shower or even a sink) and wash the eyes, making sure that the eyelids are held widely open. Wash for at least 15 minutes. If the chemical is caustic rather than acidic, the victim may not feel as much pain, since an acid causes pain due to precipitation of a protein complex, and may wish to quit earlier. An alkali or caustic chemical is more serious as it does not precipitate protein and continues to penetrate the globe, even to global rupture. It is imperative that the chemical be flushed out thoroughly.
2. Do not use an eye ointment or neutralizing agent.
3. If the chemical only came in contact with the victim's exposed skin, such as the hands, wash thoroughly until the chemical is totally removed.
4. If the chemical was in contact with the clothed portion of the body, remove the contaminated clothes as quickly as possible, protecting your own hands and body, and place the victim under a deluge shower. If the eyes were not affected initially, protect them while washing the contaminated areas. Be careful not to damage the affected skin areas by rubbing too firmly. Let the flowing water rinse the chemical off. A detergent is sometimes used, but be careful not to carry the offending chemical to other parts of the body. Be particularly careful to clean folds, crevices, creases and groin.
5. If both the eyes and portions of the body were exposed, preferably there should be a combination eyewash and deluge shower unit available. If not, take the victim to the deluge shower and tilt the head back, holding the eyelids widely open, and wash the entire body.

2.3.6.3.4. Heat Burns

First Degree (minor): Painful and red. No blisters. Skin elastic. Epidermis only. Minimal swelling.

1. Apply cold water to relieve pain and facilitate healing.
2. Avoid re-exposure, as the already injured skin can be more susceptible to further damage than normal skin.

Second Degree: Severe and painful, but no immediate tissue damage. Pale to red. Weeping blisters, vesicles. Marked swelling. Involves epidermis and dermis.

1. Immerse affected area in cold water to abate the pain.
2. Apply cold, clean cloths to the burned area.
3. Carefully blot dry.
4. Do not break blisters.
5. If legs or arms are involved, keep them elevated with respect to the trunk of the body.

Third Degree: Deep, severe burns, likely tissue damage. White, red, or black and dry and inelastic tissue. No pain, involves full thickness of skin. May involve subcutaneous tissue, muscle, and bone.

1. Do not remove burned clothing from the burned area.
2. Cover the burned area with a thick, sterile dressing or clean cloths.
3. Do not immerse an extensively burned area in cold water, because this could exacerbate the potential for shock and introduce infection. A cold pack may be used on limited areas such as the face.
4. If the hands, feet, or legs are involved, keep them elevated with respect to the trunk of the body.
5. Third-degree burns must be treated by a physician and/or hospital. They may need reconstruction, skin grafting, and prolonged care. Control of infection is mandatory.

REFERENCES

Note that many of the basic references were incorporated in the text as materials needed to plan or facilitate an effective emergency program. The following are additional references used in preparing the material.

1. **Lowery, G.G. and Lowery, R.C.,** *Handbook of Hazard Communications and OSHA Requirements,* Lewis Publishers, Chelsea, MI, 1990.
2. **Lowery, G.G. and Lowery, R.C.,** *Right-to-Know and Emergency Planning,* Lewis Publishers, Chelsea, MI, 1989.
3. **Laughlin, J.W., Ed.,** *Private Fire Protection and Detection,* ISFTA 210 International Fire Training Association, Fire Protection Publications, Oklahoma State University, Stillwater, 1979.
4. ANSI Z358.1-1989, Emergency Eyewash and Shower Equipment, American National Standards Institute, Inc., New York, 1981.
5. **Srachta, B.J.,** in *Safety and Health,* National Safety Council, Chicago, 1987, 50.
6. **Steere, N.V.,** Fire, emergency, and rescue procedures, in *CRC Handbook of Laboratory Safety,* Steere, N.V., Ed., CRC Press, Inc., Boca Raton, FL, 1971, 15.
7. ANSI Z87.1-1979. Practice for Occupational and Educational Eye and Face Protection. American National Standards Institute, Inc., New York, 1979.
8. **Schwope, A.D., Costas, P.P., Jackson, J.O., Stull, J.O., and Weitzman, D.J., Eds.,** *Guidelines for the Selection of Chemical Protective Clothing,* 3rd ed., Arthur D. Little, Inc. for U.S. EPA and U.S. Coast Guard, Cambridge, 1987.
9. **McBriarty, J.P. and Henry, N.W., Eds.,** *Performance of Protective Clothing:* Fourth Volume, American Society for Testing and Materials, Philadelphia, 1992.
10. ANSI Z88.2-1980. Practices for Respiratory Protection. American National Standards Institute, Inc., New York, 1980.
11. **Kairys, C.J.,** Hazmat protection improves with equipment documentation, in *Occupational Health and Safety,* 56, No. 12, 20, November 1987.
12. **Still, S. and Still, J.M., Jr.,** Burning issues (charts), Humana Hospital, Augusta, GA.
13. **Schmelzer L.L.,** *Emergency Procedures and Protocols,* Cancer Research Safety Workshop Workbook, Office of Research Safety, National Cancer Institute, Bethesda, MD, 1978, 106.
14. **Gröschel, D.H.M., Dwork, K.G., Wenzel, R.P., and Schiebel, L.W.,** Laboratory accidents with infectious agents, in *Laboratory Safety, Principles and Practices,* Miller, B.M., Gröschel, D.H.M., Richardson, J.H., Vesley, D., Songer, J.R., Housewright, R.D., and Barkley, W.E., Eds., American Society of Microbiology, Washington, D.C., 1986, 261.

15. **Edlich, R.F., Levesque, E., Morgan, R.F., Kenney, J.G., Silloway, K.A., and Thacker, J.G.,** Laboratory personnel as first responders, in *Laboratory Safety, Principles and Practices,* Miller, B.M., Gröschel, D.H.M., Richardson, J.H.,Vesley, D., Songer, J.R., Housewright, R.D., and Barkley, W.E., Eds., American Society of Microbiology, Washington, D.C., 1986, 279.

16. Emergency first aid guide, appendix 4, in *Laboratory Safety, Principles and Practices,* Miller, B.M., Gröschel, D.H.M., Richardson, J.H., Vesley, D., Songer, J.R., Housewright, and R.D., Barkley, W.E., Eds., American Society of Microbiology, Washington, D.C., 1986, 348.

17. *Safety in Academic Chemistry Laboratories,* 5th ed., American Chemical Society. Washington, D.C., 1985.

18. *Multimedia Standard First Aid, Student Workbook,* American Red Cross, Washington, D.C., 1981.

19. *Standard First Aid & Personal Safety,* 2nd ed., American Red Cross, Washington, D.C., 1979.

20. *Adult CPR Workbook,* American Red Cross, Washington, D.C., 1987.

21. **Hafen, B.Q. and Karren, K.J.,** *First Aid and Emergency Care Workbook,* 3rd ed., Morton Publishing, Englewood, CO, 1984.

22. **Senecal, J.A.,** Halon replacement chemicals: perspectives on the alternatives, in *Fire Technology,* 28(4), 332, November, 1992.

23. **Zurer, P.S.,** Looming ban on production of CFC's, Halon spurs switch to substitutes, *Chemical & Engineering News,* 71(46), 12, November 15, 1993.

CHAPTER 3

Laboratory Facilities—Design and Equipment

3.1. LABORATORY DESIGN

The design of a laboratory facility depends upon both function and program needs. Although there are differences among engineering, life sciences and chemistry laboratories, and within the field of chemistry (between laboratories intended for physical chemistry and polymer synthesis, to take two examples), the similarities outweigh the differences. Approximately the same amount of space is required. Certain utilities are invariably needed. Ventilation is needed to eliminate odors and vapors from the air which might have the potential to adversely affect the health of the employees, as well as provide tempered air for comfort. Provision is needed for stocking reasonable quantities of chemicals and supplies. As these are used over a period of time, chemical wastes are generated and provisions must be made for temporary storage and disposal of these according to regulatory standards. They must provide suitable work space for the laboratory workers. Many of these items, as well as others, vary only in degree. Most differences are relatively superficial and are represented primarily by the equipment which each laboratory contains and the selection of reagents used.

Not only are laboratories basically similar, but there is a growing need for "generic" laboratory spaces readily adaptable to different research programs. This is due in part to the manner in which most research is funded today. In industry, laboratory operations are generally goal oriented, i.e., they exist to develop a product or to perform basic research in a field relevant to the company's commercial interests. There is a cost-benefit factor associated with laboratory space. In the academic field, research is primarily funded by grants. These grants can be from any number of public and private sources, but, with only a moderate number of exceptions, grants are based on submission of a proposal to perform research toward a specific end to the funding agency during a stipulated period of time. At the end of this period, the grant may or may not be renewed; if not, control of the space may be turned over to another investigator. Laboratory space is too limited and too expensive (currently running in the range of $100 to $300 per square foot, dependent upon the complexity of the construction) to be allowed to remain idle. The result has been a trend to design laboratories that are relatively small, typically suitable for two to four persons to work in them simultaneously, with connections to adjacent rooms to permit growth if needed. Under these circumstances, it will be appropriate in most of this chapter to base the discussion upon a standard module. One potential result of this growing need for flexibility may be an eventual breakdown of the concept of department-owned space for research buildings, i.e., the concept of a chemistry or biology building. Eventually, facilities may be designed toward a given type of use, such as microbiology or polymer chemistry, but the users may be assigned space independently of their original discipline, based, at least in part, on current needs.

Instructional laboratories are an exception in terms of size since they are intended for continued basic programs, serving class sizes of 20 or more persons, and so normally would be somewhat larger than is needed for research programs. Also, except at advanced levels, the instructional laboratories usually do not conduct experiments or use chemicals having the same degree of risk as do research laboratories. The risk in instructional laboratories is also being reduced by the greater use of smaller quantities of chemicals because of advances in technology. However, even in the case of instructional laboratories, many of the basic safety requirements still must be incorporated in the design.

3.1.1. Engineering and Architectural Principles

The increasing cost of sophisticated laboratory space dictates a number of design considerations. It is essential that space be used to maximum advantage. Due to the necessity for mechanical services, closets, columns, wall thicknesses, halls, stairs, elevators, and restrooms, the percentage of net assignable space in even a well-designed, efficient building is generally on the order of about 65%. Due to the large number of fume hoods in a typical laboratory building and other ventilation requirements, as well as the increasingly stringent temperature and humidity constraints imposed by laboratory apparatus and computers, heating and ventilation (HVAC) systems are becoming more difficult to design. The engineer must accommodate these needs as well as the need to provide personal comfort, conserve energy, and provide low life-cycle maintenance costs. Stringent new regulatory requirements under the Americans With Disabilities Act to accommodate disabled persons in virtually every program impose costly additional constraints on accessibility and provisions for emergencies. Building designs need to be sufficiently flexible not only to suit different uses based on current technology, but should be sufficiently flexible to adopt technological innovations. For example, provision for installation of additional data, video, and voice lines in excess of current needs is almost certainly desirable. Interaction of the occupants of the building with each other, with outside services, and with other disciplines also mandates a number of design parameters. This set of parameters is the most dependent upon the specific programs using the building and will require substantial input by the users. Different disciplines perhaps require more variation in facilities for service groups than in the laboratories themselves. Typically, all of these design needs must be accommodated within a construction budget, established before the design of the building is in more than a very early conceptual stage, so the design process is a constant series of compromises. It is rare that all of the program desires (as opposed to needs) can be fully satisfied.

To the architect, a very important factor is that the building must meet all the needs in an attractive way. There is certainly nothing wrong with creating an attractive facility, in harmony with its surroundings, as long this aspect is not achieved at the expense of the basic needs of the users. Generally, the most efficient space is a cube, with no more than the minimally required penetrations of the walls and with no embellishments. No one would truly like to see this become the standard, although in the right context, such a facility could be very attractive. Buildings should fit into their environment in an aesthetic and congenial manner, but function and use factors should be preeminent in the design.

No mention has been made up to this point of health and safety design factors. They must be incorporated into virtually every other design feature. The location of a building, access to the building, the materials of construction and interior finish, size and quality of doors, width of corridors, length of corridors, number of floors, the number of square feet per floor, selection of equipment, utilities, etc. are impacted by safety and health requirements. Although it would be anticipated that architects and engineers would be thoroughly familiar with applicable safety codes and regulations, experience has shown that this is not

Figure 3.1. Standard laboratory module.

necessarily so, especially where they involve safety concepts other than those relating to fire or strength of materials. Even in these areas, the wide range of variability in interpretation of codes results in a tendency to liberally interpret the codes in favor of increasing the amount of usable space or enhancing the visual aspects of the design. It is surprising how few architectural firms maintain dedicated expertise on their permanent staff in the areas of building code compliance and health and safety. Even where such staff personnel are available, there is an inherent problem with conflict of interest. Of course, the reciprocal is also true. Most safety professionals are not artists, as many architects consider themselves, who can adequately include the aesthetic aspect in their own ideas. The eventual users may not appreciate the sole viewpoints of either of these two groups.

Since relatively few laboratories are built compared to the numbers of other types of buildings, comparatively few firms are really well prepared to design them for maximum safety, especially in terms of environmental air quality and laboratory hazards. For this reason, the eventual owners/users of a planned building should be sure to include persons knowledgeable in health and safety requirements in the group representing them to work with architects and contractors. Where this expertise is not available in-house, they should not hesitate to hire appropriate consultants to review the plans and specifications prior to soliciting bids.

Shown in Figure 3.1 is a standard laboratory module which forms the basis for much of the material in this chapter. This design, although simply a representative example, does provide a significant number of generally applicable safety features. A slightly larger variation on this design includes a central workbench down the center of the facility, but this represents an obstacle many users prefer not to have. The laboratories on either side are mirror images of this one and this alternating pattern can be repeated to fill the available space. The two side doors are operational, as shown here, and provide ready access to adjacent spaces, if needed, for the research program. Most building codes do not require more than a single exit in such a small room unless it is classified as a hazardous duty occupancy, so that if access to additional lab modules is not needed, either or both doors can be constructed as breakaway emergency exits or even made semi-permanent to allow additional bench or storage space. Where the doors are included, two readily accessible exits exist from every point within the room, even at the end of a sequence of laboratories.

Figure 3.2. A section of corridor using the standard module as a repeating feature.

In this basic 12 foot × 20 foot module, the areas where the likelihood of a violent accident are greatest are at the far end of the laboratory, away from the corridor entrance, and are well separated from stored flammable materials and other reagents. The desk area is separated from the work areas by a transparent barrier which, with the door to the laboratory proper closed, isolates the workers, when they are not actively engaged in their research, from both the possible effects of an accident and continuous exposure to the atmospheric pollutants of the laboratory. This latter factor is enhanced by the normal negative atmospheric pressure between the laboratory and the corridor, so that the air in the desk area should be virtually as clean as the corridor air. It would thus represent a safe space for employees or students to socialize, study, or have a drink or snack. The door from the corridor to the laboratory is offset at a 45° angle to minimize the loss of space within the laboratory, while at the same time avoiding the problem of a desirable outward swinging door creating an obstruction in the corridor.

Figure 3.2 provides a simplified illustration of how the modular approach can be integrated into an efficient and safe building design. This figure represents a section of a typical upper floor of a research building. Mechanical services, loading dock and receiving areas, support services, offices, conference rooms, toilets, lounges, and classrooms would be located either on other floors or further along the access corridors. Note that the fume hoods are at a back corner of the laboratory, and immediately outside is a chase to carry the exhaust duct to the roof. The location of the external chase at the juncture of two laboratories allows one chase to serve two laboratories. This external chase solves another problem if all laboratories are not originally equipped with hoods. It would be almost as cheap to go back and add a hood in this design as it would be to equip every laboratory with a hood initially.

The internal equipment is shown as the same in each laboratory, but with the exception of equipment dependent on service utilities such as water, and the fume hood exhausts, the internal arrangement is highly flexible. The individual manager can relocate virtually anything else, and with modular casework now available, there would be few restrictions on the arrangements, even in such a small module. Mention has already been made of the ability to add a fume hood later, and flammable material storage in refrigerators or flammable material storage cabinets may or may not be needed. Since the use of laboratories does change over time, the design should provide contingencies for the maximum hazard use, in terms of safety considerations, in the original construction.

The arrangement of laboratories with only a single support corridor, as shown in Figure 3.2 provides an advantageous net to gross square footage. The use of modules, arranged compactly, as these naturally permit doing, allows the architect and building owner to achieve an efficient building. The external chases lend themselves to an attractive architectural columnar appearance to the building, otherwise the absence of windows in large segments of the wall might otherwise appear too austere. The external appearance of a facility based on this design concept is shown in the section of this chapter dealing with external factors affecting ventilation.

An aspect of the design above, which may not be immediately apparent, is that such a design is especially appropriate for adding to an older facility, which was originally designed to meet less demanding standards than those of today. The newer component, situated adjacent to the original structure, and designed to meet current sophisticated research requirements, can be connected to the older one at appropriate places. By proper construction and fire separations, it would be possible to treat the old and new components as separate buildings, even though they are joined, so that it would not be necessary to renovate the older building to current construction standards. Less demanding operations, such as instructional laboratories and offices, could remain in the older component, and activities requiring additional and probably more sophisticated services, higher construction standards, etc. would be located in the new area. All of a department's operations would be in the "same" building, which has important logistical and personnel implications, and construction of an entirely new building for a department would be unnecessary. This concept is called an "infill" approach and provides some important financial savings, as it can extend the life of some older facilities. The methods of joining and maintaining separations between the two components also provide opportunities for architects to express themselves, such as making the less-expensive spaces between the two sections outside of the laboratory facility proper, into attractive communal areas.

REFERENCES

1. **Barker, J.H.**, *Designing for Safer Laboratories,* CDC Laboratory Facilities Planning Committee, Chamblee Facility, 1600 Clifton Rd., Atlanta, GA.
2. Earl Wall and Associates, Basic Program of Space Requirements, Dept. of Chemistry, VPI & SU, Laboratory Layout Studies, Blacksburg, VA, 1980.
3. **Ashbrook, P.C. and Renfrew, M.M., Eds.,** *Safe Laboratories, Principles and Practices for Design and Remodeling,* Lewis Publishers, Chelsea, MI, 1991.

3.1.2. Building Codes and Regulatory Requirements*

There are many codes and standards applicable to building construction. Many of these are incorporated by reference in the OSHA standards. A large number of the codes grew out of a concern for fire safety, and hence this general area is relatively mature. Existing health codes generally address only acute exposures and immediate toxic effects. It has been only recently that concern for long-term systemic effects has been addressed in standards, so there are fewer of them. The current OSHA laboratory standard replaces the detailed OSHA standards which were intended primarily for industrial situations and is a performance standard which requires that laboratories prepare and voluntarily comply with an industrial hygiene. The intent of the new standard is to ensure that laboratory employees are provided with at least the equivalent degree of protection as would have been provided by the general industry standards. Since most users of this Handbook may not be familiar with the OSHA General Industry Standards, there will be allusions in the text to these requirements as a reference base.

There are two specific sources from which a substantial portion of the material in this section will be derived or to which it will be compared. For a building code, the information will be drawn primarily from the BOCA (Building Officials and Code Administrators) code. This is not used universally, but is a widely accepted building code which, where applicable, provides mandatory standards. Other regional building codes are based on the same general industrial codes and recommendations of standard-setting organizations, but specific applications of these reference standards in codes for a given area may differ. Standard 45 of the National Fire Protection Association (NFPA) is specifically labeled as a laboratory safety standard. Generally, it has not been adopted as a formal legal requirement in most localities, but it does provide valuable guidance in certain areas for goals against which both existing and proposed laboratories can be measured. These two sources are primarily concerned with fire and construction safety. The materials cited are those most directly affecting the safety of building occupants or useful to persons discussing building design with architects and contractors. In addition, there will be other standards which have recently been enacted, such as the Americans With Disabilities Act, which will be superimposed on both existing facilities and, especially, new construction. This will also influence the design of laboratories. This last act is very broad in its statements, and implementation details will in many cases depend upon litigation.

Standard 45 and the BOCA code do not always agree, or at least they sometimes lead to different interpretations. The classification of the structure or building in which testing or research laboratories are operated is usually designated under BOCA as an educational or business use occupancy, although if the degree of hazard meets a number of specific criteria, it may be designated as a hazardous use facility. In NFPA Standard 45, buildings used for the purpose of instruction by six or more persons are classed as an educational occupancy. The classification is not a trivial question since it evokes a number of different design and construction constraints. A building used primarily for instruction, which might include instructional laboratories, and some testing and research laboratories might be considered primarily an educational occupancy if the research areas were separated properly from the remainder of the building. An educational occupancy is more restrictive

---

*   The basic material in this chapter concerned with building code requirements was reviewed for the 3rd edition of this handbook by Howard W. Summers, former Chief Fire Marshal (retired) for the State of Virginia. *1993 BOCA National Building Code,* copyright 1993, Building Officials and Code Administrators International, Inc., Country Club Hills, Illinois. Reprinted with permission of author. All rights reserved.

than a business occupancy, but less so than a hazard use classification. Standard 45 classifies laboratories as class A, B, or C according to the quantities of flammable and combustible liquids contained within them, with A being the most hazardous and C being the least. As discussed later, a system of ratings has been developed in the life sciences to designate laboratories according to four safety levels, with classes 1 and 2 meeting the needs of most laboratory operations, while 3 and 4 are restrictive and very restrictive, respectively. This concept, for consistency, might eventually be considered for laboratories of all types. In a later section of this handbook, such a proposed classification scheme for chemical laboratories is put forward.

3.1.2.1. Building Classification*

For the purposes of this section, the BOCA code will be used as the reference to define the classification of a building. The basic classification, therefore, will be either as an educational or business use occupancy. However, since some laboratories and ancillary spaces, such as storerooms, may meet the BOCA definitions of High Hazard Use (Group H), the following material, extracted from section 307 of the 1993 BOCA code, can be used as a guide to whether a given building or facility should be considered a Group H occupancy. The standard laboratory module, as shown in Figures 3.1 and 3.2, can meet the requirements for high hazard use in that two or more well-separated exits and the doors swing in the direction of exit travel. The doors from the modules also are offset by 45° so that the door does not swing out into the corridors when opened.

The hazard use occupancy group H is subdivided in the 1993 edition of the BOCA code into four subgroups, H-1, H-2, H-3, and H-4. Buildings which are used for the manufacturing, generation, storage, or use of hazardous materials may fall into one of these four categories, unless the amounts of materials involved are less than exempt quantities as defined by the code.

- Group H-1 generally is applied to facilities in which activities take place using materials that represent an explosive risk. In the context of laboratories, in addition to materials normally considered explosives, this includes organic peroxides, oxidizers, other highly unstable materials, and pyrophoric materials capable of *detonation*, as opposed to those which do not react as violently.
- Group H-2 includes facilities using less vigorously reacting materials as those of Group H-1, as well as flammable and combustible liquids, gases, and dusts that are a *deflagration* hazard.
- Group H-3 facility activities involve materials that represent a physical hazard due to the ability of the materials to support combustion.
- Group H-4 facilities contain materials and involve activities that present health hazards. Laboratories could be found in any of these categories in most major research facilities. Fortunately, however, most laboratories are exempt because they use relatively small quantities of these materials.

The definitions of the subgroups for peroxides, oxidizers, and reactives parallel these definitions for the use groups rather closely and so will not be repeated.

Material Safety Data Sheets, which are now required to be provided by distributors and manufacturers of commercial chemicals, give information on the characteristics of virtually

---

* The tables in this section are derived from those in the 1993 edition of the BOCA code, modified occasionally to improve clarity in the context of this handbook.

all commonly used laboratory chemicals. The definitions of explosive, flammable, combustible, and various health hazards are consistent with those provided by OSHA in CFR 29, Parts 1200, 1450, and 1910, Department of Transportation, CFR 40, Part 173, or other regulatory standards. These are discussed in detail in Chapter 4.

In the 1993 edition of the BOCA code Table 307.8(1) and Table 307.8(2) give the maximum amounts of materials in a given control area which would be exempt. These tables are provided here as Tables 3.1 and 3.2, respectively, in slightly modified form as they would apply to laboratory facilities.

The equivalent table for materials representing health hazards, Table 3.2, provides specific exemption quantities, but the limits are likely to be superseded by other applicable regulations. For example, the possession limits of a laboratory for radioactive materials are always given in Curies (or Becquerels) instead of dose rates and are limited by the terms of a facility's radioactive material license. The dose limits for sealed sources are in excess of most regulations under normal circumstances.

It is possible to have different areas in a building classified differently. If this occurs, then the requirements for each use area shall be met in those areas. Where provisions differ, the requirements providing the greater degree of safety will apply to the entire building, or a complete fire separation will be provided between the two sections. This occurs frequently when major renovations occur, such as adding a new wing to a building in the infill process. Generally, the most restrictive height and area restrictions will still apply to the entire building.

3.1.2.2. Types of Construction

There are five classifications of types of construction. Basically, without providing complete definitions, the construction materials in types 1 and 2 are considered noncombustible; type 3 may include some combustible materials, not including, however, exterior, fire, and party walls; type 4 requires the exterior walls to be of noncombustible materials while the interior structural members may be of heavy solid or laminated timber, without concealed spaces (the allowable dimensions of the wooden components are specified in the BOCA Standard), and type 5 constructions may be made of any materials permitted by the code that have appropriate fire ratings, which are dependent on a number of parameters. With the exception of type 4, these are subdivided into further subclasses. As noted in Table 3.4, for the three types of occupancy laboratories are likely to fall under, the height and area restrictions for type 5 construction would make it unlikely that a typical chemical research laboratory would use this level of construction. Therefore, in Table 3.3, where the fire resistance ratings of the various structural components in each type of construction are given, the data for type 5 construction are deleted.

In order to facilitate the use of the following tables, a number of definitions are in order (paraphrased slightly from those given in the BOCA manual):

- Fire Resistance Rating—The time in hours or fractions thereof that materials or their assemblies will resist fire exposure as determined by the fire test specified by the Code.
- Fire Separation—The distance in feet measured from the building face to the closest interior lot line, to the center line of a street or public way, or to an imaginary line between two buildings on the same property.
- Protected—Construction in which all structural members are constructed or protected in such a manner that the individual unit or the combined assemblage of all such units has the requisite fire resistance rating for its specific use or application.

Table 3.1.   Exempt Amounts of Hazardous Materials, Liquids and Chemicals Presenting a Physical Hazard (Maximum Quantities Per Control Area)[h]

Material	Class	Use Group	Storage[a] Solids lb/ft³	Storage[a] Liquid (gal or lb)	Storage[a] Gases ft³	Closed Systems[a] Solids (lb/ft³)	Closed Systems[a] Liquid (gal or lb)	Closed Systems[a] Gases ft³	Open Systems[a] Solids lb/ft³	Open Systems[a] Liquid (gal or lb)
Combustible Liquids	II IIIA IIIB	H-2 H-3 H-4	NA	120[bc] 330[bc] 13,200[cd]	NA	NA	120[b] 330[b] 13,200[d]	NA	NA	30[b] 8[b] 3,300[d]
Cryogenics Flammable or Oxidizing		H-2	NA	45[b]	NA	NA	45[b]	NA	NA	10[b]
Explosives		H-1	1[cef]	(1)[cef]	NA	1/4[e]	(1/4)[e]	NA	1/4[e]	(1/4)[e]
Flammable Gas	Gaseous Liquified	H-2	NA	NA 30[bc]	1,000[bc] NA	NA	NA 30[bc]	1,000[bc] NA	NA	NA
Flammable Liquid	1A 1B 1C	H-2	NA	30[bc] 60[bc] 90[bc]	NA	NA	30[b] 60[b] 90[b]	NA	NA	10[b] 15[b] 20[b]
Combination (1A,1B,1C)		H-2	NA	120[bcg]	NA	NA	120[bg]	NA	NA	30[bg]
Flammable Solid		H-2	125[bc]	NA	NA	25[b]	NA	NA	25[b]	NA
Organic Peroxide	UD I II III	H-1 H-2 H-3 H-3	1[ce] 5[bc] 50[bc] 125[bc]	(1)[ce] (5)[bc] (50)[bc] (125)[bc]	NA	1/4[e] 1[b] 50[b] 125[b]	(1/4)[e] (1)[b] (50)[b] (125)[b]	NA	1/4[e] 1[b] 10[b] 25[b]	(1/4)[e] (1)[b] (10)[b] (25)[b]

Table 3.1.   Exempt Amounts of Hazardous Materials, Liquids and Chemicals Presenting a Physical Hazard (Maximum Quantities Per Control Area)[h] (continued)

Material	Class	Use Group	Storage[a] Solids lb/ft³	Storage[a] Liquid (gal or lb)	Storage[a] Gases ft³	Closed Systems[a] Solids (lb/ft³)	Closed Systems[a] Liquid (gal or lb)	Closed Systems[a] Gases ft³	Open Systems[a] Solids lb/ft³	Open Systems[a] Liquid (gal or lb)
Oxidizer	4	4	1[c,e]	(1)[c,e]	NA	1/4[e]	(1/4)[e]	NA	1/4[e]	(1/4)[e]
	3	3	10[b,c]	(10)[b,c]		2[b]	(2)[b]		2[b]	(2)[b]
	2	2	250[b,c]	(250)[b,c]		250[b]	(250)[b]		250[b]	(250)[b]
	1	1	1,000[b,c]	(1,000)[b,c]		1,000[b]	(1,000)[b]		1,000[b]	(1,000)[b]
Oxidizer—Gas	Gaseous	H-2	NA	NA	1,500[b,c]	NA	NA	1,500[b,c]	NA	NA
	Liquified			15[b,c]	NA		15[b,c]	NA		
Pyrophoric		H-2	4[c,e]	(4)[c,e]	50[c,e]	1[e]	[f]	10[c,e]	0	0
Unstable (Reactive)	4	4	1[c,e]	(1)[c,e]	10[b,e]	1/4[e]	(1/4)[e]	2[c,e]	1/4[e]	(1/4)[e]
	3	3	5[b,c]	(5)[b,c]	50[b,c]	1[b]	(1)[b]	10[b,c]	1[b]	(1)[b]
	2	2	50[b,c]	(50)[b,c]	250[b,c]	50[b]	(50)[b]	250[b,c]	10[b]	(10)[b]
	1	1	125[b,c]	(125)[b,c]	750[b,c]	125[d]	(125)[d]	750[b,c]	25[d]	(25)[d]
Water Reactive	3	3	5[b,c]	(5)[b,c]	NA	5[b]	(5)[b]	NA	1[b]	(1)[b]
	2	2	50[b,c]	(50)[b,c]		50[b]	(50)[b]		10[b]	(10)[b]

NA = not applicable

a   The total quantity in use and in storage shall not exceed the amount allowed for storage.

b   The amount allowed can be increased by 100% for buildings equipped throughout with an approved sprinkler system; note c also applies and the two are cumulative.

c   The allowed amounts can be increased 100% when materials are stored in approved cabinets, fume hoods, and exhausted enclosures.

d   The permitted amounts are not limited in a building equipped throughout with an approved sprinkler system.

e   Permitted only in buildings equipped throughout with an approved sprinkler system.

f   1 lb of black sporting powder and 20 lb of smokeless powder are permitted in sprinkler or unsprinklered buildings.

g   Containing not more than the exempt amounts of Classes 1-A, 1-B, or 1-C flammable liquids.

h   Quantities in parentheses indicate quantity units in parentheses at head of columns.   1 ft = 0.028 m³, 1 lb = 0.454 kg.

Table 3.2.   Exempt Amounts of Hazardous Materials, Liquids, and Chemicals Presenting a Health Hazard (Maximum Quantities Per Control Area)[a]

Material	Storage[b]			Closed Systems[b]			Open Systems[b]	
	Solid lb[cd]	Liquid (gal or lb)[cd]	Gas ft^3	Solid lb[c]	Liquid (gal or lb)[c]	Gas ft^3	Solid lb[c]	Liquid (gal or lb)[c]
Corrosive	5,000	500	810[cd]	5,000	500	810[cd]	1,000	100
Highly Toxic	1	(1)	20[e]	1	(1)	20[e]	¼	(¼)
Irritant	5,000	500	810[cd]	5,000	500	810[cd]	1,000	100
Radioactive	25 rem - unsealed source 100 rem - sealed source			100 rem sealed sources			25 rem - sealed source	
Sensitizer	5,000	500	810[cd]	5,000	500	810[cd]	1,000	100
Toxic	500	(500)	810[cd]	500	(500)	810[cd]	125	(125)
Other health hazards	5,000	500	810[cd]	5,000	500	810[cd]	1,000	100

[a] Quantities in parentheses correspond to the units in parentheses at the heads of the columns.
[b] The total quantity in use and in storage may not exceed the amount allowed for storage.
[c] The maximum allowed amounts can be increased by 100% in buildings equipped throughout with an approved automatic sprinkler system. Note d also applies and the two allowances are cumulative.
[d] The maximum allowed amounts can be increased by 100% in buildings equipped throughout with an approved automatic sprinkler system. Note c also applies and the two allowances are cumulative.
[e] Permitted only when stored in approved exhausted gas cabinets, exhausted enclosures, or fume hoods.

Table 3.3.   Fire Resistance Ratings of Selected Structural Elements in Hours (abbreviated version of Table 602, BOCA Building Code)*

	Type of construction							
	Noncombustible					Noncombustible/combustible		
	Type 1		Type 2			Type 3		Type 4
	Protected		Protected		Unprotected	Protected	Unprotected	(Heavy timber)
Structural element	1A	1B	2A	2B	2C	3A	3B	4
Exterior walls								
Load bearing	4	3	2	1	0	2	2	2
Non-load bearing	Also must comply with Section 705.2							
	Must comply with Section 705.2							
Fire and party walls	4	3	2	2	2	2	2	2
	Also must comply with Section 707.1							
Fire separation assemblies								
Exit enclosures	2	2	2	2	2	2	2	2
Shafts (other than exits) and elevator hoistways	2	2	2	2	2	2	2	2
Fire partitions, exit access corridors	1	1	1	1	1	1	1	1
	Comply with Section 1011.4: fire retardant-treated wood permitted for types 1 and 2 if fire resistance rating 1 hour or less is required.							
Interior bearing walls, bearing partitions, columns, girders, trusses (other than roof trusses) and framing								
Supporting more than one floor	4	3	2	1	0	1	0	Section 605 governs
Supporting one floor or roof only	3	2	1.5	1	0	1	0	Section 605 governs
Structural members supporting wall	3	2	1.5	1	0	1	0	1

Table 3.3. Fire Resistance Ratings of Selected Structural Elements in Hours (abbreviated version of Table 602, BOCA Building Code) (continued)

	Type of construction							
	Noncombustible					Noncombustible/combustible		
	Type 1		Type 2			Type 3		Type 4
	Protected		Protected		Unprotected	Protected	Unprotected	(Heavy timber)
Structural element	1A	1B	2A	2B	2C	3A	3B	4
	Must not be less than supported wall; also exceptions in selected cases							
Floor construction, including beams	3	2	1.5	1	0	1	0	Section 605 governs
Roof construction, including beams, trusses and framing, arches and roof deck (15' or less in height to lowest member)	2	1.5	1	1	0	1	0	Section 605 governs
					Fire retardant-treated wood permitted for types 1 and 2 if required fire resistance rating 1 hour or less			

* This table is included only to illustrate certain design issues. Other sections of the Code will have to be considered as well, and there are possible exceptions, exemptions, or variations permitted depending upon other factors.

Table 3.4.   Height and Area Limits Use Class

Type of Construction	Business	Educational	Hazard, H-2	Hazard, H-3
Noncombustible				
1A (protected)	Not limited	Not limited	16,800 5 stories, 65 ft	33,600 7 stories, 85 ft
2B (protected)	Not limited	Not limited	14,400 3 stories, 40 ft	28,800 7 stories, 85 ft
2A (protected)	34,200 7 stories, 85 ft	34,200 5 stories, 65 ft	11,400 3 stories, 40 ft	22,800 6 stories, 75 ft
2B (protected)	22,500 5 stories, 65 ft	22,500 3 stories, 40 ft	7,500 2 stories, 30 ft	15,000 4 stories, 50 ft
2C (unprotected)	14,400 3 stories, 40 ft	14,400 2 stories, 30 ft	4,800 1 story, 20 ft	9,600 2 stories, 30 ft
Noncombustible/Combustible				
3A (protected)	19,800 4 stories, 50 ft	19,800 3 stories, 40 ft	6,600 2 stories, 30 ft	13,200 3 stories, 40 ft
3B (unprotected)	14,400 3 stories, 40 ft	14,400 2 stories, 30 ft	4,800 1 story, 20 ft	9,600 2 stories, 30 ft
4 (heavy timber)	21,600 5 stories, 65 ft	21,600 3 stories, 40 ft	7,200 2 stories, 30 ft	14,400 4 stories, 50 ft

- Walls:

    - Bearing Wall: Any wall supporting any additional vertical load in addition to its own weight.
    - Fire Wall: A fire resistance-rated wall, which may have protected openings, which is intended to restrict the spread of a fire and which is continuous from the foundation to or through the roof of a building.
    - Fire Separation Wall: Similar to a fire wall in that it is intended to restrict the spread of a fire but does not include the requirement of extending from the foundation to the roof of a building.
    - Party Wall: A fire wall on an interior lot line used for joint service between two buildings.

A number of points are immediately obvious in examining Table 3.3. In order to qualify for type 1 construction requires fire resistance ratings substantially higher than for other classes. It costs more to achieve these ratings. Ratings for load bearing walls are much higher than for walls which serve only to separate spaces. Walls which serve to prevent the spread of fire have ratings as high as or higher than bearing walls. Not given in the table is the additional requirement that fire walls must have sufficient structural stability that they will survive the collapse of the structure on either side. An important point to note is that shafts such as those that might enclose a fume hood duct, penetrating a fire separation such as a floor, must be of noncombustible material and provide a two hour fire resistance rating. The duct itself, enclosed in a shaft meeting these criteria, would not need to meet any additional fire resistance rating since it is enclosed in a shaft which does. Note that type 2C construction places no requirements of fire resistance ratings for structural components, except for those items directly related to fire separation or components associated with paths of egress. Type 3B construction is comparably tolerant for most internal components, although it is similar to types 2A and 2B for exterior walls.

3.1.2.3. Area and Height Limitations

Table 3.4, again a shorter version of one given in BOCA, gives the maximum heights, number of stories above grade, and, for buildings of only one or two stories facing on one street or public space not less than 30 feet wide, the area limitations for the classes of occupancy which are the most probable for laboratory facilities—business, educational, high hazard, and moderate hazard storage. Note that the areas are for the maximum horizontal projection of the building. There are provisions which would allow these limits to be exceeded or would allow area to be traded for additional height. For example, in Table 3.4, except for the hazard use group, the areas in the table can be increased by 200% for one- and two-story buildings and 100% for buildings with more than two stories if they are equipped throughout with an automatic fire alarm system. Similarly, for each 1% of a building's perimeter in excess of 25% fronting on a street or a vacant space, 30 feet or more wide, accessible by a legitimate (more than 18 feet wide) marked fire lane, the building areas can be increased by 2%. An automatic fire suppression system for the entire building also allows the height of a building to be increased by one additional story, or 20 more feet, again with the exception of a hazard use group building. If land values are high or space is limited, this is an option. However, the difficulty in evacuating multistory buildings argues against buildings of excessive height, especially as noted earlier for academic buildings, where the likelihood is that many of the building occupants will not be familiar with evacuation routes, coupled with the possibility that a laboratory fire may involve toxic fumes. This concept is supported in the Code by the disparity between business and educational use occupancies in Table 3.4. For example, although the areas are the same for each, one or two fewer stories are permitted for each type of construction for a building classified as an educational occupancy.

In order to erect a building more than the one or two stories for which the area figures in Table 3.4 are relevant, a penalty must be paid in terms of allowable area per floor, except for 1A and 1B construction. For 2A construction, this penalty amounts to 5% for each story above two. For all other types of construction, there is a 20% penalty for three- and four-story buildings and an additional penalty of 10% more for each story above four.

Although laboratory towers have been built, it is clear that a heavy penalty in either area per floor or in construction costs, due to building to a higher resistance rating, must be paid for erecting a code-conforming high-rise laboratory building. Since there are significant reductions in the fire resistance ratings required for structural components for 2B construction, and still more for 2C, compared to 2A, and since there are, as noted earlier, problems associated with evacuating multistory laboratory buildings, a good choice for a reasonably size laboratory facility might well be a 2B business use occupancy structure.

There are other factors and conditions that may become involved in determining the allowable area, height, etc., in addition to the ones discussed in this section. However, the intent is not to provide a course in code review, which involves much more sophisticated details than it would be possible to cover here, but to provide sufficient basic information for laboratory personnel, to allow them to understand the constraints under which the designer operates. The details of the final design must be negotiated among the architect, contractor, building official, and representatives of the owner. The participation of laboratory personnel is essential to define their program needs in the context of what is permissible under the building code. Code issues are not always clear cut, with much of the actual language subject to interpretation. Also, there are often alternative ways to provide equivalent protection so that requests to code officials for variances, based on this concept, are frequently acceptable.

There will be additional code issues addressed in many of the following sections where specific design features will be discussed in detail.

3.1.3. Laboratory Classification

There are no universal safety criteria to classify laboratories which take into account all types of risks. Standard 45 of the National Fire Prevention Association (NFPA) designates chemical laboratories of different degrees of risk, based essentially on fire safety factors, regulating the amount of solvents which each class may contain. The Centers for Disease Control has published a set of guidelines establishing a biological safety level rating system for laboratories in the life sciences and those using animals, based on a number of parameters relating to the infectiousness to humans of the organisms used in the facility. This system parallels an earlier four-level classification scheme developed for those working in recombinant DNA research. The standards associated with biological organisms are concerned with the potential risk to the public at large as well as to the laboratory workers. The Department of Agriculture regulates the importation, possession, or use of a number of nonindigenous pathogens of domesticated animals. The Drug Enforcement Agency licenses and sets standards for facilities in which controlled substances are employed to ensure that they are used safely and to guard against their loss or theft. The Nuclear Regulatory Commission (NRC) licenses agencies or individuals using radioactive materials to ensure that neither the workers nor the general public are adversely affected by the use of radiation. To obtain a license, one must demonstrate the competence to use the material safely and to be able and willing to meet an extremely detailed set of performance standards. All of these standards have been developed essentially independently and, where a regulatory agency is involved, are administered separately. In many instances laboratory operations will be affected by several sets of regulations. However, even if all of the regulatory standards were imposed simultaneously, there would still be many safety factors which would not be included. Thus, it is at least partially the responsibility of the institution or corporation to establish additional criteria to properly evaluate the degree of risk in a research program and to assign it to a suitable classified space.

Working with materials with low risk potentials will obviously be much more tolerant of poor facilities or procedures than using materials involving a high risk but not totally so. Even a small quantity of a IA flammable solvent such as ether, used in an inadequate facility, could lead to a serious accident, while the same quantity, used in a fume hood by a careful worker following sound safety procedures, could be used quite safely.

The OSHA laboratory standard mandates that performance standards be established in each facility that would ensure that the employees would be as well protected as those working in industrial situations, for which long-established general industry standards apply. This appears to bypass, at least as far as OSHA is concerned, the need for any sort of laboratory classification scheme, leaving the responsibility primarily to the local laboratory or organization. The OSHA laboratory standard does not replace the biological guidelines since the OSHA standard does not at this time include pathogens as a possible risk, nor would it supersede radiation safety standards. There is also the difficulty that research programs tend to evolve and could change the level of risk involved over a short period of time. It would be impractical to be continually shifting occupants of space as this occurred. However, the flexibility permitted by the standard laboratory module described earlier in this chapter permits easy and economical changes in a facility to modify the quality of the space for different levels of risk.

Although it is unlikely that a formal system of classifying laboratories according to a comprehensive safety standard is imminent, it does appear incumbent upon an institution or corporation to ensure that research is assigned to space suitably designed and equipped so that research can be performed with a reasonable assurance of safety. If research programs are evaluated properly, it should be possible to assign them to laboratories classified into low, moderate, substantial, and high-risk categories. This type of classification seems to be the simplest and most practical to use and has the further advantage of already being employed in life science laboratories. Before examining the features that might be incorporated into each category, which will depend somewhat upon the area of research involved, it might be well to list at least some of the parameters that should be considered in evaluating research programs.

3.1.3.1. Program-Related Factors

Evaluation of programs to permit assignment to the appropriate class of facility should depend upon several factors:

I. Materials
   A. Recognized risks
      1. Flammable
      2. Reactive
      3. Explosive
      4. Acute toxicity
      5. Known systemic or chronic health effects
         a. Carcinogens
         b. Mutagens, teratogens
         c. Affect reproduction/fertility
         d. Radiation
         e. Pathogens
         f. Affect the respiratory system
         g. Neurotoxic
         h. Known strong allergens
         i. Other known health effects
      6. Physical risks
         a. Electrical
         b. High pressure
         c. Heat and cold

        d. Sound
        e. Nonionizing radiation/light
        f. Other physical risk factors
    B. Quantities/scale of operations
    C. Procedures

  II. Information/training
    A. Health and safety training
       1. Documentation of safety and health training for laboratory managers
       2. Procedures to train new personnel
       3. Procedures to train all personnel when new materials/new procedures are used
    B. Material Safety Data Sheets available for all chemicals used
    C. General health and safety program in effect

  III. Personnel protection
    A. Exposure monitoring
    B. Personal protective equipment available
    C. Health assurance program available

The information in Part I above is, in effect, an evaluation of possible negative aspects of the program under consideration, while positive information under each of the items in Parts II and III can be used to offset, to some degree, the needs which must be met by the facility. It is preferable, however, to design-in safety rather than depending upon procedures and administrative rules.

3.1.3.2. Laboratory Class Characteristics

In the following four sections, oriented primarily toward chemistry laboratories, the reader already familiar with laboratory classification guidelines established by the Centers for Disease Control will note that in many respects the recommendations for low, moderate, substantial, and high-risk categories closely parallel those for biosafety levels one through four. It will be noted that this system will involve classifying laboratory facilities by much more than the configuration of bricks and mortar of which they are built or their contents of a single type of hazardous material, although these aspects will be important. Separate major sections in Chapter 5 are devoted to laboratories in the life sciences, animal facilities, and radiation, so reference to topics relevant to those areas will be deferred to those sections.

3.1.3.2.1. Low-Risk Facility

A low-risk facility is used for work with materials, equipment, or classes of operations, with no known or minimal risk to the workers, the general public, or to the environment. It is possible to work safely with all the necessary materials on open benches. No special protection or enclosures are needed for the equipment or operations. There is a written laboratory safety plan to which all the employees have access. Laboratory workers have been properly trained in laboratory procedures and are supervised by a trained and knowledgeable person. If there are any potential risks, the employees have been informed of them, how to detect them if they are not immediately obvious, and emergency procedures.

Although the laboratory design requirements are not stringent, features which would be difficult to change if the utilization should become one which would require a higher classification should be built to a higher level. Examples below marked with an asterisk (*) are provisions for easily cleaned and decontaminated floors and laboratory furniture and good ventilation.

Standard Practices

1. Access to the laboratory is limited at the discretion of the laboratory supervisor, as needed.
2. A program exists to ensure that reagents are stored according to compatibility.
3. An annual (or continuous) chemical inventory will be performed and information sent to a central data collection point. Outdated and obsolete chemicals will be disposed of through a chemical waste disposal program.
4. The laboratory will be maintained in an orderly fashion.
5. Although it is anticipated that the amount of hazardous chemicals used in a low risk facility will be very limited, all secondary containers containing materials incorporating more than 1.0% of a hazardous component or combination of hazardous components, which will be used more than a single work day, shall be labeled with a label listing the hazardous components (not required under the new laboratory standard, but good practice).
6. Any chemical wastes are placed in appropriate and properly identified containers for disposal through a chemical waste disposal program. Broken glass is disposed of in heavy cardboard or kraftboard boxes labeled "broken glass." Any "sharps," as defined under the blood-borne pathogen standard, will be placed in a legal container for disposal as infectious waste. Only ordinary solid, nonhazardous waste may be placed in ordinary trash containers.
7. Eating, drinking, and smoking are not permitted in the work area.
8. No food or drink can be placed in refrigeration units used in the laboratory.
9. The telephone numbers of the laboratory supervisor, any alternates, and the department head shall be posted on the outside of the laboratory door or the adjacent wall.

Special Practices

There are no special practices associated with a low-risk laboratory.

Special Safety Equipment

1. Any refrigerators or freezers shall be rated as acceptable for "Flammable Material Storage" except for ultra-low temperature units.*
2. No other special safety equipment is needed.

Laboratory Facilities

1. The floor of the laboratory is designed to be easily cleaned. Seamless floors and curved junctures to walls aid in accomplishing this.*
2. Bench tops should be resistant to the effects of acids, bases, solvents, moderate heat, and should not absorb water. The tops should have few seams or crevices to facilitate cleaning.*
3. Furniture should be designed to be sturdy and designed for convenient utilization. Storage spaces should be easily accessible.

4. Aisle spaces should be 40 to 48 inches wide and not constricted to less than 28 inches by any temporary obstacles.
5. Electrical outlets shall be three-wire outlets, with high-quality, low-resistance ground connections. Circuits should be clearly identified to correlate with labels in breaker panels.
6. The laboratory should be supplied with a sink. The plumbing shall be sized to accommodate a deluge shower and eyewash station. With average water pressure, this would normally be a 1-inch line or larger.
7. Normal building ventilation is sufficient. However, it is recommended that six air changes per hour of 100% fresh air be provided as standard.

3.1.3.2.2. Moderate-Risk Facility

A moderate-risk facility involves material, practices, and use of equipment such that improper use could pose some danger to the employees, the general public or the environment. Generally, the materials used would have health, reactivity, or flammability ratings, according to NFPA Standard 704 of 2 or less. Small quantities of materials with higher ratings might be involved in work being performed in chemical fume hoods or in closed systems. Work with special risks, such as with carcinogens, would not be performed in a moderate-risk facility. Equipment which could pose a physical hazard should have adequate safeguards or interlocks. However, in general, most operations could be safely carried out on an open work bench or without unusual precautions. The amounts of flammables kept in the laboratory meet NFPA standard 45 for Class A laboratories (or less), and when not in use are stored in either a suitable flammable material storage cabinet or other storage unit.

The person responsible for the work being performed in the laboratory is to be a competent scientist. This individual shall develop and implement a safety and health program for the facility that meets the requirements of the OSHA lab standard. The individual workers are to be fully trained in the laboratory procedures being employed and to have received special training in the risks specifically associated with the materials or work being performed. The workers are to be informed about the means available to them to detect hazardous conditions and the emergency procedures that should be followed, should an incident occur.

Standard Practices

1. Access to the laboratory work area is limited during the periods work is actively in progress, at the discretion of the laboratory supervisor.
2. A program exists to ensure that chemicals are stored properly, according to compatibility. Quantities of chemicals with hazard ratings of 3 or greater are limited to the amount needed for use in a 2-week interval, or in accordance with NFPA standard 45 for flammables, whichever is less.
3. An annual (or continuous) chemical inventory will be performed and sent to a central data collection point. Outdated and obsolete chemicals will be disposed of through a chemical waste disposal program. Ethers and other materials which degrade to unstable compounds shall be shelf dated for disposal 6 months after being opened, but no more than 12 months after purchase, even if unopened, unless processed to remove any unstable peroxides that may have formed.
4. A Material Safety Data Sheet file will be maintained for all chemicals purchased for use in the laboratory. The file will be accessible to the employees in the laboratory.

All laboratory workers shall be trained in how to interpret the information in a MSDS.

5. All secondary containers, containing materials containing more than 10% of a hazardous component or combination of hazardous components, which will be used more than a single work day shall be labeled with a label listing the hazardous components.

6. Any chemical wastes are placed in appropriate and properly identified containers for disposal through a chemical waste disposal program. Broken glass is disposed of in heavy cardboard or kraftboard boxes labeled "broken glass." Any "sharps," as defined under the blood-borne pathogen standard, will be placed in a legal container for disposal as infectious waste. Only ordinary solid, nonhazardous waste may be placed in ordinary trash containers.

7. The laboratory will be maintained in an orderly fashion.

8. No food or drink can be placed in refrigeration units used in the laboratory.

9. A placard or other warning device shall be placed on the door or on the wall immediately adjacent to the door identifying the major classes of hazards in the laboratory. (See Section 2.3.4.)

10. The telephone numbers of the laboratory supervisor, any alternates, and the department head shall be posted on the outside of the laboratory door or the adjacent wall.

Special Practices

1. Work with materials with safety and health ratings of 3 or greater in any category shall be performed in a functioning fume hood.

2. Work with substantial amounts of materials with hazard ratings of 1 or 2 shall be performed in a hood or in an assembly designed to be safe in the event of a worst-case failure.

3. Appropriate personal protective equipment shall be worn in the work area. Because eyes are critical organs very susceptible to chemical injuries or minor explosions, it is recommended that chemical splash goggles be worn whenever the work involved offers any possibility of eye injury. Wearing of contact lenses should be avoided, but if an individual must wear them for medical reasons, then that individual must wear chemical splash goggles at all times in the laboratory. A mask may be used to supplement the minimum eye protection.

Special Safety Equipment

1. Any refrigerators or freezers shall be rated as acceptable for "Flammable Material Storage" except for ultra-low temperature units.

2. A flammable material storage cabinet, either built-in or free standing, shall be used for the storage of flammable materials.

3. The laboratory shall be equipped with a fume hood.

4. The laboratory shall be equipped with an eyewash station and a deluge shower.

5. The laboratory shall be provided with one or more Class 12 ABC fire extinguishers.

6. Any special equipment mandated by the research program shall be provided.

Laboratory Facilities

1. The floor of the laboratory is designed to be easily cleaned. Seamless floors and curved junctures to walls aid in accomplishing this.

2. Bench tops should be resistant to the effects of acids, bases, solvents, and moderate heat, and should not absorb water. To facilitate cleaning, the tops should have few seams or crevices.

3. Furniture should be designed to be sturdy and designed for convenient utilization. Storage spaces should be easily accessible.
4. Aisle spaces should be 40 to 48 inches wide and shall not be constricted to less than 28 inches by any temporary obstacles.
5. Electrical outlets shall be three-wire outlets with high-quality, low-resistance ground connections. Circuits should be clearly identified to correlate with labels in breaker panels.
6. The laboratory shall be supplied with a sink. The trap should be of corrosion-resistant material. The plumbing shall be sized to accommodate the deluge shower and eyewash station. With average water pressure, this would normally be a 1-inch line or larger.
7. Six air changes per hour of 100% fresh air shall be supplied to the facility. No air shall be recirculated. The ventilation system shall be designed such that the room air balance is maintained at a small negative pressure with respect to the corridors whether the fume hood is on or off.
8. It is recommended that the facility include a separation of work spaces and desk areas as well as a second exit, as shown in the standard laboratory module, Figure 3.1 (see Section 3.1.1).

3.1.3.2.3. Substantial Risk Facility

For the two lower risk categories, it is possible to be almost completely general since they are specifically intended to be used for only limited risks. However, for both substantial risk and high-risk facilities, the nature of the risk will dictate specific safety-related aspects of the facility. Most of these can be accommodated at the substantial risk level within the standard laboratory module, appropriately modified and equipped.

The use of highly toxic (or having a seriously detrimental health characteristic, such as a potential carcinogen), highly reactive, or highly flammable chemicals or gases would mandate the work being conducted within at least a substantial risk facility. If explosives are involved, then the laboratory should be designed with this in mind. Explosion venting may be required in this instance. The location of the facility may be dictated by the need to contain or control the debris or fragments from an explosion. The level of construction may need to be enhanced to make the walls stronger to increase their explosion resistance. The use of toxic or explosive gases may require continuous air monitoring with alarms designed to alert the occupants of levels approaching an action level, which should be no higher than 50% of the level representing either a permissible exposure limit (PEL) or the lower explosive limit (LEL). The alarms must be connected to the building alarm system, which in turn should be connected to a central manned location. Highly flammable materials may require special automatic extinguisher systems, using high-speed fire detectors, such as ultraviolet light sensors coupled with dry chemical or Halon* fire suppression systems. There are, of course other risks, as tabulated in Section 3.1.3.1, which would require other precautions.

Access to a substantial risk facility should be restricted during operations and at other times to authorized personnel only at the discretion of the laboratory supervisor. The

---

* Note the discussion in Chapter 2 about the imminent phasing out of currently used chlorinated fluorocarbons due to the negative effect these materials have on the earth's ozone layer. In the context of this recommendation, alternatives of equivalent efficacy to halons should be used when available.

laboratory supervisor shall be a competent scientist, having specific knowledge and training relevant to the risks associated with the program of research in the laboratory. Each person authorized to enter the laboratory shall have received specific safety training appropriate to the work and to the materials employed. A formal, written laboratory industrial hygiene plan, including an emergency plan complying with the requirements of the OSHA laboratory standard, shall be developed and practiced at least annually. A copy of the emergency plan shall be provided to all agencies, including those outside the immediate facility, who would be called upon to respond to an incident. The emergency plan shall include a list of all personnel in the facility with business and home telephone numbers.

Standard Practices

1. Access to the laboratory is limited to authorized personnel only during operations, and to others at times and under such conditions as designated by written rules or as established by the laboratory supervisor.
2. All chemicals must be stored properly, according to compatibility. Any chemicals which pose a special hazard or risk shall be limited to the minimum quantities required to meet short-term needs of the research program, and materials not in actual use shall be stored under appropriate, safe conditions. For example, flammables not in use shall be kept in a flammable materials storage cabinet, and excess quantities of explosives should be stored in magazines, away from the immediate facility. Other materials such as drugs or radioactive materials may also require secured storage areas.
3. An annual (or continuous) chemical inventory shall be performed and sent to a central data collection point. A continuous inventory log shall be maintained of chemicals posing special risks. Outdated and obsolete chemicals will be disposed of through a chemical waste disposal program. Ethers and other materials which degrade to unstable compounds shall be shelf dated for disposal 6 months after being opened, but no more than 12 months after purchase, even if unopened, unless processed to prevent dangerous levels of the unstable peroxides from forming.
4. A Material Safety Data Sheet file will be maintained for all chemicals purchased in the laboratory for which they are available. In some cases, such as experimental compounds being tested, they are not available. Any information provided by the manufacturer will be kept in such cases. The file will be accessible to the employees in the laboratory. All laboratory workers shall be trained in how to interpret the information in a MSDS.
5. All secondary containers containing materials having more than 1% of a hazardous component or combination of hazardous components (0.1% for carcinogens), which will be used more than a single work day, shall be labeled with a label listing the hazardous components.
6. Any chemical wastes are placed in appropriate and properly identified containers for disposal through a chemical waste disposal program. Any wastes which pose a special hazard or fall under special regulations and require special handling shall be isolated and a program developed to dispose of them safely and legally. Broken glass is disposed of in heavy cardboard or kraftboard boxes labeled "broken glass." Any "sharps," as defined under the blood-borne pathogen standard, will be placed in a legal container for disposal as infectious waste. Only ordinary solid, nonhazardous waste may be placed in ordinary trash containers.
7. The laboratory will be maintained in an orderly fashion. Any spills or accidents will be promptly cleaned up and the affected area decontaminated or rendered safe.

8. No food or drink can be brought into the operational areas of the laboratory, nor can anyone smoke or apply cosmetics.
9. Any required signage or posting mandated by any regulatory agency shall be posted on the outside of the door to the entrance to the laboratory. In addition, a placard or other warning device shall be placed on the door or on the wall immediately adjacent to the door identifying any other major classes of hazards in the laboratory (see Section 2.3.4). A sign shall be placed on the door stating in prominent letters, meeting any regulatory standards, "AUTHORIZED ADMISSION ONLY."
10. The telephone numbers of the laboratory supervisor, any alternates, and the department head shall be posted on the outside of the laboratory door or the adjacent wall.

Special Practices

1. Specific policies, depending upon the nature of the hazard, shall be made part of the laboratory industrial hygiene and safety plan and scrupulously followed to minimize the risk to laboratory personnel, the general public, and the environment. Several examples of laboratory practices for various hazards are given below. This list is not intended to be comprehensive, but instead represents some of the more likely special precautions needed for a variety of types of risks.

   - All work with hazardous kinds or quantities of materials shall be performed in a fume hood or in totally enclosed systems. It may be desirable for the hood to be equipped with a permanent internal fire suppression system.
   - Work with explosives shall be limited to the minimum quantities needed. For small quantities used in a hood, an explosion barrier in the hood, with personnel wearing protective eye wear, face masks, and hand protection, may be sufficient protection. For larger quantities, the facility must be specifically designed for the research program.
   - Some gases, such as fluorine, burn with an invisible flame. Apparatus for work with such materials should be placed behind a barrier to protect against an inadvertent introduction of a hand or other part of the body, so as to prevent burns.
   - Systems containing toxic gases that would be immediately dangerous to life and health (IDLH) or gases that could pose an explosive hazard if allowed to escape, especially if they have no sensory warning properties, shall be leak tested prior to use and after any maintenance or modification which could affect the integrity of the system. Where feasible, the gas cylinders may be placed external to the facility and the gases piped into the laboratory to help minimize the quantity of gas available to an incident. Permanently installed gas sensors, capable of detecting levels of gas well below the danger limits may be needed in some cases.
   - Vacuum systems capable of imploding, resulting in substantial quantities of glass shrapnel or flying debris, shall be protected with cages or barriers or, for smaller systems, shall be wrapped in tape.
   - Systems representing other physical hazards, such as high voltage, radiation, intense laser light beams, high pressure, etc., shall be marked with appropriate signs and interlocked so as to prevent inadvertent injuries. The interlocks shall be designed to be fail safe such that no one failure of a component would render the safety interlock system inoperative.

2. Activities in which the attention of the worker is not normally engaged with laboratory operations, such as record maintenance, calculations, discussions, study, relaxation, etc., shall not be performed in the laboratory proper, but shall be performed in an area isolated from the active work area. The segregated desk area of the standard laboratory module is specifically intended to serve this purpose. Depending upon the nature of the hazard, it is usually economically feasible to make at

least a portion of the barrier separating the two sections of the laboratory transparent so that operations can be viewed, if necessary.

3. Workers in the laboratory should participate in a medical surveillance program if they actively use materials for a significant portion of their work week, which would pose a significant short- or long-term risk to their health. Employees shall be provided medical examinations if they work with any material requiring participation in a medical program by OSHA or other regulatory agencies under conditions which do not qualify for an exemption. Employees shall notify the laboratory supervisor as soon as possible of any illness that might be attributable to their work environment. Records shall be maintained of any such incident.

4. No safety feature or interlock of any equipment in the facility shall be disabled without written approval of the laboratory supervisor. Any operations which depend upon the continuing function of a critical piece of safety equipment, such as a fume hood, shall be discontinued should the equipment need to be temporarily removed from service for maintenance. Any such item of equipment out of service shall be clearly indicated with a signed "Out of Service" tag. Only the person originally signing the tag, or a specific, designated alternate, shall be authorized to remove the tag.

5. It shall be mandatory to wear any personal safety equipment required for conducting operations safely in the laboratory.

6. It is recommended that a laboratory safety committee review each new experiment planned for such a facility to determine if the experiment can be carried out safely in the facility. If the risk is such that experiments may affect the environment or the surrounding community, it is recommended that the committee include at least one layperson from the community, not currently affiliated directly or indirectly with the institution or corporation. In this context, "new" is defined as being substantially different in character, scope, or scale from any experiment previously approved for the facility.

Special Safety Equipment

1. Any refrigerators or freezers shall be rated as acceptable for "Flammable Material Storage," except for ultra-low temperature units.

2. A flammable material storage cabinet, either built-in or free standing, shall be used for the storage of flammable materials.

3. The laboratory shall be equipped with a fume hood. The fume hood should meet any specific safety requirements mandated by the nature of the research program. A discussion of hood design parameters will be found in a later section, but for high hazard use the interior of the hood and the exhaust duct should be chosen for maximum resistance to the reagents used; the blower should either be explosion-proof or, as a minimum, have nonsparking fan blades; the hood should be equipped with a velocity sensor and alarm; the interior lights should be explosion-proof, and all electrical outlets and controls should be external to the unit. It may be desirable to equip the unit with an internal automatic fire suppression system.

4. The laboratory shall be equipped with an eyewash station and a deluge shower.

5. The laboratory shall be equipped with a fire alarm system connected so as to sound throughout the building (and in a central facility manned 24 hours per day) and an appropriate fire suppression system and be provided with one or more class 12 BC, or larger, fire extinguishers, or class D units if reactive metals are in use.

6. An emergency lighting system shall be provided.

7. A first-aid kit shall be provided and maintained.

8. Any special safety equipment mandated by the research program shall be provided. For example, electrical equipment other than refrigerators may need to be designed to be explosion-safe.

Laboratory Facilities

1. The floor of the laboratory is designed to be easily cleaned. Durable, seamless floors of materials that are substantially impervious to spilled reagents are easily decontaminated, and have curved junctures to walls to aid in accomplishing this.
2. Two well-separated exit doors shall be available to the laboratory which shall swing in the direction of exit travel.
3. Bench tops should be resistant to the effects of acids, bases, solvents, and moderate heat, and should not absorb water. To facilitate cleaning, the tops should have few seams or crevices.
4. Casework should be designed to be sturdy and designed for convenient utilization. Storage spaces should be designed to meet any special requirements and should be easily accessible. It should not be necessary, for example, to stretch to reach any reagent which, if dropped, could represent a safety problem.
5. Aisle spaces should be 40 to 48 inches wide and shall not be constricted to less than 28 inches by any temporary obstacles. The aisles should lead as directly as possible toward a means of egress.
6. The organization of the facility shall be such as to reduce the likelihood of having to pass an originating or secondary hazard to evacuate the facility in the event of an emergency.
7. Electrical outlets shall be three-wire outlets with high-quality, low-resistance ground connections. Circuits should be clearly identified to correlate with labels in breaker panels. Some locations would need to be equipped with ground-fault interrupters (GFIs), such as where electrical connections are near sinks.
8. Laboratories in which the risk of electrical shock is greater than normal may also be equipped with a master "panic" manual electrical disconnect clearly marked and located in a readily accessible location.
9. The laboratory shall be supplied with a sink. The trap shall be of corrosion-resistant material. The plumbing shall be sized to accommodate the deluge shower and eyewash station. With average water pressure, this would normally be a 1-inch line or larger.
10. At least six air changes per hour (higher risk laboratories often are designed for 10 to 20) of 100% fresh air shall be supplied to the facility. No air shall be recirculated. The ventilation system shall be designed such that the room air balance is maintained at a small negative pressure with respect to the corridors whether the fume hood is on or off. Where toxic and explosive gases and fumes are present, the system is to be designed to be efficient in exhausting these fumes by locating the exhaust intakes either very near the source of fumes or near the floor (except for lighter-than-air or hot gases). Typical air flow patterns are to be such as to draw dangerous fumes away from the normal breathing zones of the laboratory's occupants.
11. The facility shall include a separation of work spaces and desk areas as well as a second exit, equivalent to the arrangement shown in the standard laboratory module, Figure 3.1 (see Section 3.1.1).

3.1.3.2.4. High-Risk Facility

A distinguishing feature of a high-risk facility is that the operations of the laboratory pose an immediate and substantial danger to the occupants, the general public, or the environment if not performed safely in a suitable facility. The users of the facility and those permitted access to it must be limited to those individuals of the highest competence, training, and character. The OSHA required laboratory safety plan must include training specifically tailored to inform the personnel in the facility of the risks to which they are exposed, the mandatory preventive safety procedures which must be followed, and the measures which must be taken in an emergency. Because it is so difficult to guarantee the degree of safety which must be met, a typical academic building would not normally be suitable, nor would most common industrial research facilities, without substantial modifications.

A second distinguishing feature of a high-risk facility is the need for isolation. If, for example, specific exceptions are permitted under the building code, then a building of use group H (hazard) shall not be located within 200 feet of the nearest wall of a building of use group A (assembly), E (education), or I (institutional). There are other restrictions under the code for specific types of risk, but each of these again define conditions of isolation. In some cases this is achieved by distance, as above. In other instances, isolation is achieved by building walls and other structural components to a higher than normal level of construction. In cases in which the level of risk is not so much physical, as is basically the concern of most building codes, but involves toxic materials or biologically pathogenic organisms, isolation can be achieved by such devices as airlocks and hermetically sealed doors. Where the risk is biological, isolation may be achieved in part by autoclaving and/or treating and disinfecting all garments, waste, and other items leaving the facility. Personnel may be required to wear self-contained, air-supplied suits while inside the facility or, in extreme cases, conduct all operations inside glove boxes or enclosures using mechanical and electrical manipulating devices. Exhaust air from such a facility may require passing through a flame to kill any active organisms. Where the risk is of this character, rather than representing a danger due to fire or explosion, it may be possible to accommodate the facility within a building of generally lower risk level as defined by the applicable building code.

It will be noted that the four sections following are similar to those for the substantial risk facility. However, there are some significant differences.

Standard Practices

1. Access to the laboratory is limited to authorized personnel only, except at times and under such conditions as designated by written rules established by the laboratory supervisor and when accompanied by an authorized individual. The doors shall be locked at all times, with a formal key (or equivalent) control program in place.
2. All chemicals must be stored properly, according to compatibility. All chemicals which pose a special hazard or risk shall be limited to the minimum quantities needed for the short-term need of the research program, and materials not in actual use shall be stored under appropriate safe conditions. For example, flammables not in use shall be kept in a flammable material storage cabinet, or excess quantities of explosives shall be stored in magazines, away from the immediate facility. Other materials such as drugs or radioactive materials may also require secured storage areas.
3. An annual (or continuous) chemical inventory shall be performed and sent to a central data collection point. A continuous inventory log shall be maintained of chemicals posing special risks, and a formal accountability program implemented. Outdated and obsolete chemicals will be promptly disposed of through a chemical waste disposal program. Ethers and other materials which degrade to unstable compounds shall be shelf dated for disposal 6 months after being opened, but no more than 12 months after purchase, even if unopened, unless processed to prevent dangerous levels of unstable peroxides from forming.

4. A Material Safety Data Sheet file will be maintained for all chemicals purchased in the laboratory for which they are available. In some instances, such as experimental compounds being tested, these are not available. Where equivalent data exist in whole or in part, this information will be made part of the MSDS file. The file will be accessible to the employees in the laboratory at all times. All laboratory workers shall be trained in how to interpret the information in a MSDS.

5. All secondary containers containing materials having more than 1% of a hazardous component or combination of hazardous components (0.1% for carcinogens), which will be used more than a single work day, shall be labeled with a label listing the hazardous components.

6. All hazardous wastes are placed in appropriate and properly identified containers for disposal through a hazardous waste disposal program. Any wastes which pose a special hazard, or fall under special regulations and require special handling (such as human blood, tissue, and other bodily fluids regulated under the blood-borne pathogens standard), shall be isolated and a program developed to dispose of them safely and legally. Normal, nontoxic waste shall be disposed of according to standard practices appropriate to such wastes, subject to any restrictions needed to prevent breaching any isolation procedures.

7. The laboratory will be maintained in an orderly fashion. Any spills or accidents will be promptly cleaned up and the affected area decontaminated or rendered safe.

8. No food or drink can be brought into the operational areas of the laboratory, nor can anyone smoke or apply cosmetics.

9. Any required signage or posting mandated by any regulatory agency shall be posted on the outside of the door to the entrance to the laboratory. In addition, a placard or other warning device shall be placed on the door or on the wall immediately adjacent to the door identifying any other major classes of hazards in the laboratory (see Section 2.3.4). A sign meeting any regulatory standards shall be placed on the door stating in prominent letters, "AUTHORIZED ADMISSION ONLY."

10. The telephone numbers of the laboratory supervisor, any alternates, and the department head shall be posted on the outside of the laboratory door or the adjacent wall.

Special Practices

1. Specific policies, depending upon the nature of the hazard, shall be made part of the OSHA-mandated laboratory safety plan and scrupulously followed to minimize the risk to laboratory personnel, the general public, and the environment. Several examples of laboratory practices for various hazards are given below. This list is not intended to be comprehensive, but instead represents some of the more likely special precautions needed for a variety of types of risks.

   • All work with hazardous kinds or quantities of materials shall be performed in a fume hood, specifically designed to provide the maximum safety for the hazard involved or in a totally enclosed system. It may be desirable for the hood or enclosed system to be equipped with a permanent internal fire suppression system. If the work involves a material which could be hazardous to the public or to the environment if released, an appropriate filtration system would probably be needed on the exhaust. If so, then a pressure sensor to measure the pressure drop across the filter would be required to ensure that the filter would be replaced as needed as the static pressure offered increases.

   • Work with explosives shall be limited to the minimum quantities needed. For small quantities used in a hood, an explosion barrier in the hood, with personnel wearing protective eye wear, face masks, and hand protection, may be sufficient protection. Note that most hoods are not designed to provide primary explosion protection. For larger quantities, the facility must be specifically designed for the research program. It is strongly recommended that a formal hazard analysis be completed, following guidelines such as those given in NFPA 49, Appendix C, if explosives are a major factor in designating the facility as a high-risk facility. During periods of maximum risk, occupancy of the facility shall be limited to essential personnel.

- Some gases, such as fluorine, burn with an invisible flame. Apparatus for work with such materials should be placed behind a barrier to protect against an inadvertent introduction of a hand or other part of the body, so as to prevent burns.
- Systems containing toxic gases that would be immediately dangerous to life and health or gases that could pose explosive or health hazard ratings of 3 or 4 (lesser ratings if they provide no physiological warning) if allowed to escape shall be leak tested prior to use and after any maintenance or modification which could affect the integrity of the system. Where feasible, the gas cylinders shall be placed external to the facility and the gases piped into the laboratory to help minimize the quantity of gas available to an incident. As few cylinders as feasible shall be maintained within a given facility, preferably three or less. Permanently installed gas sensors, capable of detecting levels of gas well below the danger limits, may be needed in some cases, such as when escaping gas provides no physiological warning signal.
- Vacuum systems, capable of imploding and resulting in substantial quantities of glass shrapnel or flying debris, shall be protected with cages or barriers, or for smaller systems, shall be wrapped in tape.
- Systems representing other physical hazards, such as high voltage, radiation, intense laser light beams, high pressure, etc., shall be marked with appropriate signs and interlocked so as to prevent inadvertent injuries. The interlocks shall be designed to be fail safe such that no one failure of a component would render the safety interlock system inoperative.

2. Activities in which the attention of the worker is not normally engaged with laboratory operations, such as record maintenance, calculations, discussions, study, relaxation, etc., shall not be performed in the laboratory proper but shall be performed in an area isolated from the active work area. The segregated desk area of the standard laboratory module is specifically intended to serve this purpose. Depending upon the nature of the hazard, it is usually economically feasible to make at least a portion of the upper half of the barrier separating the two sections of the laboratory transparent so that operations can be viewed if necessary.

3. Workers in the laboratory should participate in a medical surveillance program if they actively use materials for a significant portion of their work week which would pose a significant short- or long-term risk to their health. Employees shall be provided medical examinations if they work with any material, such as regulated carcinogens, requiring participation in a medical program by OSHA or another regulatory agency under conditions which do not qualify for an exemption. Employees shall notify the laboratory supervisor as soon as possible of any illness that might be attributable to their work environment. Records shall be maintained of any such incident as defined by the OSHA requirements for maintenance of health records.

4. No safety feature or interlock of any equipment in the facility shall be disabled without written approval of the laboratory supervisor. Any operations which depend upon the continuing function of a critical piece of safety equipment, such as a fume hood, shall be discontinued should the equipment need to be temporarily removed from service for maintenance. Any such item of equipment out of service shall be clearly identified with a signed "Out of Service" tag. Only the person originally signing the tag or a specific, designated alternate shall be authorized to remove the tag.

5. It shall be mandatory to wear any personal safety equipment required for conducting operations safely in the laboratory.
6. It is recommended that a laboratory safety committee review each new experiment planned for such a facility to determine if the experiment can be carried out safely in the facility. If the risk is such that experiments may affect the environment, or the surrounding community, it is recommended that the committee include at least one layperson from the community, not currently affiliated directly or indirectly with the institution or corporation. In this context, "new" is defined as being substantially different in character, scope, or scale from any experiment previously approved for the facility.

Special Safety Equipment

1. Any refrigerators or freezers shall be rated as acceptable for "Flammable Material Storage," except for ultra-low temperature units.
2. A flammable material storage cabinet, either built-in or free standing, shall be used for the storage of flammable materials. Any other special storage requirements, such as for locked storage cabinets or safes for drugs or radioactive materials, shall be available and used.
3. If the nature of the research program requires it, the laboratory shall be equipped with a fume hood. The fume hood shall meet any specific safety requirements mandated by the nature of the research program. A discussion of hood design parameters will be found in a later section, but for high hazard use, the interior of the hood and the exhaust duct should be chosen for maximum resistance to the reagents used; the fan should preferably be explosion-proof or, as a minimum, be equipped with nonsparking fan blades; the hood shall be equipped with a velocity sensor and alarm; the interior lights shall be explosion-proof, and all electrical outlets and controls shall be external to the unit. It may be desirable to equip the unit with an internal automatic fire suppression system.
4. The laboratory shall be equipped with an eyewash station and a deluge shower.
5. The laboratory shall be equipped with a fire alarm system connected so as to sound throughout the building (and in a central facility manned 24 hours per day) and an appropriate fire suppression system and be provided with one or more class 12 BC, or larger, fire extinguishers, or class D units if reactive metals are in use.
6. An emergency lighting system shall be provided.
7. A first-aid kit shall be provided and maintained.
8. Any special equipment mandated by the research program shall be provided. For example, electrical equipment other than refrigerators may need to be designed to be explosion-safe.
9. Any special equipment needed to maintain the required isolation for materials in the laboratory shall be provided. Examples are specially labeled waste containers, autoclaves, other decontamination equipment, or disposable clothing.

Laboratory Facilities

1. The floor of the laboratory is designed to be easily cleaned. Durable, seamless floors of materials that are substantially impervious to spilled reagents are easily de-contaminated, and have curved junctures to walls, aid in accomplishing this. The walls are to be similarly painted with a tough, substantially impervious paint (such as epoxy) to facilitate cleaning and decontamination.
2. Two well-separated exit doors shall be available to the laboratory which shall swing in the direction of exit travel.

3. Bench tops should be resistant to the effects of acids, bases, solvents, and moderate heat, and should not absorb water. To facilitate cleaning, the tops should have few seams or crevices. Although not necessarily subjected to the same level of abuse, other surfaces of the furniture should be readily cleaned or decontaminated.

4. Casework should be designed to be sturdy and designed for convenient utilization. Storage spaces should be designed to meet any special requirements and should be easily accessible. It should not be necessary, for example, to stretch to reach any reagent which, if dropped, could represent a safety problem.

5. Aisle spaces should be 40 to 48 inches wide and shall not be constricted to less than 28 inches by any temporary obstacles. The aisles shall lead as directly as possible toward a means of egress.

6. The organization of the facility shall be such as to reduce the likelihood of having to pass an originating or secondary hazard to evacuate the facility in the event of an emergency.

7. Electrical outlets shall be three-wire outlets with high-quality, low-resistance ground connections. Circuits should be clearly identified to correlate with labels in breaker panels. If the nature of the hazard generates potentially explosive or ignitable aerosols, vapors, dusts, or gases, the electrical wiring, lights, and electrical switches shall be explosion-proof. Where connections and switches are near water sources, the circuits should be equipped with ground-fault interrupters (GFIs).

8. Laboratories in which the risk of electrical shock is greater than normal may also be equipped with a master "panic" manual electrical disconnect, clearly marked and located in a readily accessible location.

9. The laboratory shall be supplied with a sink. The trap shall be of corrosion-resistant materials. The plumbing shall be sized to accommodate the deluge shower and eyewash station. With average water pressure, this would normally be a 1-inch line or larger.

10. At least six air changes per hour (higher risk laboratories often are designed for 10 to 20) of 100% fresh air shall be supplied to the facility. No air shall be recirculated. The ventilation system shall be designed so that the room air balance is maintained at a small, negative pressure with respect to the corridors, whether the fume hood is on or off.* Where toxic and explosive gases and fumes are present, the system is to be designed to be efficient in exhausting these fumes by locating the exhaust intakes either near the source of fumes or near the floor (except for lighter-than-air or hot gases). Typical air flow patterns should draw dangerous fumes away from the normal breathing zones of the laboratory's occupants.

11. The facility shall include a separation of work spaces and desk areas as well as a second exit, equivalent to the arrangement shown in the standard laboratory module, Figure 3.1 (see Section 3.1.1) unless the risk is so pronounced as to require complete separation of operational and nonoperational areas.

12. Some high-risk facilities require airlocks, changing rooms equipped with showers with "clean" and "dirty" sides, or special equipment to decontaminate materials entering or leaving the facility.

3.1.4. Access

Much of the present chapter has been spent on details directly concerning the laboratory itself. However, a laboratory is rarely an isolated structure, but is almost always a unit in

---

* This is not true for some biological laboratories or "clean rooms" where a positive pressure is maintained to reduce the likelihood of contamination of the room by external contaminants.

a larger structure. It often appears that the typical laboratory manager or employee is insufficiently aware of this. If it is necessary to dispose of some equipment, it is often simply placed outside in the hall where it is no longer of concern to him. The thought that it may reduce the corridor width to well below the required minimum width also probably does not arise. A door swinging into the hall in such a way that it may block the flow of traffic appears similarly unimportant if it preserves some additional floor or wall space within the facility. The use of the corridor as a source of make-up air often seems reasonable, yet the possibility of this permitting a fire or toxic fumes to spread from one laboratory to another or to other parts of a building is clear once it has been considered. The natural inclination for most research personnel is to concentrate one's thoughts on the operations within a laboratory since this is where virtually everything important to them takes place. The ideas presented in the previous sections relating to optimizing safety within the facility are quickly grasped and accepted by most laboratory personnel, but the importance of extending these same concepts beyond the confines of their own laboratory frequently appears to be more difficult to communicate. However, due to the inherent risks in laboratory facilities, it is critical that sufficient, safe means of egress are always available. Except for scale and specific code requirements, most of the principles used in the laboratory to allow safe evacuation extend readily to an entire building.

3.1.4.1. Exitways

An exitway consists of all components of the means of egress leading from the occupied area to the outside of the structure or to a legal place of refuge. New ADA (Americans with Disabilities Act) requirements specifically call for places of refuge where disabled persons can await assistance as part of new construction. Included as exitway components are the doors, door hardware, corridors, stairs, ramps, lobbies, and the exit discharge area. The function of the exitway is to provide a protected way of travel to a final exit from the facility to a street or open area. Elevators are not acceptable as a required means of egress.

3.1.4.1.1. Required Exits

Any required exitway is required to be maintained available at all times, unless alternate means are approved in advance by a building official,* which will provide equivalent protection. This is probably one of the most common code violations. An extremely serious violation was personally observed by the author while attending a *safety* conference at a major university. The meetings were held within a large, multistory building containing meeting rooms, dining facilities serving up to 300 persons, and offices that had all but one small, poorly marked, out-of-the-way exit blocked. This condition existed for a period of several weeks during a renovation project, during which full operations continued in the building.

Another common violation of the same type is chaining of exits for protection against theft during low-usage hours. It is common, however, for research buildings to be partially occupied at almost any time. All exits may not be required during periods of low activity, but enough legal exits must be available to serve the occupants. It is essential that

---

* A building official in this context is a person or agency specifically authorized to administer and enforce the building code applicable to the building, not a person in charge of a building or facility.

occupants know which exits remain usable, if some which are normally available are blocked during certain hours. Most of these problems arise because the persons making the decisions to eliminate or reduce the size or number of exits are not personally knowledgeable of the legal requirements and fail to check with those who do.

If sufficient legal exits cannot be maintained during renovations or at other times, the occupancy load must be reduced or perhaps sections of the building, served by the needed exits, should be closed temporarily. As a minimum, each floor of every building with an occupancy load up to 500 must have at least two exits; between 500 and 1000, three exits; and above 1000, at least four exits.

3.1.4.1.2. Exit Capacity

In designing the needed exits for a facility, it is necessary to consider: (a) the number of occupants in the building; (b) the number that *could be* in the building, if the maximum density of occupants allowed by the building code were present; or (c) the latter number, plus any persons who might have to pass through the building from another space to reach an exit. The exits must be sufficient to accommodate the largest of these three numbers. The maximum floor area allowed per occupant under BOCA is (space occupied by permanent fixtures is not counted):

Assembly without fixed seats, chairs only	7 sq. ft. net
Assembly without fixed seats, chairs and tables	15 sq. ft. net
Business areas	100 sq. ft. gross
Educational; classrooms	20 sq. ft. net
Educational; shops, etc.	50 sq. ft. net

The exit capacity from an area must be sufficient for the number of occupants of the space involved. The following table gives the width per person for the maximum permissible occupant load, expressed in inches, for the given type of usage.

Without a Fire Suppression System (required exit capacity per occupant, inches)

Use Group	Stairways	Doors, Ramps, Corridors
Assembly	0.3	0.2
Business	0.3	0.2
Educational	0.3	0.2
Hazardous	0.7	0.4
Storage	0.3	0.2

With a Full Fire Suppression System (number of occupants/unit capacity)

Use Group	Stairways	Doors, Ramps, Corridors
Assembly	0.2	0.15
Business	0.2	0.15
Educational	0.2	0.15
Hazardous	0.3	0.2
Storage	0.2	0.15

The minimum width shall be at least 44 inches (or as calculated according to the above table) for occupant loads greater than 50, or 36 inches for occupant loads of 50 or less.

3.1.4.1.3. Travel Distance

The characteristics of the routes of egress to an exit are also important, especially in as critical a facility as a laboratory building. Care should be taken, just as within the laboratory, for the distances to be as short and direct as practicable. The location of hazardous areas should be chosen to eliminate or minimize the probability of the direction of travel being toward a likely hazard during an emergency. The maximum travel distances for the use groups of interest are

Use Group	Without Fire Suppression System	With Fire Suppression System
Assembly	200 ft	250 ft
Business	200	250
Educational	200	250
Hazardous, H-2	50	100
Hazardous, H-3	100	150

As with the standard laboratory module, when a building requires more than one exit, these exits should be as remote as practical from each other. In a facility not served by a fire suppression system, the separations are to be at least half of the maximum diagonal distance of the area served. If there is an approved fire suppression system, this distance can be reduced to one fourth of the diagonal distance. They must also be arranged so that access is available from more than one direction from the area served so that it is unlikely that access from both directions will be blocked in an emergency.

If the laboratory building is classed as an assembly occupancy, there must be a main exit fronting on a street which, in addition to being large enough to serve as a sufficient discharge for all the exitways that lead to it, must also be at least large enough to accommodate half of the building's total occupant load.

It is acceptable to use an adjacent room or space as a means of egress from a room, as indicated in the standard laboratory module, if the room provides a path of egress to an exit, is not a higher hazard than the original space, and is not subject to locking. Thus, the laboratory modules should be arranged in blocks of comparable level of risk, if this concept is used to provide a second exit from each laboratory.

3.1.4.1.4. Corridors

In the introduction to this section two points were used as illustration, one maintaining the corridors free of obstruction and the second concerning the undesirability of using halls as a source of make-up air. Both of these points are intended to ensure that the corridors remain available for evacuation. A door may not swing into a hall such that it reduces the width of the corridor to less than half the legally required width, nor can the door, when fully open, protrude into the hall more than 7 inches. An obvious implication of the first provision, considering that most laboratory doors are 36 inches wide or wider, when combined with one half of the minimum legal width of 44 inches for a corridor (except for buildings occupied by 50 persons or less), would mean a minimum actual permissible width of 58 inches, unless the doors are recessed into alcoves in the connecting rooms.

Other unnecessary obstacles should be avoided as well, such as low-hanging signs, and similar devices which may protrude into a corridor, or even safety devices such as deluge showers with low hanging chains which could strike a person in the face in a partially dimmed or darkened corridor. The corridors must have a minimum of 80 inches of headroom. Door closers and stops cannot reduce this to less than 78 inches. Between the heights of 27 inches and 80 inches, objects cannot protrude into the corridor by more than 4 inches.

If the corridors were to serve as a plenum for return air, they could spread smoke and toxic fumes from the original source to other areas. Further, instead of being a protected exitway, they themselves would represent a danger. In many fires, the majority of those that fail to survive often are individuals trapped in smoke-filled corridors and stairs. Laboratories, in general, need to be kept at negative pressure with respect to the corridors, but the 200 or 300 cfm through an open door needed to maintain a negative air pressure will not violate the prohibition on the corridors as a plenum. Space above a false ceiling in a corridor can be used as a plenum, if it can be justified for the corridor to not be of rated construction or if the plenum is separated by fire resistance-rated construction. The use of spaces above false ceilings as plenums should be discouraged for other reasons, however. In recent work involving HVAC systems contaminated with microbiological contaminants, such spaces with slow-moving air have been shown to provide a favorable environment for such organisms to grow.

If a corridor serves as an exit access in the building occupancies which are being considered here, the corridor walls must be of at least 1-hour fire resistance rating. Care must be taken to construct corridor walls that are continuous to the ceiling separation to ensure this rating. Cases have been observed in which the wall was not taken above a suspended ceiling or continued into open service alcoves. Corridor floors are to have a slip-resistant surfaces.

The eventual point of exit discharge must be to a public way or a courtyard or other open space leading to a public way which is of sufficient width and depth to safely accommodate all of the occupants. On occasion, during renovation or construction projects, the areas outside the exits may not be maintained in such a way as to satisfy this condition. Such situations should be corrected promptly upon discovery.

A fairly common error that tends to creep into older buildings as renovations take place is the creation of dead end corridors. Frequently, for other design reasons, more corridors are built in originally than are actually needed. Later, as space becomes tighter or the space needs to be reconfigured, the corridors are modified in order to recoup this "wasted" space, and dead end corridors are created. These dead end corridors cannot be longer than 20 feet under most circumstances. If the corridor is of sufficient width, some of the dead end corridors can be converted into offices or other uses, as long as they conform to code requirements for the class of occupancy.

Where there would be an abrupt change in level across a corridor (or across an exit or exit discharge) of less than 12 inches, so that a stair would not be appropriate, a ramp is required to prevent persons from stumbling or tripping at the discontinuity. The ADA requirements would also mandate a ramp for at least a sufficient part of the width to accommodate disabled persons in wheel chairs or using crutches. Clearly, in such a case it would be desirable to have the ramp extend the full width of the passageway.

3.1.4.1.5. Stairs

Stairwells are also exitway components that are frequently abused. They are a means of egress providing a protected way of exiting a building. In order to provide additional ventilation or to avoid having to continually open doors, a very common practice is to use wedges of various types to permanently prop doors open. The result is to not only void the protection afforded by the required fire-rated enclosure, but to create a chimney through which fire and smoke on lower floors may rise, changing the stairway from a safety device to a potential deathtrap and providing a means for problems on lower floors to spread to higher levels.

In order to provide a protected means of egress, a required interior stair must be enclosed within a fire separation meeting the fire resistance ratings given in Table 3.3. The stair enclosure cannot be used for any other purpose, such as storage underneath the stairs or within any enclosed space under a required stair. Any doors leading into the stair enclosure must be exit doors. This also precludes creating closets underneath stairs for storage. The width of the stairs and of landings at the head, foot, and intermediate levels must meet the minimum dimensions established by the calculated required exit capacity. All doors leading onto a landing must swing in the direction of egress travel. The restrictions on reduction of width of the landings due to doors opening are the same as for corridors.

Stairways which continue beyond the floor level leading to an exit discharge onto a basement level are common. In an emergency situation, unless the stairs are interrupted at this floor, it would be very likely that persons evacuating the building would continue downward, even though there is an additional requirement that each floor level be provided with a sign indicating the number of the floor above the discharge floor, for stairways more than three floors high. The persons continuing downward might be sufficiently confused as to reenter the building on the lower level before their mistake was recognized, or in an even more serious situation cause congestion at the lower end, making access to the exit difficult or impossible for those from the lower floor. The code states that the floor level sign should be about 5 feet above the floor and readily visible whether the door is open or closed.

The treads and risers for all classes of occupancy likely to be involved in laboratory buildings shall be a minimum of 11 inches for a tread and a maximum of 7 inches and a minimum of 4 inches for a riser. The maximum variation in the actual widths for a tread or riser are to be no more than ±3/16 of an inch for adjacent steps and 3/8 inch for the maximum variation. This seems a trivial point at first glance, but the importance of it should be clear to anyone who has ever stumbled over uneven ground in the dark. As one goes up or down a flight of stairs, one quickly grows accustomed to the step configuration, and a substantial unexpected change could easily lead a person in a hurry to stumble.

A similar rationale exists for the continuance of a handrail beyond the ends of a stairway as a provision for ensuring that the person traveling the stair has something to grasp to help avoid a fall, if they cannot see that the stairs have ended. Both at the top and bottom, the handrail should turn to be parallel to the floor for at least 1 foot (plus a tread width at the bottom). There must be a handrail on both sides of a stair, and intermediate handrails must be provided so that no point over the required width is more than 30 inches from a handrail.

3.1.4.1.6. Doors

Doors are perhaps the most abused exitway component. The fire separation they are intended to provide is often defeated when they are wedged open in order to improve the ventilation or to eliminate the inconvenience of having to open them every time the passageway is used, especially if it is frequently used for moving supplies and equipment. In some instances, doors required to be shut since they represent openings into a fire resistance-rated corridor are left open simply because individuals wish to leave their office doors open to be easily accessible to persons wishing to see them, as openness and accessibility are viewed as desirable behavioral traits. The hinge assemblies of doors are often damaged when they are prevented from closing by the use of the now-ubiquitous soft drink can forced into the hinge opening. Even maintenance departments may not be aware that a required fire resistance rating is achieved by the entire door assembly, including the frame and hardware, not just by the door itself, so that a repaired door may no longer meet code specifications.

Conversely, doors required to be operable may be blocked or rendered inoperable for a variety of reasons, one of the most common being to increase security. Compact, easily portable, and salable instrumentation, especially computers and computer accessories, represents a tempting target for theft. As a result, doors which should be readily operable are fitted with unacceptable hardware to provide additional security, in many cases by the occupants themselves. In other cases, doors are blocked simply because individuals are careless and do not consider the consequences of their actions, such as locating a piece of equipment in such a way that a rarely used door cannot swing open properly.

There must be a handrail on both sides of a stair, and intermediate handrails must be provided so that no point over the required width is more than 30 inches from a handrail. Doors which are required exits must be prominently indicated as such, while doors which do not form part of a legal exitway must have signage indicating that they provide an additional means of egress, although the quality of the passages beyond may not be sufficient to meet the requirements for a protected exitway.

For the classes of occupancies being considered here, the minimum width of a door used as an exit must be at least 32 inches, and the maximum width of a single leaf of a side-hinged, swinging door must be no more than 48 inches (except for certain storage spaces). If a door is divided into sections by a vertical divider, the minimum and maximum widths apply to each section. A normally unoccupied storage space of up to 800 square feet can have a door of up to 10 feet in width. The minimum height of a door is 80 inches. If two doors are to be placed in series, as might be the case where a separation of a facility from a corridor is required to be maintained, the doors must be separated by a minimum of 7 feet.

In general, it is recommended that all doors for laboratory structures should be of the side-hinged, swinging type, opening in the direction of exit travel. For doors opening onto stairways and for an occupant load of 50 or more, or for a high hazard occupancy, doors with these characteristics are required. However, a sliding door with a side-hinged, swinging door incorporated in it of the proper dimensions is acceptable.

It must be possible to open a door coming from the normal direction of egress without using a key, and draw bolts, hooks, bars, or similar devices cannot be used. For assembly or educational occupancies, with an occupant load of 100 or more, exit doors must be fitted with panic hardware. An essential element of a door is that it cannot be too difficult to open. The opening force for most interior doors must not exceed a force of 5 pounds. To open a door that is normally power assisted must not require more than 50 pounds with the power off. Panic hardware must require no more than 15 pounds force to release, and a door not normally provided with power assistance cannot require more than 30 pounds force to initiate motion and swing to a full-open position with application of a 15-pound force. These restrictions on the force required to operate a door can easily be exceeded should the ventilation system be modified without taking this concern into account. A very moderate atmospheric pressure differential of just over 0.3 inches (water gauge) would result in a force of more than 30 pounds force on a door of the minimum acceptable size.

Addition of hoods to a laboratory, without provision of additional makeup air, could easily cause this limit to be exceeded on a more representative 3 foot by 7 foot door.

Doors opening from rooms onto corridors and into stairways and forming part of a required fire resistance-rated assembly must be rated. Most doors, such as those from offices opening on a corridor, are required to have at least a 20-minute fire rating, while doors leading from rooms of 2-hour fire resistant construction, as determined from Table 3.3, must be at least 1.5-hour fire doors, as should those entering stairways. Wired glass, one quarter inch thick, specifically labeled for such use, may be used in vision panels in 1.5-hour fire doors, provided that the dimensions do not exceed 33 inches high and 10 inches wide, with a total area of no more than 100 square inches. If the potential injuries and damage resulting from dropping chemicals as a result of being struck by a swinging door are considered, there is clearly merit in taking advantage of the provision for vision panels in laboratory, corridor, and stair doors.

Doors opening onto fire resistant-rated corridors and stairways must be self-closing or close automatically in the event of a fire. The first of these requirements usually includes offices opening directly off corridors, and is the case alluded to earlier as representing one of the most commonly violated fire regulations. For whatever reason, most individuals usually prefer to work with their office door open. Unfortunately, in an emergency evacuation, many do not remember to close their doors, and so the integrity of the fire separation is breached at these points.

3.1.4.1.7. Exit Signs, Lights, Emergency Power

The need for emergency lighting within laboratories has already been discussed independent of code issues. However, the need for lighting of exitways and identification of exits in emergencies is as critical outside the laboratory proper as it is inside. It is essential in a laboratory building that evacuation not be hindered by lack of lighting, especially in multistory buildings where stairways and corridors typically do not provide for natural lighting.

Internally illuminated exit signs are a key component of an evacuation system. In every room or space served by more than one exit, as is recommended for most laboratory rooms,* all the required means of egress must be marked with a sign with red letters on a contrasting background at least 6 inches high, with a minimum width of 3/4 of an inch for each segment making up the letters. The light intensity at the surface of other than self-luminous signs must be at least 5 foot candles.

There are self-illuminated signs, containing radioactive tritium (an isotope of hydrogen with a half-life of 12.33 years) that are acceptable, both under usual fire codes and to the NRC. The radiation from tritium is exceedingly weak (18.6 KeV beta) and, since these signs are completely sealed, no radiation can be detected from them. The transparent enclosure completely absorbs the radiation. As long as they remain sealed, they represent no hazard. However, in order to provide the required level of illumination, the amount of radioactivity in each sign is substantial. If they were to be broken, in an accident or in a fire, an individual handling them could, if the unit were broken inhale a quantity of activity substantially in excess of the permissible amount. Therefore, as a precautionary policy, radiation safety committees at a number of institutions have taken the position that these

---

*   While two exits are recommended for most laboratories in order to provide the maximum degree of safety, neither building code provisions nor OSHA regulations require two exits unless the laboratory represents a high hazard area or is occupied by more than 50 people.

signs are not permissible at their facilities. It might be well to consider the risk versus benefit whenever the use of such units is contemplated.

Note that normal glow-in-the-dark signs do not contain radioactive material. They depend upon phosphorescence, a completely different physical phenomenon, and usually remain visible for 1 to 4 hours after activation by exposure to light.

In addition to signs at the exit, it may be necessary to put up supplemental signs to assist in guiding persons to an exit, where the distance is substantial or the corridor curves or bends. If a sign incorporates an arrow, it should be difficult to modify the direction of the arrow. However, when a sign is damaged, maintenance personnel have been known to inadvertently install a replacement sign with the arrow pointing in an incorrect direction. It is well not to take anything for granted. Users of the building should verify that all exit signs indicate the correct direction to the intended final exit point.

Exit signs and means of egress must be lighted whenever a building is occupied, even if the normal source of power fails. The level of illumination at the floor level must be at least 1 foot candle. There are a number of methods in which power can be provided to emergency lighting circuits. They all must provide sufficient power to the lights and paths of egress to meet the required lighting standards for at least 1 hour, so that the building occupants will have ample time to evacuate.

For relatively small facilities, battery-powered lights, continuously connected to a charging source and which automatically come on when the power fails, are often used. Units are available which have extended useful lifetimes of 10 years or more. As with any other standby device, it is necessary to test them on a definite schedule. These battery-powered units are relatively inexpensive, currently ranging in price (in quantity) from about $40 to $150. There are battery-powered units designated as "uninterruptible power supplies" which switch over within milliseconds. These are often used to maintain power to computers or electronic equipment where a loss of power can cause data to be lost. Such units can be sized to also support emergency electrical lighting.

Standby generators are another alternative to provide energy to the required emergency lights and other equipment which may need to be supported during a power failure. A generator would be preferred over a battery system for other than small buildings to provide the fairly substantial amount of power needed. However, these generators should be checked once a month under load to ensure that they will come on within the required 10 seconds for emergency lighting and within 60 seconds for other loads. Architects often fail to consider all circumstances in designing such systems. In one case, the architect designed an excellent system, but for economic reasons, the exhaust of the generator was located immediately adjacent to the building air intake, on the premise that in an emergency the ventilation system would shut down. However, because diesel exhaust fumes were drawn into the working ventilation system during tests, this effectively prevented the scheduled operational tests from being performed until the problem was corrected. Failure to provide proper maintenance and tests can lead to embarrassing and costly incidents when outside power fails and stairways and corridors are not lighted. Academic institutions are more vulnerable than corporate facilities since, in a given building, there are more likely to be a higher percentage of individuals that are relatively unfamiliar with the evacuation routes.

A last option, but one which must be used with considerable caution, is to provide outside power from two completely separate utility power feeds. Such an arrangement can be approved by code authorities if it can be shown that it is highly unlikely that a single failure can disrupt both sources of power. For example, if the local distribution system is fed by several alternative power lines and has alternate local lines to provide power to a building, it is conceivable that local building officials would approve the system, but one must remember that entire states, and even larger regions, have suffered total power losses in recent years. Among other occurrences which could lead to such a failure are natural disasters such as hurricanes, tornadoes, blizzards, ice storms, fires, and floods. Few, if any, localities are immune to all of these. In the author's area wide-spread, lengthy power outages have occurred frequently in recent years due to ice and wind storms.

3.1.4.1.8. Other Topics

A number of other topics, related to exitways, have not been touched upon here for the same reason that has been given before. This handbook is not intended to be comprehensive. The intention is to cover those topics most meaningful to a person working in a laboratory building with enough detail so that reasonable persons can evaluate their facilities to ensure that their safety is not reduced by renovations or the actions of individuals during normal usage. The reader should also be able to follow the reasoning for many of the architectural decisions made during the planning of a facility and should be able to actively participate in the planning process. However, provisions and specifications for components such as exterior stairs, fire escapes, access to roofs, connecting floors, vestibules, and lobbies, which are all relevant topics under the general subject "means of egress," would be important to architects, but probably less so to laboratory personnel.

3.1.5. Construction and Interior Finish

Code specifications for the details of construction and the materials used as interior finish are intended to provide the degree of fire protection commensurate with the use of a building. Although, as has been noted, a laboratory building may be constructed under several different classifications, laboratory usage does imply that a certain amount of risk is involved. Therefore, care is needed to ensure that construction practices and materials used in the interior finish do not add to the risks or defeat the intended level of protection.

In addition to fire protection there are other potential hazards which may also be reduced by construction details and choices of materials. In Section 3.1.3.2, under the topic "Laboratory Facilities" for each class of laboratories, many of the features stipulated characteristics of finish materials. As a general principle, laboratory floor coverings, wall finishes, and table and bench tops should be durable, easy to clean, and resistant to the common reagents.

A normal vinyl tile floor meets many of these requirements, but the seams around each tile form cracks in which materials such as mercury and other materials can lodge. As an example, mercury can remain *in situ* in these cracks for extended periods and create a substantial mercury vapor pressure when, ostensibly, all spilled mercury may have been "cleaned up." Radioactive materials and biological agents can similarly be trapped and pose a continuing, persistent problem unless very thorough cleaning is performed on a regular basis. One argument frequently used in favor of tile floors has been that damaged or contaminated tiles can be easily and cheaply replaced. However, common vinyl asbestos tile is now included within the category of asbestos materials by the EPA and must be removed according to the procedures for removing asbestos. Replacing a tile floor, in which either the tile or the mastic affixing it to the subfloor contains asbestos fibers is expensive. Note that it is more likely that 9-inch tiles will contain asbestos than the larger, 12-inch tiles. Newly manufactured tiles do not contain asbestos, but unfortunately, the mastic often still does.

Any material used must meet required standards for fire spread and smoke generation, in addition to having the other properties for which it is selected. When a material has been selected that has both the desired properties and has the requisite ratings, the construction contract should contain peremptory language stipulating that no substitutes will be acceptable without specific approval. In too many instances, where vague language such as, "equivalent materials may be substituted," is incorporated into a contract, substitutes have been introduced which do not meet the original specifications, possibly innocently, because the supervisor on site did not realize the difference. It usually is not possible to simply look at an item and determine what its properties will be from appearances alone.

Although not particularly attractive, a plain, sealed concrete floor or one painted with a durable paint probably is the best for most laboratories, while wood and carpeting would be the worst. Many different kinds of floor finishes are available, designed to prevent slipping, generation of sparks, and resistance to corrosives or solvents. Simple concrete block walls are often used for interior partitions in buildings or, alternatively, drywall on steel studs. Both of these are relatively cheap for original construction and can be modified easily as well. Concrete block is relatively porous, which can pose decontamination problems, but it can be painted to eliminate this problem. Paints for interior surfaces are available which will provide waterproofing, resistance to corrosives and solvents, and enhance fire resistance. Where biological cleanliness is an important criteria, there are paints approved by the EPA which will inhibit the growth of biological organisms. Incidentally, the sand used in concrete blocks in many parts of the country is a source of radon, a concern to many persons.

3.1.5.1. Construction Practices

The intent of the fire code as it applies to the interior finish and to acceptable construction practices is to prevent spread of a fire from one fire area to another, i.e., to make sure that the fire walls and other fire separation assemblies are constructed in such a way and of such materials as to maintain the fire resistance rating of the structure. When architects prepare the plans and specifications, they must include documentation for all required fire resistance ratings.

It is not possible to provide total separation of fire areas and still provide for air intake, exhaust, or return air plenums, unless these are themselves separated from the surrounding spaces by required fire-resistant shaft and wall enclosures, plus properly engineered, and labeled fire dampers meeting UL 555 specifications must be installed where ducts pass through fire separations. Unless the architect, a representative of the building owner, or a contractor's inspector provides careful supervision of the workers during construction, dampers may be installed improperly, perhaps at the end of a convenient piece of ductwork, even if this happens to be in the middle of the room, far away from the fire separation wall. As noted earlier, it is possible to use the space between the ceiling of a corridor and the floor above if the space is properly separated. However, ceiling spaces used for this purpose cannot have fuel, fixed equipment, or combustible material in them. The requirement for a fire damper does not apply to ducts used for exhausting toxic fumes, as from a fume hood, since in a fire it is usually desirable for purposes of protecting both the normal building occupants and firefighting personnel for toxic materials to continue to be exhausted. This requires that ducts carrying toxic fumes be continuously enclosed within a shaft of the proper rating (normally 2-hour) to the point of exhaust.

It is essential that the integrity of the fire separation walls not be significantly diminished. For example, walls less than 8 inches thick are not to be cut into after they are constructed in order to set in cabinets or chases. Among the most common violations of the integrity of fire separations are penetrations in order to run utilities and, today, cable chases for electronic services such as video signals, data cables, and computer lines. Often, these penetrations are roughly done, leaving substantial, unfilled gaps surrounding the cables, conduits, or piping. Even in new construction, especially if the penetrations are in difficult to inspect or otherwise awkward locations, the gaps around the ducts, pipes and conduit are frequently left incompletely or poorly filled. Where retrofitting of spaces to accommodate such devices is done by maintenance and construction personnel, this deficiency is even more likely to occur if the work is inadequately supervised, since the average worker may not be aware of this requirement. Whenever unfilled spaces are found, the gaps must be filled by materials meeting fire resistance standards. If a renovation or new construction involves setting in a structural member into a hollow wall, the space around the member must be filled in for the complete thickness of the wall with approved fire stopping material.

Openings can exist in a fire wall, or else how could doors and windows exist? However, there are limits on the size of the openings—120 square feet (but no more than 25% of the length of the fire wall), except in buildings with an approved automatic sprinkler system. Larger openings (240 square feet) can exist on the first floor of a building, again with an approved automatic sprinkler system. The openings must be protected with an appropriately rated assembly, which may be a fire door. If the wall is of a 3-hour rating, the rating of the door must be 3 hours as well. For walls of 1.5- or 2-hour ratings, the doors must be 1.5 hours. Around shaft and exit enclosure walls with a fire rating of 1 hour, the door assembly must be 1 hour, also. For other fire separations with a required separation of 1 hour, the fire door need only be a 3/4-hour rated door. Unless the interior space is rated, doors to rooms such as offices opening onto a 1-hour corridor need only be rated at 20 minutes.

Fire walls shall extend completely from one rated assembly to another, such as the floor to the ceiling, extending beyond any false or dropped ceiling which may have been added. The joint must be tight.

This section has been primarily concerned with the interiors of a building, but measures are required to prevent a fire from spreading due to the exterior design of a building as well. The exterior walls must be rated to withstand the effects of fires within the building. Windows arranged vertically above each other in buildings of three or more stories for business, hazardous, or storage uses shall be separated by appropriate assemblies of at least 30 inches in height from the top of a lower window to the bottom of the one above. Although there are a number of exceptions, if the exterior wall is required to have a fire rating of 1 hour or more, then a parapet of 30 inches or greater in height above the roof is required for nonexempt structures.

3.1.5.2. Interior Finish

Materials used for interior trim or finishing must meet standards for flame spread and smoke or toxic fume generation. Materials are rated in accordance to how well they perform on tests made according to the ASTM E84 procedure, with lower numbers corresponding to the better materials. Class I materials have a rating between 0 and 25; class II, 26 to 75; and class III, 76 to 200. As far as smoke generation is concerned, materials used for interior finish must not exceed a rating of 450 as tested according to the provisions of ASTM E84. Based on these ratings, the interior finish requirements for the categories of interest are

Use Groups	Required Vertical Exits and Passageways	Corridors for Exit Access	Rooms/Enclosed Spaces
Assembly, Business, and Education	I	I	II
Hazard	I	II	III

As usual, there are numerous exceptions based on special circumstances, the most notable being (a) class III materials may be used in rooms in places of assembly if the capacity is 300 or less; (b) if there is an automatic fire suppression system, the minimum requirement for interior finish is class II; and (c) for a hazard (H) occupancy building of over two stories, class II is required.

Materials used on floors are classified differently. Often, the propensity for materials to burn is different depending upon the physical configuration. For example, a match placed on a piece of carpeting lying on the floor may smolder and go out, while a match applied to the bottom of the same piece of carpeting, mounted vertically, may result in a vigorous fire. However, most common floor coverings employed in laboratories, such as wood, vinyl, terrazzo, are exempt from being rated.

Where interior finish materials are regulated, they must be applied in such a way that they are not likely to come loose when exposed to temperatures of 200°F for up to 30 minutes. The materials must be applied directly to the surfaces of rated structural elements or to furring strips. If either the height or breadth of the resulting assembly is greater than 10 feet, the spaces between the furring strips must be fire stopped. Class II and III finish materials, less than 1/4 inch thick, must be applied directly against a noncombustible backing, treated with suitable fire-retardant material, or have been tested with the material suspended from the noncombustible backing. This seems to be a fairly minor restriction, but most of the inexpensive paneling available today from builder supply houses is either 3/16 inch or 4 mm thick. At academic institutions, many departments have their own technicians that often do the departmental remodeling as an economy move and build improper partitions of this noncomplying material. Rated paneling is available, 1/4 thick or more, which looks exactly the same on the surface. The only realistic options available to prevent violations of the code requirements is to totally prohibit purchases of building material, strictly enforce policies of no "home-built" structures, to the extent of tearing down such constructions, or to provide a source of rated material which must be used.

Roofing materials are not interior finishing materials, but also must meet standards in order to maintain adherence to classes of construction. Class 1 roofing materials are effective against a severe fire exposure and can be used on any type of construction. Class 2 materials are effective against moderate fire exposures, and Class 3 materials are effective only against light fire exposures. Class 3 is the minimum acceptable material to be used on types 2, 3, 4, and 5A construction. Typical materials meeting Class 1 requirements would be cement, slate, or similar materials, while metal sheeting or shingles would meet Class 2. Class 3 materials would be those that had been classified as such after testing by an approved testing agency.

REFERENCES

1. Occupational Exposure to Hazardous Chemicals in Laboratories, 29 CFR § 1910.1450, 55 FR 7967, March 6, 1990.
2. Standard on Fire Protection for Laboratories Using Chemicals, NFPA 45, 1989.

3. **Ashbrook, P.C. and Renfrew, M.M.,** *Safe Laboratories, Principles and Practices for Design and Remodeling,* Lewis Publishers, Inc., Chelsea, MI, 1991.
4. The BOCA Basic National Building Code/1993, 12th ed., Building Officials and Code Administrators International, Inc., Country Club Hills, IL, 1993.

3.1.6. Ventilation

Few research buildings at either corporate or academic institutions are constructed today without central air handling systems providing heating, cooling, and fresh air. Experience seems to indicate that relatively few of these are designed completely properly to provide suitably tempered air where it is needed and in the proper amounts, at all times. High energy costs mandate that the energy expended in heating or cooling the air supplied to a facility be optimally minimized. Laboratory buildings, however, have highly erratic needs for tempered air. In academic buildings, for example, when both faculty and students cease working in the laboratory to meet classes or attend to other responsibilities, fume hoods, which typically exhaust around 1000 cfm per minute, may individually be turned on and off. In a medium-sized research building containing 50 hoods, the required capacity for makeup air could theoretically vary as much as 50,000 cfm. The occupants rarely conform to a sensible daytime work regimen. In academic institutions especially, individuals are almost as likely to be working at 4:00 a.m. as at 4:00 p.m., or while the majority may be taking a Christmas vacation, there are always a few continuing work on a project that cannot be interrupted. Under such circumstances, it is very difficult to continuously provide the right amount of air all the time to every laboratory economically. Economy is the easiest parameter to forego since engineering technology is capable, at least technically, of maintaining proper ventilation under almost any circumstance, even though it may be expensive to do so. Further, the health and safety of individuals should never be compromised for economic reasons.

Most written material on laboratory ventilation concentrates almost exclusively upon fume hoods. Ventilation does play an important part in the proper performance of hoods, and they, in turn, usually have the most significant impact of any piece of laboratory equipment on the design and performance of laboratory building air handling systems. However, there are many other aspects to laboratory ventilation. Hoods will be treated as a separate topic in Section 3.2.2, and some aspects of ventilation will be deferred to that section. Those portions of hood performance which concern the general topic of space ventilation will be covered in the following material.

Active laboratory areas should be provided with 100% fresh air. No air should be recirculated. There are laboratories for which this would not necessarily be essential, but as noted earlier, the character of research conducted in a given space may change. Ventilation, which depends upon supply and exhaust plenums to the space being built into the building structure, is one of the more expensive services to provide as a retrofit. It is better to design for the most demanding requirements and use controls to modify the supply if the actual needs are less. If the active laboratory space can be adequately isolated from administrative, classroom, and service areas, the requirements for these other spaces may be met with a recirculating system, where the portion of fresh air introduced into the total air supply could be as little as 10%. For reasons associated with building air quality in nonlaboratory buildings, it is often desirable to recirculate a large fraction of a building's air, as long as sufficient fresh air is provided to accommodate the basic needs of the occupants.

The amount of fresh air to be provided to a laboratory space should depend upon the activities within the facility, but there are little data to support a given amount. Epidemiological data, gathered by OSHA, indicates that there are health risks associated with working in a laboratory. In five studies cited by OSHA involving chemical workers, although the overall mortality rate appeared to be lower among chemists than in the general population, there was some evidence that indicated additional dangers from lymphomas and leukemia, development of tumors, malignancies of the colon, cerebrovascular disease, and prostate cancer, although virtually every study indicated a lesser rate of lung cancer. The general good health might be attributable to the generally high economic and educational status of the groups being studied, which probably translates into more interest in their health and being in a position to afford to maintain it. The general consensus that it is not a good idea to smoke in laboratories could impact on the number of observed cases of lung cancer. A survey among the members of the California Association of Cytotechnologists, investigated the use of xylene in the laboratories in which they worked; of the 70 who responded, 59% felt their ventilation was inadequate, 22.6% worked where there was no exhaust system, and 43% stated that their ventilation systems had never been inspected. In several recent health hazard evaluations conducted by NIOSH, it was found that ineffective exhaust ventilation was a major contributor to the hazardous conditions. If proper procedures are employed and all operations calling for the use of a fume hood are actually performed within a hood, the general room ventilation would be expected to have relatively little bearing on the health of laboratory workers. However, sufficient hood space is not always available and, even where it is available, is not always used. Consequently, general laboratory ventilation should be sufficient to provide good quality air to the occupants.

In the absence of specific requirements there are guidelines. *Prudent Practices for Handling Hazardous Chemicals in Laboratories* recommends between 4 to 12 air changes per hour. Guidelines for animal care facilities recommend between 10 to 20. Storage facilities used for flammables are required by OSHA to have six air changes per hour and would appear to be a baseline minimum level. If the air in the room is thoroughly mixed, six air changes per hour would result in more than 98.4% of the original air being exchanged. Increasing this to seven would result in a less than 0.8% further gain at the expense of a further increase of 14.3% in the loss of tempered air. A critical consideration is whether in fact the air does become thoroughly mixed. This depends upon a great many factors, including the location of the room air intakes and exhaust outlets, the distribution of equipment and furniture, and the number, distribution, and mobility of persons in the room. At any given time, any gases or vapors present in the air will eventually diffuse and attain a fairly uniform mix, even in an unoccupied space. Substantial amounts of movement in the room will tend to redistribute air within a room more quickly, but there will still be spaces and pockets in almost any room in which, because of the configuration of the furniture and the air circulation, mixing of the air will be slow. Because thorough mixing cannot be assured, recent trends have been to specify higher exchange rates, typically 10 to 12 air changes per hour. The American Society of Heating, Refrigerating and Air Conditioning Engineers (ASHRAE) in standard 62-189 recommends 20 cfm per person of fresh air in laboratories, or 1200 cubic feet per hour. If four persons worked in the standard module, this would be approximately 5000 ft^3. A single 5-foot fume hood would exhaust this amount of air in about 5 minutes, so the make-up air for the hood would supply an ample amount to meet this criterion.

Poor design of the air intake and exhaust system can have a significant negative effect on the needed air exchange. In Figure 3.3 the results of tests of a particularly bad system

Figure 3.3.   The air entering the lab, while meeting the requirements for fresh air, does not effectively reach the breathing zone of the occupants.

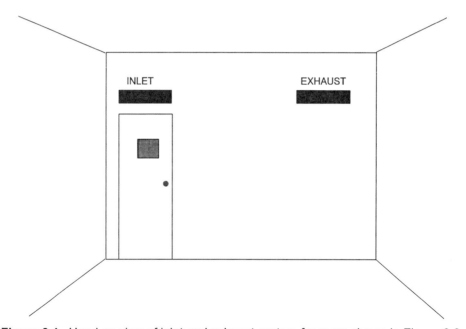

Figure 3.4. Head-on view of inlet and exhaust system for room shown in Figure 3.3.

are depicted. In this facility fresh air is delivered from a unit ventilator, mounted on a roof above a corridor and then ducted through the laboratory wall at a height of about 9 feet. Along the same wall, about 12 feet away, is an exhaust duct leading back to the roof (see Figure 3.4). Air is blown into the room horizontally toward the opposite wall. It then, supposedly, traverses the room twice and leaves through the exhaust duct or a hood.

Smoke bomb tests of this system, however, showed almost no vertical mixing of air in the room. Half an hour after the smoke was released, a clear line of demarcation about 8 feet above the floor between clear air and smoky air could still be discerned, the latter being partially replenished by exhaust air that had been recaptured and reentered the building. The occupants in this facility benefited virtually not at all from the air being introduced through the standard air intake, nor did it serve the fume hoods in the room. In order for the hoods to work properly, air had to be drawn in from either the doors or, when weather permitted, through open windows. Using the corridors as a source of air sufficient to supply even one hood violates code restrictions. Further, using open windows often results in an erratic air supply due to wind gusts, and in cold weather is clearly impractical.

Ideally, air entering the laboratory should enter gently and in such a way that the air in the breathing zones of the individuals working in the laboratory is maintained free of toxic materials and that the air flow into hoods in the room will not be interrupted or disturbed by traffic. Studies indicate that air directed toward the face of a hood or horizontally across its face will cause the most serious problems in meeting this latter condition, while air introduced through a diffuse area in the ceiling or from louvered inlets along the same wall on which the hood is situated will be affected the least by movement of personnel.

The majority of laboratory fumes and vapors are heavier than air and will preferentially drift toward the floor, although some will diffuse throughout the room air and some will be carried upward by warm air currents. Room exhausts should be located so as to efficiently pick up the fumes. Placing exhausts in the ceiling, or high up on walls is not efficient and, as in the case described previously in this section, can serve to "short-circuit" the supply of fresh air to the room. Even if high air exhaust outlets were effective, they would tend to pull noxious fumes through the occupants' breathing zones. Exhausts placed near the floor or at the rear of workbenches would prove more effective, as long as they remain unobstructed, and the direction of air movement from a source would be away from the occupant's face. Localized exhausts, using local pickups exhausting through flexible hoses, can be used to remove fumes from well-defined sources of fumes, but they must be placed close to the source. The air movement toward the nozzle is reduced to less than 10% of the original value within a distance equal to the nozzle diameter. Outside this distance, it is unlikely that a localized exhaust would be very effective in removing fumes. If all work with hazardous materials were to be done in hoods, and the hoods ran continuously, it would be possible to rely on hoods to provide the exhaust ventilation to a room. However, this is not the case. Sometimes hood sashes are closed and the hoods used to store chemicals. On other occasions, hoods are turned off while apparatus is installed, or they are off while being serviced. Therefore, the design of the air exhaust system from a laboratory must be done carefully to provide continuing replacement of fresh air in the room. The fume hood system and the supplementary exhaust system should be interlocked to ensure a stable room air balance at all times. There are advanced systems which do a very effective job of maintaining appropriate ventilation in laboratories automatically.

If there are administrative, classroom, or service areas within the same building as laboratories, the entire laboratory area should be at a modest negative pressure with respect to these spaces so that any air flow that exists will be from the nonresearch areas into the space occupied by laboratories.

It is important that the source of air for a building be as clean as possible, and that the chances for air to reenter the building be minimized. In most locations, there are preferred wind directions. In Figure 3.5A and B directional and velocity wind data are shown, averaged over a year for a typical building site. Such data can be obtained for a region from airports and weather bureaus. However, wind data are strongly affected by local terrain, other nearby buildings, trees, and other local variables (note the anomalously high

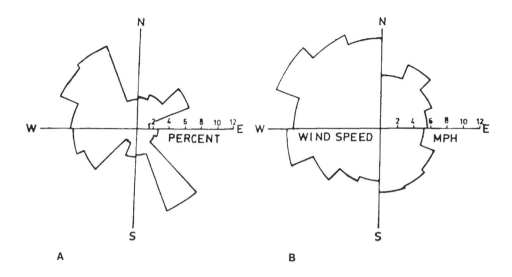

Figure 3.5.   (A) Wind direction, percent of time during year. (B) Average wind velocity vs. direction.

percentage of time the wind comes from a sharply defined southeastern direction here). Where reliable data are available or can be obtained, the air intakes should be located up-wind as much as possible with respect to the building. At this site, locating the air intakes at the northwest corner of the building would clearly be desirable, both because this is the predominant wind direction and the higher wind velocities from this direction would assist minimizing recapture of fumes. Building exhausts, again from the data, should be to the south of the air intakes. Obviously, there will be periods when this configuration will lead to the exhaust fumes being blown toward the air intakes, but other measures can be taken to also aid in the reduction of recirculation.

One situation which must be avoided is the situation shown in Figure 3.6. Here, the air intake is located in a penthouse on the lee side of a raised portion of a building. As shown in the figure, the air moving over the top of the obstruction tends to be trapped and circulate in an eddy on the downwind side of the obstruction. If fumes are swept into this volume, either from the roof above or in the space contiguous to the obstruction slightly further downwind, they will tend to remain there. If an air intake were to be located in this space, the entrapped fumes would be returned to the building.

The exhausts from the building should discharge fumes outside the building "envelope," i.e., the air volume surrounding the building where air may be recaptured. Physically, this can be done with tall individual hood exhaust ducts, or the exhausts from individual hoods can be brought to a common plenum and discharged through a common tall stack. The needed height of individual stacks often make the "tall stack" alternative a physically unattractive concept. Two "rules of thumb" are employed to estimate needed heights. For one- or two-story buildings, the stack height above the roof should be about 1.5 times the building height. For taller buildings, this "rule" would lead to very high stacks, so that a height equal to 0.5 times the buildings width is often used in such cases. We will return to the "common plenum" concept again later because there are a number of design details which are required to ensure that bringing individual exhausts to a common

Figure 3.6. Effect of building shape on wind flow in its vicinity.

duct can be safely used. In the former case, if it is desired to have the air escape from the vicinity of a building, inverted weather caps above duct outlets clearly should not be used since they would direct the air back toward the building. Updraft exhaust ducts with no weather caps are preferable, in which the outlets narrow to form a nozzle, thereby increasing the exit velocity. Since the exhaust air has a vertical velocity, it will initially continue to move upward, so that the effective height of the duct will be higher than the physical height. The gain in the effective height will depend upon a number of factors; *viz.*, the duct outlet diameter, d; the exit velocity of the gas, v; the mean wind speed, μ; the temperature difference between the exhaust gas and outside air temperatures, δT; and the absolute temperature of the gas, T. The effective height gain is given by:

$$\text{Height gain} = d[v/\mu]^{1.4}[1 + \delta T/T].$$

The following is an example of the results of using an updraft exhaust. For simplicity's sake, assume that the indoor and outdoor temperatures are the same. For a duct diameter of 8 inches, a nozzle velocity of 4900 fpm, and a mean wind speed of 700 fpm (equal to the annual average of approximately 8 mph of the site for which data were given in Figure 3.5), the height gain would be about 10 feet. Under some weather conditions, the plume would continue to rise and in others it would fall. In gusty winds, it could be blown back down upon the building roof. In any case, the effective height above the roof of about 13 feet (duct height plus height gain) would be helpful in reducing the amount recaptured by the building and obviously is far better than the alternative using weather caps, in which the fumes are always directed down toward the roof.

An examination of the equation used to determine the height gain shows that if the exit velocity could be maintained, it would be advantageous to have a larger duct diameter. For example, if several hoods could be brought to a common final exhaust duct, 2 feet in diameter, so that the exit velocity remained the same, the net gain would be 30 feet instead of 10 feet, and it would be more acceptable to have a single tall chimney rather than a forest of exhaust stacks. With this arrangement, it would be possible to be reasonably certain that the fumes would not return to the level of the air intake until the plume left the vicinity of the building for a considerably larger portion of the time.

Although some concern is usually expressed about chemical reactions due to mixing the fumes from different hoods, generally the fumes from each hood are sufficiently diluted by the air through the faces of the hoods so that the reactions in the plenum will not be a significant problem on a short-term basis. There perhaps could be long-term cumulative effects. The most serious operational problem is maintaining the balance of the system as the number of hoods exhausting into the common plenum varies. If all the hoods ran continuously, this would not be a problem, but for energy conservation, as well as other reasons, this mode of operation is not the most desirable. There are certain conditions that must be met. Each contributing hood exhaust must be kept at a negative pressure with respect to the building as a whole, so that fumes would not leak into the building through a faulty exhaust duct. In order to ensure that no fumes from the common plenum are forced back into the laboratory, the plenum must always be at a negative pressure with respect to the individual ducts, so that the plenum must be serviced by a separate blower system. It would be difficult to meet both the balance and energy conservation requirements simultaneously with a single plenum exhaust motor. Multiple motors, which go on and off line automatically, can compensate when the number of hoods which are actually on varies. Thus, a reasonably constant negative pressure differential, as determined by pressure switches, between the plenum and the individual hoods is maintained. The negative pressure in the plenum would increase the effectiveness of the individual hood exhaust fans. Some common plenum designs do not include individual fume hood fans.

A risk in such a system is that the motors serving the plenum might fail while the individual motors serving the individual hoods do not. In this event, the fumes in the common plenum would mix and the chances would be good that some fumes would be returned to the laboratories in which the hoods had been turned off. Since the hoods would be exhausting into a volume at a higher than normal pressure, the effectiveness of individual hood systems would be diminished so that the probability of fumes spilling from the hoods would increase, even for those hoods which continued to operate. If multiple motors serve the common plenum, the problems would not be serious if an individual motor failed, since the system should be designed to compensate until the motor was returned to service. However, if electrical power were to fail so that the entire plenum system were to go down, the potential would exist for serious problems within the laboratories. It is essential that such a system be provided with sufficient standby electrical power, as well as an alarm system, to permit the system to continue to serve all operations that cannot be temporarily terminated or reduced to a maintenance level. A standard close-down procedure for all the individual hoods should be developed to be implemented in such a situation.

3.1.6.1. Quality of Supplied Air

Quantity of air is important, but so is the quality. Humans and equipment work best within a fairly narrow range of temperature and humidity. The term "fresh air" implies that it is at least reasonably free of noxious fumes, but it says nothing of the temperature and humidity. In 1979, emergency building temperature regulations were imposed which required that the temperature set points be set at a minimum of 78°F in the summer and a maximum of 68°F in the winter in order to conserve energy. Although these temperatures were eminently satisfactory to some individuals, a large number were vociferously unhappy. Similarly, very low humidity in the laboratory is frequently encountered during the winter as outside air at low temperatures, containing very little moisture, is brought inside and heated, resulting in desert-like characteristics. As a result, many people develop respiratory problems. During the summer, unless sufficient moisture is wrung from the hot outside air while it is being reduced to more comfortable temperatures, the interior humidity may rise to very high levels. Persons will feel clammy and uncomfortable, since they will be unable to perspire as readily. Moist, warm air is also conducive to microbial growth, to which many individuals are allergic in varying degrees. Under very humid conditions, workers are less likely to wear personal protective equipment, such as lab coats, chemical splash goggles, and protective gloves. Unless the temperature and humidity do stay within a relatively narrow range, people become less productive and make more errors. Laboratories are not work environments where error-producing conditions should be acceptable. Although individuals differ, a comfortable temperature to a large number of people seems to lie between 68°F and 75°F, and a comfortable humidity, between 40% and 60%.

Because of all the potential negative results of having poor quality air, it is clearly desirable to have a properly designed and maintained system to make the air as conducive to comfort as possible. In order to assure this, the initial building contract should contain clauses defining explicitly the specifications for the temperature, humidity, and volume of air for each space within the building, and the contractor should be required to demonstrate that the building meets these specifications before the owner accepts the building. This is as important, from both a usage and health and safety standpoint, as any other part of the design.

A laboratory building needing 100% fresh air for space ventilation and large quantities of air discharged by each fume hood is an energy inefficient building almost by definition. Laboratory equipment in the building also is likely to be a very substantial heat load. Buildings housing laboratories, therefore, are logical candidates for energy recovery and energy management systems. In implementing an energy recovery system, care must be taken to ensure that the system is a true *energy* exchange system, where the incoming air characteristics are moderated by the air being exhausted from the building but no air is recirculated. It is difficult to manage an energy system within a building if the occupants have the capability of modifying it locally. Thus, in a managed facility, it is likely that any windows will be permanently sealed so that they cannot be opened and thus disturb the local air balance. The use of ceiling spaces as return plenums is less desirable than the use of fixed ducts, since the former permits an individual to modify the air circulation in his space by simply making an opening into the ceiling. Modifications and connections even to hard ducts need to be done by a qualified HVAC engineer, not by a local shop crew, so that the air provided remains within needed design limits for the spaces served by the system.

A building designed to meet all the requirements of a well-designed and managed building in terms of air quality is usually a "tight building" with few chances for air to leak into and out of the building. Experience has shown that such a building may lead to the "tight building syndrome,"[*] where a large fraction of the populace of an entire building appears prone to developing environmentally related illnesses, sometimes suddenly and acutely. Such problems may, on occasion, be triggered in a building by an unfamiliar odor, by the overall air quality moving out of the comfort zone, by an individual suddenly and unexplainably becoming ill, or for no apparent reason. Laboratory buildings, with

---

[*]   The term "tight building syndrome" is not synonymous with the term "sick building syndrome." NIOSH estimates that about 20% of a buildings occupants are adversely affected when a building is "sick." The term "tight building syndrome" should be applied more aptly to buildings where conditions give rise to a perception of a problem. Both of these terms are used too loosely, in the author's opinion. Conditions may exist where in a space as small as a single room an individual may become ill due to environmental conditions. To this individual, the area is "sick."

their common and frequently unpleasant, pervasive chemical odors, could be vulnerable to this problem, although occupants of such buildings may be accustomed to "strange" odors and be more willing to accept odors which are not acceptable to other personnel. Where the building occupants have little or no means of modifying their environment, i.e., they cannot open their windows, the frustration of having no control over the problem seems to exacerbate the likelihood of the perceived problem developing and worsening the impact when it does. Of course, a bad odor does not always trigger a tight building syndrome response from a building's occupants, but the inability to personally do something about it seems, at the very least, to increase a person's irascibility. Often, even after prolonged investigation, no real cause of a tight building syndrome incident is ever discovered, and it is attributed to stress or other psychological causes, especially when the problem seems to disappear when no corrective measures were taken.

Some tight building syndrome incidents represent real health problems, with individuals persistently complaining of discomfort and showing evidence of physical distress. Most commonly the symptoms are respiratory distress, headaches, fatigue, dizziness, nausea, skin and eye irritation complaints regarding odors or a chemical "taste" in their mouths. These problems are similar to allergenic reactions, and if some individuals exhibit the problems while others do not, this is typical of the widely varying sensitivity of individuals to allergens. If no other source of the problems can be located, an evaluation of the HVAC system is in order. There is a tendency by many to ascribe the blame initially to low-level volatile compounds, but microbial contaminants may actually be the real cause. Relatively few systems are as well maintained as they should be, and even fewer are cleaned routinely.

All HVAC systems include filtration systems, and some provide humidification and dehumidification functions. The filters may become dirty and the amount of fresh air may decrease to an insufficient level, or the filters may begin harboring dust mites or fungal growth. If the filters in different parts of the system become less able to pass air preferentially, the building air balance may shift so that areas which at one time had sufficient air may no longer do so. This alone can cause problems. The moisture used for humidification or the drain pans into which moisture from the dehumidifying process goes may become contaminated with biological organisms. Many individuals are allergic to dust, dust mites, fungi, and other microorganisms which would be distributed by a contaminated air supply. In severe cases, the air handling system could harbor *Legionella* organisms, which would require massive and disruptive decontamination efforts, if these were possible at all.

In order to avoid contamination of HVAC systems and to maintain a proper level of performance, a comprehensive and thorough maintenance program is required for all centralized HVAC systems. Filters should be changed or cleaned on a regular and frequent schedule. The efficiency and quality of the existing filters should be evaluated. Inefficient roll filters should be changed to a better quality, higher efficiency types. Many systems are initially equipped with filters which are not appreciably better than furnace filters used in the home, with efficiencies of as little as 25%. If the capacity of the fans can cope with higher filter efficiencies, these should be upgraded to 65% to 85%. Water employed for humidification and chillers, or cooling towers, should be checked frequently for biological growth. Condensate pans should be cleaned regularly. Decontamination of an afflicted ventilation system is difficult, time consuming, and consequently very costly. Preventive maintenance is very cost effective for ventilation systems. Fortunately laboratory areas are usually provided with 100% fresh air and do not experience the problems associated with recirculating systems, where reuse of up to 80% of the building air not only allows contaminants to build up, but allows contaminants in one area to be distributed to the entire building do not occur.

REFERENCES

1. **Li, F. et al.,** Cancer Mortality Among Chemists, in: *J. Nat. Cancer Institute.* 43, 1159–1164, 1969.

2. **Olin, R.,** Leukemia and Hodgkin's disease among Swedish chemistry graduates, in: *Lancet,* (ii) p. 916, 1976.

3. **Olin, R.,** The hazards of a chemical laboratory environment: a study of the mortality in two cohorts of Swedish chemists, in: *Amer. Ind. Hygiene Assn. J.,* 39, 557–562, 1978.

4. **Olin, R. and Ahlbom, A.,** The cancer mortality rate among Swedish chemists graduated during three decades, in: *Envir. Res.,* 22, 154–161, 1980.

5. **Hoar, S.K. and Pell, S.,** A retrospective cohort study of mortality and cancer incidence among chemists, in: *J. Occup. Med.,* 23. 485–495, 1981.

6. Comment by the California Association of Cytotechnologists re Xylene exposure re the OSHA Proposed Performance Standard for Laboratories Using Toxic Substances, 51 FR 26660, July 24, 1986.

7. U.S. Department of Health and Human Services, Centers for Disease Control, National Institute for Occupational Safety and Health, Health Evaluation Reports HETA 83-048-1347, HETA 830076-1414, HETA 81-422-1387, Cincinnati, OH.

8. *Prudent Practices for Handling Hazardous Chemicals in Laboratories.* National Academy Press, Washington, D.C., pp. 193–212, 1981.

9. The Guide for the Care and Use of Laboratory Animals, National Institute of Health Publication No. 85-23, 1985.

10. OSHA General Industry Standards, 29 CFR Part 1910, §106.(d)(4)(iv).

11. **Caplan, K.J. and Knutson, G.W.,** The effect of room air challenge on the efficiency of laboratory fume hoods, in: *ASHRAE Trans.,* 83, Part 1, 1977.

12. **Fuller, F.H. and Etchells, A.W.,** Safe operation with the 0.3m/s (60 fpm) laboratory hood, in: *ASHRAE J.,* 49, October, 1979.

13. Laboratory Ventilation for Hazard Control, National Cancer Institute Cancer Research Safety Symposium. Fort Detrick, MD, 1976.

14. **Wilson, D.J.,** Effect of stack height and exit velocity on exhaust gas dilution. *ASHRAE Handbook,* American Society of Heating, Refrigerating and Air Conditioning Engineers, Atlanta, GA, 1978.

15. **Cember, H.,** *Introduction to Health Physics,* Pergamon Press, New York, 334–339, 1969.

16. Tight building syndrome: the risks and remedies, in: *Amer. Indust. Hyg. J.,* 47, 207–213.

17. ASHRAE 90-1980, *Energy Conservation in New Building Design,* 345 E. 47th St., New York, NY 10017.

18. ANSI Z9.2-1979, Fundamentals Governing the Design and Operation of Local Exhaust Systems, American National Standards Institute, New York, NY 10018.

19. *Industrial Ventilation: A Manual of Recommended Practice,* 20th ed., American Conference of Governmental Industrial Hygienists, Cincinnati, OH, 1988.

20. **DiBerardinis, L.J., Baum, J., First, M.W., Gatwood, G.T., Groden, E., and Seth, A.K.,** *Guidelines for Laboratory Design: Health and Safety Considerations,* John Wiley & Sons, New York, 1987.

21. *Biosafety in Microbiological and Biological Laboratories,* U.S. Department of Health and Human Services, Centers for Disease Control, National Institute for Occupational Safety and Health, Publication No. (CDC) 88-8395, U.S. Government Printing Office, Washington, DC, 1988.

22. ASHRAE 62-1989, *Ventilation for Acceptable Indoor Air Quality,* American Society of Heating, Refrigerating and Air Conditioning Engineers, Atlanta, GA, 1989.

3.1.7. Electrical Systems

Electrical requirements for laboratories are relatively straightforward. The entire system must meet the NFPA Standard 70, the National Electrical Code, and must be properly inspected before being put into service. As far as the laboratory worker is concerned, the details of the service to the building are relatively unimportant, but it is important to them that there are circuits of sufficient capacity to provide enough outlets for all of the equipment in the laboratory. In most new facilities, the designers usually provide enough, but in many older facilities, where less use of electrically operated equipment was anticipated, the number of outlets is often inadequate. Many of the older electrical systems were originally designed based on two wire circuits instead of the three wire circuits that are currently required by code. Any circuits of the older type should be replaced as soon as possible, and laboratory activities should not be assigned to spaces provided with such electrical service until this is done. All circuits, whether original equipment or added later, should consist of three wires, a hot or black wire, a neutral or white wire, and a ground or green wire. The ground connection should be a high-quality, low-impedance ground (on the order of a few ohms), and all grounds on all outlets should be of comparable quality. A poor-quality ground, perhaps due to a worker failing to tighten a screw firmly, can result in a substantial difference of potential between the grounds of two outlet receptacles. This can cause significant problems in modern solid state electronic equipment which typically operates at voltages of less than 24 volts (often at 3, 5, or 6 volts). A further problem which can result from a poor ground is that leakage current through the high-impedance ground connection can develop a significant amount of localized electrical heat. This is often the source of an electrical fire, rather than a short circuit or an overloaded circuit.

Electrical circuits should be checked using a suitable instrument capable of providing a quantitative measure of the ground impedance. Commonly available, inexpensive plug-in "circuit-checkers" which indicate the condition of a circuit by a combination of lights can give a valid indication of a faulty circuit, but a "good" reading can be erroneous. If the distance to another connection is substantial, capacitive coupling of the ground wire can result in a false indication of a low impedance reading.

The female outlets, as noted, have three connections with openings located at the points of an equilateral triangle. For the common 110 volt circuit for which most common consumer and lower wattage laboratory equipment is designed, the ground connection is round, while the other two openings are rectangular. If the two rectangular openings are of different sizes, the neutral connection will be longer and the hot connection shorter. A male plug has matching prongs, and should only fit in the outlet in a single orientation. A "cheater" or adaptor can be used to allow a three-prong male plug to be inserted into a two-wire circuit, but this is not desirable. Some portable equipment, such as a drill, is available with only two connectors, but these are required now to be "double-insulated" so the external case cannot provide a circuit connection to the user. All connectors, switches, and wiring in a circuit must be rated for the maximum voltage and current they may be expected to carry. It is recommended that the female sockets be protected with ground-fault interrupters as an additional measure of protection. These devices compare the current through the two-current carrying leads, and if they are different, as would be the case where current was being diverted through an electrical short or a person, they will break the circuit in a very few milliseconds, usually before any harm occurs.

There should be enough outlets appropriately distributed in a laboratory so that it should not be necessary to use multiple outlet adapters, plugged into a single socket, or to require the use of extension cords. Where it is necessary for additional circuits to be temporarily added, the circuits should either be run in conduits or in metal cable trays, both of which should be grounded and installed by qualified technicians or electricians. Even though their use should be discouraged, extension cords will continue to be used. However, as a minimum, they must be maintained in good condition, include three wires of sufficient size to avoid overheating (preferably of 14-gauge wire or better for most common uses), and must be protected against damage. They should not be placed under stress, and should be protected against pinching, cutting, or being walked upon. Where abuse may occur, they must be protected with a physical shield sufficient to protect them from reasonably anticipated sources of damage.

Circuits must be protected by circuit breakers rated for the maximum current to be carried by the circuit. Normally, many breakers for a room or group of rooms are located together on a common breaker panel. All circuits should be identified, both within the facility and at the breaker panel, so that when required, the power supply to a given circuit may quickly and easily be disconnected. This is especially important when it is necessary to disconnect power due to an emergency. There should be no ambiguity about breaker that needs to be thrown to kill the power to a given receptacle. Laboratories with high voltage and/or high current sources should consider a master disconnect button, which anyone can use to kill all of the circuits in the facility if someone becomes connected.

Where breaker panels and electrical switches are placed in separate electrical closets or rooms, there is some question about the propriety of individual laboratory workers having access to the space. If there are electrically live parts with which individuals might accidentally come into contact, access to the spaces must only be by qualified, authorized individuals, according to code regulations. Normally, live components on breaker panel are completely covered (if not, then prompt action to replace the cover should be initiated). If, in addition, the breaker panels are segregated by a locked or otherwise secure barrier from areas containing electrically active components, then laboratory workers should be allowed to have access to the panels in order to control the circuits. However, access to these spaces must not be abused by considering them as extra "storage space." Access to the electrical panels, switches, and other electrical equipment in the space must not be blocked by extraneous objects and materials.

The location of electrical circuits and electrically operated equipment in a room should be such that they are unlikely to become wet and they should not be in an area susceptible to condensation or where a user might be in contact with moisture. As unlikely as it may appear, instances have been observed where equipment has been located and electrical circuits have been installed where water from deluge showers would inundate them. For some equipment, such as refrigerators, freezers, dehumidifiers, and air conditioning units, moisture is likely to be present due to condensation, and these equipment items must be well grounded.

3.1.7.1. Hazardous Locations

Most laboratories do not represent hazardous locations in the context of requiring special electrical wiring and fixtures, although there may be individual equipment items which may need to be treated as such. The classification of a facility as hazardous in a regulatory context, depends upon the type of materials employed in the facility and whether flammable fumes or gases, electrically conducting materials, or explosive dusts are present in the air within these facilities in the normal course of routine activities or only sporadically due to some special circumstance. Explosion proof wiring and fixtures are substantially more expensive than ordinary equivalents and should be used only if there are no acceptable alternatives. Where the need does exist, however, they definitely should be used.

The National Electrical Code defines three different categories of hazardous locations: classes I, II, and III (note that these are not the same as the classification of flammable liquids into classes I, II, and III). Class I represents locations where flammable vapors or gases may be in the air. Class II locations involve facilities where electrically conducting or combustible dusts may be found, and class III locations contain ignitable fibers. Each Class is split into two divisions, 1 and 2. In division 1 for each class, the hazardous conditions are present as a normal course of activities or are sufficiently common due to frequent maintenance or may be generated due to equipment failure, such as emission of dangerous vapors by the breakdown of electrical equipment. Division 2 includes locations which involve hazardous materials or processes similar to those in division 1, but under conditions where the hazardous gases, fumes, vapors, dusts, or fibers are normally contained, or the concentrations maintained at acceptably low levels by ventilation so that they are likely to be present only under abnormal conditions. Division 2 locations are also defined to include spaces adjacent to but normally isolated from division 1 locations from which problem materials might leak under unusual conditions.

Within class I, there is a further division by groups into A, B, C, and D, depending upon the materials employed, with the distinction between the groups being based essentially on the flammable limits in air by volume. A long list of chemicals, with the groups identified to which they may belong, is given in NFPA 497M. Information on other chemicals should be available to any researcher in the Material Safety Data Sheets which chemicals users should be maintaining, or at least to which they should have access. However, in this listing, acetylene with a flammability range of 2.5% to 81% is the only chemical listed in group A. Group B chemicals with flammability ranges between about 4% to 75%, and with flash points less than 37.8°C or 100°F include acrolein, ethylene oxide, propylene oxide, hydrogen, and manufactured gas (>30% hydrogen by volume). However, if equipment in the facility is isolated by sealing all conduits 1/2 inch or larger in diameter, according to specifications in the National Electrical Code, the first three of these may be placed in groups of lesser risk. Among these are flammable liquids with flash points above 37.8°C or 100°F, but less than 60°C or 140°F. Allyl glycidal ether and *n*-butyl glycidal ether would also be in group B, but with the same exception as to conduit sealing.

Groups C and D have flammability ranges between about 2% and 30% and 1% and 17%, respectively. A few common chemicals falling into group C are acetaldehyde, carbon monoxide, diethylether, ethylene, methylether, nitromethane, tetrahydrofuran, and triethylamine. A partial list of common group D chemicals includes acetone, benzene, cyclohexane, ethanol, gasoline, methanol, methylethyl ketone, propylene, pyridine, styrene, toluene, and xylene.

Generally, if the location uses chemicals in any of these four groups which have flash points less than 37.8°C or 100°F and otherwise meets the specifications of a class I, division 1 location, special electrical equipment would normally be required. Special electrical equipment would normally be required for class II flammables (flash points equal to or above 37.8°C or 100°F, but less than 60°C or 140°F) only if the materials are stored or handled above the flash points, while for class IIIA flammables (flash points equal to or above 60°C or 140°F, but less than 93.3°C or 200°F), special electrical equipment is needed only if there are spaces in which the temperature of the vapors may be above the flash points.

Group II includes conductive and combustible dusts, with some dusts falling into both categories. Dusts with resistivities above 10^5 ohm/cm are not considered conductive, while those with lesser resistivities are. If operations were such as to generate significant levels of dust in the ambient air in the work location, the decision as to whether special electrical equipment would be required would be based on whether a cloud of the dust in question would have an ignition sensitivity equal to or greater than 0.2 and an explosion severity equal to or greater than 0.5. Both of these are dimensionless parameters based on a comparison to a standard material, Pittsburgh seam coal.

The definitions of these two parameters are

$$\text{Ignition Sensitivity} = \frac{(P_{max} \times P)_2}{(P_{max} \times P)_1}$$

$$\text{Explosion Severity} = \frac{(T_c \times E \times M_c)_1}{(T_c \times E \times M_c)_2}$$

where P_{max} = maximum explosive pressure
P = maximum rate of pressure rise
T_c = minimum ignition temperature
E = minimum ignition energy
M_c = minimum explosive concentration
Subscript 1 = standard dust
Subscript 2 = specimen dust

There are a number of metals and their commercial grades and alloys, which could give rise to the need for special electrical equipment, some of which are listed in NFPA 497. Metals, such as zirconium, thorium, and uranium, which would be found in some special laboratories, have both low ignition temperatures, around 20°C (68°F), and low ignition energies so that work with these materials would require special precautions and safeguards.

A large number of nonconductive dusts have an ignition sensitivity of 0.2 or higher and an explosion severity of 0.5 or greater. Many agricultural products, such as grains, can form dusts in this category. Other nonconducting materials with similar characteristics would be many carbonaceous materials, chemicals, dyes, pesticides, resins, and molding compounds. A long but incomplete list is given in NFPA 497M. OSHA lists many industrial operations which could be classified as class II, and in the context of laboratory safety there are pilot or bench scale research operations which would emulate these industrial locations.

Class III locations represent hazardous locations because of easily ignitable fibers. However, it is unlikely that they would be present in sufficient concentrations to produce an ignitable mixture. Thus, no special electrical wiring requirements would normally exist. However, research facilities which generate substantial airborne quantities of fibers of cotton, synthetic materials such as rayon, wood, or similar materials should take care to avoid the potential for fire.

The special electrical equipment or wiring procedures needed to satisfy the requirements of a hazardous location are specified in the National Electrical Code, NFPA Standard 70. In general, fixtures suitable for use in hazardous locations will be rated and certified as safe by a nationally recognized testing laboratory such as Underwriters Laboratory or Factory Mutual Engineering Corporation. In some laboratory installations equipment items may be found which are nonstandard and are not listed as acceptable by any appropriate organization. In such cases, the equipment may be certified as safe if an appropriate agency charged with enforcing the provisions of the National Electrical Code so as to ensure the occupational safety of users finds the equipment in compliance with the Code. Where certification is obtained by this means, records need to be kept showing how this determination was made, from manufacturer's data or actual tests and evaluations. Special electrical equipment for use in class I locations is not necessarily gas-tight, but if an explosion does occur within it, it should be contained and quenched so that it does not propagate further.

REFERENCES

1. NFPA 70-84, National Electrical Code (latest version), National Fire Protection Association, Quincy, Mass.
2. NFPA 497-1975, *Recommended Practices for Classification of Class I Hazardous Locations for Electrical Installations in Chemical Plants,* National Fire Protection Association, Quincy, MA.
3. NFPA-497M-1983, *Manual for Classification of Gases, Vapors and Dusts for Electrical Equipment in Hazardous (Classified) Locations,* National Fire Protection Association, Quincy, MA.
4. *Prudent Practices for Handling Hazardous Chemicals in Laboratories,* National Academy Press, Washington, D.C., 1981, pp. 179–192.
5. The following ANSI Standards cover electrical apparatus for use in hazardous locations (by number only): C33.3-1973; 781-1977; 844-1978; C33.27-1973; C33.29-1972; 1002-1977; C33.97-1973. These are issued jointly with Underwriters Laboratories. Since these are updated on an irregular basis, the latest edition should be obtained when used as reference.

3.1.8. Plumbing

There are two aspects of laboratory plumbing system design, as opposed to operations, which are relevant to safety and health. The first is the capacity of the system to withstand the waste stream which the system may be called upon to handle. The second is the need to prevent the operations within the laboratory facility from feeding back into and contaminating the potable water system that supplies the building.

3.1.8.1. Sanitary System Materials

Under the standards governing the disposal of toxic and hazardous chemicals into the sanitary sewage system and into the public waters, much smaller quantities of chemicals should be going into sink drains than in the past. A very large number of the chemicals used in laboratories are now classified as hazardous waste and should be collected and disposed of according to the provisions of the Resource Conservation and Recovery Act. However, a substantial amount of acids, bases, solvents, and other chemicals still go into the sanitary system, even if from no other source than cleaning of laboratory glassware. Although the National Clean Water Act is not intended to cover small individual laboratories, or even groups of laboratories, a large research facility may add more chemicals into the sewer system than many small industries. However, there probably will be very little comparatively of any single one. Laboratory chemicals should be highly diluted in large quantities of water but, given time, can corrode ordinary plumbing systems.

Plumbing materials used to service laboratory facilities must be resistant to a large range of corrosives, physically durable, relatively easy to install and repair, and relatively cheap. Metal plumbing materials are generally unacceptable because they are vulnerable to inorganic acids. Plastics, such as are used in residential systems will not withstand many common organic solvents or will absorb other solvents and not remain dimensionally stable. Glass would serve very well but is brittle and not inexpensive in large systems. Its desirable properties make it suitable and acceptably cost effective for the components most vulnerable to the actions of chemicals—the sink trap and the fittings which connect the sink to the sanitary system. A plastic resistant to a large range of inorganic and organic waste streams dimensionally stable, durable, and relatively inexpensive is polypropylene. Other materials with comparable physical properties may be either more expensive or more difficult to install and maintain.

One characteristic not mentioned in the preceding paragraph is the necessity for the material to have established fire resistance ratings as specific structural members, i.e., tested by appropriate testing laboratories in the configurations in which they might be used. Polybutylene plumbing components are available which meet this requirement.

3.1.8.2. Backflow Prevention*

Many jurisdictions have specific legal requirements for devices preventing contamination of the potable water supply to be installed on each service line to a building's water system wherever the possibility exists that a health or pollution hazard to the waterworks system could exist. This will include any facility, such as a laboratory, where substances are handled in such a manner as to create a real or potential risk of contaminating the water supply external to the building. Although the regulations may only address the problems of cross-contamination between buildings, the risk may exist as easily within a building. For example, if by some means, contaminated water is drawn from a laboratory sink back into the potable water system, the contamination could easily express itself in the water available from the water fountain immediately outside in the hall, if the latter were to be fed from the same line supplying the faucets in the sink. The contamination would not necessarily only be close to the contaminating source, but could be anywhere in the system downstream from the origin of the pollution.

Illustrated in Figure 3.7 are two possibilities for cross-connections, the simplest being a direct connection between a contaminated system and a clean system, connected by a valve which could be opened. Obviously, this is not a desirable situation, but it can occur. The second possibility is that a pressure differential between the clean and contaminating systems may be established such that contamination may be forced or drawn into the clean system through some linkage. This linkage may not necessarily be a valve left open or a leaky valve, although these are two of the most likely sources. An example of an inadvertent cross-connection would be a heat exchanger, which has developed small leaks between the primary and secondary loops over a period of time. As long as the primary, or clean side, is at a higher pressure than the secondary, or contaminated side, any water flow would be from the clean to the contaminated side; as long as the leaks were small, they

---

* Much of the material in this section is based on the article by Paul H. Woodruff in the second edition of the *CRC Handbook of Laboratory Safety:* "Prevention of Contamination of Drinking Water Supplies."

would probably go undiscovered. If, however, the water supply on the primary side was reduced and pressure were to fall below the secondary, the flow would be reversed and the clean water supply would be polluted. This would be an example of a forced backflow problem.

A very common situation existing in a laboratory is to find a section of plastic tubing draining the effluent from a piece of apparatus, lying in the bottom of a far-from-clean sink. Perhaps the sink is being used to wash dirty glassware and the plastic tubing is under several inches of water. If the water pressure should suddenly fail, perhaps due to a reduced supply because of maintenance, coupled with the simultaneous flushing of several toilets, it would be possible for the water in the sink to be siphoned back through the system and reach the potable water supply. Any system in which a connection to the potable water supply can be flooded with contaminated water would be subject to the same type of problem. Inexpensive vacuum breakers to install on the sink faucets are available from most major laboratory supply firms.

The best correction to these problems is not always obvious. In the first example of the heat exchanger, a pressure sensor with redundancy features could be installed between the primary and secondary loops such that when the positive pressure differential falls below a stipulated value (well above the point at which the pressures are reversed), operations should be shut down and the secondary loop emptied. A better solution would be for the heat exchanger to incorporate an intermediate loop so that the secondary and primary sides could not be directly coupled.

Figure 3.8 shows simple diagrams of vacuum breakers used to prevent back-siphonage. In order to prevent backflow due to excess pressure, vacuum breakers cannot be used, and an air gap provides the most protection. Figures 3.9 and 3.10 illustrate two versions of air gaps. The second of these is very straightforward: simply do not connect anything to the water source which could become flooded. Any formal backflow or anti-siphonage device which is installed must be an approved type which has been tested by a recognized laboratory testing agency and be of satisfactory materials.

REFERENCE

Woodruff, P.H., Prevention of contamination of drinking water supplies, in CRC *Handbook of Laboratory Safety*, 2nd ed., Steere, N.V., (Ed.), CRC Press, Inc., Boca Raton, FL, 1971.

3.1.9. Other Laboratory Utilities

There are a number of other possible utilities which may be provided to a laboratory. Among these are natural gas, compressed air, distilled water, vacuum, steam, refrigerated brine, and other gases. In some instances, there are safety issues, such as limitations on the pressure available from a compressed air line and the need to incorporate provisions for pressure relief, to ensure that personnel will not be injured by explosions due to excess pressure. Often the quality of the air is more of a problem. The compressor supplying the system should be capable of supplying air which is clear of oil and moisture. Some facilities which have a large, pressurized liquid nitrogen tank at hand, use the vapors from the tank as a source of ultraclean compressed "air" to clean work surfaces.

The dangers of natural gas are well understood. This does not prevent numerous accidents each year due to gas explosions. Natural gas used to heat experimental devices

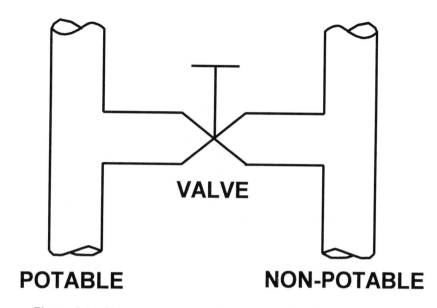

Figure 3.7. Simple cross-connection between liquid systems.

Figure 3.8. Vacuum breaker.

should not be left unattended, unless a heat-sensitive automatic cut off device is connected to the system. Otherwise, if gas service were to be interrupted, gas escaping from burners or heaters left on could easily cause a major explosion. If the gas device is set up in a properly working fume hood, the gas would be exhausted through the fume hood duct system, but as noted earlier, fumes exhausted outside a laboratory building can be recaptured under some conditions. Whenever the odor of gas is perceived, under no circumstances should electric light switches, other spark-producing electric devices, or any other possible source of ignition be operated. If the gas concentration is low enough to make it safe for personnel to enter the room, all devices connected to the gas mains in the laboratory should be immediately turned off. Any windows in the room which are not fixed shut

Figure 3.9. Minimum desirable air gap on sink.

Figure 3.10. Air gap in line subject to back pressure.

and are capable of being opened should be opened. Consideration should be given to having all nonessential personnel evacuate the building while volunteers check the remainder of the building or facility for other systems which might also be leaking. If the concentration is already high, everyone should evacuate the building at once, the gas service turned off at the service entrance to the building, and the utility service company notified. Even small gas leaks, barely perceptible by odor, should be sought out and repaired. Gas can seep slowly through cracks and seams and accumulate in confined spaces in which the air interchange is very slow and explode if an ignition source presents itself.

OSHA has specific regulations on four gases, acetylene, nitrous oxide, hydrogen, and oxygen, in the general industry standards. Generally the requirements in the OSHA regulations are those provided for by the Compressed Gas Association (GSA), and for the first two of the gases mentioned above, the OSHA standards simply refer to the appropriate pamphlet issued by the GSA. For hydrogen and oxygen the regulatory requirements are spelled out in detail in Sections 29 CFR 1910.103 and 29 CFR 1910.104, respectively. The requirements for oxygen generally pertain to bulk systems of 13,000 cubic feet or larger. Relatively few laboratory facilities would involve a system of this size.

For hydrogen, the regulations do not apply to systems of less than 400 cubic feet, so that laboratory systems involving a single typical gas cylinder would not be covered. However, in the context of the present chapter which is intended to involve the design of a building, it is entirely possible that a hydrogen gas system could be larger than 400 cubic feet and hence would be covered by the provisions of the OSHA standard. All of the requirements of the standard will not be repeated here, but among these are (1) provisions that cast iron pipes and fittings shall not be used (note that for acetylene, brass, and copper pipes may not be used); (2) that the system shall be above ground; (3) restrictions on the electrical system to be a class I system; (4) provisions that outlet openings shall be at the high points of wall and roof; and (5) requirements for explosion venting, among many others.

Every utility that is provided should be properly identified with a clear, unambiguous label. Color-coded discs with engraved name labels which screw into each service fixture are available from at least one company for most common service utilities. All fittings and connectors should be provided according to appropriate standards. For example, all gas fittings should comply with the provisions of the Compressed Gas Association Standard V-1. Every utility, if used improperly, can cause safety problems, the only difference being that some require more of an effort on the part of a user than do others.

REFERENCES

1. OSHA General Industry Standards, 29 CFR Part 1910, Sections 101–104.
2. Compressed Gas Association, Inc., 500 Fifth Avenue, New York, NY, 10036.
3. *Prudent Practices for Handling Hazardous Chemicals in Laboratories.* National Academy Press, Washington, D.C., 1981, 75–89.

3.1.10. Maintenance Factors

Every building and every piece of equipment installed in a building should be designed with the realization that each item of equipment and most building components will eventually need service. As noted at the very beginning of this chapter, the current *modus operandi* of most research facilities is to change the nature of the research in a given space fairly frequently. Thus the space itself, treated as a piece of equipment, will need relatively frequent service. It should be possible to bring any needed services to virtually any position within the laboratory. Good access to services should be provided in the initial design, making it possible for maintenance personnel to work on the equipment conveniently and safely.

All ducts, electrical circuits, and utilities need to be clearly identified with labels or other suitable means. Fume hood ducts on the roof should be identified with at least the room number where the hood is located and a hood identification if there is more than one hood in the room. It would also be desirable for the general character of the effluents from the exhaust duct to be indicated by means of a written statement or a color code. Similarly, every electrical circuit should be clearly labeled so that there would be no confusion as to which breaker controls the power to the circuit.

A significant amount of space (15 to 25%) is set aside in every major building for mechanical service rooms, electrical closets, and other building services. They are rarely seen by most of the building occupants, but they are critical to the good operation of the building. Most of the major equipment used to provide building services is either located within these spaces or the control panels are located within them. These areas need to be maintained well and not be used as a storage space. Unless a major component, such as a compressor for the air conditioning system, is out or it is necessary to turn off the water to an entire area, the operations within these rooms are usually outside the experience of the usual occupants of the building, and access to these spaces should not normally be available to them. Maintenance to most occupants of the buildings involves work either in their area or directly affecting their area. An exception is the electrical panels which control the breakers to electrical circuits. As noted earlier, there are reasons why employees other than maintenance personnel will need (sometimes urgently) to deactivate electrical circuits, and if proper segregation of the panels is done, access can be safely provided.

The building plans should show all utilities, ducts, and electrical circuits correctly as they physically exist throughout the facility. Frequently during construction projects it is necessary to make some adjustments to the original plans. Sometimes these are documented with change orders, if they are sufficiently major, but at other times they are considered minor and the building plans are not modified to show them. In principle, all changes should be reflected on the final "as built" drawings and specifications provided to the building owner. This should include not only physical changes in locations, but also any changes and substitutions made in materials as well. For example, a complicating factor in dealing with asbestos in existing buildings stems from the uncertainty as to whether asbestos was used or not, under provisions in many contracts which allowed substitution of "equivalent" materials. The resulting uncertainty has resulted in delays and additional costs in renovation projects while insulating materials are tested for asbestos. Asbestos was used in construction when it was perfectly legal and desirable to do so, and it is necessary now to know where it is located in order to remove it or treat it to render it harmless. If such information were reliably available from building plans, it would be much easier to estimate removal costs and design asbestos abatement projects.

Unfortunately, after a period of occupancy it is common for many changes to be initiated by the facility's users themselves. These are intended to serve a specific purpose and are often built without consideration for anything except this purpose in mind. As a result, maintenance may be significantly impaired. Perhaps the most common occupant-initiated changes involve electrical circuits which may or may not be installed or labeled properly. Similarly, the interior configuration of a laboratory space may be changed, which can significantly affect the distribution of air from the ventilation system serving the area. It should be institutional or corporate policy to prohibit such modifications and to require that they be removed once found. As a minimum, any such remodeling on the part of the occupants should be required to be reviewed by appropriate personnel and changed where necessary to comply with building codes and maintenance programs. They should not be allowed to be left in place where they are disruptive to good design and use practice.

Where there are unusually dangerous maintenance operations, it is necessary that appropriate provisions be made for the purpose. A good example is the need to change contaminated high efficiency particulate and aerosol (HEPA) filters, where the contaminant might be a carcinogen or another life-threatening biologically active agent. Ample access room to the exhaust duct must be provided so that bag-out procedures, where the contaminated filters are withdrawn into a sealed bag, may be done without difficulty and without risk of the agent of concern affecting the maintenance personnel or spreading to adjacent areas. Another example, representing an acute danger rather than a delayed one, would be repair to a perchloric acid ejector duct after a period of operation when the wash-down cycle failed to work. In this case, there would need to be provision for washing the interior of the duct for several hours prior to beginning maintenance operations, as well as developing contingency plans to protect the workers in the event flooding the duct was not totally effective in removing dried perchloric acid.

Maintenance operations often considered as routine, such as replacement of a fume hood motor or repair to a laboratory sink, must be considered hazardous operations if the potential exists for exposure of the workers to toxic chemicals. An example is exposure to chemicals during repairs to a hood exhaust system because of fumes from nearby exhaust ducts of operating hoods or chemical residues on the motor, fan, or other components of the unit under repair. Even maintenance activities totally unrelated to laboratory operations, such as patching a roof, can permit the workers to be exposed to possibly toxic effluents from the air handling system or fume exhausts. The converse is also true: fumes from roofing materials, paints, welding, etc. can be drawn into a building via the air intake to the accompanying discomfort of the building's occupants. Under the new 1986 OSHA Hazard Communications Standard, it is required that any individual, potentially exposed to dangerous chemicals, must be informed of the risks and of the measures needed and available to protect them from the dangerous effects. Thus, when anyone is asked to perform maintenance on a piece of laboratory apparatus which could be contaminated or is asked to work under such conditions that chemical exposures could result, it is now mandatory that the situation be evaluated for potential dangers. It may be necessary to plan for a period of time when operations can be postponed or modified to eliminate the release of toxic materials, or alternatively, the workers may be provided with appropriate protective equipment which they must wear during the maintenance operations. Since workers typically have a minimal chemical educational background, they tend to be more concerned about the dangers of exposures than an ordinary laboratory employee. It is important that an effective training program be established for them to ensure that they are informed of the actual risks to which they might be exposed, but the training should neither exaggerate or minimize these risks.

3.2. FIXED EQUIPMENT AND FURNITURE

In Section 3.1.3, requirements for furniture in the various classes of laboratories were briefly given, *viz.*, "Furniture should be designed to be sturdy and designed for convenient utilization," and "Bench tops should be resistant to the effects of acids, bases, solvents, and moderate heat, and should not absorb water." Flammable material storage cabinets were among some of the special items of equipment needed if solvents were to be used. This section will discuss these items and others in detail. The quality of the laboratory furniture should be as good as can be afforded and should be as versatile as possible, unless it is certain that usage will involve only a limited range of reagents in a few applications. Similarly, the specifications on equipment should be written to assure that it will perform well and *safely.* A refrigerator may be bought on sale for perhaps as little as one quarter of the price of a flammable material storage mode; however, the possible consequences of storing solvents in the former make it a very poor bargain in most laboratories. Even if the immediate need is not there, few of us are so certain of the future that we can be confident that in the typical 15- to 20-year life span of a refrigerator it will not become necessary to store flammable solvents in one.

3.2.1. Laboratory Furniture

Just as ordinary furniture can be bought in many grades of quality not always discernible to a casual examination, laboratory furniture also varies in quality of construction. The least expensive may look attractive and appear to offer the same features as better units, but it will not be as durable and will have to be replaced frequently. It is also likely to not be as safe. If shelves are not firmly attached, the weight of chemicals on them can cause them to collapse. If cabinet doors and drawers do not have positive catches, they can fail to close and possibly rebound, leaving them partially open. The contents in a partially open drawer may be damaged through the opening, or the protruding drawer can cause an accident to someone who does not observe the obstruction. Poorly protected work surfaces can corrode or can become contaminated and be difficult to decontaminate.

Good quality modular laboratory furniture is available today in a variety of materials and can be installed in configurations to fit almost any need. Units can be obtained precut to accommodate connections to utilities. In most cases, the utilities can be brought to the correct locations prior to installation of the furniture, which makes it simple to perform maintenance and make changes should a need arise to modify the existing configuration.

3.2.1.1. Base Units and Work Tops

Base units can be obtained in steel, wood, or plastic laminates. The steel in the steel units should be heavy gauge with a pretreatment to reduce the corrosive effects of chemicals. Painting all surfaces with a durable, baked on, chemically resistant paint finish will also help minimize chemical effects. Many individuals continue to prefer wooden laboratory furniture. Because of cost, solid wooden furniture is not an economic choice, but durable wooden (or plastic) veneer furniture is available which can meet most safety requirements. Although wood may be more absorbent to liquids than steel, it is less reactive and more resistant to a very wide range of chemicals than many materials. There may be some surface degradation, but the furniture will remain usable. There are exceptions, such as facilities using perchloric acid, where wood would not be a good choice because of the vigorous reaction of perchloric acid with organic materials.

Bench tops can be obtained of several different materials. Some are satisfactory for light duty while others can be used for almost any purpose. Again, unless it is certain that both immediate and long-range usage will permit the use of a lesser material, it is suggested that the bench tops for a new facility be selected from among the more rugged and versatile materials. Table 3.5 lists several materials provided by one vendor for laboratory bench tops, an approximate level of duty for which they appear to be suitable based on the characteristics of the materials and the vendor's own description, and a few representative applications.* Although this is only the range of materials offered by one supplier, it is a reasonable approximation of the types and ranges of materials in common use.

When using radioactive materials or perhaps some unusually toxic materials, where it would be necessary to decontaminate the work surface after a spill, the choice clearly should be limited to those materials such as stainless steel that are least likely to absorb materials. However, if the use of such materials is minimal and limited to very low levels

---

* Table derived from material presented in Fisher Scientific Laboratory Supply Catalog. Descriptive names are used instead of the trade names used in the catalog so that the materials can be compared to similar information provided by other vendors.

Table 3.5.   Work Surface Materials

Materials	Use Level	Applications
Stainless steel	Heavy	Radioisotopes, perchloric acid, solvents
Modified epoxy resin	Heavy	Most applications
Natural stone, resin impregnated	Mod. heavy	Most applications, avoid severe heat stress
Furan composition cement chemically resistant filler	Mod. heavy	Solvents, most wet chemistry operations
CaAl silicate chemically resistant filler, vinyl-coated	Moderate	Not for corrosives
Portland cement and silica mixture, vinyl-coated	Moderate	Academic laboratories, moderate duty hoods
Welded fiber, baked on resin coatings	Light	Work benches, general science facilities
Melamine laminate bonded to phenolic flakeboard	Light	Low abuse applications

of activity or concentrations (as it should be to be used on an open bench), the work surface can be protected with an absorbent paper with a chemically resistant backing. The higher cost of the premium materials could be avoided if the need does not otherwise exist.

Wherever possible, in order to avoid seams in which toxic materials could become trapped, the backsplash panel should be an integral part of the work top. Service shelves above the backsplash panel can be of lesser duty materials in most cases since they will be primarily used to store limited quantities of reagents in closed containers, intended for short-term usage.

Laboratory sinks, incorporated in the work tops, are usually made of stainless steel or materials equivalent to the top two or three other materials listed in Table 3.5, in order to provide the essential levels of resistance needed to corrosives, solvents, and other organic and inorganic materials. However, current restrictions on disposing of many chemicals into the sanitary system should reduce the burden on the laboratory sink and the remaining plumbing components.

In most instances, the tendency is to utilize all of the space underneath a work top for cabinetry and other forms of storage space. It is a good idea to leave at least one portion open and available for storage of movable carts and other items of equipment which must be left on the floor, in order to avoid reducing the aisle space below acceptable limits.

REFERENCE

1. Fisher Scientific Company, 1986 catalog, pp. 825–842.

3.2.1.2. Storage Cabinets

Facilities for storage of research materials in the laboratory should be selected with as much care as any other item of equipment. Many chemicals may be stored on ordinary shelves or cabinets, with only common sense safety provisions being necessary. Obviously, the shelves or cabinets must be sturdy enough to bear the weight of the chemicals. Storage should be such as to make it unlikely that the materials will be knocked off during the normal course of activities in the room. Shelves should not be overcrowded. It should not be necessary to strain to reach materials or to return them to their places. Incompatible chemicals should be stored well apart. Finally, the amount of storage should not be excessive, in order to restrict the amount of chemicals not in current use that would otherwise tend to accumulate within the facility. Periodically, one should "weed" the shelves of chemicals which have not been used for some time and for which no immediate use is foreseen. Older chemicals are rarely felt to be suitable for critical research needs, but can be used in less critical situations such as instructional laboratories. Although these are, as noted, common sense procedures, it is surprising how frequently many laboratories violate one or more of the rules and how many accidents occur as a result.

A number of classes of materials should be stored in units designed especially for them because of their dangerous properties, or because there are restrictions on the use of some materials which require that provisions be made for keeping them locked up. Among these are flammables, drugs, explosives, and radioactive materials. In addition, some materials, such as acids, are best stored in cabinets designed to resist the corrosive action of the materials. Most of the storage units sold for a specific purpose are intended to be free standing, but they also may be purchased to be compatible with other modular components, to be equipped with work tops, or to be used as bases for fume hoods so that work space will not be lost.

3.2.1.2.1. Flammable Material Storage

Flammable materials are among the most commonly employed chemicals in laboratories and represent one of the most significant hazards because of their ignitability characteristics. Even a small quantity when spilled can cover a surprisingly large area, and with the increased surface area provide copious amounts of vapor, which can spread even further. Since the vapors of most flammable solvents are heavier than air and tend to remain as a relatively coherent mass, they can flow for significant distances. Should the vapors encounter an ignition source and the concentration be within the flammable limits, they can ignite and possibly flash back to the original source. Everyone who was ever a boy or girl scout surely remembers the caution about using gasoline to start a fire due to that problem with gasoline. If substantial amounts of flammable solvents are openly stored on shelves or workbenches, it is possible for a small spill to quickly escalate into a large fire, possibly involving the entire facility, due to the availability of the additional fuel. Flammable material storage cabinets are primarily designed to prevent this from happening. Except for the quantities needed for the work *immediately* at hand, all of the reserves should be stored in flammable material storage cabinets, and, once the needed amounts are removed from the container, the containers should be returned to the cabinets.

Flammable liquids are divided into various classes, as given in Table 3.6. The definitions depend upon the flashpoints and in some cases the boiling points of the liquids. The flashpoint of a liquid is legally defined in terms of specific test procedures used to determine it, but conceptually it is the minimum temperature at which a liquid forms a vapor above its surface in sufficient concentration that it may be ignited. In Table 3.6, the first temperature is in degrees Celsius and the temperature in parentheses ( ) is the equivalent Fahrenheit temperature.

Table 3.6.   Definitions and Classes of Flammable and Combustible Liquids

Class	Boiling Points	Flashpoints
Flammable		
A	< 37.8 (100)	< 22.8 (73)
IB	≥ 37.8 (100)	< 22.8 (73)
IC	—	22.8(73) ≤ and < 37.8 (100)
Combustible		
II	—	37.8 (100) ≤ and < 60 (140)
IIIA	—	60 (140) ≤ and < 93.3 (200)
IIIB	—	≥ 93.3 (200)

Neither Combustible II or Combustible IIIA materials may include mixtures in which more than 99% of the volume is made up of components with flashpoints of 93.3 (200) or higher.

The OSHA Laboratory Safety Standard supersedes the OSHA General Industry Standard except for a few specific instances. However, Section 1910.106(d)(3) of the industrial standards provides guidelines on the maximum amounts of flammable liquids allowed to be stored, dependent upon class, in flammable material storage cabinets within a room and defines in Section 1910.106(d)(2) the maximum size of individual containers for the various classes of flammables. NFPA Standard 45 provides guidelines as to the maximum amounts of flammable liquids that should be allowed in the three classes of facilities A, B, and C defined in that standard. This standard has been mentioned, but the three laboratory classes have not been stressed in this volume, in favor of concepts involving wider varieties of hazards than that due to the amount of flammable materials in the laboratory alone. For the purposes of this section, the low- and moderate-risk facilities described in Sections 3.1.3.2.1 and 3.1.3.2.2, respectively, may be taken to be approximately equivalent to an NFPA Standard 45 class C facility, a substantial risk facility (Section 3.1.3.2.3) to be roughly equivalent to a class B facility, and a high risk facility (Section 3.1.3.2.4 to include a class A facility but with more restrictions than the latter would require. Note that OSHA has not adopted the restrictions of NFPA Standard 45 and does not address the issue of the total amount of flammables permitted in a laboratory area, although the amount permitted in an interior storage room is defined. The OSHA regulations regarding container sizes are based on sections of the 1969 version of the NFPA Standard 30. Before returning to the topic of flammable material storage cabinets, Table 3.7 defines the various classes of flammable and combustible liquids and the maximum container sizes permitted by OSHA for each class.

With the exception of the information for polyethylene containers taken from the 1981 NFPA Standard 30, Table 3.7 is equivalent to Table H-12 from the OSHA General Industry Standards.

There are several exceptions to Table 3.7 which would permit glass or plastic containers of no more than 1 gallon capacity to be used for class IA and class IB liquids: (a) if a metal container would be corroded by the liquid; (b) if contact with the metal would render the liquid unfit for the intended purpose; (c) if the application required the use of more than one pint of a class IA liquid or more than one quart of a class IB liquid; (d) an amount of an analytical standard of a quality not available in standard sizes needed to be maintained for a single control process in excess of 1/16 the capacity of the container sizes

Table 3.7.   Maximum Allowable Size of Containers and Portable Tanks

Container Type	Class IA	Class IB	Class IC	Class II	Class III
Glass or approved plastic	1 pt	1 qt	1 gal	1 gal	1 gal
Metal (other than DOT drums)	1 gal	5 gal	5 gal	5 gal	5 gal
Safety cans	2 gal	5 gal	5 gal	5 gal	5 gal
Metal drums (DOT specs)	60 gal	60 gal	60 gal	60 gal	60 gal
Approved portable tanks	660 gal	660 gal	660 gal	660 gal	660 gal
Polyethylene spec 34 or as authorized by DOT exemption	1 gal	5 gal	5 gal	60 gal	60 gal

allowed by the table; and (e) if the containers are intended for export outside the United States.

In section 1910.106(d)(3) of the General Industry Standards, OSHA limits the amounts of class I and class II liquids in a single flammable material storage cabinet to 60 gallons and the amount of class III liquids to 120 gallons. Thus, even if the integrity of a single storage cabinet were breached in a fire or if an accident occurred while the cabinet was open, no more than 60 gallons of class I and II liquids or 120 gallons of a class III liquid could become involved in the incident. This is not to imply that these are insignificant amounts. They are quite large and would be ample, if ignited, to create an extremely serious fire beyond the capacity of portable fire extinguishers to extinguish. Sounding an alarm and evacuation would be the proper course of action.

The purpose of a flammable material storage cabinet is to postpone the involvement of the materials within the cabinet in a fire long enough to allow persons in the immediate area to evacuate the area, or in some cases to permit the fire to be extinguished. Technically, for the cabinet to pass the 10-minute fire test according to NFPA Standard 251, the interior temperature of the cabinet should not exceed 164.1°C (325°F), all joints and seams would remain intact during the fire, and the cabinet doors would remain closed.

Most commercially available flammable material storage cabinets are made of metal and would have to meet at least the following specifications:

1. Bottom, top, and sides of at least 18-gauge sheet iron and double-walled with a 1.5-inch air space.
2. Joints shall be riveted, welded, or made tight by equally effective alternative means.
3. The cabinet door would have to be provided with a three-point lock.
4. The door sill would have to be raised at least 2 inches above the bottom of the cabinet.
5. The cabinets must be labeled in conspicuous lettering "Flammable—Keep Fire Away."

Note that there are no requirements for automatic door closers provided with a fusible link. Such a feature is clearly desirable since it ensures that the cabinet will close in the event of a fire, even if it were inadvertently left open. There are also no requirements for vent connections, but most flammable material storage cabinets have provisions for installing ventilation ducts. The value of these is not as universally accepted. If the containers within the cabinets are always tightly sealed, then there is no reason to make provisions for exhausting fumes. However, should the containers not be tightly closed, it would be possible for volatile fumes to accumulate within the cabinet that should be exhausted outside the building. Although it is likely that on some occasions containers will be returned to the cabinet improperly closed, the author's own preference is to not utilize the vents, since they are often not properly vented into an acceptable exhaust duct and volatile vapors could escape into the facility. Some also are installed using ducts made of plastic or flimsy metals which would be destroyed quickly by fire, thus reducing the protection offered by the cabinet.

Wooden flammable material storage cabinets, if properly constructed according to the provisions of Section 1910.106(d)(3)(ii)(b) of the OSHA General Industry Standards, are also acceptable. These provisions are

> "The bottom, sides, and top shall be constructed of an approved grade of plywood at least 1 inch in thickness, which shall not break down or delaminate under fire conditions. All joints shall be rabbited and shall be fastened in two directions with flathead wood screws. When more than one door is used, there shall be a rabbited overlap of not less than 1 inch. Hinges shall be mounted in such a manner as not to lose their holding capacity due to loosening or burning out of the screws when subjected to the fire test."

Although wood will eventually burn, thick sheets of plywood such as required by the standard can withstand a substantial amount of heat and is a much better insulator than metal. In 1959,* the Los Angeles Fire Department performed a number of comparative tests using various combinations of metal and wood to simulate the walls of a storage cabinet. The experimental walls were fastened to the opening of a furnace operating in the range of 704°C to 788°C (1300°F to 1450°F). A thermocouple was attached to the opposite face of the simulated wall cross-section. The experimental mockups were the following:

Metal

1. A double walled, metal structure of 18-gauge CR steel, approximately 7×10 inches with 1.5-inches air space.
2. A similar cross-section to number 1 made with a core of 5/8-inch sheetrock suspended midway in the 1.5-inch air space.
3. A similar cross-section to number 1 with a core of untreated 1/2-inch Douglas Fir plywood suspended in the air space.
4. A metal-walled structure insulated with 1 inch of 1-pound density fiberglass blanket in the 1.5-inch air space.
5. A metal-walled structure insulated with 1.5-inch of mineral rock wool (density unavailable).

Wood

1. Two layers of 1-inch Douglas Fir plywood.
2. One layer of 1-inch Douglas Fir plywood.
3. A laminated formed of 1/2-inch plywood on each side of 1/2-inch sheetrock.

---

* This material was adapted **Steere, N.V. (Ed.),** Fire-protected storage in *CRC Handbook of Laboratory Safety,* 2nd ed., CRC Press, 1971, 183–184.

Although the experimental arrangement does not simulate a storage cabinet perfectly, the temperature data as a function of time, presented in Table 3.8, do show the rate at which heat is transmitted through the various combinations. They show that the poorest of the wooden panels transmitted heat at a slower rate than the best of the metal combinations. Thus, a wooden flammable material storage cabinet, although it probably would need to be custom constructed in a laboratory facility, would appear to offer superior fire protection. Eventually, however the wood itself, even if protected by fire-retardant paints, would represent a source of fuel for a protracted fire.

3.2.1.2.2. Cabinets for Drug Storage

Security is the primary concern for the storage of controlled substances or drugs because of the potential for theft and misuse. The Drug Enforcement Agency in CFR Title 21, Parts 1301.72 through 1301.76 delineates the security requirements for Schedule I through V controlled substances, including provisions for both cabinets and storage vaults. For individuals who are practitioners (there are several different categories of practitioners, including physicians, dentists, pharmacists, and some institutional personnel, all of whom are legally allowed to dispense drugs), the requirements are simple: the storage cabinet needs to be a securely locked, substantially constructed cabinet. Most institutional programs involving drugs can be easily designed to meet this level of required security. Where the programs would fall under the requirements for nonpractitioners, Parts 1301.72 to 1301.74 describe the needed security provisions. Although these are given in considerable detail, Part 1301.71 provides that actual security requirements which offer structurally equivalent protection would be acceptable. The simplest manner in which protection can be provided for small quantities is a steel safe or cabinet. Among other requirements, these should be sufficiently durable to prevent forced entry for at least 10 minutes, should be either sufficiently heavy (750 pounds or more) or rigidly bolted to a floor or wall so that it cannot readily be carried away, and should be equipped with an alarm which will sound in a central control station manned at all times. All of these are not required for a practitioner's storage cabinet, but the implication of these provisions for tight security should serve as a guide in defining what is meant by a "securely locked, substantially constructed cabinet."

3.2.1.2.3. Storage of Radioactive Materials

Storage cabinets for radioactive materials need to meet multiple needs. As with drugs, there is the problem of security, although for a totally different reason. There is a widespread fear of radioactive materials among the general population. As a result, the public is extremely sensitive to the possibility that radioactive material could be lost or taken from the areas in which it is normally used, with the result that members of the general public could be exposed to radiation. The quantities of materials normally used in most research laboratories are often extremely small and hence would rarely cause any exposure to the public significantly above that due to natural radiation, but to address the public's concern, the rules and regulations governing the use of radioactive materials are extremely strict, including those on security. The use of radioactive materials and radiation will be dealt with at length in Chapter 5, as well as the other major concerns involving storage of radioactive materials, shielding against radiation from the original research materials and of radioactively contaminated waste. Some of the concepts which will be developed more fully there will be used here without immediate explanation. In the following material, if there are any terms which are unfamiliar, definitions or explanations will be found in Chapter 5.

Table 3.8.   Simulated Storage Cabinet Wall Configurations.
(Time [in minutes] vs. temperature data)

Sample	5		10		15		20	
	°F	°C	°F	°C	°F	°C	°F	°C
Metal								
1	430	221	500	260	510	266	650	343
2	150	66	180	82	210	99	240	116
3	130	54	140	60	160	71	170	77
4	310	154	430	221	470	243	—	—
5	270	132	433	233	466	341	—	—
Wood								
1	100	38	100	38	100	38	100	38
2	120	49	133	56	166	74	200	93
3	90	32	100	38	110	43	130	54

Title 10, Part 20.207 of the Code of Federal Regulations covers the basic security requirements very briefly, and this brevity may appear to reflect that security is not a major issue. However, directives have been issued by the NRC to emphasize the importance which the NRC places on security. All radioactive materials must be kept in a secure restricted area to which access is controlled or in a securely locked cabinet or other type of storage unit unless a qualified, authorized user is in the immediate area. This means that when an experiment is in progress involving radioactive material and the researcher leaves the laboratory, he must either lock the facility behind him or return the radioactive material to a locked storage unit, should there be no other authorized user in the laboratory unit. Users may think that the requirement for security would allow an individual to leave a facility with accessible radioactive material unattended for the few minutes it would take to go to the bathroom, obtain a soft drink or snack, but even for these short intervals the rules apply. The concern for security extends to the need to challenge an unfamiliar person in areas where radioactive materials are in place. The stranger may be a salesman, visitor, an unannounced inspector from the NRC, or, although very unlikely, someone who might take advantage of the access to the radioactive material.

If an NRC inspector were to conduct an unannounced inspection of the facility and find radioactive material in use and unattended, it would be considered a violation of the radioactive material license. Depending upon the circumstances, this could result in a fine or even a loss of the license to use radioactive materials. In an extreme case, if the institution or corporation were to be operating under a broad license and if it could be shown that the institution was not enforcing the rules, the license for the entire corporation or institution could be lost. This is a very drastic punishment which is unlikely to occur, but many well known and reputable research facilities have received very substantial fines. Clearly, then, it is essential that every laboratory using radioactive materials have a sturdy, lockable, storage cabinet (or refrigerator/freezer) in which to keep the material. Frequently, a radioactive material is incorporated in a research material having physical or chemical properties which would require specialized storage, such as flammable materials, because of these properties as well.

In addition to security, the radiation emitted by some of the isotopes used in research requires that shielding be incorporated into the storage cabinet. For the radiation from some isotopes, the choice of material used for shielding is critical because the wrong type, depending upon whether it is of a low atomic number or not, can exacerbate the radiation problem. Although almost any radioactive material can be used in some applications in the majority of most chemical and biological laboratories, a relatively small number of isotopes represent the major usage. Several of these are weak pure beta emitters and can be stored safely in any cabinet, since the betas will be completely stopped by virtually any shielding. Among these isotopes, tritium (^{3}H), ^{14}C, ^{35}S, ^{33}P, ^{45}Ca, and ^{63}Ni (used primarily as a source in gas chromatographs) are used most frequently and are among the safest to use. On the other hand, ^{32}P is also a beta emitter, but the betas are very energetic and shielding is required. As the energetic betas from this isotope are slowed down and brought to a stop in the shielding, the deceleration creates a penetrating type of electromagnetic radiation called "bremstrahlung." This effect is much more pronounced for shielding made of higher atomic number materials, such as lead, than in materials such as plastics composed primarily of carbon and hydrogen. Therefore, for this particular beta emitter, which is one of the more frequently used isotopes because of its chemical and biological properties, the shielding in the storage cabinet should be of plastic or some similar material.

Several isotopes that are commonly used also emit a penetrating electromagnetic radiation called gamma radiation. Among the more commonly used isotopes in this category are ^{125}I, ^{22}Na, ^{51}Cr, ^{65}Zn, ^{60}Co, and ^{13}Cs. The appropriate shielding material to be used in or around the storage unit in such a case would be lead.

One last type of radioactive material for which special storage criteria would be needed is in the relatively rare instances where neutron sources would be used. Neutron radiation would be found primarily in reactor facilities, but neutron sources are also used in moisture density probes in a number of research areas. If the source were to be taken from the instrument, the shielding would need to consist of a layer of plastic or paraffin, 5 to 20 cm thick, either impregnated with boron or surrounded by cadmium. Some gamma shielding might be required in an unrestricted area, which could be provided by concrete blocks.

3.2.1.2.4. Corrosive Materials

Storage cabinets can be obtained to serve either as flammable material storage cabinets or as cabinets for the storage of acids. In the latter case the shelves must be provided with protection against the effects of corrosion. One of the more economical ways of achieving the required protection is to use polyethylene shelf liners. If they are not needed or need to be replaced, they can readily be taken out or can be replaced with stock polyethylene. The interior walls can be protected from the effects of acid vapors and fumes by being painted with an acid-resistant paint. An alternative interior finish that provides excellent protection against corrosion and has recently become available is glass-reinforced cement.

3.2.1.2.5. Records Protection*

Research records are not adequately protected against fire damage in most laboratories and research facilities. Research notes, thesis materials, computer disks, and financial and academic records may be irreplaceable or more difficult to reproduce than equipment or

---

*   This section is for the most part taken directly from Steere, N.V. Ed., Fire-protected storage for records, in *CRC Handbook of Laboratory Safety,* 2nd ed., 1971, 179–180.

the building. In addition, the need exists today to be able to provide proof, at the laboratory level, that the operations of the laboratory have been in accordance with the requirements of equal employment practices and that a written industrial hygiene plan exists for the facility, as required by the OSHA Laboratory Safety Standard. This plan, as will be discussed later, requires that each laboratory have a formal performance plan to provide equal protection to the employees in the laboratory, as is provided by the specific standards in the remainder of the OSHA standard to other types of employees. Documentation will be needed to show that this is the case and this documentation might well be needed in the event of a laboratory fire.

There are several ways of providing fire-protected storage, ranging from inexpensive methods of gaining additional minutes of protection to built-in systems and construction designed to provide hours of fire protection. The potential for fire loss in laboratories calls for (1) realistic evaluation of possible losses, (2) immediate steps to minimize losses, and (3) planning for systematic long-range improvements. While various innovations and expedients may be useful for temporary or partial protection, tested and approved equipment and construction will be more economical in the long run. The laboratory standard requires an emergency evacuation plan. The requirements, of course, are based more on life safety rather than property, but in many cases the intellectual loss to researchers of all records and data could be almost as devastating as a crippling injury.

One method of providing protection for records is extremely simple. Maintain two sets of records in two different buildings. This is expensive for paper or physical records, because of the additional space required and the labor involved in the duplication of records is certainly not negligible, but it is virtually the only certain way in which valuable written records can be protected. However, as records become older and are needed only rarely, the set maintained in the laboratory can be eliminated. The amount of expensive laboratory space devoted to records can be minimized, and only that amount of space needed for records that have not been duplicated as yet need to be protected against fire damage. Fortunately, the ability to store large amounts of data on small computer disks can eliminate much of this problem. A single 3.5 inch "floppy" disk can hold up to 400 pages of written records, and it takes very little more time to prepare two disks instead of only one. Another alternative is to keep the original records on a large "hard" disk in the laboratory computer and make a copy to keep elsewhere on the floppy disks.

Computerized records can provide an alternative for a large percentage of the records, but some records, for a number of reasons, will have to be maintained as hard copy so that the problem, while reduced, still exists.

Ordinary desks and file cabinets provide almost no protection for records, since unprotected wood will burn rapidly, and metal will transmit enough heat in a few minutes to ignite or seriously char papers and other records.

Improvement of Existing Records Storage

When funds and space are limited, ordinary desks and file cabinets can be improved to provide a measure of fire-protected storage by use of special fire-retardant paints. Intumescent paints foam when heated and form an insulating barrier against heat and flames. Several such paints have been tested and are listed as fire-retardant coatings by Underwriters Laboratories, Inc. (207 East Ohio St., Chicago, Illinois, 60611) in their "Building Materials List."

Providing adequate fire protection for records will require a detailed analysis and a comprehensive plan based on determination of the value of various records, hazards to which the records are exposed, and the present protection afforded.

Acquisition of the latest version of the "Standard for the Protection of Records," National Fire Protection Association Standard No. 232, is recommended for every research department and organization. The version currently available at the time this document is being prepared was adopted in 1980 and issued in 1981. The publication details the need for management and protection of records, gives standards for fire-resistive vaults and file rooms, and describes standards for fire-resistive safes and other protective equipment. It also provides some excellent references for methods of salvaging records damaged by fire suppression efforts.

Although some large laboratories may need file-resistive record vaults and file rooms, the discussion which follows will be limited to the standards for records protection equipment. The only way to be sure that records protection equipment will meet the need for which it is purchased is to see that it bears the label of Underwriters Laboratories, Inc., or another nationally recognized testing laboratory. Chapters 4 and 5 of this handbook and Appendix A of the NFPA Standard 232 would be useful in preparing purchase order specifications.

Records Protection Equipment

Equipment should be selected to protect records from fires of the most destructive intensity and duration which could occur. If stairwells are open, if there are no automatic sprinkler systems, and if the building is of combustible construction, the probable maximum condition is complete destruction of all combustible contents and portions of the building. It is hoped that if the building is built and equipped according to the principles previously discussed in this chapter, this complete destruction will not occur. However, if firefighting efforts are seriously hampered or prevented by the effects of the chemical contents of the building, complete destruction would still be a distinct possibility. Records protection equipment in older wood-joisted or other nonfire-resistive research buildings should be rated to withstand the impact following floor collapse, as well as the fire exposure.

The method described in Appendix A-5-1.1(a) of NFPA Standard 232 for selecting records protection equipment for a building is based on estimating the combustible material to which the equipment may be exposed, assuming uniform distribution and dividing by floor area to obtain weight of combustibles per unit area. Flammable liquids are not included in this formula but may be approximated, according to Norman V. Steere, by multiplying their weights by a factor of two for the purpose of approximating their fuel contribution equivalent to ordinary combustibles. The following formula provides the fuel loading per square foot, and if the liquid fuel factor L is deleted, is the same as given in NFPA Standard 232.

$$\text{Adjusted Fuel Load } I = \frac{H}{\text{gross floor area in sq ft}}$$

Here, H is the adjusted or derated fuel load in pounds, where

H = $F - (C \times G + 0.25D)$
C = weight in pounds of books and paper in totally enclosed steel containers = volume $\times$ 28
D = weight in pounds of books and papers in partially enclosed (five-sided) steel containers = volume $\times$ 28
F = Fuel Load = $A + 2B + C + D + E + 2L$, where,

A = Weight in pounds of all wood or cellulose furniture
B = Weight in pounds of all plastic furniture
E = Weight in pounds of all free combustibles (books and paper) located on hori-
       zontal surfaces 6.5 feet or less from the floor, and in combustible containers
       and steel containers with less than five sides = volume × 28

Factor G depends upon the ratio of the factor C to that of F.
If C/F is less than 0.5, then G = 0.6; equal to or between 0.5 and 0.8; then G = 0.8 greater than 0.8, then G = 0.9

Once the factor I is calculated, then Table 3.9* may be used to determine the devices
needed to provide protected storage in a fire-resistive building, which a research building
should be. The required devices depend upon both the total combustible contents per floor
and the percentage of combustibles that are in an exposed condition on any given floor.

Since each device which provides additional protection is more expensive than the
lesser one, it can be seen from Table 3.9 that reducing the quantity of combustible contents
in a laboratory and the amount exposed will reduce the cost of records protection. Replacing or fire-retardant coating of combustible desks, filing cabinets, shelving, and containers
will also reduce the cost of providing adequate records protection.

REFERENCES

1. OSHA General Industry Standards, 29 CFR 1910, § 106(d).
2. NFPA 30, *Flammable and Combustible Liquid Code,* National Fire Protection Association,
    Quincy, MA, 1981.
3. NFPA 45, *Fire Protection for Laboratories Using Chemicals,* National Fire Protection
    Association, Quincy, MA, 1981.
4. NFPA 251, *Standard Methods of Fire Tests of Building Construction and Materials,* Na-
    tional Fire Protection Association, Quincy, MA, 1985.
5. **Steere, N.V.,** *Fire-protected storage for records and chemicals,* in *CRC Handbook of
    Laboratory Safety,* 2nd ed., Steere, N.V. (Ed.), CRC Press, Inc., Boca Raton, FL, 1971,
    179–186.
6. NFPA 232, *Standard for the Protection of Records,* National Fire Protection Association,
    Quincy, MA, 1980.
7. *Prudent Practices for Handling Hazardous Chemicals in Laboratories,* National Academy
    Press, Washington, DC, 1981, pp. 226–227.
8. Drug Enforcement Agency, Title 21 CFR Part 1301, Section 72-76.
9. Nuclear Regulatory Commission, 10 CFR Part 20 Section 207.

3.2.2. Hoods

A key item of the fixed equipment in most research laboratories other than those
employing only the least hazardous materials is a work enclosure, usually denoted, at least
for chemicals, by the common name "fume hood." The OSHA Laboratory Safety Standard, while stopping short of requiring a fume hood in each laboratory, does point out that
work with almost any hazardous substances can be done safely if done in a suitable,

---

*   This table is the same as the one provided by Steere in his article, incorporating changes in
    NFPA 232.80.

Table 3.9.   Equipment for a Fire-Resistive Building

Floor area exposed (lb per sq ft)	Noncombustible desks, filing cabinets, lockers, and other closed containers (not over 30% of combustibles exposed)	Combustible desks, filing cabinets, shelving, containers, etc.
<5	1-hour device (without impact) or file room	1-hour device (without impact) or file room
5–10	1-hour device (without impact) or file room	1-hour device (with impact) or file room
10–15	1-hour device (without impact) or file room	2-hour device or file room
15–20	1-hour device (with impact) or file room	2-hour device or file room
20–30	1-hour device (with impact) or file room	4-hour device, file room or vault
30–35	2-hour device or file room	4-hour device, file room or vault
35–45	2-hour device or file room	6-hour vault or file room
45–50	4-hour device, file room or vault	6-hour vault or file room
50–60	4-hour device, file room or vault	6-hour device or file room with no combustible near door

properly functioning, hood. Some laboratory facilities have been constructed recently with no open bench space, with all work within the facility being done within hoods. Others are being built or planned around the same concept. It has been suggested that a reasonable rule of thumb to be used in designing laboratories is to provide one hood for every two research personnel. However, since the activities in a laboratory vary so widely, there can be no absolute standard for a suitable number for a given laboratory other than that there should be sufficient hoods available so that no work which should be performed in a hood need be done on an open bench. In the standard laboratory module described in Section 3.1.1, only one hood is shown. However, additional hoods could be added as needed, with these being placed as far back in the laboratory as possible in each case. The air supply intakes to the room should be designed to permit location of these additional hoods so that the flow of air into the hoods will be minimally disturbed. In general, the air speed at the face of the hood must be somewhat less than the face velocity of the hood, in the range of 20 to 30 fpm.

The purpose of a hood is to capture, retain, and ultimately discharge any noxious or hazardous vapors, fumes, dusts, and microorganisms generated within it. It is not intended to capture contaminants generated elsewhere in the room, although since a hood exhausts so much of the laboratory air when operating, it does serve as a major component of the ventilation system for the room. A few specially designed hoods are intended to confine moderate explosions, but most are not. Unless it is designed, built, used, and maintained properly a hood will not perform its intended function.

Some individuals have been observed to be so hypnotized by the concept of a hood that they continue to use hoods which are not functioning, still counting on them to provide a normal level of protection. It actually has been necessary on occasion to padlock the sashes of hoods closed to prevent this. Unless a hood is fully functional, it should not be used. The OSHA Laboratory Safety Standard recommends that hoods be equipped with a monitor to determine if the air is moving through the face of the hood properly or not. New hoods should be equipped with velocity detectors, and older hoods should be retrofitted. The cost of the sensors is not exorbitant.

There are many different hood designs, such as chemical or biological, even within a single category. Some configurations perform better than others, while the desirability of some features depend upon individual preferences. A good quality general-purpose hood should be able to withstand corrosion, be easily decontaminated (especially for some uses), be suitable for the use of flammable materials, and be capable of withstanding the effects of a fire for a reasonable period of time, sufficient to either allow an attempt to put out the fire or to initiate an orderly evacuation. The design should be such as to minimize the possibility of initiating a fire or explosion. Since research programs change frequently, it would be highly desirable to install a hood used for heavy risk applications rather than attempt to match the specifications solely to current usage. As indicated above, a hood not performing its function adequately represents a risk to the user, rather than acting as a safety device.

3.2.2.1. Factors Affecting Performance

This has been covered briefly before in discussing the location of the hood within a laboratory and the effects of doors, windows, air supply inlets, and traffic. Although there have been a number of advances in fume hood designs and standardization of commercial units since the article "Hoods for Science Laboratories" by Harold Horowitz, S.A. Heider and Caldwell N. Dugan appeared in the 2nd edition of the *Handbook of Laboratory Safety*, (pp. 154–165), much of the material is still relevant. In addition, Norman V. Steere had a good article on pp. 141–149 of the same handbook, while portions of the article "Air Conditioning for Laboratories" provided by the American Society of Heating, Refrigeration, and Air Conditioning Engineers, Inc. (pp. 166–174), also provided excellent information on factors affecting fume hood performance. Much of the material found in these earlier articles will be recognized in the material to follow, revised, combined, and edited to meet current regulations or guidelines. Some material is used without changes, identified as quotations. In the next few sections, most of the material is oriented toward what has become the typical hood, the "conventional" hood with a vertical sash, with air entering the hood through the face of the hood below the sash. Other types of hoods and some of the parameters affecting their performance will be discussed in Section 3.2.2.2.

3.2.2.1.1. Face Velocity

One of the major factors affecting the performance of a hood is the face velocity: "Satisfactory performance of a fume hood requires that the airflow past the opening into the enclosure occur within minimum and maximum limits, since both are of great importance. They vary somewhat depending upon a number of factors relating to the design of the particular hood, its location in the laboratory, and the degree of hazard of the experiments; therefore, the limits must be selected with judgment. The minimum face velocity must be great enough to ensure that the direction of air movement at any point in the area of the open face of the hood will always be into the hood. It is desirable that the lowest possible amount of air be exhausted, consistent with safety requirements, because of the economic advantage of reducing losses of heated air in winter and cooled air in summer. Factors influencing the minimum face velocity are numerous.

"The upper limit of air velocity for a fume hood is related to the aerodynamic flow pattern created by the air stream flowing past the person in front of the hood, past the experiment itself, and out the exhaust opening. The person standing in front of the hood serves as a barrier to the air stream, and when the air velocity reaches a certain point, a low pressure or partial vacuum is created directly in front of him. The low-pressure zone extends into the fume hood increasingly as the face velocity is increased. A velocity can be reached where the low-pressure zone may extend into the area of the fume hood occupied by the experimental apparatus. When this condition is reached, the fumes generated by the experiment will fill the low-pressure zone and may contact the researchers skin, or be inhaled unless the worker is protected by a portable shield or a horizontal-sliding sash," or the vertical sash is lowered well below the individuals face.

"Within the two limiting extremes of face velocity, the choice of face velocity is based on the desire to reduce to a minimum the total flow of air being exhausted and to maintain a safe environment." There is no uniform agreement on what the minimum safe velocity is, with recommended minimums from various sources ranging from 60 to 150 fpm. Tests have shown that although the face velocity can be reasonably uniform across the opening to the enclosure, the amount of spillage may vary widely from point to point, depending in part upon the external factors listed earlier, but also upon the internal design, setup of the hood, and the manner in which it is used. It has been shown that factors such as the distance of the fume source from the face of the hood and the height of the sash opening (for a vertically sliding sash) can affect the performance significantly. The presence of a heat source within the hood can cause the spillage to increase. A depressed area of the base (to retain spilled liquids) can modify the air flow pattern into the hood. Generally, however, the influence of these factors on the spillage from the hood diminishes as the face velocity increases. As a result of these factors, although hoods can be operated safely at 60 fpm, they are likely to be more vulnerable to operating conditions. An individual walking at about 2/3 of a mile per hour (a comfortable walking speed is on the order of 2 or 3 mph) would exceed this speed. Too rapid an arm movement can cause pulses of spillage. Unless every other factor is optimally adjusted to prevent spillage, 60 fpm provides little safety margin. Although the actual operating velocity may be adjusted to a speed such as 100 fpm, an operating velocity recommended by many sources, it is recommended that the system should be designed with a capacity to provide a somewhat greater speed, perhaps 125 fpm. This permits some error in the actual performance as compared to the design value, so that an operating value of 100 fpm can definitely be obtained. In addition, as the performance of the hood decreases with time due to losses in fan effectiveness from dirt in the exhaust duct or stretching of the drive belt from the motor to the fan and other factors, the design level capacity of 125 fpm provides some slack in the maintenance required. The air velocity can be adjusted in many hoods by a damper in the exhaust duct. In others, variable motor speed controls are provided. In both cases the air balance to the room can be disturbed. It is preferable that adjustments be made at the fan location by maintenance personnel who should at the same time check the room air balance to ensure that it remains about the same and slightly negative with respect to the corridor.

Conventional hoods can also be designed to incorporate devices to vary the volume of air discharged as the sash height changes. These devices, called appropriately variable air volume (VAV) devices, usually work by adjusting a damper in the exhaust duct. There is a variety of ways in which the changing air volume requirements are detected and adjustments made. There are several advantages to incorporating such a system in laboratory hoods. Among them is that the flow variation in the flow rate across the work surface and experimental apparatus can be substantially reduced. Another, and probably the advantage most important to designers, is that by reducing the volume of air needed, the loss of tempered air from the room can be reduced. If the room supply system is coupled with the hood exhaust system to maintain an overall balance, significant energy savings are possible. Note that energy conservation is a primary reason for another type of hood called the auxiliary air or add-air type. This system can serve somewhat the same purpose. A refinement to the VAV system is to include a proximity detector in the hood so that it can detect whether a worker is standing in front of the hood or not. A time interval can be established so that if no one is using the hood for a fixed time, the volume of air can be automatically reduced.

Most good quality commercial hoods today incorporate air foil features at the top, bottom, and sides to ensure an aerodynamic flow of air into the hood, to minimize eddy currents, and to provide for maximum operating efficiency. Many are available which provide a reasonably constant volume of air through the hood, but split between an opening below the sash and a path over the top of the sash, inexpensively eliminating excessive velocities at smaller sash openings (please refer to Section 3.2.2.2).

A common error which frequently leads to low face velocities and poor performance is for the scientist to personally select the exhaust motor from the catalog. Each hood installation needs to be configured by a ventilation engineer. The length and diameter of the duct, the number of bends and turns, the type of fan, and the termination of the duct all will influence the size of the blower motor required. Few research personnel are qualified to correctly select the size fan required. A short review of the factors involved in the calculations in any edition of *Industrial Ventilation, A Manual of Recommended Practice* of the American Council of Governmental Industrial Hygienists should be sufficient to convince most to defer to ventilation engineers to design the exhaust system and specify the components.

As noted earlier, individuals will continue to use hoods that are not working because they appear that they should be working. The air speed of about 1 mph is so slight that unless there are visible fumes, it is not readily apparent whether the hood is working or not. A light found on the outside of many hoods usually only indicates that power is being supplied to the motor. It does not provide a positive indication of air flow. In at least one instance, a perchloric acid hood was checked and found to have zero face velocity. When the fan on the roof was examined, the motor was running, but the blades had corroded so badly that they had fallen off and were strewn across the roof. A positive velocity sensor of some type should be incorporated into every hood, as recommended in the Laboratory Safety Standard. It should provide a visual and audible alarm when the air velocity falls below a predetermined safe level and give an indication should the sensor itself fail. Note that this conforms also with the intent of the ADA requirements for disabled persons. The loudness of the alarm and the brightness of the visual indicator may have to be specially selected for persons with corresponding disabilities.

Hoods should be checked upon installation, when any maintenance is done upon them, when there is any modification to the room which could affect airflow patterns in the vicinity of the hood, when any significant maintenance or modifications are made to the building HVAC system, and on a regularly scheduled basis. This last should be done at least annually and preferably quarterly or semiannually. The anemometers or velometers employed for testing the performance of the hoods should be calibrated before use. Measurements should be made at a number of places over the sash opening to assure that the velocity does not vary by more than ±25% at any point. For a conventional hood, it may be desirable to mark the sash heights on the side of the hood opening at the positions where the face velocity drops below the selected acceptable level and above a level of 150 fpm, where turbulence due to a person standing in front of the hood could lead to spillage. Smoke generators are recommended to supplement the air speed measurements to check for spillage under various configurations of face velocity, baffle positions, and experimental apparatus positioning.

3.2.2.1.2. Dimensions

"The depth of a hood is the most important dimension with respect to satisfactory operation. In general the deeper the hood, the more satisfactory it will be in providing uniform suction across the open face. The depth is also of great importance with respect to the size of apparatus and experimental setups that can be accommodated." A deeper unit will also allow fume generating apparatus to be placed farther back in the hood. "Generally, the lower the height of a fume hood (and the clear opening), the more satisfactory will be its airflow characteristics. A very tall hood with the exhaust duct connection near the top," as most are, "will present the greatest risk of escape for fumes because of the difficulty of securing uniform airflow at the base."

3.2.2.1.3. Baffles

"The function of the baffles in a fume hood is to distribute the suction of the exhaust duct in such a way that uniform airflow through the face of the hood will result. The design and position of baffles and their openings are critical to satisfactory performance. Opinions on proper design of baffles vary widely among experts on the subject." Most commercial hoods are provided with a single baffle which has adjustable slots at the top and bottom. The baffle and the slots should be adjusted after installation in the laboratory to provide the most uniform airflow across the face of the hood. Larger hoods may have more slots in the baffle to aid in adjusting the airflow.

3.2.2.1.4. Construction Materials

The three previous factors were concerned with the ability to capture and retain noxious and dangerous fumes and vapors within a stream of air passing through a hood. This section is concerned with the physical ability of a hood to contain and to withstand the corrosive actions of the materials used in it. In principle, the selection of the materials used in the fabrication of the hood should depend upon the types of chemicals intended to be used in the research program. However, hoods are major fixed pieces of equipment that would be difficult and expensive to change as the nature of a research program changes, so, as has been the general tenor of the recommendations in this chapter, the hood should be selected to be as versatile as funds will permit. If a hood is intended for a dedicated use which is not expected to change for an extended period, then it is only common sense not to spend more than necessary. The materials used for the walls and lining, base, sash, and some of the interior fittings will be discussed separately.

Walls and Lining

"A great many materials have been used for fume hood construction. Most of them have proven to be satisfactory when used correctly and properly selected with regard to the requirements of the experiments." As noted above, "it makes good sense to use the least expensive material that will do the job. Unfortunately, however, materials which are most versatile tend to be the most expensive." In the following list, the materials discussed are the ones normally available from commercial vendors. It is rare for hoods to be fabricated by the user. The list is approximately in order of versatility.

Transite: Transite is a material in which asbestos fibers are bonded with a resin and until recently was probably the most popular lining material for general-purpose, heavy-duty hoods, Although it may become discolored with use, it is highly resistant to a large number of chemicals. In recent years, because of the concern for the health effects of airborne asbestos fibers, its use has decreased rapidly. Although the asbestos fibers are tightly bonded in the transite, some older hoods have been observed in which the fibers had become friable and likely to be become detached. When transite is broken or cut, asbestos fibers may again become airborne and pose a potential health hazard to anyone in the vicinity at the time. As older hoods using transite are taken from service, they should not be replaced with the same material; even if hoods using this material remain available. The linings of hoods using transite constitute an asbestos hazard when they are removed and disposed of. Their disposition must be done in conformance with EPA and OSHA regulations involving asbestos handling.

Stainless steel: Stainless steel may be attacked by some chemicals but type 316 stainless or equivalent is commonly used for the lining of perchloric acid hoods and in radioisotope hoods which need to be easily decontaminated. Because of its vulnerability to some chemicals and its relatively high cost, it is not recommended for general purpose fume hood use. Among the problem chemicals for stainless steel are acids and compounds containing halides.

Fiberglass-reinforced polyester: This material is popular for lining general-purpose hoods and is highly resistant to a large number of materials. However, as with stainless steel, there are some materials which may cause some problems with heavy usage. Care should be used in selecting this material to avoid applications involving chemicals for which it is not suited. Among those chemicals for which it is suitable for limited service are acetone, ammonium hydroxide, benzene, and hydrofluoric acid.

Smooth surfaced, glass reinforced cement: At least one major firm has substituted this material for transite as a general-purpose hood lining material. It is a relatively new material, but if it proves to be sufficiently durable, it will probably become a popular choice.

Epoxy resin: This material is comparable to fiberglass in versatility and is affected by some chemicals. If a wide range of chemicals are to be used, advice should be sought from the vendor to ascertain if there are any significant problems. Among the chemicals for which it may be unsatisfactory are benzene, fatty acids, concentrated hydrochloric acid, and nitric acid.

Polyvinyl chloride (PVC): PVC offers good protection for a wide range of chemicals but is affected by some, including liquid ammonia, amyl acetate, aniline, benzene, benzaldehyde, bromine, carbon disulfide, carbon tetrachloride, chloroform, ether, fluorine, nitric acid, and fuming sulfuric acid.

Plastic-coated steel: Plastic (usually vinyl) clad steel linings are used for limited or light duty applications for materials which will not adversely affect the vinyl cladding.

Cold rolled steel: This material is intended for light duty liner applications only. However, carefully prepared and treated to resist corrosion, or vinyl clad as for epoxy resin, it is the most popular material for the external components of fume hoods.

Fume Hood Bases

Hoods which need to be decontaminated frequently, such as radioisotope hoods, often come with a stainless steel base surface which forms an integral part of the hood interior. Other models are available with integrated bases of different materials. However, many hoods are sold without bases and they can be selected separately. Base units and work tops are discussed in Section 3.2.1.1 and the same selection criteria discussed there apply here. The most popular material used for fume hood bases is a molded, modified epoxy resin. Since the material is molded, it is easy to incorporate a shallow depression in the work surface to function as a water tight pan to catch and retain spills.

Sashes

The most popular sash configuration is the vertically sliding type. The sash must be counterweighted, especially if the sash window is made of heavy glass. In order to avoid a "guillotine" effect should the counterweight cable break, a safety device must be incorporated in the sash. As for transparent materials used for the sash window, only a few are used in good quality, currently available commercial fume hood models.

Laminated safety glass: This is probably the best material for a sash because of the resistive properties of glass to most chemicals and the safety features provided by laminating tempered glass. It is not as effective as is tempered glass for higher temperature applications (good only up to about 70°C). In the event of a moderate explosion, it is possible that the sash will remain within the frame. If the sash remains essentially intact, the employees in the laboratory will continue to be afforded some protection against fires within the hood and escape hazardous fumes.

Tempered glass: Although tempered glass breaks into fragments that are not sharp in the event of an explosion, the result will be a loss of glass from the sash and the loss of protection afforded by the enclosure. Tempered glass will withstand more heat (up to about 200°C) than laminated glass.

Clear, high impact PVC: PVC will provide impact protection comparable to laminated safety glass, but will not be as resistant to the effects of chemicals in the hood.

Plexiglass: Plexiglass is not as durable as the other materials and is primarily provided in economy models.

Polycarbonate: Polycarbonate sashes are recommended when heavy use of hydrofluoric acid is involved. Polycarbonate plastics offer good physical strength.

Internal Fixtures

The lights in the hood should be shielded from the hood body in, as a minimum, a vapor-proof enclosure. These can also be selected as explosion-proof units should the usage include substantial amounts of very flammable materials. In fact, if a large portion of the work in the hood is expected to involve very volatile flammable materials, the hood should be designed completely as an explosion-proof unit. In any case, all electrical outlets and switches should be located outside of the hood on the vertical fascia panels. Similarly, utilities should be provided by remotely controlled valves in the side panels operated by handles outside the hood. The connections to the utilities inside the hood chemically resistant and should be clearly identifiable as to the utility provided.

3.2.2.2. Types of Chemical Fume Hoods

There are several different types of fume hoods: (1) conventional hood, vertical sash, (2) conventional hood, horizontal sash, (3) bypass hood, (4) auxiliary air hood, (5) walk-in hood, and (6) self-contained hood. The differences in types 1, 4, and 6 are especially important in terms of the amount of tempered air lost during operations, while 1, 2, and 3 differ primarily in the airflow patterns through the sash openings. Figures 3.11 to 3.16 illustrate each of these types and the air currents through them during typical operations. In addition, there are specialty fume hoods for perchloric acid and radioisotopes, which will be treated separately. All of the hoods discussed in this section will be updraft units, where the exhaust portal is at the top of the hood, with of course the exception of the self-contained type.

One class of hood which will not be discussed here is the canopy hood. These have their uses, where it is desired to capture and exhaust hot fumes carried upward by convection currents until they come close enough to the canopy so that the fumes become entrained within the hood. The speed of the air movement in the vicinity of the hood face, due to the air flowing through the canopy, falls off very rapidly to about 7.5% at a distance equal to the effective size of the canopy opening. If the canopy is at a reasonable distance away from the bench top, the airflow at the work surface due to the hood will be on the order of the average air movement speed within the room, or less. A further disadvantage would be that the fumes, if drawn upward, would pass through a worker's breathing zone. For these reasons, canopy hoods are not recommended as general usage laboratory fume hoods.

3.2.2.2.1. Conventional Fume Hood

A conventional fume hood with the sash open and closed is illustrated in Figure 3.11. Note that a section of the internal baffle essentially remains in contact with the sash at all opening positions. It is equipped with a vertically opening sash and an interior baffle so arranged so that some of the air sweeps the base of the hood and is directed up behind the baffle to the exhaust opening. There can be additional slots in the baffle, perhaps one in the middle and one near the top, through which air can pass to provide more uniform airflow. The remaining air passes through the hood interior and is directed into the exhaust portal over the top of the interior baffle. The volume of air through the hood is relatively constant, although some losses occur as the sash opening becomes smaller. A more important consequence of the decreasing sash opening is the increasingly high velocity of the air through the narrowing opening. The increased air speed could disturb operation of the experiment within the hood, and the effects of turbulence around apparatus sitting on the bottom of the hood would be increased. A person representing an external obstruction could also give rise to increased turbulence so that spillage from the hood would occur, although with the bottom of the sash well below face height, the possibility of toxic fumes entering directly into the breathing zone would not be high.

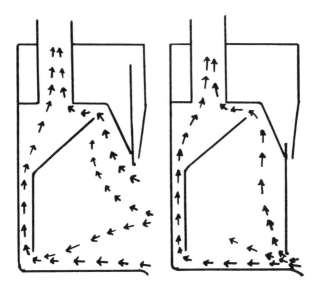

Figure 3.11. Conventional hood, sash open and closed.

Figure 3.12. Conventional hood, horizontal sash.

3.2.2.2.2. Conventional Hood, Horizontal Sash

Some hoods are available with two or three horizontally sliding sashes instead of a vertical sash (Figure 3.12). With three sashes, the maximum hood opening that is possible would be greater than with two. It would also be possible with three sashes to have more flexibility where the sash opening would be located. Since it would be possible to have the center section of the hood opening blocked, and the fume generating apparatus placed behind it, fumes entrained in the eddy currents would have more difficulty reaching a person working in the position shown in Figure 3.13 than with a similar vertical sash hood. The section of sash between the operator and the experiment would also afford some protection from a moderate explosion or chemicals ejected from a runaway reaction.

Figure 3.13. Protection offered by overlapping sashes.

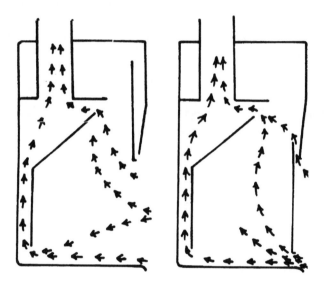

Figure 3.14. Bypass hood, sash open and closed.

3.2.2.2.3. Bypass Hood

The bypass hood (Figure 3.14) is designed so that a portion of the air entering the face of the hood may pass over the top of the sash opening as well as below it. This has two consequences. The first is that the air velocity near the work surface remains reasonably constant, so that excessive air speeds will not occur which could be detrimental to delicate apparatus or experiments. The second is that there is less static pressure and hence less frictional resistance to the flow of air than with the conventional hood, so that the volume of air through the hood remains more nearly the same at different sash heights, permitting better control of the laboratory air balance. This is probably the best basic choice for a laboratory fume hood that is not equipped with a VAV device.

3.2.2.2.4. Auxiliary Air Hoods

This type of hood is somewhat controversial. The original comments of Horowitz et. al. which appeared in the 1974 edition of this handbook are still echoed in today's literature.

"The auxiliary air hood attempts to reduce air-conditioning requirements by providing a separate supply of air that has not been cooled and dehumidified in the summer or fully heated in the winter. The supply of air for such a hood may be drawn from outside or from the service chases within the building, which are, in turn supplied by air from attic or mechanical equipment rooms. Such hoods can substantially reduce the air-conditioning equipment capacity required to make up losses through fume hoods; operating costs can likewise be reduced. However, there are a number of disadvantages to such a hood. One type of auxiliary air-supply hood discharges untreated air just in front of the hood, usually at the head. A scientist working at the hood must work in unconditioned air. The disadvantages are obvious, and the annoyance of scientists has been evidenced by their very human attempts to invent means of foiling the intended mode of operation. One such effort consists of securing cardboard over the outlets with adhesive tape, thus closing or reducing the auxiliary supply.* Attempts to rectify this problem by partially cooling or heating the air supply, depending upon the season, substantially reduce any economic advantage of this type of hood. Another type of hood introduces the auxiliary air within the hood enclosure and is inherently unsafe because the face velocity is reduced below the rate necessary to capture fumes."

There have been changes in hood design since the article quoted above was published which permits the supplementary air to be brought down outside the sash, but not in such a way that the operator will be standing in the airflow, as shown in Figure 3.15. Specifications on this type of hood usually permit up to 70% of the air to be provided by the auxiliary air supply with at least 85% of the supplied air passing through the hood face. Untempered air, especially during the winter, when the outside temperature may be very cold, would mandate that this supplied air be heated to at least 10° to 15°C, which does eliminate much of the savings for this type of hood. The installation and configuration of this type of hood is critical and proper operation is difficult to maintain. Most authorities still discourage selection of an auxiliary air hood.

3.2.2.2.5. Walk-In Hood

Walk-in hoods (Figure 3.16) are hoods which usually rest directly upon the floor or on a pad resting directly upon the floor. They are designed to accommodate tall apparatus which will not fit in a standard hood sitting upon a base unit or work bench. Because their height is usually somewhat out of proportion to their width, the airflow characteristics may not be as favorable for avoiding spillage as with a standard hood, as noted in the earlier section on the effect of dimensions on hood performance. Because their height would

---

* It should be noted that here is a good example where the users would continue to use a hood which they themselves have caused to function improperly. There is a serious attitude problem among many, although by no means a majority, of scientific workers which puts expediency and comfort ahead of safety.

Figure 3.15. Auxiliary or add-air hood.

Figure 3.16. Typical walk-in hood.

require an abnormally long sash travel, these hoods are often provided with dual sashes, each of which would cover half the opening or with a single sash which would come down only about halfway, with swinging doors being used to provide access to the lower portion.

These doors may terminate a few inches above the floor to ensure that there is always some airflow through the entire length of the vertical space. As with other types of hoods, they may be obtained in different configurations, such as bypass or auxiliary air units.

3.2.2.2.6. Self-Contained Hoods

These are listed last because they are the least desirable alternative as chemical fume hoods. A self-contained chemical unit should not be confused with a biological safety cabinet, which is often also self-contained. Some of the newer designs give a reasonable approximation to a conventional hood in performance, although they are not recommended for general usage. However, there are circumstances where they can be used relatively safely. Good quality units can cost as much as a conventional hood, so their use is not mandated upon the basis of purchase price. Their use is usually contemplated when a hood is needed, but no exhaust duct is accessible and it is impractical to install one. The commercial units available are generally intended for use in histology and cytology procedures involving materials such as xylene, formalin, toluene, alcohol, etc. The newer units provide for a choice of filters which makes them suitable also for applications other than organics.

A typical self-contained unit pulls room air through the face of the unit over the work surface and through a filter selected for the material intended to be used in the hood. The fan unit does not become contaminated because the air is filtered before it reaches it. A typical filter will absorb several pounds of the solvent before it becomes saturated and must be replaced. For average use, the filters may last 1- to 2-years, at which time they must be replaced. A major problem with most of the units available is the inability to tell when the filter has become saturated. At least one vendor has solved this problem by placing a material at the back of the filter which reacts with the solvent when it is close to no longer being absorbed by the filter and then emits a pungent odor. The same manufacturer also provides an electronic detector for solvent vapors. The unit provided by this same vendor probably comes about as close to the performance of a regular fume hood as any self-contained unit on the market, but it is also comparably expensive. The face velocity is set at 80 fpm, which is comparable to that of a standard hood, although it is in the range in which spillage of fumes from the hood can occur unless the hood is used with care (fume generators, 20 cm from the sash opening or better) and located where external air movement problems will be minimal. However, even this relatively well performing unit is no real substitute for an installed, ducted, fume-hood which should be the choice whenever possible. The physical design is similar to a standard conventional hood.

3.2.2.2.7. Other Modes of Exhaust

The trend is to do more laboratory work in hoods, but the use of hoods is a substantial burden on the energy budget of a building. An alternative is to use spot ventilation. Many laboratory supply houses provide small exhausters in which an adjustable inlet can be placed very close to well-defined spot sources of noxious fumes. The fumes are then typically discharged into a fume hood. If the work is repetitive enough to warrant setting up a permanent spot exhaust, then the designs shown in Figures 3.17 and 3.18 might be usefully employed for limited risk work. In the first design, a repetitive operation involving pipetting into test tubes or sample vials could be done perhaps virtually at the face of

Figure 3.17. Local fixed exhaust.

Figure 3.18. Rear plenum exhaust system.

the cowl-shaped spot exhauster. With this physical relationship, the air intake should be very effective in exhausting any aerosols or fumes which might be generated. Note that the airflow would be, as with a normal fume hood, away from the research worker.

The second design would be more effective for a tray-type operation. Here, one or more narrow slots, with air being drawn through them at a relatively high inlet velocity, are placed at the rear or, occasionally, to the side of the workbench close to the level of the bench top. This design takes advantage of the fact that most solvents are heavier than air. Air flowing across the surface would entrain the vapors from the tray and exhaust them through the slots at the rear of the workbench. The fumes would not have an opportunity to rise into the worker's breathing area, being pulled back and away from the worker. A modified version of this system is used for silk screening. In this version the entire circumference has either an aerodynamic slot around the edge or the last several inches close to the edge of the table top are perforated with hundreds of holes, through which in both cases air is pulled down prior to being exhausted. There are consumer range tops for cooking built on this principle that work very well. Smoke from the food being prepared rarely rises more than an inch or two above the cook-top surface before being captured and exhausted. Chemical fumes should act in a similar manner.

3.2.2.2.8. Perchloric Acid Hood

Individuals working with perchloric acid and perchlorates must be trained in procedures necessary to conduct their research with maximum safety. These are extremely dangerous materials.

There is an extensive section later on the problems of working with perchlorates and perchloric acid; this section will be restricted to a discussion of the critical factors applicable to the proper performance of perchloric acid hoods. It is sufficient to say at this point that perchloric hoods are designed to avoid accumulation of precipitates from perchloric acid or to avoid perchloric acid from coming into contact with materials with which it may react vigorously and explosively. Hoods designed for hot perchloric acid use should not be used for research using other types of materials. The hood should be prominently labeled with a sign stipulating that it is for perchloric acid work only. Exhaust systems serving hoods used with heated perchloric acid should not be manifolded into a common exhaust plenum.

For conventional hoods and their variants which have been covered in Sections 3.2.2.2.1 to 3.2.2.2.6, the discussion of appropriate ducts and exhaust fans has been deferred to separate sections. However perchloric fume hood systems are uniquely dangerous and will be treated as an integrated topic.

Perchloric hoods are usually constructed with an integral liner of a single piece of stainless steel, such as 316 stainless, which will resist the effects of the acid, although PVC can also be used as a liner. The liner should have coved corners and as few seams as possible to allow ease of decontamination. In order to avoid buildup of perchloric precipitates in the hood and duct system, a hood intended to be used for perchloric acid work must be equipped with a rinse system which will make it possible to thoroughly flush the interior of the hood and ductwork with water. This may be done with a manual control system or by an automatic system that will come on and rinse the system for 20–30 minutes at the end of a work session. A combination of an automatic system which can be bypassed for additional rinses is preferable so the researcher may choose to clean the system if necessary.

The ductwork should also be stainless steel or PVC. Caution must be exercised to ensure that workers on installation do not use standard organic caulks to seal the joints. Organic materials, when contaminated with perchloric acid, are highly flammable and dangerous, and such joints will also tend to leak, allowing perchloric acid to escape outside the ductwork. Under circumstances allowing this errant material to be exposed to heat or to receive a sudden shock, the result could be a fire or an explosion in the space outside the duct. The most desirable procedure for stainless steel ducts is to weld the sections of stainless steel ducts together. This will require heliarc welding, which is a relatively expensive procedure compared to welding ordinary steel ductwork. Some fluorinated hydrocarbon materials can be used as sealants if welding is not feasible.

It is recommended that the interior fittings of a perchloric acid hood should be non-sparking and the lights should be explosion-proof. This concept should be extended to any apparatus placed in the hood. With the dangers already represented by perchloric acid, there should be no contributory factors that could initiate an explosion. PVC ductwork can be employed instead of stainless steel, but it would be much less likely to remain intact in the event of a significant fire exposure. However, if the ductwork is enclosed within a 2-hour fire-rated chase, as it usually should be, this would not be a serious drawback.

The ductwork for a perchloric acid hood should have as few bends as possible and be taken to the roof in the shortest, most direct *vertical* path. No horizontal runs should be permitted, and even slopes of less than 70° to 80° should be avoided wherever possible. For aesthetic reasons, architects prefer to place exhaust ducts away from the edge of a building so that they cannot be seen easily. As a result, horizontal runs of 100 feet or more of perchloric fume hood exhaust ducts have been observed in some older designs. Even if a wash-down mechanism were incorporated in the design, it would be unlikely to come into contact and clean the upper portion of the duct in the horizontal section. In one instance in which a perchloric hood was installed below grade in the basement of a building, the exhaust duct first was run horizontally for approximately 75 feet under the floor of an adjacent section of the building and an exhaust fan was installed in this horizontal run. The duct then ran vertically for three floors. When this was discovered, the horizontal section of the ductwork beyond the fan had corroded through and perchloric acid crystals were observed on the external surface of the duct and on the ground below it. This became a major and costly removal project.

As a minimum, the blades and any other portion of the exhaust fan coming into contact with the perchloric fumes should be coated with PVC, Teflon, or another approved material that will resist the effects of the perchloric acid. An induction exhaust fan, where none of the fumes actually pass through any part of the motor or fan is recommended. Under no circumstances should the exhaust fumes be directed down upon the roof to be absorbed in the roofing material. The contaminated roofing material could itself constitute a danger. The exhaust point should be well above the roof (at least 10 to 15 feet) to avoid the fumes readily reaching any portion of the roof prior to dilution by the outside air. The wash-down mechanism should be capable of cleaning the entire duct from the point of exhaust all the way back to the hood. The washdown system plumbing should automatically drain when shut off to avoid rupturing the supply lines due to freezing in the winter. The rinse water may be permitted to drain directly into the sanitary system where it will be quickly diluted.

Normally, in this handbook, specific manufacturers and brand names are avoided but Labconco makes an excellent perchloric acid ejector duct with all of these desirable features that exhausts the perchloric fumes at a point about 10 feet above the roof level. It is not inexpensive but performs exceptionally well.

Maintenance personnel, as well as the laboratory employees, should be trained in the dangers inherent in the material and the potential for injury represented by any residual material in crevices or other places where perchloric acid or by-products might accumulate. This training should be done in a positive way to instruct individuals how to work with such material properly, and not in such a way as to unduly frighten anyone.

3.2.2.2.9. Radioisotope Fume Hood

Radioisotopes are frequently used in the life sciences and in nuclear medicine in diagnostic applications. Only rarely are the amounts employed large enough to be of immediate danger to the laboratory worker, if used properly. In addition, many of the more commonly used radioisotopes emit low energy beta radiation only, which will not penetrate the skin. However, not all radioactive materials have this favorable property and many even relatively safe materials may cause delayed injury, perhaps 20 years later, if ingested and inhaled. The word "may" is not to be construed here in any definite sense of "they will cause injury." It is intended to indicate only that the possibility of a health effect may be increased. At very low levels, there is no direct evidence of either immediate or delayed injury, although there has been an enormous number of studies trying to resolve the issue. Most studies, seeking to prove the point either way, have typically been controversial and not accepted universally. The possibility of health effects mentioned above is based on a conservative linear extrapolation from known detrimental effects at high levels to possible detrimental effects at the much lower levels usually encountered in research. However, because of the possibility of a finite risk and because of public concern, it is public policy that laboratory use of radioactive materials be stringently controlled to minimize exposures. A carefully monitored license is required. There is a separate major section in Chapter 5 on radiation safety.

A radioisotope fume hood is designed to minimize risks of exposure to the laboratory worker by making it easier to maintain the hood in an uncontaminated condition. The liner is usually made of a single piece of stainless steel, as with perchloric acid units and for the same reason, for ease of decontamination. There should be a minimal number of seams or hard to clean areas. The major classes of research employing radioisotopes are often intended to retain the compounds containing radioisotopes in the end product, since a frequent purpose in employing radioactive materials is as a tracer. Therefore, although there will be some radioactive fumes generated, there may be less than with some other dangerous materials. Even so, it is usually recommended that the ductwork for a radioactive fume hood be of stainless steel since this is easily decontaminated.

Where relatively high levels of radioactive materials are used or where the levels of fumes generated could be substantial, it may be necessary to install an absorbent filter or for particulate fumes, a two-stage high efficiency particulate air (HEPA) filter unit, which will filter out 99.97% of all particles 0.3 microns in size or larger, in line in the exhaust duct to ensure that the legal minimum concentrations of activity can be maintained at the point where the fumes are discharged to the outside. The most appropriate location for this filter is at the exit portal of the hood, since this will prevent any of the ductwork from being contaminated and the location will make it convenient to service the filter. Where a HEPA filter is used in a fume hood involving chemicals, it is essential that a device to monitor the air velocity through the face of the hood be installed. The air passages in a HEPA filter used in a chemical system where particulates are generated will soon become clogged, increasing the difficulty of drawing air through the filter so that the efficacy of the fan unit will fall off rapidly. A less expensive prefilter will significantly extend the life of the HEPA filter. A device to measure the pressure drop across the filter unit may be used to monitor the condition of the filter, but this will not necessarily measure the velocity of the airflow through the hood and duct. Gaseous radioactive materials will not be stopped by a HEPA filter. An activated charcoal or alumina absorbent filter should be used for these materials.

Although the stainless steel liner is relatively easy to clean, a number of measures make it easier to keep it in an uncontaminated condition. Among these are the use of trays to contain any spills which might occur and the use of plastic-backed absorbent paper (which may be discarded as waste) on the work surfaces. Many of the procedures used in the life sciences, where the preponderance of radioactive materials used in research laboratories are employed, generate aerosols, so in cleaning the interior of the hood, attention needs to be paid to the interior walls and sash surfaces.

The property of emitting radiation which makes radioisotopes useful in research and potentially dangerous is also the property which makes their use relatively easy to control by research personnel who conscientiously follow good laboratory practice. The work surface, the hands and clothing of the persons performing the work, and the tools and equipment employed in the work can be easily checked for contamination by the use of appropriate instrumentation. It is unfortunate that the dispersion of many dangerous chemical and biological agents in the laboratory cannot be monitored and controlled so readily.

The only other unique feature in a radioactive fume hood may be the need of the base to support shielding materials. The base may need to be stronger than usual to support the concentrated weight of the lead shielding which may be needed to protect the workers from radiation, such as when synthesizing a compound, where relatively large amounts may be employed and substantial amounts of shielding may be required. Most of the time, only small quantities of radioactive materials are in use at a single time, so personnel shielding needs would be small. However, substantial amounts of lead may be needed to provide adequate shielding against background radiation for the sensitive detectors used to detect minute traces of the experimental radioisotopes in the material being studied.

Fume hoods used for radioactive materials should be marked "RADIOISOTOPE HOOD" and in addition should be labeled with a "CAUTION—RADIOACTIVE MATERIALS" sign bearing the standard radiation symbol. The isotopes being used should be identified on the label. Under some circumstances, additional signs may be needed.

3.2.2.2.10. Carcinogen Fume Hood

Clearly, work with a carcinogen mandates a high-quality fume hood. The features discussed in the previous two sections, which minimize spaces for materials to be trapped and facilitate decontamination, are strongly recommended. In the OSHA General Industry Standards, Subpart Z-Occupational Health and Environmental Control, many of the specific standards for the carcinogens regulated in this section contain the following standard paragraph relating to fume hoods:

> "'Laboratory type hood' is a device enclosed on three sides and the top and bottom, designed and maintained so as to draw air inward at an average linear face velocity of 150 feet per minute with a minimum of 125 feet per minute; designed, constructed and maintained in such a way that an operation involving (name of regulated carcinogen) within the hood does not require the insertion of any portion of any employees' body other than his hands and arms."

3.2.2.3. Exhaust Ducts

Exhaust ducts are necessary to take the fumes from the hood to the point at which the fumes are to be exhausted. For the purpose of this section, it will be assumed that the duct will exhaust directly to the outdoors, rather than to a plenum (see Section 3.1.6).

It is recommended that if it is desired to manifold more than one hood into a common duct that this practice be limited to hoods within the same room. Otherwise, it is less likely that individuals using different hoods would be aware of each other's activities, and one might make changes which would affect the performance of hoods in the other room. For example, a pressure differential might be established between one laboratory and another, so that fumes could be exchanged between the two areas. In addition, maintenance of the air balance may be made more difficult, and there may be problems in complying with fire codes, if fire walls are penetrated.

3.2.2.3.1. Materials

Many of the comments in Section 3.2.2.1 regarding materials are relevant here as well. At one time, transite was a very popular duct material, but it is no longer recommended due to the concerns regarding asbestos. The joints between sections required cutting by the maintenance personnel preparing or installing the ductwork, with a consequent release of asbestos airborne fibers which could be inhaled. Stainless steel is used for special types of applications, such as for perchloric acid systems, but is not universally suitable for all chemicals. PVC can be used for many applications, is easy to install and custom fit, and is comparatively inexpensive. Steel ductwork coated with a chemically resistant material, such as an epoxy coating, is popular because it is relatively inexpensive and is especially adaptable for custom installations.

3.2.2.3.2. Dimensions[*]

There are numerous sources of frictional losses for the airflow through the hood and ductwork to be overcome by the exhaust fan. The air entering the hood must be accelerated from a minimal velocity in the room to the velocity within the duct. Due to physical factors, there is always some turbulence created during this process, so that the pressure difference created by the fan must be sufficient to provide the desired airspeed in the duct and overcome the losses due to turbulence. In the specifications for a hood, the hood static pressure data provided are a direct measure of the total of the energy needed for acceleration and to overcome turbulence losses. The static pressure will be proportional to the square of the velocity of the air entering the hood face.

There are significant losses along even a straight, relatively smooth section of duct due to air friction. This is due to the energy required to maintain a velocity gradient ranging from the essentially stagnant air in contact with the walls to the rapidly moving air near the center of the duct. For example, for a nominal 10-inch internal diameter PVC duct, through which 1500 cfm of air is passing, each 10-feet section contributes about 0.1 inch, water gauge, pressure loss. This also varies rapidly with the volume of air movement; by increasing the speed by one third, the pressure loss is increased by about two thirds. The diameter of the duct is also important. For the same volume of air, 1500 cfm, the losses in a nominal 8-inch I.D. duct will be more than three times greater than in the 10-inch duct, while the losses in a 12 inch duct will only be about 40% of the amount in a 10-inch duct. In general, the total amount of duct friction in a round duct varies directly proportionally to the length, inversely to the diameter, and is proportional to the square of the velocity of the air moving through it. The equations below give an approximate value for the skin resistance of a duct.

$$wg = \frac{2LV^2}{CDT} \text{ for circular ducts}$$

---

[*]   For a more thorough discussion of the material covered briefly in this section, the reader is referred to ANSI Z9.2 or in the ACGIH, *Industrial Ventilation, A Manual of Recommended Practice,* for an even more complete treatment.

where wg   = pressure loss in inches water gauge
    L   = length of duct in feet
    V   = velocity of air in feet per second
    D   = diameter of the duct in inches
    T   = absolute temperature on Fahrenheit scale (460 + °F)
    C   = constant = 55 for new steel ducts and 45 for older steel ducts

Similarly for a rectangular duct, sides A and B:

$$wg = \frac{LV^2(A+B)}{CABT} \text{ for rectangular ducts}$$

A shock loss occurs whenever there is a sudden change in the air velocity, caused by a change in the direction of the air or a change in the diameter of the duct. Every bend in the duct dramatically decreases the efficacy of the fan motor. The sharper the bend, generally the more severe the loss becomes. This loss may be estimated from the following equation:

$$wg = \frac{0.1188V^2k}{T}$$

where k varies as follows:

Mean Radius of Bead/Duct Diameter or Width (width is dimension of side measured along radius of bend)	Circular Ducts k	Rectangular Ducts k
Right angle elbow	—	1.25
0.50	0.75	0.95
0.75	0.38	0.33
1.00	0.25	0.17
1.5	0.17	0.09
2.00	0.15	0.08
3.00	0.13	0.07
6.00	0.10	0.05

The last equation also can be used to estimate the loss in changing duct sizes by substitution of k_c for k in the equation. The ratio given below is the ratio of the smaller flow area to the larger:

Ratio	0.1	0.2	0.3	0.4	0.5	0.6	0.7	0.8	0.9	1.0
k_c	1.25	1.20	1.15	1.05	0.95	0.80	0.65	0.45	0.25	0

Similarly, if a secondary branch joins another duct, the angle at which it joins critically affects the fractional velocity pressure loss of the air through the secondary branch, ranging from a very small percentage at shallow angles to over 40% at 60°. Branches should not enter at right angles or opposite each other.

It has already been pointed out that a deflecting weather cap is inappropriate for a fume hood since it is highly undesirable for the exhaust fumes to be deflected back toward the

Figure 3.19. Exhaust duct rain shield.

roof. It would also cause significant pressure losses. The design shown in Figure 3.19 would cause minimal pressure losses and yet would protect almost as well against rain, unless the rain were falling virtually straight down. Unless the fan were off, the normal speed of the discharged air would keep any rain from entering the duct as well.

3.2.2.3.3. Fan Selection

Centrifugal fans are the most commonly used type of fume hood exhaust fans. Within this category are several different variants, the choice depending upon the requirements of the individual installation (Figure 3.20).

Forward curved or squirrel cage fans are primarily suited for relatively low pressure applications, from 0 to 2 or 3 inches water gauge. They typically have a relatively large number of small blades, set close together around the periphery of the wheel, each blade being curved forward in the direction of wheel rotation. Because the spaces between the blades in this type of fan are small, they are prone to collecting dust and becoming clogged.

Paddlewheel or radial blade fans are used where high pressures of 15 inches water gauge or more are required. The radial blades, typically six, are relatively heavy and resist corrosion and abrasion well. Since they have large spaces between the small number of blades, they are the least likely to become clogged. Fans with backward curved blades are best for medium pressures, from approximately 1 to 8 inches water gauge. This type of fan resembles the forward curved type in that the blades are placed around the periphery of the wheel, but they usually have fewer blades, typically less than 16, but more than radial blade fans. This type of fan operates more efficiently than the other two types.

The inlets to the fans should be designed to take maximum advantage of the fan's performance capabilities. Either the duct should feed directly into the fan intake or be brought into it with a smooth bend to the duct. Connections which require the air to make a right angle turn will significantly and adversely affect the performance of the fan.

Figure 3.20. (Top) Types of centrifugal fan blade designs; (bottom) exhaust duct fan connections.

For general purpose fume hoods, as a minimum, the fan blades should be nonsparking. Usually this is achieved by using coated aluminum or stainless steel fan impellers. If the fume hood is certain to be used heavily for highly flammable solvents, the motor should be selected to be explosion-proof as well. If the exposure to corrosion is expected to be severe, materials with good corrosion resistance, such as PVC or fiberglass reinforced polyester materials should be selected for the fans. If, for reasons of economy, ordinary steel fans are used, they should be coated with Teflon, or if the corrosion problem is somewhat less severe, PVC or polypropylene will serve.

For flexibility, the fan should be driven by a belt from the motor, since within limits, this will allow the speed of the fan to be changed to make up losses in efficiencies in the exhaust system over a period of time, or to make planned design changes. If a deliberate change is to be made in the amount of air discharged through a given duct, the ramifications of the change on the room air balance and the performance of other hoods in the facility should be considered in advance.

REFERENCES

1.   **Horowitz, H., Heider S.A. and Dugan, C.N.,** Hoods for science laboratories, in *CRC Handbook of Laboratory Safety,* 2nd ed., Steere, N.V. (ed.), CRC Press. Boca Raton, FL, 1971, 154–165.

2.   American Society of Heating, Refrigerating, and Air Conditioning Engineers, *Applications Handbook*, Chap. 15, 1978.

3.   Committee on Industrial Ventilation, *Industrial Ventilation, Section 4, Hood Design Data,* 15th ed., American Conference of Governmental Industrial Hygienists, Cincinnati, OH, 1978.

4.   *Industrial Ventilation, A Manual of Recommended Practices*, 20th ed., American Conference of Governmental Industrial Hygienists, Cincinnati, OH, 1988.

5. **Fuller, F.H. and Etchells, A.W.,** Safe operation with the 0.3 m/s (60 fpm) laboratory hood, *Am. Soc. of Heating, Refrigerating, and Air Conditioning Eng. J.,* October 1979, p. 49.

6. **Caplan, K.J. and Knutson, G.W.,** The effect of room air challenge on the efficiency of laboratory fume hoods, ASHRAE Trans. 83, Part 1, 1977.

7. **Steere, N.V.,** Ventilation of laboratory operations, in *CRC Handbook of Laboratory Safety,* 2nd ed., Steere, N.V. (Ed.), CRC Press, Inc. Boca Raton, FL, 1971, pp. 141–149.

8. *Useful Tables for Engineers and Steam Users,* 8th ed., Babcock & Wilcox Co., 1963.

9. *Laboratory Chemical Fume Hoods Standards, Manual 232.1*, United States Department of Agriculture, Administrative Services Division, December, 1981.

10. ANSI Z9.2-1979, *Fundamentals Governing the Design and Operation of Local Exhaust Systems,* American National Standards Institute, Inc., New York, NY.

11. ASHRAE 110, *Method of Ting Performance of Laboratory Fume Hoods,* American Society of Heating, Refrigerating, and Air Conditioning Engineers, Fairfax, VA, 1985.

12. *Corrosion Resistance Chart*, Sheldons Manufacturing Corp., 1400 Sheldon Drive, Elgin, IL, 60120.

3.2.2.4. Biological Safety Cabinets

The primary safety role of a biological safety cabinet is intended to protect the laboratory worker from particulates and aerosols generated by microbiological manipulations. A second important goal is to provide an uncontaminated work area for the work in progress. To accomplish both of these aims, the designers of these cabinets depend upon the ability of HEPA filters to remove the contaminants from air passing through them. Since HEPA filters are ineffective against gaseous chemicals, biological safety cabinets are generally not intended to be used for protection against this type of chemical hazard. In general, they are not really intended for other chemicals either since chemicals other than gases can quickly clog the passages of the HEPA filter. One type, in which the air is totally exhausted after a single pass through the work area, may be used to a limited degree for chemical applications.

There are basically three classes of biological safety cabinets, classes I, II, and III. Class II units are divided into two main subclasses, IIA and IIB. There are several versions of the latter of these two.

3.2.2.4.1. Class I Cabinet

The class I cabinet, illustrated schematically in Figure 3.21, is essentially a variation on the chemical fume hood, resembling closely the self-contained type. As with a chemical fume hood, worker protection is provided by air flowing inward through the work opening to prevent escape of biologically active airborne agents. Unlike a chemical hood, the view screen, equivalent to the sash in a chemical fume hood, is usually designed to provide a fixed opening, typically 8 to 10 inches high (20 to 25 cm). The view screen is generally made so that it can swing upward to permit putting equipment into the unit, although some units have an air-lock type door on the end of the unit to serve the same purpose.

The airflow through the face may range between 75 and 100 fpm. An aerodynamic shape at the entrance or suction slots near the edge to ensure the flow of air is inward aids in the performance of the cabinet, while a baffle at the rear provides that some of the inlet air will flow directly across the work surface as the remainder is drawn upward through the

Figure 3.21. Class I biosafety cabinet.

cabinet to the exhaust portal. It is primarily the directional flow of air which ensures that the agents of concern stay within the cabinet.

As with the chemical fume hood, the performance of the cabinet can be adversely affected by a number of factors. The velocity of air movement within the room can result in a degradation of the performance of the hood, as can too-rapid movements by the worker, location of the work too close to the entrance, or perhaps the effect of a thermal source, e.g., a Bunsen burner, within the unit. The effectiveness of the unit can be enhanced by placing a panel over the opening in which are cut arm holes. The airflow through these arm holes will be much higher than the designed air speeds through the front, especially with the worker using them, but the relatively small gaps normally should keep the turbulence caused by the higher speeds through the portals from causing materials to escape from the cabinet and creating a problem for the user.

A class I cabinet, used properly, can provide excellent protection for the research worker, but it does not provide any protection for the active work area within the cabinet, since the air flowing into the cabinet is "dirty" air, i.e., ambient air from the room that has not been specifically cleaned. For chemical work, this was not discussed since it is rarely a concern of the chemists, but with biological research, elimination of contamination of research materials is likely to be important.

The air from the cabinet is usually exhausted through a HEPA filter placed above the exit portal and before the exhaust fan. If the discharged air does not contain any chemical agent or other material which could be expected to pass through a HEPA filter, it is not absolutely essential that the exhaust be to the outdoors. Absorbent filters, such as that used in the self-contained chemical hood, can be used to supplement the HEPA filter if there are possible chemical effluents. However, a HEPA filter can begin to leak, and hence it may be desirable to take the exhaust to the outside if the contaminants can cause problems to the workers. It is not as critical that the entire ductwork be maintained at a negative pressure as with a chemical hood, so the exhaust fan can be integrated into the cabinet if necessary. The exhaust can be into a chemical air discharge system as well.

A unit sometimes confused with a class I biological safety cabinet is a horizontal laminar flow cabinet or work table. This type of unit serves precisely the opposite function of a class I cabinet. Clean air which has been HEPA filtered is blown across the work surface *toward* the worker so that the research or product materials are protected against contamination, but the worker is not protected at all. Such a unit is unsuited for microbiological work, except for applications which would cause no harm to the user.

3.2.2.4.2. Class II Cabinets

Class II cabinets provide protection for both the researcher and for the research materials within the cabinet. Figure 3.22 shows an idealized version of a class II cabinet. In this design, room air is drawn in the front opening, but instead of passing over the work surface air is pulled down into a plenum by a fan unit under the work surface. Then the output of the fan is passed up through a channel at the rear of the cabinet into a space between two HEPA filters. A portion of the air is exhausted at this point through one of the filters and the other portion passes through the second filter, where it is then directed down as clean air to the work surface. This stream of air is considered to be a laminar flow stream that resists encroaching air and remains clean, so that the work surface is in a clean environment. Part of this air enters the intake air grill at the front of the cabinet and the remainder goes through an exhaust grill at the rear of the unit. Both of these airstreams continue going through the cycle. The result is that "dirty" room air or air that has passed over the work surface, and hence is also "dirty," is restricted to the air on its way to the filters. The work surface is in a clean air space, and pathogens are blocked from escaping into the room by the inward flow of air at the work opening and by being removed by the HEPA exhaust filter. As noted in the beginning of this section there are several variants on this concept.

Class IIA Biosafety Cabinets

Class IIA units are very similar to the basic class II biosafety cabinet just described. In order to keep the work area free of room air and air that has been contaminated by the work, the plenum containing the blower and the channel through which air is delivered to the HEPA filters under a positive pressure must be carefully sealed to be leak tight. The class IIA unit is designed so that about 30% of the air is exhausted from the cabinet and about 70% recirculated each cycle. Thus, the make-up air through the front must provide an amount of air equal to that exhausted, or 30%. The air being supplied through the front opening must be carefully balanced with the amount of air being exhausted. If too small an amount of air is exhausted to the outside, the air in the working volume could be at a positive pressure and pathogens could be forced out into the operator's area. If too small an amount of make-up air is provided at the face, the pressure in the working volume could become negative, allowing the work area to be contaminated by room air. This type of unit is especially sensitive to anything in the work area that could perturb the laminar flow of air. Examples would be equipment blocking portions of the duct, rapid arm motions, and gas flames.

In consequence of both a smaller front opening, 8 to 10 inches, and the fact that only 30% of the air is being discharged at any one time, the amount of tempered air needed to be provided for a laboratory using this type of cabinet is much less than for a chemical

Figure 3.22. Class II biosafety cabinet.

fume hood, typically in the range of 250 cfm for a 4-foot unit as compared to close to 1000 cfm for the same size chemical hood. It could even be more favorable if the bio-safety cabinet exhaust were to be discharged into the room, in which case no additional tempered air would need to be supplied to the room other than that needed to provide the recommended amount for personnel comfort and well-being.

Class IIB Biosafety Cabinets

Class IIB biological safety cabinets differ in several ways from class IIA cabinets. Two major differences are the amount of air recirculated (a much smaller proportion for a type B unit than for a type A) and that the air in the plenums surrounding the work area is filtered or clean air rather than contaminated air.

Shown in Figure 3.23 is a generalized drawing of a representative class IIB biosafety cabinet, designed originally by the National Cancer Institute. As with other class II units, air entering the cabinet is immediately drawn into an air intake. Also entering this intake is filtered air from above, aiding in blocking the entrance of room air further into the working volume. In this unit, 100 fpm of room air is provided as supply air. The air drawn through the front grill is drawn through HEPA filters below the work surfaces by blowers situated below the filters. The resulting clean air is forced up plenums on each end of the unit and into a diffuser area above the work area. The clean air is circulated into the work zone through a diffuser panel at a reduced speed of 50 fpm. Most of this air, 70%, is exhausted as contaminated air through two rows of slots at the rear of the cabinet, through a HEPA filter, while 30% is recirculated through the front grill. Blowers, external to the cabinet, which are typically a part of the laboratory system, provide the exhaust pressure.

Figure 3.23. Class IIB biosafety cabinet.

Another version of the class IIB biosafety cabinet is the total exhaust unit shown in Figure 3.24. All of the air entering the cabinet makes only one pass through the cabinet before being discharged through a HEPA filter. One hundred (100) fpm of room air enters through the front opening, providing protection for the operator against pathogens escaping from the cabinet, and is drawn down into a plenum below the work surface by a blower. All of this air is exhausted. None is recirculated. In order to provide clean air for the work zone and to block the entrance of room air into the cabinet interior, air is drawn in from above the cabinet by another blower through a HEPA filter into the work zone. Part of this air passes through the inlet opening at the front of the cabinet and part through a slot at the rear as contaminated air. All of the air in the exhaust plenum is discharged through another HEPA filter.

Since none of the air that has passed through the work zone is recirculated, at least in principle this type of cabinet could be used for moderate chemical applications. However, care would have to be taken to ensure that the exhaust filter would not suffer from loading of the filter by chemicals. The different rates of loading of the inlet and exhaust filters are a problem for all types of class II cabinets, but the more complicated airflow systems of class IIB biosafety cabinets, where two or more blowers are involved, exacerbate the problems.

Table 3.9 provides a summary of the characteristics of the three types of class II biosafety cabinets.

Class II biosafety cabinets are intended for low- to moderate-risk hazards. As a minimum, they should be required to meet the National Safety Foundation (NSF) Standard 49 for class II (laminar flow) biohazard cabinetry. The working enclosures and plenums through which air moves should be constructed of materials that are easy to decontaminate, such as stainless steel or a durable plastic.

Figure 3.24. Total exhaust-biosafety cabinet.

Table 3.10.   Basic Characteristics of Class II Biosafety Cabinets

	Type		
	2A	2B	2B (total exhaust)
Inlet air speed	75 fpm	100 fpm	100 fpm (minimum)
Fraction of air recirculated	70%	30%	0%
Positive pressure plenums	Contaminated	Filtered	Contaminated, but isolated
Exhaust air for a typical 4-foot unit	225 cfm	500 cfm	700 cfm

3.2.2.4.3. Class III Cabinets (Glove Boxes)

Class III units are totally sealed cabinets in which the user performs the manipulations with the research materials inside the chamber, using attached arm length, impermeable gloves. Materials are usually placed in the cabinet prior to work beginning, or cycled into the interior through an air lock. Materials taken from the chamber are usually taken out through a double-door autoclave, through a second air lock where decontamination procedures may be carried out, or through a chemical dunk tank. These units are intended to be used for high-risk materials. Protection for the work is simply that provided by preliminary cleaning of the interior.

Because most are totally sealed, glove boxes typically have minimal ventilation requirements. At least one vendor, however, Baker Co. of Sanford, Maine, makes a class III unit in which air enters through a HEPA filter and is exhausted through ultra-high-efficiency HEPA filters (99.999% efficient for particle removal) in tandem. These filters, which exceed the requirements for normal HEPA filters by better than a factor of 10, are especially tested to ensure their performance. The cabinets should be kept under slight negative pressure, so that if there are any leaks at all, the leakage would be from the outside to the inside.

3.2.2.4.4. HEPA Filters

An accepted definition of a HEPA filter is

"A throwaway extended-pleated-medium dry-type filter with (1) a rigid casing enclosing the full depth of the pleats, (2) a minimum particle removal efficiency of 99.97% for thermally generated monodispersed DOP smoke particles with a diameter of 0.3 microns, and (3) a maximum pressure drop of 1 inch water gage when clean and operated at its rated airflow capacity."

A properly functioning HEPA filter is an essential component in both class I and class II units and may be important in some class III biosafety cabinets.

HEPA filters used in class II biosafety cabinets meet the requirements of Underwriters Laboratory's UL 586 and mil-spec MIL-F-51068. As stated in the definition, they are made by folding a continuous sheet of filter paper back and forth over corrugated separators. The separators provide strength to the assembly and form air passages between the pleats. The filter paper is made up of submicron fibers in a matrix of larger, 1 to 5 micron fibers with a small admixture of organic binder. It is a fragile material and subject to puncture or cracking if abused. The filter is usually glued to the edge of the frame by a glue that hardens, thus making this connection a highly vulnerable stress point. The frame is seated on a gasket and this seat is also vulnerable to leaks. If there are any questions about the integrity of the HEPA filter, operations should cease until it has been tested and recertified.

A standard filter assembly of 24 × 24 inches and 5-7/8 inches deep has a rated capacity of 500 cfm. This filter will contain about 110 sq. ft. of filter paper. Operating at rated capacity, the air will be moving through the filter at about 5 fpm and the speed of the air leaving the face will be about 125 fpm. The filter will offer, in clean condition, a pressure drop across the filter of no more than 1 inch water gauge. When the pressure drop reaches 2 inches water gauge, it is usually time to replace the filter.

Because the desired laminar air velocity inside a class II unit is somewhat less than 125 fpm (in the unit designed by the National Cancer Institute, described in Section 3.2.2.4.2, the downward design velocity is 50 fpm), most HEPA filters operate at well below their rated capacity. This results in a significantly increased lifetime. The resistance decreases inversely proportionally to the velocity of the air through the filter, and the amount of loading on the filter will decrease directly as the volume of air passing through the filter. The net effect is to greatly extend the life of the filter unit.

3.2.2.4.5. Installation, Maintenance, and Certification

The guidelines governing the most favorable location of laminar flow cabinets are essentially the same as for a chemical fume hood. Place them in the far end of a laboratory, in a low traffic area, and where there are no drafts. The environment of the hood should be checked to ensure that air speeds in the neighborhood of the cabinet opening are small compared to the face velocities of the cabinet.

The installation of laminar flow cabinets is a specialized skill. An examination of a cabinet shows that the components are carefully sealed together, with a large number of bolts being used to maintain sufficient pressure on the seals between the individual sections of the cabinet to ensure that they fit tightly together. Shipping or even moving the cabinet across the room may be sufficient to break some of the critical seals. Demonstration that the unit's seals are intact and that it meets all specifications at the factory simply shows that it was in good condition at that time. It does not demonstrate that it is still in a similar condition after installation. The equipment needed to test the integrity of the cabinets is expensive and requires skills and training which relatively few laboratories, institutions, or corporations have in-house. However, to be sure that a cabinet is providing the protection for the operator, product, or both for which it was designed the cabinet needs to be tested and certified at installation, after any relocation, and periodically, such as either annually or after 1000 hours of use.

A purchase order for a cabinet should include provision and funds for testing and certification of the unit after delivery and setup but before final payment is made. Few, if any, vendors routinely make provision for this service or have their own personnel to do it. In order to avoid a potential conflict of interest, this should be arranged by the purchaser with an independent contractor. The cabinet vendor should be willing to delay payment of their invoice for a reasonable period to allow this to be done. This provision needs to be made in the purchase contract so that no misunderstanding will occur. There are training programs that teach how to perform the tests properly and there are firms or individual consultants available that will perform the tests at a reasonable fee.

It has been noted that it is not *essential* that the cabinets be exhausted outdoors, but there are several reasons why it is highly desirable. The most obvious reason is that no system is absolutely foolproof. A cabinet may be checked and certified in the morning, and an accident may occur during the afternoon which could cause a seal to fail or cause the HEPA filter to begin to leak. If the cabinet is exhausted outside the facility, no pathogens should be released into the facility. Decontamination may be required for formaldehyde, or a comparably undesirable chemical, that should not be released into the laboratory. Some research may involve chemical carcinogens, radioisotopes, or other materials not eliminated by the HEPA filters and should not be discharged indoors. In fact, it may be desirable to incinerate some organisms by passing them through a flame before evacuating them even to the outside. Finally, class IIB cabinets typically do not provide an integral exhaust blower to discharge air from the unit, but depend upon the facility to provide the exhaust fan. If the cabinet is connected to the system to provide suction on the exhaust portal, there is no reason for the system to exhaust back into the laboratory or building. Where cabinets are exhausted to the roof, the height of the exhaust stack should be sufficient to ensure that the effluents are discharged above head height of any maintenance personnel who may be present on the roof. No weather cap should be provided for these exhaust ducts.

The removal and replacement of contaminated HEPA filters should be performed by trained personnel who take precautions against any exposure to themselves and to avoid contaminating the facility. This will require planning ahead of time to provide sufficient access to the filters. For filters anticipated to be contaminated with human pathogens (or animal, where animal exposures are a matter of concern), provision for isolation bag-out procedures should be made in advance. In general, HEPA filters should be disposed of as contaminated biological waste, preferably by incineration.

Cabinets should be decontaminated periodically. A UV tube placed inside the cabinet will aid in disinfecting the surface, but will not significantly decontaminate the air passing through it. Cleaning after each day's use or at the end of a sequence of operations with a weak solution of household bleach is recommended, but other materials such as quaternary ammonium compounds may serve as well, be less irritating to the user, and cause less corrosion. There are procedures recommended by the National Cancer Institute and available on a slide cassette package for decontaminating cabinets. Basically, this consists of sealing the cabinet and vaporizing an amount of dry paraformaldehyde in it sufficient to provide a concentration throughout the cabinet of 7 to 8.5 mg/m^3. The cabinet should remain sealed for about 4 hours. The temperature should be between 20° and 25°C and the humidity above 70%. The paraformaldehyde should then be exhausted to the outdoors and the cabinet ventilated for at least 8 hours. It should be noted that since this procedure was recommended, questions about the carcinogenicity of formaldehyde have arisen. Because of this carcinogenic property, the OSHA 8-hour time weighted average (TWA) has been lowered to 0.75 ppm, with an action level set at 0.5 ppm. A short-term (15 minute) limit of 2 ppm has also been set. The ACGIH recommended limits are even lower, 0.3 ppm as a ceiling limit. Persons performing the decontamination must not exceed the occupational limits. There is no question that some individuals are irritated by levels somewhat lower than the levels now permitted.

REFERENCES

1. National Sanitation Foundation, Standard No. 49, *Class II (Laminar Flow) Biohazard Cabinetry,* 1983.
2. **Burchsted, C.A., Kahn, J.E. and Fuller, A.B., Eds.,** *Nuclear Air Cleaning Handbook,* Oak Ridge National Laboratory, ERDA 76-21.
3. **McGarrity, G.J. and Coriell, L.L.,** Modified laminar flow biological safety cabinet, Applied Microbiology, 28(4), 647–650, October 1974.
4. **Coriell, L.L. and McGarrity, G.J.,** Biohazard hood to prevent infection during microbiological procedures, Applied Microbiology, 16(12), 1895–1900, December 1968.
5. Selecting a Biological Safety Cabinet, (slide cassette) NAC no. 00709 and no. 01006, National Audio Visual Center (GSA), Washington, D.C., 1976.
6. *Effective Use of the Laminar Flow Biological Safety Cabinet,* (slide/cassette) NAC no. 00971 and no. 003087, National Audio Visual Center (GSA), Washington, D.C., 1976.
7. *Certification of Class II (Laminar Flow) Biological Safety Cabinets,* (slide/cassette) NAC no. 003134 and no. 009771, National Audio Visual Center (GSA), Washington, D.C., 1976.
8. *Formaldehyde Decontamination of the Laminar Flow Biological Safety Cabinet,* (slide/cassette) NAC no. 005137 and no. 003148, National Audio Visual Center (GSA), Washington, D.C., 1976.

3.2.3. Built-In Safety Equipment

In a very real sense, the hoods and safety cabinets that have been the subjects of the last several sections can be considered built-in safety equipment. However, the items to be discussed in the following sections are equipment which comes to mind for most persons in the context of fixed safety equipment.

3.2.3.1. Eyewash Stations

One of the most devastating injuries a person can suffer is loss of eyesight. There are a number of protective measures which should be taken in the laboratory to prevent eye injury. However, should all of these measures fail and chemicals enter the eye, an effective eyewash station is an essential item of fixed equipment that should be immediately available. Although superseded by the Laboratory Safety Standard, OSHA does require in Section 1910.151(c) of the General Industry Standards that:

"Where the eyes or body may be exposed to injurious corrosive materials, suitable facilities for quick drenching or flushing of the eyes and body shall be provided within the work area for immediate emergency use."

There are no fixed standards on the maximum acceptable distance of travel to reach an eyewash station. It has been recommended by the National Safety Council that the travel distance should be no more than 25 feet or that the station take no more than 15 seconds to reach. A more realistic criterion should probably be that no one should have to open a door to reach an eyewash station, and the distance *and* time of travel should be no greater than those recommended. No one in pain and possibly blinded should have to overcome any additional impediments or obstacles in seeking relief. An eyewash station should be centrally placed in a laboratory along a normal path of egress or in an otherwise equally logical location in a given facility.

Small squeeze bottles containing a pint, or perhaps at most a quart, of water are not acceptable as eyewash devices. The basic problem is lack of volume. As a minimum, eyes suffering even a light chemical burn need to be flooded with potable water for 15 to 20 minutes. The second problem is that the water in the bottle may become contaminated. Where water lines are not available, eyewash units connected to pressurized portable containers of water are acceptable substitutes if they contain sufficient amounts of water, preferably 15 to 20 gallons.

Eyewash stations should provide an ample amount of water, at least 0.4 gallons (1.5 liters) per minute, at a relatively low pressure, 25 psi or less, in such a manner as to flood both eyes with aerated potable water. If the unit is intended to cover the entire face, then to meet ANSI 358.1 requirements, 3 gallons or 11.4 liters per minute would be needed. The most common type, with two nozzles facing upward and aimed slightly inward toward each other is probably the best overall design. An alternative consisting of a spray nozzle, connected to a flexible hose, is not a bad supplement, but it should not be the only eyewashing device available. An individual alone may be in too much pain to do much more than hold his face in flowing water and certainly could not simultaneously manipulate a hose and use his hands to hold his eyelids open.

Turning on the eyewash should require minimal manual dexterity. Any number of mechanisms to turn them on are possible but perhaps the most popular is a simple paddle that the injured person can push aside. The eyewash should remain on continuously with no additional effort after the initial activation, but if an automatic cutoff is provided, it should not activate for at least 15 minutes or until 40 to 50 gallons of water have been delivered. Many eyewash stations are mounted as part of the plumbing over a sink. This is convenient but not essential. In the following section on deluge showers, it will be pointed out that in the case of eye injuries, safety is more important than spilled water, which can be mopped up, so that floor drains are not required for safety shower installations. Preferably, eyewash stations and deluge showers should be installed as a package since it is likely that if the eyes and face have been exposed to chemicals, other portions of the body quite possibly may have been contaminated as well.

A major problem with most eyewash stations and deluge showers is that they are usually connected to the cold water line. Typically, tap water temperatures are in the 60° to 70°F (15.5° to 21°C) temperature range, but in colder climates can be much less during the winter. Water at temperatures in the 50s or lower can be painful itself, and in extreme cases can cause the injured person to go into shock. Although relatively few eyewash installations are capable of conveniently providing it, lukewarm water between 90° and 95°F (32° to 35°C) would be ideal.

A permanently installed eyewash station is an essential component in or very near every laboratory, but, if one is not available, any source of water, provided it is not too hot or extremely cold, should be used in an emergency. A sink faucet, a shower, even a large basin of water in which the injured person could immerse their eyes is adequate.

All eyewash stations should be checked under full flow conditions on a definite schedule. Any deficiencies should be corrected immediately.

Brief instructions on how to activate and use an eyewash fountain should be placed immediately adjacent to the unit in addition to simplified instructions on how to help the patient keep their eyes fully open so that the water will be able to reach the injured tissues.

3.2.3.2. Safety (Deluge) Showers

According to ANSI 358.1, safety showers should also be placed so that it should not be necessary to travel more than 100 feet (30.5 meters) to reach one in the event of an accident, nor should there be any obstacles in the way. This is a substantial distance, and consideration should be given to reducing it. However, showers often are placed in hallways, usually where they can service more than one laboratory. This also serves to avoid creating a massive water flood in a crowded laboratory in favor of a much more easily cleaned hallway. This latter point is not an insignificant advantage since it may be difficult to locate a shower in a small facility so that water will not splash into sophisticated and easily damaged equipment. Nevertheless, the potential of a serious life-threatening or maiming injury in a chemical spill is sufficiently likely so that the concern for personal safety should override this and other factors. The design of the laboratory should take into account the prevention of damage of equipment by the shower. In addition, units located in public corridors are more subject to vandalism than those inside rooms; therefore, it is recommended that each laboratory be individually equipped with a combination safety shower and eyewash station. As noted in the previous section, accidents involving facial splashes are also likely to involve other parts of the body. Clearly, if the units are separated, it is not practical to travel from one to the other when both are needed. Both should be in the same location.

The demands of a safety shower on the water supply are much more severe than for an eyewash fountain. Their alternate name, deluge shower, is not idly applied. The water supply should be able to provide up to 30 gallons (114 liters) per minute. For normally available water pressures, this would usually indicate that the shower water supply be at least a 1- inch line. The problem of temperature is, again, a serious one, but even more so because of the greater area of the body involved. Usually the showers are connected to a cold water supply line. In colder climates, the stress of inundating the entire body with cold water at temperatures perhaps in the 50° to 60°F (10° to 15.5°C) range may be sufficiently severe to cause a person to go into shock. Ideally, water temperatures should be around 90° to 95°F (32° to 35°C) but not over 100°F (38°C).

Instructions for activating and deactivating a shower should be prominently posted near the shower. The mechanism for turning a shower on can be a paddle, which is simply pushed out of the way, or a chain which can be pulled. Both are simple and require minimal physical control or the manual dexterity that may be important in this type of an emergency. The shower should continue to run until it is deliberately turned off, although it may be equipped with an automatic cutoff, especially if it is installed where there is no drain, but it should deliver at least 50 to 60 gallons before any automatic cutoff activates. Showers should be checked at least once a year, which is conveniently done by catching the flow in a large funnel connected to a fire hose which is discharged into a 55-gallon drum. This provides both a rate and volume check simultaneously, while avoiding creating a mess to be cleaned up.

It should not be necessary to point out that safety showers should not be located near any source of electricity with which the flowing water from the shower could come into contact, but numerous instances have been observed where this has occurred. Usually it has been due to the laboratory workers themselves moving portable equipment too near the shower, but occasionally one is found improperly installed by maintenance personnel or because of changes made by renovation crews.

As noted, drains are not essential, although if it is feasible to make them available, they are desirable. Without a drain, a sizable mess will be created which will have to be mopped up when a shower is used, but this should be sufficiently rare that it may not justify the cost of installing additional drains to accommodate the safety showers, especially if a retrofit is necessary.

REFERENCES

1. ANSI Z358.1, *American National Standard for Emergency Eyewash and Shower Equipment,* American National Standards Institute, Inc., New York, NY, 1991.
2. **Stearns, J.G.,** Safety showers, in *CRC Handbook of Laboratory Safety,* 2nd ed., Steere, N.V., CRC Press, Inc., Boca Raton, FL, 1971, pp. 121–123.
3. **Srachta, B.J.,** Safety showers and eyewashes, in *Safety and Health,* National Safety Council, Chicago, IL, August 1987, pp. 50–51.

3.2.3.3. Fire Suppression Systems

Many laboratory buildings have automatic fire suppression systems, but there are still some that do not, even though data show that the most common type, a water sprinkler system, is effective well over 90% of the time in either extinguishing a fire or controlling it until firefighting personnel arrive. In a very high percentage of the cases in which the sprinkler was not effective, the reason the system did not succeed was due to either poor installation or human error. Some scientists do not want a fire suppression system in their laboratories because they feel that it will lead to problems. The possibility that a fire in one area will cause the entire system throughout the building to activate is one basis for concern, although in most systems only those sprinkler heads in the vicinity of a fire activate. In over one third of the fires in which the sprinkler system was successful, only one sprinkler head was activated in controlling the fire. There is some concern that water will react with chemicals in the laboratory, will spread the fire due to burning solvents being carried away by excess water, or will damage sensitive equipment. If there are problem chemicals, there are alternative fire suppression systems that do not use water and will not damage even delicate equipment.

The potential for fires is higher in most research facilities than in a typical building because of the variety and character of the chemicals employed. Often, much of the equipment is home-built or temporarily rigged and has not been checked for safety by an accredited testing laboratory. Many of the operations involve heat. Some type of fire-suppression system is recommended for laboratory buildings, if for no other reason than that the presence of one in a building can lead to substantial savings for insurance costs. We are in a time of insurance premiums rapidly escalating at a rate that does not appear to show signs of abating. Unless the building is insured for its full replacement value, including the cost of its contents, the increased cost of construction, and the high cost of sophisticated research equipment, it is likely that insurance will pay only a fraction of the actual losses. Further, insurance cannot replace in most cases the intellectual properties lost.

The time to stop a fire is when it is very small. This is the purpose of portable fire extinguishers, which laboratory personnel can use before a fire becomes too large to control. Unfortunately, most laboratories are not staffed at all hours, while some heat-generating equipment such as stills or heat baths may be left functioning continuously. A sprinkler system serves essentially the same purpose as a person with an extinguisher 24 hours a day, 365 days a year. It can put water, carbon dioxide, or other fire suppression media directly onto a fire still small enough to be easily extinguished. There are fires which expand so rapidly that a fire suppression system will be overwhelmed but, as noted above, statistically the number of fires in this category are very small.

OSHA does not mandate automatic fire extinguishing systems, but the General Industry Standards, Subpart L, Sections 1910.155-165, do provide regulations covering the essential requirements which installed systems must meet.

The principles of an automatic fire suppression system are straightforward, but designing an actual system requires substantial engineering skills to be sure that the system will adequately and efficiently serve its purpose. The following brief sections on built-in fire suppression systems are intended only to supply basic information on the essential features of the various types of fire suppression systems to provide some insight in this critical area to concerned laboratory facility managers.

3.2.3.3.1. Water Sprinkler Systems

There are several essential components of an automatic water sprinkler system. The supply system to provide water to the sprinkler system must have sufficient pressure to deliver water to the units highest up in the building. There must be a control valve to serve the suppression system. A distribution system must carry the water to the spaces to be protected by the system, when needed, to respond to the presence of a fire. In each area where the sprinkler system is to provide protection, a carefully engineered pattern of sprinkler heads is required to distribute the water to ensure that complete coverage is obtained. The sprinkler heads themselves must be selected to meet the needs of the location. Generally, the heads themselves incorporate the heat sensing devices, in the form of fusible links, which will cause them to activate. The design of the system must take into account the maximum normal temperature that will be reached in the vicinity of the sprinkler heads and specify an operating temperature for the fusable links a reasonable amount above this temperature. In some systems, the sensing devices are separate from the sprinkler heads. Finally, when a sprinkler system is activated, a sensing device is needed which will transmit an alarm to the occupants of the building, preferably to a manned location which can immediately summon firefighters, or to a fire station directly. Any automatic water sprinkler system should be installed according to the NFPA Standard 13, as most recently amended.

The water supply system may be from a water works system, a gravity tank, or a pressure tank. NFPA Standard 13 provides specifications for each of these to ensure that they provide sufficient water volume and pressure to supply a sprinkler system. This standard provides for seven different categories of occupancy classifications ranging from light hazard to extra hazard. These levels are only for the purpose of designing an appropriate sprinkler system and do not correspond directly to the occupancy classes under the BOCA Building Code. The extra hazard, group 2, represents a facility where, in part, there are moderate to substantial amounts of flammable and combustible liquids. As a minimum, for a light hazard occupancy, the water supply must provide for a residual pressure of 15 psi at the level of the highest sprinkler and a duration of 30 minutes for the volume of water needed for the system. The requirements on duration and the amount of water needed are higher for other classifications. A gravity tank supply should be at least 35 feet above the level of the highest sprinkler head to provide an adequate pressure, while a pressure tank of sufficient size, filled two thirds, must be pressurized to at least 75 psi.

There must be at least one fire department connection through which a fire department can pump water into the sprinkler system. When a large number of sprinkler heads have become involved in a fire, or the fire department has connected to the same supply to service their own hoses, the flow of water through the sprinkler pipes may be reduced so that the required volume and pressures cannot be met. In such cases the fire department can boost the pressure and volume through this connection, preferably from separate water mains not being used in their own fire suppression efforts.

There are two different types of water sprinkler systems, wet-pipe and dry-pipe. The lines leading to the sprinkler heads are full of water in a wet-pipe system so that water will be discharged immediately from an open sprinkler head, while in a dry-pipe system, the lines are full of air under pressure instead of water. This latter type of unit should be used where the temperature is not maintained above freezing at all times. Both types of systems include a main control valve which is designed not only to supply water to the sprinkler heads but also to provide a mechanism to cause an alarm to sound. The valves also usually provide a visual indication of whether they are open or closed. Except during maintenance, they should be in the open position. There are a number of design features for these valves which are beyond the scope of this document, except to note that some are intended to avoid false alarms due to surges or variations in the water supply pressure to the system.

A variation of the dry-pipe system is the deluge system. The sprinkler heads are continuously open and water is prevented from entering the system by a deluge valve. When a fire is detected the valve to the water supply opens and water flows into the system and out of all the sprinkler heads. This not only wets the immediate area of a small fire, but the entire area to which the fire may spread. This type of system is usually chosen when the contents of the space are unusually hazardous. A variation on this system is the pre-action system in which the sprinkler heads are not open, but when a fire is detected the deluge valve opens and water is supplied to the sprinkler heads. The water entering the system causes an alarm to be sounded. When the heat causes the sprinkler heads to fuse, or open, water is discharged onto the fire.

After the control valve is activated, water is distributed through one or more vertical risers to portions of the system. Smaller cross-mains are connected to the risers which then service several still smaller branch lines. The sprinkler heads are connected to these branch lines. For laboratories, both the piping and sprinkler heads should be especially selected to prevent corrosion.

Sprinkler heads are very simple and rugged devices in which a valve is kept tightly closed by lever arms held in place by fusable links or other devices which fuse or open when they are heated to a predetermined level. When this occurs the valve opens and water is discharged. There are many different versions of this simple device. Figure 3.25 shows a generalized drawing of an upright design.

In most cases, the water flow pattern is designed to be similar to that shown in Figure 3.26 to obtain a uniform overlapping distribution. The water is intended to be in the form of a fine spray. In some instances, where there are expected to be strong vertically rising convective airflow currents, the design may be modified to provide larger droplets which would be more likely to overcome the upward moving air.

Figure 3.25. Upright sprinkler head.

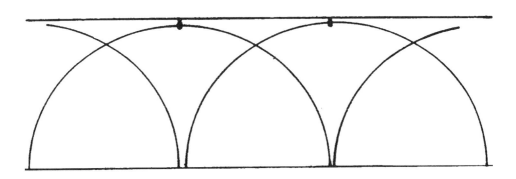

Figure 3.26. Flow pattern from sprinkler system.

The orifice within the sprinkler head is usually 1/2 or 17/32 of an inch in diameter. Depending upon the design and the level of risk of the occupancy, a sprinkler head with an orifice of these sizes is intended to protect between 90 and 130 square feet for the higher levels of risk.

The temperature at which the fusable link or other device fuses depends upon the normal maximum temperature to which it may be exposed. For maximum ceiling temperatures in the vicinity of 100°F or 38°C, which would be comparable to those found in most laboratories, the fusable links should be selected to open with a temperature rating in the range of 135° to 170°F (57° to 77°C).

Water normally is not dangerous to humans, and a water sprinkler system is comparatively inexpensive and easy to maintain. Major advantages of water sprinkler systems are due to the physical properties of water which has a high heat capacity and a large heat of vaporization. It provides cooling and wetting of the fuel, which aids in putting out the fire, and once the fire is subdued, the presence of water discourages re-ignition of the fire. Water may cause considerable damage to equipment and can react with many materials.

3.2.3.3.2. Halon Systems

Halon is a generic name for halogenated agents used in fire suppression systems. There are two major agents of this type in common use, Halon 1301 and Halon 1211. The different numbers in the nomenclature correspond to the numbers of carbon atoms, fluorine atoms, chlorine atoms and bromine atoms in that order. If there were a fifth number in the designation, it would refer to the number of iodine atoms. Halon 1301 is the one most commonly used in occupied spaces while Halon 1211 applications are typically in storage areas or other areas which are rarely or sporadically occupied.

There is clear evidence that the class of chemicals to which the Halon's belong are causing a depletion of the ozone layer in the upper atmosphere. As a result there has been an international agreement to phase out the production of these materials by 1996. Systems already in place can continue to be used but replacement of the materials, in the event the system is activated will become increasingly difficult. Alternatives, are being sought, which do not contain the critical elements, bromine and chlorine, that cause the ozone depletion, but which have similar fire suppression capabilities. The following material is still included in this edition because of the large number of existing systems. The current favorite as a replacement, CF_3I, is characterized as a "drop-in" replacement for Halon-1301 so some of the discussion will still apply. There are still toxicity, corrosion, and ozone-depletion potential questions to be resolved.

Halon 1301 Systems

Halon 1301 is a colorless, odorless gas that is noncorrosive and is electrically nonconductive. It is relatively nontoxic, and at concentrations up to at least 7% by volume in air, personnel can perform a normal evacuation of an area without significant risk. It is especially useful for fires in normally occupied spaces; where flammable and combustible liquids are present; and where electrical devices such as computers, data terminals, and electronic instrumentation and control equipment are present. These latter characteristics make it very useful for laboratory applications.

The design of Halon 1301, bromotrifluoromethane ($CBrF_3$) systems is covered in NFPA Standard 12A. This standard also contains much useful background material on the characteristics of the material, and is the major reference for this section.

There are two basic types of systems, one providing total flooding of an area and one designed for local application of the extinguishing material directly to the fire. Fixed systems will consist of a container of liquified compressed gas, pressurized with nitrogen, detectors, distribution piping, nozzles, manual releases, and a control panel incorporating an alarm. The control panel may also provide circuitry to cut off ventilation systems, and close windows, louvers and other openings to the space.

It is generally accepted that Halon agents interrupt the chemistry of a fire to extinguish it. Since fires involving the vapor phase of a material are most susceptible to contact with the extinguishing agent, Halon systems work best on this type of fire. In the case of a solvent fire for example, if the fire can be detected quickly enough and Halon 1301 applied immediately to the fire, substantial fires can be put out in seconds. Gaseous systems such as Halon work well on liquid fires, in part also, because the temperature of the liquid generally will not exceed the boiling point of the liquid, so that if the availability of oxygen is eliminated, the liquid itself will not be hot enough to reignite the fire. Other types of fires, such as fires on the surfaces of solid materials will take longer, while deep seated fires may not be appropriate for this agent, although if a sufficient concentration can be maintained long enough, even deep seated fires can be extinguished.

In a total flooding system, the required concentration of the extinguishing agent will depend upon what fuel is present and if it is sufficient to only extinguish the fire or if it is desirable to "inert" the space, i.e., provide a sufficient concentration of Halon 1301 in the atmosphere so that the flammability range for the fuel-air combination will be non-existent at the ambient temperature. The latter concentrations are substantially greater than required for extinguishing the fire. The concentrations needed at a temperature of 77°F or 25°C.[*] to extinguish fires for a number of common flammable liquids are given in Table 3.10, as well as design concentrations to provide a reasonable safety factor, and required inerting concentrations. Note that the design concentration was chosen to be at least 5%.

Note that if ethylene were the primary fuel present, a higher concentration of the extinguishing agent than the recommended 7% would be needed. This would not preclude the use of Halon 1301, but it would mandate an effective and rapid evacuation plan. The fact that only flammable liquids or gases are given in Table 3.10 is not to infer that Halon 1301 is only suitable for such materials. If a fire is confined to the surface of a solid material, chances are that Halon will be effective, i.e. a 5% concentration could possibly put it out if the fuel is allowed to "soak" in the design concentration for 10 or more minutes and if the fuel temperature falls below the level required for spontaneous ignition after the Halon is dissipated. Deep-seated fires, which in this context are defined as ones which will not be put out at a concentration of 5% with a soaking time of 10 minutes, may still be extinguished at higher concentrations, longer soaking times, or a combination of both extended time and increased concentrations. Some of the advantages of Halon 1301 diminish if higher concentrations and longer times are required and some disadvantages arise. Most of these negative factors are related to health factors.

Halon 1301 as a pure material appears to have very low toxicity. At concentrations of Halon 1301 below 7%, little if any effect on humans has been noted for test periods up to 30 minutes. In order to provide a reasonable safety margin, exposures of up to 15 minutes are permissible below 7% concentrations by volume. Some individuals have experienced mild effects on the central nervous system, such as dizziness and a tingling sensation in the extremities at concentrations between 7 and 10%, so to be conservative at these concentrations, exposures should be limited to 1 minute. Above 10%, the effects on the central nervous system are more pronounced; between 10 and 15%, exposures should be limited to 15 seconds. There should be no exposures to personnel above 15%. Although a number of other physical effects have been investigated, the only other significant effect noticed in tests on animals has been to the cardiovascular system, where the heart has been made abnormally sensitive to elevated levels of adrenaline, as might be present during the stress of a fire, leading to possible cardiac arrhythmia. All of the observed effects have been shown to be transitory, disappearing after the exposure ceased.

The decomposition products of Halon 1301 are dangerous in sufficient concentrations. The chief ones are the acids of the halide components, HF, HBr, and Br_2. Fluorine is too reactive to be present in substantial amounts alone. Small amounts of carbonyl fluoride and carbonyl bromide (COF_2 and $COBr_2$), respectively, have been found as well. Because of the potential for these toxic materials being present after the intervention of a Halon 1301 system in a fire, fire personnel and others entering the area of a fire should exercise caution. Where the fire involves substantial amounts of solid fuel, the use of a positive pressure, air-supplied breathing apparatus is recommended to search the area for injured personnel.

---

[*]   These data are derived from NFPA 12A, Tables A-2-3.2.2 and A-2-3.2.3.

Table 3.11.   Halon 1301 Characteristics (values in volume percent in air at 77°F [25°C] at 1 atmosphere)

Fuel	Extinguishing Concentrations	Design Concentrations	Inerting Design Concentrations
Acetone	3.3	5.0	7.6
Benzene	3.3	5.0	5.0
Ethanol	3.8	5.0	11.1
Ethylene	6.8	8.2	13.2
Methane	3.1	5.0	7.7
n-Heptane	4.1	5.0	6.9
Propane	4.3	5.2	6.7

The disadvantages alluded to above, when discussing the use of Halon 1301 for deep-seated fires in solid materials, are the risks associated with the required higher concentrations of Halon 1301 due to the increased generation of decomposition by-products with more extinguishing agent present, especially if an individual is injured or unconscious and is unable to evacuate quickly. Cost is a further and significant disadvantage. Halon 1301 is expensive and most systems are designed to completely release the supply of the agent. As a result, fixed Halon systems are usually installed where they will be cost effective, such as in large computer installations, where other expensive configurations of sensitive electronic devices are in use, or where substantial amounts of solvents are stored or in use.

To take advantage of the excellent extinguishing properties of Halon 1301, especially for very rapidly spreading fires from burning flammable liquids, a quick-acting fire detector should be selected, such as one that detects either ultraviolet or infrared light from the flames. Although precautions have to be taken for both these types of detectors to avoid false activation of the system, the rapid response capability of these units means that the extinguishing agent can be applied to the fire almost instantaneously.

Local application systems work precisely as do the fixed systems, with the exception that the agent is applied directly to the immediate area of the fire. The primary need is for the concentration to remain sufficiently high in the area of the fire long enough to ensure extinction of the fire and to permit sufficient time to elapse to allow cooling of the fuel to avoid spontaneous reignition once the concentration decreases below the level required to extinguish a fire. While in a fixed system it is possible to calculate the amount of Halon 1301 needed with some precision, each local application is a unique situation and must be evaluated individually.

As with other common types of fire suppression systems, Halon 1301 is ineffective for fires involving a number of materials including reactive metals, metal hydrides, and materials not requiring the presence of air to burn, such as gunpowder, some organic peroxides and hydrazine.

Although Halon 1301, or it is hoped, its eventual replacement it is not ideal for every laboratory facility, it does offer advantages over many other systems. Its toxicity is low. It is very effective on many common types of laboratory fires. It acts rapidly, which is especially critical in minimizing the spread of a fire in a laboratory situation. It will not damage the increasingly sensitive equipment found in modern laboratories. Its cost and the propensity to be set off by minor flames, if ultraviolet or infrared sensors are used, are its major disadvantages.

Halon 1211 Systems

The material in the preceding section is generally relevant to Halon 1211 bromochloro-difluoromethane ($CBrClF_2$) systems as far as effectiveness is concerned. There are some minor differences. Instead of being odorless, it has a faintly sweet smell, but otherwise many of the physical properties are similar. Given in Table 3.11 are the extinguishing characteristics for Halon 1211. Again, the two materials are very similar, as can be seen by a comparison of the data in Table 3.11 to the corresponding information in Table 3.10 for Halon 1301.

The fact that the design characteristics are not very different could be taken to imply that the two materials could be used virtually interchangeably. However, the onset of toxic effects on personnel for the most part do not become significant for Halon 1301 until concentrations of 7% or higher are reached, a level above most of the required design concentrations. For Halon 1211, the onset of problems, such as dizziness, become definite within a few minutes at exposures above 4%, below the required extinguishing concentrations. For this reason, the use of Halon 1211 is not approved for occupied spaces or for normally unoccupied spaces where an evacuation time of more than 30 seconds would be required.

Even though there are restrictions on its use, because of its excellent fire extinguishing properties Halon 1211 is frequently used in flammable storerooms, especially small stock rooms where self-contained units can be simply hung from the ceiling. Most portable Halon extinguishers are filled with Halon 1211.

3.2.3.3.3. Carbon Dioxide Systems

Carbon dioxide is a colorless, odorless, nonconductive, chemically inert gas. It is commonly used in small portable fire extinguishers for putting out class B fires, i.e., fires that involve solvents, petroleum products, grease, and gases. It can also be used for class C fires where the nonconductivity of the extinguishing agent is important. It is not as effective against fires involving ordinary combustible class A materials, such as wood and paper. Carbon dioxide is also used in total flooding systems. The characteristics of carbon dioxide as an extinguishing agent are included in NFPA Standard 12, which also provides information on the requirements of an approved system.

A major distinction between the use of carbon dioxide as an extinguishing agent and the Halon systems described in the last two sections is that the extinguishing mode for carbon dioxide is primarily simple smothering of the burning fuel, with no chemical action involved. There is little cooling action, with an effectiveness for carbon dioxide of about one tenth that of an equivalent amount of water. As a result, once the carbon dioxide has dissipated, the possibility of reignition exists if there are any sufficiently hot areas still present.

Table 3.12 gives the minimum extinguishing concentrations by volume and the design concentrations to provide a margin of safety. These should be compared to the equivalent data in Tables 3.10 and 3.11.

It is clear that far higher concentrations are required than for Halon systems, so high in fact that an individual trapped in a space flooded by the minimum recommended amount of 34% in the table would quickly become unconscious due to lack of oxygen, since in such a case the oxygen concentration would fall to 13.8%.

When a fixed system is activated, a major portion of the rapidly expanding gas will become carbon dioxide vapor, while the remainder will become very fine particles of dry

Table 3.12.   Halon 1211 Characteristics (values in volume percent in air at 77°F
[25°C] at 1 atmosphere).

Fuel	Extinguishing Concentrations	Design Concentrations	Inerting Design Concentrations
Acetone	3.6	5.0	NA
Benzene	2.9	5.0	5.0
Ethanol	4.2	5.0	NA
Ethylene	7.2	8.6	13.2
Methane	3.5	5.0	10.9
n-Heptane	4.1	5.0	NA
Propane	4.8	5.8	7.7

ice. There will also be condensed water vapor due to the cooling action of the expanding gas. As a result, visibility may be limited and individuals trapped in an area may have difficulty finding their way out. An area equipped with an automatic carbon dioxide extinguishing system should be posted with warning signs, such as:

WARNING
AREA EQUIPPED WITH CARBON DIOXIDE FIRE SUPPRESSION SYSTEM
EVACUATE IMMEDIATELY
WHEN ALARM SOUNDS

Pre-discharge alarms are essential as is personnel training where a carbon dioxide system is installed; individuals in adjacent spaces where the gas could flow as well as those working in the immediate area protected by the system should be included in the training. Any aisles providing a path of egress should be amply wide and kept clear at all times. Doors should swing in the direction of exit travel. Any automatic door closing systems, used to minimize leakage of the suppressant from the fire area, should be equipped with delay circuits. Self-contained, positive pressure breathing apparatus should be maintained nearby to make it safe for personnel to conduct rescue efforts for individuals that may be trapped in the area. Firefighting and rescue forces entering an area where a total flooding system has been triggered should exercise caution and, unless they are confident that the gas has been dissipated, should wear breathing apparatus.

Because of the risk to personnel, carbon dioxide systems should not be used in most laboratory situations, although they could be used in storage areas or where electrical and electronic devices are employed. It does not leave a residue and will not damage equipment.

3.2.3.3.4. Dry Chemical Systems

Dry chemical fire extinguishing systems use a variety of dry powders as the firefighting agent. They can be stored in pressurized containers and discharged, when needed, very much like the water and gaseous materials previously covered. Most of the agents are primarily effective against fires involving solvents, greases, and gases and can be used around active electrical circuits and electrical equipment, since the chemicals are noncon-

Table 3.13.   Carbon Dioxide Characteristics (values in volume percent in air)

Fuel	Extinguishing Concentrations	Design Concentrations
Acetone	27	34
Benzene	31	37
Ethanol	36	43
Ethylene	41	49
Methane	25	34
Propane	30	36

ductive. They do leave a residue, which can be a problem for delicate electronic equipment. In general, however, the residue can be readily brushed off surfaces and vacuumed or swept up. There are multipurpose formulations which can be used on ordinary combustibles such as wood and paper. Most dry chemical formulations use monoammonium phosphate, sodium bicarbonate, potassium chloride, or potassium carbonate as their fire extinguishing agent.

Dry chemical agents are nontoxic, but personnel in the area when they are being discharged may experience respiratory difficulties and vision problems due to the copious amounts of powder in the air.

The primary extinguishing mechanism is similar to that of the halogenated agents. They disrupt the chemistry of a fire so that it will not propagate. Other mechanisms include reducing the oxygen concentration in the flame zone, heat absorption by the chemical agent, and for liquid fires reducing the amount of vapor entering the air from the liquid by reducing the amount of energy radiated by the flame reaching the surface and causing evaporation. The action of the dry chemicals is very rapid, as with the halogenated agents, which make them desirable where it is essential to prevent a fire from spreading. If the problem with the residue is acceptable, a dry chemical system would appear to be a good choice for laboratories, for chemical stock rooms, and for storerooms.

Monoammonium phosphate ($NH_4H_2PO_4$) is described as a multipurpose agent. As with other common dry chemical agents, it is effective on class B and C fires, but unlike most others, which are not recommended for class A fires, this compound does provide a mechanism for putting out fires in ordinary combustible materials. When heated, it decomposes and forms a solid residue that adheres to the heated fuel surface and excludes oxygen, thus eliminating an essential component for a fire except in the case of those materials which will sustain combustion without additional oxygen. Because of the formation of the solid residue adhering to equipment, it is less desirable for laboratory applications, but it would appear to be a good choice for a stockroom where materials are often stored in their original paper and wood shipping containers.

3.2.3.3.5. Foam Systems

Foam systems are usually intended to be used to extinguish fires involving flammable liquids rather than as general purpose fire extinguishing agents, although some versions are useful on class A fires. The foam is intended to float on the surface of the liquid and extinguish the fire by excluding air and by cooling the fuel and hot objects that may have been heated by the action of the flames. It also prevents reignition by suppressing the production of flammable vapors which could come into contact with heated solid objects. Most foam systems are not intended to be used on three-dimensional fires, i.e., fires involving solid materials of a significant height. There are high expansion foam systems which are designed to smother a fire by flooding the area with a foam layer about 2 feet thick, that can be used in storage areas.

To be effective, the foams must remain intact and not mix with the burning liquids. Some foams work well on ordinary liquid hydrocarbons but readily mix with and lose their effectiveness on polar liquids such as alcohols, acetone, and ketones. However, there are different formulations effective for each type of liquid, and at least one, a synthetic alcohol foam, may be used on either one.

Unless the laboratory is a very specialized one utilizing large quantities of flammable liquids, it is unlikely that a foam fire suppression system would be selected. Foam-generating equipment could be useful in large facilities storing significant amounts of flammable liquids.

3.2.3.4. Fire Detection and Alarm Systems

Every laboratory facility should be equipped with at least a manually activated alarm system, although an automatic system is preferable since it will continue to function when the facility is unoccupied. This is a special advantage in academic institutions since there are periods when the population of the campus is very low. In many cases, depending upon local code requirements and the occupancy classification, an automatic fire alarm system may be required rather than optional. Every component of a system should be approved by Underwriters Laboratories, Inc., the Factory Mutual System, or other nationally recognized accrediting and testing organizations.

3.2.3.4.1. Detectors

The first essential component in any system is a device used to detect and initiate an alarm. In a manual fire alarm system this essential device is an individual who will recognize the probability of a fire. This is not a trivial point. If a fire is behind closed doors or in a concealed chase or plenum, it may not be readily apparent. Persons in buildings where the likelihood of a fire exists should be familiar with the normal state of affairs and be able to recognize discrepancies. For example, there are fairly common situations in chemistry laboratories and operations such as welding where visible smoke and fumes may be generated. Certainly no fire alarm should be sounded in such situations. However, where smoke is observed under unusual circumstances, it should be investigated. When starting to enter a room and the door or doorknob is warm, it can be reasonably assumed that a fire is burning on the other side of the door. Under suspicious circumstances which cannot readily be clarified (the door should not be opened, since this could provide a fresh supply of air to the fire and result in a possibly violent increase in the vigor of the fire), it is usually desirable to activate an alarm.

In a manual fire alarm system, the device used to activate the alarm in most cases is a pull box or pull station. The mechanism is very simple—pulling the switch either makes or breaks an electrical circuit which in turn causes an alarm to sound. As will be discussed more fully in the next section, the alarm may sound only in the individual building, or the alarm could initiate a signal at a remote location as well.

In most research buildings, vandalism is rarely a problem. If local circumstances are such that this is not the case, there are protective measures which can be taken. The use of a glass cover, to be broken by a small hammer, is to be avoided since the person pulling the alarm can be cut by the broken glass. An alternative that has proven very effective is a cover for the pull station which sounds a local alarm when removed. Usually this frightens away a person intending to initiate a false alarm.

In an automatic system, devices are used that depend upon physical phenomena uniquely characteristic of fires. These can be classified under four categories: heat, visible products of combustion, invisible products of combustion, and electromagnetic energy output.

Heat is the most obvious choice of a characteristic by which a fire can be automatically recognized. In the section on fire suppression systems, the fusable links in the sprinkler heads represented one type of heat detector. Alloys can be developed that will have reproducible melting points. When the temperature at the detector site exceeds the melting point of the alloy, contacts are allowed to move so that the device can either make or break a circuit, just as with a manual alarm system. There are plastics which can perform in the same manner. Fixed temperature systems are very stable and not prone to false alarms, but are slow to respond. There are several other versions of these fixed temperature detectors, including bimetallic strips, where the differential rate of expansion of two different metals causes the strip to flex or bend to either make or break the contact. Others depend upon the expansion of liquids.

One version of a heat-sensing detector that has recently had wide application and depends upon the properties of materials changing with temperature is cables which can be run in cable trays or conduit along with data and video cables to detect excessive heat or fires within the chases. These serve to isolate within the length of each section of the heat-sensitive cable where a problem has occurred or perhaps even provide warning of an impending problem. Two wires in the cable are normally insulated from each other, but heat causes the insulator to change so that a short develops between the wires. Transmittal of data and linkage of computers is such an important part of modern technology that the use of these cables as part of a heat-sensing system is highly recommended.

Rate-of-rise heat sensors react more quickly to rapid changes in temperature than the fixed temperature units. Most of these units use the thermal expansion of an enclosed volume of air to activate a pair of electrical contacts. They include a small vent to allow some air to leak slowly in and out to avoid false alarms due to slow changes in the ambient temperature or due to changes in barometric temperatures. The vent is designed to be too small to accommodate rapid changes in air volume due to a sudden rise in temperature. Some units that use thermocouples to detect temperature changes take advantage of the property of two dissimilar metals in contact with each other generating a small electric current when a temperature difference is established between one junction and another. A rapid change in the electric current can be detected and used to initiate an alarm, while slow changes in the ambient temperature can be discriminated against electronically.

Smoke is the obvious visible product of combustion that usually accompanies a fire, although the smoke-generating properties of fires vary greatly. In recent years the use of local smoke detectors has become very widespread. Smoke detectors include two components, a light source and a light sensor. Because of the latter component, an alternate name is "photoelectric" detector. In most units of this type, the detector is usually shielded from the light source. The detector is activated by the light being scattered into the sensor by the smoke which enters into the unit. The location of the unit should be such that there is some air movement in the area which will cause smoke to be carried into the detector. They should not be placed in areas of stagnant air.

A major failing of smoke detectors is that they can be triggered by extraneous light scattering materials. These can be aerosols, such as generated by spray cans or vapor from showers, dust from sweeping, maintenance operations such as welding or construction activities, or insects. They can be set off accidentally by cooking, persons smoking near them, or, unfortunately, by individuals deliberately introducing smoke into them. Some of these deficiencies can be avoided by careful location of the detectors or by training in which maintenance personnel are instructed to cover the units when working nearby. The design should include a very fine grid over the passages by which smoke enters the unit so that only the smallest insects can enter them, or they can be constructed with dual chambers which will require both chambers to provide a positive indication. Pest strips on the outside can be used to kill any insect crawling into them. Finally, a frequent cleaning (at least twice per year) program will assist in reducing false activations.

The actions recommended above can keep the number of false alarms to a reasonably acceptable level, but deliberately initiated false alarms are a more difficult problem. Fortunately, at corporate facilities and in the academic and service buildings at academic institutions, deliberate alarms are not a major problem, although they are a significant problem in dormitories at academic institutions. One of the few recourses available is to use verification circuits to test the detector a brief time later (30 seconds, for example) after the initial triggering event. If the device has cleared, then the alarm will not be activated, while if it has not, an alarm will be sounded.

Ionization detectors are alternatives to smoke detectors. In this type of detector, a very weak radiation source emitting beta particles is placed near two plates, one of which is charged positively and the other negatively. As the beta particles pass through the air between the plates they create ion pairs which are collected by the plates, creating a weak electric current. When ionized combustion products enter the air space between the plates, they neutralize the ionization current and cause it to decrease or cease. The lack of current then triggers an alarm circuit. Some of the same problems, such as cooking or sources of ionized particles from some laboratory operations, can also trigger these units and cause false alarms. Both this type of detector and the smoke detector have good sensitivity.

Sensors can be built to detect the light from a fire. However, in occupied spaces it would not be feasible to use the visible light region between about 4000 and 7000 Angstroms. Sensors can be designed to detect energy generated by the fire in the ultraviolet region below 4000 Angstroms and in the infrared region above 7700 Angstroms.

Other sources of ultraviolet and infrared light must be prevented from entering these two types of light sensors. Welding generates ultraviolet light. Lightning reflected from a polished floor has been known to trigger an ultraviolet sensor. For infrared detectors, fairly sophisticated filters are needed to eliminate the background infrared radiation. Electronic filtering circuits can be used to help systems employing both types of these detectors avoid triggering false alarms.

3.2.3.4.2. Automatic Alarm Systems

These systems can range from being very simple to very sophisticated. In the simplest possible system, an alarm triggered by any of the automatic sensors in the previous section will sound in the building and nowhere else. It will continue alarming until shut off. At the other extreme, the system will not only sound a local alarm and initiate a number of local measures to promote life safety and to confine and extinguish the fire, but, will also send a large amount of useful information to a central location where firefighting resources are available. No matter what level system is present, the control panel should be located at the most convenient point of entrance for firefighting personnel so that they can quickly gain as much information as possible from the indicators on the panel. It should not be hidden away in a location difficult to reach, such as in a locked electrical closet.

A significant improvement in a basic system is for the building to be divided into zones with a separate module in the control panel for each zone. A zone will normally include a number of detectors within a fairly compact, contiguous area of a building. Firefighters reporting to the scene can tell from the panel in which zone the alarm had been initiated. A map of the building, indicating the boundaries of the individual zones, should be located in the immediate area of the control panel. Although the individual zone where the alarm was initiated will be identified on the panel, there still would have been only a single alarm which would have sounded throughout the building. If provision has been made to transmit the signal to a remote location, this type of system can directly transmit the zone information to the emergency personnel, or it can be limited to a single indication of a problem in the building. In some systems, this signal may go instead to a locally manned station, perhaps a security area manned 24 hours per day, which will in turn notify a fire safety crew.

The alarm signal may be any number of different devices, such as a bell, horn, or a recorded voice message. It should be different from any other similar signal. For example, in an academic institution there should be no confusion between the fire alarm signal and class bells. In addition, the requirements for the disabled need to be met. In the case of an alarm signal, it is necessary to make provision for individuals with a hearing handicap by means of a visual signal, usually a strobe light. ADA now requires that installed alarms provide sound levels that are at least 15 decibels above ambient sound levels, or 5 decibels above the maximum sound level for a 60 second duration with a maximum level of 120 decibels. Visual alarms must be of the xenon strobe type, clear or nominal white in color. The minimum acceptable intensity is 75 candela, with a pulse duration of a maximum of 0.2 seconds and a maximum duty cycle of 40%. The flash rate must be between 1 and 3 Hertz.

In addition to initiating an alarm in the building that may or may not also be sent to a central station, frequently the control system will be capable of initiating a number of other actions. The system can activate mechanisms to close fire doors, turn off ventilation systems, close dampers, send elevators to the ground floor and lock them out of service, and activate emergency lights for the evacuation routes.

The advent of small but increasingly powerful computers shows promise for revolutionizing automatic fire alarm systems. If enough contact points are available and with the proper interface to the computer, a multiplexing system would make it possible to identify the individual detector in alarm at a remote computer console. With enough memory, it would be possible to display a map of the building on the screen locating the alarm source and, if the information has been put into a database, list the physical conditions or the types and amounts of hazardous materials expected to be present. Under current OSHA and EPA regulations, firefighting groups must be notified of hazardous materials they may encounter at facilities in their jurisdiction. In addition, they must be provided with Material Safety Data Sheets for these materials, so that they may be informed of the risks to which they may be exposed and the appropriate emergency measures they may need to take.

The required hazardous material information is primarily being provided to the fire departments as printed material at the present. However, generic databases containing thousands of Material Safety Data Sheets recently have become available on optical disks (CD-ROM) which provide encyclopedic amounts of data. Small computers have also developed the capability of placing information on these disks so that locally specific information can be added. It should not be too long before companies selling computerized alarm systems will incorporate these most recent advances into their products and be able to provide even small organizations with extremely sophisticated information resources to aid firefighters in coping with fires in research buildings and other complex facilities.

Often fire alarm system needs can be accommodated on computer or communication systems installed for other purposes. For example, several companies sell small systems, primarily designed for energy management which are designed to accept inputs from other types of systems such as security and fire safety, and they provide software packages to support them.

The cost of laboratory facilities is so great and the progress in science is so rapid that no competitive corporate or academic research organization can afford the loss of a major facility. Even if insurance is available to cover the physical losses, how can the value of intellectual properties, perhaps irreplaceable, be determined? It takes a minimum of 2 to 3 years to construct a major facility. What would the experimental research personnel, formerly housed in the lost facility, do in the meantime? What would be the position of granting agencies that might be supporting the research? The cost of a good fire safety system, including appropriate detectors, an alarm system, and a fire suppression system, is inexpensive compared to the potential losses including, in addition to those cited, injuries to personnel.

REFERENCES

1. NFPA 11, *Low Expansion Foam and Combined Agent Systems,* National Fire Protection Association, Quincy, MA.
2. NFPA 11A, *Medium and High Expansion Foam Systems,* National Fire Protection Association, Quincy, MA.
3. NFPA 12-1985, *Carbon Dioxide Extinguishing Systems,* National Fire Protection Association, Quincy, MA.
4. NFPA 12A, *Halon 1301 Fire Extinguishing Systems,* National Fire Protection Association, Quincy, MA.
5. NFPA 11-1983, *Low Expansion Foam and Combined Agent Systems,* National Fire Protection Association, Quincy, MA.
6. NFPA 15-1985, *Water Spray Fixed Systems for Fire Protection,* National Fire Protection Association, Quincy, MA.
7. NFPA 13, *Sprinkler Systems, Installation of,* National Fire Protection Association, Quincy, MA.
8. NFPA 17, *Dry Chemical Extinguishing Systems,* National Fire Protection Association, Quincy, MA.
9. IFSTA, *Private Fire Protection and Detection,* Laughlin, J.W., Ed., International Fire Training Association, Fire Protection Publications, Oklahoma State University, Stillwater, OK, 1979.
10. OSHA, *General Industry Standards,* 29 CFR 1910, Subpart L, Section 155-165.
11. **Senecal, J.A.,** Halon replacement chemicals: perspective on the alternatives, *Fire Technology,* 28(4), 332, November, 1992.
12. **Hall, J.,** The U.S. experience with sprinklers: who has them? how well do they work?, *NFPA Journal* 87(6), 44, November/December, 1993.
13. **Zurer, P.S.,** Looming ban on production of CFC's, halons spurs switch to substitutes, *Chemical & Engineering News,* 71, 46, November 15, 1993.
14. **Cummings, R.B. and Jaeger, T.W.,** ADA sets a new standard for accessibility, *NFPA Journal,* 87(3), 43 May/June, 1993.
15. Americans With Disabilities Act, FR July 26, 1991.

3.2.4. Other Fixed Equipment

Buildings need to be equipped with two separate elevator systems—one for personnel and one for freight. In order to make spaces available to the disabled, many buildings without elevators are being retrofitted with them, and others are having former freight elevators converted to be used for passengers. Because it is often too expensive or there may be no reasonable way to install two separate elevators, many of these elevators are being used for both purposes. This is not desirable, especially in research facilities where hazardous materials are in use. Passenger elevators should not be used to transport hazardous materials, nor should passengers, other than those essential to managing the materials, use a freight elevator at the same time it is being used to move hazardous items between floors.

Both types of elevators should be equipped with a means to signal to a manned location in the event of an emergency, preferably by telephone (unfortunately, the experience with misuse of elevator telephones in some locations has not been positive). In the event of a fire in a building, elevators should automatically be sent to the ground floor (or the floor representing the normal entrance level to the building), and should be constrained to remain there unless fire or other emergency personnel override the interlock. The ADA also affects elevators. If elevators are employed to make areas of a building accessible to the disabled, other provisions must be made to evacuate them since elevators are to be made nonusable under normal procedures.

3.3. CHEMICAL STORAGE ROOMS

The OSHA requirements for inside storage rooms for chemicals are given in 29 CFR Part 1910.106(d)(4) and are based on NFPA Standard 30-1969. This latter standard has been amended since 1969, but at this time, the OSHA requirements have not been changed to reflect the later changes. The BOCA building code is based on the changed standards. The major difference in the more recent NFPA standard involves changes in the ventilation requirements, although there are minor differences elsewhere. In a few instances, additional safety precautions which are generally accepted practices have been added as recommendations in the following material.

3.3.1. Capacity

The amount of flammable and combustible liquids permitted in inside storage rooms depends upon the type of construction and whether an automatic fire suppression system is installed or not. The permissible amounts under both the OSHA and more recent NFPA standards are the same and are given in Table 3.13.

It should be noted that these limits are generous for most operations, permitting between 300 and 5000 gallons of flammable and combustible liquids overall in an inside storage room. However, NFPA Standard 30 at the current time does not permit more than 660 gallons of 1A liquids, 1375 gallons of 1B liquids, 2750 gallons of 1C liquids, and 4125 gallons of class II liquids, so these limits should not be exceeded individually. Excess storage should be avoided, regardless of the legal maximums permitted. For other materials which normally will be kept in the same storage room, the OSHA standard permits other materials to be stored in the same space, provided that they create no fire hazard to the flammable and combustible liquids. Materials which react with water must not be stored in the same room.

Table 3.14.   Storage in Inside Rooms

Fire Protection Provided	Fire Resistance (hours)	Maximum Size (ft²)	Total Allowable Quantities (gal/ft² of floor area)
Yes	2	500	10
No	2	500	4
Yes	1	150	5
No	1	150	2

Note: 1 square foot = 0.0929 square meter. 1 gallon = 3.785 liter.

3.3.2. Construction Features

In the earlier material (Sections 3.1.2.1 to 3.1.2.3) on the building code, the factors that govern the classification of the storage space as to the degree of hazard and the implications of the classification were covered in detail. However, there are minimum construction standards which are specified by OSHA. These basic requirements are

1. Inside storage rooms shall be constructed to meet the fire-resistive ratings for their use. Doors are to be approved self-closing fire doors. Windows opening on the room, exposing other parts of the building or other properties, are to be protected according to NFPA Standard 80, Standard for Fire Doors and Windows.
2. The room must be liquid-tight where the floor meets the walls. Spilled liquids are to be prevented from running into adjacent rooms by one of three methods: (a) at least a 4-inch (10.16 cm) high sill or ramp at the opening; (b) the floors can be recessed at least 4 inches; or (c) an open-grated trench can be cut in the floor within the room which drains to a safe location.
3. If the room is to be used for class I liquids, the wiring must be adequate for class I, division 2, hazardous locations. If only class II and class III liquids are to be stored in the room, the wiring need only meet standards for general use.
4. Shelves, racks, scuffboards, and floor overlay may be made of wood if it is at least 1 inch (nominal) thick.

3.3.3. Ventilation

1. Every inside storage room must be provided with ventilation, either gravity or mechanical. Either type system must be capable of six complete air changes per hour according to the OSHA standard. Instead of six air changes per hour, NFPA 30 specifies an exhaust ventilation rate of 1 cfm per square foot of floor space (0.305 m³/min/m², but not less than 150 cfm, 45.7 m³/min). For a ceiling height of 10 feet (3.05 meters), the two requirements are the same.
2. If a mechanical system is used to provide the ventilation, OSHA requires that it be controlled by a switch on the outside of the door to the room. The lights in the room are to be operated from the same switch, and if class I liquids are dispensed within the room, a pilot light must be installed adjacent to the operating switch. In order to accommodate hearing handicapped users, a strobe alarm light is recommended as well.

3. If the ventilation is provided by a gravity system, both the intake air inlets and the exhaust air outlet must be on the exterior of the building.
4. Exhaust air should be taken from a point no more than 1 foot (0.3048 meters) from the floor and exhausted directly to the roof of the building. The air intake in the room should be on the opposite side of the room from the exhaust. Since, in general, the vapors are heavier than air, this design is intended to sweep the floor clean of vapors before they accumulate and pose a hazard. The aisles of the room should be such as not to block this sweeping action. If ducts are used, they must meet requirements of NFPA 91, Standard for the Installation of Blower and Exhaust Systems, for Dust, Stock, and Vapor Removal or Conveying, and should not be used for any other purpose. It would be preferable for the air intakes to be on the upper portion of the building and upwind from the most prevalent wind direction.

3.3.4. Fire Safety

1. There shall be no smoking or open flames in a flammable or combustible material storage room. A prominent **FLAMMABLE MATERIAL STORAGE/NO SMOKING** sign should be posted on the outside of the door to the facility.
2. At least one 12 B or larger portable fire extinguisher must be located outside the door to a flammable material storage area, no more than 10 feet from the door.
3. Any fire suppression system installed in the storage room must meet the standards of 29 CFR Parts 1910.155-165.
4. There shall be at least one clear aisle at least 3 feet (0.9144 meters) wide in every flammable and combustible material storage room. No container should be more than 12 feet (3.66 meters) from an aisle. Containers of 30 gallons (113.5 liters) capacity or larger must not be stacked more than one layer high.

REFERENCES

1. NFPA 30, *Flammable and Combustible Liquids Code,* National Fire Protection Association, Quincy, MA.
2. *Prudent Practices for Handling Hazardous Chemicals in Laboratories,* National Academy Press, Washington, D.C., 1981, 218–222.
3. ANSI Z9.2, *Fundamentals Governing the Design and Operation of Local Exhaust Systems,* American National Standards Institute, Inc., New York, NY, 1979.
4. NFPA 91, *Standard for the Installation of Blower and Exhaust Systems for Dust, Stock, and Vapor Removal or Conveying,* American National Standards Institute, Inc., New York, NY.

3.4. MOVABLE EQUIPMENT

Many items of movable equipment represent special safety problems in the laboratory. In some instances it is the probability of initiating an explosion or fire that is the major concern. In other cases, the major problem may be generation of toxic fumes and in still others the equipment is inherently dangerous due to the physical injuries which improper maintenance or use may cause.

Laboratory equipment often includes electric motors, switches, relays, or other spark-producing devices. In the presence of vapors from flammable materials, a spark can initiate a fire if the concentration of the vapor is between the upper and lower flammable limits and the temperature is above the flash point at which the given vapor can be ignited. Motors used in laboratory equipment should be induction motors which are nonsparking. Series-wound motors with graphite brushes, such as those used in many home appliances, should not be used in laboratory equipment, and appliances such as hot plates, vacuum cleaners, blenders, and power tools designed for the home should not be brought into laboratories where flammable liquids are actively used. Switches and contacts for electrical controls should be located in flammable vapor-free areas wherever possible. Equipment should be purchased which is designed to minimize the possibility of flammable vapors entering internal spaces where sparks may occur or where the vapors may come into direct contact with heating elements.

Electric shock is another hazard common to many pieces of laboratory equipment. Any electrically powered item of laboratory equipment which is subject to spillage of chemicals or water or exhibits signs of excessive wear should be used carefully. All equipment should be provided with three-wire power cords (some tools may be double-insulated as an acceptable alternative), which should be replaced if the insulation is cracked or frayed. Metallic parts of the equipment should be grounded separately if necessary. Care must be taken to ensure that any ground is, in fact, a good one. An alligator clip on a water pipe is not sufficient. A poor ground connection can generate a high temperature if sufficient current passes through the high resistance contact, and instead of being a safety feature actually represents a fire hazard. The potential difference between two poorly grounded pieces of electronic can be enough to damage sensitive electronic components. Portable equipment can also be connected through ground-fault interrupters, devices which detect a diversion through an alternate path (such as a person) to the standard connecting wiring and which, upon detection, shut down the circuit within a very few milliseconds.

A large variety of devices are left on continuously or for long periods are left operating while unattended. Any device which could overheat to a degree that it could result in a fire within the facility should be equipped with redundant controls, heat sensors, or overload protection which will cause the equipment to shut off if excess heat is generated or to fail in such a way as to minimize heat generation. Although much concern about fires stems from the presence of flammable vapors, many fires start in overheated ordinary combustible materials, with flammable and combustible liquids becoming involved at a later stage.

Many items of equipment result in significant amounts of fumes being generated in laboratory operations. Many of these are typically being used outside of fume hoods. In general, these items of equipment, if used properly, do not cause fumes to be generated at levels exceeding or even approaching the allowable limits of exposure established by OSHA. However, these levels are subject to revision as more information becomes available, usually being lowered, and there are individuals of more than average sensitivity for whom even the original limits may be too high. Under these circumstances, it would appear prudent for laboratory managers to adopt as an informal policy within their facility a chemical policy similar to that used to minimize exposure to radiation. This policy, which is intended to achieve exposures As Low As Reasonably Achievable (ALARA), would appear to be a good working policy as well as a good basis for equipment design.

In the following sections, attention is directed to a number of specific items of equipment for which explicit problems may arise.

3.4.1. Refrigeration Equipment

Refrigeration units represent a hazard as an item of laboratory equipment for a number of reasons. For example, improper use of laboratory refrigerators for food to be consumed by the laboratory workers is a continuing problem but is readily solvable by firm enforcement of policies defining acceptable practices by managers. Under no circumstances should laboratory refrigerators and freezers, used for toxic chemicals and pathogenic biological agents, ever be permitted to be used for such a purpose. However, the major problem with refrigeration units is the tightly sealed space within them.

The confined space within refrigeration units permits vapors from improperly sealed containers to accumulate. In some instances, the vapors may be toxic, and an individual peering in to find the container desired has the potential to breathe in fumes which may substantially exceed acceptable safe levels. Unless the material has a distinctive or offensive odor, it may not even occur to the person using the unit that a problem may exist. It may be desirable to have more than one refrigerator in a laboratory, one of which would be designated and prominently labeled for the storage of dangerous materials only, which would encourage users to be especially careful in using it and would encourage them to make sure that everything placed in it was properly sealed. Unfortunately, both space and funds are often limited in laboratories, and unless the laboratory manager is careful to maintain a strict policy on segregation of storage, the possibility exists, as storage needs increase, that the unrated refrigerator could be improperly used. Careful training in how to seal containers placed in refrigerators and freezers and insistence that these procedures be followed should be a part of every laboratory's management program.

Beakers, flasks, and bottles covered with aluminum foil or plastic wrap are unacceptable for storage of volatile materials in a refrigeration unit. Corks and glass stoppers also may not form a good seal. Screw-cap tops with a seal inside are much better, when screwed on firmly; however, no type of top is foolproof when used in haste.

The problem of confined flammable vapors is fortunately, one for which an engineering solution has been developed and is commercially available, since the consequences of an accumulation of flammable vapors in a normal refrigeration unit are potentially life threatening. The vapors of flammable liquids may be ignited by sparks (or other heat sources), but some liquids are more easily ignited than others over a wider range of concentrations. Data for some flammable liquids are given in Table 3.14.

The materials in Table 3.14 were selected as examples because they have a wide range of concentrations in which their vapors could be ignited and flash points[*] well under temperatures found in most household refrigerators and freezers. The reason this information is important is that the interiors of household refrigerators contain a number of electrical contacts which could generate sparks to ignite the contained vapors. Among these are the light switch, temperature control, defrost heater (in "frost-free" models), and fan. Many frost-free models also have a drain which could allow the vapors to reach the space occupied by the compressor. Models are available in which all of these sparking devices have been eliminated, modified to be explosion-proof (such as the compressor), or moved to a safer location outside the refrigeration unit.

The confined vapors in a refrigerator or freezer, if ignited, can create a major explosion and fire within a facility. Anyone standing in front of an exploding unit would be in very real danger of being seriously injured or even losing their life. It is highly likely that any

---

[*]   The flash point is the lowest temperature at which a liquid gives off vapor in a high enough concentration to form an ignitable air-vapor mixture above the surface of the liquid.

Table 3.15.   Flammability Characteristics of Some Common Solvent Vapors

Chemical	Flash Point °F (°C)	Ignition Temperature °F (°C)	Flammable Limit (% by volume in air)	
			Lower	Upper
Carbon disulfide	−22 (−30)	176 (80)	1.3	50
Diethylether	−49 (−45)	320 (160)	1.9	36
Ethylene oxide	<−18 (<0)	804 (429)	3.6	100
Ethyl nitrite	−31 (−35)	194 (90)	4.0	50
Propylene oxide	−35 (−37)	840 (449)	2.8	37
Vinyl ethyl ether	−50 (−46)	395 (202)	1.7	28

containers of flammable liquids nearby not directly involved in the explosion would be broken and contribute fuel to the fire which would probably follow the explosion. The freely flowing flammable liquids could spread to other stored materials within the laboratory. In an unprotected (by sprinklers or other fire suppression system) facility or in an older structure, the final result could be the destruction of an entire building. There could also be other toxic or hazardous materials within the refrigerator or laboratory which could be spread by the incident even if the fire did not spread. In at least one instance, the refrigerator which blew up was in a very active radiochemical laboratory. An entire floor of the building had to be decontaminated at a considerable expense in time and money.

Refrigerators which have been commercially modified to be safe for the storage of flammable materials are designated as "Flammable Material Storage Units" and meet NFPA Standard 56C and Underwriters Laboratories, Inc. standards. These units are not explosion-proof. They have only had components removed which could cause sparks within the interior of the unit. Only refrigerators or freezers intended to be used in hazardous locations where a spark inside or outside the units could cause a fire or an explosion are designated as explosion proof. The electrical power wiring to this latter class must be installed in conformance with "Commercial Refrigerator/Freezer for Hazardous Locations" class 1, groups C and D code requirements. Although in theory it is possible to modify an ordinary refrigerator to be acceptable, in practice it is difficult to be sure that it has been done properly. It is strongly recommended that all laboratory refrigerator and freezer units be purchased already modified to be safe for storage of flammables, with the exception of ultra-low temperature units which operate at temperatures lower than the flash points of any commonly used flammable liquids.

The costs of "Flammable-Material Storage" refrigerators and freezers are usually two to four times higher than comparable products to be used in the home. As a result, many individuals object to a blanket policy requiring the purchase of these safer units, especially those who are not using the refrigerators in their laboratories to store solvents or other flammable liquids. There are several reasons to override these objections. The most important are based on the exceptionally long useful life of refrigerators. Only rarely do they last less than 10 years, and many continue to work well for more than 20 years. Few research programs endure for comparable periods and few individuals remain in the same position as long. Thus, although assurances can be given and signs can be placed on doors of the units forbidding the use of the refrigerator in question for the storage of flammables, there is no feasible means of guaranteeing that they will not be used at some time during the useful life of the refrigerator or freezer for the storage of flammables. Although the initial cost is high, over the total life of a unit it is an extremely inexpensive price to pay to totally eliminate a major source of fire and explosions.

Ordinary sized refrigerators and freezers, as well as combination units, are available from a number of sources. As the size of the units become larger, however, only a few suppliers offer flammable material safe units and then usually as ordinary refrigerators modified at the factory to be explosion safe for confined vapors at an additional cost.

There are legitimate reasons to make a few exceptions. If the use is for a basic departmental function which would never entail the use of flammable liquids and the department is a stable, established discipline or research field, then there is reason to accept ordinary units to avoid paying the additional costs involved. Refrigeration units to be placed permanently in isolated, normally unoccupied locations also might be candidates for quality consumer units. Refrigerators to be used only for the storage of food and beverages for the convenience of the employees should also be permitted, but they should *not* be allowed to be placed directly within the working area of a laboratory.

Large walk-in refrigerators and freezers, or cold rooms, pose an additional problem with condensation of water vapor on the equipment when electrically operated equipment is placed inside them, due to the very high humidity usually present. Care should be taken to avoid shorts and electrical shocks to personnel. All of the equipment should be well grounded and any electrical cords should be insulated with waterproof insulation. A recommended precaution would be to have all of the electrical sockets in the interior wired with ground-fault interrupters or to require that any equipment used inside must be connected through one.

REFERENCE

1. NFPA 325M-1977, *Fire Hazard Properties of Flammable Liquids, Gases, and Volatile Solids,* National Fire Protection Association, Inc., Quincy, MA.

3.4.2. Ovens

Electrically heated ovens are another device found in many laboratories for which the problem of ignition of flammable fumes may exist. The ovens are used for baking or curing materials, out-gassing, removing water from samples, drying glassware, or in some cases providing a controlled, elevated temperature for an experiment. Very few are provided with any provision for preventing any of the materials evaporated from the samples from entering the laboratory. Ovens should be designed so that any fumes generated in the interior do not have an opportunity to come into contact with the heating elements or any spark-producing control components. A single pass-through design in which air is drawn in, heated, and then exhausted is a relatively safe design as long as nothing in the oven impedes the flow of air. Most ovens intended for home kitchens are not constructed in this manner and should not be used in the laboratory environment.

Every oven should be equipped with a backup thermostat or temperature controller to either control the unit should the primary one fail or shut the oven down. If the secondary controller permits the oven to continue to operate after the primary device fails, it should provide a warning that the failure has occurred so that the researcher can make the decision as to whether or not to continue the operation. In any event, as soon as practicable and before beginning a new run, the oven should be repaired. No unit with only a single thermostat should be used for long, unattended programs.

Because most laboratory ovens do exhaust directly to the laboratory, they should not be used to heat any material from which a toxic vapor or gas would be expected to evolve, unless provisions are made to exhaust the fumes outdoors, as would be done with a fume hood. Since the exhausted gases would be warmer than the ambient air, they should rise. This is one instance in which a canopy hood placed over an oven, or at the least the oven exhaust, would be a satisfactory choice for a hood.

Ovens can be purchased which are suitable for heating materials which contain flammable liquids. One commercial model, designed to be used for small amounts of solvents, purges the interior with several complete air changes prior to turning the heat on in order to remove any residual gas which may be present. It also automatically turns off the heat if either the exhaust fan fails or the temperature rises above the maximum temperature for which the unit is designed. The door to the unit has an explosion venting latch, which allows it to blow open in an explosion. However, a recommended feature of ovens used for solvents are explosion vents on the rear of the unit, so that any explosion would be vented away from the laboratory and its occupants.

Where asbestos has been used as insulation in laboratory ovens, some concern has been voiced about the potential exposures of service personnel performing maintenance on the units. Although under normal circumstances such operations will involve minimal contact with the insulation, it is desirable to purchase units which use other insulating materials. Eventually, the ovens containing asbestos will have to be removed from service, and the asbestos insulation would represent a possibly expensive disposal problem. The oven could not simply be taken to the typical local municipal landfill.

3.4.3. Heating Baths

Heating baths are used to heat containers partially immersed in them and to maintain them at a stable temperature, on some occasions for extended periods. Heating baths should be equipped, as in the case of ovens, with redundant heat controls or automatic cutouts should the temperature regulating circuits fail. The material used in the bath may be flammable, and excessive temperatures could result in a fire.

A number of materials are used in heating baths. Water can be used up to about 180°F (82°C). Mineral oil and glycerine are used up to about 300°F (about 150°C). Paraffin is employed in the range up to about 400°F (about 200°C). These last three materials are flammable, although the NFPA rating of each is 1, on a scale of 0 to 4. Silicone oils are recommended at temperatures up to about 570°F (about 300°C). These are also moderately flammable and are more expensive than organic oils. In a 1957 tabulation of materials in use for heating baths, R. Egly listed tetracresyl silicate as an expensive material, but one which had very good characteristics. It was listed as nontoxic, noncorrosive, fire resistant, and suitable for use from near room temperatures to approximately 750°F (about 400°C).

Heating baths should be in durable, nonbreakable containers and set up on a firm support so that they will not be likely to tip over. They should not be placed near flammable and combustible material, including wood and paper which, if exposed to continuing heat over a sufficient period of time, could reach kindling temperatures, or near sources of water (particularly deluge showers) which could cause the bath liquid to splatter violently from the container. In most cases the bath temperatures are high enough to cause severe burns. If it is necessary to move the full container, it should be done while the liquid is cool, again to avoid the risk of burns.

If the container itself does not include a heating element, any immersion heater should be insulated to avoid the potential of electrical shock and should include a cut-out device if the temperature exceeds the set point. Alternatively, a second temperature sensor should be placed in the heat bath to act as a circuit breaker to cut off power to the heater if its thermostat fails. The thermostat should always be set well below the flash point of the heating liquid in use. A thermometer placed in the bath at all times it is in use is recommended to provide a visual indication of the actual temperature of the bath. Digital controllers normally provide this additional information.

REFERENCE

1. *Techniques in Modern Chemistry,* Vol. 3, Part 2, Weissberger, (Ed.) Interscience, New York, NY, 1957, p. 152.

3.4.4. Stills

Individual stills are frequently set up in laboratories to provide distilled water to the facility and are usually left running for extended periods unattended. The concern here, as with many of the other devices discussed in this section, is the possibility of over-heating with the subsequent initiation of a fire. Some units use two water sources, pre-treated water for the boiler, from which various impurities are removed, and ordinary tap water for cooling, while some use tap water for both. The still should be equipped with an automatic cutoff should it overheat, in case of either the water pressure fails or the boiler becomes dry. Both the water supply and heat should cut off if the collector bottle becomes full; if the power fails, a valve should shut the water supplies off.

3.4.5. Kjeldahl Systems

Kjeldahl units and other digester and distillation units used for nitrogen determinations and trace element analyses can be sources of potentially unacceptable fume levels if not vented properly. Larger units are constructed so several digestions can take place simultaneously, and thus substantial amounts of corrosive fumes can be generated. The fumes in such units are usually drawn through a manifold to a discharge point at one end of the unit. At this point, the fumes are either exhausted by an integral blower or drawn to an aspirator where they are diluted and condensed by the water spray and disposed of into the sanitary system. Because the blower is integral to the unit, the exhaust duct downstream from the unit would be at a positive pressure. If the duct were to corrode, fumes could leak into the surrounding spaces. The total volume of fumes disposed of into the sanitary system would normally be small enough to be well diluted in the sanitary waste stream, so the aspirator method has some advantages. An unexpected hazard associated with Kjeldahl units, reported in the August 10, 1992 issue of *Chemical & Engineering News,* was the discovery of significant amounts of mercury (20 milliliters) in the condensation tubes while the unit was being cleaned. As the individual reporting this finding, David Lewis, Chemist for the City of Lompoc, CA, noted, the possibility exists for workers using a mercury sulfate catalyst to have a long-term exposure to mercury vapor.

3.4.6. Autoclaves

Pressurized sterilizing chambers or autoclaves are used primarily in the life sciences. Glassware, instruments, gloves, liquids in bottles, biological waste, dressings, and other materials are sterilized in them by steam under pressure, typically at a pressure of a little under 2 atm, at temperatures of up to 275°F (135°C). Since they are heated pressure vessels, they should be checked periodically to ensure that the seals to the closures are in good condition, and they should be equipped with safety devices to prevent excessive temperatures and pressures. There are a number of potential problems associated with their use. The requirements for treating infectious waste, under the bloodborne pathogen standard and other federal and state regulations, will increase the amount of materials processed through steam sterilizer units, with more accidents occurring as a consequence. All users should be thoroughly trained in safe techniques and acceptable practices.

Fortunately, most autoclaves are designed so that they cannot be opened while the chamber is under pressure. However, the materials inside will still be very warm, and removing them too hastily or forgetting to wear insulating gloves would very likely cause the item being handled to be dropped. In some cases this would only cause a loss of sterility in the dropped material, but in other cases a bottle containing a liquid would probably be broken.

Liquids placed inside in sealed bottles may explode, and liquids in ordinary glass bottles instead of Pyrex containers designed for the temperatures and pressures may rupture. If the unit is set to exhaust rapidly, as might be done for instrument sterilization, boiling may take place in bottles of liquids, with a subsequent loss of liquid into the autoclave. Flammable liquids or chemicals which could become unstable at the temperatures reached in the autoclave should not be run through the sterilizing cycle.

Operating instructions and a list of good safety practices should be posted near any autoclave for ready reference.

REFERENCE

1. **B.M. Miller,** et al. (Eds.), *Laboratory Safety Practices,* American Society for Microbiology, Washington, D.C., 1986.

3.5. ANIMAL LABORATORIES—SPECIAL REQUIREMENTS*,**

Much of the requirements on such physical facilities are identical to those found in the material in Chapter 5 for microbiological laboratories. The floors, walls, casework, and equipment should be designed to be easily decontaminated. Storage facilities should be available to store medically regulated waste (current term for infectious waste), pending collection and disposal. Ventilation recommendations can be met in more than one way, either as air changes per hour or as volume per animal housed. Generally, animal facilities are provided with 10 to 20 air changes per hour. Current recommendations for human-occupied spaces have increased in recent years to 10 to 12 air changes per hour, so laboratory spaces can be designed toward a common goal. Animal holding facilities are usually designed with both temperature and humidity controls, while human spaces often do not include the latter in the design. The comfort levels for both temperature and humidity for animals are essentially the same as for humans, although certain species have

---

*   Much of this material was prepared by David M. Moore, D.V.M., for the third edition of the handbook. With the new bloodborne pathogen standard and infectious waste regulations, a small amount of additional material has been added.

** Additional information will be given as an Appendix to the end of the material dealing with animal facilities from the Centers for Disease Control Guidelines for Vertebrate Animal Bio-safety.

more stringent requirements. Typically, 30 to 70% relative humidity is desirable while temperatures of 18° to 26°C (≈64° to 79°F) will be satisfactory.

Workers who work with items possibly contaminated with diseases communicable to humans, including tissue, fluids, fecal materials, and equipment which has come into contact with any of these, should be offered appropriate immunizations, if safe effective vaccines are available. Tetanus shots are recommended for all who work with animals, while those who work with wild animals should be offered rabies vaccinations. A pre-employment medical examination is mandatory and should include medical and work histories. Periodic examinations may be desirable and should be considered. Any worker who may come into contact with human or primate tissue, blood, and fluids must receive training to meet the standards of the OSHA bloodborne pathogen standard and be offered shots for Hepatitis-B.

3.5.1. Fixed Equipment in Animal Holding Facilities

In addition to items such as fume hoods and casework, for which the requirements will be similar to those described previously in this chapter, there are some special sanitation equipment items.

3.5.1.1. Cage Washers

Most facilities have equipment for sanitization of cages and cage racks. There are three major types of mechanical cage washers:

1. Rack washer—This unit can hold one or more cage racks or racks containing cage boxes. It can have a single entrance or two entrances, allowing movement of cages/racks from a "dirty" processing area through to a "clean" area, with the areas separated by a wall. Spray arms in the unit direct water at high pressure on all sides, and the unit reaches the recommended temperature of 180°F for a minimum of 3 minutes, which is sufficient to destroy pathogenic (disease causing) microorganisms. It can handle a larger number of cages/racks than a cabinet washer and is thus less labor intensive.
2. Cabinet washer—The cabinet washer has smaller internal dimensions than the rack washer and can accommodate cages, but not racks. For larger, heavier cages (i.e., rabbit, nonhuman primate, dogs), this unit proves to be more labor intensive and more time consuming. Racks would require some other method for sanitization: steam generators, high-pressure spray units, or chemical disinfection.
3. Tunnel washer—"Shoebox" rodent cages, water bottles, cage pans, and other small equipment can be placed on this unit's conveyor belt. It is more efficient than the cabinet washer for sanitizing small items.

These units generate quite a bit of heat, and the ventilation for the cage wash area should be adjusted accordingly.

3.5.1.2. Autoclaves

Autoclaves were described earlier, but are especially important in animal facilities. Autoclaves provide support for animal surgical facilities and may be used in barrier facilities to sterilize food, bedding, water and water bottles, cages, and other equipment prior to entry into the barrier. Steam autoclaves can potentially dull sharp surgical instruments, and the heat can reduce vitamin levels in feed. Special autoclavable diets are manufactured with higher levels of vitamins to assure that appropriate levels remain after autoclaving. Ethylene oxide sterilizers are used for materials damaged by the temperatures in steam autoclaves (i.e., surgical instruments, plastic tubing and catheters, electronic devices). However, ethylene oxide gas poses a health risk for humans, and materials should be allowed to "off gas" for 24 to 48 hours before coming in contact with animals. Care should be taken in designing a gas scavenging system, and safety standard operating procedures should be established for use of the ethylene oxide sterilizer. Steam sterilization is one of the two universally accepted methods of treating medically regulated waste. The use of autoclaves for tissue and related materials is not as desirable as it might be, since the end product can still decompose and be objectionable when placed in landfills. Autoclaving is more practical for labware, syringes, and other "sharps." However, the current climate is to render these types of items unrecognizable due to adverse publicity engendered by hospital waste washing up on beaches in recent years.

3.5.1.3. Incinerator

Most facilities dispose of solid wastes and animal carcasses by incineration. This is the second universally accepted method of treating medically regulated wastes. States are adopting very stringent operating conditions on incinerators employed for this purpose. Mixed waste, which contains animal material, chemicals meeting EPA regulatory criteria, or radioactive materials may or may not be incinerated according to the characteristics of the waste. The reader is referred to the section in Chapter 4, written by Mr. Lawrence Doucet, P.E., on the various alternatives which currently exist for processing medically regulated waste.

REFERENCE

1. NCI, *Chemical Carcinogen Hazards in Animal Research Facilities,* Office of Biohazard Safety, National Cancer Institute, Bethesda, MD, 1979, 15–17.

3.5.2. Equipment for Animal Laboratories and Holding Areas

Caging for small and large laboratory animal species may be either fixed or movable. Sanitization of fixed caging and the room environment is less easily accomplished than is sanitization of movable racks and cages. Fixed caging might also provide safe haven for vermin and reduces the flexibility of use of that holding room.

Animal caging is designed for the convenience of the investigator and the husbandry staff, but more importantly, for the comfort, safety, and well-being of the animal. *The Guide for the Care and Use of Laboratory Animals* provides recommendations for caging materials and cage sizes for a variety of species. Adherence to these recommendations will also assure compliance with the Federal Animal Welfare Act requirements regarding cage sizes. A description of various caging systems is given by Hessler and Moreland (1984).

REFERENCES

1. **Hessler, J.R. and Moreland A.F.,** Design and management of animal facilities, in *Laboratory Animal Medicine,* Fox, J.G. et al. (Eds.), Academic Press, Orlando, 1984, pp. 517–521.
2. ILAR. *The Guide for the Care and Use of Laboratory Animals.* NIH Publication 85-23, NIH Bethesda, MD, 1985.

APPENDIX: VERTEBRATE ANIMAL SAFETY BIOSAFETY LEVEL CRITERIA

Most of the following material is taken directly from the referenced material. It may be abridged or modified very slightly to correct minor grammatical or typographical errors.

Introduction

Management must provide facilities and staff and adopt procedures that reasonably assure appropriate levels of environmental quality, safety, and care for work in which experimental animals are used. Animal care facilities, practices, and quality of animal care must meet all applicable standards, such as provided in *Guide for the Care and Use of Laboratory Animals,* (HHS Publ. No. 86-23, Rev. 1985), and Laboratory Animal Welfare Regulations, 9 CFR, Subchapter A, Parts 1, 2, and 3. Facilities for laboratory animals should be physically separate from other activities such as animal production and quarantine, clinical laboratories, and, especially, facilities that provide patient care. The facilities should be such as to aid cleaning and housekeeping. Floor drains should be installed only on the basis of clearly defined needs. A "clean hall/dirty hall" layout is very useful in reducing cross-contamination. If floor drains are installed, the drain trap should always contain water. Following are specific facility standards and practices recommended for use when vertebrate animals are used.

Animal Biosafety Level 1

A. Standard Practices

1. Doors to animal rooms open inward, are self-closing, and are kept closed when experimental animals are present.
2. Work surfaces are decontaminated after use or after any spill of viable materials.
3. Eating, drinking, smoking, and storing food for human use are not permitted in animal rooms.
4. Personnel must wash their hands after handling cultures and animals and before leaving the animal rooms.
5. All procedures are carefully performed to minimize the creation of aerosols.
6. An insect and rodent control program is in effect.

B. Special Practices

1. Bedding materials from animal cages are removed in such a manner as to minimize the creation of aerosols and disposed of in compliance with applicable institutional or local requirements.

2. Cages are washed manually or in a cage washer. Temperature of the final rinse water in a mechanical washer should be 180°F.
3. The wearing of laboratory coats, gowns, or uniforms in the animal room is recommended. It is further recommended that laboratory coats worn in the animal room not be worn in other areas.

C. Containment Equipment

Special containment equipment is not required for animals infected with agents assigned to Biosafety Level 1.

D. Animal Facilities

1. The animal facility is designed and constructed to facilitate cleaning and housekeeping.
2. A hand-washing sink is available in the animal facility.
3. If the animal facility has windows that open, they are fitted with fly screens.
4. It is recommended, but not required, that the direction of airflow in the animal facility is inward and that exhaust air is discharged to the outside without being recirculated to other rooms.

Animal Biosafety Level 2

A. Standard Practices

1. Doors to animal rooms open inward, are self-closing, and are kept closed when infected animals are present.
2. Work surfaces are decontaminated after use or spills of viable materials.
3. Eating, drinking, smoking, and storing of food for human consumption are not permitted in animal rooms.
4. Personnel must wash their hands after handling cultures and animals and before leaving the animal room.
5. All procedures are carefully performed to minimize the creation of aerosols.
6. An insect and rodent control program is in effect.

B. Special Practices

1. Cages are decontaminated, preferably by autoclaving, before they are cleaned and washed.
2. Surgical-type masks are worn by all personnel entering animal rooms housing nonhuman primates.
3. Laboratory coats, gowns, or uniforms are worn while in the animal room. This protective clothing is removed before leaving the animal facility.
4. The laboratory or animal facility director limits access to the animal room to personnel who have been advised of the potential hazards and who need to enter the room for program or service purposes when work is in progress. In general, persons who may be at increased risk of acquiring infection or for whom infection might be unusually hazardous are not allowed in the animal room.

5. The laboratory or animal facility director establishes policies and procedures whereby only persons who have been advised of the potential hazard and meet any specific requirements (e.g., for immunization) may enter the animal room.

6. When the infectious agent(s) in use in the animal room requires special entry provisions (e.g., vaccination), a hazard warning sign incorporating the universal biohazard symbol is posted on the access door to the animal room. The hazard warning sign identifies the infectious agent, lists the name and telephone number of the animal facility supervisor or other responsible person(s), and indicates the special requirements for entering the animal room.

7. Special care is taken to avoid skin contamination with infectious materials; gloves should be worn when handling infected animals and when skin contact with infectious materials is unavoidable.

8. All wastes from the animal room are appropriately decontaminated - preferably by autoclaving—before disposal. Infected animal carcasses are incinerated after being transported from the animal room in leakproof, covered containers.

9. Hypodermic needles and syringes are used only for the parenteral injection or aspiration of fluids from laboratory animals and diaphragm bottles. Only needle-locking syringes or disposable needle syringe units (i.e., the needle is integral to the syringe) are used for the injection or aspiration of infectious fluids. Needles should not be bent, sheared, replaced in the sheath or guard, or removed from the syringe following use. The needle and syringe should be promptly placed in a puncture-resistant container and decontaminated, preferably by autoclaving, before discard or reuse.

10. If floor drains are provided, the drain traps are always filled with water or a suitable disinfectant.

11. When appropriate, considering the agents handled, baseline serum samples from animal care and other at-risk personnel are collected and stored. Additional serum samples may be collected periodically, depending on the agents handled or the function of the facility.

C. Containment Equipment

Biological safety cabinets, other physical containment devices, and/or personal protective devices (e.g., respirators, face shields) are used whenever procedures with a high potential for creating aerosols are conducted. These procedures include necropsy of infected animals, harvesting of infected tissues or fluids from animals or eggs, intranasal inoculation of animals, and manipulations of high concentrations or large volumes of infectious materials.

D. Animal Facilities

1. The animal facility is designed and constructed to facilitate cleaning and housekeeping.

2. A hand-washing sink is available in the room where infected animals are housed.

3. If the animal facility has windows that open, they are fitted with fly screens.

4. It is recommended, but not required, that the direction of airflow in the animal facility is inward and that the exhaust air is discharged to the outside without being recirculated to other rooms.

5. An autoclave which can be used for decontaminating infectious laboratory waste is available in the building with the animal facility.

Animal Biosafety Level 3

A. Standard Practices

1. Doors to animal rooms open inward, are self-closing, and are kept closed when work with infected animals is in progress.
2. Work surfaces are decontaminated after use or spills of viable materials.
3. Eating, drinking, smoking, and storing of food for human consumption are not permitted in animal rooms.
4. Personnel must wash their hands after handling cultures and animals and before leaving the animal room.
5. All procedures are carefully performed to minimize the creation of aerosols.
6. An insect and rodent control program is in effect.

B. Special Practices

1. Cages are autoclaved before bedding is removed and before they are cleaned and washed.
2. Surgical-type masks or other respiratory protection devices (e.g., respirators) are worn by personnel entering rooms housing animals infected with agents assigned to Biosafety Level 3.
3. Wrap-around or solid front gowns or uniforms are worn by personnel entering the animal room. Front-button laboratory coats are unsuitable. Protective gowns must remain in the animal room and must be decontaminated before being laundered.
4. The laboratory or other responsible person restricts access to the animal room to personnel who have been advised of the potential hazard and who need to enter the room for program or service purposes when infected animals are present. In general, persons who may be at increased risk of acquiring infection or for whom infection might be unusually hazardous are not allowed in the animal room.
5. The laboratory or other responsible person establishes policies and procedures whereby only persons who have been advised of the potential hazards and meet any specific requirements (e.g., for immunization) may enter the animal room.
6. Hazard warning signs, incorporating the universal biohazard symbol, are posted on access doors to animal rooms containing animals infected with agents assigned to Biosafety Level 3. The hazard warning sign should identify the agent(s) in use, list the name and telephone number of the animal room supervisor or other responsible person(s), and indicate any special conditions of entry into the animal room (e.g., the need for immunizations or respirators).
7. Personnel must wear gloves when handling infected animals. Gloves are removed aseptically and autoclaved with other animal room wastes before being disposed of or reused.
8. All wastes from the animal room are autoclaved before disposal. All animal carcasses are incinerated. Dead animals are transported from the animal room to the incinerator in leakproof, covered containers.
9. Hypodermic needles and syringes are used only for gavage, parenteral injection, or aspiration of fluids from laboratory animals and diaphragm bottles. Only needle-locking syringes or disposable needle syringe units (i.e., the needle is integral to the syringe) are used. Needles should not be bent, sheared, replaced in the sheath or guard, or removed from the syringe following use. The needle and syringe should be promptly placed in a puncture-resistant container and decontaminated, preferably

by autoclaving, before discard or reuse. Whenever possible, cannulas should be used instead of sharp needles (e.g., gavage).

10. If floor drains are provided, the drain traps are always filled with water or a suitable disinfectant.

11. If vacuum lines are provided, they are protected with HEPA filters and liquid disinfectant traps.

12. Boots, shoe covers, or other protective footwear and disinfectant footbaths are available and used when indicated.

C. Containment Equipment

1. Personal protective clothing and equipment and/or other physical containment devices are used for all procedures and manipulations of infectious materials or infected animals.

2. The risk of infectious aerosols from infected animals or their bedding can be reduced if animals are housed in partial containment caging systems, such as open cages placed in ventilated enclosures (e.g., laminar flow cabinets), solid wall and bottom cages covered by filter bonnets, or other equivalent primary containment systems.

D. Animal Facilities

1. The animal facility is designed and constructed to facilitate cleaning and housekeeping and is separated from areas which are open to unrestricted personnel traffic within the building. Passage through two sets of doors is the basic requirement for entry into the animal room from access corridors or other contiguous areas. Physical separation of the animal room from access corridors or other activities may also be provided by a double-door clothes changing room (showers may be included), air lock, or other access facility which requires passage through two sets of doors before entering the animal room.

2. The interior surfaces of walls, floors, and ceilings are water resistant so that they may be easily cleaned. Penetrations of these surfaces are sealed or capable of being sealed to facilitate fumigation or space decontamination.

3. A foot, elbow, or automatically operated hand-washing sink is provided near each animal room exit door.

4. Windows in the animal room are closed and sealed.

5. Animal room doors are self-closing and are kept closed when infected animals are present.

6. An autoclave for decontaminating wastes is available, preferably within the animal room. Materials to be autoclaved outside the animal room are transported in a covered, leakproof container.

7. An exhaust air ventilation system is provided. This system creates directional airflow that draws air into the animal room through the entry area. The building exhaust can be used for this purpose if the exhaust air is not recirculated to any other area of the building, is discharged to the outside, and is dispersed away from occupied areas and air intakes. Personnel must verify that the direction of the airflow into the animal room is proper. The exhaust air from the animal room that does not pass through biological safety cabinets or other primary containment equipment can be discharged to the outside without being filtered or otherwise treated.

8. The HEPA filtered exhaust from class I or class II biological safety cabinets or other primary containment devices is discharged directly to the outside or through the

building exhaust system. Exhaust air from these primary containment devices may be recirculated within the animal room if the cabinet is tested and certified at least every 12 months. If the HEPA filtered exhaust air from class I or class II biological safety cabinets is discharged to the outside through the building exhaust system, it is connected to this system in a manner (e.g., thimble unit connection) that avoids any interference with the air balance of the cabinets or building exhaust system.

Animal Biosafety Level 4

A. Standard Practices

1. Doors to animal rooms open inward and are self-closing, and are kept closed when work with infected animals is in progress.
2. Work surfaces are decontaminated after use or spills of viable materials.
3. Eating, drinking, smoking, and storing of food for human consumption are not permitted in animal rooms.
4. All procedures are carefully performed to minimize the creation of aerosols.
5. An insect and rodent control program is in effect.
6. Cages are autoclaved before bedding is removed and before they are cleaned and washed.

B. Special Practices

1. Only persons whose entry into the facility or individual animal rooms is required for program or support purposes are authorized to enter. Persons who may be at increased risk of acquiring infection or for whom infection might be unusually hazardous are not allowed in the animal facility. Persons at increased risk may include children, pregnant women, and persons who are immunodeficient or immunosuppressed. The supervisor has the final responsibility for assessing each circumstance and determining who may enter or work in the laboratory. Access to the facility is limited by secure, locked doors; accessibility is controlled by the animal facility supervisor, biohazards control officer, or other person(s) responsible for the physical security of the facility. Before entering, persons are advised of the potential biohazards and instructed as to appropriate safeguards. Personnel comply with the instructions and all other applicable entry and exit precautions. Practical and effective protocols for emergency situations are developed.
2. Personnel enter and leave the facility only through the clothing change and shower rooms. Personnel shower each time they leave the facility. Head covers are provided to personnel who do not wash their hair during the exit shower. Except in an emergency, personnel do not enter or leave the facility through the air locks.
3. Street clothing is removed in the outer clothing change room and kept there. Complete laboratory clothing, including undergarments, pants and shirts or jumpsuits, shoes, and gloves, are provided and used by all personnel entering the facility. When exiting, personnel remove laboratory clothing and store it in a locker or hamper in the inner changing room before entering the shower area.
4. When infectious materials or infected animals are present in the animal rooms, a hazard warning sign incorporating the universal biohazard symbol is posted on all access doors. The sign identifies the agent in use, lists the name and telephone number of the animal facility supervisor or other responsible person(s), and indicates

any special conditions of entry into the animal room (e.g., the need for immunizations and respirators).

5. Supplies and materials to be taken into the facility enter by way of the double-door autoclave, fumigation chamber, or air lock, which is appropriately decontaminated between each use. After securing the outer doors, personnel inside the facility retrieve the materials by opening the interior doors of the autoclave, fumigation chamber, or air lock. This inner door is secured after materials are brought into the facility.

6. Materials (e.g., plants, animals, clothing) not related to the experiment are not permitted in the facility.

7. Hypodermic needles and syringes are used only for gavage, parenteral injection, or aspiration of fluids from laboratory animals and diaphragm bottles. Only needle-locking syringes or disposable needle syringe units (i.e., the needle is an integral part of the unit) are used. Needles should not be bent, sheared, replaced in the sheath or guard, or removed from the syringe following use. The needle and syringe should be promptly placed in a puncture-resistant container and decontaminated, preferably by autoclaving, before discard or reuse. Whenever possible, cannulas should be used instead of sharp needles (e.g., gavage).

8. A system is developed and is operational for the reporting of animal facility accidents and exposures, employee absenteeism, and for the medical surveillance of potential laboratory-associated illnesses. An essential adjunct to such a reporting-surveillance system is the availability of a facility for the quarantine, isolation, and medical care of persons with potential or known laboratory-associated illnesses.

9. Baseline serum samples are collected and stored for all laboratory and other at-risk personnel. Additional serum specimens may be collected periodically, depending on the agents handled or the function of the laboratory.

C. Containment Equipment

Laboratory animals, infected with agents assigned to Biosafety Level 4, are housed in the class III biological safety cabinet or in partial containment caging systems (e.g., open cages placed in ventilated enclosures, solid wall and bottom cages covered with filter bonnets, or other equivalent primary containment systems), in specially designed areas in which all personnel are required to wear one-piece positive pressure suits ventilated with a life support system. Animal work with viral agents that require Biosafety Level 4 secondary containment, and for which highly effective vaccines are available and used, may be conducted with partial containment cages and without the one-piece positive pressure personnel suit if the facility has been decontaminated, if no concurrent experiments are being done in the facility which require Biosafety Level 4 primary and secondary containment, and if all other standard and special practices are followed.

D. Animal Facility

1. The animal rooms are located in a separate building or in a clearly demarcated and isolated zone within a building. Outer and inner changing rooms separated by a shower are provided for personnel entering and leaving the facility. A double-door autoclave, fumigation chamber, or ventilated air lock is provided for passage of materials, supplies, or equipment which are not brought into the facility through the changing room.

2. Walls, floors, and ceilings of the facility are constructed to form a sealed internal shell which facilitates fumigation and is animal- and insect-proof. The internal surfaces of this shell are resistant to liquids and chemicals, thus facilitating cleaning and decontamination of the area. All penetrations in these structures are sealed.

3. Internal facility appurtenances, e.g., light fixtures, air ducts, and utility pipes, are arranged to minimize the horizontal surface area on which dust can settle.

4. A foot, elbow, or automatically operated hand-washing sink is provided near the door of each animal room within the facility.

5. If there is a central vacuum system, it does not serve areas outside the facility. The vacuum system has in-line HEPA filters placed as near as is practical to each use point or service cock. Filters are installed to permit in-place decontamination and replacement. Other liquid and gas services for the facility are protected by devices that prevent backflow.

6. External animal facility doors are self-closing and self-locking.

7. Any windows must be resistant to breakage and sealed.

8. A double-door autoclave is provided for decontaminating materials that leave the facility. The autoclave door which opens to the area external to the facility is automatically controlled so that it can be opened only after the autoclave "sterilization" cycle is completed.

9. A pass-through dunk tank, fumigation chamber, or an equivalent decontamination method is provided so that materials and equipment that cannot be decontaminated in the autoclave can be safely removed from the facility.

10. Liquid effluents from laboratory sinks, cabinets, floors, and autoclave chambers are decontaminated by heat treatment before being discharged. Liquid wastes from shower rooms and toilets may be decontaminated with chemical disinfectants or by heat in the liquid waste decontamination system. The procedure used for heat decontamination of liquid wastes must be evaluated mechanically and biologically by using a recording thermometer and an indicator microorganism with a defined heat susceptibility pattern. If liquid wastes from the shower rooms are decontaminated with chemical disinfectants, the chemicals used must have documented efficiency against the target or indicator microorganisms.

11. An individual supply and exhaust air ventilation system is provided. The system maintains pressure differentials, and directional airflow is required to assure inflow from areas outside of the facility toward areas of highest potential risk within the facility. Manometers are provided to sense pressure differentials between adjacent areas that are maintained at different pressure levels. The manometers sound an alarm when a system malfunctions. The supply and exhaust airflow is interlocked to assure inward (or zero) airflow at all times.

12. The exhaust air from the facility is filtered by HEPA filters and discharged to the outside so that it is dispersed away from occupied buildings and air intakes. Within the facility, the filters are located as near to the laboratories as is practical in order to reduce the length of potentially contaminated air ducts. The filter chambers are designed to allow *in situ* decontamination before filters are removed and to facilitate certification testing after they are replaced. Coarse filters are provided for treatment of air supplied to the facility in order to increase the lifetime of the HEPA filters.

13. The treated exhaust air from class I or class II biological safety cabinets can be discharged into the animal room environment or to the outside through the facility exhaust system. If exhaust air from class I or class II biological safety cabinets is discharged into the animal room, the cabinets are tested and certified at 6-month intervals. The treated exhaust air from class III biological safety cabinets is discharged without recirculation via the facility exhaust air system. If the treated exhaust air from any of these cabinets is discharged through the facility exhaust air

system, it is connected to this system in a manner that avoids any interference with the air balance, the cabinets, or the facility exhaust air system.

14. A specially designed suit area may be provided in the facility. Personnel who enter this area wear a one-piece positive pressure suit that is ventilated by a life support system. The life support system is provided with alarms and emergency backup breathing air tanks. Entry to this area is through an air lock fitted with airtight doors. A chemical shower is provided to decontaminate the surface of the suit before the worker leaves the area. The exhaust air from the area in which the suit is used is filtered by two sets of HEPA filters installed in series. A duplicate filtration unit and exhaust fan are provided. An automatically starting emergency power source is provided. The air pressure within the suit area is lower than that of any adjacent area. Emergency lighting and communication systems are provided. All penetrations into the inner shell of the suit area are sealed. A double-door autoclave is provided for decontaminating waste materials to be removed from the suit area.

Additional information on microbiological laboratories can be found in Chapter 5, Section 5.5.9. Information on the infectious agents can be found in Chapter 5, Section 5.5.10.

REFERENCE

1. **Richardson, J.H. and Barkley, W.E., Eds.,** *Biosafety in Microbiological and Biomedical Laboratories,* 2nd ed., U.S. Department of Health and Human Services, Public Health Service Centers for Disease Control and National Institutes of Health, U.S. Government Printing Office, Washington, D.C., May 1988.

CHAPTER 4

Laboratory Operations

4.1. GENERAL CONSIDERATIONS

The attitude of laboratory personnel toward safety is the most important factor affecting the safe conduct of research. It is more important than the quality of the equipment, regulations, managerial policies, and the inherent risks associated with the materials being employed and the operations being conducted. If the attitude of everyone in the laboratory is positive, and this attitude is clearly supported by either the corporation or the academic institution, then it is highly probable that a strong effort will be made for the research program to be conducted safely. Conscientious individuals will try to follow the standards of behavior established by their organization to ensure for themselves that their operations are as safe as possible, and will attempt to comply with regulations and policies which have been established for their protection. On the other hand, no matter how strict management policies are and how many regulations have been established, individuals with an attitude that safety concerns are not important and that nothing will ever happen to them will manage, somehow, to circumvent any inconvenient restrictions. Occupational Safety and Health Administration (OSHA), in its new laboratory standard, recognized the importance of the local laboratory manager by placing the responsibility for developing and implementing a sound safety plan squarely on this individual.

Rarely do you have as black and white a situation as implied by the two extremes in the preceding paragraph. No one is so careful that they avoid taking any risks, nor is any one totally unconcerned about their own safety. The goal should be to avoid taking unreasonable risks, and it is the responsibility of laboratory managers to establish, by policy and example, reasonable standards of conduct to ensure that this goal is met.

Generally, a safe laboratory operation is usually a well-run operation. For example, labeling of secondary containers of reagents is not only a good safety practice to avoid accidental reactions leading to injuries, but it serves to prevent errors which could negatively affect the research program as well.

The failure of a laboratory manager to establish the right atmosphere of safety and to enforce established safety and health policies can render the manager vulnerable to litigation on the part of an injured employee, especially if it can be shown that the failure was due to willful negligence. The OSHA Laboratory Standard does require a written hygiene plan for the laboratory facility, but if written policies were not available, if a reasonable individual can be shown to have been likely to have anticipated a problem, and if due care to protect an employee under the individual's supervision was not exercised, a civil court suit against the manager by the injured party could very well be successful. On the other hand, an employee (at least in an academic institution), who deliberately does not comply with safety precautions of which they have been informed and which are normally expected to be followed, may weaken his case due to contributory negligence to the extent that the suit would not succeed or the award be substantially diminished. In the corporate world, there are workman's compensation laws that govern situations in which an employee is injured and usually provide for compensation to the employee regardless of who is at fault, although there are differences in coverage depending upon many factors in the different states. The whole concept of liability is constantly being modified by court actions. However, for financial as well as ethical reasons, the prudent manager or employer should be sure that the research programs for which the manager is responsible are conducted according to good safety practices, as defined by laws and regulations, corporate and institutional policies, and reasonableness.

It is symbolic of our society that this chapter, intended to provide guidelines to assist in making laboratory operations safer, should start with such a strong legal tone. Formal safety standards have been established because of concerns for the rights of individuals and society, due to abuses by the very small minority that may place results ahead of the well-being of the persons involved. Individuals are no longer willing to accept what they believe to be excessive risks on behalf of their employer and are willing to go to court to protect themselves, to the extent that this prerogative is at risk of being abused. However, even without the need for laws and regulations, such a chapter in a book on laboratory safety would still be needed to provide guidelines to research personnel on how to avoid or minimize the risks associated with the conduct of research.

Much has been made of the professional expertise, experience, and judgment of scientists which should allow them to be the best judge of the safety program needed in their research. In chemistry laboratories in the academic world, however, where competent, enlightened scientists should be found, it has been estimated that the accident rate is 10 to 50 times higher than that in industrial laboratories. The broad range in the estimate is attributed to the reluctance of academic personnel, particularly students, to report accidents. The disparity between the two situations may be explained by the greater likelihood in industry that scientists will be required to do a careful hazard analysis and follow strict safety precautions. Expertise is often confined to the scientific object of the research program. Very few scientists have taken formal courses in safety, health, and toxicology. Most of the relevant safety articles are published in journals devoted to topics outside of their major field of interest. They are likely to have no better judgment or common sense, on average, than any comparably well-educated and intelligent group. They may, in fact, because of the intensity of their interest in a very narrow field, have only a limited awareness of information extraneous to those interests which would assist them in making safety decisions. In the academic area, many profess to be concerned that academic freedom could be abridged by rules imposed from the outside. Academic freedom, however, should not be confused with issues governing the health and safety of individuals and the environment transcend this desirable concept.

There are legitimate concerns that research laboratories may become overregulated by too-specific a set of rules, since they do not fit the standard mold for which the original OSHA and other regulatory standards were designed. Instead of working with a few chemicals, a single laboratory may work with hundreds over the course of time, often for limited periods. Safety and health information may be extremely limited or nonexistent for newly synthesized substances or for many of the materials with which a scientific investigator may work. In general, research laboratory safety and health policies should not be regulated on a chemical by chemical basis except for specific, known serious risks, but this does not mean that otherwise there should be no safety rules. Health and safety programs should be based on well-defined general policies, sufficiently broad in scope, conservatively designed to encompass any reasonable hazard to laboratory personnel. They should be administered uniformly as institutional or corporate policies, tempered by local circumstances, to assure that all laboratory workers, including students, are equitably treated.

4.2. OSHA LABORATORY SAFETY STANDARD

The OSHA Laboratory Safety Standard, 29 CFR Part 1910.1450, addresses the issue of local responsibility by requiring that each laboratory develop an individual chemical hygiene plan as part of an overall organizational plan, so individuals responsible for the laboratories must take time to consider the safety factors applicable to their work. It must be written to ensure that it is available to all the employees so documentation will exist that the effort has been made. The new standard is a performance plan, superseding the General Industry Standards for working with chemicals with a few exceptions, which reduces the number of explicit requirements to a very few. It also replaces, for laboratory operations, the Hazard Communication Standard, 29 CFR Part 1910.1200. This second standard addresses many of the same issues. The details of many of the topics found in the following sections, such as a discussion of the contents of Material Safety Data Sheets, definitions of toxic, acutely toxic, etc., are given in later sections of this chapter in order that the general provisions of the Laboratory Safety Standard not be obscured at this point by a profusion of details.

The entire Laboratory Safety Standard, as published in the *Federal Register,* is only about nine pages long. Although it is a performance standard, with few explicit requirements, it does not relieve the laboratory manager of any safety responsibility. It simply leaves up to that individual, supported by the organization, the method for creating a safety program at least as effective for the laboratory's employees as would have the General Industry Standard. The next several sections will deal with the requirements of the OSHA *Occupational Exposure to Hazardous Chemicals in Laboratories* standard, to use its official title. Information which facilitates compliance with these requirements represents the bulk of the first through fourth chapters of this book.

4.2.1. The Chemical Laboratory

The standard applies only to laboratory use of chemicals and their hazards. The definition of hazard is very broad—"a hazardous chemical means one for which there is statistically significant evidence based on at least one study conducted in accordance with established scientific principles that acute or chronic heath effects may occur in exposed employees. The term 'health hazard' includes chemicals which are carcinogens, toxic or highly toxic agents, reproductive toxins, irritants, corrosives, sensitizers, hepatoxins, nephrotoxins, agents which act on the hematopoietic systems, and agents which damage the lungs, skin, eyes or mucous membranes." The standard also mentions physical hazards for materials that are flammable, combustible, compressed gases, explosives, oxidizers, organic peroxides, pyrophoric, reactive or unstable, or water reactive. Not all uses of chemicals with these properties are covered by the standard but only those uses which occur in a "laboratory" on a "laboratory scale." Note that the list of hazardous properties does not include radioactive, ionizing and nonionizing radiation, or contagious. Operations involving these types of hazards are covered under other standards or regulated by other agencies. The definitions are somewhat circular but the standard is clearly intended to exclude workplaces where the intent is to produce commercial quantities of a substance or where procedures are part of a production process or which simulate a production process. A laboratory is where small quantities of hazardous chemicals are used on a nonproduction basis. Laboratory-scale operations are those in which containers used in the work are designed to be safely and easily manipulated by one person. Also, a laboratory uses a variety of chemicals and procedures. The scale is such that standard laboratory practices and equipment can be used to minimize the exposure to the chemical hazards. The utilization of chemicals with similar hazardous properties in a nonlaboratory environment falls under the OSHA hazard communication standard.

4.2.2. Chemical Hygiene Plan

A key component of the OSHA standard is the Chemical Hygiene Plan (CHP). This is an explicit requirement for laboratory activities that conform to the definitions given in the preceding section. The facility must develop and carry out a written CHP which satisfies several criteria. The first three are generalizations but are nevertheless essential. It is not required in these three sections to define how one is to accomplish them.

1. Capable of protecting employees from health hazards associated with hazardous chemicals in the laboratory.
2. Capable of keeping exposure levels below the Permissible Exposure Levels (PELs) as listed in the General Industry Standards, 29 CFR 1910, Subpart Z.
3. The CHP shall be readily available to employees, employee representatives, and on request to OSHA.

The remaining elements of the plan are more much more explicit in their requirements. The standard states "The Chemical Hygiene Plan shall include each of the following elements and shall indicate specific measures that the employer will take to ensure laboratory employee protection."

4. Standard operating procedures to be followed when working with hazardous chemicals.
5. Criteria the employer will use to select and implement measures to reduce employee exposures. This covers engineering controls, personal protective equipment, and hygiene practices. Control measures to reduce exposures to extremely hazardous chemicals are considered especially important.
6. Fume hoods and other protective equipment must be functioning properly and a program must exist to ensure that this is so.
7. Employee information and training must be provided.
8. Defining a program to determine the need for and procedures for a pre-initiation approval process for some operations.
9. Provisions for medical consultation and medical surveillance for employees when conditions exist in which exposures in excess of the PELs or action levels may have occurred or may routinely occur.
10. Designation of personnel responsible for implementation of the CHP, to include designation of a chemical hygiene officer (CHO) and, if appropriate, a chemical hygiene committee. Most organizations with a variety of laboratories would normally choose to form such a committee.
11. Special provisions for additional protection for work with particularly hazardous materials such as carcinogens, reproductive toxins, and acutely toxic substances.

If the scientific worker, for whom this handbook is intended, follows the recommendations in this handbook, the requirements to meet the desired outcome of the standard will be met, but a written plan is required. The next section will define what must be covered by the plan to meet the 11 requirements listed above. The topics will not be covered in the order in the list.

4.2.2.1. Goals

The introduction to the plan should succinctly state that the organization, for the specific laboratory plan, is committed to providing a program that reduces exposure of employees to hazardous chemicals to below acceptable limits by (1) Providing them with adequate facilities for their work; (2) provision of appropriate engineering controls or, if that is not feasible for good reasons, with personal protective equipment; (3) providing them, in a timely manner, with appropriate training in procedures which they are to follow, access to information about the chemicals with which they are working, the risks associated with the chemicals, how to recognize hazards which may arise, and emergency responses; (4) providing medical consultation and surveillance as needed; (5) providing ready access to the plan; and (6) monitoring the continuing efficacy of the plan.

4.2.2.2. Organization

The organization responsible for implementation of the plan, including key individuals by title, should be identified, along with a brief description of the responsibility assigned to each. An organizational chart should be provided with the following positions (or groups) identified:

A. The senior person in the organization who is charged with the overall responsibility for safety and health programs in the organization. This position should be at a sufficiently high level to ensure that the program receives adequate support.

B. The organization under the executive authority charged with actual implementation of the plan. Normally this would consist of the environmental health and safety department and the chemical hygiene committee.

C. The CHO for the organization. This person could be the head of the health and safety department or the chairperson of the chemical hygiene committee. However, neither of these persons would normally be able to devote full time to this work and it is a critical, full-time position. The responsibility may be delegated to a another person, most probably in the health and safety department. The chemical hygiene committee should function to define policies and provide oversight of the program, while the health and safety staff should provide the daily operational support. The duties of the CHO should include:

1. Assist the individual laboratory managers to develop their own chemical hygiene programs. They should not be, and indeed are not likely to be, sufficiently familiar with the operations of individual laboratories to be expected to write the plans for specific laboratories. They should provide a template or format for the persons locally responsible for a specific facility.

2. They should develop a "train-the-trainer" program to assist the local managers in providing the appropriate training for their personnel.

3. The CHO should develop a CHP covering the entire organization, containing basic policies for chemical procurement, storage, handling, disposal, facility standards, basic training, availability of Material Safety Data Sheets and other chemical information, personal protective equipment guidelines, emergency planning for the organization, and auditing and inspection protocols.

4. They should conduct, or have done under their supervision, laboratory inspections of equipment, specifically including fume hoods and other fixed safety equipment, maintenance and housekeeping, chemical storage, and compliance with the organization- and laboratory-specific safety plans.

5. They should see that a medical consultation and surveillance program is available to the employees in the event of overexposure conditions and conduct environmental monitoring as required to support this program.

D. The local laboratory management line of authority. This could be one or more persons, dependent upon the size of the operation, with one individual designated as the senior person to whom responsibility ultimately devolves. The latter individual is responsible for seeing that the facility develop a CHP for that facility. A recommended approach would be to make the laboratory plan a second part of a document of which the organization's CHP would be the first. This would serve two purposes: every employee would have access to the policies of the organization, and would eliminate repetitious and possibly conflicting interpretations of these broad, basic policy areas. The laboratory management might choose to designate a laboratory hygiene officer, if the number of employees is large enough, to perform some of the following responsibilities and to liaise with the organizational CHO. Regardless of how it is done, the local laboratory management has the responsibility to:

1. See that the physical facilities are adequate and in good working order.
2. See that maintenance and housekeeping are satisfactory.
3. Develop and implement safe standard operating procedures for the activities conducted within the facility. These should be written and maintained in a suitable form to which the employees would have ready access.
4. Conduct training programs or see that training programs are provided to the employees to inform them of the contents and location of the CHP for the facility, the location and means of accessing chemical information, such as Material Safety Data Sheets, the standard operating procedures for the facility, the risks associated with the chemicals, warning characteristics of the chemicals in use, including possible symptoms indicating overexposures or adverse reactions, emergency response or evacuation plans, and availability of the medical program.
5. Ensure that chemicals are stored, handled, and disposed of properly.
6. Conduct in-house inspections of the facility, conduct, or have conducted, inventories of the chemical holdings and make sure that suitable personal protective equipment is available and employed as needed.

E. As discussed in Chapter 1, the employee is the one ultimately responsible for complying with safety policies, in this instance, as contained in the CHP and standard operating procedures. They have the responsibility for developing good personal safety habits.

4.2.2.3. Training and Information Program

The CHP must contain a description of the organization's information and training program. The training and educational programs are to be made available at the time of the employee's initial assignment of exposure situations. Refresher training is to be provided at a frequency determined by the employer. The information to be provided to the employees must include:

1. The contents of the laboratory safety standard. Since the standard, including its appendices, is quite short, this may be accomplished by including a copy as an appendix to the CHP.

2. The location and availability of the organization's and laboratory's CHP. This is most easily accomplished by maintaining a master copy of the basic CHP for the organization at a central location, such as the environmental health and safety department, with copies of the laboratory CHP in the individual laboratories. The latter should include the basic plan. Access to a computerized information system is becoming widely available in many commercial laboratory organizations and larger academic institutions. The basic unit can be part of this information system and be available to anyone with access to the system at any time. A computer version has a distinct advantage in that it can be updated at any time without distribution of many hard paper copies.

3. The OSHA PELs or action levels for the chemicals in use in the employee's work area. The entire list of PELs can be made an appendix to the CHP to satisfy this requirement rather than having to modify this information whenever a new material is brought into the facility. Not every chemical has an established PEL or action level, but the American Conference of Governmental Industrial Hygienists (ACGIH) publishes a more comprehensive list, updated annually, and the National Institutes of Occupational Safety and Health (NIOSH) also publishes lists of recommended exposure limits, and these must be made available in the absence of OSHA PELs. The three sets of levels do not always agree. Where they differ, the OSHA PELs and action levels are the legally applicable limits. Copies of the ACGIH and NIOSH limits are available as published documents and can be provided as reference material, available in the workplace.

4. The location and availability of reference material on the hazards, safe handling, storage, and disposal of the chemicals found in the laboratory. Note that OSHA uses the word "found," not the phrase "in use." For laboratories that have accumulated a large inventory of rarely used materials, this alone is an excellent reason to dispose of excess and obsolete materials. The minimum means of complying with this requirement is to maintain a file of the Material Safety Data Sheets (MSDSs) provided by the manufacturer of the chemicals. The MSDSs will satisfy the previous requirement for PELs or other recommended exposure levels since they include this information. As will be discussed later, maintaining an up-to-date copy of MSDSs in every laboratory is very difficult, but computer versions of these data are available which can serve as an alternative. MSDSs should be supplemented by other compilations of data. One weakness in the MSDS system is that in order to avoid liability due to recommending a less than necessary level of care, many manufacturers have gone to the other extreme and recommend very conservative measures. Manuals such as *The Merck Manual and Properties of Industrial Chemicals* by Sax would be good supplements to the MSDS data. Labels on commercial chemicals provide much information. The standard requires that these labels not be defaced or removed. All of this material need not be in each laboratory, but the employee must be told where it is and how to obtain access to it. This access should be convenient.

5. Indicators and symptoms associated with exposure to chemicals used in the laboratory.

All of the above is basic information which can be provided as part of the basic plan for the organization, if the employees know where the material is and have reasonable means to obtain access to it. Some organizations accomplish this by computers, and as the use of computers approaches universality, this is likely to become the favored approach.

The required training program must include the following elements:

1. The employees must be informed of the methods used to detect releases or the presence of hazardous chemicals in the workplace. Some of these are available to the employee directly, such as information concerning warning properties of the

chemicals (odor, visual indicators) or symptoms which might be experienced (irritation, nausea, or dizziness). Other means of detecting materials which may be used would include fixed alarms, such as gas monitors, or environmental monitoring by safety and health support staff. Among equipment which might be available would be detector tubes, ambient gas meters, passive dosimeters, and sophisticated devices such as portable infrared, atomic absorption, or gas chromatograph instruments. Detection methods which are available and might be employed should be listed in the CHP. Where access to these methods is through nonlaboratory personnel, the training should include how to obtain the required aid and the telephone numbers of support personnel. Some of this material, such as the environmental monitoring services, should be in the organization's basic plan, but the indicators such as odor or the presence of local fixed gas monitors should be part of the laboratory's own plan.

2. The chemical and physical hazards of the chemicals in the workplace. This is almost the same as the basic information on PELs and MSDSs listed in the previous section. Those requirements basically defined limits of exposure and the sources of data. This requirement provides that the employees be given chemically specific hazard information on the chemicals in their work area. It is most important that the chemicals in actual use are the principal ones for which this information is provided. However, generic hazard information by class for chemicals present but not in use should be provided as well. There is always the potential for an accident involving chemicals not in current use. The employees must be informed that they are not to deface or remove the labels on commercial containers of chemicals, since they represent a primary source of information. It is not required by the standard, but the requirement from the Hazard Communication Standard 29 CFR 1910.1200, that secondary containers intended for use beyond a single work shift should be labeled, is highly recommended.

3. The employees must receive training on the measures they can take to protect themselves. The content of this training should be made part of the CHP for each individual laboratory. Among these measures are

   a. Work practices specific to the laboratory. These include the standard operating and administrative procedures developed so that the work can be carried out safely and efficiently.
   b. Emergency procedures. This can include a wide variety of measures, including how to put out a small fire, how to evacuate an area (including primary and secondary escape routes), steps to take to bring a reaction under control if time permits, how to relieve pressure on pressurized equipment, how to clean up minor spills, how to report larger spills and secure help in responding to them, how to use personal protective equipment available to them, first aid, and close-down procedures in the event of a fume hood failure or failure of any other item of protective equipment. Means of initiating a general evacuation from a facility, or the building in which the laboratory facility is located must exist and should be identified in this section.

4. The details of the CHP applicable to their area, including the basic organizational plan.

The items listed above are for normal laboratory work. If there are some operations which require prior approval by a more senior individual or external group, then these must be included in the training program as well. This need not be a special and possibly more hazardous laboratory evolution, although that is the primary intent of this requirement, but it could represent the purchase of selected items of equipment which must meet certain standards of performance, such as refrigeration units, fume hoods, heating devices, storage cabinetry for flammables, certain classes of chemicals such as carcinogens, etc.

Additional training is also needed for working with extremely hazardous materials. The training must include:

1. Where the work must be done. An area must be designated. This can be an isolated suite of laboratories with controlled access or an area as small as a fume hood explicitly defined as the area where the work is to be done.
2. The use of special containment devices, such as hoods or fully contained glove boxes.
3. Standard operating procedures for the work with the material, including use of appropriate personal protective equipment.
4. Means of safe removal and disposal of contaminated material.
5. Procedures to decontaminate the work area.

4.2.2.4. Medical Program

The CHP must define the means by which the facility will comply with the medical requirements of the standard. In most cases this procedure should be the same for all laboratories within an organization, so the means should be spelled out in the basic plan. There are four specific requirements:

1. Employees working with hazardous chemicals must be provided an opportunity to have a medical examination, and follow-up examinations if necessary, under any of the following circumstances:

   a. The employee develops any signs or symptoms associated with the chemicals to which they may have been exposed in the laboratory.
   b. For specific substances regulated by OSHA, e.g., formaldehyde, for which exposure monitoring and medical surveillance requirements exist in the standard for that substance, the employee must be offered the prescribed medical surveillance program if environmental monitoring shows a routine exposure level above the action level (or PEL, if an action level is not specified).
   c. An incident occurs such as a spill, leak, or explosion and there is a likelihood that the employee might have received an exposure to a hazardous substance; the employee must be offered an opportunity for a medical consultation. The consultation is for the purpose of determining if a medical examination is needed.

2. "All medical examinations and consultations shall be performed by or under the direct supervision of a licensed physician and shall be provided without cost to the employee, without loss of pay, and at a reasonable time and place."
3. The employer must provide the following information to the referral physician, if available:

   a. The identity of the hazardous chemical(s) to which the employee may have been exposed.
   b. A description of the conditions under which the exposure occurred, including quantitative exposure data.
   c. A description of the signs and symptoms of exposure the employee is experiencing, if any.

4. The examining physician must provide a written opinion to the *employer* in a timely manner which shall include or conform to the following requirements:

   a. Any recommendation for further medical follow-up.
   b. The results of the examination and any associated tests.
   c. Any medical condition (not limited to the ones that may have resulted from the exposure) revealed in the course of the examination which may place the employee at increased risk as a result of exposure to a hazardous chemical found in the workplace.
   d. A statement that the employee has been informed of the results of the consultation or medical examination and any medical condition that may require further examination or treatment by the physician.*
   e. The written opinion *shall not* reveal (to the employer) specific findings or diagnoses unrelated to occupational exposure. This obviously is to protect the employee's privacy rights.

5. OSHA does not include the use of respirators under the medical program, but it is closely related since under the General Industry Standard 29 CFR 1910.134 the ability to use a respirator depends upon the employee's health. A basic requirement is the ability of the employee to pass a pulmonary function test, but the employee must not have any other health problems which would preclude the use of respirators if they are needed or required to protect the employee. A statement must be included in the CHP that the organization has a respirator protection program which meets the requirements of the general industry standards. This program should be a written one and included in the employee's training.

4.2.2.5. Laboratory Produced Chemicals

A characteristic of a research laboratory is that chemicals are produced or synthesized in the course of the research. If the composition of the chemical is known and it is a hazardous material, all of the training requirements and other provisions of the standard apply. If the composition is not known, it shall be assumed to be hazardous and, with the exception of the requirements for MSDSs and similar information sources, the provisions of the CHP apply. If the chemical is produced for a user outside the laboratory, the provisions of the Hazard Communication Standard (29 CFR 1910.1200) apply, including the requirement for providing an MSDS and proper labeling of the material. Compliance with these requirements will be the responsibility of the individual laboratory and a commitment to this compliance should be in the laboratory CHP.

4.2.2.6. Record Keeping

The employer must commit to establishing and maintaining for each employee an accurate record of any measurements taken to monitor employee exposure and any medical consultations and examinations, including tests or written opinions required by the standard.

---

* At the author's institution, when the medical surveillance program began several years ago, 22% first-time participants had significant untreated health problems of which, they stated, they were not aware. Very few of these were related to occupational exposures, but some did require adjustments in their duties.

Further, the employer shall assure that such records will be kept, transferred, and made available in accordance with 29 CFR 1910.20.

4.2.2.7. Summary

The sections immediately preceding this one detail the requirements of the OSHA Laboratory Standard and suggest general means by which an organization and/or laboratory can comply with it. Appendix A of the standard provides many recommendations of how compliance can be achieved. These recommendations are not mandatory. The standard is a performance standard which allows a great deal of flexibility on the part of the employer. As noted earlier, the bulk of this handbook (with the exception of Chapter 5, which covers laboratories generally working with materials not covered by the laboratory standard) is designed to provide specific information on how to achieve the appropriate level of performance in all facets of laboratory safety, including design and outfitting of facilities, covered primarily in Chapter 3. The remainder of this chapter starts from the point of an assumption that an adequate facility is available and proceeds from that point the very beginning of planning a program to be done in the facility.

4.3. OPERATIONAL PLANNING

A typical research proposal submitted to a funding source goes into great detail on the significance of the proposed research, the approach to be taken, and the results sought. Typically, the proposal always provides a thorough justification for the technical manpower and equipment resources needed to carry out the planned program. The hazards which will be encountered, if these are sufficiently dangerous or unusual, and the means by which they will be controlled are likely to receive much less attention. Unless the research involves very stringently regulated materials, the reviewer of the proposal often must take on faith, if the question arises at all, that a basic infrastructure has been established to ensure that the research can be carried out safely and in compliance with contemporary regulatory standards. This situation does show signs of changing in some areas, such as when human subjects are involved. The needed infrastructure does not just happen, it requires careful planning. It is the intent of this chapter to provide essential information to guide planning for safe operations in the laboratory.

The first order of priority, after authority to proceed on a specific program is obtained, is to order all of the essential items of equipment which will be needed. Orders for major items of equipment frequently take extended periods to be processed and delivered, 9 to 10 weeks being as short an interval as might reasonably be expected, especially in a facility supported by public funds, encumbered by an abundance of bureaucratic requirements. If installation is required, such as when an additional hood is needed, this period could be extended for months since the installation will have to be carefully planned to ensure, among other things, that the air handling system has sufficient capacity and that fire code requirements can be met, especially if the ductwork must penetrate multiple floors. Scheduling and pricing of the actual work cannot be done in such instances without working plans. This delay may be critical when the work is scheduled to be completed within a fixed contract period with annual renewals depending upon progress, as are many academic research contracts.

If new employees need to be hired, a number of factors must be considered in addition to technical skills. As noted earlier, attitude is extremely important. A research laboratory is not the place for a casual attitude toward safety. Skills and experience are, of course, important, but a vital consideration should be a compatible personality. It is critical in any group effort for personnel to be able to work together. It is not necessary to be "popular," but it is important for individuals to be receptive to the ideas of others and tolerant of differences in points of view. A group of persons working under the stress of strained relationships is likely to be an unproductive and unsafe group. Obviously, it is important that an individual to be hired is safety conscious and willing to comply with the employer's safety policies. A principal investigator needs to establish a clear line of authority for the laboratory personnel, both for day to day operations and for emergencies. These may not be the same. The individual trained to manage the scientific aspects of the research may not have as appropriate a background to handle an emergency situation as would a senior technician who might have received special training in safety areas such as chemical spill control or emergency first aid. Where there is the possibility of ambiguity, responsibility for various duties needs to be clearly assigned, especially those associated with safety. It would be well, for example, to designate a relatively senior person as the laboratory chemical hygiene officer (LCHO) and if necessary provide access to additional safety training to that individual. This individual could be responsible, under the laboratory CHP for such items as safety orientation of new employees and safety training of all employees when new materials or procedures are incorporated into the laboratory operations. They might be asked to perform a hazard analysis of any new laboratory operations and to secure any authorizations or clearances which might be needed. It could be this individual's duty to assign other persons the responsibility of being sure that chemicals are shelved according to compatibility and to maintain safety items such as first aid kit supplies, personal protective equipment, spill kit materials, Material Safety Data Sheets, or maintenance of equipment in safe condition. The LCHO and/or the laboratory supervisor needs to act as liaison with the safety department to provide access to any new information which might affect the laboratory's operations. A knowledgeable employee, who could be the LCHO, needs to be designated as the person responsible for the safe disposal of hazardous materials. This individual needs to be responsible for seeing that all surplus and waste materials are properly identified and segregated if waste materials are combined into common containers. Individuals handling hazardous waste must receive training in the risks associated with that responsibility.

The emergency planning required under the OSHA Laboratory Safety Standard requires an effective emergency plan to be developed for each individual laboratory which is consistent with and integrated into the plan for the entire building and that of the corporation or institution. It needs to take into account procedures for temporarily interrupting the research operations or for automating uninterruptible operations to allow employee evacuation during an emergency. This plan should be reviewed periodically to ensure that it is still appropriate. As has been noted many times, research programs, especially in academic institutions, change rapidly, not only in the materials in use and the operations being conducted, but also in the participating personnel. Evacuation plans need to be tested periodically to ensure that they are effective. It was remarked upon earlier that the transient nature of a building's population in the academic environment creates difficulties in ensuring participation of all of the occupants. Drills held at least once a year should include enough "permanent" occupants to help those who are less familiar with the evacuation procedure.

Every aspect of the laboratory operations should be evaluated to see if it could be made more efficient and safer. Purchasing of reagents, for example, should be reviewed to see how much is actually needed on hand at a given time.

If all chemicals are ordered early in the program and the program needs shift, a substantial investment in surplus chemicals could result. Today, where disposal of waste chemicals has become such a major legal issue, the cost of disposing of surplus chemicals may exceed the original costs. The quality of partial containers of chemicals may have become dubious, and the initial investment in the excess will represent a drain on the currently available funds. Anticipation of needs is critical, especially where equipment is involved. As noted earlier, delivery of essential items of equipment may be delayed for extended periods. The temptation is to "make-do" with equipment not specifically designed to meet the actual needs, with serious safety implications being involved on occasion.

The regulations, and the information on which they are based, change so frequently that it is unreasonable to expect every purchaser to be able to keep up with the current regulations. Further, the entire body of relevant information regarding laboratory safety has become so extensive and so complex that again it is unlikely that a single individual can be sufficiently knowledgeable to adequately consider every factor. For example, the review of the purchase of a fume hood is usually not based so much on the characteristics of the hood, but on the installation. Has the location been reviewed for availability of sufficient make-up air? Has the path of the exhaust duct been selected and has the exhaust blower been sized appropriately? Will fire separations have to be penetrated? The flagging of the purchase will permit these questions to be answered, and if they had not been considered, someone needs to ensure that they are before the order is processed. It is highly likely that the order will have to be modified if these factors have not been addressed, and it is highly desirable that specifications be changed prior to ordering unsuitable equipment.

An evaluation of the potential exposures of individuals to hazardous materials should be made as soon as possible. It may be necessary to consider selectively placing individuals in work assignments, although one has to be very careful in such cases to avoid triggering charges of discrimination based on factors such as gender or disability. Still, if there are known risks, for example, of teratogenic effects from a chemical, it would certainly be surprising if an expectant mother did not have some concerns about working in an area where it was in use, even if the levels were well below the acceptable OSHA limits for the average worker. Any work regimen would need to be fully discussed between the individual and the supervisor in such a case and be based on knowledge, not speculation. Often, once the exposure potential or lack of one is clearly understood, concerns may disappear. Failure to consider the employee's rights to a working environment free of recognized hazards could lead to a complaint to OSHA or another regulatory agency, which could, in extreme cases, cause the program to be interrupted pending resolution of the safety issues.

Prior planning is needed, especially in facilities in which students are expected to be working. Legal safety standards usually have been designed for employees, and although many graduate students and students on work-study programs receive stipends for their efforts, they may not be considered or treated as "real" employees by others in the work area. They typically have less experience and a different purpose in being in the laboratory than do permanent personnel. The pressures associated with completing the various hurdles of a degree program, especially those accompanying completing a research program for a thesis or dissertation within a tight schedule, often lead to students working long hours, going without enough sleep, and eating odd diets. The result may be working without adequate supervision and being affected by factors that could cause impaired judgment. The laboratory safety program should take these factors into account and make a special effort to see that these younger persons understand the goals of the safety program, as it bears upon the operations of the laboratory, and the need to comply with the safety polices of the organization and the laboratory.

4.3.1. Quantities

The recommendation that volumes of reagents kept on hand be kept to the minimum needed for a reasonably short working period is found in virtually every laboratory safety manual. However, a visit to almost any laboratory will reveal many bottles and other types of containers accumulating substantial layers of dust. Many of the more recently acquired reagents very likely will be duplicates of these older materials. There must be good reasons for this apparently needless duplication.

It would appear to make a great deal of sense to order what you need and replace it when it appears that more will be needed. There are at least three reasons why this common sense approach is so rarely followed, two of which are attributable to factors in the purchasing process:

1. It takes time to process an order. Unless a central stores facility maintains a stock of chemicals at the research facility, the processing of a requisition, receipt of an order by the vendor, and delivery are unlikely to take less than 1 month, unless an alternative buying process has been established, such as a blanket order system or previously cleared requisitions for low-value purchases. Under these circumstances, a purchaser tends to buy more than is currently needed in order to avoid having to order frequently and to avoid delays in receipt of the needed material.
2. Chemical costs decrease rapidly with the size of the container. For example, for one grade of sulfuric acid, the following pricing schedule has been established by one major vendor (note that these have been normalized to set the price per liter of the smallest size to equal 1).

Container Size	Cost/Liter
One liter, each	1.000
6 × 1 liter, case	0.558
4 liter, each	0.526
4 × 4 liter, case	0.359
10 liter	0.303
20 liter	0.225

Obviously, if the volume of usage justifies the purchase, the largest size is the most economical to buy. However, there are several reasons why such a purchase is probably unwise. It increases the potential risk, as in this example, to have more material than is actually needed, and storage space will have to be found for the excess material. If it is not used relatively quickly, the quality may become suspect, and users will be reluctant to use it in their research programs. The cost of disposal of any eventual surplus material is likely to wipe out any initial economic gain from buying in volume, unless the surplus can be used by someone with less critical applications.

3. In addition to the two reasons given above, sometimes a researcher wants to be sure of the consistency of the reagent, so he buys enough for his needs from one lot. However, some chemical firms will, upon request, set aside an amount of a given lot and maintain it at their regional warehouse to accommodate a larger user.

An examination of the purchases of the various kinds of research reagents by most university or corporate research facilities will probably reveal that a relatively small proportion of them are bought in substantial quantities. At the author's institution, fewer than 75 of the more than 1200 different chemicals purchased during a typical year exceeded an amount of 50 kg. Where this is true, it would appear feasible to set up a central stores for at least a limited list of chemicals. Stocking of these stores areas should probably emphasize the middle ranges of sizes. If, in the example given above, multiple-case lots of 4-liter containers were the primary sizes purchased from the vendor for stocking, most of the cost savings of volume purchases could be passed on to the local purchaser. Smaller sizes would have the advantage of being likely to be completely emptied, thus eliminating the cost of waste disposal completely for these containers, but forcing the users to buy small sizes could lead to buyer resistance because of a perceived inconvenience. It might be desirable to restrict the purchases of larger sizes to those who can establish a need or, for those items for which it is feasible, disburse chemicals from drums into smaller containers by stores workers.

Except for the high-volume materials, most remaining chemicals are bought in relatively small quantities to meet specific needs of individual programs. Some chemicals pose unusual hazards, such as ethers that degrade over a short period of time. It is desirable to keep track of which group is ordering them and where they are to be found. A central stores area would make a convenient distribution center for these special materials and would facilitate maintenance of records of their use.

Bar code technology has now made it possible to conveniently mark every container received and distributed from a stores area with a unique identification code which includes the name of the chemical, the date received, the quantity, and the recipient. The last of these can be tied to a specific facility, and a specific laboratory within the facility. The availability of powerful desktop computers now makes it possible, with appropriate software that is commercially available, to establish a tracking program for every container from the point of purchase to its final disposition. Networking software even makes it possible to have more than one point of receipt and still accomplish the same task. There are obvious implications with such a program to enable volume purchasing, control of total amounts on hand, and disposal of chemicals approaching dates at which point they may no longer be safe to retain. Bar code technology, using reading devices to scan the container codes into "notebook" size computers that are almost as powerful as the desktop units, makes it possible to quickly inventory all of the containers in a laboratory and to keep track of chemical containers if they are transferred from one laboratory to another.

On April 22, 1987, the EPA Community Right-to-Know standard (40 CFR, Part 370) became law (also known as SARA Title III), requiring users of hazardous materials to inform nearby communities when they had significant holdings of any of several hundred hazardous chemicals. The definition of significant holdings varies from 1 pound (0.454 kg) to 10,000 pounds (4539 kg), depending upon the chemical. Where the amount exceeds another, usually larger, threshold, the law requires that emergency planning programs be established. It is also required to report within 60 days any time these two levels are exceeded. Clearly, it is desirable to maintain amounts in storage less than the trip-point levels.

There are several exemptions to the Community Right-to-Know standard, one of which provides important relief to laboratories from the inventory and reporting provisions of the standard. The EPA, in its final rule, provided an exemption for "any substance to the extent that it is used in a research laboratory or a hospital or other medical facility under the direct supervision of a technically qualified individual." The research laboratory exemption applies only to the chemicals being used in the laboratory, not the laboratory itself, under the direction of a person meeting the specified criteria. Basically the same limitation which qualified a laboratory-scale operation under the laboratory standard applies here. The exemption does not apply to pilot plant-scale operations or production-like programs. The difficulty of preparing the reports required under SARA Title III make this exemption extremely useful. It is surprising how many of the chemicals can be found within individual laboratories in excess of the reportable or emergency planning thresholds, and if the total number of laboratories in a larger research organization is considered, it would be very difficult to comply. Implementation of a chemical tracking program, using the bar coding concept and suitable software, will make it possible to comply with the law, should the exemption be removed.

Although the exemption is very useful as a practical matter, it is somewhat troubling to have to depend upon since the risks that evoked passage of the Right-to-Know act are real. Many universities and industrial research facilities are located in smaller towns and may represent a significant chemical release risk to the community, perhaps the largest. The individual containers are small, but if a fire involved a large chemical building, the total amount of chemicals released into the air and perhaps running off in the water being used to fight the fire could be very large. Not only could there be a large quantity of chemicals involved, the release would be very complex because of the very large variety present, with the toxicity of the release being impossible to predict. In a large release from a burning chemical, one could find oneself facing the problem of evacuating thousands of students and employees from an academic research building within a very short time. The available emergency resources could be overwhelmed easily. This scenario for laboratory organizations may be the most pressing factor in developing a chemical management program. It is recommended that, where research-oriented firms and institutions do represent a significant environmental hazard, a representative participate on the local emergency planning committees established under the EPA standard, even if technically exempt.

In summary, it is desirable to order and maintain in stock as small amounts of chemicals as practicable in order (a) to minimize the risks in the event of an incident, (b) to reduce the overall expense by reducing the amount requiring disposal as hazardous waste, and (c) to minimize the problem of complying, at least in spirit, with the Community Right-to-Know standard. However, in order to encourage a laboratory manager to buy and stock smaller containers, purchasing procedures need to be established to conveniently provide smaller sizes at a reasonable cost.

4.3.2. Sources

One of the more difficult tasks associated with the purchase of equipment and materials meeting acceptable safety standards is to do so in a system which requires acceptance of the low bid. Many of the safety standards or guidelines are minimal standards, which it is often desirable to exceed. Usually chemicals from any major company or distributor will be acceptable, but the same is not necessarily true of equipment. In order to obtain the quality desired, purchase specifications must be carefully written to include significant differences which will eliminate marginally acceptable items. In some cases, it is virtually impossible to write such a specification, and it is necessary to include a performance criteria. This often requires considerable effort on the part of the purchaser. As an example, chemical splash goggles are sold by many companies, at prices that differ by an order of magnitude or more. All of these will usually meet ANSI Standard Z-87 for protective eyewear, but many are sufficiently uncomfortable or fog up so rapidly that they will not be worn. Thorough comparative testing under actual laboratory conditions will identify a handful of the available models that offer superior performance. With documented data, it is usually possible to obtain permission of the purchasing department to limit purchases to sources meeting acceptably high safety and performance criteria, rather than minimal standards. This applies not only to smaller items but also to major ones, such as fume hoods. Where there is a significant difference in quality which will enhance the performance and/or the safety of any unit at a reasonable price, a cooperative effort should be made by the purchaser, the Purchasing Department, and the Safety Department to obtain needed items from these sources.

4.3.3. Material Safety Data Sheets

The federal government enacted a hazard communication standard in 1984. Chemical manufacturers, importers, and distributors were required to comply with the standard by November 25, 1985, and affected employers by May 25, 1986. Originally, the standard applied only to Standard Industrial Code Classifications 20 through 39. After September 23, 1987, it has been required that Material Safety Data Sheets (MSDS) be provided to nonmanufacturing employees and distributors with the next shipment of chemicals to these groups. As of May 23, 1988, all employers in the nonmanufacturing sector must have been in compliance with all provisions of the standard. The laboratory safety standard specifically mentions that at least MSDSs need to be available to laboratory employees. Many states have enacted similar standards which extended the coverage within their own jurisdiction. Some specifically extended coverage to public employees, which included individuals at public universities and colleges.

Under the hazard communication standard, chemical manufacturers and importers must obtain or develop a MSDS for each hazardous chemical they produce or import. These MSDSs must reflect the latest scientific data. New information must be added to the MSDS within 3 months after it has become available. The manufacturer or importer must provide an MSDS to a purchaser the first time a given item is purchased and an updated version after the information becomes available. A distributor of chemicals must provide MSDSs to their customers.

The MSDSs can be in different formats as long as the essential information is included, although a standard format may be adopted. The minimal information to be provided, which must be in English, is

1. The identity of the chemical as used on the label of the container.

   a. For a single substance, the chemical name and other common names.
   b. Mixtures tested as a whole: The chemical and common names of all ingredients which contribute to known hazards, and common names of the mixture itself.
   c. Mixtures untested as a whole: Chemical and common names of all ingredients which are health hazards and which are in concentrations of 1% or more, or carcinogens in concentrations of 0.1% or more. Carcinogens are defined to be those established as such in the latest editions of (a) National Toxicology Program (NTP) *Annual Report on Carcinogens*, (b) International Agency for Research on Cancer (IARC) Monographs, or (c) 29 CFR Part 1910, Subpart Z, "Toxic and Hazardous Substances," OSHA.

   If any of the ingredients which do not exceed the concentration limits in the previous paragraph could be released from the mixture such that they could exceed an established OSHA PEL, or an ACGIH threshold level value, or could represent an occupational health hazard, their chemical and common names must be given as well. The same information is also required for any ingredient in the mixture which poses a physical hazard (as opposed to a health hazard).

2. Physical and chemical characteristics of the hazardous chemicals.
3. Physical hazards of the hazardous chemical, specifically including the potential for fire, explosion, and reactivity.
4. Known acute and chronic health effects and related health information.* This information is to include signs and symptoms of exposure and any medical conditions which are generally recognized as being aggravated by exposure to the chemical.
5. Primary routes of entry into the body.
6. Exposure limits data.
7. If the hazardous material is considered a carcinogen by OSHA, IARC, or the NTP (see 1.c above).
8. Precautions for safe handling, including protective measures during repair and maintenance of apparatus employed in using the equipment and procedures for cleanup of spills and leaks.
9. Relevant engineering controls, work practices, or personal protective equipment.
10. Emergency and first aid procedures.
11. Date of MSDS preparation or latest revision.
12. Name, address, and telephone number of the entity responsible for preparing and distributing the MSDS.

Although this list appears straightforward, the MSDSs provided by different companies vary significantly in quality. Many are incomplete, perhaps not always due to lack of information. There are generic sources of MSDSs which are prepared by firms independently of the original manufacturers and are possibly more free of bias. On June 3, 1993, the American National Standards Institute approved a voluntary consensus standard for MSDSs developed by the Chemical Manufacturers Association in an effort to provide more uniformity in the documents. Several industries claimed that there were problems with the new form which were not fully considered. As a result, at the time of this writing, no consensus standard has been adopted.

Provision of a MSDS at the time of the initial purchase of a chemical is a responsibility of the chemical vendor, and if the vendor fails to provide it, it is the responsibility of the purchaser to take the necessary steps to require the vendor to do so. A typical MSDS can be up to several pages long, and a comprehensive file of hundreds of these, which might be required in a typical laboratory, or thousands if the file is maintained at a central location in an organization, will be bulky and difficult to maintain.

Both the distributor of a chemical and the purchaser have a major problem in complying with the requirement that a MSDS be provided to the user, where purchasing authority is widely distributed, as it often is on a university campus. Many institutions permit direct delivery to the actual location ordering a given material, while in others there is a central receiving point. In the former situation, a chemical vendor may supply an MSDS to the first purchaser of a chemical at the institution, but subsequent purchasers may not receive one, because they did not receive a copy of the first one sent to the initial purchaser. Where all the separate purchasers of a chemical are part of the same institution and located within contiguous confines of a single site, it is probable that the vendor technically can meet the legal requirement of furnishing an MSDS to the institution as an entity by providing a single MSDS to the individual laboratory first ordering a substance. In a large research institution, this would result in a very incomplete distribution of MSDSs. Designation of a single department, such as the safety department, to receive all MSDSs from the

---

* It is possible that in the near future, usage will substitute "immediate" for "acute" and "delayed" for "chronic."

chemical vendors and to establish a master file of them, with perhaps some partial or complete duplicate files at other locations, will partially alleviate the problem. These files would need to be in places that are easily accessible to the users for a large portion of the day in order to approximate compliance with the requirement of being readily available to the employees. The laboratory standard does not contain the language "readily accessible," but only requires that the employees know where they are being held by their employer. However, the organization should still make arrangements to facilitate access. Unless the information as to which unit actually ordered the material accompanies the MSDS, it would be impossible to distribute them further internally, unless an individual department requests a specific MSDS which they wished to maintain in their local file. However, unless each department received a notice of the receipt of any revised MSDS and took the initiative to upgrade their own files, the local files would soon become obsolete. This could lead to possible liability problems if an employee assumed that the local files were current.

Some chemical manufacturers or distributors have avoided the entire problem, as far as they are concerned, by sending an entire set of MSDSs for all of their products to corporations or institutions with whom they do a substantial business. It is then up to the university or corporation to decide how to distribute them properly to comply with the regulatory requirement that any needed MSDS be readily available to employees.

If all chemicals are delivered to a central receiving location, a fairly straightforward, but labor intensive, solution to the problem exists. A master file of all MSDSs can be maintained at the central receiving location, as well as a list for each chemical of all departments or other definable administrative units which have previously ordered the material. If a department is not on the latter list for a given chemical, then a copy of the MSDS can be made and sent along with the material when it is delivered. A revised MSDS would be sent to every department listed as having the specific chemical in their possession. It would require that a copy of every purchase order and/or invoice were sent to the department maintaining the file in order to maintain the departmental lists. Although this sounds relatively easy, the amount of record maintenance required and the time spent in checking the files would be substantial. For a major research institution, the amount and variety of materials ordered, coupled with the large number of independent administrative units, would probably mandate at least a full-time equivalent clerical employee for the program.

Computer technology has provided solutions to all or part of the management problem for distribution of MSDSs within complex organizations in which the variety of chemicals is numbered in the hundreds or thousands, instead of a few.

Although hard copy compilations of MSDSs are available either in print or on microfiche, the most flexible approach is to obtain access to an on-line source of MSDSs or subscribe to a vendor that will send an updated CD-ROM disk on a quarterly basis (this meets the 3-month update requirement). There are several firms which provide one or the other of these services. Some of the same firms also fulfill a requirement for the users of hazardous materials that they have access to a 24-hour emergency assistance number in the event of an accident. On-line services typically charge for accessing their services on a per-call basis, although some provide a limited amount of free time each month. Access to a server computer housing the CD-ROM data base through modems or a network is a useful service, but the user must be sure that license requirements are met. If license agreements are available, then access can in principle be made available to every user of chemicals in a facility with access (rapidly becoming the norm) to a computer or terminal 24 hours per day through a network or modem. Both of these means can provide access to a very large MSDS data base that is current and reliable. Providers of generic data bases do assume the liability of ensuring that their information is correct, and this factor contributes in part to the relatively high cost of computer MSDS data bases. The other major reason is the substantial amount of research needed to keep up with all of the current published material available.

4.3.4. Purchase of Regulated Items

There are a number of classes of items for which purchases must be carefully monitored for compliance with safety and security regulations. Several of these can be purchased only if a license is held by the individual or by the corporation or institution. There are many restrictions, in addition, on the transportation of hazardous materials. Usually, the purchaser will expect the vendor to be responsible for meeting these shipping requirements. However, there will be occasions when the institution or corporation will initiate a shipment. It is recommended that a subscription to a hazardous materials transportation regulatory advisory service be taken out by anyone who ships any hazardous material frequently, due to the relatively rapid changes in shipping regulations. Such information is also rapidly becoming available from on-line or CD-ROM computer services. Some are being provided by the regulatory agencies themselves.

4.3.4.1. Radioisotopes

The purchase of radioactive materials, with certain exceptions, is generally restricted to those persons who are licensed to own and use the materials under one of the sections of Title 10, Code of Federal Regulations, usually Part 30. In this context, the word "person" is used quite broadly. In Part 30, which provides the rules for domestic licensing of by-product material, "person" is defined as: "Any individual, corporation, partnership, firm, association, trust, public or private institution, group, Government agency other than the Commission or Department..., any State, any foreign government or nation or any political subdivision of any such government or nation, or other entity; and any legal successor, representative, agent or agency of the foregoing." Clearly, virtually any assemblage of persons can be licensed to own and use radioactive by-product materials, if they can fulfill the licensing conditions provided by Part 30 and have an approved radiation management program meeting the standards of Part 20. In approximately half of the states, the oversight function to ensure compliance with the standard is done by the state rather than the Nuclear Regulatory Commission (NCR). These are "agreement states."

There are a few more definitions which will be useful. The federal regulations in Part 30 usually apply only to "by-product material." This refers to... "radioactive materials, other than special nuclear material, yielded in or made radioactive by exposure to the radiation incident to the process of producing or utilizing special nuclear material." The NRC definition of special nuclear material is lengthy, but essentially it means plutonium or uranium enriched in the fissionable isotopes U-233 or U-235. There are naturally occurring radioactive materials which are mostly unregulated and there are radioactive materials made radioactive by using accelerators. These latter materials are regulated by the states independently, not by the NRC. Exposure to some natural radioactive materials, such as radon, are regulated under some circumstances.

There are a number of classes of radioactive materials which do not require a license. If the amount is less than the exempt quantity for a given material, as listed in Paragraph 30.71, Schedule B of the regulations, a license is not required. The amount meeting this criteria is given in Table 4.1 for a few of the radioisotopes most commonly used in research. The units are in microcuries where 1 microcurie is equal to 37,000 disintegrations

Table 4.1.   Exempt Quantities of Some of the Most Often Used Radioisotopes

Isotope	Quantity (μCi)	Isotope	Quantity (μCi)
Calcium 45	10	Iodine 131	1
Carbon 14	100	Iron 59	10
Cesium 137	10	Mercury 203	10
Cobalt 60	1	Molybdenum 99	100
Chromium 51	1000	Nickel 63	10
Hydrogen 3	1000	Phosphorus 32	10
Iodine 125	1	Sulfur 35	100

per second, since this is the way they appear in the regulations. A set of units different from these has been recommended by the International Commission on Radiological Protection. In the International System of Units (SI units), the unit of activity is the Becquerel (Bq) and is equal to 1 disintegration per second. A microcurie, therefore, equals 37,000 Bq.

There are a number of other classes described in paragraphs 30.15-20 of 10 CFR, in which the persons purchasing certain items containing radioactive materials are exempt from having a license, although the original manufacturer must have had a specific license to allow production of the units. Among these are self-luminous devices and gas and aerosol detectors.

The amounts in Table 4.1 are very small and are usually exceeded in most research applications. For practical research using radioactive materials, it is necessary to obtain a license; a discussion of this will be deferred to Chapter 5. However, assuming that a license has been obtained and a radiation safety program has been established satisfying the NRC (or its equivalent in an agreement state; henceforth, when the NRC is mentioned, it will be understood to include this addendum), there are still formal steps to go through in purchasing and receiving radioactive materials.

In a research institution, it is common practice to establish a license to cover all users of radiation at the institution. This is called a broad license and provides limits on the total amount of each isotope that can be in possession of the licensee at any specific time. These limits are normally chosen by the institution and approved by the NRC. If there are several separate users, as is usually the case, the sum of all their holdings for each isotope, including unused material, material in use, and material as waste, must not exceed these limits. Since each individual user cannot keep track of the holdings of other independent users, it is essential that all purchase orders, as well as all waste materials, be passed by or through a radiation safety specialist, whose responsibility (among many others) is to ensure that the license limits are not violated. Adherence to this and all other radiation safety regulations is essential. At one time the primary threat in the event of a violation was the possibility of suspension of a license. This was such a severe penalty that it was invoked very infrequently. In recent years, substantial fines have been levied against universities and other users who violate the regulations and the terms of their licenses. On March 12, 1987, a city attorney filed 179 *criminal* charges against a major university within the city's jurisdiction and 10 individual members of its faculty for violations of the state standards. This established a major precedent. More recently, another university reached a settlement with the surrounding community to conduct a $1,300,000 study of the possible dispersion of radioactive materials into the community in addition to a substantial fine, because of their management of the use of radioactive materials. As this is being written, a major study on the use of radioactive materials in "research" on possibly unsuspecting or involuntary participants shortly after World War II is underway after release of hitherto secret papers. Even at this late date, there may be significant repercussions.

Unless a vendor has a valid copy of the license for a person ordering radioisotopes, they are not allowed to fill an order. Since the radiation safety specialist is such a key person in the process in any event, it should also be this person's responsibility to maintain current copies of licenses, including any amendments, in the hands of prospective suppliers of radioactive materials. At many facilities, the radiation safety specialist has been assigned virtually all responsibility for ordering and receipt of radioactive materials. Title 10 CFR, Part 20.205, requires each licensee to establish safe procedures for receipt and opening of radioactive packages. Although mistakes are rare, they do happen, so it is highly desirable that the radiation safety specialist directly receive each package of radioactive materials, check that its paperwork is correct, check the external radiation levels, and check the containers for damage. It has happened that all of the paperwork conformed to the expected material, but the wrong material or the wrong amounts of the ordered material were shipped. Where it is impossible for the radiation specialist to receive all packages, provision needs to be made for temporary secure storage of packages until they can be checked.

Many radioactive materials are used in the form of labeled compounds, often prepared specifically to order. In some of these, the half-life of the isotope used in the compound is short so that procedures need to be established to ensure prompt handling and delivery to the user. In other cases, the compound itself will deteriorate at ordinary temperatures. These packages are usually shipped packed in dry ice and must be delivered immediately upon receipt or stored temporarily in a freezer until delivery. If it is necessary to ship radioactive material, the material must be packaged according to Title 49, CFR 173. Again, the radiation safety specialist is the individual who normally would be expected to be familiar with all current standards affecting shipment and be able to arrange for transportation according to the regulations.

4.3.4.2. Controlled Substances (Drugs)

The purchase, storage, and use of many narcotic, hallucinogenic, stimulant, or depressive drugs are regulated under Title 21, Code of Federal Regulations, Part 1300 to the end. In addition, these substances are usually regulated by state law, which in many cases is much more stringent than federal law. The controlled substances covered by the Controlled Substances Act are divided into five schedules; Schedule I substances have no accepted medical use in the United States, have a high potential for abuse, and are the most tightly controlled, while Schedule V substances contain limited quantities of some narcotics with limited risk. In this case, the Drug Enforcement Agency (DEA), which is the federal agency regulating the use of these substances, does not permit a broad agency license, but requires a single responsible individual in each functionally independent facility to obtain a separate license which spells out which Schedules of controlled substances are permissible for the facility to possess. This individual can permit others to use the controlled substance under his direction or to issue it to specific persons for whom he will take the responsibility, but there is no required equivalent to the radiation safety officer to monitor programs internally. Thus, the individual license holder is responsible for ordering, receiving, and maintaining an accurate current inventory for the drugs used in his laboratory.

One institutional responsibility that should be assigned to an individual or department is monitoring the expiration dates of licenses. Although the DEA has a program which should remind each licensee in ample time that their license is on the verge of expiring, experience has shown that the program has not been entirely successful. An individual should maintain a file of all licenses held by employees of the institution and take appropriate steps to see that applications for renewals are filed in a timely manner to avoid purchasing of controlled materials on expired licenses. In organizations that have a pharmacy or pharmacists on their staff, the senior pharmacist would be the logical person to perform these limited regulatory roles.

Packages containing controlled substances must be marked and sealed in accordance with the provisions of the Controlled Substances Act when being shipped. Every parcel containing these sensitive materials must be placed within a plain outer container or securely wrapped in plain paper through which no markings indicating the nature of the contents can be seen. No markings of any kind are permitted on the parcel which would reveal the nature of the contents.

4.3.4.3. Etiologic Agents

Hazardous biological agents are classified as "etiologic agents." An etiologic agent is more specifically defined as (1) a viable microorganism, or its toxin, which is listed in 42 CFR 72.3 or (2) which causes or may cause severe, disabling, or fatal human disease. The importation or subsequent receipt of etiologic agents and vectors of human diseases is subject to the regulations of the Public Health Service, given in Title 42, Section 71.156. The Centers for Disease Control (CDC) issues the necessary permits authorizing the importation or receipt of regulated materials and specifies the conditions under which the agent or vector is shipped, handled, and used. The interstate shipment of indigenous etiologic agents, diagnostic specimens, and biological products is subject to applicable packaging, labeling, and shipping requirements of the Interstate Shipment of Etiologic Agents (42 CFR Part 72). Packaging and labeling requirements are illustrated in Figure 4.1.

In addition to the regulations of the Public Health Service, the Department of Transportation has additional regulations in Title 49, CFR Section 173.386-388. Shipments are limited to 50 milliliters or 50 grams in a passenger carrying airplane or rail car, and 4 liters or 4 kilograms in cargo aircraft. The U.S. Postal Service provides regulations covering the mailability of biological materials in the Domestic Mail Manual, Section 124.38. All of these provide explicit instructions on how etiologic agents can be shipped. There are additional restrictions for international shipments, covered by the International Mail Manual. The ability to make foreign shipments is restricted to laboratories, by approval of the General Manager, International Mail Classification Division, USPS Headquarters, Washington, D.C. 20260-5365.

Whether a person or laboratory purchases a given etiologic agent should depend upon the facilities available for the research program, the training and experience of the laboratory employees, and the type and scale of the operations to be conducted. If, as discussed in that material, the etiologic agent is one that would require the planned operations to be conducted in a laboratory meeting Biological Safety Standard level 3 or 4, the purchase should require the prior approval of the institutional biosafety committee. Operations and classification of Microbiological and Biomedical laboratories will be covered in some detail in Chapter 5, Section 3.

There are comparable restrictions for the importation, possession, use, or interstate shipment of certain pathogens of domestic livestock and poultry, administered by the U.S. Department of Agriculture.

Figure 4.1. Packaging of etiologic agents.

For additional information regarding etiologic agents of human diseases and related materials, write to:

Centers for Disease Control
Attention: Office of Biosafety
1600 Clifton Road, N.E.
Atlanta, GA 30333
or call (404) 329-3883

For additional information regarding animal pathogens, write:

Chief Staff Veterinarian
Organisms and Vectors
Veterinary Services
Animal and Plant Health Inspection Service
U.S. Department of Agriculture
Federal Building Room 810
Hyattsville, MD 20782
or call (301) 436-8017

4.3.4.4. Carcinogens

There are no restrictions on ordering known carcinogenic materials. However, for the carcinogenic materials covered by the regulations in 29 CFR Part 1910, Subpart Z, their

Table 4.2.   Regulated Carcinogenic Materials

Asbestos, tremolite, actinolite, and anthrophyllite	Benzidine 4-Aminodiphenyl	Vinyl chloride Inorganic arsenic
4-Nitrobiphenyl	Ethyleneimine	Coke oven emissions
α-Naphthylamine	Formaldehyde	1,2-dibromo-3-chloro-
Methyl chloromethyl ether	β-Propiolactone	propane
3,3'-Dichlorobenzidine (and its salts)	2-Acetylaminofluorene 4-Dimethylaminoazo	Acrylonitrile Ethylene oxide
Bis (chloromethyl) ether	benzene	Benzene
α- and β-Naphthylamine	N-Nitrosodimethylamine	Cadmium

Note: Coke oven emissions on this list are the result of processes and not chemicals to be used in the laboratory.

purchase for research should be limited to individuals who formally commit themselves to complying with the terms and conditions of the standards. As noted earlier, although the laboratory standard does preempt most of the usual OSHA standards, where there exist specific regulations for individual materials, these regulations still apply. To ensure that this is done, every requisition for purchase of one of the regulated carcinogens should be referred to the institutional safety department for review. This will normally involve a review of the research protocols to ascertain if the use is liable to meet any criteria exempting the proposed program from some of the more stringent and often expensive requirements. If the program does not appear to qualify for exemptions, then the investigator and the safety reviewer should go through each of the requirements under the standard to confirm that they can be met. Although this will seem excessive to some users, it serves not only to protect the employees, but also to minimize the potential for litigation for the research director and the academic institution or corporation.

There are a number of known carcinogenic materials, and the list is growing as the necessary studies of suspected carcinogens are completed. It is recommended that purchases of these be limited and exposures minimized as much as possible to promote the safety of everyone exposed to the materials and in consideration of potential future regulatory restrictions. As discussed in Section 4.3.3, for the purpose of the MSDSs, a listing as a carcinogen by either the NTP, or the IARC is sufficient to be considered as one for the purpose of the Laboratory Safety Standard or the Hazard Communication Standard. However, the only carcinogens specifically regulated as such in 29 CFR Part 1910, Subpart Z are those for which individual regulatory standards have been issued. Materials currently regulated as carcinogenic by OSHA are given in Table 4.2.

In addition to the list in Table 4.2, a so-called "California" list of chemicals has been developed as a result of passage of Proposition 65 in California in 1986, which requires that the governor of the state publish each year a list of chemicals known to cause cancer or reproductive toxicity. The California list as of June 3, 1993, is given in Table 4.3.

4.3.4.5. Explosives

A substantial amount of laboratory research involves materials considered, in the legal sense of the term, as explosives rather than simply chemicals which can explode under appropriate conditions. The term "explosive" in this relatively narrow sense is defined as any material determined to be within the scope of Title 18, United States Code, Chapter 40,

Table 4.3A. "California" List of Carcinogens (continued)

Chemical	CAS Number
A-*a*-C (2-amino-9H-pyrido[2,3-b]indole	26148685
Acetaldehyde	75070
Acetamide	60355
Acetochlor	34256821
2-Acetylaminofluorene	53963
Acifluorfen	62476599
Acryiamide	79061
Acrylonitrile	107131
Actinomycin D	50760
Adriamycin (doxorubicin hydrochloride)	23214928
AF-2,[2-(2-furyl)-3-(5-nitro-2-furyl)]acrylamide	3688537
Aflatoxins	—
Alachlor	15972608
Alcoholic beverages, when associated with alcohol abuse	—
Aldrin	309002
Allyl chloride	107051
2-Aminoanthraquinone	107951
p-Aminoazobenzene	60093
o-Aminoazotoluene	97563
4-Aminobiphenyl (4-aminodiphenyl)	92671
3-Amino-9-ethylcarbazole hydrochloride	6109973
1-Amino-2-methylanthraquinone	82280
4-Amino-5-(5-nitro-2-furyl)-1,3,4-thiadiazole	712685
Amitrole	61825
Analgesic mixtures containing phenacetin	—
Aniline	62533
o-Anisidine	90040
o-Anisidine hydrochloride	134292
Antimony oxide (antimony trioxide)	1309644
Aramite	140578
Arsenic (inorganic arsenic compounds)	—
Asbestos	1332214
Auramine	492808
Azaserine	115026
Azathioprine	446866
Azacitidine	320672
Azobenzene	103333
Benz[a]anthracene	56553
Benzene	71432
Benzidine [and its salts]	92875
Benzidine-based dyes	
Benzo[*b*]fluoranthene	205992
Benzo[*j*]fluoranthene	205823
Benzo[*k*]fluoranthene	207089

Table 4.3A.   "California" List of Carcinogens (continued)

Chemical	CAS Number
Benzofuran	271896
Benzo[a]pyrene	50328
Benzotrichloride	98077
Benzyl chloride	100447
Benzyl violet 4B	1694093
Beryllium and beryllium compounds	—
Betal quid with tobacco	—
bis(2-chloroethyl) ether	111444
N,N-bis(2-chloroethyl)-2-naphthylamine (chlornapazine)	494031
Bischloroethyl nitrosurea (BCNU) (carmustine)	154938
Bis(chloromethyl) ether	542881
Bitumens, extracts of steam-refined and air-refined	—
Bracken, fern	—
Bromodichloromethane	75274
Broform	75252
1,3-Butadiene	106990
1,4-Buanediol dimethanesufonate (busulfan)	55981
Butylated hydrooxyanisole	25013165
β-Butyrolactone	3068880
Cadmium and cadmium compounds	—
Captafol	2425061
Captan	133062
Carbon tetrachloride	56235
Carbon-black extracts	—
Ceramic fibers (airborne particles of respirable size)	—
Certain combined chemotherapy for lymphomas	—
Chlorambucil	305033
Chloramphenicol	56757
Chlordane	57749
Chlordecone (kepone)	143500
Chlordimiform	6164983
Chlorendic acid	115286
Chlorinated paraffins (average chain length, C212, approximately 60% chlorine by weight)	108171262
Chlorodibromethane	124481
Chloroethane (ethyl chloride)	75003
1-(2-Chloroethyl)-3-cyclohexyl-1-nitrosourea (CCHU) (lomustine)	13010474
1-(2-Chloroethyl)-3(4-methylcyclohexyl)-1-nitrosourea (methyl-CCNU)	13909096
Chloroform	67663
Chloromethyl methyl ether (technical grade)	107302
4-Chloro-o-phenylenediamine	95830
p-Chloro-o-toluidine	95692
Chlorothalonil	1897456
Chlorozotocin	54749905

Table 4.3A. "California" List of Carcinogens (continued)

Chemical	CAS Number
Chromium (hexavalent compounds)	—
Chrysene	218019
C.I. Acid Red 114	6459945
C.I. Basic Red 9 monohydrochloride	569619
Ciclosporin (Cyclosporin A; Cyclosporin)	59865133
	79217600
Cionnamyl anthranilate	87296
Cisplatin	15663271
Citrus Red No. 2	6358538
Cobalt metal powder	7440484
Cobalt [II] oxide	1307966
Coke oven emissions	—
Conjugated estrogens	—
Creosotes	—
p-Cresidine	120718
Cupferron	135206
Cycesin	14901087
Cyclophosphamide (anhydrous)	50180
Cyclophosphamide (hydrated)	6055192
D&C Orange No. 17	3468631
D&C Red No. 8	2092560
D&C Red No. 9	5160021
D7C Red No. 19	81889
Dicarbazine	4342034
Daminozide	1596845
Dantron (chrysazin; 1,8-dihydroxyanthraquinone)	117102
Daunomycin	20830813
DDD (Dichlorodiphenyldichloroethane)	72548
DDE (Dichlorodiphenyldichloroethylene)	72559
DDT (Dichlorodiphenyltrichloroethane)	50293
DDVP (Dichlorvos)	62737
N,N'-Diacetylbenzidine	613354
2,4-Diaminoanisole	615054
2,4-Diaminoanisole sulfate	39156417
4,4'-Diaminodiphenyl ether (4,4'-Oxydianiline)	101804
2,4-Diaminotoluene	95807
Diaminotoluene (mixed)	—
Dibenz[a,h]acridine	226368
Dibenz[a,j]acridine	224420
Dibenz[a,h]anthracene	53703
7H-Dibenzo[c,g]carbazole	194592
Dibenzo[a,e]pyrene	192654
Dibenzo[a,h]pyrene	189640
Dibenzo[a,i]pyrene	189559
Dibenzo[a,l]pyrene	191300

Table 4.3A.   "California" List of Carcinogens (continued)

Chemical	CAS Number
1,2-Dibromo-3-chloropropane (DBCP)	96128
p-Dichlorobenzene	106467
3,3'-Dichlorobenzidine	91941
1,4-Dichloro-2-butene	764410
3,3'-Dichloro-4,4'-diaminodiphenyl ether	28434868
1,1-Dichloroethane	75343
Dichloromethane (methylene chloride)	75092
1,2-Dichloropropane	78875
1,3-Dichloropropene	542756
Dieldrin	60571
Dienestrol	84173
Diepoxybutane	1464535
Diesel engine exhaust	—
Di(2-ethylhexyl)phthlate	117817
1,2-Diethylhydrazine	1615801
Diethyl sulfate	64675
Diethylstilbestrol	56531
Diglycidyl resorcinol ether (DGRE)	101906
Dihydrosafrole	94586
Diisopropyl sulfate	2973106
3,3'-Dimethoxybenzidine (o-dianisidine)	119904
3,3'-Dimethoxybenzidine dihydrochloride (o-dianisidine-dihydrochloride)	20325400
Dimethyl sulfate	77781
4-Dimethylaminoazobenzene	60117
trans-2-[(Dimethylamino)methylimino]-5-[2-(5-nitro-2-furyl)vinyl]-1,3,4-oxadiazole	55738540
7,12-Dimethylbenz(a)anthracene	557976
3,3'-Dimethylbenzidine (o-Tolidine)	119937
3,3'-Dimethylbenzidine hydrochloride	612828
Dimethyl carbamoyl chloride	79447
1,1-Dimethylhydrazine (UDMH)	57147
1,2-Dimethylhydrazine	540738
Dimethylvinylchloride	513371
1,6-Dinitropyrene	42397648
1,8-Dinitropyrene	42397659
2,4-Dinitrotoluene	121142
1,4-Dioxane	123911
Diphenylhydantoin (phenytoin)	57410
Diphenylhydantoin (phenytoin), sodium salt	630933
Direct Black 38 (technical grade)	1937377
Direct Blue 6 (technical grade)	2602462
Direct Brown 95 (technical grade)	16071866
Disperse Blue	2475458
Epichlorohydrin	106898
Erionite	12510428

Table 4.3A. **"California" List of Carcinogens (continued)**

Chemical	CAS Number
Estradiol 17β	50282
Estrone	53167
Ethinylestradiol	57636
Ethyl acrylate	140885
Ethyl methanesulfonate	62500
Ethyl-4,4'-dichlorobenzilate	510156
Ethylene dibromide	106934
Ethylene dichloride (1,2-dichloroethane)	107062
Ethylene oxide	75218
Ethylene thiourea	96457
Ethyleneamine	151564
Folpet	133073
Formaldehyde (gas)	50000
2-(2-Formylhdrazino)-4-(5-nitro-2-furyl)thiazole	3570750
Furazolidone	67458
Furmecyclox	60568050
Gasoline engine exhaust (condensates/extracts)	—
Glasswool fibers (airborne particles of respirable size)	—
Glu-P-1 (2-amino-6-methyldipyrido[1,2-a:3',2'-d]imidazole)	6770114
Glu-P-1 (2-aminodipyrido[1,2-a:3',2'-d]imidazole)	67730103
Glycidaldehyde	765344
Glycidol	556525
Grissofulvin	126078
Gyromitrin (acetylaldehyde methylfomylhydrazone)	16568028
HC Blue 1	2784943
Heptachlorhexachloro	76448
Heptachlor epoxide	1024573
Hexachlorobenzene	118741
Hexachlorocyclohexane (technical grade)	—
Hexachlorodibenzodioxin	34465468
Hexachloroethane	67721
Hexamethylphosphoramide	680319
Hydrazine	302012
Hydrazine sulfate	10034932
Hydrazobenzene (1,2-diphenylhydrazine)	122667
Indeno [1,2,3-cf]pyrene	193395
IQ (2-amino-3-methlimidazo[4,5-f]quinoline)	76180966
Iron dextran complex	9004664
Isosafrole	120581
Lacrofen	77501634
Lasiocarpine	303344
Lead acetate	301042
Lead and lead compounds	—
Lead phosphate	7446277
Lead subacetate	1335326

Table 4.3A.   "California" List of Carcinogens (continued)

Chemical	CAS Number
Lindane and other hexachlorocyclohexane isomers	—
Mancozeb	8018017
Maneb	12427382
Me-A-α-C (2-amino-3-methyl-9H-pyrido[2,3-b]indole)	68006837
Medroxyprogesterone acetate	71589
Melphalan	148823
Merphalan	531760
Mestranol	72333
8-Methoxypsorlen with ultraviolet therapy	298817
3-Methoxypsorlen with ultraviolet therapy	484208
2-Methylaziridine (propyleneimine)	75558
Methylazoxymentanol	590965
Methylazoxymentanol acetate	592621
3-Methylcholanthrene	54695
5-Methylchrysene	3697243
4,4'-Methylene bis(2-chloroaniline)	101144
4,4'-Methylene bis(N,N-dimethyl)benzeneamine	101611
4,4'-Methylene bis(2-methylaniline)	838880
4,4'-Methylenedianiline	101779
4,4'-Methylenedianiline dihydrochloride	13552448
Methylhydrazine and its salts	—
Methyl iodide	74884
Methyl methanesulfonate	66273
2-Methyl-1-nitroanthraquinone (of uncertain purity)	129157
N-Methyl-N'-nitro-N-nitrosoguanidine	70257
N-Methylolacrylmide	924425
Methylthiouracil	56042
Metiram	9006422
Metronidazole	443481
Michler's ketone	90948
Mirex	2385855
Mitomycin C	50077
Monocrotaline	315220
5-(Morpholinomethyl)-3-[(5-nitro-furfurylidene)-amino]-2-oxalolidinone	139913
Mustard gas	505602
Nafenopin	3771195
1-Naphthylamine	134327
2-Naphthylamine	91598
Nickel and certain nickel compounds	—
Nickel carbonyl	1346393
Nickel refinery dust from the pyrometallurgical process	—
Nickel subsulfide	12035722
Nididazole	61574
Nitrilotriacetic acid	139139
Nitrilotriacetic acid, trisodium salt monohydrate	18662538

Table 4.3A. "California" List of Carcinogens (continued)

Chemical	CAS Number
5-Nitroacenaphthene	602879
5-Nitro-o-anidisidine	99592
o-Nitroanisole	91236
4-Nitrobiphenyl	92933
6-Nitrochrysene	7490/6028
Nitrofen (technical grade)	1836755
2-Nitrofluorene	607578
Nitrofurazone	59870
1-[(3-Nitrofurfurylidene)-amino]-2-imidazolidinone	555840
N-[4-(5-Nitro-2-furyl)-2-thiazlyl]acetamide	531828
Nitrogen mustard (mechlolethamine)	51752
Nitrogen mustard hydrochloride (mechlolethamine hydrochloride)	5867
Nitrogen mustard N-oxide	126852
Nitrogen mustard N-oxide hydrochloride	302705
2-Nitropropane	79469
1-Nitropyrene	5522430
4-Nitropyrene	57835924
N-Nitrosodi-n-butylamine	924163
N-Nitrosodiethanolamine	1116547
N-Nitrosodiethylamine	55185
N-Nitrosodimethylamine	62759
p-Nitrosodiphenylamine	156105
N-Nitrosodiphenylamine	86306
N-Nitrosodi-n-propylamine	621647
N-nitroso-N-ethylurea	759739
3-(N-N-Nitrosomethylamine)propionitrile	60153493
4-(N-N-Nitrosomethylamino)-1-(3-pyridyl)1-butanone	64091914
N-Nitrosomethylethylamine	10595956
N-Nitroso-N-methylurea	684935
N-Nitroso-N-methylurethane	615512
N-Nitrosomethylvinylamine	4549400
N-Nitrosomorpholine	59892
N-Nitrosonornicotine	16543558
N-Nitrosopiperidine	100754
N-Nitrosopyrrolidine	930552
N-Nitrososarcosine	13256229
Norethistrone	68224
Ocratoxin A	303479
Oil Orange SS	2646175
Oral contraceptives, combined	—
Oral contraceptives, sequential	—
Oxadiazon	434071
Oxymetholone	434071
Panfuran S	—
Pentachlorophenol	87865

Table 4.3A. "California" List of Carcinogens (continued)

Chemical	CAS Number
Phenacetin	62442
Phenazopyridine	94780
Phenazopyridine hydrochloride	136403
Phenesterin	3546109
Phenobarbital	50066
Phenoxybenzamine	59961
Phenoxybenzamine hydrochloride	63923
Phenyl glycidal ether	122601
Phenylhydrazine and its salts	—
o-Phenylphenate, sodium	132274
Poligeenan	5397398
Polybrominated biphenyls	—
Polychlorinated biphenyls	—
Polychlorinated biphenyls (containing 60% or more chlorine by molecular weight)	—
Polychlorinated dibenzo-p-dioxins	—
Polychlorinated dibenzofurans	—
Ponceau MX	3761533
Ponceau 3R	3564098
Potassium bromate	7758012
Procarbazine	671169
Procarbazine hydrochloride	366701
Progesterone	57830
1,3-Propane sulfone	1120714
β-Propiolactone	57578
Propylene oxide	75569
Propylthiouracil	51525
Radionuclides	—
Reserpine	50555
Residual (heavy) fuel oils	—
Saccharin	81072
Saccharin, sodium	128449
Safrole	94597
Selenium sulfide	7446346
Shale oils	58308349
Silica, crystalline (airborne particles of respirable size)	—
Soots, tars, and mineral oils (untreated and mildly treated oils and used engine oils)	—
Sterigmatocystin	10048132
Streptozotocin	18883664
Styrene oxide	98093
Sulfalate	95067
Talc containing asbestoform fibers	—
Testosterone and its esters	58220
2,3,7,8-Tetrachlorodibenzo-p-dioxin (TCDD)	1746014
1,1,2,2-Tetrachloroethane	79345

Table 4.3A. "California" List of Carcinogens (continued)

Chemical	CAS Number
Tetrachloroethylene (p-chloroethylene)	127184
p,α,α,α-Tetrachlorotoluene	5216251
Tetranitromethane	509148
Thioacetamide	62555
4,4'-Thiodianiline	139651
Thiourea	62566
Thorium dioxide	1314201
Tobacco, oral use of smokeless products	—
Tobacco smoke	—
Toluene diisocyanate	26471625
o-Toluidine	95534
o-Toluidine hydrochloride	636215
p-Toluidine	106490
Toxaphene (polychlorinated camphenes)	8001352
Treosulfan	299752
Trichlormethine (trimustine hydrochloride)	817094
2,4,6-Trichlorophenol	88062
1,2,3-Trichloropropane	96184
Triphenyltin hydroxide	76879
Trichloroethylene	79016
Tris(aziridinyl)-p-benzoquinone (triazoquinone)	68768
Tris (1-aziridinyl)phosphine sulfide (thiotepa)	52244
Tris (2-chloroethyl) phosphate	115968
Tris (2,3-dibromopropyl) phosphate	126727
Trp-P-1 (Tryptophan-P-1)	62450060
Trp-P-2 (Tryptophan-P-2)	62450071
Trypan blue (commercial grade)	72571
Unleaded gasoline (wholly vaporized)	—
Uracil mustard	66751
Urethane (ethyl carbamate)	51796
Vinyl bromide	593602
Vinyl chloride	75014
4-Vinyl-1-cyclohexene diepoxide (vinyl cyclohexene dioxide)	106876
Vinyl trichloride (1,1,2-Trichloroethane)	106876
2,6-Xylidine (2,6-Dimethylaniline)	87627
Zineb	12122677

Table 4.3B. Reproductive Toxicants (continued)

Chemical	CAS Number
Acetohydroxamic acid	546883
Actinomycin D	50760
All-trans retinoic acid	302794
Alprazolam	28981977
Amikacin sulfate	3983555

Table 4.3B.   Reproductive Toxicants (continued)

Chemical	CAS Number
Aminoglutethimide	125848
Aminoglyosides	—
Aminopterin	54626
Angiotensin converting enzyme (ACE inhibitors)	—
Anisindione	117373
Aspirin (NOTE: It is especially important not to use aspirin during the last 3 months of pregnancy unless specifically directed to do so by a physician because it may cause problems in the unborn child or complications during delivery.)	50782
Barbiturates	—
Benomyl	17804352
Benzphetamine hydrochloride	5411223
Benzodiazepines	—
Bischloroethyl nitrosurea (BCNU) (carmustine)	154938
Bromoxynil	1689845
Butabarbital sodium	143817
1,4-Butanediol dimethylsulfonate (busulfan)	55981
Carbon disulfide	75150
Carbon monoxide	630080
Carboplatin	41575944
Chenodiol	474259
Chlorcyclizine hydrochloride	1620219
Clorambucil	305033
Chlordecone (kepone)	143500
Chlordiazepoxide	58253
Chlordiazepoxide hydrochloride	438415
1-(2-Chloroethyl)-3-cyclohexyl-1-nitrosourea (CCNU) (lomustine)	13010474
Clomiphene citrate	50419
Chlorazepate dipotassium	57109907
Cocaine	50362
Colchicine	64868
Conjugated estrogens	—
Cyanazine	21725462
Cycloheximide	66819
Cyclophosphamide (anhydrous)	50180
Cyclophosphamide (hydrated)	6055192
Cyhexatin	13121705
Cytarabine	147944
Danazol	17230885
Daunorubicin hydrochloride	23541506
Demeclocycline hydrochloride (internal use)	64733
Diazepam	439145
Dicumarol	66762
Diethylstilbestrol (DES)	56531

Table 4.3B.   Reproductive Toxicants (continued)

Chemical	CAS Number
Dinocap	39300453
Dinoseb	88857
Diphenylhydantoin (phenytoin)	57410
Doxycycline (internal use)	564250
Doxycycline calcium (internal use)	94088854
Doxycycline hyclate (internal use)	24390145
Doxycycline monohydrate (internal use)	17086281
Ergotamine tartrate	379793
Ethyl alcohol in alcoholic beverages	—
Ethylene glycol monoethyl ether	110805
Ethylene glycol monomethyl ether	109864
Ethylene glycol monoethyl ether acetate	111159
Ethylene glycol monomethyl ether acetate	110496
Ethylene thiourea	96457
Etoposide	33419420
Etratinate	54350480
Fluorouracil	51218
Fluoxymesterone	76437
Flurazepam hydrochloride	1172185
Flutamide	13311847
Halazepam	23092173
Hexachlorobenzene	118741
Ifosfamide	3778732
Iodine-131	24267569
Isotretinoin	4759482
Lead	—
Lithium carbonate	554132
Lithium citrate	919164
Lorazapam	846491
Lovastatin	75330755
Medroxyprogesterone acetate	71589
Megestrol acetate	595335
Melphalan	148823
Menotropins	9002680
Meprobamate	57534
Mercaptopurine	6112761
Mercury and mercury compounds	—
Methacycline hydrochloride	3963959
Methimazole	60560
Methotrexate	59052
Methotrexate sodium	15475566
Methyl bromide	74839
Methyl mercury	—
Methyltestosterone	58184

Table 4.3B.   Reproductive Toxicants (continued)

Chemical	CAS Number
Midazolam hydrochloride	59467968
Minocycline hydrochloride (internal use)	13614987
Misoprostol	62015398
Mitoxantrone hydrochloride	70476823
Nafgarelin acetate	86220420
Neomycin sulfate (internal use)	1405103
Netilmicin sulfate	56391572
Nicotine	54115
Nitrogen mustard (mechlorethamine)	51752
Nitrogen mustard hydrochloride (mechlorathamine hydrochloride)	55867
Norethisterone (norethindrone)	68224
Norethisterone acetate (norethindrone acetate)	51989
Norethisterone (norethindrone)/ethinyl estradiol	68224/57636
Norethisterone (norethindrone)/mestranol	68224/72333
Norgrestrel	6533002
Oxazepam	604751
Oxytetracycline (internal use)	79572
Oxytetracycline hydrochloride (internal use)	2058460
Paramethadione	115671
Penicillamine	52675
Pentobarbital sodium	57330
Phenacemide	63989
Phenprocoumon	435972
Pipobroman	54911
Plicamycin	18378897
Polychlorinated biphenyls	—
Procarbazine hydrochloride	366701
Propylthiouracil	51525
Retinol/retinyl esters, when in daily dosages in excess of 10,000 IU, or 3000 retinol equivalents. (NOTE: Retinol/retinyl esters are required and essential for maintenance of normal reproductive function. The recommended daily level during pregnancy is 8000 IU.)	—
Ribavirin	36791045
Secobarbital sodium	309433
Streptomycin sulfate	3810740
Tamoxifen citrate	54965241
Temazepam	846504
Testosterone cyoionate	846504
Testosterone enanthate	315377
2,3,7,8-Tetrachlorodibenzo-*p*-dioxin (TCDD)	1746016
Tetracycline (internal use)	—
Thalidomide	50351
Thioguanine	154427
Tobacco smoke (primary)	—

Table 4.3B.   Reproductive Toxicants (continued)

Chemical	CAS Number
Tobramycin sulfate	49842071
Toluene	108883
Triazolam	28911015
Trilostane	13647353
Uracil mustard	66751
Urofollitropin	26995915
Valproate (valproic acid)	99661
Vinblastine sulfate	143679
Vincristine sulfate	2068782
Warfarin	81812

Table 4.3C.   Female Reproductive Toxicity

Chemical	CAS Number
Aminopterin	54626
Anabolic steroids	—
Carbon disulfide	75150
Cocaine	50362
Cyclophosphamide (anhydrous)	50180
Cyclophosphamide (hydrated)	6055192
Ethylene oxide	75218
Lead	—
Tobacco smoke (primary)	—
Uracil mustard	66751

"Importation, Manufacture, Distribution and Storage of Explosive Materials," and any material classified as an explosive by the Department of Transportation in the Hazardous Material Regulations (Title 49 CFR, Parts 100-199). A list of the materials that are within the scope of Title 18, United States Code, Chapter 40 is published periodically by the Bureau of Alcohol, Tobacco, and Firearms, U.S. Department of the Treasury.

Classification of explosives by the U.S. Department of Transportation, Title 49 CFR, Chapter I is as follows (the wording follows that used by OSHA in Title 29 CFR Part 1-910.109(a)(3)):

1. Class A—"Possessing detonating or otherwise maximum hazard, such as dynamite, nitroglycerin, picric acid, lead azide, fulminate of mercury, black powder, blasting powder, blasting caps, and detonating primers."
2. Class B—"Possessing flammable hazard, such as propellant explosives (including some smokeless propellants), photographic flash powders, and some special fireworks."
3. Class C—"Includes certain types of manufactured articles which contain Class A or Class B explosives, or both, as components but in restricted quantities."

Table 4.3D. Male Reproductive Toxicity

Chemical	CAS Number
Anabolic steroids	—
Benomyl	17804352
Carbon disulfide	75150
Colchicine	64868
Cyclophosphamide (anhydrous)	50180
Cyclophosphamide (hydrated)	6055192
1,2-Dibromo-3-chloropropane (DBCP)	96128
m-Dintrobenzene	99650
o-Dintrobenzene	528290
p-Dintrobenzene	100254
Dinoseb	88857
Ethylene glycol monoethyl ether	110805
Ethylene glycol monomethyl ether	109864
Ethylene glycol monoethyl ether acetate	111159
Ethylene glycol monomethyl ether acetate	110496
Lead	—
Nitrofurantoin	67209
Tobacco smoke (primary)	—
Uracil mustard	66751

4. Forbidden or Not Acceptable Explosives—"Explosives which are forbidden or not acceptable for transportation by common carriers by rail freight, rail express, highway, or water in accordance with the regulations of the U.S. Department of Transportation, 49 CFR, Chapter 1."

Some activities involving explosives require a federal license or permit under Title XI, 18 Unites States Code, Chapter 40. Those activities not covered in these regulations are covered by NFPA 495, where the latter standard has been adopted as a legal requirement. Under NFPA 495, no explosive materials shall be sold or transferred in any way to a person without a valid permit to have them, and no one is to conduct any operations involving explosives without an appropriate permit. Laboratories that are engaged in research with an explosive would require a "Permit to Use" under NFPA 495, unless test blasts are involved, in which case an additional "Permit to Blast" would be required.

The OSHA standards are based on the 1970 version of NFPA 495 and differ in some respects from the current version. However, they still require stringent safety precautions which must be followed by any research facility employing or investigating explosive materials. In consequence, any acquisition of explosive materials should be internally reviewed to ensure the ability on the part of the intended recipient to provide adequate facilities and safeguards to comply with the standards.

4.3.4.6. Equipment

In Chapter 3 a large number of items of equipment were discussed in terms of the requirements which should be met in order to ensure that the equipment meets appropriate safety standards. These standards change as technological improvements occur, as new information becomes available, and as new or revised regulatory standards become effective. Acquisition of many items of equipment affecting safe operations of a facility should be internally reviewed, in part to ensure that the equipment being acquired meets current safety specifications, and in part to alert units within the organization that need to know of the purchase. In the latter case, it may be necessary to confirm that sufficient attention has been given to the installation. Among those items of equipment which need to be routinely reviewed, usually by the safety department, sometimes by buildings and grounds, and if separate from the latter, planning and engineering, are

- Biological safety cabinets
- Chemical fume hoods
- Electron microscopes
- X-ray equipment
- Equipment containing
  radiation sources

- Lasers
- Safety equipment
- Water stills
- Refrigeration equipment

4.3.5. Free Materials

It is tempting to accept free materials. However, with the advent of regulations providing for the safe and environmentally responsible disposal of hazardous chemicals under the Resource Conservation and Recovery Act (RCRA), the acceptance of free materials poses the risk that the "free" material may eventually generate unexpected and substantial disposal costs. In many cases, however, where the study of an experimental material is involved, free materials are the basis for the intended research program, but the amount proffered often is substantially more than is actually needed. The excess amount is typically unusable for other purposes and will require legal disposal. Unfortunately, a few organizations have shifted disposal costs to the recipient organization by giving inexperienced persons unwanted surplus or out-of-date materials. Even if this is not the case, when normal purchasing procedures are bypassed, it could increase the possibility of unrecorded dangerous materials arriving on site. All of these negative possibilities can be avoided by the adoption of a policy similar to the following regulating the receipt of free materials:

1. The amount of free material which may be accepted by an individual, laboratory, or other administrative unit to be used in a program of research must be limited to the amount which is likely to be actually needed in the proposed program.
2. The donor must agree in writing to accept the return of any unused amounts or pay for the legal and safe disposal of the material. The recipient may agree to waive this requirement if they are prepared to pay for the disposal from their own funds.
3. If the utilization or storage of the free material is likely to pose any substantive risk to personnel or property, the safety department and the risk management (insurance) office must be informed prior to completion of the agreement to accept the material, to allow time for a determination of whether the risks are acceptable or not, if adequate facilities are available, and to ensure that the proposed research has been reviewed for health and safety implications.

In some cases, undesirable materials may be a part of or within an item of equipment. In an example at the author's institution, an engineering department accepted a large "free" transformer, which eventually turned out to be unusable. The transformer contained 474 gallons of high-purity polychlorinated biphenyl (PCB) insulating oil. The total disposal cost of the transformer and its contents was well over $10,000 because of the PCBs. Some states have passed laws or published policies requiring any state agency to audit any property the agency might acquire, either as a purchase or as a gift, for evidence of any prior hazardous waste dumping.

These policies may appear unnecessarily formal to many individuals. However, the cost of disposing of hazardous materials and the rate of increase of these costs have reached the point that hazardous waste disposal is becoming a major problem for many corporations and institutions. Most cannot afford to incur further costs by accepting free materials without consideration of the future obligations which the gift may engender.

REFERENCES

1.   **Bilsom, R.E.,** *Torts Among the Ivy: Some Aspects of the Civil Liability of Universities,* University of Saskatchewan, Saskatoon, Saskatchewan, Canada.
2.   **Gerlovich, J.A. and Downs, G.E., Eds.,** Legal liability, in *Better Science Through Safety,* Iowa State University Press, Ames, IA, 1981, 17.
3.   **Kaufman, J.,** Laboratory Safety Workshop, Curry College, Milton, MA.
4.   Occupational Safety and Health Administration, Hazard Communication; Final Rule 29 CFR Parts 1910, 1915, 1917, 1918, 1926, and 1928, *Fed. Reg.,* 52(163), 31852, 1987.
5.   Occupational Health Services, Inc., New York, NY.
6.   Nuclear Regulatory Commission, Title 10, Code of Federal Regulations; Parts 20, 30 and 33, Washington, D.C., 1988.
7.   Food and Drug Administration, Title 21, Part 1300, Washington, D.C.
8.   U.S. Public Health Service, Title 42, Parts 71.156 and 72, Washington, D.C.
9.   Department of Transportation, Title 49, Code of Federal Regulations, Sections 173.386 to 173.388, Washington, D.C.
10. Occupational Safety and Health Administration, Title 29, Code of Federal Regulations, Subpart Z, Washington, D.C.
11. Title 18, United States Code, Chapter 40, Importation, Manufacture, Distribution, and Storage of Explosive Material, 1988.
12. Department of Transportation, Title 49, Code of Federal Regulations, Parts 100 to 199, Washington, D.C., 1988.
13. Occupational Safety and Health Administration, Title 29, Code of Federal Regulations, Part 1910, §109, Washington, D.C., 1988.
14. Code for the Manufacture, Transportation, Storage and Use of Explosive Materials, NFPA 495, National Fire Protection Association, Quincy, MA, 1982.

4.4. PURCHASING OF ANIMALS*

4.4.1. Introduction

Nothing can sabotage an animal-related experiment more quickly than the use of animals with latent (hidden) or overt signs of disease. The use of healthy, unstressed animals is critical to obtaining quality experimental results. The following information will aid in the selection of appropriate vendors and provide criteria for selecting "clean" animals.

---

*   This section was written by David M. Moore, D.V.M.

4.4.1.1. Selection Criteria for Rodents and Rabbits

Animals purchased from commercial vendors could have latent bacterial, viral, or rasitic diseases which can radically affect experimental results.[1-3] To avoid the use of infected ("dirty") animals as well as the contamination and subsequent infection of "clean" animals in a facility, selection criteria for the purchase of animals should be established.

Commercial vendors should be required to submit copies of their monthly or semiannual quality assurance health testing reports for review prior to purchase. Testing includes serological evaluation for the presence/absence of microbial pathogens (disease-causing bacteria), and evaluation for internal and external parasites. Small[4] lists the organisms which should *not* be present in mice, rats, hamsters, guinea pigs, and rabbits prior to purchase.

Since the vendor animal health report represents a snapshot in time and does not guarantee avoidance of subsequent contamination in the vendor's facility or during transport to the research facility, it would be wise to test a representative sample of the animals after receipt while they are housed in the quarantine area. Should serological testing reveal a latent viral infection, the entire group of animals should be destroyed and the quarantine area thoroughly disinfected. Studies using inbred strains of rats and mice can be decimated by genetic impurity. Vendors also test inbred strains for genetic purity, and a copy of their most recent test report can be requested prior to animal purchase.

4.4.1.2. Laws Affecting Animal Purchasing

The Federal Animal Welfare Act (PL 89-544) governs the purchase of dogs and cats for use in research and teaching. Under this law, these species may be purchased only from (1) a dealer licensed with the U.S. Department of Agriculture Animal, Plant Health Inspection Service (USDA/APHIS), (2) a commercial breeder licensed with USDA/APHIS, (3) local or county animal shelters (recognizing that, at this time, 13 states prohibit the sale of pound animals for research or teaching), or (4) another research institution which has obtained the animals from any of these sources.

Research facilities must keep the following records on dogs and cats, usually on the USDA Individual Health Certificate and Identification Form (VS Form 18-1):

1. The name, address, and license number of the dealer from whom the animal was purchased.
2. The date of acquisition of each live dog and cat.
3. The official USDA tag number or tattoo assigned to each animal.
4. A description that includes species, sex, date of birth or approximate age, color and distinctive markings, and breed or type.
5. Any identification number assigned to that animal by the research facility.

4.4.1.3. Transportation of Animals

Transport of animals from the vendor to the research facility is also regulated by the Animal Welfare Act. Regulations on ambient temperature limits might prohibit shipment of animals on days that are too cold or too hot. Since most shipments are by common carrier (airplane, truck), one cannot be sure that animals were not subject to environmental stressors (heat or cold) while in loading areas or during transport. Heat stress can cause debilitation or death in rodents and other species. Additionally, animal transport boxes from several vendors might be shipped together, with the potential for cross-contamination of "clean" animals by "dirty" animals during shipment. Some vendors ship animals in their own environmentally controlled vehicles to preclude this problem, but this service is not available to all parts of the country.

4.4.1.4. Additional Laws Affecting Animal Purchase

Some states, such as California, prohibit entry of certain species (i.e., gerbils, ferrets) into the state without appropriate approval forms. Contact the state veterinarian in your state to ascertain whether similar regulations/restrictions exist.

REFERENCES (SECTIONS 4.4 TO 4.4.1.4)

1. **Bhatt, P.**, Virus infections of laboratory rodents, *Lab. Anim.*, 9(3), 43, 1980.
2. **Orcutt, R.P.**, Bacterial diseases: agents, pathology, diagnosis, and effects on research, *Lab. Anim.*, 9(3), 28, 1980.
3. **Hsu, C.K.**, Parasitic diseases: how to monitor them and their effects on research, *Lab. Anim.*, 9(3), 48, 1980.
4. **Small, J.D.**, Rodent and lagomorph health surveillance-quality assurance, in *Laboratory Animal Medicine*, Fox, J.G. et al., Eds., Academic Press, Orlando, FL, 1984, p. 709.
5. **Smith, K.P., Hoffman, H.A., and Crowell, J.S.**, Genetic quality control in inbred strains of laboratory rodents, *Lab. Anim.*, 11(7), 16, 1982.
6. **Weisbroth, S.H.**, The impact of infectious disease on rodent genetic stocks, *Lab. Anim.*, 13(1), 25, 1984.

4.5. STORAGE

Laboratory storage practices may enhance or diminish overall laboratory safety. There are many factors to be considered in addition to those concerned with flammable materials, touched upon in Chapter 3 in the design and selection of facilities and equipment for those specific substances. Among these are the amount, location, and organization of the stored chemicals. The types of vessels in which they are contained and the information on the container labels are important. Some types of materials represent special hazards for which specific protective measures may be indicated or perhaps are mandatory due to regulations. The following sections will address these topics.

4.5.1. Compatible Chemical Storage

Many laboratories, if not the majority, find it easiest to store a large portion of their chemicals alphabetically, although there are often partial exceptions, even in these facilities. The most frequently used flammable liquids, for example, are frequently placed in a common area. Other less frequently used flammable liquids may still be stored on shelves with other chemicals. Comparable usage may occur for acids, bases or other heavily used reagents. For the most part, however, general chemical storage in a laboratory is often done without regard to compatibility. Determination of compatibility does require some effort, as reflected by Table 4.4, adapted from material in the *Journal of Hazardous Materials*.

Table 4.4.   Compatible Chemical Groups

Number	Chemical Group	Do Not Store With Group Numbers
1	Inorganic acids	2–8, 10, 11, 13, 14, 16–19, 21–23
2	Organic acids	1, 3, 4, 7, 14, 16–19, 22
3	Caustics	1, 2, 5, 7, 8, 13–18, 20, 22, 23
4	Amines and alkanolamines	1, 2, 5, 7, 8, 13–18, 23
5	Halogenated compounds	1, 3, 4, 11, 14, 17
6	Alcohols, glycols, glycol ethers	1, 7, 14, 16, 20, 23
7	Aldehydes	1–4, 6, 8, 15–17, 19, 20, 23
8	Ketones	1, 3, 4, 7, 19, 20
9	Saturated hydrocarbons	20
10	Aromatic hydrocarbons	1, 20
11	Olefins	1, 5, 20
12	Petroleum oils	20
13	Esters	1, 3, 4, 19, 20
14	Monomers, polymerizable esters	1–6, 15, 16, 19–21, 23
15	Phenols	3, 4, 7, 14, 16, 19, 20
16	Alkylene oxides	1–4, 6, 7, 14, 15, 17–19, 23
17	Cyanohydrins	1–5, 7, 16, 19, 23
18	Nitriles	1–4, 16, 23
19	Ammonia	1–2, 7, 8, 13–17, 20, 23
20	Halogens	3, 6–15, 19, 21, 22
21	Ethers	1, 14, 20
22	Elemental phosphorus	1–3, 20
23	Acid anhydrides	1, 3, 4, 6, 7, 14, 16–19

There are longer and more detailed lists of compatible chemicals which could be used to determine appropriate storage. However, even the system represented by Table 4.4 may be too elaborate to encourage individuals to use it. Several of the major chemical firms have developed less complicated systems using a color code to define the groups which should be stored together. Unfortunately, although there are some similarities, the schemes of the different companies are not wholly compatible. The color code systems of two major chemical companies are shown in Figure 4.2, for comparison with each other and with Table 4.4. The color codes used to define the major groups for chemical vendors are prominently incorporated into the label. This makes it easy to decide where to place a new container and return it after use. Although providing less selectivity in segregating different materials than a system listing individual chemicals that should not be stored together, or a list of classes of materials, as in Table 4.4, the ease of use should make color coding acceptable to most laboratory managers and employees.

Both of these systems provide an alternative for the color blind. In the Fisher system, the first letter of the color is displayed prominently in the color bar, while the Mallinckrodt system spells the color out.

Figure 4.2. Chemical label color code schemes for two vendors.

Other firms have similar systems, which also vary slightly in detail. Usually they agree for chemicals coded red, blue, yellow, and white, but differ in the way they denote exceptions or indicate a material for which there are few or no storage problems. For example, the J.T. Baker Company, one of the pioneers in developing a color-coded storage system, uses striped colors to denote exceptions and orange to denote chemicals which may be stored in the general storage area.

Until an accident occurs in which the contents of the different containers come into contact with each other, failure to store materials according to compatibility may have few, if any, repercussions. In some instances, mixing of spilled chemicals will result in no reaction or relatively nonviolent reactions which can be controlled easily. However, in other cases, the result could be the insidious release of deadly toxic fumes or perhaps a violent explosion or fire. In any event, failure to store chemicals according to their properties is too much of a risk to personnel, to property, and possibly even to the intellectual value of accumulated research data files that may represent the product of years of effort. It is inexpensive insurance to make the relatively modest effort to segregate chemicals according to a color code system or even to follow a more complicated program using groups such as those defined in Table 4.4.

4.5.2. Labeling

There are two important types of labels in laboratories. The labels on commercial containers are usually extremely comprehensive, providing not only information on the nature, amount, and quality of the product but also a very large amount of safety-related data. Typically, a commercial label will readily meet the requirements of the hazard communication standard. On the other hand, labels placed on secondary containers in the laboratory by employees may be something such as "soln. A" or even less. This may be sufficient if all of the material is to be promptly used by the individual placing the label on the container, but otherwise it is not. In most instances, secondary containers of hazardous chemicals should be marked with labels identifying the chemical in the container and providing basic hazard warnings. The secondary label should be affixed before the container is put into use.

Although, as noted earlier, the laboratory standard preempts the hazard communication standard for laboratory employees, in order to provide equivalent protection, it is difficult to imagine an alternative labeling procedure which would be equivalent without meeting the requirements of the hazard communication standard. Under the latter standard, the container label containing the hazard information must be in English as the primary language. As long as an English version is on the label, the same information may be provided in other languages as well to meet the needs of the personnel in the area. The intent of the labeling requirements under the hazard communication standard is primarily to protect the immediate users of the material by ensuring that they have access to the identity of the material with which they are working. However, superficial and uninformative labels cause a major problem for the legal disposal of containers possibly holding hazardous chemicals. It is expensive to dispose of unknown materials under RCRA rules. This is a serious problem in academia, due to the rapid turnover in research personnel, especially students as they finish their degree programs. Full compliance with the hazard communication standard should greatly alleviate this problem.

Laboratory employees should familiarize themselves with the commercial labels on the chemical containers that they are using. The labels will reflect the information available to the manufacturer at the time the material was packaged. The information will not be as current as that provided by an up-to-date MSDS. These are revised as often as significant new information becomes available. However, in most cases the information for a specific chemical will change sufficiently slowly so that the information on the label may be used with reasonable confidence.

An examination of typical chemical labels will reveal that the following safety-related information is included on the labels of virtually all manufacturers:

- Name of compound
- CAS number
- Impurities, other components
- Flash point (if applicable)
- NFPA hazard diamond
- Risk descriptor (Danger, Warning, Caution)

- Risk descriptive statement
- Storage color code
- Handling advice
- Recommended fire extinguisher class
- First aid emergency medical advice
- UN number

Many labels contain additional useful information. The following items also are found frequently on chemical container labels:

- DOT symbols
- Hazard ratings (may be different from NFPA ratings for laboratory applications)
- Recommended protective equipment (some do this in words, some by means of stylized pictographs)
- Target organs
- Symptoms of exposure
- Spill response procedures
- Location for dating receipt of material
- Bar code
- Space for local information

4.5.3. Regulated Materials

Many regulated materials require special storage facilities, primarily for security reasons, although the security requirements frequently are based upon safety or environmental concerns. The following materials were discussed in Section 4.3.4 concerning restrictions on their purchase:

- Explosives
- Controlled substances (drugs)
- Radioisotopes
- Etiologic agents
- Carcinogens

This list is approximately in order of the security required, although there is considerable overlap.

4.5.3.1. Explosives

Explosives pose the most immediate danger to individuals of any of the items on the list in the preceding section. OSHA covers the storage requirements for all Class A, B, and C explosives, any special industrial explosives, and any newly developed and hence unclassified explosives in 29 CFR Part 1910.109(c). All of these materials must be kept in magazines which meet the specified requirements in the same section. As noted earlier, the OSHA standard is based on a 1970 version of NFPA Standard 495, although it does include some amendments added in 1978. Where the OSHA requirements do not cover a specific point, the most recent version of NFPA 495 may be consulted. Some state and local jurisdictions may have adopted later versions as part of their codes or may have added special regulations of their own.

The OSHA storage requirements differentiate between storage of less than 50 pounds ($\approx$22.7 kg) and more than 50 pounds. Some large research projects may require more than 50 pounds to be available at a given time, but most programs typically involve much smaller amounts, often on the order of a few grams. Class 1 magazines which are required for the larger quantities have structural requirements appropriate to room size spaces, while Class 2 magazines, appropriate for smaller operations, may be mounted on wheels and are mobile. This section will be limited to a discussion of Class 2 magazines, which are the most suitable for typical laboratory-scale use of explosives.

Normally, explosive magazines are expected to be located outdoors. However, a Class 2 magazine may be permitted within a building (the OSHA standard specifically references warehouses and wholesale and retail establishments) when it is located on a floor which has an entrance at outside grade level and the magazine is not located more than 10 feet from the entrance. Also, it is not normally expected to have two magazines within the same building, but if one is used solely to store no more than 5000 blasting caps and it is at least 10 feet from the other, it is permissible to do so.

Class 2 magazines may be constructed primarily of wood or metal and will normally be a combination of the two. The wood of the bottom, sides, and cover of a primarily wooden magazine must be of 2-inch hardwood. The corners must be well braced and the magazine must be covered with sheet metal of not less than 20-gauge thickness. In order to avoid contact of the stored explosive with metal, any exposed nails in the interior must be well countersunk.

Primarily metal magazines must have their bottom, top, and cover constructed of sheet metal, and lined on the interior with 3/8 inch plywood or the equivalent. The edges of the metal covers must overlap the sides by at least 1 inch. Both metal and wood magazines must be lockable. The covers must be attached securely with substantial strap hinges.

Class 2 magazines must be painted red and labeled on each of the four sides and the top with white letters at least 3 inches high:

EXPLOSIVES—KEEP FIRE AWAY

When Class 2 magazines are kept inside a building, they must be equipped with substantial wheels or casters to make it possible to remove them from the building in the event of a fire. When necessary due to climate, Class 2 magazines must be ventilated.

The general OSHA standards for the method of storage of explosives within magazines are generally intended to apply to larger units, but a number are appropriate for smaller Class 2 units. Among the more relevant are

1. Packages must be stored lying flat with the top side up. Piles of packages or containers must be stable.
2. The oldest material of a given type of explosive must be used first to minimize the risks associated with instability upon aging.
3. Packing and unpacking of explosives must not be done within 50 feet of a magazine.
4. Except for metal tools to cut open fiberboard boxes, all other tools must be nonsparking.
5. The magazine must not be used for the storage of metal tools or for other general storage.
6. Smoking, matches, open flames, spark-producing devices, firearms (other than those in the possession of guards), and combustible materials must not be permitted within 50 feet of the magazine.
7. A competent individual must be charged with the care of the magazine at all times and is responsible for enforcement of all safety precautions. It is most important that access to the explosives in the magazine be limited to those who have demonstrated or documented experience in the safe use and handling of the explosive materials stored in the magazine. It is also important that an accurate record of the contents be maintained. This record should include an identification of each separate package in the magazine, when it was placed inside, the contents of the package, and the date and amounts of any materials that have been removed. The log should also identify individuals by name, not initials, who have been permitted to remove explosives. The log should be audited periodically by an independent person.

In addition to the materials specifically classified as explosives, there are many chemicals which, under appropriate conditions, can act as explosives, i.e., react or decompose very rapidly, accompanied by a large release of energy. The violence of an explosively rapid reaction is largely dependent upon the gas pressures produced in the reaction, enhanced by any thermal energies produced.

The best precaution in working with potential explosives is to minimize the amounts actually present. In part, this may be achieved by maintaining proper inventory control in order to dispose of chemicals which tend to form unstable materials with age such as isopropyl ether, in which peroxides form, or perchloric acid or picric acid, which become dangerous if they are allowed to dry.

Other safety measures which should be taken in storing these laboratory materials are

1. Keep the minimum quantities needed in a cool, dry area protected from heat and shock.
2. The materials should be segregated during storage from materials with which they could react, as well as flammables, corrosives, and other chemicals which are likely to interact with each other.
3. Potentially explosive materials should be stored and used in an area posted with a sign in prominent letters:

CAUTION
POTENTIAL EXPLOSIVE HAZARD

4. If the material is being kept because of its potentially explosive properties, it should be treated as an explosive of the appropriate class and kept in a magazine or the equivalent.
5. Make sure that all occupants of the laboratory are aware of the potential risks and are trained in emergency procedures, including evacuation procedures, fire containment, and emergency first aid for physical injuries that might result from an explosion.

As noted, many laboratory chemicals may cause an explosion under appropriate conditions. Table 4.5 represents a brief list of chemicals which have a reactivity rating of 3 or 4 (mostly 4) according to the NFPA No. 704M system for the identification of hazardous chemicals and are sensitive to shock, heat, and/or friction. Materials on this list and those with comparable properties should always be treated with extreme care.

4.5.3.2. Controlled Substances (Drugs)

It was pointed out in Section 3.2.1.2.2 that the major concern where controlled substances are involved in research is security or control of the materials. This is particularly important for those materials which are usable by individuals for personal use or for sale as narcotics. The storage unit must be sufficiently strong to prevent forced entry for at least 10 minutes or more and must be either sufficiently heavy (750 pounds or more) or be rigidly bolted to the floor or wall to prevent the entire storage unit from being carried away. This is relatively easy to provide, but, as with explosives, it is equally important that a complete current record be maintained of all materials in the storage cabinet. The log should contain the following information: the date and amounts of each substance placed in the storage cabinet, and the dates and amounts disbursed from it. The amounts must be accurately quantified. The name, not initials, of the person to whom the materials are issued should be recorded. Distribution of the material should be under either the direct control of the principal investigator identified as the holder of the license allowing possession and use of controlled substances, or an alternate designated by that person in writing.

Unfortunately, because human nature is involved, narcotics intended for research purposes are occasionally abused by those persons who may legally possess them. Provision should be made for a periodic audit of the utilization of controlled substances by each licensee to ensure that any such behavior be quickly detected, for the licensee's own protection and for that of the organization.

Table 4.5.   Highly Reactive Shock/Heat Sensitive Materials

Ammonium perchlorate	Dibenzoyl peroxide
Ammonium permanganate	Diisopropyl peroxydicarbonate
Anhydrous perchloric acid	Dinitrobenzene (ortho)
Butyl hydroperoxide	Ethyl methyl ketone peroxide
Butyl perbenzoate	Ethyl nitrate
t-Butyl peroxyacetate	Hydroxylamine
t-Butyl peroxypivalate	Peroxyacetic acid
1-Chloro-2,4-dinitrobenzene	Picric acid
Cumene hydroperoxide	Trinitrobenzene
Diacetyl peroxide	Trinitrotoluene

4.5.3.3. Radioisotopes

Radioactive by-product materials are probably the most closely regulated research materials in wide use. It is rare for any organization involved to a significant degree in research, especially in the life sciences, not to have a license to use radioactive materials. The possession of the license and use of radioactive material under the provisions of the license commits the organization to establishing a radiation safety committee and to designate a radiation safety officer (RSO). Among the duties of this officer, under the terms of the license, are to make sure that each person authorized to have or use radioisotopes or sealed sources of radiation have no more than they are authorized to possess at one time, and that the total amount of each radioisotope does not exceed the overall limits provided for under the organization's broad license.* It is necessary for each authorized user to make sure that all radioactive materials in their possession are securely stored. Normally, unlike controlled substances, most radioisotopes have no practical use outside the research laboratory so that the strength of the storage facilities is not comparably demanding, but whenever a legitimate user is not physically present, even for short intervals, the radioactive material must be under lock and key. Failure to provide this security would probably result in a serious citation if discovered by a NRC inspector, with the organization being fined. It is necessary to keep track of the materials used, although the type of use makes it difficult to do so with high accuracy. For example, if the use of ^{14}C results in generation of carbon dioxide, a somewhat uncertain amount may escape into the exhaust system in a fume hood, while some of the remainder may be retained within the experimental materials and another part in the apparatus.

Many radioactive materials used in the life sciences are incorporated in materials that require refrigeration to prolong their usefulness. Not only must a refrigerator or freezer used for radioactive storage be lockable, but care must be taken to avoid contaminating the unit and its contents by spillage or leakage of material from a container due to freezing of the contents. In such a case, the lost material usually will be trapped in the ice or frost

---

* Most larger organizations and many smaller ones have a "broad" license to use radioactive materials covering the entire organization, or at least that part of it at a single location. Although it is possible to have individual users directly licensed by the NRC, it is still required, in the latter case, to establish a committee and safety officer to monitor the use of radioisotopes.

within a freezer. The trapped material could represent a personnel problem or the possibility of an uncontrolled release of radioactivity to the sanitary system when the freezer is defrosted.

Not only must the experimental radioactive materials be stored properly, but all of the waste products which could contain any residual active materials must be kept and stored within the laboratory until they can be disposed of safely, usually by a radiation safety specialist. The temporary storage of radioactive waste gives rise to the possibility of accidental removal of the radioactive waste as ordinary trash. Any trash containers containing radioactive waste should be distinctively marked, and custodial personnel should receive special training to recognize the containers and instructions not to combine the contents with ordinary trash. Laboratory personnel must be sure to place radioactive waste and potentially contaminated waste in these special receptacles. Written labels or signs are not sufficient to prevent accidental losses of radioactive waste in this way. An estimate of the amount of activity going into the waste containers needs to be maintained. If, by chance, a waste can is disposed of inappropriately, it will be necessary to make a reasonable estimate of the amount lost in order to judge the potential risk to the public. Of course, an effort would be expected of the organization to recover the lost radioactive material.

The inability to account for the radioactive material in the possession of a licensee is likely to be taken as a serious event by the NRC or its local surrogate in an agreement state. Each loss of radioactive material, if in such quantity and under such circumstances as to potentially pose a hazard to persons in unrestricted areas, must be reported immediately by telephone and telegraph to the Director of the NRC for the region in which the facility is located. More details on the reporting process will be provided in Chapter 5. Even if the amounts lost do not appear to pose a significant risk, operational procedures for the laboratory in question should be reviewed by the internal radiation safety committee, and the NRC notified as an item of information. Operations of facilities where continuing problems of accountability occur, even for minor problems, should be carefully monitored because they can lead to a loss of credibility for the oversight program for the entire institution or commercial research laboratory. A pattern which appears to reflect poor governance of the facilities using radiation, with a subsequent failure of the radiation safety program to take prompt effective corrective actions has resulted in the NRC imposing substantial fines in recent years.

Another security issue that has been discussed during recent NRC inspections has been the question of ensuring that only authorized persons have reason to be in the facilities where radioactive materials are in use. Most radioisotope labs do not use radioactive materials in such quantities that they meet the requirements for restricted areas under 10 CFR Part 20. Usually when personnel are at work, the laboratory doors are not locked. It is not incorporated in the standards, but it has been made clear that it would be desirable for holders of NRC licenses to challenge the presence of any person not explicitly known to the employees as being in the area appropriately. In industry, most employees in a facility wear badges identifying them, but in academic institutions where most scientists also teach, such a policy would make it difficult for them to maintain free access to them by their students. NRC inspections are always unannounced, so all security and compliance measures must be current and ready to withstand an inspection at any time.

Many laboratories conduct operations which do not require the use of radioactive materials in the same space as those in which radioactive materials are employed. In many of these instances different personnel are employed in the two programs. Nonusers of radioisotopes should be made sufficiently aware of the procedures required for the safe, legal use of radiation so that they will neither inadvertently violate any safety requirements for the use of by-product materials, nor misunderstand any actions of the employees involved with radiation. While the latter persons are present, they can and must take precautions to avoid exposing the other persons in the laboratory unnecessarily to radiation, but when they are not present the nonusers need to be aware of the areas where radioactive sources and waste are stored and areas they should avoid if there is any possibility of contamination. Any area containing radioactive materials should be clearly marked with signs bearing the radiation symbol and the label:

CAUTION
RADIOACTIVE MATERIALS

4.5.3.4. Etiologic Agents

Laboratories which employ etiologic agents in their research in addition to chemicals are different in a number of ways from the ordinary chemical laboratories, primarily because the agents which are involved are often infectious to humans. There are data to support that there does exist a higher rate of incidence of the diseases associated with the organisms which are involved with the research, but data also exist which show that there are virtually no secondary infections from the primary laboratory infections. There is a wide variation in the level of professionalism among the employees in laboratories employing biological agents, so it is extremely important that the laboratory manager and all employees set and maintain high standards of safety.

There is less agreement in this area than most as to the actual risks posed by a number of operations and specific etiologic agents. In many cases, the scale of the operation determines the level of safety required, and in other cases there are other factors such as the availability of a safe, effective vaccine. The CDC has published a safety guide to assist the laboratory director in making appropriate decisions as to the level of biosafety precautions required and has defined sets of standard practices, special practices, containment equipment, and laboratory equipment appropriate for each level. A condensed version of the salient features of these guidelines will be found in Section 5.3. Note that the word "guidelines" was used. The CDC manual is a set of recommended practices that do not have the force of law; however, they do form a body of information which has been carefully reviewed by a large number of highly trained, experienced, professional individuals. It is strongly recommended that they be incorporated into a formal biosafety program at any academic or corporate research institution that has a substantial level of research in the life sciences.

Although a biosafety committee is normally not required by law, unlike for radiation safety, the creation of one to monitor the use of etiologic agents within an organization is highly desirable. For specific areas such as human subjects, recombinant DNA, and research involving the use of animals, regulatory committees are required. If a committee is provided with an appropriate charge, it can be of considerable help in expenditures to provide safety resources and services for the biological laboratories as well as providing a uniform set of safety and performance standards for the laboratories.

In the context of this section any biologically active materials which could result in human infections should be stored and used in such a way as to preclude or minimize the probability of infections for all laboratory employees, as well as any other employees, such as custodial workers and maintenance personnel, and visitors to the laboratory. This can best be done by limiting access to the facility to those who have a specific need to be present, by keeping materials in active use within the appropriate enclosure, by keeping materials not currently required in storage, and by decontaminating work surfaces frequently. Strict controls are recommended in the guidelines for the two most restrictive biological safety levels. Specific regulations are dictated by the OSHA bloodborne pathogen standard for exposures to human blood, tissue, mucus, and other bodily fluids. This standard is intended to control exposure to the HIV (AIDS) and hepatitis B viruses, but the same regulations could be used for any human pathogen. Due to an increase in tuberculosis as a secondary infection related to the spread of AIDS, it is likely that there will be additional regulations in which tuberculosis will be targeted. Infectious waste regulations also arise from the concern about AIDS but apply to any human pathogen. There will be separate sections later in this chapter on both the bloodborne pathogen and infectious waste issues.

4.5.3.5. Carcinogens

Cancer in its many forms is the second largest cause of death in the United States and most other developed countries. Anything which increases the likelihood of initiating the eventual onset of cancer, therefore, is a matter of significant concern to those working with such materials. Although there are a few specific materials which have been known or suspected for some time to be carcinogenic agents, the long latency period for many materials, ranging from as few as 5 years to as many as 30, has delayed identification of causal relationships in many cases. OSHA has in the past regulated possible carcinogens on a case by case basis and continues to do so. Chemically specific regulations listed in Subpart Z of the OSHA regulations pertaining to exposure levels and medical surveillance requirements are not preempted by the laboratory safety standard. This is the basis of the list in Section 4.3.4.4. As a result of the difficulty of generating a standard for each chemical and taking the standard through the entire regulatory process, OSHA proposed and has adopted a generic carcinogen standard, 29 CFR 1910.101-152. In this standard, which was intended to shorten the regulatory process, they proposed standards and procedures for evaluating human epidemiological data and animal studies so that materials which met these standards would then be considered to be carcinogenic. At this time, disagreements on the validity of the criteria have resulted in no new materials being considered carcinogens under the new standard, with the only ones currently regulated as carcinogens being materials which have individually gone through the standard setting process.

Before a research program is begun using any of the currently regulated carcinogens, or any which may become regulated in the future by either the individual standard-setting process or the generic standard, there are a number of factors which should be considered:

1. Are there suitable alternative, noncarcinogenic materials which could be used?
2. Are there acceptable permissible exposure limits and ceiling limits below which it would be acceptable to use the material, with resultant exemptions from some restrictions upon the use of the material?
3. If the answer to question 2 is no, then can the provisions of the standard be met? An examination of the standards for the currently regulated carcinogens needs to be done prior to acquiring the material.

It would be inappropriate in this section, which is concerned primarily with storage (and security) of materials, to pursue the subject of research with carcinogens further at this time. However, unless there is a firm commitment to compliance with all regulations pertaining to carcinogenic research, it makes very little sense to acquire a carcinogenic chemical, especially if it is regulated or appears likely to become regulated, so that it cannot be used without complying with a perhaps onerous set of restrictions. The responsibility of assuring that it is not used and the potential liability if someone exposed to the material were to eventually develop cancer would appear to preclude acquiring the material unless a definite research program exists of sufficient merit which would justify the risk and the compliance effort.

If a program mandating the use of a carcinogen does exist, the following recommendations should be considered as a basis for minimizing the risks to the employee. The materials should be kept in either a secure cabinet in sealed, unbreakable containers or in a sealed, protected system. A current inventory of the quantities held must be maintained. The containers should be marked:

DANGER, CONTAINS _____
CANCER HAZARD

Containers for a potential carcinogen in which the evidence of carcinogenicity is based on limited but suggestive evidence should be labeled:

POTENTIAL CANCER HAZARD

The laboratory standard has modified the area control requirements to simply stating that an area must be designated as the area in which the work is to be done, and suggests that this area could be as small as a fume hood. By also stipulating that the laboratory and the organization describe the additional protective measures needed to work with extremely dangerous materials, with carcinogens being used as an example of such a material, it is implied that special care needs to be taken by anyone in the area where this work is to be done. Therefore, it would not be inappropriate for these areas to be identified by signs such as:

DANGER, _____
CANCER HAZARD
AUTHORIZED USERS ONLY

Under the laboratory safety standard additional safety measures are expected to be provided to any employee working with carcinogens. As with other dangerous materials they must be informed of the potential risk and procedures to be followed in an emergency. An emergency plan specific to the area and the research program must be developed in order to prevent exposures of the normal occupants of the area and to prevent the accidental exposures of others outside the immediate research facility. The training program should include everyone who may work within the designated area and handle the materials, including persons who may only stock the shelves and those individuals performing custodial and maintenance services, unless all such services are performed only with operations suspended and in the presence of a qualified laboratory employee who is fully trained.

REFERENCES

1.   *J. Hazardous Materials,* 1, 334, 1975.
2.   Allied Fisher Scientific, Fairlawn, NJ, 1986.
3.   LabGuard Safety Label System, Mallinkrodt, Paris, KY, 1986.
4.   Baker Saf-T-Data Guide, J.T. Baker Chemical Company, Phillipsburg, NJ, 1987.

5. *Chemical Labeling Standard*, ANSI Z129.1 (pending), American National Standards Institute, New York, 1987.

6. Occupational Safety and Health Administration, General Industry Standards, Title 29, Code of Federal Regulations, Part 1910.109(c).

7. Occupational Safety and Health Administration, General Industry Standards, Title 29, Code of Federal Regulations, Part 1910.1200.

8. Occupational Safety and Health Administration, General Industry Standards, Title 29, Code of Federal Regulations, Part 1910.1450.

9. *Hazardous Chemical Data*, NFPA 49, National Fire Protection Association, Quincy, MA.

10. Nuclear Regulatory Commission, Title 10, Code of Federal Regulations, Part 20, §207, Washington, D.C.

11. Biosafety in Microbiological and Biomedical Laboratories, HHS Publication No. 88-8395, Dept. of Health and Human Services, Washington, D.C., 1988.

12. **Favero, M.S.,** Biological hazards in the laboratory, in *Proc. Institute on Critical Issues in Health Laboratory Practices,* Richardson, J.H., Schoenfeld, E.S., Tulis, J.J., and Wagener, W.W., Eds., DuPont, Wilmington, DE, 1985.

13. **Pike, R.M.,** Laboratory-associated infections: incidence, fatalities, cases, and prevention, *Annu. Rev. Microbiol.,* 33, 41, 1979.

14. Occupational Safety and Health Administration, General Industry Standards, Title 29, Code of Federal Regulations, Part 1910.101 to 1910.152, Washington, D.C.

4.5.4. Ethers

Ethers represent a class of materials which can become more dangerous with prolonged storage because they tend to form explosive peroxides with age. Exposure to light and air enhance the formation of the peroxides. A partially empty container increases the amount of air available, and hence the rate at which peroxides will form in the container. It is preferable, therefore, to use small containers which can be completely emptied, rather than take the amounts needed for immediate use from a larger container over a period of time unless the rate of use is sufficiently high so that peroxides will have a minimal time in which to form.

The following material is taken from the second edition of this handbook from an article by Norman V. Steere, "Control of Peroxides in Ethers." It has been edited slightly to conform with the format of the current edition. The sections on detection and estimation of peroxides and removal of peroxides have been substantially shortened, in line with the philosophy espoused elsewhere in this section to keep on hand only amounts that will be quickly used.

Ethyl ether, isopropyl ether, tetrahydrofuran, and many other ethers tend to absorb and react with oxygen from the air to form unstable peroxides which may detonate with extreme violence when they become concentrated by evaporation or distillation, when combined with other compounds that give a detonatable mixture, or when disturbed by unusual heat, shock or friction. Peroxides formed in compounds by autoxidation have caused many laboratory accidents, including unexpected explosions of the residue of solvents after distillation, and have caused a number of hazardous disposal operations. Some of the incidents of discovery and disposal of peroxides in ethers have been reported in the literature, some in personal communications, and some in the newspapers. An "empty" 250-cc bottle which had held ethyl ether exploded when the ground glass stopper was replaced (without injury), another explosion cost a graduate student the total sight of one eye and most of the sight of the other, and a third explosion killed a research chemist when he attempted to unscrew the cap from an old bottle of isopropyl ether.

Appropriate action to prevent injuries from peroxides in ethers depends on knowledge about formation, detection, and removal of peroxides, adequate labeling and inventory procedures, personal protective equipment, suitable disposal methods, and knowledge about formation, detection, and removal of peroxides.

4.5.4.1. Formation of Peroxides

Peroxides may form in freshly distilled and undistilled and unstablized ethers within less than 2 weeks, and it has been reported that peroxide formation began in tetrahydrofuran after 3 days and in ethyl ether after 8 days. Exposure to air, as in opened and partially emptied containers, accelerates the formation of peroxides in ethers, and while the effect of exposure to light does not seem to be fully understood, it is generally recommended that ethers which will form peroxides should be stored in full, air-tight, amber glass bottles, preferably in the dark.

Although ethyl ether is frequently stored under refrigeration, there is no evidence that refrigerated storage will prevent formation of peroxides, and leaks can result in explosive mixtures in refrigerators since the flash point of ethyl ether is $-45°C$ (49°F).

The storage time required for peroxides as H_2O_2 to increase from 0.5 to 5 ppm has been reported to be less than 2 months for a tin-plate container, 6 months for an aluminum container, and over 17 months for a glass container. The same report stated that peroxide content was not appreciably accelerated at temperatures about 11°C (20°F) above room temperature. Davis has reported the formation of peroxides in olefins, aromatic and saturated hydrocarbons, and ethers particularly, with initial formation of an alkyl hydroperoxide which can condense on standing or in the presence of a drying agent to yield further peroxidic products. Davis refers to reports that the hydroperoxides initially formed (e.g., from isopropyl ether and tetrahydrofuran) may condense further, particularly in the presence of drying agents, to give polymeric peroxides and that cyclic peroxides have been isolated from isopropyl ether.

The literature contains an extensive report on autoxidation of ethyl ether.

Isopropyl ether seems unusually susceptible to peroxidation and there are reports that a half-filled 500-mL bottle of isopropyl ether peroxidized despite being kept over a wad of iron wool. Although it may be possible to stabilize isopropyl ether in other ways, the absence of a stabilizer may not always be obvious from the appearance of a sample, so that even opening a container of isopropyl of uncertain vintage to test for peroxides can be hazardous. Noller comments that "neither hydrogen peroxide, hydroperoxide nor the hydroxyalkyl peroxide are as violently explosive as the peroxidic residues from oxidized ether."

4.5.4.2. Detection and Estimation of Peroxides

Appreciable quantities of crystalline solids have been reported as gross evidence of formation of peroxides, and a case is known in which peroxides were evidenced by a quantity of viscous liquid in the bottom of the glass bottle of ether. If similar viscous liquids or crystalline solids are observed in ethers, no further tests are recommended, since in four disposals of such material there were explosions when the bottles were broken.

The potassium iodide method of testing for peroxides in ethyl ether as described by the Manufacturing Chemists Association is as follows:

> Add 1 cc of a freshly-prepared 10% solution of potassium iodide to 10 cc of ethyl ether in a 25 cc glass-stoppered cylinder of colorless glass protected from light; when viewed transversely against a white background, no color is seen in either liquid.

If a yellow color appears when 9 cc of ethyl ether are shaken with 1 cc of a saturated solution of potassium iodide, according to one organization, there is more than 0.005% peroxide and the ether should be discarded.

4.5.4.3. Inhibition of Peroxides

No single method seems to be suitable for inhibiting formation in all types of ethers, although storage and handling under an inert atmosphere would be a generally useful precaution.

Some of the materials which have been used to stabilize ethers and inhibit formation of peroxides include the addition of 0.001% of hydroquinone or diphenylamine, polyhydroxyl-phenols, aminophenols, and arylamines. Addition of 0.0001 g of pyrogallol in 100 cc ether was reported to prevent peroxide formation over a period of 2 years. Water will not prevent the formation of peroxides in ethers, and iron, lead, and aluminum will not inhibit the peroxidation of isopropyl ether, although iron does act as an inhibitor in ethyl ether. Dowex-1$^\copyright$ has been reported effective for inhibiting peroxide formation in ethyl ether, 100 parts per million (ppm) of 1-naphthol for isopropyl ether, hydroquinone for tetrahydro-furan, and stannous chloride or ferrous sulfate for dioxane. Substituted stilbene-quinones have been patented as a stabilizer against oxidative deterioration of ethers and other compounds.

4.5.4.4. Removal of Peroxides

Reagents which have been used for removing hydroperoxides from solvents are reported to include sodium sulfite, sodium bisulfite, stannous chloride, lithium tetrahydroalanate (caution: use of this material has caused fires), zinc and acid, sodium and alcohol, copper-zinc couple, potassium permanganate, silver hydroxide, and lead dioxide.

Decomposition of ether peroxides with ferrous sulfate is a commonly used method; 40 g of 30% ferrous sulfate solution in water is added to each liter of solvent. Caution is indicated since the reaction may be vigorous if the solvent contains a high concentration of peroxide.

Reduction of alkylidene or dialkylperoxides is more difficult, but reduction by zinc dissolving in acetic or hydrochloric acid, sodium dissolving in alcohol (not ease of ignition of hydrogen), or the copper-zinc couple might be used for purifying solvents containing these peroxides.

Addition of 1 part of 23% sodium hydroxide to 10 parts of ethyl ether or tetrahydro-furan will remove peroxides completely after agitation for 30 minutes; sodium hydroxide pellets reduced but did not remove the peroxide contents of tetrahydrofuran after 2 days. Addition of 30% of chloroform to tetrahydrofuran inhibited peroxide formation until the eighth day with only slight change during 15 succeeding days of tests; although sodium hydroxide could not be added because it reacts violently with chloroform, the peroxides were removed by agitation with 1% aqueous sodium borohydride for 15 minutes (with no attempt made to measure temperature rise or evolution of hydrogen).

A simple method for removing peroxides from high-quality ether samples without need for distillation apparatus or appreciable loss of ether consists of percolating the solvent through a column of Dowex-1$^©$ ion exchange resin. A column of alumina was used to remove peroxides and traces of water from ethyl ether, butyl ether, dioxane, and petroleum fractions and for removing peroxides from tetrahydrofuran, decahydronapthalene (decalin), 1,2,3,4-tetrahydronaphalene (tetralin), cumene and isopropyl ether.

Because they have a limited shelf life, as noted earlier, ethers should be bought in the smallest practicable containers appropriate to the rate of usage within the facility, preferably in 500-mL containers. In Section 4.3.1 it was acknowledged that buying in small sizes does invoke a heavy financial penalty, but some of this price disadvantage can be eliminated by buying in case lots or having a central stock room buy in multiple case lots. It also eliminates much of the need for frequent checking of the conditions of the contents of a large container. No matter what size container is purchased, each container should be dated when it is received and placed in stock. For isopropyl and diethyl ethers, it is recommended that even unopened containers be disposed of after 1 year, while opened containers should be discarded 6 months after they are first used, if they have not been tested periodically during the interval. Opened containers should be tested after 1 month and continue to be tested until emptied, or at frequent intervals. If only modest amounts of peroxides are found, the ethers can be decontaminated. If an alumina column is used, the contaminated alumina can be treated with an aqueous solution of ferrous sulfate and discarded as chemical waste. However, it may be difficult to get a hazardous waste disposal firm to accept it. They are very reluctant to accept any waste material if there is any possibility of an explosion. The concern about possible explosions, the cost of the manpower involved in the needed frequent checking, and the possibility of having to pay a premium price for disposal of excess materials as a potential explosive should obviate the argument of a lower unit cost for the larger sizes. The implementation of the computerized chemical tracking program, previously mentioned, places this responsibility at the point of receipt rather than in the individual laboratory. The information can be used by those responsible for disposing of waste and surplus chemicals to retrieve these materials before a target date so that the commercial disposal firms will accept the materials for transportation.

Table 4.6 contains a brief list of some the more common chemicals in which peroxides may form.

Table 4.6.   Some Materials Tending to Form Peroxides

Acrolein	Cyclooctene	p-Dioxane
Aldehydes	Decahydronaphthalene	Divinyl ether
Allyl ethyl ether	Diacetylene	Ethyl methyl ether
Allyl phenyl ether	Dibutyl ether	Methyl acetylene
Benzyl ether	Dicyclopentadiene	o-Methylanisole
Benzoyl n-butyl ether	Diethylene glycol	m-Methylphenetole
Bromophenetole	Diethylene glycol diethyl ether	Phenetole
Butadiene	Diethylene glycol mono-o-butyl ether	Tetrahydrofuran
p-Chloroanisole	Diethyl ether	Tetrahydronaphthalene
Cumene	Dimethyl ether	Vinyl acetate
Cyclohexene	Dimethyl isopropyl ether	Vinylidene chloride

ACKNOWLEDGMENT

[From original article]. Reprinted with permission from *The Chemistry of Ether Linkage*, 1967, Saul Patai, Editor, Interscience Publishers, Inc., Division of John Wiley & Sons, Inc.

REFERENCES (SECTION 4.5.4)

1. **Douglas, I.B.,** *J. Chem. Educ.*, 40, 469, 1963.
2. **Steere, N.V.,** Control of hazards from peroxides in ethers, *J. Chem. Ed.*, 41, A575, 1964.
3. Accident Case History 603, Manufacturing Chemists Association (reported in part in Reference 2).
4. **Fleck, E.,** Merck, Sharp & Dohme Company memo, Rahway, NJ, May 11, 1960.
5. **Noller, C.R.,** *Chemistry of Organic Compounds,* W.B. Saunders, Philadelphia, 1951.
6. **Rosin, J.,** *Reagent Chemicals and Standards*, 4th ed., D Van Nostrand, Princeton, NJ, 1961.
7. **Brubaker A.R.,** personal communication, December 4, 1964.
8. **Davies A.G.,** Explosion hazards of autoxidized solvents, *J. R. Inst. Chem.*, 386, 1956.
9. **Lindgren, G.,** Autoxidation of diethyl ether and its inhibition by diphenylamine, *Acta Chir. Scand.*, 94, 110, 1946.
10. **Pajaczkowski, A.** personal communication, September 29, 1964.
11. **Dugan, P.R.,** *Anal. Chem.*, 33, 1630, 1961.
12. **Dugan, P.R.,** *Ind. Eng. Chem.*, 56, 37, 1964.
13. **Feinstein, R.N.,** Simple method for removal of peroxides from diethyl ether, *J. Org. Chem.*, 24, 1172, 1969.
14. **Kirk, R.E. and Othmer, D.F., Eds.,** *Encyclopedia of Chemical Technology,* Vol. 5, New York, 1950, 142, 871.
15. **Kirk, R.E. and Othmer, D.F., Eds.,** *Encyclopedia of Chemical Technology*, Vol. 6, New York, 1950, 1006.
16. **Jones, D.G.,** British Patent 699,179, *Chem. Abstr.*, 49, 32, 1953.
17. **Moffett, R.B. and Aspergren, B.D.,** Tetrohydrofuran can cause fire when used as solvent for LiAlH, *Chem. Eng. News,* 32, 4328, 1954.
18. Chemical Safety Data Sheet—SD 29, Ethyl Ether, Manufacturing Chemists Association, Washington, D.C., 1956.
19. **Dasler, W. and Bauer, C.D.,** Removal of peroxides from ether, *Ind. Eng. Chem. Anal. Ed.*, 18, 52, 1946.
20. **Birnbaum, E.R.,** personal communication, August 11, 1964.
21. Manuals of Techniques, Inc., Beverly, MA, 1964.
22. **Ramsey, J.B. and Aldridge, F.T.,** Removal of peroxides from ethers with cerous hydroxide, *J. Am. Chem. Soc.*, 77, 2561, 1955.
23. **Steere, N.V.,** Control of peroxides in ethers, in *CRC Handbook of Laboratory Safety*, 2nd ed., Steere, N.V., Ed., CRC Press, Boca Raton, FL, 1971, 250.
24. **Mateles, R.I.,** Cumene hydroperoxide explosion, Letter to the editor, *Chem. Eng. News.,* 71(22), 4, 1993.
25. **Dussault, P.,** Potential hazard with 2-methoxyprop-2-yl hydroperoxide, Letter to the editor, *Chem. Eng. News.*, 71(32), 2, 1993.

4.5.5. Perchloric Acid[*]

4.5.5.1. Hazards of Perchloric Acid

Hazards of perchloric acid are described in MCA SD-11 as follows:

1. Perchloric acid is strong, and contact with the skin, eyes, or respiratory tract will produce severe burns.
2. Perchloric acid is a colorless, fuming, oily liquid. When cold its properties are those of a strong acid, but when hot, the acid acts as a strong oxidizing agent.
3. Aqueous perchloric acid can cause violent explosions if misused, or when in concentrations greater than the normal commercial strength (72%).
4. Anhydrous perchloric acid is unstable even at room temperatures and ultimately decomposes spontaneously with a violent explosion. Contact with oxidizable material can cause an immediate explosion.

The following are listed among the causes of fires and explosions involving perchloric acid:

1. The instability of aqueous or of pure anhydrous perchloric acid under various conditions.
2. The dehydration of aqueous acid by contact with dehydrating agents such as concentrated sulfuric acid, phosphorous pentoxide, or acetic anhydride.
3. The reaction of perchloric acid with other substances to form unstable materials.

Combustible materials, such as sawdust, excelsior, wood, paper, burlap bags, cotton waste, rags, grease, oil, and most organic compounds, contaminated with perchloric acid solution are highly flammable and dangerous. Such materials may explode on heating, in contact with flame or by impact or friction, or they may ignite spontaneously.

4.5.5.2. Perchloric Acid Storage

Within the laboratory: The maximum advisable amount of acid stored in the main laboratory should be no more than two 8 pound (3.6 kg) bottles. A 450-gram (1 pound) bottle should be sufficient for individual use. Storage of perchloric acid should be in a fume hood set aside for perchloric acid use. The acid should be inspected monthly for discoloration; if any is noted, the acid should be discarded.

Outside of the laboratory: Perchloric acid should be stored on an epoxy-painted metal shelf, preferably in a metal cabinet away from organic materials and flammable compounds. Discolored acid should be discarded.

---

[*]   The material in this section is taken from the article by K. Everett and F. A. Graf, Jr., handling perchloric acid and perchlorates, in the *CRC Handbook of Laboratory Safety,* 2nd ed., Steere, N.V., Ed., CRC Press, Boca Raton, FL, 1971. A revised version of this article will be found in Section 4.6.2.4.

4.5.5.3. Storage of Anhydrous Perchloric Acid

The anhydrous acid is much more dangerous than the concentrated 72% acid. If stored for more than 10 days, the acid is likely to develop a discoloration and be capable of spontaneously exploding. A principal hazard in the use of the anhydrous acid is the breakage of containers. Anhydrous perchloric acid will explode in contact with wood, paper, carbon, and organic solvents. Anhydrous perchloric acid should only be made as required and should never be stored.

The building in which perchloric acid is stored, as with the building in which it is used, should present few opportunities for the perchloric acid to encounter organic materials. It would be desirable for the floors to be of epoxy-painted concrete. The bench tops should be of materials which will not react with the perchloric acid, such as epoxy composites. Painted metal furniture should be used instead of wood. None of the furniture should be screwed into the floor, forming openings into which acid could enter and form dangerous metal perchlorates, which could then explode upon the bolt being removed. In general, the furniture in the laboratory should have as few seams in which the acid could enter and become dry and more dangerous as possible.

REFERENCE*

1. Chemical Safety Data Sheet—SD-11, Perchloric Acid, Manufacturing Chemists' Association, Washington, DC.

4.5.6. Flammable Liquids

Should a fire or an explosion occur in a laboratory, a major concern is to reduce the amount of fuel available to support a fire. Many solvents commonly used in laboratories are highly flammable, and even a small quantity involved in the fire could have the capacity to significantly increase the probability of the fire spreading.

The OSHA General Industry Standard, in Section 1910.106(d)(3), provides restrictions on the maximum amounts of flammable liquids allowed to be stored, dependent upon class, in flammable material storage cabinets within a room, and defines in Section 1910.106(d)(2) the maximum size of individual containers for the various classes of flammables. NFPA Standard 45 provides guidelines as to the maximum amounts of flammable liquids that should be allowed in the three classes of facilities (A, B, and C) defined in that standard. This standard has been mentioned but the three laboratory classes have not been stressed in this volume, in favor of a four-level standard, paralleling legally defined classes of risk for biological facilities, and involving wider classes of hazards than that due to the amount of flammable materials in the laboratory alone. For the purposes of this section, the low- and moderate-risk facilities described in Sections 3.1.3.2.1 and 3.1.3.2.2, respectively, may be taken to be approximately equivalent to an NFPA Standard 45 Class C facility, a substantial-risk facility (Section 3.1.3.2.3) to be roughly equivalent to a Class B facility, and a high-risk facility (Section 3.1.3.2.4) to include a Class A

---

* The user should refer to the MSDS accompanying the material upon receipt or a generic one provided by a commercial service.

facility, but with more restrictions than the latter would require. Note that OSHA has not adopted the restrictions of NFPA Standard 45 and does not address the issue of the total amount of flammables permitted in a laboratory area, although the amount permitted in an interior storage room is defined. The OSHA regulations regarding container sizes are based on sections of the 1969 version of the NFPA Standard 30. Before returning to the topic of flammable material storage cabinets, Tables 4.7 and 4.8 define the various classes of flammable and combustible liquids and the maximum container sizes permitted by OSHA for each class. The third table provides the maximum amount of flammable liquids permitted in the three classes of laboratories defined by NFPA 45 (*outside* flammable material storage cabinets and not in safety containers), if this latter standard were to be adopted.

The definitions depend upon the flashpoints, and in some cases, the boiling points of the liquids. The flashpoint of a liquid is legally defined in terms of specific test procedures used to determine it, but conceptually is the minimum temperature at which a liquid forms a vapor above its surface in sufficient concentration that it may be ignited.

In Table 4.7, the first temperature is in degrees Celsius and the temperature in parentheses ( ) is the equivalent Fahrenheit temperature. Note that neither Combustible II or Combustible IIIA materials include mixtures in which more than 99% of the volume is made up of components with flashpoints of 93.3°C (200°F) or higher.

Table 4.8 is equivalent to Table H-12 from the OSHA General Industry Standards. Polyethylene containers have become widely available and are included in later versions of NFPA 30, so information on this type of container has been appended to the end of the table.

Several exceptions to Table 4.8 permit glass or plastic containers of no more than 1-gallon capacity to be used for Class IA and Class IB liquids: (1) if a metal container would be corroded by the liquid, (2) if contact with the metal would render the liquid unfit for the intended purpose, (3) if the application required the use of more than one pint of a Class IA liquid or more than one quart of a Class IB liquid, (4) an amount of an analytical standard of a quality not available in standard sizes needed to be maintained for a single control process in excess of one sixteenth the capacity of the container sizes allowed by the table, and (5) if the containers are intended for export outside the U.S.

Table 4.9 is derived from data in NFPA Standard 45 and gives the amounts permitted by that standard in laboratories, excluding quantities in storage cabinets and safety cans. The reader is referred to the original standard for the total amounts that would be permitted in laboratories.

In the author's opinion, the amounts permitted by Standard 45 would be excessive for most laboratories, especially academic research facilities. Under these limits, the standard laboratory module, used as the basis for most of this chapter, would be permitted to have 80 gallons of mixed flammables or 40 gallons of Class IA flammables (Table 4.10) in un-protected storage cabinets or containers, if the facility were rated as a high hazard operation. Except in unusual cases where some very active laboratories could use such large quantities in a sufficiently short period to justify their presence, flammable materials in these amounts, other than those contained in research apparatus, should be kept in protected storage.

In Section 1910.106(d)(3) of the General Industry Standards, OSHA limits the amounts of Class I and Class II liquids in a single flammable material storage cabinet to 60 gallons and the amount of Class III liquids to 120 gallons. Thus, even if the integrity of a single

Table 4.7.   Definitions, Classes of Flammable and Combustible Liquids

	°C (°F)	
Class	Boiling Points	Flash Points
Flammable IA	< 37.8 (100)	< 22.8 (73)
Flammable IB	≥ 37.8 (100)	< 22.8 (73)
Flammable IC	—	22.8 (73) ≤ and < 37.8 (100)
Combustible II	—	37.8 (100) ≤ and < 60 (140)
Combustible IIIA	—	60 (140) ≤ and < 93.3 (200)
Combustible IIIB	—	≥ 93.3 (200)

Table 4.8.   Maximum Allowable Size of Containers and Portable Tanks

Container Type	IA	IB	IC	II	III
Glass or approved plastic	1 pt	1 qt	1 gal	1 gal	1 gal
Metal (other than DOT drums)	1 gal	5 gal	5 gal	5 gal	5 gal
Safety cans	2 gal	5 gal	5 gal	5 gal	5 gal
Metal drums (DOT specs.)	60 gal	60 gal	60 gal	60 gal	60 gal
Approved portable tanks	660 gal	660 gal	660 gal	660 gal	660 gal
Polyethylene spec. 34 or as authorized by DOT exemption	1 gal	5 gal	5 gal	60 gal	60 gal

storage cabinet were broached in a fire, or if an accident occurred while the cabinet was open, no more than 60 gallons of Class I and II liquids or 120 gallons of a Class III liquid could become involved in the incident. Even these amounts would add an enormous amount of fuel to a fire.

4.5.7. Refrigeration Storage

Two of the most dangerous storage units in any laboratory is the ordinary refrigerator, and to a somewhat lesser extent, the freezer. This is primarily due to the storage of flammable materials within them, although there are also problems due to individuals using them as a place to store food which they have brought to the laboratory for their own consumption, as well as for their intended function. Refrigerators intended for the storage of laboratory chemicals and biological materials should not be used for personal items, particularly food and beverages.

Table 4.11 lists the flash points for a number of common solvents with flash points below or close to the normal operating temperature (about 38°F or 3.3°C) of a common refrigerator. Also given are the flammable limits in percent by volume in air for these same solvents. Most of these evaporate rapidly so that they quickly reach equilibrium concentrations in a small confined space. Some of the most dangerous are acetaldehyde, carbon disulfide, diethyl ether, and ethylene amine, which have broad explosion limits.

Table 4.9.   Amounts of Flammable Liquids Permitted in Laboratory Units Outside
Flammable Material Storage Units and Safety Cans (in gal) [*,**,***]

Laboratory Class	Flammable/ Combustible Liquid Class	Max. Quantity (per 100 ft^2)	Max. Quantity per Laboratory Unit	
			Unsprinklered	Sprinklered
A	I	10	300	300
	I, II, and IIIA	20	400	400
B	I	5	150	150
	I, II, and IIIA	10	200	200
C	I	2	75	75
	I, II, and IIIA	4	100	200

[*]   Class A units cannot be used as instructional laboratories.
[**]   Instructional laboratory units are permitted only half of the amounts given in the table.
[***] The total amounts cannot exceed the amounts permitted using the limits per 100 square feet, including the areas of offices and other nonresearch areas that are part of the laboratory unit.

Table 4.10.   A Brief List of Some Common Class IA Solvents

Acetaldehyde	Furan	Methyl sulfide
2-Chloropropane	Methyl acetate	N-Pentane
Collodian	Isoprene	Pentene
Ethyl ether	Ligroine	l-propylamine
Ethanthiol	Methylamine	Propylene oxide
Ethylamine	2-Methylbutane	Petroleum ether
Ethyl vinyl ether	Methyl formate	Trimethylamine

Storage of flammable materials in refrigerators or in other confined spaces in which the vapors can be trapped and which also contain sources of ignition represents a potential explosion hazard. Carelessly closed containers, e.g., screw caps that are not firmly tightened or beakers containing solvents covered only with aluminum foil or plastic wrap, will allow vapors to escape from the container and, given sufficient time, build up in the confined space until they may reach a concentration in excess of the lower flammable limit. A spark may then cause ignition, and because the reaction is temporarily constrained, very high pressures can build up until the refrigerator door latch fails and a powerful explosion ensues. Many such cases have been documented, and in most cases, workers in the vicinity in front of the refrigeration unit likely would have sustained serious if not fatal injuries. Fortunately, it appears in many cases that the propensity of laboratory workers to place improperly sealed containers in refrigerators is greatest at the end of the work day, when they may be in a hurry to leave. This pattern, coupled with the materials remaining undisturbed for extended periods of time after normal working hours, tends to make night hours the most likely time for a refrigerator explosion to occur.

A normal refrigerator has many sources of ignition—the thermostat, interior light, the light switch on the door, the defrost heater, the defrost control switch, the compressor unit,

Table 4.11.   Flammability Characteristics of Some Common Solvents

| Chemical | Flashpoint (°C) | Flammable Limits (%) | |
		Lower (%)	Upper (%)
Acetaldehyde	−37.8	4	60
Acetone	−17.8	2.6	12.8
Benzene	−11.1	1.3	7.1
Carbon disulfide	−30.0	1.3	50
Cyclohexane	−20.0	1.3	8
Diethyl ether	−45.0	1.9	36
Ethyl acetate	−4	2.0	11.5
Ethyleneimine	−11	3.6	46
Gasoline (approximate)	−38	1.4	7.4
n-Heptane	−3.9	1.05	6.7
n-Hexane	−21.7	1.1	7.5
Methyl acetate	−10	3.1	16
Methyl ethyl ketone	−6.1	1.8	10
Pentane	−40.0	1.5	7.8
Toluene	4.4	1.2	7.1

and the fan. Most of these are located within the space being maintained cool, but self-defrosting units contain an internal drain that can permit the internal vapors to flow into the compressor space below the usable space.

It is possible to modify a normal home refrigeration unit to remove the internal sources of ignition, but unless it is done by a person who knows precisely what to do and does it very carefully, the result may not be as safe as one initially designed to be used for flammable material storage. The liability which could result from the failure to prevent an explosion by an improperly locally modified unit makes this an imprudent economy measure. In most cases, refrigeration units need only be rendered safe for prevention of ignition by components of the refrigerator themselves, i.e., be designated as safe for the storage of flammable materials, instead of meeting standards for total explosion safety, which would permit them to be operated in locations where flammable vapors and gases exist outside the refrigeration units. The additional cost of the latter units plus the cost of making the proper electrical explosion proof connections are unnecessary expenses for most laboratories.

In virtually all cases refrigeration units which operate as "ultra-lo" units in which the internal temperatures are of the order of −60°C to −120°C need not be flammable material storage units or explosion safe. Note that none of the materials in Table 4.11 have flash points in this range.

As noted earlier, refrigerators last for as many as 20 to even 30 years. It is not feasible to accept assurances by laboratory managerial personnel that no flammable materials will ever be placed in an ordinary refrigerator, because neither the individual making the promise nor the program for which the refrigerator is purchased is likely to occupy the same laboratory space for such an extended period. It is also not reasonable to depend upon marking laboratory refrigerators, no matter how prominently, as not to be used for flammable material storage and count on compliance with the restriction. If there is an ordinary consumer-quality refrigerator in the laboratory, it is virtually certain that someone will eventually use it improperly. Therefore, it is recommended that all refrigerators to be used in laboratory areas be required to be initially constructed for flammable material storage, and bear an appropriate label on the front that it meets such standards. Exceptions should be very few, and restricted to those programs that are, in fact, essential to the basic operations of a stable department, and not depend upon the program of an individual or a limited number of persons constituting a temporary research group. Older units not meeting the standards for flammable material storage should be phased out as rapidly as possible, or moved from laboratories in which usage of flammables is a normal activity to noncritical areas where current usage is not likely to involve storage of flammable liquids, and replaced with a suitable refrigerator or freezer. This procedure will enable all laboratories to be equipped with safe units relatively economically over a period of time and improve the general level of safety at the same time.

REFERENCES

1. Occupational Safety and Health Agency, Title 29 Code of Federal Regulations, Section 106(d)(3), General Industry Standards, Washington DC, 1988.
2. Flammable and Combustible Liquids Code, NFPA-30, National Fire Protection Association, Quincy, MA, 1981.
3. Standard on Protection for Laboratories Using Chemicals, NFPA-45, National Fire Protection Association, Quincy, MA, 1982.
4. Flammability Ratings of Flammable Liquids, Fisher Chemical Company, Chicago, IL.
5. **De Roo, J.L.,** The Safe Use of Refrigerator and Freezer Appliances for Storage of Flammable Materials, Union Carbide, South Charleston, WV.
6. **Langan, J.P.,** Questions and Answers on Explosion-Proof Refrigerators, Kelmore, NJ.
7. Properties of Common Flammable and Toxic Solvents, Division of Industrial Hygiene, New York State Department of Labor, Albany.

4.5.8. Gas Cylinders

Gas cylinders are used for many purposes in the research laboratory. Most individuals probably think of gas cylinders in the context of the standard industrial gas cylinder, which is 23 centimeters (9 inches) in diameter and 140 centimeters (55 inches) high. However, there are many other sizes available. All of these cylinders can represent a significant hazard. A standard cylinder weighing about 64 kilograms (140 pounds) often contains gas at pressures of about 21 mega-Pascals or 3000 pounds per square inch or even more. Should the valve connection on top of the cylinder be broken off, such a cylinder would correspond to a rocket capable of punching a hole through most laboratory walls and would represent a major danger to all occupants in any area where such an incident occurred. The contents of cylinders also frequently represent inherent hazards. These pressure-independent hazards associated with the contents include flammability, toxicity, corrosiveness, excessive reactivity, and potential asphyxiation if the volume of air displaced by the contents of the cylinder is sufficient. Obviously, measures need to be taken to ensure that the integrity of the cylinder is totally maintained. Compressed gas cylinders can be used safely, if due care is taken with them and with the accessories and systems with which they may be combined.

A compressed gas is defined by the federal Department of Transportation (DOT) as "any material or mixture having in the container either an absolute pressure greater than 276 kPa (40 lbf/in.^{2}) at 21°C (69.8°F), or an absolute pressure greater than 717 kPa (104 lbf/in.^{2}) at 54°C (129.2°F) or both, or any liquid flammable material having a Reid vapor pressure greater than 276 kPa (40 lbf/in.^{2}) at 38°C (100.4°F)."

The actual pressure in a cylinder will depend upon the type of gas in the cylinder and the physical state of the contents. Cylinders containing gases which are gaseous at all pressures practicable for the cylinder, such as nitrogen or helium, will have a pressure which reflects the amount of material in the cylinder, while those that are in equilibrium with a liquid phase, such as ammonia, carbon dioxide, or propane, will be at the pressure of the vapor as long as any of the material remains in the liquid phase, provided that the critical temperature is not exceeded. The weight of the cylinder in excess of its empty weight is used to measure the amount of gas in the cylinder for the latter type of gas.

Compressed gas cylinders usually will be stamped near the top of the cylinder with the DOT code appropriate to the specification under which the cylinder was manufactured and the pressure rating at 21°C, usually in lbf/in^2. The last date on which the cylinder was tested usually will be stamped near the upper end of the cylinder. In most cases, the test interval for a steel cylinder is 10 years. It is the responsibility of the company distributing the cylinders to be sure that they are within the appropriate test span. Unfortunately, it appears that this is not always done. A spot-check of a large group of cylinders at one facility revealed that more than 10% were significantly beyond the required test date.

Very few laboratory facilities have the capability to refill cylinders, nor in most cases is the use heavy enough to make it economical to acquire the capacity to so safely. Most users of compressed-gas cylinders have an arrangement with a vendor to periodically replace empty cylinders with full ones of the same type on a regular basis, paying a demurrage charge on the number which they maintain in use. As a result, the ones on hand are continually changing and care must be taken to ensure that the identity of the gas in the cylinder is known. Color codes are unreliable, especially for the caps, since these are always taken off in use. Tags attached to the cap are not appropriate for the same reason. A name stenciled on the side of the cylinder or an adhesive label placed on the side of the cylinder would form the basis of a satisfactory system, but the contents of the replacement cylinders should be confirmed upon each delivery. A cylinder for which the contents are not certain should not be accepted. Any units for which the identification label or stencil has become defaced should also be returned to the supplier. Unidentified cylinders are very difficult and expensive to dispose of through most commercial waste disposal firms. Special problems are posed by those that are still pressurized but have physically damaged or corroded valves.

OSHA at one time had specific standards for the inspection of compressed gas cylinders and for the safety relief devices for the cylinders. However, these standards were revoked as of February 10, 1984. OSHA still does have general standards on gas cylinders incorporated into Section 1910.101(a) requiring that visual and other inspections shall be conducted as prescribed in the hazardous materials regulations of the Department of Transportation (49 CFR Parts 171-179 and 14 CFR Part 103). Where those regulations are not applicable, visual and other inspections shall be conducted in accordance with Compressed Gas Association (CGA) pamphlets C-6-1968 and C-8-1968. Section 1910.252 of the OSHA standards, which concerns welding, cutting, and brazing, also contains regulations for compressed gas cylinders used in these operations. Users should also seek further guidance from the most current edition of these publications and from the most current version of the CGA publication "Safety Relief Device Standards."

4.5.8.1. Bulk Storage

Most building codes have restrictions on the location and arrangement of bulk storage facilities of compressed gas cylinders. Storage areas for gas cylinders should not intrude on a required path of egress, such as stairs and hallways, or be in an outside area where occupants evacuating a building would be required to pass them or congregate. Cylinders in storage containing oxidizing agents should be separated by at least 25 feet from those containing reducing agents, or these two different types should be separated by a fire wall at least 5 feet high with at least a 30-minute fire rating. If possible, flammable gas cylinders should be stored separately from other cylinders, even those containing inert gases; in the event of a fire, their contribution of additional fuel would not increase the possibility of other nearby cylinders rupturing. Nominally empty cylinders should not be stored in the same location as full ones and all "empty" cylinders should be clearly marked as empty. If separate storage is provided, a simple marking of a cylinder with the letters "MT" is usually sufficient. Empty oxidizing- and reducing-agent cylinders should be separated as if they were full, since they should never be emptied to less than about 172 kPa (25 lbf/in^2), in order to prevent contamination of the interior of the cylinders.

In addition to any code restrictions, there are a number of common sense guidelines which should be followed in providing a storage location. They should be stored in well-ventilated areas, and air should be able to circulate freely around them, so that any leakage of gases will be quickly dissipated. The area in which they are stored should not be damp, and the surface on which they sit should be dry to minimize corrosion of the steel. Outdoor storage areas should be protected from the weather, and cylinders should not be stored in areas where heavy objects might fall on them, nor in areas with heavy vehicular traffic where they might be struck by a moving vehicle. Indoor storage areas should be of fire-resistant construction. Since many compressed gases are heavier than air, the storage location should preferably be above grade, and should not readily connect to spaces where it would be likely for escaping gases to flow and collect. Cylinder caps should always be on the cylinders while they are in storage and at any time they are moved. Signs listing all gases stored at the location should be prominently displayed at the storage area. All cylinders containing a specific variety should be grouped together.

Compressed gas cylinders should be handled carefully when moving them to and from the bulk storage area and during normal use. Although they appear sturdy, the cylinders are designed as shipping containers for the gases within them and hence are designed to be as light as possible, consistent with reasonable margins of safety for pressure and physical handling. Cylinders should always be moved using a transporting device, e.g., strapped to a common hand dolly or using any number of commercial units especially designed to transport cylinders. They should never be moved by supporting the valve cap with one hand while rotating and rolling the base of the cylinder along the ground with the other. There are also manual handtrucks specially designed for moving heavy objects up and down stairs. If a freight elevator is not available when moving a cylinder from one floor to the next, one of these devices should be employed. A cylinder should never be hoisted by attaching a cable to the cylinder cap.

Except those cylinders designed to hold toxic gases, most cylinders incorporate a rupture disk as an overpressure safety device, which will melt at the relatively low temperatures of between 70° and 95° (158° and 203°). Because of these safety devices, the temperature in an area where cylinders are stored should not exceed 52°C (125°F), nor should the cylinders be exposed to localized heating.

Cylinders should be stored in an upright position, with the valve end up, never on their side. The storage area should contain facilities which would make it possible to firmly secure the cylinders in an upright position and prevent them from falling or being knocked over. Parallel bars with space between them sufficient to accommodate a cylinder, with a chain holding them in place make a secure storage area, although stored compactly in a mass with a chain drawn tightly around the entire group also would be satisfactory. In the latter case, however, the length of the chain holding them upright should be adjustable so that if the number stored decreases substantially, the chain will not become so slack as to permit cylinders to fall over.

4.5.8.2. Laboratory Storage

Storage of cylinders in a laboratory at a given time should be restricted to those in actual use or attached to a system ready for use. If this is not feasible, the actual numbers of cylinders present should be maintained at an absolute minimum. No cylinder should be in a laboratory which is not securely fastened to a support so that it cannot fall over. No free-standing cylinder should be allowed to be present in a laboratory, even a nominally "empty" cylinder. There is always a possibility that the empty cylinder has been mis-labeled. A good rule of thumb is to treat a compressed gas cylinder as you would a gun; unless confirmed otherwise, always assume that it is loaded and treat it as such.

As long as cylinders in a laboratory are not connected to a system and potentially in use, the rules in the preceding section (4.5.8.1. Bulk Storage) should apply. Such restrictions as the physical separation of oxidizing and reducing agents should not be abrogated unless circumstances are appropriate.

Additional information on operations involving gas cylinders will be provided in Section 4.6.2.8.

REFERENCES

1. *Prudent Practices for Handling Hazardous Chemicals in Laboratories*, National Academy Press, Washington, D.C., 1981, 75, 221.
2. Specialty Gas Data Sheets, Air Products and Chemicals, Emmaus, PA.
3. *Safety In Academic Laboratories,* 5th ed., American Chemical Society, Washington, D.C., 1991.
4. **Pinney, G.,** Compressed gas cylinders and cylinder regulators, in *CRC Handbook of Laboratory Safety,* 2nd ed., Steere, N.V., Ed., CRC Press, Boca Raton, FL, 1971, 565.
5. *Safety Relief Device Standards—Cylinders for Compressed Gases,* Compressed Gas Association, New York, 1965.
6. *Standard Compressed Gas Cylinder Valve Outlet and inlet Connections,* Vol. 1, Compressed Gas Association, New York, 1965.

4.5.9. Animal Food and Supply Storage[*]

4.5.9.1. Animal Food

The Federal Animal Welfare Act (PL 89-544) requires that animal food be stored in facilities which protect it from infestation or contamination by vermin (wild rodents, birds,

---

[*] This section was written by David M. Moore, D.V.M.

and insects). Food can be stored in individual animal rooms in vermin-proof containers with lids, such as plastic garbage containers. Ideally, bulk-food shipments should be stored in a room or warehouse where the temperature can be maintained at less than 70°F and the relative humidity at 50% or less. The room should have doors that prevent the entry of rodents or birds. Vermin control is important since wild rodents, birds, and insects can contaminate stored feed with bacteria, viruses, or parasites which could adversely affect laboratory animal health. Pesticides should not be used to control vermin in this area while food supplies are present; contamination of food with pesticides can seriously affect experimental results in animals. Boric acid powder can be placed along the walls to control cockroaches, without the negative experimental impact of organophosphate insecticides.

Most commercial laboratory diets contain preservatives and stabilizers which maintain nutrient quality in the diet for up to 6 months. However, diets containing vitamin C (i.e., guinea pig chow and nonhuman primate chow) have a limited shelf life of 90 days because of the instability of vitamin C in the diet. Feed sacks are coded at the manufacturer with the date of milling, and this date should be recorded upon receipt of the shipment, recognizing that the food should be used within 90 days after the milling date to avoid vitamin C deficiency problems in guinea pigs and monkeys. Most facilities use the "first-in, first-out" method of warehousing feed, stacking the pallets/bags such that the oldest feed is most accessible for transport to the individual animal rooms.

The ingredients in purified or chemically defined diets are not as stable as those in most commercial diets, and the NIH Guide for the Care and Use of Laboratory Animals recommends that these diets be stored at 39°F or colder.

Diets which contain potential or known hazardous compounds (carcinogens, mutagens) should not be stored in the same area as control diets.

4.5.9.2. Supply Storage

Potentially hazardous compounds such as detergents, chemical disinfectants, and insecticides should not be stored in the same area with bulk-feed stores to prevent contaminating the latter. The storage area should be clean and orderly, with appropriate precautions taken to keep it free of vermin.

4.5.9.3. Animal Carcass Storage

Animal carcasses that are not immediately incinerated should be kept refrigerated at 44°F or lower. Those to be kept for an extended period should be frozen. Refrigerator units should not be used to store food if used for carcass retention.

REFERENCES

1. Guide for the Care and Use of Laboratory Animals, NIH Publ. 85-23, U.S. Department of Health and Human Services, Washington, D.C., 1985, 22.
2. Animal Welfare Act (PL 89-544), Title 9, Subchapter A, Subpart B, Section 3.25—Facilities, General, Paragraph C—Storage.
3. **Hessler, J.R. and Moreland, A.F.,** Design and management of animal facilities, in *Laboratory Animal Medicine,* Fox, J.G. et al., Eds., Academic Press, Orlando, FL, 1984, 509.

4.6. HANDLING AND USE OF CHEMICALS:
LABORATORY OPERATIONS

Laboratory personnel work in a *potentially* extremely hazardous and unforgiving environment. The substances with which they work may be toxic, flammable, explosive, carcinogenic, pathogenic, or radioactive, to mention only a few unpleasant possibilities. The hazards may cause an immediate or acute reaction, or the effects may be delayed for several years. A worker may be lulled into a false sense of security because of the seeming safety of a material, according to current knowledge, but eventually evidence may develop that continued exposure may cause unexpected or cumulative and irreversible effects.

The equipment in the facility, if used improperly or if it becomes defective, could represent physical hazards that could result in serious injuries or death. Electric shock, cuts, explosions due to rupture of high pressure systems (or implosion of large vacuum systems), exposure to cryogenic materials, excessive levels of exposure to ionizing and nonionizing radiation, heat, mechanical injuries due to moving systems, equipment or supplies simply falling on a person, among many other possibilities, may occur in a laboratory environment.

Of course, the laboratory environment is not the only place an injury can occur. Scoffers who do not put a high probability on the possibility of an accident happening in the laboratory often point out that they could have been injured while driving to work. Certainly this is possible, as is being struck by lightning or any number of other possibilities. If we were to brood about all the things that could happen each day, we might choose to not get up each morning. However, it is necessary that we do so and that some risks be taken. It is prudent, however, to follow practices which will minimize the risks. Most of us would not choose to drive with faulty brakes, deliberately drive on the wrong side of the road, or through red lights and stop signs, nor would most of us deliberately violate similar common sense rules governing practices which would lead to ill health. It is impossible to achieve absolute safety, but in the presence of hazards, it is only reasonable to take those steps which will efficiently and effectively reduce the risks to acceptable levels. Laboratory workers should follow ALARA principles (using the parlance of radiation safety) and reduce the risks to a level as low as is reasonably achievable.

Laboratory operations are so varied that it would be totally impractical to attempt to exhaustively cover the topic. There are, however, some basic considerations which should be used to enhance the safety of laboratory operations. Some of these are common sense and some have been made mandatory by regulatory requirements, because some safety-related practices are too important to be left to choice. Prior to addressing specific topics, the following list of simple rules, if followed religiously, would dramatically reduce the number of laboratory accidents, or would diminish the consequences of those that do occur.

1.   Plan the work carefully. At the beginning of an extended project, formally analyze the proposed program for possible hazards and consider the consequences of possible failures or errors. Ask a colleague to review the hazard analysis with you. Being too close to a subject often leads to overlooking potential problems.
2.   Make sure the right equipment is available and in good condition. All too often, makeshift equipment or deteriorated equipment is the cause of an accident. Rarely is it worth the risk to take chances. Most persons with more than a few years of experience can think of a number of examples where this has proven true, sometimes tragically.
3.   Make sure all systems are assembled in a stable and solid manner, making sure that accommodations for the specific limitations or failure modes of the individual components are factored into the operation of the total system.

4. If the release of a toxic or hazardous substance may occur, the work should be done in a fume hood appropriately designed for the operation. Use the fume hood so as to maximize its effectiveness.

5. Use an explosion shield, other protective enclosure, and/or personal protection equipment such as goggles and a face mask if there is a possibility of a violent reaction. Do not overlook the possibility that scaling up a process will change the safety parameters.

6. Chemicals should be handled carefully at all times, using appropriate containers and carrying devices. Open containers should be closed after use, and unneeded reagents should be returned to storage.

7. Do not hurry unnecessarily or compromise on safety. Take the time to do things properly, e.g., label temporary containers as they are employed. Many accidents are due to unnecessary haste or the use of "shortcuts."

8. Follow good safety practices with electrical circuits and equipment. Avoid use of extension cords, multiple plugs, and devices to defeat the need to use three-wire connectors.

9. Do not work alone. As a minimum, a second person should be aware of an individual working alone in a laboratory and definite arrangements should be made for periodic checks. Excessively long working hours increase the likelihood of mistakes due to fatigue.

10. Follow good housekeeping practices. Maintain the work area in an orderly fashion.

11. Do not set up equipment so as to block means of egress from the work area. Consider the activities of others sharing the facility with you in establishing your own work space.

12. Conscientiously use any protective equipment required and wear appropriate clothing.

13. Make sure that you are familiar with and conscientiously follow all safety and emergency procedures.

14. The work area in a laboratory is not a restaurant or a place to socialize. Coffee and meal breaks should be taken at a desk outside the active work area or in a lounge set aside for the purpose.

15. Anyone indulging in horseplay or practical jokes within a laboratory should be excluded from the facility.

16. Never work while under the influence of drugs or alcohol.

REFERENCE

1. *Safety in Academic Laboratories,* 5th ed., American Chemical Society, Washington, D.C., 1991.

4.6.1. Physical Laboratory Conditions

Many of the points to be made in this section were alluded to in Chapter 3, where the factors that should be considered in laboratory design were discussed. If the layout of the laboratory is similar to that of the standard laboratory module shown in Figure 3.1, and repeated here as Figure 4.3, many safety practices which depend upon the physical configuration of the facility will almost automatically follow. However, in many cases, laboratories are often placed in structures originally designed for other purposes and ill adapted for the intended use. Even in this latter case, safety can be significantly enhanced by following a few straightforward guidelines as closely as the available space permits.

Figure 4.3. Standard laboratory module.

4.6.1.1. Organization of the Laboratory

The basic premise in laying out the interior design of a laboratory facility or allocating space for the various activities within an existing facility is to separate areas of high risk from those of low risk as much as possible, and to place high-risk operations where there will be the least traffic and the least probability of blocking escape from the laboratory in case of an accident. Escape routes should, wherever possible, lead from high- to low-risk areas. A high-risk component may not always be obvious. For example, storage of chemicals in appropriate cabinets does not represent a high risk under most circumstances, but if left open, a flammable material storage cabinet along a path of egress can become a major danger if the liquids stored inside become involved in a fire. If the configuration of the laboratory permits, the laboratory furniture should be selected to permit two alternative evacuation paths from any point in the room. One of these, constituting a secondary escape path, may not necessarily lead directly from a high- to a low-hazard area, but even a poor alternative is better than none at all.

Fume hoods are intended to be used to house activities that should not be done on an open bench because of the potential hazard which the activities represent, usually the generation of noxious fumes. The ability of fume hoods to capture and retain fumes generated within them is especially vulnerable to air movement, either due to traffic or other factors such as the location of air system ducts, windows, doors, or fans. Clearly, they should be located, as in the standard laboratory module, in a remote portion of the laboratory selected for low traffic and minimal air movement. Other fume generating apparatus, such as Kjeldahl units, should also be placed in out of the way places where errant air motion will not result in dispersion of the fumes generated into more heavily occupied areas of the room. A point that needs to be considered is the work habits of laboratory employees. Data on the possible health effects of long-term exposures to the vapors from most laboratory chemicals is relatively scant, although there are beginning to be more epidemiological data for various modes of exposures, indicating some general problems. There are very little data on the synergistic effects of combinations of general laboratory chemicals. It is known, however, that the sensitivity of individuals to materials varies widely. In the spirit of ALARA, as applied to chemical usage, when evidence to the contrary is missing, a conservative approach is recommended to limit exposures. Research personnel should spend no more time than is essential to the work in progress in areas where the generation and concentration of chemical vapors is likely to be high (although certainly not above the PEL or action levels established by OSHA), as compared to other spaces in the facility. The practice of allowing or requiring laboratory technicians to have desks in the work area either for convenience or so that they can keep an eye on the work in progress should be discontinued for both health and safety reasons. The standard module provides for this oversight safely by making the barrier between the desk area and the laboratory proper transparent.

The location of the various items of equipment in a laboratory should depend upon a number of factors, such as frequency of use, distances to be traveled, and the need to transport chemicals to and from the work location and the storage cabinets. The distances traveled to and from the most heavily used apparatus should be minimized, as should the frequency and distances involved in the use of chemical reagents. Specialized work, such as the use of radioactive materials, should be isolated from the other activities in the laboratory, especially if only some of the laboratory's employees are involved in the activity while others are not.

Dangerous apparatus should be placed in areas in which protection can be afforded to the maximum number of laboratory employees. For example, a temporary glass system containing a highly reactive material under pressure might be located to one side of the laboratory with explosive barriers placed on either side of the system so that if the system did rupture, the flying particles of glass would be directed toward a normally unoccupied area of the laboratory. Of course, if the probability of an explosion is significant, instead of being only a remote possibility, the work should take place in a laboratory built with the proper explosion venting and explosion-resistant barriers.

Safety showers and deluge showers should be conveniently located within the facility so that the approach to them is uncomplicated and unlikely to be blocked. They should also be located close to the primary entrance to the laboratory so that persons rendering assistance to an injured individual should not have to enter the laboratory any farther than necessary, to reduce the possibility of having to enter a contaminated area. Note that in the standard laboratory module, both shower and eyewash station are located at the end of the workbench closest to the entrance and immediately adjacent to a secondary entrance. Also note that the portable fire extinguisher is on the wall behind the workbench, near the fume hood and close to the combination eyewash station and deluge shower, nearest the areas where it might be needed and useful. A first aid kit should be maintained in the desk area.

4.6.1.2. Eating, Studying, and Other Social Activities

Unless provision is made for acceptable alternatives or extremely tight discipline is maintained in the laboratory, it should be recognized that the work area of the laboratory will be used for eating, studying, and social activities. Such activities, however, should not take place, and as a minimum the laboratory should be clearly marked where these activities will not be allowed under any circumstances. The NRC, for example considers failure to prevent eating, drinking, and smoking in the active work area a citable offense. In recognition of the need to control these activities, the standard laboratory module does provide a convenient and acceptable location within the laboratory for eating, record and labbook maintenance, studying, and social activities by placing a desk space immediately inside the laboratory facility, separated from the rest of the laboratory by a partition. The location of this office class space also provides direct access to the exitway corridor and to the reminder of the building without having to reenter the laboratory in the event of an emergency. The upper half of the partition separating the laboratory and desk spaces in the laboratory module is intended to be transparent, made either of tempered safety glass or a plastic material such as Lexan™. This makes it possible to keep an eye on laboratory operations while taking a coffee or lunch break in safety, or to perform any other desired activity without interfering with laboratory operations or being disturbed by others still in the laboratory. Since the laboratory, in most instances, should be at a modest negative pressure with respect to the corridor, the passage of vapors from the laboratory into the desk area should be inhibited, reducing the routine exposure to the laboratory atmosphere to those at their desks significantly. The low speed of the air through the door, on the order of 10 to 20 fpm, will still make it possible for traffic through the door to carry some laboratory odors into the office compartment. The two doors in sequence also serve an additional safety function, representing a simplified air lock, separating the corridor from the laboratory, thus adding some stability to the HVAC demands within the room. If they are closed following an evacuation, they would provide an additional barrier to any fire or noxious gases spreading from the laboratory to the remainder of the building.

4.6.1.3. Maintenance

Topics generally overlooked in laboratory safety are safety factors involved in providing needed maintenance and custodial services. Access to equipment needing service must be provided to service personnel under conditions which make it possible for them to perform their work safely. Generally, equipment maintenance in the laboratory by support personnel should be coordinated by an individual who is familiar with current and recent research programs and can advise the workers who arrive on the scene of possible risks in handling the various components. An example is maintenance on fume hood exhaust fans. Instances are known of workers servicing fume hood exhausts who suffered severe reactions to contaminants on the equipment and the roof in the vicinity of the exhaust duct, even though the hood was not in use at the time. It is not enough to warn the maintenance department upon the initial request for services. Direct information needs to be provided to the service persons on the scene. Some materials can remain a problem for extended periods of time. In such cases, the workers need to be protected while doing the work, using appropriate items of protective clothing such as gloves, respirators, coveralls, and goggles or full-face respirators, depending upon the level of risk. The levels will depend upon the contaminants that might exist due to work in the laboratory, and these risks should be evaluated conservatively.

Where fume hood exhausts are brought to the roof through individual ducts, the area in which maintenance is needed may be surrounded by exhaust ducts still in use, since in most cases it is impractical to shut off operations for an entire building or even a significant portion of one because it is too disruptive to the research programs. Therefore, it is probably desirable to have a standard personnel protective equipment package for the maintenance workers to use, consisting of half- or full-face respirators providing protection against solvents, particulates, and inorganic acids, chemically resistant coveralls, and gloves selected to provide a broad spectrum of protection against chemicals. Requiring personnel to wear these may appear to be excessively cautious but, as noted earlier, there have been instances where unanticipated severe and long-lasting health effects have occurred. Not all employees who work on the roof of a chemistry building may be employees of the organization. Outside contractors also are used to do a variety of maintenance duties and, under the hazard communication standard, they must be apprised of the risks to which they might be exposed. The mix of materials exhausted through ducts is typically so complex that meeting this requirement is difficult, if not impossible. Recommending to them to wear equivalent protection should fulfill the spirit of the standard.

Fume hood maintenance is one of the more active areas in which maintenance personnel have concerns and where both support and laboratory personnel need to assume responsibility for seeing that the work is properly coordinated. Some simple suggestions that have been found useful are to ensure that each exhaust duct on the roof is properly labeled with the room location of the hood itself. Workers have been known to turn off power to motors on hoods in active use. Where hoods are dedicated to special uses which represent unusual hazards, such as radioactive materials, perchloric acid, exceptionally toxic gases, or any other especially unusual risk, the duct should also be labeled with the application involved or a color code employed to identify these unusual risks. The latter program would alert maintenance personnel to definitely contact the laboratory from which the duct came before working in the vicinity of the duct. Power to the motors on the roof should also be provided in such a way as to ensure that the workers on the roof can completely control the circuits while working to avoid accidental activation of the circuits from the laboratory. However, should the exhaust motor be turned off by maintenance workers without prior notification of laboratory personnel, an alarm should sound in the laboratory, warning that the hood is not functional. A tagging and lock-out procedure should also be employed during the maintenance operation.

Once hoods are removed from service to perform maintenance, they should not be returned to use until it is verified that they are performing according to required standards. It is easy to erroneously wire a three-phase motor so that the fan rotates opposite to the desired direction. Belts may need to be tightened or a pulley size changed to achieve the proper face velocity.

Fume hoods have been used to illustrate some of the problems that can arise from lack of coordination of maintenance and laboratory personnel, but there are many other possible problems. Explosions can occur if gas service is turned off without everyone being aware of it and they leave gas jets open, flooding a facility with gas when service is restored. Stills can overheat if condenser water supplies are interrupted. Electrical service to an area should be discontinued and restored only with full prior notification to all persons that might be affected. Today, with the large amount of computer automation being used, interrupting the power to the facility may disrupt the entire operation and bring down the wrath of the scientist on the worker who caused the problem. Individual laboratory technicians or students often modify their facility without informing the groups responsible for maintenance, thereby raising the possibility of an injury to an unsuspecting service-person, or a repair which will not be based on an accurate assessment of the conditions which could affect his work.

4.6.1.4. Housekeeping

Another maintenance issue is what reasonably should be expected of custodians? Experience has shown that there is a tremendous variation in the level of expectations and wishes among laboratory supervisors. There are those who do not wish custodians to enter their laboratory at all, while there are those who have no qualms in asking them to clean up a hazardous chemical spill. Most safety and laboratory personnel would agree that the latter is asking too much, while most would also agree that, if they wish, facility personnel should be allowed to take care of their own housekeeping, as long as reasonable standards of cleanliness are maintained. Most laboratory groups, however, fall somewhere between these two extremes.

The salary levels of most custodial positions are usually among the lowest in most organizations and limit the skill levels one can expect from the persons filling the positions. Unfortunately, literacy rates are less than average and, in many cases, it certainly would be unrealistic to expect a significant level of technical training which would permit an understanding of the problems that custodial employees might encounter in a laboratory. As a result, custodial workers are often afraid of the laboratory environment. However, alternative positions are also usually hard to find for these employees, so they frequently are very concerned about losing their jobs. Most cannot afford to do so. As a result of these conflicting pressures, they may attempt to do things they really do not understand and are afraid to ask about and may make mistakes in consequence. It is the responsibility of the laboratory supervisor, working with custodial management, to carefully establish the areas of responsibility for the custodians in the laboratory.

Among things a custodian can reasonably be expected to do in most laboratories are:

1. Clean and maintain the floor area.
2. Dispose of ordinary trash. However, if other than ordinary solid waste is generated in the laboratory, it should be placed in distinctively shaped and/or colored containers. If the custodians are still expected to handle it, then the circumstances and procedures should be carefully delineated and training given.
3. Wash windows. If they are expected to wash bench tops or other laboratory furniture, it should be only when additional supervision is provided by laboratory personnel.

Among items which they should not be expected to do are

1. Clean up chemical spills. They are not trained to do it according to established regulatory guides nor to do it in such a way as to ensure that they do not expose themselves to the potential injury.
2. Dispose of broken glass, syringes, or "empty reagent containers." These items can be disposed of by them if they are carefully prepared by the laboratory workers in advance. For example, broken glass can be disposed of if it is placed in a sturdy kraftboard box, sealed, and labeled as "broken glass." Other items, such as syringes, are now considered to be regulated medical wastes in most instances, as, for example, if they fall under the provisions of the bloodborne pathogen standard. Syringes are to be placed in a leak- and puncture-proof container specifically intended for them. The needles are not to be sheared, broken, or removed in any way. The containers are to be rendered biologically safe and disposed of by techniques which will be discussed in detail in the section on infectious waste disposal. Custodians are not to handle these wastes. Empty reagent bottles should be triple rinsed and then placed in a box labeled "triple rinsed reagent containers." All of the restrictions on glassware should be thoroughly explained to the custodial employees. There are specific requirements under the bloodborne pathogen standard that this be done. If custodians believe that they are being asked to handle unsafe waste, they should ask their supervisors to intercede for them.
3. Handle special wastes in any way, including radioactive materials, chemical wastes, or contaminated biological materials. All of these require special handling by specialists and precautions must be taken to ensure that these materials are not accidentally collected by custodians. The custodians should be given awareness training to ensure that they have sufficient knowledge to allow them to recognize these special wastes.

4. Clean the work surfaces and equipment in the laboratory, except in special circumstances and under the direct supervision of a responsible laboratory employee.
Even in this case, a preparatory program should have been carried out in advance by
laboratory personnel to remove or secure items which could be dangerous in the area
being cleaned.

Housekeeping also means maintaining the laboratory in a reasonably organized fashion
on a day-to-day basis. This is the responsibility of all laboratory personnel, but individuals
will follow the laboratory manager's own performance as a guide. Reagents not in use
should be returned to proper storage. Secondary containers should be labeled according to
the requirements of the hazard communications standard. Glassware should be cleaned and
put away. Trash should not be allowed to accumulate. Equipment should not be allowed
to encroach upon aisles. Cables and temporary electrical extensions should not become a
tripping hazard. Periodically refrigerators and other storage units should be gone through
and cleaned out. An audit of materials should be made periodically to dispose of old,
degraded, and obsolete materials before they become a hazard. Chemicals stored inappropriately outside of their hazard class should be returned to their proper locations.
Bottles heavily covered with dust, indicating a lack of use for an extended period, are likely
to remain unused and should be eliminated. No one should expect a busy laboratory to be
spotless but neither should it be a disaster area. Unless a concerted effort is made, eventually housekeeping problems tend to slowly accumulate. An effective mechanism used by
the author to combat this erosion of order was to schedule a quarterly "field day" during
which all personnel, including faculty, staff, and students, ceased research and returned
everything to reasonable order. This rarely took more than a few hours and furthered a
sense of cooperation between the various groups of people.

4.6.1.5. Signs and Symbols

Many situations exist in which a person entering an area needs to be made aware that
a hazard exists in the area or needs to know of restrictions placed on persons entering the
area. In addition, there are signs which are intended to provide information to individuals
in an emergency. There are literally hundreds of specialized safety signs and symbols
which can be purchased for the laboratory. Given below is a partial list of some of the
more important ones, along with a brief description of the types of applications for which
they would be needed. In many cases the signs in this list are mandated by regulatory
requirements, while in other cases they represent common safety practices. In most cases,
the hazard signs will be prefaced by a risk descriptor, defining the level of risk represented
in the specific instance. The three cautionary words in normal use, in decreasing order of
risk are **DANGER**, **WARNING**, and **CAUTION**.

1. **AREA UNSAFE FOR OCCUPANCY**—This is used to indicate a contaminated
   area or an area otherwise rendered unsafe, temporarily or otherwise, for normal use.
2. **AIRBORNE RADIOACTIVITY AREA**—Some applications involving radioactive
   materials result in the generation of airborne radioactive materials in excess of those
   permitted by the standards of the NRC, or of the equivalent state agency in an
   agreement state. Should such an operation exist, the boundaries of the room,
   enclosure, or operating area where the airborne material may exist must be posted
   with this sign. The legend will be accompanied by the standard radiation symbol
   shown in Figure 4.4.

Figure 4.4. Radiation symbol.

Figure 4.5. Biohazard symbol.

3. **ASBESTOS**—Asbestos is still used in a number of products employed in laboratories. If there is the potential while using these products that asbestos fibers may become airborne, the area needs to be marked with a sign:

CAUTION
ASBESTOS-CONTAINING MATERIAL PRESENT

Note that a sign such as this does not say that there are asbestos fibers in the air. The sign is intended to alert people that their actions could result in the generation of airborne asbestos fibers. If there is a risk that airborne asbestos fibers may be present, the appropriate department needs to be notified (usually either physical plant or health and safety) to correct the problem.

4. **AUTHORIZED ADMISSION ONLY**—This sign may accompany many other signs or it may stand alone in restricting access to an area to those who have legitimate reasons to be there, or who are aware of the risks within the area to which they may be exposed.

5. **BIOLOGICAL HAZARD**—The sign will be accompanied by the standard biological hazard symbol shown in Figure 4.5, indicating that an agent which may prove infections to human beings is present within the area.

6. **CARCINOGENIC AGENT**—The laboratory safety standard requires that areas in which carcinogenic agents are in use be designated as such. This can be done with a sign such as:

CANCER-SUSPECT AGENT
AUTHORIZED PERSONNEL ONLY

Where the agent might be unusually dangerous, the agent would be specified and any special protective measures needed would be appended.

7. **CHEMICAL SPLASH GOGGLES REQUIRED WHILE WORK IN PROGRESS**—It is recommended that this sign (Figure 4.6) be used at the entrances to all active laboratories where chemicals are employed and are actively being used. In order to enforce the requirement, care must be taken to select goggles which resist fogging, do not become oppressively warm while being worn at comfortable room conditions, and do not exert uncomfortable pressure on the face. They should also accommodate wearing normal size prescription glasses at the same time. Many goggles which meet the minimum regulatory standard based on ANSI Z87.1 for

Figure 4.6. Eye protection pictograph.

impact protection do not meet all of these practical considerations, but there are several that do. When work is not in progress, or when a person is in an area well separated from the active work area, it may be permissible, for reasons of comfort, to allow goggles to be removed.*

8. **CRYOGENIC LIQUIDS**—All containers which contain cryogenic liquids, most commonly liquid nitrogen (as in the example below) but also other gases maintained at very low temperatures, should be prominently labeled:

CAUTION
LIQUID NITROGEN

The container, usually a large flask with walls separated by a vacuum called a dewar, will also usually be labeled with the cautionary information:

FRAGILE CONTAINER UNDER VACUUM
MAY IMPLODE

9. **EMERGENCY INFORMATION SIGNS**—Prominent signs, such as those shown in Figure 4.7, should be posted near the safety device mentioned to aid in locating them in an emergency. Symbols can be used in place of or in addition to some of these.

10. **EXPLOSIVES**—If explosives are stored in Class 1 magazines, or in outdoor Class 2 magazines, the property most be posted with signs stating **EXPLOSIVES—KEEP OFF**. Class 2 magazines must have labels on all sides except the bottom in letters at least 3 inches high, **EXPLOSIVES—KEEP FIRE AWAY**.

11. **FLAMMABLE MATERIALS**—Cabinets containing flammable materials and areas or rooms where flammable materials are stored or used must be posted with this sign, which may also be indicated by the symbol shown in Figure 4.8. This sign should always be accompanied by the **NO SMOKING** sign, which may be augmented by a standard no smoking symbol.

12. **HIGH VOLTAGE DANGER**—Spaces which contain accessible high voltage panels, such as switch rooms and electrical closets, should be locked and provided with these signs to warn persons lacking training and experience in working with high voltage circuits not to enter. Equipment containing high voltage circuits should also bear this warning label.

13. **HYDROGEN—FLAMMABLE GAS, NO SMOKING OR OPEN FLAMES**— This sign must be posted in all areas where hydrogen is used or stored.

---

*   A point is intended to be made here, which will not be repeated for reasons of brevity, that signs and rules must take into account human factors. Otherwise, they are likely to be ignored and weaken compliance with other rules.

Figure 4.7. Various emergency signs.

Figure 4.8. Flammable and **no smoking** signs.

14. **INTERLOCKS ON**—Equipment with internal hazards, such as X-ray diffraction cameras, or areas in which the space is rendered unsafe to enter by the presence of a hazard, are often provided with a fail safe circuit, or interlock, which will turn off the equipment creating the problem if the circuit is broken. The sign provides a warning that the interlock is on to prevent access to the hazard.

15. **LASERS**—Labeling of lasers should follow 21 CFR 1040, the Federal Laser Product Performance Standard. The spaces in which lasers are located should also have a similar warning at the entrance. The label will depend upon the class of laser

involved. All of the labels will include a stylized sunburst symbol, with a tail extending to the left (see Chapter 5). The signal word **CAUTION** is to be used with Class II and IIIA laser systems while the signal word **DANGER** is to be used for all Class IIIB and Class IV systems.

16. **MACHINE GUARDS IN PLACE**—OSHA requires that many machines, such as vacuum pumps or shop equipment, be provided with guards over the moving parts. Signs should be posted near these machines to remind employees not to use the equipment if the guards are not in place.

17. **MICROWAVES**—This sign must be posted in any area where it is possible to exceed the current occupationally legal limit of exposure to microwave electromagnetic radiation.

18. **NO EATING, DRINKING, SMOKING, OR APPLYING COSMETICS**—This sign should be posted whereever toxic materials are used, in the working areas of wet chemistry laboratories, or in biological laboratories using pathogenic substances.

19. **NO SMOKING**—A **NO SMOKING** sign, as shown in Figure 4.8 must be posted wherever flammables are in use; where there is a risk of explosion due to the presence of explosives or from gases, vapors, or dusts; and where toxic materials are in use.

20. **RADIATION AREA**—Areas where the radiation exceeds a level established by the NRC must be posted with this sign (Figure 4.9). If the level exceeds a higher level set by the NRC, the area must be posted with a **HIGH RADIATION AREA** sign. Most of these areas will be within an area posted with a **RESTRICTED AREA—AUTHORIZED ADMISSION ONLY** sign. Specifics on these requirements will be given in Chapter 5. Signs defining radiation areas should be used to post areas where radioactive materials are stored unless radiation levels equal or exceed the stipulated limits. Areas where radioactive materials are stored should be posted with a **CAUTION—RADIOACTIVE MATERIALS** sign.

21. **RADIOACTIVE WASTE**—This is not a sign specifically required by the NRC but is recommended to denote areas within a laboratory where radioactive waste is temporarily stored prior to being removed for permanent disposal, in order to help avoid accidental removal of radioactive waste as part of normal laboratory waste. Much radioactive waste resembles ordinary trash, such as paper.

22. **REFRIGERATOR (FREEZER) NOT TO BE USED FOR STORAGE OF FLAMMABLES**—All refrigerators or freezers not meeting the standards permitting the storage of flammable materials (see Section 4.5.7) must be marked with this sign.

23. **REFRIGERATORS NOT TO BE USED FOR STORAGE OF FOOD TO BE USED FOR HUMAN CONSUMPTION**—Laboratory refrigeration units used for the storage of chemicals and biological materials must be posted with this sign to prevent the use of units to store lunches and other food.

24. **RESPIRATORY PROTECTIVE EQUIPMENT REQUIRED**—Wherever airborne pollutants are present which exceed the PELs established by OSHA, respiratory protection is required (see Figure 4.10). In many cases, ACGIH threshold limit values (TLVs) are lower than the OSHA PELs and respiratory protection is recommended when the levels approach these lower limits. It is recommended that in most cases an action level of half or less of the TLV values be set to accommodate in part the different sensitivity of individuals to materials.

25. **SAFETY GLASSES REQUIRED**—This sign is to be posted wherever there is a risk of eye injury due primarily to impact.

26. **TOXIC GAS**—Areas where toxic gases are used or stored must be posted with this warning sign.

27. **ULTRAVIOLET LIGHT, EYE PROTECTION REQUIRED**—This sign should be posted wherever there is a risk of eye injury due to ultraviolet light emission.

Figure 4.9. Basic radiation area sign.

Figure 4.10. Respirator needed pictograph.

There are many other signs and symbols identifying hazards or denoting specific requirements to aid in reducing a specific risk. The following generic signs are representative of many of these.

28. **(SPECIFIC ITEM) PERSONAL PROTECTIVE EQUIPMENT RE-QUIRED**—Many other risks exist which would require specific items of protective equipment. Where these items are needed, the area should be appropriately posted.
29. **(SPECIFIC) TOXIC OR HAZARDOUS MATERIAL**—There have been a number of materials that pose known risks, and the areas in which these materials are used should be posted with an appropriate sign.
30. **(SPECIFIC) WASTE CHEMICAL ONLY**—Disposal of waste chemicals according to RCRA standards requires that wastes be identifiable, in some cases by class only, but in most cases it is desirable that wastes not be mixed. Posting of areas or containers with this sign where several waste streams of different character will aid in legal disposal of waste.

4.6.2. Working Procedures

This section can only touch upon the broad topic of safe laboratory working procedures because of the immense scope of the subject. Some of the more common areas which offer the potential for mishaps will be covered in some detail, but undoubtedly there will be areas that are considered comparably important by many that will be touched upon lightly or not at all. Sections will be devoted to a few of the more hazardous chemicals to illustrate the precautions that need to be taken when working with such materials. In addition to physical hazards, such as fire and explosions, health risks will be emphasized, since in many cases these are more insidious and less often recognized by many laboratory workers. The next several sections will be concerned primarily with physical hazards and the latter part of the chapter will be devoted to short- and long-term aspects of laboratory operations on workers' health.

4.6.2.1. Protection Against Explosions

Unusually careful planning must take place whenever there is any reason to suspect that work to be undertaken may involve the risk of an explosion. However, not all potentially explosive situations are recognized in advance. Letters from experimenters describing work in which unexpected explosions occurred can be found in a substantial proportion of issues of *Chemical and Engineering News.* Because these incidents were unanticipated, sufficient protective measures often were not employed; consequently, injuries which could have been avoided are reported in these letters. Explosions may occur under a variety of conditions, the most obvious being a runaway or exceedingly violent chemical reaction. Other situations could include the ignition of escaping gases or vapors, ignition of confined vapors with the subsequent rupture of the containment vessel, rupture of a system due to overpressure caused by other mechanisms, or a violent implosion of a large vessel operating below atmospheric pressure. Partial confinement within a hood can enhance the dangerous effects of an explosion; areas in front of the open face may be damaged more severely than if the explosion were not confined.

Injuries can occur due to the shock wave from a detonation (if the release of energy occurs at supersonic speeds) or deflagration (if the energy release occurs at subsonic speeds). Most laboratory reactions belong to the latter class. Hearing loss may result if the shock wave causes a substantial overpressure on the eardrums. According to Table C-3.1(a) of NFPA 45, Appendix C, the equivalent of as little as 1 gram of TNT can rupture the eardrum of a person within 0.75 m ($\approx$2.4 ft), while 10 g is likely to rupture the ears of 50% of persons within 67 cm (2.2 ft) of the expression. The shock wave, as a wave, can "go around" barriers or be reflected and reach areas that would be shielded from direct line-of-sight interactions. Injuries can occur due to the heat or flames from the explosion. Fume hood materials should be selected to contain fires occurring within them. However, if the sash is severely damaged, flames or burning material can escape through the front opening and the flames may spread to other fuels in the vicinity. Due to this possibility, flammable materials should not be stored in close proximity to fume hoods. Respiratory injuries can occur due to inhalation of fumes and reaction products. However, the most serious hazard is flying debris, including fragments of the containment vessel, other parts of the experimental apparatus, or nearby materials or unreacted chemicals which can inflict physical injuries. The risk of the latter type of injuries can be reduced by eliminating the possibility of line-of-sight or single-ricochet paths for missiles from likely sources of an explosion to workers or to equipment which could be damaged and result in secondary harmful events. The possibility of extraneous material becoming involved in an explosion is a powerful argument in favor of not using a hood as a storage area, especially in experimental activities. The reflected shock wave can act in much the same way as a piece of physical debris in causing damage external to a fume hood. Overreaction of a worker or involuntary reflexive actions to even minor explosions can also lead to quite serious secondary injurious incidents.

In addition to immediate injuries, an insufficiently contained explosion can lead to fires or cause damage sufficient to wipe out expensive apparatus, destroy months or years of research effort, or even destroy an entire facility. Conservative precautionary measures to reduce the likelihood of these repercussions are worthwhile from this aspect alone. Ordinary fume hoods offer a fair amount of protection to the sides and rear of the hood, if they are of good quality with substantial walls. However, most fume hoods are not intended to provide really significant explosion protection against a major explosion for a user standing immediately in front of the hood, although sash materials are usually designed not to contribute to the hazard. This is accomplished by having the sash material made of either laminated or tempered glass so that the broken sash will not cut persons standing in front of the hood. A hood with a three-section horizontal sash, where the user stands behind the central section, provides superior protection to the more common vertical sash hood. If the work to be done involves a known explosive risk, certainly a hood specifically designed to contain any anticipated explosion, or to provide safe explosion venting should be employed.

For the majority of laboratories equipped only with ordinary fume hoods, supplementary measures should be taken to minimize the type of risks described above if a careful analysis of the planned operation reveals any likelihood of a potential explosion.

A simple way to reduce the potential risks is to minimize the amount of material involved in the experiment. The smallest amount sufficient to achieve the desired result should be used. The trend toward microscale experimentation supports this option. Care should be taken in scaling up from a preliminary trial run in which minimal quantities were employed. Increasing the amount of material in use could significantly change the physical parameters so that insufficient energy removal, inadequate capacity for the reaction products, or excessive pressures could develop in the scaled-up version of the work and lead to a dangerously unsafe condition.

A number of other measures can be taken to enhance the protection of workers against explosions. Provision of barriers is a straightforward measure. The selection of an appropriate barrier will depend upon the circumstances. A variety of factors should be considered.

The strength of the barrier material is clearly an important factor. Tests have been made of many materials commonly used in laboratory protective barriers and available either in commercial units or readily amenable for fabrication of custom shielding. Table 4.12 is adapted from a study by Smith in which each material tested was 0.25 inches or 6.4 millimeters thick. The relative susceptibility to fracture was measured by either the ASTM D 256 test method or by dropping balls from various heights. It required 12 to 16 foot pounds of energy to fracture the polycarbonate material in the ASTM D 256 test. Additional protection can be obtained by increasing the thickness of the materials used in fabricating the shield, approximately proportional to the thickness. An equal thickness of steel would have a relative effectiveness on this scale of about 40. Resistance to fracture is not the only consideration. Wired glass, for example, may represent an additional hazard due to the presence of the wire, if shattered. Ordinary glass should not be used due to the danger of cuts from the flying debris. Methyl methacrylate is not suitable where high temperatures may occur. However, sheets of methyl methacrylate are commonly available at moderate cost and can readily be fabricated into custom shields. Polycarbonate obviously offers considerable strength, but can be damaged by organic solvents. Steel is resistant to both heat and solvents, but does not offer the desired transparency. However, there are alternatives to this deficiency, such as mirrors, optical devices, or closed-circuit television. Remotely controlled manipulating devices can be used to control apparatus behind any shield material.

The simplest types of supplementary protection suitable for moderate risks are commercial shields which are available from most laboratory supply firms. Shields usually found in catalogs are curved sheets of transparent material, most commonly polycarbonate, weighted at the bottom to increase their stability. Since these are free standing, they often will not remain upright in explosive incidents and, if the explosion is severe enough, may actually be hurled through the air and cause injury themselves. Since the scale of an explosion cannot always be accurately estimated, it would be desirable to secure these shields firmly to the work surface. For small-scale reactions, they offer a worthwhile degree of added protection. The shields should be located so as to provide the maximum

Table 4.12. Shock Tests on Transparent Shields

Material	Thickness		Drop Ball		ASTM D 256
	mm	in.	kg/m	in./lb	
Double-strength glass	3.2	0.125	446	25	
Laminated glass	6.4	0.25	1,964	110	
Plate glass	6.4	0.25	1,964	110	
Wired glass	6.4	0.25	2,000	112	
Tempered glass	6.4	0.25	10,393	582	
Methyl methacrylate	6.4	0.25	19,400	1086	0.4 to 0.5
Polycarbonate	6.4	0.25			12.0 to 16.0

protection against flying debris, chemicals, or, as noted earlier, external shock wave interactions. Individuals in the laboratory should be trained to use the shields correctly and not to move or modify them to improve their convenience in performing tasks, if these changes could reduce the level of protection.

For larger-scale experiments, especially if the operation is to continue for a significant length of time, the small commercial shields should be replaced with custom-fabricated shields. Because of its availability in a wide range of thicknesses, relatively high impact resistance, moderate cost, and ease of fabrication, methyl methacrylate is a convenient material to use in constructing custom transparent shields. The same essential design considerations that apply to commercial shields also apply to custom shields: (1) select the material that has the appropriate mechanical and physical properties and (2) place the shield to provide the maximum protection against flying debris and other external effects and to minimize direct injuries to personnel and secondary events. As the size or hazard of the experiment increases, more consideration should be given to relocating the work to a more appropriate containment facility, preferably one specifically designed and engineered to limit damage should an explosion occur. If it is essential that the work be done in a specific location not explicitly designed for work with potentially explosive situations, it is strongly recommended that the design of the proposed renovation or modification be done by a qualified engineer and fabricated by professionals, not graduate students.

There are differences in the cost of different types of shielding materials and in the approach that is taken to provide adequate protection. A cost-benefit analysis is always appropriate in selecting or designing any experimental system. On the other hand, selecting too inexpensive an approach can be a false economy. The person making the decision may be the one injured or killed if the protection is insufficient. There is always the question of liability if others are injured and, finally, there is always the ethical question of what should have been done if one only did what was required to be done and a person was injured as a result,

4.6.2.1.1. Personal Protective Equipment

Personal protective equipment will be covered in more detail in Chapter 6, but an important aspect of enhancing the safety of workers where the potential for explosions exists is to provide and require the use of protective equipment. Every laboratory worker in a facility where any potentially injurious chemical is being used should always wear protective eyewear. Common spectacles with side shields are not nearly as effective as properly selected chemical safety goggles in protecting workers' eyes. The latter not only fit all around the eyes and thus protect against direct impact, but also protect the eyes against flying liquids. Since a properly fitting pair of goggles offers a snug fit to the face, it generally provides superior protection from lateral impacts which could knock off an ordinary pair of safety spectacles. Additional protection to the eyes, face, and throat should be provided, as circumstances warrant, by the use of face masks in addition to goggles. A mask which protests the throat is preferable to one which does not, due to the vulnerability of the carotid arteries on each side of the neck.

In addition to eye protection, sturdy laboratory smocks, preferably made of a flame-retardant or flame-proof material, should be worn. Short-sleeved shirts, T-shirts, shorts, or sandals should not be allowed in any laboratory where the potential for exposure of the skin to chemicals exists or where even minor explosions can occur. Sturdy gloves or gauntlets, selected for the immediate requirement of manual dexterity, should be worn if manual manipulation of apparatus or materials with the potential for explosion is needed. Although the work should be done in a specially equipped facility if the potential for a major explosion exists, in principle, even bullet-proof vests could be used in certain situations. The potential for hearing damage can be greatly reduced if research personnel use good quality earmuffs. If these are to be issued, the workers should be provided an audiometric hearing evaluation beforehand to determine if a hearing loss has actually occurred subsequent to an accident.

4.6.2.1.2. Summary

Laboratory explosions occur frequently in situations in which they are least expected. Simple caution should dictate a conservative estimate of the probabilities of such an event and encourage the use of appropriate preventive measures in all laboratory work. It is far better to prevent an explosion than to attempt to confine one or reduce the severity of the damage resulting from one. However, appropriate barriers and personal protective devices can aid in reducing the seriousness of consequences when explosions do occur, if these ameliorative measures are used properly. Training all personnel to understand the risks associated with their work, coupled with encouragement by management and the cooperation of all laboratory workers to follow safe procedures, can significantly reduce the risks of explosions.

REFERENCES

1. Fire Protection for Laboratories Using Chemicals, Appendix C., NFPA Standard 45, National Fire Protection Association, Quincy, MA.
2. **Smith, D.T.,** Shields and barricades for chemical laboratory operations, in *Handbook of Laboratory Safety,* 2nd ed,, Steere, N.V., Ed., CRC Press, Boca Raton, FL, 1971, 113.

4.6.2.2. Corrosive Chemicals

The definition of corrosive chemicals is very broad. However, in the sense that the action of the chemical will result in an immediate, acute erosive effect on tissue as well as other materials, strong acids and bases, dehydrating agents, and oxidizing agents are commonly considered to be corrosive materials. These terms may not be mutually exclusive.

Accidents with corrosive materials in which the material may splash on the body are very common in the laboratory. The eyes are particularly vulnerable to injury, and injuries to the respiratory system may range from moderate irritation to severe injury. Skin injuries may be very slow to heal. Ingestion can cause immediate injury to the mouth, throat, and stomach, and in severe cases can lead to death. Some common household chemicals that are equivalent to laboratory corrosives, such as drain cleaners, are common causes of child fatalities. Work with strongly corrosive materials should be done in a fume hood. Personal protective equipment, such as gloves, laboratory aprons, and chemical splash goggles should be used and, if the possibility of inhalation is significant, appropriate respirators fitted with specific cartridges for the type of materials being used. Every laboratory, especially those using these materials, should be individually equipped with deluge showers and eyewash fountain combinations.

Corrosive chemicals should always be handled with the greatest care. Where available, they should be purchased in containers coated with a protective plastic film so that if they are dropped the probable result should be at most a leak through the film instead of a potentially dangerous splashing of the chemical onto the skin of the person transporting the material and possibly others. Although there is a surcharge in most cases for ordering containers with protective films, the additional protection afforded is substantial. Even with the film, the material leaking out of the container can still represent a nasty mess which must be cleaned up promptly.

Safety carriers are available to use in transporting containers of dangerous chemicals. These should be used whenever the chemical containers are unprotected, breakable bottles. Unfortunately, it is unlikely that they will be used for the movement of chemicals a few feet within a laboratory, which is much more common than transporting materials from one laboratory to another, or from the stockroom to the laboratory. Just as most car accidents occur within a few miles of home, most chemical accidents occur where the workers spend most of their time. The use of protected containers can largely ameliorate the seriousness of these accidents, since it does not require an extra effort to use them.

When it is necessary to move chemical containers a significant distance, the use of safety carriers, even for protected containers, is recommended. Although the result is likely to be much less serious when using coated bottles, accidents can still occur and it is always desirable not to have to clean up after one. If several bottles are to be moved at once, the bottles should be moved on a low cart, with a substantial rim around the edge, so that if the cart is struck the chemicals will be likely to stay on it, or if not, have only a short distance to fall.

Wherever possible chemicals should be moved from one floor to another on a freight elevator rather than carrying them manually up and down a flight of stairs. If only a passenger elevator is available, the use of the elevator should be delayed until no passengers are using it or passengers desiring to use it should be courteously asked to wait until the chemicals have been moved. Although elevator accidents are rare, should a dangerous material be released in one while the passengers were trapped inside, the results could be catastrophic. The fumes could easily cause serious injury before it would be possible to leave the elevator.

Laboratory personnel should be trained to quickly limit the area affected by a spill, and the necessary supplies to enable them to do so should be immediately available within the laboratory. Kits containing absorbent pillows, neutral absorbent materials, or neutralizing materials are commercially available and will enable knowledgeable persons to safely contain most small accidents, such as spills of the contents of a single bottle of reagent, until a final cleanup in which the materials in the container as well as those used to contain it can be disposed of, usually as hazardous waste.

If the average person who has taken chemistry class retains one safety rule, it is probably the one about always adding acid to water, never the reverse (usually brought up in the context of sulfuric acid, which is a strong dehydrating agent). This precaution is taken to avoid splashing of acid on the body due to the localized generation of heat as the two substances mix.

There are a number of other basic safety procedures involving corrosives. Keep the container sizes and quantities on hand as small as possible, consistent with the rate of use. Store each class by themselves. Keep containers not in use in storage, and store the containers either in separate cabinets or on low shelves rather than high ones. Remember that reactions involving these substances will usually generate substantial heat, so that closed containers in the area of a spill involving these highly reactive materials could become hot and rupture due to excessive pressures.

Some brief comments about some of the classes of corrosives are given below. In the sections which follow, some of these as well as others will be covered in more detail.

1. **Strong Acids**—Concentrated strong acids can cause severe and painful burns. The pain is due in part to the formation of a protein layer which resists further penetration of the acid. In general, inorganic acids are more dangerous than organic acids, although the latter can cause deep-seated burns on extended contact with the skin. Leakage from containers and material remaining on the outside of the containers following a sloppy transfer can cause corrosion of the shelving and, if the acids are stored with materials with which they may react, accidents can result if the chemicals come into contact.

2. **Strong Alkalis**—Alkali metal hydroxides are very dangerous when allowed to come into contact with tissue. The contact with the skin is likely to be less painful than a comparable exposure to acid because the protective protein barrier is not formed. Damage may extend to greater depths as a result of the lesser pain because the injured person may not be as aware of the seriousness of the incident. Any area exposed to a strong alkaline material should be flooded with water for at least 15 minutes or longer. This is especially important in eyes since the result of an exposure can be a rupture of the global structure of the eye.

3. **Nonmetal Chlorides**—Compounds such as phosphorous trichloride and corresponding bromides react violently with water and are a common cause of laboratory accidents.

4. **Dehydrating Agents**—Strong dehydrating agents such as sulfuric acid, sodium hydroxide, phosphorous pentoxide, calcium oxide, and glacial acetic acid can cause severe burns to the eyes because of their strong affinity to water. When they are added to water too rapidly, violent reactions accompanied by spattering can occur.

5. **Halogens**—Halogens are corrosive on contact with the skin, eyes, and the linings of the respiratory system as well as being toxic. Because they are gases, they pose a greater danger, especially by inhalation, of coming into contact with sensitive tissue.

4.6.2.3. High-Energy Oxidizers

Oxidizing agents such as chlorates, perchlorates, peroxides, nitric acid, nitrates, nitrites, and permanganates represent a significant hazard in the laboratory because of their propensity under appropriate conditions to undergo vigorous reactions when they come into contact with easily oxidized materials, such as metal powders and organic materials such as wood, paper, and other organic compounds. Elements from group 7A of the Periodic Table, fluorine, chlorine, bromine, and iodine—react similarly to oxygen and are also classified as oxidizing agents.

Most oxidizing materials increase the rate at which they decompose and release oxygen with temperature. The rate of decomposition of hydrogen peroxide goes up by a factor of about 1.5 with each 10°F (5.6°C). Because of this ability to furnish increasing amounts of oxygen with temperature, the reaction rate of most oxidizing agents is significantly enhanced with increasing temperature and concentrations. The hazard associated with these agents therefore increases as well. For example, cold perchloric acid at a concentration of 70% or less has little oxidizing power, while at concentrations above 73% has significant oxidizing power at room temperatures, which increases still further at higher concentrations. Hot, concentrated perchloric acid is a very strong oxidizing agent. Containers of oxidizing agents may explode if they are involved in a fire within a laboratory.

The quantities of strong oxidizing agents stored within the laboratory should be minimized, and their containers should be rigorously segregated from materials with which they could react. The containers should be protected glass, with inert stoppers instead of rubber or cork.

Quantities of potentially vigorously reacting materials, such as strong oxidizing agents, used in a given research operation or evolution should be kept to the minimal quantities needed. The work should always be performed in a hood, with appropriate safety features (such as the wash-down system, recommended for research involving hot perchloric acid digestions). Oxidizing agents should be heated with glass-heating mantles or sand baths. The use of personal protection devices, including sturdy gloves and eye protection which provides both chemical splash and impact protection, should be mandatory. If the potential for explosions is determined to be significant, the operator as well as others within the facility should be protected with explosion barriers. If the risk is sufficiently high, the research should be performed in an isolated facility especially designed for the program, which would include explosion venting and explosion-resistant construction.

The following are brief comments regarding the hazardous properties of a number of representative common, powerful oxidizing reagents. Most form explosive mixtures with combustible materials, organic substances, or easily oxidizable materials and most yield toxic products of combustion. Current MSDSs will provide additional data on each of these materials.

Ammonium Perchlorate (NH_4ClO_4)—Similar in explosion sensitivity to picric acid. Explosive when mixed with organic powders or dusts. Highly sensitive to shock and friction when mixed with powdered metals, carbonaceous materials, and sulfur.

Ammonium Permanganate (NH_4MnO_4)—May become shock sensitive at 60°C (140°F) and may explode at higher temperatures. Avoid contact with readily oxidizable, organic, or flammable materials.

Barium Peroxide (BaO_2)—Combinations of this compound and organic materials are sensitive to friction and contact with small quantities of water.

Bromine (Br)—Highly reactive material. Causes serious burns to tissue; toxic; when inhaled can cause serious damage to respiratory system.

Calcium Chlorate ($Ca(ClO_3)_2$)—Explosive mixtures are ignitable by heat and friction.

Calcium Hypochlorite ($Ca(ClO)_2$)—Ignites easily when in contact with organic and combustible material. Chlorine evolved at room temperatures when mixed with acids.

Chlorine Trifluoride (ClF_3)—Vapor at room temperature and is dangerously reactive. Most combustible materials ignite spontaneously on contact. This is an exception to the use of glass containers. The material reacts strongly with silica, glass, and asbestos. Extremely toxic. Causes severe burns to tissue.

Chromium Anhydride or Chromic Acid (CrO_3)—Ignites on contact with acetic acid and alcohol and may react sufficiently vigorously with other organic materials to ignite.

Dibenzoyl Peroxide (($C_6H_5CO)_2O_2$)—Extremely explosion sensitive to shock, heat. and friction. Comparatively low toxicity.

Fluorine (F_2)—Extremely reactive gas, reacting vigorously with most oxidizable materials at normal room temperatures, often vigorously enough to ignite. Causes severe burns to tissue. Severe danger of damage to respiratory tract.

Hydrogen Peroxide (H_2O_2)—Commercial products usually sold inhibited against decomposition. At concentrations between 35 and 52%, shares properties of hazards of other oxidizing agents associated with coming into contact with easily oxidizable materials, but may violently decompose when coming into contact with many common metals and their salts, e.g., brass, bronze, chromium, copper, iron, lead, manganese, silver, etc. At higher concentrations, most combustible materials will ignite on contact. Mixing of organics with concentrated hydrogen peroxides may create very sensitive explosive combinations.

Magnesium Perchlorate ($Mg(ClO_4)_2$)—Sensitive to ignition by heat or friction.

Nitric Acid (HNO_3)—Explosively reactive with carbides, hydrogen sulfide, metallic powders, turpentine. Causes severe burns to tissue.

Nitrogen Peroxide (in equilibrium with nitrogen dioxide) (N_2O_4; NO_2)—May cause fire on contact with clothes and other combustible materials. Reactions with other fuels and chlorinated hydrocarbons may be violent. Vapors are life threatening at very low concentrations. Severely dangerous to tissue.

Nitrogen Trioxide (N_2O_3)—May cause fire on contact with combustible materials. Very damaging to tissue, especially respiratory tract, where fatal pulmonary edema may result, although onset of symptoms may be delayed for several hours.

Perchloric Acid ($HClO_4$)—Very dangerous oxidizing agent at high concentrations and elevated temperatures (see Section 4.6.2.4).

Potassium Bromate ($KBrO_3$)—Sensitive to ignition by heat or friction. Relatively moderate health hazard.

Potassium Chlorate ($KClO_3$)—Similar explosive properties to potassium bromate but is toxic and fumes liberated by combustion are toxic.

Potassium Perchlorate ($KClO_4$)—Similar to potassium chlorate above. Yields toxic fumes in fires and is irritant to eyes, skin, respiratory system.

Potassium Peroxide (K_2O_3)—Reacts vigorously with water. Mixtures with combustible, organic, or easily oxidizable materials are explosive. They ignite easily with heat, friction, or small quantities of water. Toxic, if ingested.

Propyl Nitrate (normal) ($CH_3(CH_2)_2NO_3$)—Very dangerous material. Forms explosive mixtures with air. Very wide flammable limits (2% to 100%), Flash point = 20°C (68°F); very low energy required for ignition comparable with acetylene and hydrogen. Vapors are heavier than air and may travel some distance to ignition source and flash back. Material itself is toxic by either inhalation or ingestion and combustion products highly toxic.

Sodium Chlorate ($NaClO_3$)—Properties similar to potassium chlorate above.

Sodium Chlorite ($NaClO_2$)—Releases explosive, extremely poisonous chlorine dioxide gas upon contact with acid.

Sodium Perchlorate ($NaClO_4$)—Properties similar to potassium perchlorate above.

Sodium Peroxide (Na_2O_2)—Properties similar to potassium hydroxide above.

REFERENCES

1. Hazardous Chemical Data, NFPA 49, National Fire Protection Association, Quincy, MA.
2. **Armour, M.A., Browne, L.M., and Weir, G.L.,** *Hazardous Chemicals, Information and Disposal Guide,* 2nd ed., University of Alberta, Edmonton, Canada, 1984.
3. **Lewis, R.J., Sr.,** *Sax's Dangerous Properties of Industrial Materials*, 8th ed., Van Nostrand Reinhold, New York, 1993.

4.6.2.4. Perchloric Acid*

4.6.2.4.1. Introduction

Considerable interest has been taken in the explosion hazards to be encountered in the use of perchloric acid since a mixture of perchloric acid and acetic anhydride exploded in a Los Angeles factory in 1947, killing 15, injuring 400, and causing $2 million in damage. On a smaller scale, Robinson reported a detonation of 3 g of a perchlorate salt of a rhodium-polyamine complex undergoing an evaporation step in a rotary evaporator.[1] A violent explosion destroyed the evaporator, smashed a lab jack, cracked the bench top, and chipped walls over 15 feet away. Fortunately, this happened in an empty laboratory. Literature surveys reveal that descriptions of explosions in laboratories using perchloric acid have been reported over a period of more than a century.

The most detailed available account of the chemistry of perchloric acid and a reference highly recommended to everyone who will be working with perchlorates is given by J.S. Shumacher in the American Chemical Society monograph *Perchlorates, Their Properties, Manufacture, and Uses.*[2] Cummings and Pearson have reviewed the thermal decomposition and the thermochemistry, and in a later report Pearson reviewed the physical properties and inorganic chemistry of perchloric acid.[3,4] Shorter, but very useful, reviews of the chemistry of perchloric acid also have been published[5,6] and several accidents which have occurred in France have been reviewed.[7] A review of the circumstances leading to the Los Angeles explosion and of five laboratory incidents involving perchloric acid has been published,[8] and a résumé of five serious accidents has been reported.[9] A summary of a number of accidents reported in the literature is appended to this section.

4.6.2.4.2. General Discussion

Perhaps the most disturbing features of accidents involving perchloric acid are (1) the severity of the accidents and (2) the persons involved are, in the majority of cases, experienced workers. Harris concludes that the basic cause of accidents involving perchloric acid is contact with organic material or the accidental formation of the anhydrous acid.[8] Smith[6] emphasizes the hazard of allowing strong reducing agents to come into contact with concentrated (72%) perchloric acid (Table 4.13).

---

* This material is taken from an updated (by Graf) version of an article by K. Everett and F.A. Graf, Jr., which appeared in the second edition of the *Handbook of Laboratory Safety*. The remaining portion of this article forms part of Section 4.5.6.

Properties of Perchloric Acid Solution

A review of the latest MSDSs prepared to meet the OSHA hazard communication and laboratory safety standards by each manufacturer is strongly recommended.

The important physical and chemical properties of perchloric acid are that it is a clear liquid like water, it has no odor, the boiling point (of constant boiling mixture at atmospheric pressure) is 203°C (397°F), and while under a high vacuum a 73.6% composition can be produced. Perchloric acid can be dangerously reactive. At ordinary temperatures 72% perchloric acid solution reacts as a strong, nonoxidizing acid. At elevated temperatures (approximately at 160°C, 320°F), it is an exceedingly strong and active oxidizing agent, as well as a strong dehydrating reagent. Contact with combustible material at elevated temperatures may cause fire or explosion.

Relative Oxidizing Power of Perchloric Acid

Although no data are known to describe the change in the oxidizing power of perchloric acid with temperature and/or concentration, some observations describe the phenomenon sufficiently.

Cold perchloric acid, 70% or weaker, is not considered to have significant oxidizing power. The oxidizing power, however, increases rapidly as the concentration increases above 70%. Acid of 73%+ (which gives off fumes in even relatively dry air) is a fairly good oxidizer at room temperature. The monohydrate of perchloric acid (85% acid strength and a solid) is indeed a very good oxidizer at room temperature, as it will even react with gum rubber, whereas the 73% acid does not.

Temperature increases will also increase the oxidizing power of perchloric acid solutions; therefore, hot, strong perchloric acid solutions are very powerful oxidizing agents.

Explosive Reactions

A chemical explosion is the result of a very rapid increase in volume due to the evolution of gas or vapor, the reaction normally being exothermic. The force of the shock wave is governed by the rate at which the reaction takes place. The point is made by Burton and Praill[5] that, apart from the thermochemical aspects of explosions, it is the velocity of the decomposition which determines whether or not a reaction is explosive, and that the power of the explosive is governed largely by the pressure of the gases produced in the decomposition. Where the temperature of the explosion is several thousand degrees celsius, the power of the explosion is increased further by the thermal expansion of the gases.

When considering the hazards involved in the use of perchloric acid this point should be clearly recognized: many of the reported serious laboratory accidents involved only small quantities (<1 g) of reactant (see Appendix included within this section).

The System: Perchloric acid–Acetic Anhydride–Acetic Acid

Both Shumacher[2] and Burton and Praill[5] examined the perchloric acid-acetic anhydride system in some detail, the former reproducing a triangular diagram for the system. Burton and Praill quote the equation:

$$2.5Ac_2O + HClO_4 \cdot 2.5H_2O = 5AcOH + HClO_4 + 18.4 \text{ kcal}$$

showing that there is a considerable evolution of heat when mixing the reagents, and that if excess acetic anhydride is present the solution may be considered to be a solution of anhydrous perchloric acid in acetic acid. The most explosive mixture is given as the one in which complete combustion occurs.

$$CH_3COOH + HClO_4 = 2CO_2 + 2H_2O + HCl$$

These authors give the explosion temperature as 2400°C and calculate that 1 gram of the mixture instantaneously produces about 7 liters of gas at the explosion temperature. Finally, they state that for this system, all the investigated explosions have been due to the use of potentially dangerous mixtures, together with faulty equipment or technique.

In 1961 in a university laboratory, a young metallurgy student lost the sight of both eyes when an explosion took place while he was preparing an acetic anhydride–perchloric acid–water electropolishing mixture. Bartlett and Turner[12] investigated the process and discovered that the reagents should be mixed in the following order:

1. Add the perchloric acid to the acetic anhydride.
2. Add water to the mixture, slowly, and in small portions.

The Use of Magnesium Perchlorate as a Drying Agent

Several cases are on record in which magnesium perchlorate (anhydrone) has exploded while being used as a desiccant (Appendix 1). Smith[6] regards the use of magnesium perchlorate for the drying of alcohol vapors as permissible, but Burton and Praill[5] warn that if magnesium perchlorate is to be used for drying organic liquids, the purity of the drying agent should be determined since the preparation may have left traces of free perchloric acid in the salt. Explosions involving magnesium perchlorate may have been caused by the formation of perchloric esters in the system. It should be noted that methyl and ethyl perchlorate are violently explosive compounds.

Miscellaneous Reactions

Burton and Praill[5] state that it is impossible to overemphasize that the simple alkyl esters are extremely dangerous. They note that many documented explosions resulted from the standard method of determining perchlorates or potassium as potassium perchlorate during which an ethyl alcohol extraction is used. The same authors state that most organic perchlorate salts, with the exception of the diazonium salts, are safe unless they are overheated or detonated, but mention that pyridine perchlorate can be detonated by percussion.

Reducing Agents

In general, mixtures of strong reducing agents and concentrated 72% perchloric acid should be regarded as very hazardous[6] (see Appendix 1).

Safe Handling of Perchlorates

Shumacher[2] states that:

> Perchlorates appear to fall into two broad categories: those (1) more or (2) less sensitive to heat and shock. Included in the group of those qualitatively less sensitive are pure ammonium perchlorate, the alkali metal perchlorates, the alkaline earth perchlorates, and perchloryl fluoride. Among the more sensitive compounds are the pure inorganic nitrogenous perchlorates, the heavy metal perchlorates, fluorine perchlorate, the organic perchlorate salts, the perchlorate esters, and mixtures of any perchlorates with organic substances, finely divided metals, or sulfur. Any attempt to establish a more precise order of the degree of hazard to be expected from any given perchlorate seems unwarranted on the basis of available data. Each perchlorate system must be separately (and cautiously) evaluated.
>
> There do not appear to be any uniform recommendations for the safe handling of perchlorates which are generally applicable. A number of heavy metal and organic perchlorates, as well as hydrazine perchlorate (hydrazinium diperchlorate) and fluorine perchlorate, are extremely sensitive and must be handled with great caution as initiating explosives. Mixtures of any perchlorates with oxidizable substances are also highly explosive and must be treated accordingly. For all of these, it is essential to avoid friction, heating, sparks, or shock from any source (as well as heavy metal contamination), and to provide suitable isolation, barricades, and protective clothing for personnel. However, the more common ammonium, alkali metal, and alkaline earth perchlorates are considerably less hazardous.

Synthesis of new inorganic or organic perchlorates should only be undertaken by an experienced, cautious investigator who is familiar with the literature.

A simple test to evaluate impact sensitivity can be conducted by placing a crystal or two of the perchlorates on a steel block and striking with a hammer. The degree of noise and relative impact to produce an explosion can be roughly correlated with the impact sensitivity.

A simple thermal stability test can be conducted by placing a crystal or two on a hot plate and observing the time to create a violent decomposition reaction. A gram of the material can be heated slowly in a loosely closed vial for a more exact determination of thermal stability.

Recommendations for the Safe Handling of Perchloric Acid

Several organizations have drawn up recommendations for the safe handling of perchloric acid, among them the Association of Official Agricultural Chemists, the Factory Mutual Engineering Division, and the Association of Casualty and Surety Companies. Graf updated these recommendations in a paper.[13] The recommendations from these and other sources are combined and summarized below.[*]

Floors—Perchloric acid should be handled in a masonry building with concrete or tile floors. Handling acid on wooden floors is dangerous, especially after the acid has dried. The wooden floor will then become sensitive to ignition by friction.

---

[*] Note that since the original preparation of this article, commercial equipment specifically designed to be used with perchloric acid has become generally available—specifically hoods and ejector ducts to exhaust perchloric fumes well above roof level.

Table 4.13.   Materials Compatible With and Resistant to 72% Perchloric Acid

Compatible	
Material	**Compatibility**
Elastomers	
Gum rubber	Each batch must be tested to determine compatibility
Vitons[a]	Slight swelling only
Metals and alloys	
Tantalum	Excellent
Titanium (chemically pure grade)	Excellent
Zirconium	
Niobium	Excellent
Hastelloy C*	Excellent
Plastics	Slight corrosion rate
Polyvinyl chloride	
Teflon[a]	
Polyethylene	
Polypropylene	
Kel-F[b]	
Vinylidine fluoride	
Saran[c]	
Epoxies	
Others	
Glass	
Glass-lined steel	
Alumina	
Fluorolube	

Incompatible
Plastics
Polyamide (nylon)
Modacrylic ester, Dynel (35–85%) acrylonitrile
Polyester (dacron)
Bakelite
Lucite
Micarta
Cellulose-based lacquers
Metals
Copper
Copper alloys (brass, bronze, etc.) for very shock-sensitive perchlorate salts
Aluminum (dissolves at room temperature)
High nickel alloys (dissolves)
Others
Cotton
Wool
Wood
Glycerin-lead oxide (letharge)

[a] du Pont trademark; [b] 3M Company trademark; [c] Dow Chemical Corporation trademark.

For conventional wooden wall construction, which is [not] desirable, it is highly recommended that a 6-inch concrete curb be provided for the walls to rest on. In this way, acid seepage under the wall is minimized.

Concrete of course, is not resistant to acids, and thus should be covered. Epoxy paints in general are resistant to room temperature perchloric acid spills; however, epoxy paint will peel off concrete if pools of water stand for several days. Therefore, the floor should have a gentle slope to a drain and contain no low spots.

No equipment of any kind should ever be bolted to a floor by using bolts that screw into the floor. Perchlorates can enter and form hazardous metallic perchlorates that can be initiate a detonation when the bolt is removed. Studs, firmly and permanently set into the floor to which the equipment can be bolted, are far safer. The nuts can then be flushed with water and sawed off with a hacksaw under a constant water spray or remove that equipment.

Laboratory benches—Laboratory benches should be constructed of resistant materials and not wood to prevent acid absorption, especially at the bottom surface which rests on the floor and would be subject to the greatest exposure from acid spills. Bench tops of resistant and nonabsorbent materials such as chemical stoneware, tile, epoxy composites, and polyethylene are recommended.

Shelves and cabinets—Shelves and cabinets of epoxy-painted steel are highly recommended over wood.

Heating source—Hot plates (electric), electrically or steam-heated sand baths, or a steam bath are recommended for heating perchloric acid. Direct flame heating or oil baths should not be used.

Vacuum source—Smith describes a simple apparatus, using a water aspirator or pump, for drawing fumes from a reaction vessel. The use of this apparatus is to be commended in that the contamination of the fume cupboard duct with a dust/perchloric acid layer is avoided and the vapors are drawn into water and discharged safely to the drain.* A similar apparatus is marketed for carrying out Kjeldahl digestions. Vacuum pumps from which all traces of petroleum lubricants have been flushed and refilled with halocarbon, Kel-F, or fluorolube are recommended.

Silverman and First have described a self-contained unit which has been developed and field tested for collecting and disposing of chemical fumes, mists, and gases.[15] It is portable and compact and, when assembled, only requires connection to an electrical receptacle and water tap to be completely operational. Although originally designed for use in filter-type radiochemical laboratory hoods for safe disposal of perchloric acid fumes arising from acid digestions, it may be used as a substitute for a permanent hood in a variety of [low risk] locations. Collection and disposal of the acid at the source of emission is the guiding design principle.

The conventional vacuum pump has been used both in the laboratory and in a pilot plant handling large quantities of 72% perchloric acid. The pumps were protected by the use of well-designed cold traps and desiccant columns as well as maintaining the practice of changing petroleum-based oil daily. The desiccant was also routinely changed, along with the frequent thawing and removal of the cold trap contents.

---

* For an individual laboratory, such a practice should pose no problem. However, if the facility is big enough, eventually such practices may become regulated, especially if the water treatment works is marginal. The same concern may eventually become involved in maintaining air quality near facilities with a large number of hood exhausts.

Glassware. The hazards that may ensue if an apparatus cracks or breaks due to thermal or mechanical shock are sufficient to make it desirable that quartz apparatus be considered, especially as it is necessary in many experiments to chill rapidly from the boiling point.

Glass-to-glass unions, lubricated with 72% perchloric acid, seal well and prevent joint freezing arising from the use of silicon lubricants. Rubber stoppers, tubes, or stopcocks should not be used with perchloric acid due to incompatibility.

Stirrers. Pneumatically driven stirrers are recommended rather than the electric motor type. Repeated exposure of the motor windings to perchloric acid vapor could result in a fire, unless the motor is an explosion-proof type, which would be unlikely.

Sundry items—The choice of tongs for handling hot flasks and beakers containing perchloric acid mixtures should be given due thought. Since the use of radioactive materials has become commonplace, much thought has been put into the design of indirect handling equipment. The cheap, commonly used crucible tongs are most unsuitable for picking up laboratory glassware. If possible, tongs with a modified jaw design should be used to ensure that a safe grip is obtained.

Handling 73% or Less Perchloric Acid in the Laboratory

1. Use chemical splash and impact-rated goggles for eye protection whenever the acid is handled.
2. Always transfer acid over a sink in order to catch any spills and afford a ready means of disposal.
3. In wet combustions with perchloric acid, treat the sample first with nitric acid to destroy easily oxidizable matter.
4. Any procedure involving heating of perchloric acid must be conducted in a perchloric acid fume hood, with the sash down.
5. No organic materials should be stored in the perchloric acid hood.
6. Do not allow perchloric acid to come into contact with strong dehydrating agents (concentrated sulfuric acid, anhydrous phosphorous pentoxide, etc.).
7. Perchloric acid should be used only in standard analytical procedures from well-recognized analytical texts. (This does not apply to analytical research workers.)
8. Keep the quantities of perchloric acid handled at the bare minimum for safety.

Handling 73 to 85% Perchloric Acid in the Laboratory

Use the same precautions as above.

Handling Anhydrous Perchloric Acid (Greater than 85%)

1. Only experienced research workers should handle anhydrous perchloric acid. These workers must be thoroughly familiar with the literature on the acid.
2. A safety shield must be used to protect against a possible explosion, and the acid must be used in an appropriate hood with a minimum of equipment present. No extraneous chemicals should be present in the hood.
3. A second person should be informed of the intended use of anhydrous acid and be in the same room with the research worker using this extremely strong oxidizer.
4. Safety goggles, face shield, thick gauntlets, and a rubber apron must be worn.
5. Only freshly prepared acid should be used.

6. Do not make any more anhydrous perchloric acid than is required for a single day's work.
7. Dispose of any unused anhydrous acid at the end of each day by dilution and neutralization.
8. Contact of the anhydrous acid with organic materials will usually result in an explosion.
9. Any discoloration of the anhydrous acid requires its immediate disposal.

Acid Disposal

Spills—Perchloric acid spilled on the floor or bench top represents a hazard. It should not be mopped up, nor should dry combustibles be used to soak up the acid. The spilled acid should first be neutralized and then soaked up with rags or paper towel. The contaminated rags and paper towel must be kept wet to prevent combustion upon drying. They should be placed in a plastic bag and sealed and then placed in a flammable waste disposal can. If the spill can be rinsed down a chemical drain, neutralization of the wetted area is recommended followed by additional rinsing. Other recommendations in the literature include wearing a face shield and gloves while working on the spill. Cover the spill with a weak solution of sodium thiosulfate, and then transfer the slurry into a large container of water, where it should be neutralized with soda ash. After neutralization, it can be drained into the sewer, accompanied by abundant water.

Disposal—Stir the acid into cold water until the concentration is less than 5%; follow by neutralization with aqueous sodium hydroxide; then dispose of the resulting mixture in the sanitary system, accompanied by abundant water. Larger quantities in the original unopened containers may be acceptable to a commercial hazardous waste vendor. If it is potentially explosive, the best option available is to hire a firm specializing in disposal of exceptionally hazardous materials. This will be expensive.

Dismantling an Exhaust Ventilation System Suspected of Contamination with Perchlorates

Dismantling a laboratory exhaust system contaminated with shock-sensitive perchlorates is a hazardous operation, as evidenced by published and unpublished case histories. The procedures used by one university to reduce the hazards were described by Peter A. Breysse in the *Occupational Health Newsletter* (15[2, 3] 1, 1966) published by the Environmental Health Division, Department of Preventive Medicine, University of Washington. The problem, procedures, and confirmation of perchlorate contamination were reported as follows.

A short time ago, the manager of maintenance and operations was requested to dismantle and relocate six laboratory exhaust systems. The possibility of perchloric acid contamination of these systems was considered. An investigation indicated that several laboratories serviced by the exhaust systems were utilizing or had in the past used perchloric acid for wet ashing of tissues. Furthermore, the exhaust hoods were constructed with sharp corners and cracks, permitting the accumulation of contaminants not readily noticed or easily removed. The ducts were made of ceramic material and contained numerous joints as well as a number of elbows, areas conducive to perchlorate buildup. Organic compounds were also used to pack the duct joints, as an adhesive for the flexible connectors, and as a sealing compound for the fan.

Recognizing the potential dangers of dismantling these systems, the following procedures were established and successfully carried out:

1. It was deemed desirable to dismantle the systems on the weekend when occupancy would be at a minimum.
2. The entire system was washed for 12 hours, just prior to dismantling, by introducing a fine water spray within the hoods, with the fans operating.
3. The fans were then hosed down.
4. Fan mounting bolts and connectors were carefully removed. Nonsparking tools were and should be used throughout.
5. The fans were immediately removed to the outdoors. As an added precaution during removal the fans were covered with a wet blanket.
6. After all the fans were taken outdoors, one fan at a time was placed behind a steel shield for protection during dismantling. This fan was again washed down.
7. Plate bolts were evenly loosened to remove the plate without binding. If a fan puller is necessary, it should be nonsparking.
8. All disassembled parts were washed and cleaned. The gasket material contained on the flanges was scraped off with a wooden scraper.
9. Ordinarily the ceramic ducts would be removed by breaking them apart with a sledge hammer. In this instance, the ducts were washed down again just prior to and during dismantling. A high-speed saw was used to remove the ductwork.

One of the flexible connectors and a piece of duct-joint sealing compound were collected and taken to the laboratory for examination. Qualitative analysis by X-ray fluorescence and chemical tests indicated the presence of perchlorates in both samples. While these procedures for dismantling and decontamination seem unduly severe, the uncertainty requires that they be followed.

One of the problems in conducting such an operation is providing appropriate liability insurance. This question should be considered before doing such work in-house, and contracts for work done by outside vendors should have provision of liability insurance as a specific requirement.

Some additional words of caution include: (1) do not unscrew any nuts or bolts—cut them off after water flushing, (2) do not produce any impact upon the hood parts, and (3) do not cause any friction between parts of the hood during dismantling.

Summary and Conclusions

The use of perchloric acid is becoming increasingly widespread and the properties of both the acid and its derivatives make it likely that the trend will continue. Safety hazards associated with the use of perchloric acid may be reduced, provided that its hazardous properties are clearly recognized, the purpose of the acid in a process is fully understood, and measures are taken to avoid known possibilities.

It is clear, however, that no one should attempt to use perchloric acid who is not fully conversant with the chemistry of the material, who has not made a careful appraisal of operating conditions and techniques, and who exhibits an unsafe attitude about his work.

Appendix 1: Some Accidents Involving Perchloric Acid

1. Explosions may occur when 72% perchloric acid is used to determine chromium in steel, apparently due to the formation of mixtures of perchloric acid vapor and

hydrogen. These vapor mixtures can be exploded by the catalytic action of steel particles.[2]

2. Two workers are reported to have dried 11,000 samples of alkali-washed hydrocarbon gas with magnesium perchlorate over a period of 7 years without accident. However, one sample containing butyl fluoride caused a purple discoloration of the magnesium perchlorate, with the subsequent explosion of the latter.[2]

3. A worker using magnesium perchlorate to dry argon reported an explosion and warned that warming and contact with oxidizable substances should be avoided.[2]

4. An explosion was reported when anhydrous magnesium perchlorate used in drying unsaturated hydrocarbons was heated to 220°C.[2]

5. An explosive reaction takes place between perchloric acid and bismuth or certain of its alloys, especially during electrolytic polishing.[2,5]

6. Several explosions reported as having occurred during the determination of potassium as the perchlorate are probably attributable to heating in the presence of concentrated perchloric acid and traces of alcohol. An incident in a French laboratory is typical: an experienced worker in the course of a separation of sodium and potassium removed a platinum crucible containing a few decigrams of material and continued the heating on a small gas flame. An explosion pulverized the crucible, a piece of platinum entering the eye of the chemist.[7]

7. A violent explosion took place in an exhaust duct from a laboratory hood in which perchloric acid solution was being fumed over a gas plate. It blew out windows, bulged the exterior walls, lifted the roof, and extensively damaged equipment and supplies. Some time prior to the explosion, the hood had been used for the analysis of miscellaneous materials. The explosion apparently originated in deposits of perchloric acid and organic material in the hood and duct.[8]

8. A chemist was drying alcohol off a small anode over a Bunsen burner in a hood reserved for tests involving perchloric acid. An explosion tore the exhaust duct from the hood, bent a portion of the ductwork near the fan, and blew out many panes of window glass.[8]

9. An employee dropped a 7-pound (3.2 kg) bottle of perchloric acid solution on a concrete floor. The liquid was taken up with sawdust and placed in a covered, metal waste can. Four hours later, a light explosion blew open the hinged cover of the can. A flash fire opened three sprinklers which promptly extinguished the fire.[8]

10. A 7-pound bottle of perchloric acid solution broke while an employee was unpacking a case containing three bottles. The spilled acid instantly set the wood floor on fire, but it was put out quickly with a soda-acid extinguisher.[8]

11. At a malleable iron foundry, perchloric acid had been used for about 4 years in the laboratory for the determination of the silicon contents of iron samples. A cast iron, wash-sink drain at the bench used for this purpose had corroded and the leaking acid had soaked into the wood flooring, which was later ignited while a lead joint was being poured. This fire was extinguished and part of the wood flooring was removed. Later in the day, at a point slightly removed from the location of the first fire, a similar fire occurred when hot lead was again spilled. This time the fire flashed with explosive violence into the exhaust hood and stack above the workbench. Laboratory equipment and records were wetted down extensively and damaged.

12. A stone table of a fume hood was patched with a glycerin cement and several years later, when the hood was being removed, the table exploded when a worker

struck the stone with a chisel. The hood had been used for digestions with perchloric acid and, presumably, acid spills had not been properly cleaned up.[9]

13. A conventional chemical hood normally used for other chemical reactions, including distillation and ashing of organic materials, was also used during the same time for perchloric acid digestion. During a routine ashing procedure, the hot gases went up the 12-inch tubular transit exhaust duct and one of a series of explosions occurred that tore the duct apart at several angles and on the horizontal runs.[9]

14. During routine maintenance involving partial dismantling of the exhaust blower on a perchloric acid ventilating system, a detonation followed a light blow with a hammer on a chisel held against the fan at or near the seal between the rear cover plate and the fan casing. The intensity of the explosion was such that it was heard 4 miles away. Of the three employees in the vicinity, one sustained face lacerations and slight eye injury; the second suffered loss of four fingers on one hand and possible loss of sight in one eye; the third was fatally injured with the 6-inch chisel entering below his left nostril and embedded in the brain.[9]

15. A 6-pound (2.7 kilograms) bottle of perchloric acid broke and ran over a fairly large area of a wooden laboratory floor. It was cleaned up, but some ran down over wooden joists. Several years later a bottle of sulfuric acid was spilled in this same location and fire broke out immediately in the floor and the joists.[9]

16. A chemist reached for a body of perchloric acid stored on a window sill above a steam radiator. The bottle struck the radiator, broke, and the acid flowed over the hot coils. Within a few minutes the wooden floor beneath the radiator burst into flame.[9]

17. An explosion occurred when an attempt was made to destroy benzyl celluloses by boiling with perchloric acid.[11]

18. An explosion occurred as anhydrous perchloric acid was being prepared via sulfuric acid dehydration and extraction with methylene chloride when a stopper was removed from the separatory flask.[14]

19. A rat carcass was dissolved in nitric acid, the fat skimmed off, and perchloric acid added. The mixture was heated to dryness and touched, setting off an explosion that cracked the fume hood and nearly blew out the sash.[16]

20. A perchlorate-doped polyacetylene film was prepared and stored under argon in a sealed vessel. Two weeks later, the film detonated when the vessel top was being removed. Earlier safety testing failed to show any reaction to flame or impact.[17]

21. Perchlorate-doped polyacetylene samples combusted violently in the oxygen atmosphere of a Schoniger flask.[18]

22. An explosion occurred in a fume hood upon ether drying of a second crop of crystals of hexamine chromium (111) perchlorate that were washed with absolute ethanol and anhydrous diethylether. Following aspiration of the ether wash, the ether damp filter cake was agitated with a glass stirring rod and the mass detonated.[19]

23. Perchlorate-doped, highly conducting polythiophene Pt-ClO_4 exhibits excellent ambient stability, but the film should not be heated above 100°C. Touching an extremely dry Pt-ClO_4 film (kept in a desiccator over P_2O_5) with tweezers might cause an explosion of the film.[20]

24. Some samples of rare earth organic fluoride were re-ashed with perchloric, sulfuric, and nitric acid in 1-liter beakers. One of the beakers started foaming, turned yellow, and then exploded. The surface of the hot plate was bent downward, and the imprint of the beaker was left in the metal surface of the hot plate by the force of the explosion.[1]

25. Drying an acetonitrile adduct of neodymium perchlorate at 80°C in vacuum apparently produces a compound that can detonate on mechanical contact.[21]

ACKNOWLEDGMENT

This section was based on and developed from a safety assessment prepared for use at the University of Leeds, England, and is reprinted with permission, as well as the work of Graf and associates at Thiokol Chemical Corporation, Brigham City, Utah.

REFERENCES

1. **Robinson, W.R.,** Perchlorate salts of metal ion complexes: potential explosives, *J. Chem. Ed.*, 62(11), 1001, 1985.
2. **Shumacher, J.C.,** *Perchlorates: Their Properties, Manufacture and Uses,* American Chemical Society Monogr. Ser. No. 146, Reinhold, New York, 1960.
3. Perchloric Acid: A Review of its Thermal Decomposition and Thermochemistry, Rocket Propulsion Establishment Tech. Note No. 224, Ministry of Aviation, London, 1963.
4. Perchloric Acid: A Review of the Physical and Inorganic Chemistry, Rocket Propulsion Establishment Tech. Note No. 352, Ministry of Aviation, London, 1965.
5. **Burton, H. and Praill, P.F.G.,** Perchloric acid and some organic perchlorates, *Analyst*, 80, 16, 1955.
6. **Smith, G.F.,** The dualistic and versatile reaction properties of perchloric acid, *Analyst*, 80, 16, 1955.
7. **Moureu, H. and Munsch, H.,** Sur Quelches Accidents Causes par la Manipulation de l'Acide Perchloriques et des Perchlorates, *Arch. Mal. Prof. Med. Trav. Secur. Soc.,* 12, 157, 1951.
8. **Harris, E.M.,** Perchloric acid fires, *Chem. Eng.*, 56(1), 116, 1949.
9. **Schleffler, G.L.,** University Health Service, University of Minnesota, private communication to safety officer, Leeds University, London.
10. **Jacob, K.D, Brabson, J.A., and Stein, C.,** *J. Assoc. Off. Agric. Chem.,* 43, 171, 1960.
11. **Sutcliffe, G.R.,** *J. Textile Ind.,* 41, 196T, 1950.
12. **Bartlett, R.K. and Turner, H.S.,** An unappreciated hazard in the preparation of electropolishings. *Chem. Ind.,* November 20, 1965, pp. 1933–1934.
13. **Graf, F.A.,** Safe handling of perchloric acid, *Chem. Eng. Prog.,* 62(10), 109, 1966.
14. Explosion Involving Anhydrous Perchloric Acid, letter WLC-151/63, Thiokol Chemical Corporation, April 11, 1963.
15. **Silverman, L. and First, M.W.,** Portable laboratory scrubber for perchloric acid, *Ind. Hyg. J.,* 463, 1962.
16. **Muse, L.A.,** Letter to the editor, *Chem. Eng. News,* 51(6), 29, 1973.
17. **Elsenbanner, R.L. and Miller, G.G.,** Letter to the editor, *Chem. Eng. News,* 63(25), 4, 1985.
18. **Varyn, M.E., Schenoff, J.B., and Dabkowski, G.M.,** Letter to the editor, *Chem. Eng. News,* 63(27), 4, 1985.
19. **Pennington, B.E.,** Letter to the editor, *Chem. Eng. News,* 62(32), 55, 1982.
20. **Osterholm, J. and Passiniemi, P.,** Letter to the editor, *Chem. Eng. News,* 64(6), 2, 1986.
21. **Raymond, K.N.,** Letter to the editor, *Chem. Eng. News,* 63(27), 4, 1985.
22. **Everett, K. and Graf, F.A., Jr.,** Handling perchloric acid and perchlorates, in *CRC Handbook of Laboratory Safety,* 2nd ed., Steere, N.V., Ed., CRC Press, Boca Raton, FL, 1971, 265.

4.6.2.5. Ethers

Most of the concern in working with ethers is due to the problem of peroxide formation. Section 4.5.4 already dealt with this at some length, discussing this problem in the context of storage of ethers. If ethers are bought in small sizes, so that the containers in use on a given day are emptied, the concerns of peroxide formation in partially full containers in storage should not arise. Peroxides can form in unopened containers as well, although due to the absence of excess available air in the restricted empty space above the liquid level, the rate should be much slower. However, the inventory of unopened containers in a laboratory should be maintained at a reasonable level so that no ethers should be kept for extended periods. All containers should be stamped with the date received and a schedule of disposal or testing established. If a partially empty container is kept, a target date for testing for peroxides should be placed on the container, as well as subsequent dates for the tests to be repeated. A centralized computer tracking program for chemicals could take this responsibility from the individual laboratory managers and largely eliminate the problem of outdated ethers. The effort needed to maintain a reliable program in the laboratory to check for the presence of peroxides could almost certainly be used to greater advantage on the basic research program. Unused portions of ethers should not be returned to the original container. Small quantities can be allowed to evaporate in a fume hood.

Because of the explosion risk associated with peroxides, older containers of ethers are not normally accepted as part of an ordinary hazardous waste shipment by a commercial chemical waste disposal firm. They would have to be disposed of as unstable explosive materials which is much more expensive than if the materials were not outdated. Any savings in buying the "large economy size" would be more than compensated for by the additional disposal costs. Attempting to treat the ethers to remove the peroxides or to dispose of them by laboratory personnel carries with it the risk of an explosion and the subsequent liability for injuries.

Although the major risk usually associated with ethers is, as noted, the problem of peroxides, they also pose additional problems because of their properties as flammable solvents. Many of them have lower explosion limits, in the range of 0.7 to 3%, and flash points at room temperature or, in several cases, much lower. Ethyl ether, the material that most frequently comes to mind when "ether," otherwise unspecified, is mentioned, has a lower explosion limit of 1.85%, an upper explosion limit of 36%, and a flash point of $-45°C$ ($-49°F$). The vapors of most ethers are heavier than air, and hence can flow a considerable distance to a source of ignition and flash back. Because of their flammable characteristics, many ethers, especially ethyl ether, placed in improperly sealed containers in an ordinary refrigerator or freezer release vapors which represent a potential "bomb" that can be ignited by sparks within the confined space and explode with sufficient force to seriously injure or even kill someone in the vicinity. A fire is virtually certain to result should such an incident occur which, depending upon the type of construction and the availability of fuel in the laboratory area, could destroy an entire building. Because ethers, as well as other flammable solvents, are used so frequently in laboratory research, all refrigeration units in wet chemistry laboratory environments should be purchased without ignition sources within the confined spaces and with explosion-proof compressors, i.e., they must meet standards for "flammable material storage," even if the current program does not involve any of these materials. At some time during the effective lifetime of most refrigeration units, it is likely that flammable liquids will be used in the facility where they are located.

A spill should be promptly cleaned up using either a commercial solvent spill kit material to absorb the liquid or a preparation of equal parts of soda ash, sand, and clay cat litter, which has been recommended as an absorbent. Since the lower explosion limit concentrations are so low for so many of the commonly used ethers, all ignition sources should be promptly turned off following a spill and all except essential personnel required to leave the area. The personnel performing the clean up should wear half-face respirators equipped with organic cartridges. The resulting waste mixture from the clean up can be placed in a fume hood temporarily until removed from the laboratory for disposal as a hazardous waste.

Class B fire extinguishing agents are to be used to combat ether fires as well as other fires involving flammable liquids. Usually the most effective are dry chemical extinguishers which interrupt the chemistry of a fire, while carbon dioxide units can be used to smother small fires. Portable Halon extinguishers are also usable if the fire is such that the fuel has time to cool before the concentration of the halogenated agents falls below the critical concentration at which it is effective (the current extinguishing agents used in most of these, Halon-1301 and Halon-1211, damage the ozone in the atmosphere and will be replaced). It is worth pointing out once again that unless there is a reasonable chance of putting out a fire with portable extinguishers, it is preferable to initiate an evacuation as quickly as possible and to make sure that everyone can safely leave the building, rather than engage in a futile attempt to put out the fire.

In general, the toxicity of ethers is low to moderate, although this generalization should be confirmed for each different material to be used by information obtained either from the container label or the MSDS. Prolonged exposure to some ethers has been known to cause liver damage. Many have anesthetic properties and are capable of causing drowsiness and eventual unconsciousness. In extreme cases of exposure, death can result. Four ethylene glycol ethers—2-methoxyethanol, 2-ethoxyethanol, 2-methoxyethanol acetate, and 2-ethoxyethanol acetate—have been identified as causing fetal developmental problems in several animal species, including fetal malformations and resorption and testicular damage. Studies have also shown adverse hematologic effects and behavioral problems in the offspring of animals.

4.6.2.6. Flammable Solvents

Much of the concern in the literature is centered on the flammable characteristics of flammable liquids, much of which has already been addressed in Section 4.5.7. This characteristic will be treated in more detail in the following section. There are many other issues relating to the health effects of these solvents which also need to be considered.

4.6.2.6.1. Flammable Hazards

The fire hazard associated with flammable liquids should be more appropriately associated with the vapors from the liquid. It is the characteristics of the latter which determine the seriousness of the risk posed by a given solvent. In previous sections on storage of liquids, two of the more important properties of flammable liquids were mentioned in terms of defining the classes to which a given solvent might belong. Definitions of Class 1A, 1B, 1C, 2A, 3IA and 3B were based on the boiling point and flash point of the solvents. The formal definition of these two terms will be repeated below, as well as three other important parameters relevant to fire safety, the ignition temperature and the upper and lower explosion limits.

Boiling point (bp)—This is the temperature at which the vapor of the liquid is in equilibrium with atmospheric pressure (defined at standard atmospheric pressure of 760 mm of mercury).

Flash point (fp)—This is the minimum temperature at which a liquid gives off vapor in sufficient concentration to form an ignitable mixture with air near the surface of a liquid. The experimental values for this quantity are defined in terms of specific test procedures which are based on certain physical properties of the liquid.

Ignition (autoignition) temperature—This is the minimum temperature which will initiate a self-sustained combustion independent of the heat source.

Lower explosion (or flammable) limit (lel)—This is the minimum concentration by volume percent in air below which a flame will not be propagated in the presence of an ignition source.

Upper explosion (or flammable) limit (uel)—This is the maximum concentration by volume percent of the vapor from a flammable liquid above which a flame will not be propagated in the presence of an ignition source.

For a fire to occur involving a flammable liquid, three conditions must be met: (1) the concentration of the vapor must be between the upper and lower flammable limits; (2) an oxidizing material must be available, usually the air in the room; and (3) a source of ignition must be present. The management strategy is usually to either maintain the concentration of the vapors below the lower flammable limit by ventilation (such as setting the experiment up in an efficient fume hood) or to eliminate sources of ignition. The latter is easier and more certain because the ventilation patterns even in a hood may be uneven or be disturbed so that the concentrations may locally fall into the flammable range. While some materials may require an open flame to ignite, it is much easier to ignite others. For example, the ignition temperature for carbon disulfide is low enough (80°C or 176°F) that contact with the surface of a light bulb may ignite it.

In order to work safely with flammable liquids there should be no sources of ignition in the vicinity, either as part of the experimental system or simply nearby. Use non-sparking equipment. When pouring a flammable liquid from one metal container to another, both of the containers should be grounded, as the flowing liquid can itself generate a static spark. Flammable materials should be heated with safe heating mantles (such as a steam mantle), heating baths, or explosion-proof heating equipment. Many ovens used in laboratories are not safe for heating flammables because the vapors can reach the heating element, or either the controls or thermostat may cause a spark. Any spark-emitting motors should be removed from the area. Flammable materials should never be stored in an ordinary refrigerator or freezer because they have numerous ignition sources in the confined volume. Placing the entire system in a hood where the flammable vapors will be immediately exhausted aids in limiting the possibility of the vapors coming into contact with an ignition source.

The vapors of flammable liquids are heavier than air and will flow for a considerable distance away from the source. Should they encounter an ignition source while the concentration is in the flammable range, a flame may be initiated and flash back all the way to the source. In at least one instance a fire resulted when a research worker walked by a fume hood carrying an open beaker of a low ignition point volatile solvent. There was an open flame in the hood, and when the fumes were pulled into the hood, the fumes ignited and flashed back to the container, which was immediately dropped, with the result that the feet and lower legs of the worker and the entire floor of the laboratory became engulfed in flame. Fortunately, a fire blanket and fire extinguisher were immediately available, so that the worker escaped with only minor burns and the fire was extinguished before anything else in the laboratory became involved. This incident clearly illustrates the need to consider all possibilities of fire when using solvents, and that a hood does not totally isolate a hazardous operation.

REFERENCES

1. **Armour, M.A., Browne, L.M., and Weir, G.L.,** *Hazardous Chemicals Information and Disposal Guide,* 2nd ed., University of Alberta, Edmonton, Canada, 1984.
2. *Federal Register,* F.R. 10586, April 2, 1987.
3. **Langan, J.P.,** Questions and Answers on Explosion-Proof Refrigerators, Kelmore, Inc., Newark, NJ.
4. *Prudent Practices for Handling Hazardous Chemicals in Laboratories,* National Academy Press, Washington D.C., 1981, 57.
5. **Lewis, R.J.,** *Sax's Dangerous Properties of Industrial Materials,* 8th ed., Van Nostrand Reinhold Company, New York, NY, 1993.
6. Hazardous Chemicals Data, NFPA 49, National Fire Protection Association, Quincy, MA.

4.6.2.7. Reactive Metals

Lithium, potassium, and sodium are three metals that react vigorously with moisture (lithium to a lesser extent than the other two, except in powdered form or in contact with hot water) as well as with many other substances. In the reaction with water, the corresponding hydroxide is formed along with hydrogen gas, which will ignite. Lithium and sodium should be stored under mineral oil or other hydrocarbon liquids that are free of oxygen and moisture. It is specifically recommended that potassium be stored under dry xylene.

In many ways the chemical hazards of the three metals are similar. All three form explosive mixtures with a number of halogenated hydrocarbons; all three react vigorously or explosively with some metal halides, although potassium is significantly worst in this respect, and the reaction of all three in forming a mercury amalgam is violent. They all react vigorously with oxidizing materials. Potassium can form the peroxide and superoxide when stored under oil at room temperature and may explode violently when cut or handled. Sodium reacts explosively with aqueous solutions of sulfuric and hydrochloric acids. The literature provides a number of other potentially dangerously violent or vigorous reactions for each of these three materials. It is not the intent here to list all of the potentially dangerous reactions that may occur, but to point out that there are many possibilities for incidents to happen. No one should plan to work with these materials without carefully evaluating the chemistry involved for potential hazards. The materials should be treated with the care which their properties demand at all times.

In one instance, a very old can of sodium was determined to have "completely" reacted to form sodium hydroxide, and the worker decided to flush it with water to dispose of the residue. The bottom 2 inches were still sodium metal and consequently the can exploded. The only entrance to the laboratory was blocked by an ensuing fire so that the occupants had to escape through windows. Fortunately, they were on the first floor and the windows were not blocked. The latter point is worth noting because shortly before this incident, bars over the windows to prevent break-ins had been removed at the insistence of the organization's safety department. Unsubstantiated assumptions or misplaced priorities are a major cause of injuries.

Since they all react vigorously with moisture, care should be taken to avoid skin and eye contact, which could result in burns from the evolved heat and direct action of the hydroxides. The materials should always be used in a hood, and, as a minimum, gloves and chemical splash and impact-resistant goggles should be worn while working. If the risk of a violent reaction cannot be excluded, additional protection such as an explosion barrier or a face mask should be considered.

If a fire should result, appropriate class D fire extinguishers should be available within the laboratory. These vary somewhat in their contents, which usually are specific for a given material. Appropriate material suitable for extinguishing fires involving these reactive materials are dry graphite, soda ash, and powdered sodium chloride. Other materials which might be used in these class D units as extinguishing agents are pulverized coke, pitch, vermiculite, talc, and sand. These materials will usually be mixed with various combinations of low melting fluxing salts, resinous materials, and alkali-metal salts which, in combination with the other material in the extinguisher, form a crust to smother the fire. Water (obviously), carbon dioxide, or halogenated units should not be used. There are a number of other reactive metals, which while not as active chemically as these three, once ignited, require class D fire extinguishing agents as well. These include magnesium, thorium, titanium, uranium, and zirconium. Other materials for which class D units should be used include metal alkyds and hydrides and organometallic compounds.

As with many other materials, one of the major considerations in using these reactive metals is the problem associated with disposal of unneeded surplus quantities or waste materials. There are suggestions in the literature for treating waste for each of these three materials. For example, small amounts of potassium residues from an experiment should be treated promptly by reacting them with *tert*-butyl alcohol because of the danger that they will explode (even if the potassium is stored properly). This is appropriate for small quantities in the laboratory, but disposal of substantial quantities of unwanted material is a different matter. Recycling or transfer to another operation needing the material should be investigated, but disposal by local treatment should be avoided. There are limits to local treatment permitted under the RCRA, beyond which a permit is required to become a treatment facility. In addition, there are safety and liability risks associated with processing dangerously reactive materials which must be considered. Reactive metals are among those materials that require special handling by commercial waste disposal firms. Because special procedures are required, the cost of disposal is much higher than for routine chemical waste. Quantities purchased and kept in stock should be limited.

REFERENCES

1. Hazardous Chemicals Data, NFPA 49, National Fire Protection Association, Quincy, MA.
2. **Armour, M.A., Browne, L.M., and Weir, G.L.,** *Hazardous Chemicals Information and Disposal Guide,* 2nd ed., University of Alberta, Edmonton, Canada, 1984.
3. **Lewis, R.J.,** *Sax's Dangerous Properties of Industrial Materials,* 8th ed., Van Nostrand Reinhold Company, New York, NY, 1993.
4. **Sax, N.I. and Lewis, R.J., Sr.,** *Rapid Guide to Hazardous Chemicals in the Workplace,* Van Nostrand Reinhold Company, New York, NY, 1986.
5. Portable Fire Extinguishers, NFPA-10, National Fire Protection Association, Quincy, MA.

4.6.2.8. Mercury

Mercury and its compounds are widely used in the laboratory. As metallic mercury, it is often used in instruments and laboratory apparatus; in the latter application especially it is responsible for one of the more common types of laboratory accidents, mercury spills. Thermometers containing mercury are frequently broken; mercury is often spilled in working with mercury diffusion pumps or is lost when cleaning the cold traps associated with high vacuum systems in which mercury pumps are used. Over a period of time the small amounts of mercury lost each time can add up to a substantial amount. In one instance the cold traps were always cleaned over a sink. After a few years of this, a large amount of mercury accumulated in the sink trap and finally eroded the metal and spilled on the floor. Over 15 pounds of mercury were recovered. Mercury is frequently ejected from simple manometers consisting of mercury in plastic tubing, connected to a system under vacuum. As an example of the consequences of this last type of accident, an employee, working in a room previously used for years as an undergraduate biology laboratory, was diagnosed as having a fairly severe case of mercury poisoning, although he did not use mercury. Upon investigation, more than 50 pounds of mercury were retrieved from under the wooden floorboards in the room. Although the instructors had "cleaned up the mercury" when spills had occurred, over the years a substantial amount had obviously not been recovered, but had worked its way through the cracks in the floor. No measurements of the airborne concentration were made at the time, but the normal equilibrium vapor pressure of mercury in air at normal room temperatures is between 100 and 200 times the current permissible levels of mercury in the workplace. It should also be noted that the vapor pressure rises rapidly with temperature. At the temperature of boiling water at standard atmospheric pressure, the vapor pressure is more than 225 times higher than at 20°C (68°F) (0.273 mm Hg) and reaches 1 mm Hg at 126.2°C (259.2°F). Clearly, mercury always should be heated in a functioning fume hood instead of an open bench.

Mercury poisoning has been known to affect many individuals, among them such prominent scientists as Pascal and Faraday and also workers in various industries such as those exposed to mercury as an occupational hazard while using mercuric nitrate in the hat industry in making felt. A frequently cited example of the effects of mercury poisoning is the "Mad Hatter" in Lewis Carroll's *Alice in Wonderland.* In the 1950s, many Japanese in a small fishing village suffered serious permanent damage to their central nervous system and, in many cases, death or permanent disability due to eating fish containing methyl mercury as a result of the industrial discharge of mercury compounds into the sea near their village. In the 1970s, fish taken from some of the common waters of Canada and the U.S. were found to be contaminated with mercury, and fishing was banned in some areas. Some commercial swordfish and tuna were found to be contaminated with mercury and had to be withdrawn from the market. In at least one instance fish in a river in the southeastern United States were found to be contaminated by mercury-containing waste from a chemical plant. Eventually the plant closed down its operations because the expense of modifying its operation to eliminate the discharge would have been too high. Mercury compounds were at one time used as fungicides; individuals died from mistakenly eating seed corn treated with these materials. Mercury, once it enters the ecology, is slow to biodegrade. As a result of the dangers inherent in materials containing mercury, the use of mercury for most agricultural purposes has been banned, and dumping wastes containing mercury compounds in such a way as to be able to contaminate the environment is no longer permitted. Currently it is difficult to dispose of compounds containing mercury. Waste mercury metal, on the other hand, can be recycled.

Elemental mercury is probably not absorbed significantly in the gastrointestinal tract, but many of its compounds are. Poisoning due to inhalation and absorption of mercury vapors results in a number of symptoms. Among these are personality and physiological changes such as nervousness, insomnia, irritability, depression, memory loss, fatigue, and headaches. Physical effects may be manifested as tremors of the hands and general unsteadiness. Prolonged exposure may result in loosening of the teeth and excessive salivation. Kidney damage may result. In some cases the effects are reversible if the exposure ceases, but as noted in the previous paragraph, ingestion of some organic mercury compounds may be cumulative and result in irreversible damage to the central nervous system. Alkyl mercury compounds have very high toxicity. Aryl compounds, and specifically phenyl compounds, are much less toxic (in the latter case comparable to metallic mercury) and therapeutic compounds of mercury are less toxic still. In the case of the Minimata Bay exposure in Japan it was found that the fetus was especially vulnerable to the exposure. Mercury passes readily through the placenta from mother to child. In recognition of the seriousness of the potential toxic effects the permissible ceiling exposure level for metallic mercury is currently set by OSHA at 0.01 mg/M^3, and for alkyl organo-mercury compounds, an 8-hour time weighted average of 0.01 mg/M^3 has been established by OSHA with a ceiling average of four (current ACGIH recommendation is three) times that level.

The following four paragraphs are taken directly from the article by Steere from the second edition of this handbook.

Absorption of Mercury Vapor

The metabolism of inhaled mercury in the rat was reported in 1962 by Hayes and Rothstein in studies using a special exposure unit, [203]Hg as a tracer, and whole body counting techniques which allowed frequent estimation of the body burden. They found that 86% of the mercury was cleared from inhaled air, deposited in the lung, oxidized to the ionic form, and absorbed by the blood within a matter of hours. They report a study by Clarkson in 1961 which showed mercury in the blood to be bound 50% to the red blood cells and 50% to serum protein, with less than 0.1% filterable.

Distribution of Mercury in the Body

In their study of the metabolism of inhaled mercury in the rat, Hayes and Rothstein found after exposures of 5 hours to a mercury level of 1.4 mg/M^3 that the metal was generally distributed in the body, but became highly localized in the kidney with an accumulation after 15 days of 70% or more of the body burden, or 150 times as high a concentration of the metal on a per gram basis as the other tissues. Assuming filterable mercury at 0.1%, they calculate the maximum amount of mercury that could be filtered by the kidneys in 24 hours after exposure at 0.86% of the body burden, and in comparison with the actual deposition of 16.7% of the body burden conclude that the extraordinary affinity of the kidney for mercury is not related to filtration phenomena.

Excretion of Mercury

In the rat inhalation studies, about 30% of the body burden of mercury cleared from the body rapidly with a half-time of 2 days, associated with a rapid fecal excretion, and the rest moves more slowly with a half-time of 20 days, with about equal rates of fecal and urinary excretion. The study showed that excreta during 15 days accounted for 42% of the initial body burden, and suggested that an additional reduction of 14% of initial body burden may in large part be accounted for by loss of mercury in the form of vapor from the animals.

Pharmacology of Mercury

The pharmacology of mercury and other heavy metals was reviewed in 1961 by Passow, Rothstein, and Clarkson in an article which mentions that mercury in the body has high affinity for sulfhydryl groups, chloride ions, and amines or simple amino acids, and that mercury inhibits urease, invertase, and other enzymes carrying SH groups. Mercury can also block glucose uptake by erythrocytes and muscle, produce K^+ loss from the cells of all species, cause lesions of the central nervous system, and influence bioelectric phenomena by altering transmembrane potentials and by blocking nerve conduction.

Mercury is dense (specific gravity of just under 13.6 at 4°C (39.2°F)) and has a high surface tension and low viscosity. As a result, it tends to break up into small droplets when it is poured or spilled. Anyone that has tried to pick up small droplets using a stiff piece of paper can attest to the appropriateness of the alternate name "quicksilver." As the droplets are disturbed as, for example, when walked upon on the laboratory floor, they tend to break up into smaller and smaller droplets, eventually becoming too small to see. In a laboratory where mercury has been in use for an extended period of time, it is instructive to run a pen knife in the cracks in a tile floor or in the seams where cabinets and bench tops fit together. Invariably, small droplets of mercury will be found.

Although a thin film of oxide will form on the skin of mercury droplets, it is very fragile and will break. Similarly, sprinkling flowers of sulfur on the location of a mercury spill has been suggested as a control measure but the surface film which forms also apparently is very fragile and will allow the mercury underneath the film to be readily exposed.

4.6.2.8.1. Control Measures

Exposure Reduction

One of the simplest control measures is to reduce the amount of mercury used in instrumentation and equipment. Mercury thermometers can be replaced with alternatives. Vacuum gauges can be used to replace manometers, and oil diffusion pumps can replace mercury diffusion pumps.

Wherever possible, work with mercury should be done in a fume hood, preferably one that has a depressed surface, so that a lip will aid in preventing mercury spills from reaching the floor, and with a seamless interior, as recommended for radiological work and work with perchloric acid. As noted earlier, heating mercury causes it to emit fumes at concentration levels two to three orders of magnitude above the PELs. Heating of mercury should never be done on the open bench.

The general restrictions on eating, drinking, and smoking in the laboratory should be strictly enforced in laboratories where mercury compounds are commonly used. Depending upon the level of use and the availability of fume hoods, the use of personal protective equipment is recommended in addition to the use of goggles and laboratory aprons. Respiratory protection, consisting of half-face respirators fitted with a cartridge which will absorb mercury, should be considered if there is any potential for mercury exposure. A number of mercury compounds are absorbed through the skin and are strong allergens. The skin in such cases should be protected with gloves covering the forearms as well as the hands.

Tile is a commonly used material for the floors in a laboratory because it is "easy" to maintain and inexpensive to install. Ease of maintenance is not the case for a tile floor in a laboratory using mercury, because of the propensity of the extremely small (20 microns or less) mercury droplets to collect in the cracks. A seamless vinyl or poured epoxy floor should be used instead, with the joints of the floor with the wall being curved or "coved." Similarly, the bench top should be curved where it joins the back panel. Existing tile floors, especially the smaller 9 inch × 9 inch size, frequently represent a maintenance problem since a large proportion contain asbestos, as may the mastic holding them to the floor. When these tiles need replacing, the work must be done in conformance with EPA and OSHA standards and can be very costly.

Monitoring

The fumes from mercury provide no direct sensory evidence that it is present. Where use is substantial, monitoring measures should be readily available. The least expensive means to detect mercury levels is a detector tube in which a given volume of air is pulled through a glass tube containing a material that undergoes a color change when mercury comes into contact with the material. Normally the air is drawn through by means of a hand-operated pump. This method provides a reasonable accuracy, but any finding within approximately 25% of the PEL should be considered to be sufficient warning of a possible overexposure of personnel working in the area. Each measurement requires a new tube. A popular instrument used to provide a direct reading of mercury concentrations in the air is a hand-held atomic absorption spectrometer in which air contaminated with mercury is drawn through the instrument and the degree of interference with ultraviolet light of the wavelength corresponding to a characteristic line of the mercury spectrum is translated into a numerical reading of the concentration of mercury in the air. The calibration of these instruments must be carefully maintained. Another instrument which provides an accurate, rapid reading of the level of mercury in the air depends upon the property of mercury to amalgamate with a thin gold film. This latter type of instrument is probably the most accurate and reliable but is also the most expensive. Unless the work with mercury compounds is quite heavy, the laboratory may be unable to afford either of the last two types of monitoring devices. In such a case the safety department should be provided with an instrument to be used at all locations within an organization.

Spill Control Measures

Large globules of mercury can be cleaned up mechanically by carefully brushing them onto a dustpan or a stiff piece of paper. Another simple device is the use of a small mechanical hand-held pump to suck the globules into a small container. This is a tedious procedure limited to small spills and, of course, to droplets big enough to be seen. Bulk mercury recovered by these procedures can be recovered and purified for reuse.

Mercury spill kits available commercially usually include a small pump, sponges impregnated with a material to absorb mercury and which can be used to wipe up the area of a small spill, and a quantity of an absorbent powder that reacts with mercury to form a harmless amalgam. The latter can be spread on cracks and seams in the floor and furniture and is effective in collecting mercury from otherwise inaccessible places. After leaving the material on the floor or contaminated surfaces for several hours in order to allow the amalgam to form, the powder can be swept or brushed up and the waste material disposed of as a hazardous waste.

Ordinary vacuum cleaners *MUST NOT* be used to used to clean up a mercury spill. An ordinary vacuum filter bag will not stop an appreciable fraction of small particles in the region of several microns or less and, more importantly, the mercury globules pulled into the bag will be broken up into even finer droplets and spewed out of the vacuum's exhaust into the air, substantially increasing the surface area of mercury exposed to the air and enhancing the rate at which mercury vapor will be generated. There are commercial vacuum cleaners which are specifically designed to pick up mercury, however. One such unit, sold by Nilfisk of America, Inc., which also makes specialized units for other toxic materials, first draws the mercury into a centrifugal separator and collects the bulk of the material into an airtight plastic bottle. The contaminated air is then passed into a collection bag which collects bulk solid waste and then through an activated charcoal filter. Additional filters (some optional) follow the charcoal filter collector. This unit can be used to clean up virtually any spill alone but can be used with other control measures to ensure a complete clean up of the spilled material. The hose in the Nilfisk unit vacuum cleaner has an especially smooth surface to prevent mercury particles from adhering to the inside of the hose. As with most specialized units, the vacuum cleaner is not inexpensive. In some instances, a special purpose mercury vacuum cleaner is virtually indispensable for use on a spill on a porous, rough material such as carpeting. The use of carpeting as a laboratory floor covering is very rare but, in at least one instance where this was done, a large area of the carpet was thoroughly contaminated by an extensive mercury spill.

Ventilation

The ventilation system in a laboratory using mercury or mercury compounds should conform to the general recommendation that wet chemistry laboratories involving any hazardous material be provided with 100% fresh air instead of having a portion of the air recirculated. Local ventilation systems, such as the exhausts of mechanical pumps servicing mercury diffusion pumps, should be collected with a local exhaust system and discharged into the fume hood exhaust system in the room or to a separate exhaust duct provided to service such units. The mercury vapor is much heavier than air so it is important that the room exhausts be placed near the floor or at the back of the workbench to collect as much of the vapors as possible.

Medical Surveillance

It is recommended, as a minimum, that permanent employees working with mercury or mercury compounds be provided with periodic physical examinations with a test protocol selected specifically for mercury poisoning. Women who may be pregnant should be especially careful and encouraged to participate in the medical surveillance program if they cannot avoid exposure entirely.

REFERENCES

1. **Steere, N.V.,** Mercury vapor hazards and control measures, in *CRC Handbook of Laboratory Safety*, 2nd ed., Steere, N.V., Ed., CRC Press, Boca Raton, FL, 1971, 334.
2. **Vostal. J.J. and Clarkham, T.W.,** Mercury as an environmental hazard, *J. Occupational Medicine,* 15, 649, 1973.
3. **Armour, M.A., Browne, L.M., and Weir, G.L.,** *Hazardous Chemicals Information and Disposal Guide*, 2nd ed., University of Alberta, Edmonton, Canada, 1984.

4.6.2.9. Hydrofluoric Acid

Anhydrous hydrofluoric acid (HF) (CAS 7664-39-3) is a clear, colorless liquid. Because it boils at 19.5°C (67.1°F) and has a high vapor pressure, it must be kept in pressure containers. It is miscible in water, and lower concentration aqueous solutions are available commercially. It is an extremely dangerous material and all forms, including vapors and solutions, can cause severe, slow-healing, burns to tissue. At concentrations of less than 50%, the burns may not be felt immediately, and at 20% the effects may not be noticed for several hours. At higher concentrations, the burning sensation will become noticeable much more quickly, in a matter of minutes or less. Fluoride ions readily penetrate skin and tissue and, in extreme cases, may result in necrosis of the subcutaneous tissue which eventually may become gangrenous. If the penetration is sufficiently deep, decalcification of the bones may result. The current OSHA PEL 8-hour time weighted average to HF is set at 3 ppm (2.5 mg/M^3), which also is the ceiling TLV currently recommended by the ACGIH. Chronic exposure to even lower levels may irritate the respiratory system and cause problems to the bones. Even brief exposures to high levels of the vapors may cause severe damage to the respiratory system, although the sharp, irritating odor of the acid will usually provide a warning to assist in avoiding inhalation in normal use. Contact with the eyes could result in blindness. If eye exposure occurs, it is urgent to flush the eyes as quickly as possible. Every laboratory using hydrofluoric acid should have both an eyewash station and deluge shower within the laboratory. Dilute solutions and vapors may be absorbed by clothing and held in contact with the skin, which will probably not result in an immediate sensation of pain as a warning but eventually may lead to skin ulcers which, again, may take some time to heal. A generalization might be made here about absorbent clothing. In many instances, as in this case, absorbent clothing which can retain toxic materials and maintain them in close contact with the skin may be worse than no protection at all, changing the exposure from a transient phenomena to a persistent one. This is not always a problem, but it should be kept in mind as a possibility when choosing protective apparel. All work with hydrofluoric acid should be done in a fume hood.

Hydrofluoric acid attacks glass, concrete, and many metals (especially cast iron). It also attacks carbonaceous natural materials such as woody materials, animal products such as leather, and other natural materials used in the laboratory such as rubber. Lead, platinum, wax, polyethylene, polypropylene, polymethylpentane, and Teflon will resist the corrosive action of the acid. In contact with metals with which it will react, hydrogen gas is liberated and hence the danger exists of a spark or flame resulting in an explosion in areas where this may occur.

The following article, "Treatment of Severe Hydrofluoric Acid Exposures," is taken with permission from *Journal of Occupational Medicine*, 25(12), 861, 1983.

4.6.2.9.1. Treatment of Severe Hydrofluoric Acid Exposures

Hydrogen fluoride (HF) or its aqueous solution is capable of producing severe damage to the lungs, skin, eyes, and other soft tissues from brief acute exposures. It can also produce changes in bone, teeth, muscle, and connective tissue in chronic overexposures. These chronic changes are well described in recent monographs[1,2] and will not be discussed here.

DuPont experience with hydrofluoric acid exposures has resulted in previous publications on treatment of the resulting burns.[3,4] More recent experience has led to revisions in treatment methods that appear to have improved results. This report will update the recommendations of the 1965 and 1966 reports.

The treatment methods we have found to be the most successful are summarized.

Principles

The two basic tenets of the treatment program we have developed are the following:

1. Speed is vital in initiating both first aid and definitive treatment. The exposed individual must remove contaminated clothing immediately, get under a safety shower, and begin flushing exposed skin areas with water.
2. Evaluation and monitoring must always extend to the entire individual. Delayed systemic poisoning may appear after relatively small areas of skin have been burned.[5]

Calcium glutonate 2.5% gel has proved a distinct advance in local treatment. It is packaged in 25-gram tubes, which may be given to the patient with instructions to rub it on the burned areas whenever pain recurs. This method was described in 1974 by Browne[6] and has gained worldwide acceptance.

Eye exposures should also be treated with calcium gluconate by irrigation with a 1% solution of calcium gluconate in physiologic saline solution, followed by 1% calcium gluconate in saline as eye drops. For infiltration of deeply burned skin, 10% calcium gluconate is diluted with an equal amount of physiologic saline solution to yield a 5% solution. For inhalation exposures, 2.5% to 3% solution is given by nebulizer with 100% oxygen, preferably by intermittent positive pressure breathing (IPPB). Calcium gluconate is also added to the intravenous solution for management of extensive skin or inhalation exposures to prevent potentially lethal hypocalcemia.[7]

Skin Burns

Skin contacted by concentrated HF vapor or aqueous solution rapidly becomes erythematous with a white or gray color at the surface due to the coagulation of tissue. Treatment begins with immediate copious washing with water for five minutes. All clothing must be removed. Leather gloves and shoes cannot be decontaminated and must be destroyed.

All burned areas should have calcium gluconate 2.5% gel (HF burn jelly) applied to them as a first-aid measure. The application and massaging into the skin of HF burn jelly should be continued for three to four days, four to six times daily. Care should be taken to see that the personnel who apply the jelly, especially on the initial application, wear rubber gloves to prevent skin contamination with HF and the development of hand burns.

The attending physician should carefully estimate the total area of skin involved and the depth of the burns. Burn areas greater than 50 to 100 cm^2 should be of great concern and the patient should be hospitalized immediately. Involvement of areas greater than 100 to 150 cm^2 probably will require the facilities of an intensive care unit. All patients with extensive skin burns will have some fume inhalation unless the individual was wearing a proper respiratory protection apparatus. Each patient's pulmonary status must be carefully and rapidly evaluated.

Blood for a complete blood cell (CBC) count and for determining liver function, blood urea nitrogen or creatinine, electrolytes, serum calcium, magnesium, and fluoride levels should be drawn immediately. Intravenous fluid therapy should be initiated as soon as possible. Twenty milliliters of calcium gluconate should be added to the first liter of intravenous solution to raise the amount of available calcium in the blood. Frequent measurement of serum calcium and magnesium levels should guide further administration. The electrocardiogram should be monitored continuously for prolongation of the QT interval.[7,8]

The maintenance of an adequate amount of calcium ion in the blood and tissue is essential in the treatment of extensive HF exposures, whether the exposures occurred by skin, inhalation, or ingestion. Fluoride ion is rapidly absorbed and depletes the tissue calcium rapidly. Blood calcium must be frequently and carefully monitored and sufficient calcium given intravenously, as gluconate, to maintain the calcium level at or slightly above the upper limit of normal. This will encourage movement of the ion into areas of lesser concentration in the extracellular fluid and then into the tissues.

Deep burns with HF solutions of concentrations greater than 20% may be treated by the use of subcutaneous infiltration of calcium gluconate. Ampules of 10% calcium gluconate should be diluted with physiologic saline solution to obtain a 5% concentration of calcium gluconate. The solution may be infiltrated under the burned area using a 25-gauge × 1½-inch or 24-gauge × 3½-inch spinal needle, if the area is extensive. Infiltration should be avoided on the digits and undertaken with great care on the hands, feet, and face. The administration of calcium gluconate in concentrations above 5% tends to produce severe irritation of the tissues that may cause keloids and intractable scarring. Burns with solutions of HF less than 20% do not usually require infiltration.

System therapy in the severely burned patient is generally supportive: maintaining electrolyte balance and carefully monitoring the patient for signs of renal or hepatic toxicity and maintaining respiratory and cardiac functions. ECG monitoring is necessary for the early detection of arrhythmias that may result from the alterations in calcium metabolism and for observing prolongation of the QT interval.

The use of adequate doses of corticosteroids to maintain the blood pressure and exert an anti-inflammatory action in the injured area is strongly advised. Hydrocortisone sodium succinate (Solu-Cortef) 500 mg, four to six times a day in the intravenous bottle, is usually adequate.

Antibiotics by the systemic or local route are usually unnecessary. Our observation has been that HF burns seldom become infected. Since the tissue is digested, enzymes are unnecessary. One can use silver-sulfadiazine cream for local treatment after the initial several days of treatment with calcium glutonate or infiltration.

Inhalation Therapy

Persons exposed to HF by inhalation should immediately be given 100% oxygen by mask or catheter. As soon as possible, they should be given 2.5% to 3% calcium gluconate solution by inhalation, preferably by IPPB utilizing a nebulizer alone. The patient should be carefully watched for edema of the upper airway with respiratory obstruction, and the airway should be maintained by tracheostomy or endotracheal intubation if necessary. All those with a history of exposure who experience respiratory irritation should be immediately admitted to an intensive care unit and carefully watched for 24 to 48 hours. Delayed pulmonary edema is likely in patients with burns of the skin of the face or neck.

If pulmonary edema develops, the patient should be placed on IPPB with positive end-expiratory pressure (PEEP). The administration of respiratory care should be very closely supervised, including the continuing administration of calcium gluconate by inhalation. The general supportive principles given in the discussion of skin burns should be observed with the addition of the use of antibiotics to prevent pulmonary infections and the use of arterial blood gases for control of respiratory therapy.

Toxicity from pulmonary or dermal absorption of fluoride ion may develop rapidly in the liver and kidneys and may require more energetic measures of control, up to and including hemodialysis, if the blood urea nitrogen and potassium levels rise. Supportive care is necessary for all organ systems.[7]

Pulmonary reaction to HF continues for up to three weeks, and the patients must be closely monitored. The great solubility of HF in water will limit mild exposures to the bronchial tree, but severe exposures have been found on autopsy to result in bronchiolar and alveolar damage.

Radiological studies and pulmonary function tests are a great assistance in evaluation of long-term effects of Hf exposure. Dyspnea during mild or heavy physical activity may occur as long as seven to nine months after exposure. The need for corticosteroid administration on every third day may be necessary periodically for as long as three months to control inflammatory episodes.

Eye Exposure

The cornea and conjunctiva can be extensively damaged by vapor exposure alone. More severe damage is caused by liquid HF, which rapidly penetrates the outer layer of epithelium of the eye.

Immediate rapid copious washing of the eyes should be followed by ice packs. The ice pack should be used until a medical facility is reached. Here the eyes should be washed thoroughly with 1% calcium gluconate in normal saline for five to 10 minutes; thereafter, calcium gluconate in normal saline should be instilled every two to three hours for 48 to 72 hours. No oils or ointments should be used. Inflammation may be decreased by the use of corticosteroid solutions for ophthalmic use.

The cornea may cloud rapidly and obscure vision when exposed to HF. Desquamation may occur in four to 24 hours. Recovery usually takes four to five days; scarring may occur, as may corneal perforation. An ophthalmologist should always be consulted and should be instructed that long-term monitoring may be required. Surgical procedures may be needed to correct defects caused by HF exposure. Severe exposures to anhydrous HF have led to destruction of the eye and denucleation to protect surrounding tissues.

4.6.2.9.2. Summary

In all HF exposures, time is of the utmost importance. One must help the patient wash with copious amounts of water immediately to minimize the extent and depth of burns, and then minimize the time needed to institute adequate medical treatment. The amount of tissue injury depends upon the concentration and amount of HF encountered and the duration of the exposure. Low concentrations may lead to delayed effects.

REFERENCES

1. **Proctor, N.H. and Hughes, J.P.,** *Chemical Hazards of the Workplace,* J.B. Lippincott Co., Philadelphia, 1978, 290.
2. **Knight, A.L.,** *Occupational Medicine: Principles and Practical Applications,* Zenz, C. (Ed.), Year Book Medical Publishers, Inc., Chicago, 1975, 649.
3. **Wetherhold, J.M. and Shepherd, F.P.,** Treatment of hydrofluoric acid burns, *J. Occup. Med.,* 7, 193, 1965.

4. **Reinhardt, C.F., Hume, W.G., and Linch, A.L. et al.,** Hydrofluoric acid burn treatment, *Am. Ind. Hyg. Assoc. J.,* 27, 166, 1966.
5. **Gosselin, R.D., Hodge, H.C., Smith, R.P. et al.,** *Clinical Toxicology of Commercial Products: Acute Poisoning,* 4th ed., Williams & Wilkins Co., Baltimore, 1976, 159.
6. **Browne, T.D.,** The treatment of hydrofluoric acid burns, *J. Occup. Med.,* 24, 80, 1974.
7. **Tepperman, P.B.,** Fatality due to acute systemic fluoride poisoning following a hydrofluoric acid skin burn, *J. Occup. Med.,* 22, 691, 1980.
8. **Abukurah, A.R., Moser, A.M., Baird, C.L. et al.,** Acute sodium fluoride poisoning, *JAMA,* 222, 816, 1972.
9. **Trevino, M.A., Herrmann, G.H., and Sproul, W.L.,** Treatment of severe hydrofluoric acid exposures, *J. Occup. Med.,* 25(12), 861, 1983.

4.6.2.10. Hydrogen Cyanide

Hydrogen cyanide (HCN) (CAS 74-90-8), also called hydrocyanic acid or prussic acid, is an extremely dangerous chemical that is toxic by ingestion, inhalation, or by absorption through the skin. The current OSHA 8-hour PEL to the vapors from this chemical is 10 ppm, as is the current ACGIH ceiling limit (with a cautionary note that skin absorption could be a contributory hazard). The material has a characteristic odor of bitter almonds, but the odor is not usually considered to be sufficiently strong to be an adequate warning of the presence of the vapors at or above the PEL. Not only is HCN toxic, it has a very low flash point, −17.8°C (0°F), a lower explosion limit of 6%, and an upper explosion limit of 41%, so that it also represents a serious fire and explosion hazard. It has a boiling point of 26°C (79°F), so that it is normally contained in cylinders in the laboratory. Heating of the liquid material in a pressure-tight vessel to temperatures above 115°C (239°F) can lead to a violent, heat-generating reaction. The material is usually stabilized with the addition of a small amount (0.1%) of acid, usually phosphoric acid, although sulfuric acid is sometimes used. Samples stored more than 90 days may become unstable.

Hydrogen cyanide can polymerize explosively when amines, hydroxides, acetaldehyde, or metal cyanides are added to the liquid material, and it also may do so above 184°C (363°F).

Although there will be variations among individuals, a concentration of 270 ppm in air is usually considered fatal to humans. A few breaths above this level may cause nearly instantaneous collapse and respiratory failure. Exposures at lesser levels may be tolerated for varying periods, e.g., 18 to 36 ppm may be tolerated for several hours before the onset of symptoms. Initial symptoms of exposure to HCN include headache, vertigo, confusion, weakness, or fatigue. Nausea and vomiting may occur. The respiratory rate usually increases initially and then decreases until eventually it becomes slow and labored, finally ceasing. The symptoms reflect the mechanism by which the toxic action occurs. The chemical acts to inhibit the transfer of oxygen from the blood to tissue cells by combining with the enzymes associated with cellular respiration. If the cyanide can be removed, the transfer of oxidation will resume. On average, absorption of 50 to 100 mg of HCN, directly by ingestion or through the skin as well as by inhalation can be fatal.

Treatment of a person poisoned by HCN is based on the introduction of methemoglobin into the bloodstream to interact with the cyanide ions to form cyanmethemoglobin. In any area where HCN is being used, a special emergency kit should be provided, containing an ample supply of ampules of amyl nitrite, a solution of 1% sodium thiosulfate solution, and an oxygen cylinder accompanied by a facepiece and tubing to permit administering the oxygen. This kit should be labeled **FOR HCN EMERGENCIES ONLY**. Sodium nitrite might also be kept in the kit if there is someone available qualified to administer drugs intravenously. Introduction of sodium nitrate directly into the bloodstream has been suggested as a means to increase the rate of conversion of cyanide to the thiocyanate, which is less toxic. Treatment should begin as soon as possible after an acute exposure and after recognition of the symptoms in less intense exposures. If the exposure has been due to contamination in the air in the area, the patient should be removed from the area (if the source of vapor is from a cylinder, the valve on the cylinder should be closed). Any contaminated clothes should be removed and the skin flooded with water.

If the patient is not breathing, resuscitation should be begun. As soon as the patient is breathing, an open amyl nitrite ampule should be held under the patient's nose for 15 seconds per minute, with oxygen being administered during the remaining 45 seconds. Medical aid should be called for immediately. If there is a person available qualified to administer drugs intravenously, injection of sodium nitrite while administering amyl nitrite should prove beneficial. Subsequent intravenous injection of sodium thiosulfate also has been suggested as an ameliorative action. Rescue squad teams usually have at least one member qualified to administer drugs while under the direction (by radio) of an emergency room physician. If the patient has swallowed HCN, the recommended treatment is to get the patient to swallow one pint of the sodium thiosulfate solution, followed by soapy water or mustard water to induce vomiting. Vomiting should not be induced in an unconscious patient. Application of amyl nitrite may restore consciousness.

All work with HCN must be done in a fume hood, operating with a face velocity of at least 100 fpm and with the apparatus set well back from the face of the hood to ensure that all vapors will be captured and discharged by the exhaust system. The hood should comply in every respect to recommended good practices for the location, design, and operations of hoods in Chapter 3. The hood should have its own individual duct to the roof. If the work is a continuing program, the exhaust duct should be labeled **DANGER, DO NOT SERVICE OR WORK IN THE VICINITY WHILE UNIT IS OPERATING**.

Protective gloves and chemical splash goggles should be worn while working with HCN. No one should work alone with this material. The laboratory entrance(s) should be posted with a warning sign **DANGER, HCN, AUTHORIZED PERSONNEL ONLY**. As noted above, care needs to be taken to be sure that persons outside the area, such as workers on the roof, are not inadvertently exposed. All work with HCN should be done in trays or other shallow containers of sufficient capacity to retain any spill from the apparatus.

Cleaning up of spills represents a serious problem with a chemical as dangerous as HCN, so extra care should be taken to avoid accidents with the material. If a spill occurs outside a hood, the laboratory should be evacuated as quickly as possible. All ignition sources and valves to cylinders of HCN should be turned off. If any individuals are splashed with the compound in an accident, they should immediately remove their contaminated clothes and step under a nearby deluge shower, preferably located in a space outside the laboratory in which the accident occurred. The occupants of the latter space should be warned of the accident and encouraged to evacuate as well. If the spill is substantial, the evacuation of either all the contiguous spaces or the entire building might be considered. The evacuation of additional spaces is especially important in facilities in which the reentry of fumes exhausted from the building is known to be a problem. Medical observation and care for any exposed persons should be obtained as soon as possible. It would be desirable to have sufficient self-contained escape-type breathing devices on hand to equip every occupant of the laboratory. Unless laboratory personnel have received specific training in handling hazard material incidents, the nearest hazardous material emergency response center should be called for assistance for a substantial spill.

Hydrogen cyanide is categorized as a chemical which is immediately dangerous to life and health (IDLH). As such, clean up of spills should be handled very carefully. Individuals performing the clean up should wear a self-contained, positive pressure breathing apparatus, equipped with a full facepiece, rubber or neoprene gloves, and chemically protective outerwear. A type C supplied air respirator unit operated at a positive pressure can be used as well, but a self-contained unit should be available as a backup. Anyone asked to wear this equipment must have received prior training in the proper use of the equipment. The material can be cleaned up using absorbent pillows or other absorbent materials. Waste should be placed in double heavy-duty plastic bags, which are then tightly closed by twisting the top, folding the top over and wrapping it securely with duct tape. The sealed plastic bags should then be placed in heavy plastic containers or steel drums which can be tightly sealed. Waste material should not be placed in fume hoods to evaporate or be disposed of in drains. In the latter case the possibility of fumes collecting in sections of the drain piping and reentering a building thorough a dry sink trap is too great. The waste should not go to a normal landfill. Incineration is the preferred means of disposal.

A leaking cylinder which cannot be readily repaired should be taken to a remote location where the gas in the cylinder can be released safely. Cylinders which are damaged but not leaking should be returned to the vendor for disposal wherever possible. Disposal of gas cylinders by commercial hazardous waste firms can be very expensive.

REFERENCES

1. Occupational Health Guidelines for Chemical Hazards, Mackison, F.W., Stricoff, R.S., and Partridge, Jr., L.J., (Eds.), U.S. Department of Health and Human Services and U.S. Department of Labor. DHHS (NIOSH) Pub. No. 81-123, 1981.
2. *Prudent Practices for Handling Hazardous Chemicals in Laboratories,* National Academy Press, Washington, D.C., 1981, 45–47, 133–134.
3. **Chen, K.K. and Rose, C.L.,** Nitrite and thiosulfate therapy in cyanide poisoning, *JAMA,* 149, 113, 1952.
4. **Hirsch, F.G.,** Cyanide poisoning, *Arch. Environ. Health,* 8, 622, 1964.

4.6.2.11. Fluorine Gas

Fluorine (CAS no. 7782-41-4) is an extremely reactive gas which reacts violently with a wide variety of materials, a representative sample of which are most oxidizable substances, most organic matter, silicon-containing compounds, metals, halogens, halogen acids, carbon, natural gas, water, polyethylene, acetylides, carbides, and liquid air. Many of these reactions will initiate at very low temperatures. Because it will react with so many materials, extreme care must be taken when working with fluorine. The work area should be very well ventilated and free of combustible materials which would act as fuel in the event of a fire. A written hazard analysis should be prepared for the research program prior to beginning work and an emergency contingency plan developed as a part of the laboratory industrial hygiene plan required by the OSHA laboratory standard. Written standard operating procedures are required, and employees fully trained in the nature of the risks and protective measures necessary to avoid injury.

The OSHA PEL is 0.1 ppm or 0.2 mg/M^3. However, the 1993-94 ACGIH TWA levels are 10 times higher with short-term exposure limits (STEL) another factor of 2 higher. An exposure to 25 ppm for 5 minutes has caused severe eye symptoms. The LC_{50} (50% lethal concentration) for a 1-hour exposure for rats and mice is 185 and 150 ppm, respectively. It is highly irritating to tissue.

Fluorine will react with brass, iron, aluminum, and copper to form a protective metallic fluoride film. Circulating a dilute mixture of fluorine gas and inert gases through a system of these metals will passivate the surfaces and render them safe to use, provided the film remains intact. However, it is recommended that an inert gas be circulated through any fluorine system before the fluorine is introduced.

All systems containing fluorine should be checked frequently for leaks. Filter paper moistened with potassium iodide can be used to perform the tests. The paper will change color when any escaping gas comes in contact with it.

Work with systems using fluorine should be done within a fume hood. The research worker should be protected by an explosive shield. The worker should also wear protective goggles and a face mask. Unless the cylinder valve is operated through a remote control device, the user should wear sturdy gloves with extended cuffs to protect his hands and arms while manipulating the valve. A protective apron should be worn as well. However, all of these may give only limited protection in the event of an accident since fluorine may react with many common items of personnel protective gear.

Self-contained escape breathing apparatus should be available for all occupants of a laboratory in which fluorine is in active use. In the event of an accident, immediate evacuation of the area should take place, being sure to close doors as personnel leave to isolate the problem as much as possible. Evacuation of nearby areas should be considered or, depending upon the scale of the accident, perhaps the entire building, especially, as noted elsewhere, if there are known problems with exhausted materials reentering the building. No remedial measures should be attempted under most circumstances; the incident should be allowed to proceed until the fluorine supply is exhausted. Firefighting efforts should be aimed at preventing a fire from spreading. Applying water directly to the leak could intensify the fire. Obviously, it would be desirable to use smaller cylinders, if practicable, for the research program to reduce the scale of any incident.

Cylinders with valves that cannot be dislodged without application of sufficient force to damage the valve or the connection to the cylinder should be returned to the vendor for repair rather than take a chance on a massive rupture and release of the contents of the cylinder. Ordinary maintenance personnel should never be asked to attempt to free the valve. Although the consequences here would be exacerbated by the extremely hazardous properties of the contents of the cylinders, the same recommendation would apply to virtually any cylinder containing a substantial volume of gas under high pressure. There are firms that specialize in handling dangerous situations such as peroxides, explosives, highly reactive materials, and damaged cylinders.

REFERENCES

1. Occupational Health Guidelines for Chemical Hazards, Mackison, F.W., Stricoff, R.S., and Partridge, Jr., L.J., (Eds.), U.S. Department of Health and Human Services and U.S. Department of Labor. DHHS (NIOSH) Pub. No. 81-123, 1981.
2. Hazardous Chemical Data., NFPA-49, National Fire Protection Association, Quincy, MA.
3. Threshold Limit Values for Chemical Substances and Physical Agents and Biological Exposure Indices, American Conference of Governmental Industrial Hygienists, Cincinnati, OH, 1993–1994.

4.6.2.12. General Safety For Hazardous Gas Research

The previous section was devoted to fluorine as an example of an exceptionally hazardous gas. However, many others commonly used in the laboratory pose comparable risks, some because of their own toxicity or that of gases or vapors evolved as a consequence of their decomposition, others because of their flammable or explosive properties or due to reactions with other chemicals, or a combination of all of these characteristics. Even relatively innocuous gases such as nitrogen, carbon dioxide, argon, helium, and krypton can be a simple asphyxiant if they displace sufficient air, leaving the oxygen content substantially below the normal percentage. If the content of these inert gases approaches one third or higher, symptoms of oxygen deprivation begin to occur, and at concentrations of around 75%, persons will survive for only a brief period. Any gas under high pressure in a cylinder poses a problem if the cylinder is mishandled so as to rupture the containment of the gas. In such a case the cylinder can represent an uncontrolled missile with deadly consequences for anyone in the vicinity. The ability of escaping gas to move readily throughout a volume greatly enhances the likelihood that a flammable gas will encounter a source of ignition. This problem is shared by the vapors of many volatile liquids. Many gases are heavier than air and may collect in depressions or areas with little air movement, representing a danger to unsuspecting persons. Gases often do not have a distinctive odor or are not sufficiently irritating to warn of their presence, and some that do, such as hydrogen sulfide, act to desensitize the sense of smell at levels which would be dangerous.

Cylinders connected to systems in the laboratory must always be strapped firmly to a support to ensure that they do not fall over. Not only is there a risk of breaking the connection on the cylinder side of the regulator valve, with the concomitant risk of the cylinder becoming a missile, but the connection to the system also may be broken so that gas will escape from the low pressure side of the regulator. If the amount of gas to be stored in the laboratory is substantial, it may be preferable to pipe the gas in from a remote outside storage area with control valves located both outside and within the laboratory.

Where the explosion risk is substantial, the facility may need to be designed with explosion venting so that the force of any explosion and the resulting flying debris can be released in a relatively safe direction, minimizing the risk to the occupants. Systems for smaller-scale operations should be placed within a hood, perhaps one specially designed to provide partial protection against explosions. Explosion shields are available to aid in protecting the research worker. Wherever the possibility of an explosion exists, laboratory personnel should wear impact-resistant goggles, possibly supplemented by a face mask to protect the lower part of the face and throat. Gauntlets should be worn if there is risk to the hands and forearms while conducting experimental evolutions.

Not all work with high pressures involves gas cylinders. Reaction bombs are commercially available which operate at pressures up to 2000 psi (13.8 MPa) and temperatures up to 350°C (662°F). An error on the part of the research worker could permit these design parameters to be exceeded, and although these units are designed with a substantial safety factor, the failure of one of them could lead to disastrous consequences. Not only is there the immediate danger of injury due to flying components of the system, there is the risk of reactions involving reagents from broken bottles. These secondary events could escalate the consequences far beyond the original scope of the incident. Any device in which the potential for a high-pressure accident exists should be set up in a hood to provide some explosion protection, and explosion barriers should be used to provide additional protection to the occupants of the room. Personnel working in this type of research should be especially careful to not work alone. They should wear goggles, a face mask, and sturdy gloves to protect their hands and forearms.

Systems involving toxic gases should be adequately ventilated. If possible, the systems should be set up totally within a fume hood. Large walk-in hoods often are used for this purpose. All systems should be carefully leak-tested prior to introduction of toxic materials into the system, periodically thereafter, and after any maintenance or modifications to the system which could affect its integrity.

Many gases are potentially so dangerous that access to the laboratory should be limited to essential personnel that are authorized to be present. When working with such materials, no one should work alone. It may be desirable to have one of the persons somewhat removed from the immediate area of operations, but a second person should be within the working area. Entrances to a high hazard gas research facility should be marked with a **DANGER, SPECIFIC AGENT, AUTHORIZED PERSONNEL ONLY** or a comparable warning sign. Hood exhausts also should bear a comparable warning legend. In some cases, automatic alarm sensors have been developed to detect the presence of gas levels approaching dangerous levels. It is recommended that warning trip points on these devices be set at no more than 50% of either the OSHA PEL or the ACGIH TLV value, whichever is lower. If an automatic sensing device is available, circuitry can be devised to activate a valve to cut off the gas supply as well as to provide a warning. The latter is especially important if the operation is left unattended and no signal is transmitted to a manned location, rendering an alarm ineffective. The growing application of programmed personal computers, or laboratory computer workstations dedicated to experimental control has increased the amount of sophisticated experimentation that can be automated.

In any laboratory involving the use of highly hazardous materials, an emergency plan should be developed in advance of initiation of any major project or any major modification to an ongoing project based on a thorough hazard analysis. Because of the special problems associated with gases, the emergency contingency plan should make provision for rapid evacuation of the immediate laboratory using short-duration, self-contained breathing apparatus and provision for initiation of evacuation from other areas of the facility. Provision of detailed information to emergency response groups is now required under the Community Right-to-Know law for many hazardous substances when the amount involved exceeds a prescribed threshold amount. It is also recommended that employees involved in any research involving hazardous materials participate in a medical surveillance program, consisting of a comprehensive prior screening exam, acquisition of a serum sample for comparison with a sample following a possible incident, and a complete medical history, so that baseline information on individuals will be available to medical personnel called upon to treat personnel that may have been exposed, as for example in this area, to toxic gases. If there are specific organs or bodily functions that might be affected, the examination may need to include special tests in these areas.

The most common problem associated with the use of gas cylinders is the problem of leaks. The cylinder can develop leaks at any of four points, assuming that no gross rupture of the cylinder wall itself has occurred: the valve threads, valve safety device threads, the valve stem, and the valve outlet. Repairs of leaks in the first two of these would require repairs done at high pressure and are not to be attempted in most laboratory facilities. It may be possible that some adjustments are incorporated into the design of the cylinder to allow stopping leaks in the latter two areas. In either case, it is best to contact the manufacturer for advice before attempting any repair. Forced freeing of a "frozen" or corroded valve should not be attempted.

Leaks involving cylinders containing corrosive materials may increase in size with time as the corrosive gas interacts with atmospheric moisture and erodes the opening. Removal of leaking cylinders containing corrosives from an occupied facility to a remote location should be done as quickly as possible. The vendor should be called for advice or assistance. If time permits, a solution would be to slowly exhaust the leaking cylinder into a neutralizing material. If the protective cap has corroded into position so that no bleeder hose can be attached, two heavy-duty plastic bags can be placed over the leaking end of the cylinder and the gases conducted from the bags through a hose into a drum of neutralizing solution. If the leak is too large for easy handling, a commercial, state, or local hazardous material response group should be called upon for assistance. The wearing of protective suits and self-contained breathing apparatus may be required, which requires special training.

Leaks involving toxic and flammable gases pose the risk of personal injury to individuals attempting repairs, and repairs should not be attempted by local personnel unless it is certain that they can be done safely. Evacuation is usually recommended. Where flammable gas leaks are concerned, all ignition sources must be turned off prior to evacuating a facility, if time permits. It is foolish to risk life or personal injury unless by doing so it may be possible to save others.

In cases where the melting point or boiling point of the contained gas is sufficiently high, the leak rate may be substantially reduced or virtually stopped by putting the body of the cylinder into a cooling bath while deciding upon the appropriate corrective measures or waiting for assistance. This should not be done if the contained material will react with the coolant.

REFERENCES

1. Handling and Storage of Flammable Gases, Air Products and Chemicals, Inc., Allentown, PA.
2. *Prudent Practices for Handling Hazardous Chemicals in Laboratories,* National Academy Press, Washington, D.C., 1981, 75-101, 221, 223–224, 229.

4.6.2.12.1. Some Hazardous Gases

The properties of a number of hazardous gases will be given in this section, along with a few brief comments on noteworthy problems associated with these materials. The common chemical name, formula, and CAS number will be given for each substance, followed by a definition of the primary hazard class(es) represented by the material, the boiling point as degrees Celsius (degrees Fahrenheit);[*] the explosive range (lower explosive limit [lel]—upper explosive limit [uel]) in percent by volume in air (NA where not applicable or not available); the vapor specific gravity referred to air as 1; OSHA 8 hour time weighted average PEL in parts per million[**] and finally, salient comments about the material.

Acetylene—C_2H_2; 74-86-2; explosive, flammable, asphyxiant; bp = $-84.0°C$ [$-119°F$]; er = 2–82%; sp g = 0.91; PEL = 2500 ppm (10% of lel). Acetylene in cylinders is usually dissolved in acetone and is relatively safe to handle, but purified acetylene has a very low ignition energy and a relatively low minimum ignition temperature, 300°C [571°F]. It forms explosive mixtures with air over a wide range. Utilization of acetylene should be at a pressure of 15 psi gauge or less. Piping for acetylene systems should be steel, wrought iron, malleable iron, or copper alloys containing less than 65% copper. May form explosive compounds with silver, mercury, and unalloyed copper.

Ammonia—NH_3; 7664-41-7; flammable gas, Causes tissue burns, strong respiratory irritant; bp = $-33.3°C$ [$-28°F$]; er = 15–28%; sp g = 0.6; PEL = 50 ppm. May cause

---

[*]   In most cases, the celsius temperature is from the literature, while the Fahrenheit temperature is calculated from the former.

[**] The ceiling value is given if the level should not exceed this value at any time or for a limited period. If a level has not been set by OSHA and an ACGIH TLV value is available, the latter will be given.

severe injury to respiratory system and eyes, common 35% laboratory solution can cause severe skin burns. High concentrations may cause temporary blindness. Baseline physical should stress respiratory system and eyes. Skin should be examined for existing disorders. Tests should include pulmonary function and chest X-ray. Should wear self-contained breathing apparatus and rubber shoe covers when cleaning up a spill (by dilution with ample amounts of water and mop to a drain). On EPA list of extremely hazardous substances, 40 CFR Section 302.

Arsine—AsH_3; 7784-42-1; deadly poison by inhalation, fire and explosion hazard; bp $= -62.5°C$ [$-144.5°F$]; er = NA; sp g = 2.66; PEL = 0.05 ppm. Recognized carcinogen; causes pulmonary edema; primarily poisonous due to interaction with hemoglobin, causes anemia; early symptoms are headache, dizziness, nausea, and vomiting. Severe exposures result in kidney damage, delirium, coma, and possible death. In increased use due to applications to semiconductor research. Should be used in a sealed system, or the system should be set up in an efficient fume hood. Leaking cylinders can be handled by allowing the leaking gas to interact with a 15% aqueous solution of sodium hydroxide.[8] The arsine forms a water soluble precipitate. After neutralization with sulfuric acid, the precipitate is filtered out to be disposed of as hazardous waste, and the neutral solution can be disposed of into the drain, diluted by large amounts of water. Individuals doing this should wear self-contained breathing apparatus and chemically resistant gloves. Arsine will ignite in contact with chlorine and undergoes violent oxidation by fuming nitric acid. On EPA list of extremely hazardous substances, 40 CFR Section 302.

Boron Trifluoride—BF_3; 7637-07-2; inhalation poison, irritant, nonflammable; bp $= -100°C$ [$-148°F$]; er = NA; sp g = 2.3; ceiling PEL = 1 ppm. Irritating to eyes and respiratory system. Animals have been shown to have kidney damage after high exposures. Baseline physical should stress respiratory system, eyes, and kidneys. Tests should include a chest X-ray, pulmonary function test. Cylinders with a slow leak can be allowed to leak into an efficient fume hood for disposal, or seal the cylinder and return to the vendor. Produces a thick white smoke by interaction with moisture in humid air, but otherwise insufficient data on warning properties. On EPA list of extremely hazardous substances, 40 CFR Section 302.

1,3-Butadiene—C_4H_6; 106-99-0; flammable, irritant to eyes and mucous membranes, suspect carcinogen; bp $= -4.5°C$ [$23.9°F$]; er = 2–11.5%; sp g = 1.9; PEL = 1000 ppm (ACGIH = 10 ppm, suspect human carcinogen). Narcotic at high concentrations. May form peroxides on exposure to air, at high temperatures, may self-polymerize exothermally, forms explosive mixtures with air. Participation in a medical surveillance program recommended for users. On EPA list of extremely hazardous substances, 40 CFR Section 302. On "California" list as a carcinogen.

Carbon Dioxide—CO_2; 124-38-9; nonflammable, asphyxiant; sublimes at $-78.5°C$ [$-109°F$]; er = NA; sp g = 1.53; PEL = 5000 ppm. Very common laboratory gas, also used as "dry ice." Causes problems primarily by displacement of air. At 5% concentration, respiratory volume quadrupled. Heart rate and blood pressure increases reported at 7.6%. At 11%, unconsciousness typically occurs in 1 minute or less. No warning other than symptoms, dizziness, headaches, shortness of breath, and weakness because it is colorless and odorless.

Carbon Disulfide—CS_2; 75-15-0; flammable liquid (vapor pressure 400 mm at 28°C [82.4°F], poisonous; bp $= 46.5°C$ [$115°F$]; er = 1.3-50%, flash point $= -30°C$ [$-22°F$] and ignition temperature is only 90°C [194°F]; sp g of vapor = 2.64; PEL = 20 ppm. Central nervous system poison. Extended exposure can cause permanent damage to the CNS in severe cases. Numerous other physiological problems to heart, kidneys, liver, stomach. Has strong narcotic and anesthetic properties. Poisonous if inhaled or ingested, or with prolonged contact with skin. Baseline physical should stress central and peripheral nervous systems, cardiovascular system, kidneys, liver, eyes, and skin. Tests should include urinalysis for kidney function, liver panel, electrocardiogram, and ophthalmic exam. Vapors can be ignited by contact with an incandescent light bulb. Air-carbon disulfide mixture can explode in the presence of rust. Do not pour down a sink. Do not use where electrical sparks possible. Use nonsparking tools. Explosion-proof wiring, fixtures not necessarily effective to arrest flame. Use dry chemical or carbon dioxide to fight fires. On EPA list of extremely hazardous chemicals, 40 CFR Section 302. On "California" list as a chemical with reproductive toxicity.

Carbon Monoxide—CO; 630-08-0; flammable, poisonous, experimental teratogen; bp = $-191.1°C$ [$-311.9°F$]; er = 12.5–74%, sp g = 0.9678; PEL = 50 ppm. Carbon monoxide combines highly preferentially with hemoglobin to the exclusion of oxygen and hence is a chemical asphyxiant. Pregnant women and smokers more susceptible to risk. On "California" list as a chemical with reproductive toxicity. Baseline physical recommended. Medical history taken to discover history of problems involving heart, cerebrovascular disease, anemia, and thyroid toxicosis. A complete blood count should be taken. Insidious poisonous gas. Does not have adequate warning properties, odorless and nonirritating. Contact with strong oxidizers may cause fires and explosions. Dangerous fire hazard.

Chlorine—Cl_2; 7782-50-5; nonflammable gas but supports combustion of other materials, toxic; bp = $-34.5°C$ [$-30.1°F$]; er = NA; sp g = 2.49; ceiling PEL = 1 ppm. Forms explosive mixtures with flammable gases and vapors. Reacts explosively with many chemicals and materials such as acetylene, ether, ammonia gas, natural gas, hydrogen, hydrocarbons, and powdered metals. Many incompatibles; carefully review the literature and MSDS before working with this material. Systems should be set up in an efficient fume hood. Strong odor, noticeable well below acute danger levels. Inhalation can cause severe damage to lungs. Baseline physical exam should stress eyes, respiratory tract, cardiac system, teeth, and skin. Pulmonary function test and chest X-ray recommended. Warning of the presence of gas well below PEL due to odor and irritating properties. On EPA list of extremely hazardous chemicals, 40 CFR Section 302.

Cyanogen—C_2H_2; 460-19-5; highly flammable, toxic; bp = $-21°C$ [$-5.8°F$], er = 6.6–32%, sp g = 1.8; ACGIH TLV = 10 ppm. Odor resembles almonds, poisonous by inhalation and skin; symptoms are headache, dizziness, nausea, vomiting, and rapid pulse; severe exposures lead to unconsciousness, convulsions, and death; strong eye irritant; fire hazard when exposed to oxidizers, flame, sparks. Reacts with fluorine (ignites) and oxygen (combination of liquid cyanogen and liquid oxygen will explode).

Cyanogen Chloride—CClN; 506-77-4; nonflammable, inhalant poison, eye irritant; bp = $13.1°C$ [$55.6°F$]; er = NA; sp g = 1.98; ACGIH ceiling level = 0.3 ppm. Toxic properties similar to hydrogen cyanide; heat causes it to decompose and emits highly toxic and corrosive fumes.

Diazomethane—CH_2NO_2; 334-88-3; explosion hazard, irritant to respiratory system, eyes, suspect carcinogen; bp = $-23°C$ [$-9.4°F$]; er = NA; sp g = 1.4; OSHA PEL = 0.2 ppm. One of the most dangerous chemicals used in chemical laboratories. Strong allergen, irritating to eyes and to respiratory system. May sensitize as well as irritate. Recommend initial medical history, full chest X-ray, and pulmonary function tests to individuals planning to work with this material. Explosively sensitive to shock, heat (about 100°C, 212°F), exposure to rough surfaces (e.g., ground glass joints), alkali metals, calcium sulfate. Do not store, prepare fresh when needed. Always use in an efficient fume hood, use an explosive shield, wear impact and chemical splash-resistant goggles, possibly supplemented by a face mask. Respiratory protection recommended if levels may approach PEL.

Diborane—B_2H_6; 19287-45-7; flammable, poison, suspect carcinogen; bp = $-92.5°C$ [$-134.5°F$]; er = 0.9–98% (ignites at temperatures of 38–52°C [100–125°F] or less in humid air); sp g = 0.96; OSHA PEL = 0.1 ppm. Respiratory irritant, causes pulmonary edema. Strong irritant to skin, eyes, other tissues. Keep cool and away from oxidizing agents. Baseline physical should stress lungs, nervous system, liver, kidneys, eyes. Pulmonary function test and chest X-ray recommended. Reacts with aluminum and lithium to form hydrides which may explode in air. Use only in an efficient fume hood. Use an explosion shield, impact and chemical splash protecting goggles, supplemented with a mask, and hand and forearm protection. On EPA list of extremely hazardous chemicals, 40 CFR Section 302.

Dimethyl Ether—C_2H_6O; 115-10-6; flammable, inhalation and skin irritant, narcotic properties; bp = $-23.7°C$ [$-10.7°F$]; er = 3.4–27%; sp g = 1.62; OSHA PEL = NA. Explosion hazard when exposed to flames, sparks, forms peroxides, sensitive to heat.

Ethylene—C_2H_4; 74-85-1; simple asphyxiant, flammable, poisonous to plants; bp = $-103.9°C$ [$-155°F$]; er = 2.7–36%; sp g = 0.98; OSHA PEL = NA. Dangerous when exposed to heat and flames.

Ethylene Oxide—C_2H_4O; 75-21-8; flammable, strong irritant, inhalation, and oral poison, carcinogen, bp = 10.7°C [51.3°F]; er = 3–100%; sp g = 1.52; OSHA PEL = 1 ppm. Regulated carcinogen, 29 CFR Section 1910.1047. Increases rate of miscarriages. Very commonly used as sterilizing agent. Should be used in a sealed system and exhausted outdoors. Personnel exposures should be monitored. It is a very strong irritant to eyes, skin, and to respiratory tract. Can cause pulmonary edema of respiratory system at high levels. Forms explosive mixtures with air. Very reactive when in contact with alkali metal hydroxides, iron and aluminum oxides, and anhydrous chlorides of iron, aluminum, and tin. Also reacts with acids, bases, ammonia, copper, potassium, mercaptans, and potassium perchlorate. On EPA list of extremely hazardous chemicals, 40 CFR Section 302. On "California" list as a carcinogen, and a chemical with reproductive toxicity for women.

Formaldehyde—CH_2O; 50-00-0; irritant to eyes, skin, respiratory system, oral poison, allergen, OSHA regulated carcinogen, flammable; bp = $-19.4°C$ [$-3°F$]; er = 7–73%; sp g = slightly greater than 1; OSHA PEL = 0.75 ppm. Previous physical properties are for the gas. It is normally sold as an aqueous solution of 37 to 52% formaldehyde by weight. Other solvents are also used. Liquid formaldehyde when heated can evolve the gas which will burn. Ingestion of the solution will cause stomach pain, nausea, and vomiting and can result in loss of consciousness. Severe eye irritant. Extended exposures can cause skin and bronchial problems. The gas is on the EPA list of extremely hazardous materials, 40 CFR Section 302. As a gas, is on the "California" list as a carcinogen

Hydrogen—H_2; 1333-74-0; flammable, explosive, asphyxiant; bp = $-252.78°C$ [$-422.99°F$]; er = 4.1–74.2%; sp g = 0.0695; OSHA PEL = NA. Gaseous hydrogen systems of 400 cubic feet (11.35 cubic meters) and containers of liquid hydrogen of more than 150 liters (39.63 gallons) are regulated by OSHA, 29 CFR Section 1910.103. Systems to be used for hydrogen should be purged with inert gas prior to use. Consideration should be given for incorporation of safety systems required for larger systems, depending on the research program and facilities available. Hydrogen will burn with virtually an invisible flame. Care should be used in approaching a suspected hydrogen flame; holding a piece of paper in front of you is recommended to detect the flame. Because it is so light, it does tend to escape from rooms where leaks occur.

Hydrogen Chloride (Anhydrous)—HCl; 7647-01-0; nonflammable, toxic gas by all routes of exposure and intake; bp = $-84.8°C$ [$-121°F$]; er = NA; sp g = 1.27; OSHA ceiling PEL = 5 ppm. Gas combines with moisture to become corrosive to eyes, skin, and respiratory system. Baseline physical exam should stress respiratory system, eyes, and skin.

Pulmonary test and chest X-ray recommended. Exposure to airborne concentrations above 1500 ppm can be fatal in a few minutes. Irritating properties detectable at about PEL. Should only be used in an efficient fume hood. Workers should wear gas-tight goggles, acid, resistant aprons, gloves, and outerwear.

Hydrogen Fluoride—HF; 7664-39-3; strong irritant and corrosive to eyes, respiratory system, internal tissue, skin, via contact, inhalation, ingestion, noncombustible; bp = 19.5°C [67.2°F]; er = NA; sp g = 0.7; OSHA PEL = 3 ppm. Readily dissolves in water to form hydrofluoric acid. Prolonged exposure can cause bone changes. Baseline physical should stress eyes, respiratory tract, kidneys, central nervous system, skin, and skeletal system. Special tests should include urinalysis, pelvic X-ray (use shielding to protect genitals as much as possible), and an ophthalmic examination. Provides warning at levels near PEL due to irritant effects. Gas on EPA list of extremely hazardous chemicals, 40 CFR Section 302. For other comments see Section 4.6.2.9 Hydrofluoric Acid.

Hydrogen Selenide—H_2Se; 7783-07-5; flammable, very toxic via inhalation, also toxic by contact with eyes and skin; bp = −41.3°C [−42°F]; er = NA; sp g = 2.1; OSHA PEL = 0.05 ppm. Very offensive odor, but threshold for detection above dangerous levels. Recommend initial medical screening prior to work with this substance for existing respiratory problems and impaired liver function. Causes irritation of eyes, nose, throat, and lungs. Symptoms of exposure include nausea, vomiting, followed by a metallic taste, garlic odor to breath, dizziness, fatigue. Can cause eye, liver, spleen, and lung damage. Use only in an efficient fume hood. Additional respiratory and eye protection recommended if potential exposure problem. Contact with acids, halogenated hydrocarbons, oxidizers, water may result in fire or explosion. On EPA list of extremely hazardous chemicals, 40 CFR Section 302.

Hydrogen Sulfide—H_2S; 7783-06-4; flammable, irritant, asphyxiant; bp = −60°C [−76°F]; er = 4.3–46%; sp g = 1.19; OSHA ceiling PEL = 20 ppm (single 10 minute peak PEL = 50 ppm). Strong odor of rotten eggs, but sense of smell desensitized by gas after short interval (minutes) at high levels. Preliminary medical exam recommended prior to work with this material, with stress on eyes and lungs, to include chest X-ray and pulmonary function test. Severe eye and respiratory irritant at moderate concentrations. Rapidly acting systemic poison which causes respiratory paralysis at high levels. Exposures at 1000 to 2000 ppm can cause immediate death. Prolonged exposure above 50 ppm can damage eyes and cause respiratory problems. Susceptibility increases with repeated exposures. Symptoms caused by low concentrations are headache, fatigue, insomnia, irritability, gastrointestinal problems. Highly reactive with strong nitric acid and strong oxidizing agents. On EPA list of extremely hazardous chemicals, 40 CFR 302.

Methane—CH_4; 74-82-8, flammable, simple asphyxiant; bp = −161.4°C [−258.6°F]; er = 5–15%; sp g = 0.52; OSHA PEL = NA. Keep away from sources of ignition.

Methyl Acetylene—C_3H_4; 74-99-7; flammable, anesthetic; bp = −23°C [−10°F]; er = 1.7–11.7; sp g = 1.4; OSHA PEL = 1000 ppm. Sweet odor. Overexposure causes drowsiness. Reactions with chlorine and strong oxidizing agents may result in fires or explosions. It forms very shock-sensitive compounds with copper. Equipment components containing more than 67% copper should not come into contact with the compound.

Methyl Acetylene Propadiene Mixture (MAPP)—C_3H_4 isomers; No CAS No.; flammable, anesthetic; bp = −34.5°C [−30°F]; er = 3.4–10.8%; sp g = 1.5; OSHA PEL = 1000 ppm. Foul odor. Reactivity similar to methyl acetylene (MA). Overexposure can cause drowsiness and unconsciousness with MA. Odor detectable well below PEL. Will attack some plastics, films, and rubber.

Methyl Bromide—CH_3Br; 74-83-9; fumigant, toxic by inhalation, contact with eyes and skin, and ingestion; cumulative poison; bp = 3.6°C [38.4°F]; er = 13.5–14.5% (requires high energy ignition source); sp g = 3.3; OSHA ceiling PEL = 20 ppm (skin). Severe respiratory irritant, neurotoxin, narcotic at high concentrations. Symptoms of overexposure include headache, visual disturbances (blurred or double vision), nausea, vomiting, in some cases vertigo, tremors of the hand, and in more severe cases convulsions may occur. Persistent depression, anxiety, hallucinations, inability to concentrate, vertigo may follow severe exposures. Kidney damage may occur. Contact with skin can cause skin rash or blisters. Complete preuse physical recommended stressing nervous system, lung function, skin condition. Chest X-ray and pulmonary test recommended. Possible carcinogen. Contact with aluminum and strong oxidizers can lead to fires or explosions. On EPA list of extremely dangerous chemicals, 40 CFR Section 302. On "California" list as a chemical with reproductive toxicity.

Methyl Chloride—CH_3Cl; 74-87-3 flammable, Moderate irritant, suspected carcinogen, poison; bp = $-23.7°C$ [$-10.7°F$]; er = 8.1–17%; sp g =1.78; OSHA PEL = 100 ppm (ceiling = 200 ppm), peak = 300 ppm for 5 minutes in any 3-hour period). Dangerous fire hazard from heat, flame oxidizers. Prolonged exposures can cause psychological problems due to damage to central nervous system. Also can damage liver, kidneys, bone marrow, cardiovascular system. Faint, sweetish odor does not provide adequate warning of overexposure. Use only in an efficient fume hood.

Methyl Mercaptan—CH_4S, 74-93-1; flammable, inhalant poison, possible carcinogen; bp = 5.95°C [42.7°F]; er = 3.9–21.8%; sp g = 1.66; OSHA ceiling PEL = 10 ppm. Warning odor of rotten cabbage. Reacts vigorously with oxidizing agents. Will decompose to emit toxic and flammable vapors when reacts with water, steam, and acids.

Nitric Oxide—NO; 10102-43-9; noncombustible, strong oxidizing agent, inhalant poison; bp = $-152°C$ [$-241°F$]; er = NA, sp g = 1.0; OSHA PEL = 25 ppm. Nitric oxide causes narcosis in animals. Causes drowsiness. Sharp, sweet odor provides warning well below dangerous levels. Changes to nitrogen dioxide in air (see following material), although conversion is slow at low concentrations. Reacts vigorously with reducing agents. Will attack some plastics, films, rubber. On EPA list of extremely hazardous chemicals, 40 CFR Section 302.

Nitrogen Dioxide—NO_2; 10102-44-0; poisonous by inhalation; bp = $-21°C$ [9.8°F]; er = NA; sp g = 2.83; OSHA ceiling PEL = 5 ppm. Brown pungent gas. Odor detection threshold approximately the same as the OSHA PEL. Strong respiratory irritant. Exposure to 100 ppm for 1 hour will normally cause pulmonary edema and possibly death; 25 ppm will cause chest pain and respiratory irritation. Onset of symptoms may be delayed. Recovery from an overexposure may be slow and may result in permanent lung damage. Recommend a baseline physical examination, with emphasis on respiratory and cardiovascular systems. Special tests recommended are chest X-ray, pulmonary function test, and electrocardiogram. Reacts vigorously with chlorinated hydrocarbons, ammonia, carbon disulfide, combustible materials, possibly resulting in fires and explosions. High-temperature glassblowing operations may generate significant levels of this material. On EPA list of extremely hazardous chemicals, 40 CFR Section 302.

Nitrogen Trifluoride—NF_3; 7783-54-2; poisonous by inhalation; bp = $-12°C$ [$-200°F$]; er = NA; sp g = 2.5; OSHA PEL = 10 ppm. May affect the capacity of the blood to carry oxygen; reacts vigorously with reducing agents. Baseline physical should stress examination of blood, cardiovascular, nervous systems, and liver and kidney function. A complete blood count should be taken. Persons with a history of blood disorders should take special care to avoid exposures.

Oxygen Difluoride—OF_2; 7783-41-7; poison via inhalation, strong respiratory irritant, noncombustible, strong oxidizing agent; bp = $-145°C$ [$-229°F$]; er = NA; sp g = 1.86; OSHA PEL = 0.05. Foul odor, but detection threshold too high to provide adequate warning. Sense of smell fatigues rapidly. Inhalation at less than 1 ppm causes severe headaches. Very corrosive to tissue. Strong irritant to respiratory system, kidneys, internal genitalia. Baseline medical exam recommended, with particular emphasis on affected systems. Use only in efficient hood system.

Ozone—O_3; 10028-15-6; strong irritant to eyes and respiratory system; bp = $-112°C$ [$-169.6°F$]; er = NA; sp g = 1.65; OSHA PEL = 0.1 ppm. Sharp, distinctive odor; odor detection level about the same as the PEL. Affects central nervous system. May be mutagen. Powerful oxidizing agent for both organic and inorganic oxidizable materials. Some reaction products are very explosive. Baseline physical recommended with emphasis on heart and lungs. Chest X-ray and pulmonary function test recommended. On EPA list of extremely hazardous materials, 40 CFR Section 302.

Phosgene—CCl_2O; 75-44-5; inhalant poison, nonflammable; bp = 8.2°C [46.7°F]; er = NA; sp g = 3.4; OSHA PEL = 0.1 ppm. Odor of "moldy hay." Sense of smell desensitized quickly, irritating properties well above PEL, so does not provide adequate warning. Must be used carefully in an efficient fume hood. Paper soaked in a 10% mixture of equal parts of *p*-dimethylaminobenzaldehyde and colorless diphenylamine in alcohol or carbon tetrachloride, then dried, makes a good color indicator. Color changes from yellow to deep orange at about the maximum allowable concentration. Severe respiratory irritant, but irritation does not manifest itself at once, even at dangerous levels. Decomposes in presence of moisture in lungs to HCl and CO. Baseline medical examination recommended, with emphasis on respiratory system. Chest X-ray and pulmonary function test recommended. On EPA list of extremely hazardous chemicals, 40 CFR, Section 302.

Phosphine—PH_3; 7803-51-2; flammable, inhalation poison; bp = $-87.8°C$ [$-126°F$]; er = 1%–?; sp g = 1.17; OSHA PEL = 0.3 ppm. Fishy odor detectable well below PEL. Inhalation is a severe pulmonary irritant and systemic poison. Results of severe overexposure are chest pains, weakness, lung damage, and in some cases coma and death. Persons with prior history of respiratory problems should not work with this material without precautions. Baseline physical should include pulmonary function test. On EPA list of extremely hazardous chemicals, 40 CFR Section 302.

Propane—C_3H_8; 74-98-6; flammable, asphyxiant; bp = $-42.1°C$ [$-43.7°F$]; er = 2.3–9.5; sp g = 1.6; OSHA PEL = 1000 ppm. High exposures (100,000 ppm) caused dizziness after a few minutes. Odorless, nonirritating, so commercially sold propane usually has a foul-smelling odorant added as a warning device. Reacts vigorously with strong oxidizing agents. Explosion hazard from heat and flames.

Propylene—C_3H_6; 115-07-1; flammable, simple asphyxiant; bp = $-47.7°C$ [$-53.9°F$]; er = 2–11.1%; sp g = 1.5; OSHA PEL = simple asphyxiant. Dangerous when exposed to heat and flames. Can react vigorously with oxidizing agents.

Silane—SiH_4; 7803-62-5; flammable, respiratory irritant; bp = $-112°C$ [$-169.6°F$]; er = NA; sp g = NA; ACGIH TLV = 5 ppm. Moderate respiratory irritant. Repulsive odor. Easily ignites in air. Reacts vigorously with chlorine, bromine, covalent chlorides.

Stibine—SbH_3; 7803-52-3; flammable, poison by inhalation; bp = $-17°C$ [1°F]; er = NA; sp g = 4.34; OSHA PEL = 0.1 ppm. Odor similar to hydrogen sulfide but data not available as to whether the odor is an adequate warning of the PEL. Toxic hemolytic agent, which causes injury to liver and kidneys. Probable lung irritant. Symptoms of overexposure may be delayed for up to 2 days and would include nausea, headache, vomiting, weakness, and back and abdominal pain. Death would result from renal failure and pulmonary edema. Baseline medical exam should stress blood, liver, and kidneys. Special tests should include complete blood count, urine analysis, and liver panel.

Sulfur Dioxide—SO_2; 7446-09-5; nonflammable, strong respiratory, eye, skin irritant; bp = $-10.05°C$ [13.9°F]; er = NA; sp g = 2.26; OSHA PEL = 5 ppm. Sharp, irritating odor. Detectable well below OSHA PEL. Reacts rapidly with moisture to form corrosive sulfurous acid (H_2SO_3). Mostly absorbed in upper respiratory tract. High concentrations can cause pulmonary edema and respiratory paralysis. Reacts vigorously with water, with some powdered metals, and with alkali metals such as sodium and potassium. Baseline medical exam should emphasize eyes and respiratory tract. Recommend chest X-ray and pulmonary function tests. Some individuals (10 to 20% of young adults) may be hypersensitive to the material. On EPA list of extremely hazardous chemicals, 40 CFR Section 302.

Sulfur Tetrafluoride—SF_4; 7783-60-0; powerful irritant, poisonous by inhalation; bp = $-40°C$ ($-40.2°F$); sp g = NA; ACGIH ceiling TLV = 0.1 ppm. Reacts with water, steam, and acids to produce toxic and corrosive fumes. On EPA list of extremely dangerous chemicals, 40 CFR Section 302.

Trifluoromonobromomethane (Halon 1301)—$CBrF_3$; 75-63-8; affects heart at high levels; bp = $-57.8°C$ [$-72°F$]; er = NA; sp g = 5; OSHA PEL = 1000. Popular fire extinguishing agent used for solvent fires and for delicate electronic equipment because of its nondamaging properties. Currently production discontinued due to effects of this gas on earth's ozone layer. At design use range no observed effects but at high concentrations can cause cardiac arrhythmias. Individuals with heart problems should use with caution. Can emit dangerous gases and vapors on decomposition by heat.

Vinyl Chloride—C_2H_3Cl; 75-01-4; flammable, dangerous irritant, carcinogen; bp = $-13.9°C$ [7°F]; er = 4–33%; sp g = 2.15; OSHA PEL = 1 ppm, ceiling = 5 ppm. Regulated under 29 CFR Part 1910.1017. Vinyl chloride *monomer* has been shown to cause a rare liver cancer, angiosarcoma. Latency period 20 years or more. Participation in medical surveillance program required by OSHA standard if use meets prescribed conditions. Dangerous irritant to respiratory system, skin, eyes, mucous membranes. Reacts vigorously with oxidizers. Decomposes when heated to generate phosgene. On "California" list as a carcinogen.

REFERENCES

1. Occupational Health Guidelines for Chemical Hazards, DHHS (NIOSH) Pub. No. 81-123, 1981. Mackison, F.W., Stricoff, R.S., and Partridge, L.J., Jr., (Eds.), U.S. Department of Health and Human Services and U.S. Department of Labor, Washington, D.C.
2. **Sax, N.I. and Lewis, R.J., Sr.,** *Rapid Guide to Hazardous Chemicals in the Workplace,* Van Nostrand Reinhold Company, New York, 1986.
3. **Armour, M.A., Browne, L.M., and Weir, G.L.,** *Hazardous Chemicals Information and Disposal Guide,* 2nd ed., University of Alberta, Edmonton, Canada, 1984.
4. **Lewis, R.J.,** *Sax's Dangerous Properties of Industrial Materials,* 8th ed., Van Nostrand Reinhold Company, New York, NY, 1993.
5. Threshold Limit Values for Chemical Substances and Physical Agents and Biological Exposure Indices, American Conference of Governmental Industrial Hygienists, Cincinnati, OH, 1993.
6. OSHA, General Industry Standards, Subpart Z, Occupational Health and Environmental Control, 1910.1000-1500.
7. Hazardous Chemical Data, NFPA-49, National Fire Prevention Association, Quincy, MA.
8. *Prudent Practices for Handling Hazardous Chemicals in Laboratories,* National Academy Press, Washington, D.C., 1981.
9. *NIOSH Pocket Guide to Chemical Hazards*, Hazardous Materials Publishing Co., Inc., Kutztown, PA, 1991.

4.6.2.13. Cryogenic Safety*

Cryogenics may be defined as low temperature technology or the science of very low temperatures. To distinguish between cryogenics and refrigeration, a commonly used measure is to consider any temperature lower than −73.3°C (−100°F) as cryogenic. Although there is some controversy about this distinction and some who insist that only those areas within a few degrees of absolute zero may be considered as cryogenic, the broader definition will be used here.

Low temperatures in the cryogenic area are primarily achieved by the liquefaction of gases, and there are more than 25 which are currently in use in the cryogenic area. However, the seven gases which account for the greatest volume of use and applications in research and industry are helium, hydrogen, nitrogen, fluorine, argon, oxygen, and methane (natural gas).

Cryogenics is being applied to a wide variety of research areas, a few of which are food processing and refrigeration, rocket propulsion fuels, spacecraft life support systems, space simulation, microbiology, medicine, surgery, electronics, data processing, and metalworking. (Recent advances in high temperature superconductors will further increase the use of liquid nitrogen.)

Cryogenic fluids (liquified gases) are characterized by extreme low temperatures, ranging from a boiling point of −78.5°C (−109°F) for carbon dioxide to −269.9°C (−453.8°F) for an isotope of helium, helium-3. Another common property is the large ratio of expansion in volume from liquid to gas, from approximately 553 to 1 for carbon dioxide to 1438 to 1 for neon. Table 4.14 contains a more complete summary of the properties of cryogenic fluids.

4.6.2.13.1. Hazards

There are four principal areas of hazard relating to the use of cryogenic fluids or in cryogenic systems. These are flammability, high pressure gas, materials, and personnel. All categories of hazard are usually present in a system concurrently and must be considered when introducing a cryogenic system or process.

The flammability hazard is obvious when gases such as hydrogen, methane, and acetylene are considered. However, the fire hazard may be greatly increased when gases normally thought to be non-flammable are used. The presence of oxygen will greatly increase the flammability of ordinary combustibles and may even cause some noncombustible materials like carbon steel to burn readily under the right conditions. Liquified inert gases such as liquid nitrogen or liquid helium are capable, under the right conditions, of condensing oxygen from the atmosphere and causing oxygen enrichment or entrapment in unsuspected areas. Extremely cold metal surfaces are also capable of condensing oxygen from the atmosphere.

The high pressure gas hazard is always present when cryogenic fluids are used or stored. Since the liquified gases are usually stored at or near their boiling point, there is always some gas present in the container. The large expansion ratio from liquid to gas provides a source for the build-up of high pressures due to the evaporation of the liquid. The rate

---

*   This section, except for a short amount of material appended at the end, is taken directly from the article, "Cryogenic Safety," by Spencer,[15] with minor editing in the second edition of the handbook.

Table 4.14.   Properties of Cryogenic Fluids

Gas	Normal Boiling Point °C	°K	Volume Expansion to Gas	Flammable	Toxic	Odor
Helium-3	−269.9	3.2	757 to 1	No	No	No
Helium-4	−268.9	4.2	757 to 1	No	No	No
Hydrogen	−252.7	20.4	851 to 1	Yes	No	No
Deuterium	−249.5	23.6	—	Yes	Radioactive	No
Tritium	−248.0	25.1	—	Yes	Radioactive	No
Neon	−245.9	27.2	1438 to 1	No	No	No
Nitrogen	−195.8	77.3	696 to 1	No	No	No
Carbon monoxide	−192.0	81.1	—	Yes	Yes	No
Fluorine	−187.0	86.0	888 to 1	No	Yes	Sharp
Argon	−185.7	87.4	847 to 1	No	No	No
Oxygen	−183.0	90.1	860 to 1	No	No	No
Methane	−161.4	111.7	578 to 1	Yes	No	No
Krypton	−151.8	121.3	700 to 1	No	No	No
Tetrafluromethane	−128	145	—	No	Yes	No
Ozone	−111.9	161.3	—	Yes	Yes	Yes
Xenon	−109.1	164.0	573 to 1	No	No	No
Ethylene	−103.8	169.3	—	Yes	No	Sweet
Boron trifluoride	−100.3	172.7	—	No	Yes	Pungent
Nitrous oxide	−89.5	183.6	666 to 1	No	No	Sweet
Ethane	−88.3	184.8	—	Yes	No	No
Hydrogen chloride	−85.0	188.0	—	No	Yes	Pungent
Acetylene	−84.0	189.1	—	Yes	Yes	Garlic
Fluoroform	−84.0	189.1	—	No	No	No
1,1-Difluoroethylene	−83.0	190.0	—	Yes	No	Faint ether
Chlorotrifluoromethane	−81.4	191.6	—	No	Yes	Mild
Carbon dioxide	−78.5	194.6	553 to 1	No	Yes	Slightly pungent

of expansion will vary, depending on the characteristics of the fluid, container design, insulating materials, and environmental conditions of the atmosphere. Container capacity must include an allowance for that portion which will be in the gaseous state. These same factors must also be considered in the design of the transfer lines and piping systems.

Materials must be carefully selected for cryogenic service because of the drastic changes in the properties of materials when they are exposed to extreme low temperatures. Materials which are normally ductile at atmospheric temperatures may become extremely brittle when subjected to temperatures in the cryogenic range, while other materials may improve their properties of ductility. The American Society of Mechanical Engineers' *Boiler and Pressure Vessel Code, Section VIII—Unfired Pressure Vessels* may be used as a specific guide to the selection of materials to be used in cryogenic service. Some metals which are suitable for cryogenic temperatures are stainless steel (300 series and other austenitic series), copper, brass, bronze, monel, and aluminum. Non-metal materials which perform satisfactorily in low temperature service are Dacron, Teflon, Kel-F, asbestos impregnated with Teflon, Mylar, and nylon. Once the materials are selected, the method of joining them must receive careful consideration to ensure that the desired performance is preserved by using the proper soldering, brazing, or welding techniques or materials. Finally, chemical reactivity between the fluid or gas and the storage containers and equipment must be studied. Wood or asphalt saturated with oxygen has been known to literally explode when subjected to mechanical shock. When properties of materials which are being considered for cryogenic use are unknown or not to be found in the known guides, experimental evaluation should be performed before the materials are used in the system.

Personnel hazards exist in several areas where cryogenic systems are in use. Exposure of personnel to the hazards of fire, high pressure gas, and material failures previously discussed must be avoided. Of prime concern is bodily contact with the extreme low temperatures involved. Very brief contact with fluids or materials at cryogenic temperatures is capable of causing burns similar to thermal burns from high temperature contacts. Prolonged contact with these temperatures will cause embrittlement of the exposed members because of the high water content of the human body. The eyes are especially vulnerable to this type of exposure, so that eye protection is necessary.

While a number of the gases in the cryogenic range are not toxic, they are all capable of causing asphyxiation by displacing the air necessary for the support of life. Even oxygen may have harmful physiological effects if prolonged breathing of pure oxygen takes place.

There is no fine line of distinction between the four categories of hazards, and they must be considered collectively and individually in the design and operation of cryogenic systems.

4.6.2.13.2. General Precautions

Personnel should be thoroughly instructed and trained in the nature of the hazards and the proper steps to avoid them. This should include emergency procedures, operation of equipment, safety devices, knowledge of the properties of the materials used, and personal protective equipment required.

Equipment and systems should be kept scrupulously clean and contaminating materials avoided that may create a hazardous condition upon contact with the cryogenic fluids or gases used in the system. This is particularly important when working with liquid or gaseous oxygen.

Mixing of gases or fluids should be strictly controlled to prevent the formation of flammable or explosive mixtures. As the primary defense against fire or explosion, extreme care should be taken to avoid contamination of a fuel with an oxidant, or the contamination of an oxidant by a fuel.

As a further prevention when flammable gases are being used, potential ignition sources must be carefully controlled. Work areas, rooms, chambers, or laboratories should be suitably monitored to automatically warn personnel when a dangerous condition is developing. When practical, it would be advisable to provide for the cryogenic equipment to be shut down automatically as well as to sound a warning alarm.

When there is a possibility of personal contact with a cryogenic fluid, full face protection, an impervious apron or coat, cuffless trousers, and high-topped shoes should be worn.

Watches, rings, bracelets, or other jewelry should not be permitted when personnel are working with cryogenic fluids. Basically, personnel should avoid wearing anything capable of trapping or holding a cryogenic fluid in close proximity to the flesh. Gloves may or may not be worn, but if they are necessary in order to handle containers or cold metal parts of the system, they should be impervious and sufficiently large to be easily tossed off the hand in case of a spill. A more desirable arrangement would be hand protection of the potholder type.

When toxic gases are being used, suitable respiratory protective equipment should be readily available to all personnel. They should thoroughly know the location and use of this equipment.

4.6.2.13.3. Storage

Storage of cryogenic fluids is usually in a well insulated container designed to minimize loss of product due to boil-off.

The most common container for cryogenic fluids is a double-walled, evacuated container known as a Dewar flask, made of either metal or glass. The glass container is similar in construction and appearance to the ordinary "Thermos" bottle. Generally, the lower portion will have a metal base which serves as a stand. Exposed glass portions of the container should be taped to minimize the flying glass hazard if the container should break or implode.

Metal containers are generally used for larger quantities of cryogenic fluids and usually have a capacity of 10 to 100 liters (2.6 to 26 gallons). These containers are also of double-walled evacuated construction and usually contain some absorbent material in the evacuated space. The inner container is usually spherical because that shape has been found to be the most efficient in use. Both the metal and glass Dewars should be kept covered with a loose-fitting cap to prevent air or moisture from entering, and to allow built-up pressure to escape.

Larger capacity storage vessels are basically the same double-walled containers, but the evacuated space is generally filled with powdered or layered insulated material. For economic reasons, the containers are usually cylindrical with dished ends, approximating the shape of a sphere, which is expensive to build. Containers must be constructed to withstand the weights and pressures that will be encountered and adequately vented to permit the escape of evaporated gas. Containers also should be equipped with rupture discs on both inner and outer vessels to release pressure if the safety relief valves should fail.

Cryogenic fluids with boiling points below that of liquid nitrogen (particularly liquid helium and hydrogen) require specially constructed and insulated containers to prevent rapid loss of product from evaporation. These are special Dewar containers which are actually two containers, one inside the other. The liquid helium or hydrogen is contained in the inner vessel, and the outer vessel contains liquid nitrogen which acts as a heat shield to prevent heat from radiating into the inner vessel. The inner neck as shown in the illustration, should be kept closed with a loose fitting, non-threaded brass plug which prevents air or moisture from entering the container, yet is loose enough to vent any pressure which may have developed. The liquid nitrogen fill and vent lines should be connected by a length of gum rubber tubing with a slit approximately 2.54 cm (1 inch) long near the center of the tubing. This prevents the entry of air and moisture, while the slit will permit release of the gas pressure. Piping or transfer lines should be double-walled evacuated pipes to prevent the loss of product during transfer.

Most suppliers are now using a special fitting to be used in the shipment of Dewar vessels. Also, an automatic pressure relief valve and a manual valve prevent entry of moisture and air, which will form an ice plug. The liquid helium fill (inner neck) should be reamed out before and after transfer and at least twice daily. Reaming should be performed with a hollow copper rod, with a marker or stop to prevent damaging the bottom of the inner container. Some newer style Dewar vessels are equipped with a pressure relief valve and pressure gauge for the inner vessel.

Transfer of liquids from the metal Dewar vessels should be accomplished with special transfer tubes or pumps designed for the particular application. Since the inner vessel is mainly supported by the neck, tilting to pour the liquid may damage the container, shortening its life or creating a hazard due to container failure at a later date. Piping or transfer lines should be so constructed that it is not possible for fluids to become trapped between valves or closed sections of the line. Evaporation of the liquid in a section of line may result in pressure build-up and eventual explosion. If it is not possible to empty all lines, they must be equipped with safety relief valves and rupture discs. When venting storage containers and lines, proper consideration must be given to the properties of the gas being vented. Venting should be to the outdoors to prevent an accumulation of flammable, toxic, or inert gas in the work area.

ACKNOWLEDGMENT

Reprinted with permission from the *Journal of the American Society of Safety Engineers,* Vol. 11(8), 15–19, August 1963, Chicago, IL.

ADDENDUM TO SECTION 4.6.2.13 (1988)

There are some applications in which Dewars with wide mouths are used, such as storage of certain biological materials. These come with a loosely fitting cap to prevent absorption of air and moisture into the liquid nitrogen, the refrigerant most frequently used in these Dewars. As briefly alluded to in the preceding article, a buildup of oxygen in liquid nitrogen containers over a period of time can become a problem if care is not taken to keep the cap on or to change the entire volume occasionally. If the liquid takes on a blue tint, it is contaminated with oxygen and should be replaced. The contaminated liquid should be treated as a dangerous, potentially explosive material. Most users fill Dewars from larger ones, usually by pressurizing the larger one with nitrogen from a cylinder, thereby forcing the liquid into the smaller one. In order to not waste liquid nitrogen by evaporation in a warm container, neither of the two Dewars are usually allowed to become totally empty, again leading to possible oxygen contamination. If these practices are continued for a sufficiently long time, the oxygen content of the cryogenic liquid may become dangerously high.

There are two relatively common ways to maintain a supply of liquid nitrogen at a facility, one being to have a large reservoir of several thousand liters capacity from which individual users fill their smaller Dewars. The boil-off from a large reservoir can be used to provide a supply of ultraclean "air" to laboratories to clean surfaces. Liquid nitrogen is also usually available, if reasonably close to a distributor, in 160-liter pressurized containers delivered directly to the laboratory. In either case, the quantities actually needed for most small laboratories can be obtained frequently enough to avoid having an elaborate piping and control system from a large central reservoir, with the associated problems of avoiding blockage of the system by ice plugs. There are, of course, applications for which automatically controlled systems are necessary that provide safety relief and warning devices.

REFERENCES

1.   *Cryogenics,* Marsh & McLendon Inc., Chicago, IL, 1962.
2.   *Industrial Gas Data,* Air Reduction Sales Co., Acton, MA.
3.   *Matheson Gas Data Book*, 47th ed., The Matheson Co., Inc. East Rutherford, NJ, 1961.
4.   *Precautions and Safe Practices for Handling Liquid Hydrogen,* Linde Company, New York, 1960.
5.   *Precautions and Safe Practices for Handling Liquified Atmospheric Gases,* Linde Company, New York, 1960.
6.   **Braidech, M.M.,** *Hazards/Safety Considerations in Cryogenic (Super Cold) Operations,* Conference of Special Risk Underwriters, New York, 1961.
7.   **Hoare, Jackson, and Kurti,** *Experimental Cryophysics,* Butterworths, London, 1963.
8.   **MacDonald, D.K.C.,** *Near Zero, An Introduction to Low Temperature Physics,* Anchor Books, Doubleday & Co., Inc., New York, NY, 1961.
9.   **Neary, R.M.,** *Handling Cryogenic Fluids,* Linde Company, New York, 1960.
10. **Scott, R.B.,** *Cryogenic Engineering,* D. Van Nostrand Company, Inc., Princeton, NJ, 1959.
11. **Timmerhaus, K.D., Ed.,** *Advances in Cryogenic Engineering,* Vol. 7, Plenum Press, New York, 1961.
12. **Vance, R.W., and Duke, W.M., Eds.,** *Applied Cryogenic Engineering,* John Wiley & Sons, Inc., New York, 1962.
13. **Zenner, G.H.,** Safety engineering as applied to the handling of liquified atmospheric gases, in *Advances in Cryogenics Engineering,* 6, Plenum Press, New York, 1960.
14. Cryogenic Safety. A Summary Report of the Cryogenic Safety Conference, Air Products, Inc., Allentown, PA, 1959.
15. **Spencer, E.W.,** Cryogenic safety, in *CRC Handbook of Laboratory Safety,* 2nd ed., Steere, N.V., Ed., CRC Press, Boca Raton, FL, 1971.

4.6.2.14. Cold Traps[*]

Cold traps are used in instrumentation and elsewhere to prevent the introduction of vapors or liquids into a measuring instrument from a system, or from a measuring instrument (such as a Mcleod gauge) into the system. A cold trap provides a very low-temperature surface on which such molecules can condense and improves pump-down (the achievable vacuum) by one or two magnitudes.

However, cold traps improperly employed can impair accuracy, destroy instruments or systems, and be a physical hazard. For example, many of the slush mixtures used in cold traps are toxic or explosive hazards, and this is not indicated in the literature.

The authors became aware of the deficiencies in tunnel instrumentation, where it was necessary to measure pressures in the micron to 760-torr[**] region. The instrumentation system used Stratham gauges for ambient pressure down to 100 to 150 torr or about 2 to

---

[*]   This section is taken from the article "Cold Traps," by Kaufman and Kaufman[4] in the second edition of this handbook.

[**] The torr is equal to a pressure of 1 mm Hg at standard condition.

3 psia and NRC alphatron gauges for pressures to 5×10^{-2} torr. To prevent calibration shifts and contamination of the NRC transducers by oil fumes from the vacuum pump and possible wind tunnel contaminants, a cold trap was placed in the line.

The cold trap was filled with liquid nitrogen, and the valve to the tunnel line shut off. When the valve was opened, cold gas shot out, shown by the condensation; the overpressure developed in the system destroyed the stratham straingage bridge, although it was not sufficient to rupture the transducer diaphragm. As no satisfactory explanation was forthcoming, a glass cold trap was procured and set up in a dummy system. The cause of the phenomenon soon became apparent: air in the trap and system lines was liquified in the trap. When the valve was opened, this liquid air was being blown into the warmer lines by atmospheric pressure: The resultant volatilization of liquid into gas was practically an explosion.

Nevertheless, cold traps are often the only satisfactory means of removing contaminants, although in ordinary experimental work the charcoal trap is occasionally acceptable. A charcoal trap will remove oil and condensable vapors so that pressures to 10^{-8} torr or better may be secured, but it presents a serious restriction on pumping speed.

The errors introduced by the water vapor, when measuring low pressures, depend on the vacuum gauge used. The presence of water vapor also affects the magnitude of vacuum that can be achieved. The equilibrium point of a dry-ice-acetone slush is $-78°C$ ($-108.4°F$), which although sufficient to trap mercury vapor effectively, does not remove water vapor; a temperature of at least $-100°C$ ($-148°F$) is required to eliminate water vapor or, alternatively, exposure to anhydrous phosphorous pentoxide (P_2O_5). This material is usually rejected for field use because of possible biological, fire, and explosive hazards: in absorbing water it produces heat and reacts vigorously with reducing materials.

Slush mixtures using liquid air and liquid oxygen were considered and dropped, either because of the explosive hazard or toxicity of the vapors or because they were not cold enough. Table 4.15 lists many common thermal transfer and coolant fluids with their hazards and limitations.

4.6.2.14.1. Virtual Leaks

If the cold trap is chilled too soon after the evacuation of the system begins, gases trapped will later evaporate when the pressure reaches a sufficiently low value. The evaporation of the refrigerated and trapped gases is not rapid enough to be evacuated by the system, but is enough to degrade the vacuum, producing symptoms very similar to those of a leak.

To avoid these virtual leaks, keep the trap warm until a vacuum of about 10^{-2} torr is obtained. The tip of the trap is then cooled until ultimate vacuum is reached, at which time the trap may be immersed in the coolant to full depth.

4.6.2.14.2. Safety Precautions

If liquid nitrogen is the coolant, liquid air can condense in the trap, inviting explosion. Liquid air, comprising a combination primarily of oxygen and nitrogen, is warmer than liquid nitrogen. Depending on the nitrogen content, air liquifies anywhere from $-190°C$ ($-310°F$) ($5°C$ warmer than liquid nitrogen) to $-183°C$ ($-297.4°F$) (liquid oxygen). If liquid nitrogen is used, the trap should be charged only after the system is pumped down lest a considerable amount of liquid oxygen condenses, creating a major hazard.

Table 4.15.   Thermal Transfer Fluids[a] Used with Instrumentation-Type Cold Traps

Element[b]	Temperature[c]		Hazard[d]			Remarks
	°C	°F	Inhalation Toxicity	Skin Toxicity	Explosive or Fire	
Glycerine, 70% by weight Pressure water 30%	-38.9	-38.0	None	Slight	Slight	Vapor 2.5 × 10⁻³ torr at 50°C (122°F)
Ethyl alcohol, dry ice	-78	-108	Moderate	Slight	Dangerous	—
Ethylene glycol 52.5% by volume, water 47.5%	-40	-40	None	Slight	Slight	—
Chloroform, dry ice	-63.5	-82	Extreme	Slight	Slight	Vapor pressure 100 torr at 22°C (71.6°F)
Liquid SO_2	-75.5	-103	Extreme	Extreme	None	Very dangerous
Methanol (methyl alcohol), dry ice	-78	-108	Slight	Slight	Dangerous	Vapor pressure 100 torr at 22°C (71.6°F) Ingestion very dangerous
Acetone, dry ice	-78	-108	Moderate	Slight	Dangerous	Vapor pressure 400 torr at 30°C (102.2°F)
Methyl bromide, dry ice	-78	-108	Extreme	Extreme	Moderate	Very dangerous
Fluorotrichloromethane (Freon 11), dry ice	-78	-108	Slight	Slight	None	—
Methylene chloride, dry ice	-78	-108	Moderate	Moderate	Slight	Very dangerous to eyes Vapor pressure 380 torr at 22°C (71.6°F)
Calcium chloride	-42	-44	Slight	Slight	None	—
Ethyl methyl ketone	-78	-108	Moderate	Moderate	Dangerous	—

a   Transfer fluids will freeze solid and become colder if subject to temperatures lower than their freezing point. A slush mixture is secured by lowering temperature, such as by the introduction of limited quantities of dry ice, until the mixture is quasi-frozen.

b   These materials are often sold under trade names (listed in the *CRC Handbook of Chemistry and Physics*). In general, any combination of elements shown was selected for the coldest mixture obtainable.

c   If the refrigerant is dry ice, the transfer fluid will not go below -78°C (-18.4°F), the temperature of solid CO_2.

d   The consensus is that many of these liquids, while hazardous at room temperature, are not hazardous when cooled, since their evaporation at low temperatures is fairly low. For utmost safety those noted to be dangerous should not be employed unless venting or other special precautions are taken. For greater detail, see Reference 2.

Handle any liquid gas carefully; at its extremely low temperature, it can produce an effect on the skin similar to a burn. Moreover, liquified gases spilled on a surface tend to cover it completely and intimately, and therefore cool a large area.

The evaporation products of these liquids are also extremely cold and can produce burns. Delicate tissues, such as those of the eyes, can be damaged by an exposure to these cold gases which is too brief to affect the skin of the hands or face.

Eyes should be protected with a face shield or safety goggles (safety spectacles without side shields do not give adequate protection). Gloves should be worn when handling anything that is or may have been in contact with the liquid; asbestos gloves are recommended (this is no longer true because of the concern about airborne asbestos fibers from products containing asbestos. Gloves made of an artificial material such as Kevlar™ or Zetex™ are recommended as an alternative), but leather gloves may be used. The gloves must fit loosely so that they can be thrown off quickly if liquid should spill or splash into them. When handling liquids in open containers, high-top shoes should be worn with trousers (cuffless if possible) worn outside them.

Stand clear of boiling and splashing liquids and its issuing gas. Boiling and splashing always occurs when charging a warm container or when inserting objects into the liquid. Always perform these operations *slowly* to minimize boiling and splashing.

Should any liquified gases used in a cold trap contact the skin or eyes, immediately flood that area of the body with large quantities of unheated water and then apply cold compresses. Whenever handling liquified gases, be sure there is a hose or a large open container of water nearby, reserved for this purpose. If the skin is blistered, or if there is any chance that the eyes have been affected, take the patient immediately to a physician for treatment (call for emergency medical aid; normally rescue squads can be in immediate contact with an emergency room physician by radio).

Oxygen is removed from the air by liquid nitrogen exposed to the atmosphere in an open Dewar. Store and use liquid nitrogen only in a well ventilated place; owing to evaporation of nitrogen gas and condensation of oxygen gas, the percentage of oxygen in a confined space can become dangerously low. When the oxygen concentration in the air becomes sufficiently low, a person loses consciousness without warning symptoms and will die if not rescued. The oxygen content of the air must never be allowed to fall below 16%.

The appearance of a blue tint in liquid nitrogen is a direct indication of its contamination by oxygen, and it should be disposed of, using all the precautions generally used with liquid oxygen. Liquid nitrogen heavily contaminated with oxygen has severe explosive capabilities. In addition, an uninsulated line used to charge Dewars will condense liquid air; liquid air dripping off the line and revaporizing causes an explosive hazard during the charging operation.

If the cold trap mixture is allowed to freeze, and the cold trap becomes rigid, slight movement in other parts of the apparatus could result in breakage of the trap or other glassware.

If a gas trap has to be lifted out of the Dewar cold bath for inspection, it will be difficult to reinsert into the slush. Therefore, it is preferable to use a liquid that will not freeze at $-78.5°C$.

ACKNOWLEDGMENT

Reprinted with permission of Rimbach Publications, Pittsburgh, Pennsylvania, from *Instrument and Control Systems,* Vol. 36, pp. 109–111, July 1963.

REFERENCES

1. **Strong, J., Neher, H.V., Whitford, A.E., Cartwright, C.H., and Hayward, R.,** *Procedures in Experimental Physics,* Prentice-Hall, Inc., New York, 1938.
2. **Sax, N.I.,** *Dangerous Properties of Industrial Materials*, Reinhold Publishing Co., New York, 1961.
3. **Dushman, L.,** *Scientific Foundation of Vacuum Techniques,* John Wiley & Sons, New York, 1962.
4. **Kaufman, A.B. and Kaufman, E.N.,** Cold traps, in *CRC Handbook of Laboratory Safety,* 2nd ed., Steere, N.V., Ed., CRC Press, Boca Raton, FL, 1971, 510.

4.6.2.15. Care and Use of Electrical Systems

Some of the problems associated with electrical systems have been covered in previous sections, such as 3.1.7. There may be some unavoidable repetition in this section, which will be primarily concerned with the safe use of electricity rather than the characteristics of individual items, although there will be a brief discussion of generic problems associated with the design of equipment. Most of the hazards associated with the use of electricity stem from electrical shock, resistive heating, and ignition of flammables, and most of the actual incidents occur because of a failure to anticipate all of the ways in which these hazards may be evoked in a laboratory situation. This lack of appreciation of the possible hazards may be reflected in the original choice of suitably safe electrical equipment or improper installation of the equipment. In some instances the choice of equipment may simply involve a continued use of equipment on hand under conditions for which it is no longer suitable, so that safety specifications are not really considered. Often, this is due to a familiarity with the existing resources rather than a deliberate choice.

Part of the problem is a feeling that questionable electrical practices routinely followed at home as well as in the laboratory are actually safe unless you do something "really bad," such as standing in water while in contact with an electrically active wire, or a similar feeling about wiring, "Just hook it up with the extension cord bought on sale at the department store." When asked about a number of similar practices involving multiple connections to a single outlet or the use of extension cords, most people will answer that they know that they should not do it, but see no real harm.

Two major electrical factors need to be considered in the choice of most electrical items of equipment. The equipment needs to be selected so that it will not provide a source of ignition to flammable materials, and it should be chosen so as to minimize the possibility of personnel coming into contact with electrically live components. This latter problem will be addressed first.

4.6.2.15.1. Electrical Shock

OSHA has included the relevant safety portions of the National Electrical Code in 29 CFR 1910, Subpart S. This regulatory standard, as are many other sections of the OSHA regulations, is primarily oriented toward industrial applications, but it does speak directly to the problem of preventing individuals from coming into contact with electricity. Live parts of electrical equipment operating at 50 volts or more must be guarded against accidental contact. Indoor installations that contain circuits operating at 600 volts or more, and accessible to electrically untrained persons, must have the active components within metal enclosures or be located within a space controlled by a lock. The higher voltage equipment also must be marked with appropriate warning signs. Access points to spaces in which exposed electrically live parts are present must be marked with conspicuous warning signs which forbid unqualified persons to enter.

The effects of electricity on a person depend upon the current level and, of course, on physiological factors unique to the individual. Table 4.16 gives typical effects of various current levels for 60-Hz currents for an average person in good health.

Several things affect the results of an individual incident. The duration of the current is important. In general, the degree of injury is proportional to the length of time the body is part of the electrical circuit. A suggested threshold is a product of time and energy of 0.25 watt-seconds for an objectionable level. The voltage is important because, for a given resistance R, the current I through a circuit element is directly proportional to the applied voltage V.

$$I = V/R \tag{1}$$

If the contact resistance to the body is lowered so that the total body resistance to the flow of current is low, then even a modest applied voltage can affect the body. The condition of the skin can dramatically alter the contact resistance. Damp, sweaty hands may have a contact resistance which will be some orders of magnitude lower than dry skin. The skin condition is more important for low voltage contacts than for those involving high voltages, since in the latter case the skin and contact resistance break down very rapidly. The remaining resistance is the inherent resistance of the body between the points of contact, which is on the order of 500 to 1000 ohms. As can be noted in Table 4.16, the difference in a barely noticeable shock and a potentially deadly one is only a factor of 100. For an individual with cardiac problems, the threshold for a potentially life-threatening exposure may be even lower. The major danger to the heart is that it will go into ventricular fibrillation due to small currents flowing through it. In most cases, once the heart goes into ventricular fibrillation, death follows within a few minutes.

Even if an individual survives a shock episode, there may be immediate and long-term destruction of tissue, nerves, and muscle due to heat generated by the current flowing through the body. The heat generated is basically resistive heating such as would be generated in heating coils in a small space heater, with the exception that the resistive elements are the tissues and bones in the body. The power, P, or heat is given by

$$P = I^2R \tag{2}$$

The scope of the effects of external electrical burns is usually immediately apparent, but the total effect of internal burns may become manifest later on by losses of important body functions due to the destruction of critical internal organs, including portions of the nervous system.

Several means are available to prevent individuals from coming into contact with electricity, in addition to exclusion of unqualified personnel from space, as mentioned in the introduction to this section. These include insulation, grounding, good wiring practices, and mechanical devices. Before addressing these latter options, it might be well to briefly discuss the concept of a qualified person.

Certainly a licensed electrician would in most cases be a qualified person, and a totally inexperienced person would just as clearly not be a qualified individual. There is no clear definition of the training required to be "qualified" to perform laboratory electrical and electronic maintenance. As a minimum, such training should include instruction in the

Table 4.16.   Effects of Electrical Current in the Human Body

Current (milliamperes)	Reaction
1	Perception level, a faint tingle.
5	Slight shock felt; disturbing but not painful. Average person can let go. However, vigorous involuntary reactions to shocks in this range can cause accidents.
6 to 25 (women) 9 to 30 (men)	Painful shock, muscular control is lost. Called freezing or "let-go" range.*
50 to 150	Extreme pain, respiratory arrest, severe muscular contractions, individual normally cannot let go unless knocked away by muscle action. Death is possible.
1,000 to 4,300	Ventricular fibrillation (the rhythmic pumping action of the heart ceases). Muscular contraction and nerve damage occur. Death is most likely.
10,000-	Cardiac arrest, severe burns, and probable death.

*   The person may be thrown away from the contact if the extensor muscles are excited by the shock.

consequences of electrical shock, basic training in wiring color codes (so as to recognize correct leads), familiarity with the significance of ratings of switches, wiring, breakers, etc., simple good wiring practices, and recognition of problems and poor practices, such as frayed wiring, wires underfoot, wires in moisture, overloaded circuits, use of too small or wrong type of conductors, poor grounding procedures, and improper defeat of protective interlocks. This would not make an individual a licensed electrician, which requires extensive training and experience, but would reduce the number of common electrical errors.

Insulation is an obvious means of protecting an individual against shocks. In general, good wiring insulation is the most critical, particularly that of extension cords, which are often abused. Insulation must be appropriate for the environment, which may involve extremes of temperature or exposure to corrosive vapors or solvents. The insulation itself may need to be protected by a metal outer sheath, or the wire may need to be installed in conduit.

As it ages, insulation may become brittle and develop fine cracks through which moisture may seep and provide a conductive path to another component or to a person who simply touches the wire at the point of failure. Many plastic or rubber insulating materials will soften with heat and, if draped over a metal support, may eventually allow the wire to come into contact with the metal, thus rendering the metal electrically active, if it does not first cause other problems. Extension cords, as noted above, are particularly susceptible to abuse. They are often carelessly strewn across the floor or furniture. On the floor, they may be walked upon, equipment may be rolled across them, or they may become pinched between items of furniture. Extension cords should only be used as temporary expedients, but if they are used, they should be treated as any other circuit wiring, put out of harm's way, and properly supported on insulators separated by distances not to exceed 10 feet. Defective extension cords with badly deteriorated insulation should be discarded.

Insulation is not used solely to protect wiring. Insulation in the form of panels supporting printed circuits may break if excessive force is applied. If an arc temporarily flashes across an insulating surface, a carbonized conducting path may be permanently established on the surface which could render an external component such as a chassis mounting screw "hot." Care should be taken with all electrical equipment, especially older items, to ensure that the integrity of the insulation has been maintained.

Proper grounding of equipment is another requirement to ensure that components are not electrically live. Most equipment for use with 120-volt circuits comes with a three-wire power cord, which requires a mating female connector at the power source, many of which are designed so that the neutral, hot, and ground connections can be readily identified and matched. The ground wire, which is either green or perhaps green with yellow stripes, is always connected to the female socket which accommodates the round prong on the male connector. The neutral circuit wire which normally completes the circuit for the equipment is usually white or gray. The socket and corresponding male connector are often wider than the connections for the hot wire. The hot wire is usually covered with a black insulator, although red may also be used. Where there are both red and black wires, usually both will be hot wires. Some equipment is double insulated, and does not have the third ground wire in the power connector. Usually, these will have a polarized connector, so that the neutral and hot wires will be properly oriented. Older circuits unfortunately do not always provide the proper connections and should be replaced. If this is not feasible, the third wire on the power connector, if one is present, should be directly connected to a good quality ground.

Auto-transformers, which may be used to supply variable voltages to heating devices, may be connected in such a way that either outlet line may be high with respect to ground. They should be purchased with a switch which breaks the connection of both outlet sockets to the power input line, or they should be rewired with a double-pole power switch to accomplish this.

The quality of all ground connections (and of all connections) needs to be good. This is often taken for granted, but the connection may vibrate or work itself loose, or a careless worker may fail to tighten a connection. In such cases, a significant difference of potential may arise between two different items of equipment. This can be enough to give rise to a discernible shock for a person coming into contact simultaneously with both pieces of equipment, and in some cases can cause damage to the equipment if they are interconnected. A careful researcher should have the electrical circuits checked for the resistance to ground of all the wiring in his facility. A ground with a resistance of 100 ohms will be at a difference of 10 volts with respect to ground if a current of 100 milliamperes were to flow through the ground connection. Good quality grounds with resistances of a few ohms are easily achievable with care.

Adapters or "cheaters" can be used to allow power cords with three wires to be connected to wiring providing only two wires, or used to avoid connecting the third wires to ground. There are only a few exceptions which would make this an acceptable practice; one is where an alternative direct connection to ground is provided for the equipment. Another would be in a very few occasions in which even the difference of a few millivolts between the separate ground connections would affect the experimental data signals. In the latter case, ground connections can be made directly between the components with a single connection being made to the building ground. Except in clearly defined situations where their use is clearly made safe, these adapters should not be used.

Simple devices such as fuses, circuit breakers, and ground fault interrupters are available to cut off equipment when they overload or short out or an imbalance develops between the input and output current from a device or circuit. More sophisticated devices can also be used to determine a problem, such as a redundant heat detector used to deactivate a circuit serving a still, condenser, or heat bath should the temperature become too high.

Fuses and circuit breakers are the simplest devices used to shut off a circuit drawing too much current. A fuse inserted in one of the circuit legs functions by melting at a predetermined current limit, and the breaker by mechanically opening the circuit. The latter device is more flexible in that it can be reset while the fuse must be replaced. A ground fault interrupter (GFI), on the other hand, specifically can protect an individual who comes into contact with a live component. The individual's body and the wires become parallel circuits through which a fraction of the total current flows. The amount through the body makes the two normal halves of the circuit out of balance, which the GFI detects, and causes it to break the circuit. A GFI can detect a difference on the order of 5 milliamperes and can break the circuit in as little as 25 milliseconds. A review of Table 4.16 would show that the contact might be barely noticeable but would cause no direct harm because of the short duration of the current flow. Although GFIs are generally used in the construction industry, they would serve a useful purpose in laboratories, such as, where moisture would be a problem. Any laboratory containing equipment operating at high voltages should have each electrical outlet protected by a GFI, supplemented by a master disconnect switch in an obvious and easily accessible space. It is critical to remove the current source in as short a time as possible.

The best defenses against electrical shock injuries are good work practices, as invoked by using good judgment and exercising care appropriate to the risk. The basic principles embodied in the OSHA lock-out, tag-out provisions of their electrical standard should be followed. These are basic common sense. Maintenance of electrical equipment or wiring should be done only with the system deenergized unless it is essential that the circuit be active for the required maintenance. In the latter case specific care is taken to come into contact (if necessary) with only one side of a circuit, so that the circuit cannot be completed through the body. Procedures should be followed to confirm that power to the system has been disabled and remains so during the duration of the maintenance activity or, alternatively, if the circuit must remain powered, that a second person is available to disable the circuit and assist in the event of an incident. Formal lock-out procedures are recommended where high voltage circuits are involved. The tools used to perform maintenance should be in good condition. Barriers may be needed to isolate live circuits in the maintenance area. Good judgment should be used to determine safe distances, to not use metal ladders, etc., where it would be possible to contact a hot circuit. In some cases, it might be necessary to use rubber gloves and gauntlets, insulating mats, and hard hats certified for electrical protection.

A relatively simple protective stratagem which should be followed by anyone working with or handling live electrical circuits is to remove all conducting jewelry, specifically items on or near the hands such as rings, watches, and bracelets or necklaces which may dangle from the neck and complete a circuit to the neck. If work activities involving direct contact with electrical components are infrequent, removing these items at the time may suffice. If the work is a normal activity, the practice of avoiding wearing metallic objects should be routine to avoid having to remember to remove them. Avoiding the use of conductive items in the vicinity of electricity should extend to any object which might come into contact with the circuit. Many tools must be metallic, but any tool used in electrical work should have insulating materials in those areas normally in contact with the hands.

Interlocks should never be bypassed by the average laboratory worker. If it becomes necessary, the decision to do so must be done with the knowledge of all persons who might be affected by the decision. Bypassing an interlock should not be a decision permitted for an inexperienced graduate student or new employee. Whenever an interlock is bypassed, a definite procedure, requiring positive confirmation that all personnel are no longer at risk, must be adopted and in place. This may involve actual locks, for which only the responsible person has a key, tags which cannot be removed without deliberately breaking a seal, an alarm, or a combination of the above. However, no preventive procedure should depend upon the continued functioning of a single device, such as a microswitch, which may fail in such a way as to defeat the alarm or interlock.

Each circuit should be clearly identified and labeled to correspond to a circuit breaker in a service panel. Access to these service panels should be provided to most laboratory employees, but they should not be permitted to remove the panel covers protecting the wiring. No closet containing an electrical service panel should be allowed to be used for a storage closet by laboratory personnel. Access to the panel should not be blocked by extraneous items, and accidental contact with the wires should not be possible.

One of the most effective safety practices, as well as one highly conducive to productivity, is a definite scheduled program of preventive maintenance. Each item of equipment should be periodically removed from service, carefully inspected and calibrated, any faults or indications of deterioration repaired, and tagged with the date of review and the name of the maintenance person, if more than one technician could have been responsible. A permanent file or maintenance log on each major item of equipment is useful for identifying trends or weak components.

Finally, in a facility in which electrical injuries are a reasonable possibility, it is strongly recommended that at least some permanent personnel be trained in CPR and the measures to be taken should a person receive a severe shock. Individuals should also be trained to effect a rescue without themselves becoming a casualty. If, for example, a live wire is lying across a person and the circuit cannot be readily broken, they should be instructed to find a meter stick or some other insulated device to lift the energized wire from the victim, or use rubber gloves or other insulator in attempting to loosen a person from a circuit.

4.6.2.15.2. Resistive Heating

This is one of the two major electrical sources of ignition of flammable materials in a laboratory; the other is sparks. Electrical heating can occur in a number of ways—poor connections, undersized wiring or electrical components (or alternatively, overloaded wiring or components), or inadequate ventilation of equipment.

Equation 2 in the previous section shows that the power or heat released at a given point in a circuit is directly proportional to the resistance at that point. A current of 100 milliamperes through a connection with a resistance of 0.1 ohms would generate a localized power dissipation at that point of only 10 milliwatts, while a poor connection of 1000 ohms resistance would result in a localized power dissipation of 100 watts. The former would normally cause no problems, while the latter might raise the local temperature enough to exceed the ignition temperature of materials in the vicinity. Poor or loose connections have, in fact, caused many fires due to just such localized heating. An alligator clip used to attach a grounding wire is a good example of a potentially poor connection. Similarly, a contact which has been degraded by a chemical, a wire that has been insecurely screwed down, or the expansion and contraction of a wire such as aluminum may in time result in this kind of problem.

Overheating of switches, fixtures, and other electrical components due to electrical overloads can be avoided very simply by reading the electrical specifications for the component, usually printed or embossed on the item, and complying with the limitations. If a switch is rated to carry 7 amperes at 120 volts, it will not survive indefinitely in a circuit in which it is carrying 30 amperes.

Each size or gauge wire is designed to carry a maximum amount of current. This is based on the voltage drop per unit length and the amount of power dissipated in the wire. The voltage drop should not exceed 2 to 5% due to wiring resistance. A 5% drop in a 120-

volt circuit supplying an item of equipment would mean an actual voltage at the connection to the equipment of only 114 volts. The heat developed in an overloaded circuit may heat the wiring to a point where the insulation may fail or in extreme cases actually catch on fire. Even moderate overheating, continued long enough, will probably cause an eventual breakdown in the insulation. In addition, any energy dissipated in the wiring is wasted energy.

Table 4.17 gives the maximum current for copper wire of various sizes, the resistance, voltage drop, and power loss per 50 ft (about 15.25 m) of line (the latter two values computed for a wire carrying the maximum rated current). Most inexpensive extension cords purchased at a department store are made of either 16 or 18 gauge wire. As can be seen from the table, inexpensive extension cords do not carry sufficient current to be useful for providing power to more than a few instruments at most, when properly used. Over-loading them will cause a larger voltage drop and power dissipation (heating) in the wire. Although extension cords made of wire which is too small will probably not immediately fail in most applications, they are not suitable for continued use. The lower available voltages can result in damage to equipment or failure of relays in control circuits, as their magnetic fields become weaker.

Additional electrical load is a problem for extension cords and for the permanent wiring as well, if multiple outlet plugs are used in a socket. Virtually every safety professional has at least one photograph or slide of several multiple plugs plugged into each other, all drawing current from a single socket. The result will usually be an overheated fixture and wiring, as well as a lower voltage at the plug. This process can continue all the way back to service panels and to the power supply to the facility. Examples of breaker panels almost too hot to touch are, unfortunately, fairly common. Low overall supply voltages are becoming common in older facilities as the larger electrical loads of laboratories replace the lesser loads of classrooms in many academic institutions. Table 4.18 lists a number of common laboratory devices with the current and power requirements for representative units. Note that some of these normally require 208- to 240-volt circuits. Some may also require connection to three-phase current, which if not done correctly can result in an accident or, at best, poor performance (a relatively common mistake is to wire a three-phase motor so that it operates backwards). A fume blower miswired in this way would result in minimal exhaust velocity.

Devices with resistive heating elements, such as furnaces, heat guns, hot plates, and ovens should be configured in such a way that personnel cannot come into contact with an electrically active element, nor should volatile solvents be used in the proximity of such devices (or in them, as in an oven) where the temperature will exceed the ignition temperature of the solvent.

4.6.2.15.3. Spark Ignition Sources

Induction motors should be used in most laboratory applications instead of series wound electric motors, which generate sparks from the contacts of the carbon brushes. Sealed explosion-proof motors can also be used but are expensive. It is especially important to use nonsparking motors in equipment which result in substantial amounts of vapor, such as blenders, evaporators, or stirrers. Equivalent ordinary household equipment or other items such as vacuum cleaners, drills, rotary saws, or other power equipment are not suitable for use in laboratories where solvents are in use. Blowers used in fume exhaust systems should at least have nonsparking fan blades, but in critical situations with easily ignitable vapors being exhausted, it may be worth the additional cost of a fully explosion-proof blower unit.

Table 4.17.   Electrical Characteristics of Wire per 50 Feet

Wire Size	Resistance (Ω)	Maximum Amperes	Voltage Drop (V)	Power Loss (W)
18	0.3318	7	4.6	32
16	0.2087	10	4.2	42
14	0.1310	15	3.9	59
12	0.0825	20	3.3	61
10	0.0518	30	3.1	93
8	0.0329	40	2.6	105
6	0.0205	55	2.3	124
4	0.0129	70	1.81	127
3	0.0103	80	1.64	131
2	0.00809	95	1.53	138
1	0.00645	110	1.42	156
0	0.00510	125	1.27	159

Any device in which an electrically live circuit makes and breaks, as in a thermostat, an on-off switch, or other control mechanism, is a potential source of ignition for flammable gases or vapors. Special care should be taken to eliminate such ignition sources in equipment in which the vapors may become confined—as already discussed for refrigerators and freezers, but also possible in other equipment such as blenders, mixers, and ovens—or the units should not be permitted to be used with or in the vicinity of materials which emit potentially flammable vapors.

REFERENCES

1.   OSHA, General Industry Standards, Subpart S-Electrical 29 CFR 1910.301-339.
2.   National Electrical Safety Electrical Code, ANSI C2, American National Standards Institute, Inc., New York.
3.   *Prudent Practices for Handling Hazardous Chemicals in Laboratories,* National Academy Press, Washington, D.C., 1981.
4.   **Lockwood, G.T.,** Protective lockout and tagging of equipment, in *CRC Handbook of Laboratory Safety,* 2nd ed., Steere, N.V., Ed., CRC Press, Inc., Boca Raton, FL, 1971, 511.
5.   **Electronic Industries Association,** Grounding electronic equipment, in *CRC Handbook of Laboratory Safety,* 2nd ed., Steere, N.V., Ed., CRC Press, Inc., Boca Raton, FL, 1971, 516.
6.   **Dalziel, C.F.,** Deleterious effects of electrical shock, in *CRC Handbook of Laboratory Safety,* 2nd ed., Steere, N.V., Ed., CRC Press, Inc., Boca Raton, FL, 1971, 521.
7.   **Ehrenkranz, T.E., and G.W. Marsischky,** Electrical equipment, wiring, and safety procedures, in *CRC Handbook of Laboratory Safety,* 2nd ed., Steere, N.V., Ed., CRC Press, Inc., Boca Raton, FL, 1971, 528.
8.   **Ehrenkranz, T.E.,** Explosion-proof electrical equipment, in *CRC Handbook of Laboratory Safety,* 2nd ed., Steere, N.V., Ed., CRC Press, Inc., Boca Raton, FL, 1971, 540.

Table 4.18.   Electrical Requirements of Some Common Laboratory Devices

Instrument	Current (amperes)	Power (watts)
Balance (electronic)	0.1–0.5	12–60
Biological safety cabinet	15	1,800
Blender	3–15	400–1,800
Centrifuge	3–30	400–6,000
Chromatograph	15	1,800
Computer (PC)	2–4	400–600
Freeze dryer	20	4,500
Fume hood blower	5–15	600–1,800
Furnace/oven	3–15	500–3,000
Heat gun	8–16	1,000–2,000
Heat mantle	0.4–5	50–600
Hot plate	4–12	450–1,400
Kjeldahl digester	15–35	1,800–4,500
Refrigerator/freezer	2–10	250–1,200
Stills	8–30	1,000–5,000
Sterilizer	12–50	1,400–12,000
Vacuum pump (backing)	4–20	500–2,500
Vacuum pump (diffusion)	4	500

9.   Controlling Electrical Hazards, U.S. Department of Labor, Occupational Safety and Health Administration, Washington, D.C., OSHA-3075, 1983.

10.   Electrical Hazard, U.S. Department of the Interior, National Mine Health and Safety Academy, Safety Manual No. 9.

4.6.2.16. Glassware

It is recommended that most laboratory glassware be made of borosilicate glass. Pyrex™ is one of the most familiar brand names under which this type of glass is known. Borosilicate glass has many of the desirable characteristics of pure silica glass. It has good chemical durability and a low temperature expansion coefficient. Exceptions to this recommendation are reagent bottles, stirring, tubing, and any other glass item where the value of a low temperature expansion glass is not needed.

Breakage is the major safety concern with glassware. Bottles often break when dropped. The seriousness of this type of accident can be reduced by purchasing reagents in glass containers coated with plastic. This largely eliminates the danger of cutting oneself on the sharp edges since the broken container normally will be held more or less together by the film, and it should reduce the amount of liquid leaking so that the problem of harm due to splashing liquids is also minimized. Highly corrosive materials which for any reason are not in these safety containers should always be transported in rubber or plastic carriers designed for the purpose.

Laboratory systems routinely are made of glass flasks, other glass components, and glass tubing. These systems may explode if the internal pressures become too high or implode if the system is evacuated. Many circumstances may give rise to overpressure, a reaction which escalates beyond that anticipated, an explosion, a too-violent reaction, or excessive heat, among others. When possible, spherical containers should be used where pressurized systems (over- or underpressure) are constructed, or if not, then the reduced strength of nonspherical containers should be compensated for by using thicker glass. If the system is to be evacuated, it is common practice to tape the glass containers to reduce the problem of flying glass should the container implode. Flying glass fragments with razor-sharp edges are extremely dangerous. Large systems, especially, should be contained in a surrounding wire cage for the same reason. Glass systems can be set up within a fume hood and protected with explosion barriers as additional protection.

Glass under strain is very vulnerable to sharp blows, but strains can also release spontaneously, especially with rising temperature. In some instances, the release will cause the glass under stress to crack. Systems should be put together to minimize additional mechanical strains placed on the components. The strains in glass can also release spontaneously for no apparent cause. In one instance in which the author was personally involved, a gallon jug of a hazardous chemical was delivered to the secretaries in a research-oriented department. No one was at the laboratory to receive it so it was left in the departmental mail room. About 3 hours later, the jug spontaneously split, the chemical ran out, dissolved the plastic packaging protective material, ran into the carpet and the floor underneath, and dripped through into an office area below, damaging the floor there as well. The material was toxic, so evacuation was necessary; the cleanup required working in protective clothing and the use of respiratory protection. The direct cost to repair the damage was on the order of $2000 to 3000, with an additional comparable cost in manpower. Fortunately, only a handful of persons were in the building at the time, and no one suffered an acute exposure.

Glass systems under either positive or negative pressure are very dangerous due to the inability of the glass to withstand impact. The force on the surface of an evacuated container may be very large. For example, the net force on an evacuated spherical container 10 cm (4 inches) in diameter is equal to the force equivalent to the weight of an 81-kg (179-lb) mass. There is essentially no difference in this force for a good vacuum and a poor one. The force exerted on a container with a poor vacuum equivalent to 1 mm of mercury is within 0.13% of the force on a container with a vacuum several orders of magnitude better. The forces on high pressure systems can, of course, be much higher than at atmospheric pressures. Pressures in gas cylinders can be 200 times atmospheric pressure or more, and hence never should be attached to most glass research systems except through a regulator valve. The existing forces on pressurized glass systems make them unusually vulnerable to other factors, such as a sharp blow, which they might otherwise withstand. A warning sign should be placed on systems under pressure to alert personnel of the unusual danger. A wire cage, mentioned earlier, to contain flying glass also can protect the glass from external blows. Additional protection can be provided by setting glass systems up in fume hoods and keeping the sashes closed. Walk-in hoods can be used for larger systems.

Anything that weakens the strength of the glass in a system under pressure increases the risk of either an implosion or explosion. For example, some laboratories use metal evaporating systems under a large bell jar to deposit metal films. A current is passed through a small crucible or "boat," raising its temperature to a sufficient level to evaporate a metal placed within it. In order to get a good deposition of the metal, a good vacuum needs to be established. Although there usually is not sufficient air in the bell jar to transfer heat by either conduction or convection, substantial radiative transfer does occur. In order to get good evaporation, in most cases the metal must be heated to a brilliant white-hot temperature. This corresponds to a temperature on the order of 1500°C (2700°F) or better, well above the point at which borosilicate glass begins to soften. If the temperature is maintained at this level for a substantial length of time, the strength of the glass bell jar can be impaired with a resulting implosion. Metal-film evaporating systems should always be enclosed with a sturdy wire cage during use.

Broken glassware offers additional opportunities for injury due to cuts from the sharp fragments. Washing glassware offers the opportunity for cutting oneself on a broken flask or beaker. Accidentally shattering a glass container while holding it can also represent an opportunity for injury. Cleaning up broken glass must be done carefully. Note that broken glassware placed into ordinary trash represents an unnecessary and unacceptable hazard to an unsuspecting maintenance worker. Custodial workers routinely transfer trash to plastic garbage bags which offer no protection against broken glass. All broken glassware should be placed in a sturdy kraftboard box, taped closed, with the words **BROKEN GLASS** written prominently on the outside before being placed beside, not in, the ordinary trash containers.

Working with glass tubing offers a number of opportunities for injury. A section of tubing can be cut to length by scribing the tubing with a triangular file for approximately one third of its circumference, then wrapping it with cloth and exerting sideways pressure on the opposite side of the tubing with the thumbs placed a bit to each side of the filed line. The tubing should snap cleanly. If it does not, the file cut should be deepened and the procedure repeated. Multiple file marks may increase the chance of an irregular break which may have sharp edges. Fire polishing of the tubing should be done in every case to avoid cuts from any such edges.

Inserting glass tubing into stoppers or flexible plastic or rubber hoses is the source of one of the most common laboratory accidents with glass. Usually the cause is trying to force the tubing into the hole with no hand protection. Serious cuts can occur in such cases. The worker should either wear leather gloves or protect the hand holding the tubing by wrapping the tubing with a cloth and wrapping cloth around the hand holding the flexible tubing or stopper.

For flexible tubing, the hole size is predetermined by selecting the plastic or rubber tubing to have an internal diameter just slightly smaller than the glass tubing to which it is to be mated. To cut a hole in a rubber stopper, a well-sharpened borer just a size smaller than one which will slide over the glass tubing should be used. The stopper should be held only with the fingers. Wrapping the entire hand around it to get a "better grip" offers an opportunity to cut that hand with the borer. Holding the stopper in that manner should not be necessary. The borer should be lubricated with either water or glycerol. The hole should be cut by exerting a steady pressure on the borer while using a smooth rotary motion to slice through the stopper. The index finger should be placed along the barrel of the borer and close to the stopper. In the beginning, this will limit the distance the borer will travel if a slip occurs, while it will limit the distance the borer will extend beyond the stopper when penetration occurs.

Once a clean, well-shaped hole is prepared in a stopper, or if plastic or rubber tubing is to be attached to a section of glass tubing, the following steps should be taken:

1. Lubricate the glass tubing with stopcock grease, or other suitable lubricant. Use leather gloves to protect the hands or, as mentioned earlier, protect the hand holding the glass by wrapping a cloth around the glass tubing, and protect the other hand by wrapping it with a piece of toweling or other bulky cloth.
2. Hold the glass tubing an inch or two from the end.

3. Insert the glass into the hole, using moderate inward pressure and a slight twisting motion. Neither the forward pressure nor the twisting force should be excessive to avoid breaking the glass.
4. Continue the process in step 3 until the glass has been inserted to the desired or needed distance.

If a cut occurs, despite all the precautions, bleeding can be controlled in most cases by placing pressure directly on the wound, preferably by placing a clean dressing over the wound and holding it tightly in place while seeking first aid.

4.6.2.16.1. Other Safety Problems

It is relatively easy to heat and bend pieces of glass tubing to achieve a desired shape or to draw out a piece of tubing to diminish the internal diameter of the tube. Working glass to create complicated shapes for laboratory applications, however, is best left to the professional glassblower. In order for the shapes to be free of strains which could cause the glass to fail under stress, all of the components should be properly annealed, which requires experience and the correct equipment to do properly.

There are two special health problems associated with fabricating glass, other than the danger of cutting oneself. When a glassblower is configuring a glass component, he frequently employs a glass lathe which has some asbestos components. This asbestos tends to become quite friable with time, and individuals working near the lathe can be exposed to airborne asbestos fibers. Alternative insulating materials, such as Zetex™, should be used or the glass worker and nearby personnel should wear appropriate respiratory protection. The use of localized exhaust to remove the asbestos fibers is usually not practical since the moving air will cause problems with the work.

The second special problem is primarily associated with the construction of relatively large components of silica glass. The temperature at which fused silica can be conveniently worked is about 1580°C (2876°F), while ordinary borosilicate glass can be worked at a temperature nearly 500°C (900°F) lower. The formation of nitrogen dioxide from the air is significant at the higher temperatures involved in fabricating silica components.

Both of these problems can be alleviated by ensuring that the glassblowing shop or work area is well ventilated. Because heat rises, canopy hoods placed directly a few feet above the workstations will be effective in trapping the gases, fumes, and even particulates entrained in the rising hot air. This is a major exception to the previous recommendation against canopy hoods.

Glassware Cleaning—Cleaning of glassware in laboratories is an essential part of laboratory procedures. In most cases a simple cleaning with soap and water is sufficient, but in some cases chemical cleaning is necessary. The use of strong chemical agents such as sulfuric acid, perchloric acid, chromic acid, nitric acid, etc. should not be used unless the need specifically exists. When employing these strong reagents, protective gloves, eye protection, chemically resistant aprons, and possibly respiratory protection is recommended.

When washing glassware with soap or detergent and water in a sink, under normal circumstances household gloves (such as Playtex HandSaver® gloves) usually be sufficient, although gloves made of different materials may be indicated depending upon the materials originally in the items being cleaned. Scouring powder and brushes might be needed for stubborn residues. Glassware may break while being washed so care must be taken to avoid cuts. A soft mat in the bottom of the sink will reduce the chance of this occurring.

In the life sciences contact with contaminated glassware could allow an individual to contract a contagious disease. Biologically contaminated glassware should be autoclaved and or submerged in a sterilizing bath before further cleaning. This is a requirement under the OSHA bloodborne pathogen standard for items which may be contaminated with human blood, tissue, or other bodily fluids. In some larger laboratories glassware is washed in a central location, and in such facilities it is the responsibility of the individuals using the glassware to ensure that workers in the washing facility are protected from materials still in or on the containers and utensils.

REFERENCES

1. **Smith, G.P.,** Glass, in *CRC Handbook of Laboratory Safety*, 2nd ed., Steere, N.V., Ed., CRC Press, Boca Raton, FL, 1971, 544.
2. *Guide for Safety in Academic Chemistry Laboratories*, 5th ed., American Chemical Society, Washington, D.C., 1990, 9.
3. **Tucker, B.,** Acid cleaning of glassware, in *CRC Handbook of Laboratory Safety*, 2nd ed., Steere, N.V., Ed., CRC Press, Boca Raton, FL, 1971, 557.
4. Occupational Exposure to Bloodborne Pathogens, 29 CFR Part 1910.1030, FR 58, 235, December 6, 1991, 64175.

*4.6.2.17. Distillation Units**

4.6.2.17.1. The Safe Operation of Laboratory Distillations Overnight

Introduction

Scientific laboratories differ greatly in the extent to which laboratory distillations are permitted to run unattended during the evening hours. Most scientists consider the familiar "distilled water" still to be free of trouble. Many laboratories have had such stills operating day and night for years with the very minimum of attention. However, other distilling units are often regarded as hazardous either because of their construction or because the material being distilled is toxic or flammable.

If these other stills are examined carefully, the sources of potential trouble can be eliminated, and these stills too can be operated into or through the evening hours. The location and surroundings, at least, of "trouble-free" water stills need to be reviewed too in many laboratories. A serious fire occurred in the author's institution due to a water still overheating and setting some nearby combustible materials on fire.

Failure of the Cooling Water

Most distillation units operate with water-cooled condensers; therefore, it is essential for safe operation that the water supply be dependable. However, water pressures change:

---

* Section 4.6.2.17 is a slightly edited version of an article that appeared in the *CRC Handbook of Laboratory Safety*, 2nd edition.

1. There are decreases in water pressure which may be caused by failure of a pump, by partial blocking of the mains, by increased consumption by other laboratories on the same water main, by water main breaks, by maintenance personnel working on the system, lawn sprinkling, or water department use.
2. There are increases in water pressure which may occur when auxiliary pumps are being switched on, usually at the beginning of the working day; when other laboratories turn off their water valves, usually at the end of the day; or at other times during the day or evening when water system pressures increase because of decreased consumption by many or major users.

Either type of pressure variation can cause trouble. Too low a pressure results in an inadequate water flow to the cooling condenser, thereby allowing distillate vapors to escape. Too high a pressure can cause the tubing connections to the condensers to expand and burst, which is doubly undesirable: the laboratory is flooded, and vapors may be released since there is no flow of cooling water to the condenser.

It is not difficult to protect against both hazards if the following four-point approach is used:

1. Install a simple pressure regulator in the water main ahead of the valve that is used to adjust the flow to the condenser. The regulator can be adjusted to maintain an intermediate water pressure; it will then minimize the effect of both increases and decreases in the supply pressure.
2. Protect the pressure regulator with a suitable water filter, which will prevent particles of rust and other foreign materials from affecting its operation. Commercial filters with replaceable cartridges made of cotton or polypropylene twine are readily available.
3. Monitor the flow of water flowing from the cooling condenser to the drain. Monitoring this emergent water flow ensures that water has flowed through the condenser. Thus, it is inherently a more reliable approach than monitoring the main pressure.
4. Install a solenoid valve in the water supply line and connect it electrically to the water-flow monitor so that electrical power to this valve and to the still-pot heater will be turned off when the water-flow monitor responds to a water failure.

A number of water-flow monitoring devices have been described in the literature. A modification of Houghton's device is shown in Figure 4.11A, made of glass, and in 4.11B, made of metal. This design has the advantages that it imposes no back-pressure on the apparatus being monitored; it possesses a short time delay so that adjustments in the flow can be readily made; and its flow-restricting capillary can be readily inspected. [A number of commercial units are now available in virtually any general scientific supply catalog.]

Figure 4.12 shows a commercially available water-flow monitor based on this same principle. The commercial unit has a built in alarm plus a number of electrical receptacles, several of which are automatically turned off when the cooling water fails. The four-point approach recommended above is shown in Figure 4.13, items A, B, C, and D.

Failure of the Electrical Supply

It is important that the line voltage used for the distillation remain relatively constant since even moderate changes will affect the rate of distillation. However, electric voltages, like water pressures, change:

Figure 4.11 Cooling water-flow monitors (1A glass, 1B metal). When the flow ceases, the column drains through the capillary, and the mercury contact or the pressure switch opens.

Figure 4.12 Water-flow monitor equipped with alarm and convenience options.

Figure 4.13 Distillation column showing (A) water pressure regulator, (B) water filter, (C) emergent water-flow monitor, (D) solenoid valve in water line, and (EE′) (FF′) capacitance-actuated controllers for monitoring still-pot and still-head thermometers. (Glass equipment: Courtesy of Ace Glass Co., Vineland, New Jersey.)

1. There are decreases in line voltage due to increased consumption by other laboratories, usually early in the day. This will be most noticeable if the lines are inadequate. (Voltage levels from the power distributor may drop significantly due to excess demand placed on the generating and distribution system. This normally occurs during extremely hot or cold weather when either air-conditioning or space-heating demand loads the system. Under heavy demand situations the power company may deliberately drop the supplied voltage by 5%).
2. There are increases in line-voltage when additional generating equipment is turned on, or when other laboratories turn off their electrically-operated equipment at the end of the day.

The following two-point approach will guard against difficulties from these variations.

1. As with water pressure, there is a convenient device for overcoming source fluctuations: the constant-voltage transformer. A 500-, 1000-, or 2000-watt unit would be suitable. Use the size which is appropriate to the electrical load. Constant-voltage transformers work best when nearly fully loaded Other types of voltage stabilizing

devices are also commercially available. Although bulky, the insurance transformer provides against fluctuations in line voltage greatly outweighs any inconvenience. These units are inherently reliable and trouble free.

2. A constant-voltage transformer cannot cope with line voltage failure. Such failure in itself is not serious: the distillation merely stops. However, when the power is turned on again, "bumping" may well take place in the still-pot. If "bumping" is a serious problem, one should consider wiring a suitable relay into the still-pot heating circuit. Failure of the electrical supply should cause the relay to "drop-out" and stay out until it is manually reset. An alternative that is now available at a reasonable cost is the uninterruptible power supply. These devices switch over to a battery operated supply until the power failure is cleared. The capacities available vary, but there are sizes which would carry even a heat-pot for periods from a few minutes to perhaps an hour. Most power interruptions are relatively short. Many laboratory facilities also now incorporate emergency generators in their design for critical applications. An alternative way to achieve this same result is to plug the solenoid valve referred to previously into the water-flow monitor provided. A failure of the electric supply will cause the solenoid valve to close, shutting off the flow of cooling water and hence causing the water-flow monitor to de-energize its outlets. Since the solenoid valve is connected to one of these outlets, both it and the water flow-monitor are unable to reset themselves should the power failure be corrected. This approach has the advantage of having an inherent time delay—it will not be affected by momentary electrical cessations.

Termination of Distillation[*]

A question that often arises in deciding to leave an unattended distillation running during the night is how to stop the distillation if it is not to run until morning. There are several methods; the choice depends upon the nature of the distillation.

Timer Terminations. If the distillation is similar to the distillation of a solvent, the still-pot and the still-head temperature will not change greatly during the course of a distillation. Consequently, if, as has been recommended, a constant-voltage transformer is used to guard against variations against line voltage, the distillation rate will be reasonably constant. With experience, one can, therefore, estimate rather closely the amount of distillate that should accumulate during the night, and the distillation can be safely terminated with an electric timer. Such timers are familiar tools in most laboratories.

Temperature Terminations. If the boiling range of the material is wide rather than narrow, the rate of distillation will not remain constant. In this case, it will be better to terminate the distillation at a predetermined temperature rather than at a predetermined time. This is a common situation, not only with distillations left running for days on end but also with one-day runs that are not quite finished by "quitting time."

The scientist can readily terminate such a distillation through the use of a capacitance-actuated controller which sense the position of the mercury in a thermometer located either in the still-pot or in the still-hood. This is shown in Figures 4.13 and 4.14, EE′ and FF′. When the still-pot heater is plugged into the controller, the heat to the still-pot will be cut automatically when the temperature rises to the desired value. (A limit switch in the controller, which must be manually reset, keeps the power from turning on again when the temperature drops.)

---

[*]   Automated control of experiments is routinely within the capability of mini- or microcomputers. Control, however, is still based on the same type of measurable parameters.

Figure 4.14. Distillation column with still-head temperature being monitored for over temperature or under temperature by FF', and still-pot being monitored for over-temperature by either EE' or GG'. Note: If FF' is not used for safety monitoring, it may be used to control the reflux ratio automatically by turning the reflux timer off whenever the still-head temperature exceeds the set point.

Weight and Volume Terminations. If one prefers to terminate the distillation on the basis of the weight of the volume of the distillate, several other approaches may be used:

1. If the system need not be airtight and the product receiver can be supported on a pan balance, movement of the balance can be used to open an electrical circuit and terminate the distillation. When the balance moves, the heater is turned off (Figure 4.15). The weights used on the balance should allow for the force used to actuate the switch. This approach can be modified if the system must be airtight; flexible tubing can be run from the condenser to the product receiver mounted on the balance. It may be necessary to run two flexible lines—one for liquid flow, the other for vent. Care should be used in selecting the tubing used. It must not be attacked by the materials being distilled.

2. If a volume termination is preferred, there are several choices—these include the use of mercury contacts, capacitance-actuated controllers, and optical means of detecting liquid level. Figure 4.16 shows three examples schematically.

3. It is quite likely that automatic fraction collecting devices that have been developed for chromatographic purposes can be adapted for use with distillation columns. Termination of the distillation would be done on the basis of the number of fractions collected of a fixed volume.

Possible Overheating of the Still-Pot

Once the foregoing factors have been taken into consideration, the more subtle hazards should receive attention. Laboratory distillations involving many liters of raw material are not likely to run dry overnight. However, it is conceivable that sooner or later this unlikely situation might develop because of some oversight. The distilling apparatus can be monitored by an over-temperature or an under-temperature approach.

1a. If a heating mantle is used with the still-pot, one can make use of the thermocouple that is embedded in most heating mantles. This can be attached to a suitable millivoltmeter equipped with an adjustable over-temperature sensing pointer (Figure 4.14, CC'). These meters, known as meter-relays are commercially available. They should be ordered complete with auxiliary relays capable of interrupting heater currents of several amperes.

1b. The still-pot itself can be monitored for over-temperature by the use of a thermoregulator plus an electronic relay or by a thermometer with a capacitance-actuated controller similar to that described above, but for monitoring rather than for termination (Figure 4.14, EE').

2. In some cases it is convenient to monitor the still-head temperature against a decrease in temperature. A decrease should occur if the liquid in the still-pot becomes quite small in volume; the *rate* of distillation will then decrease greatly and the still-pot temperature will drop. This monitoring of the still-head temperature can also be accomplished readily with either the thermocouple plus meter-relay or with the still-head thermometer plus a capacitance-actuated controller (Figure 4.14 FF').

Breakage and Physical Separation

In any safety discussion it is important to consider the possible breakage of the glass equipment. Such breakage is highly unpredictable since it may result either from residual internal stresses in the glass, from stresses generated by improper external supports, or from an accidental blow.

Figure 4.15. Termination of distillation by weight of product.

Figure 4.16. Termination of distillation by volume of product.

Internal stresses in the glass apparatus may arise from inexperienced glass blowing or from careless commercial manufacture. *It is strongly recommended that all parts of the glass distillation equipment be carefully annealed and be checked for residual stress by means of polarized light.* Commercial polarized light units are available for this purpose, or a stress-analysis unit can be improvised from two sheets of Polaroid and a light source. Some large research laboratories inspect with polarized light all glass apparatus which they receive.

The method used to support the distillation column is of greatest importance in minimizing stress. Thought must be given to the design of the supporting framework, the method of clamping, and the design and location of the individual clamps.

A complete distillation column is usually rather tall and narrow. The column, head, and receivers can therefore be supported by a single support rod. The use of a rigid support rod and heavy-duty clamps and clamp holders is important in order to prevent one part of the apparatus from shifting with respect to the other parts. It is recommended that the rod be

SPRING
CLAMP
on each
joint

Figure 4.17. Ball-socket links in transfer lines can flex and hence eliminate any stress present.

at least 1.90 cm (¾ inch) in diameter and securely fastened to a wall. The following procedures are suggested:

1. The column itself is first firmly clamped either at its top, bottom, or at some enlarged portion. The clamp used must be strong enough to carry the full weight of the column and head.
2. A second clamp is then placed on the column. This is tightened only slightly; it is intended as a guide against lateral movement.
3. When the still-pot is attached to the bottom of the column, it can be held in place by a cradle supported by springs. The still-pot will then be free to align itself with the column.
4. The distillation head is placed on the top of the column and is held loosely in position by a clamp which allows it to move vertically, but largely restricts lateral movement.
5. The receivers are then supported with care in order to prevent them from putting strain on the distillation head. Careful support of this part as well as of the rest of the apparatus is essential in order to prevent physical parting of the ground joints.
6. If the apparatus is extensive, or if the receivers when loaded will be heavy, then it may be desirable to introduce semi-flexible connecting links between the column and the receivers. One way of doing this is shown in Figure 4.17. The links illustrated are easily fabricated from ground-glass ball-and-socket joints.

In passing, mention should be made of an alternative method of supporting distillation equipment: namely, to mount the assembly on a rigid panel made of metal or other noncombustible material. In this case, a large number of supporting brackets may be used to clamp all parts of the assembly. A glassblower may seal the parts together after they are supported, thus eliminating any residual stresses due to the clamping.

Finally, breakage due to accidental blows should be prevented (1) by locating the apparatus in a corner of the laboratory, out of the main line of traffic, and (2) by the use of adequate safety shields. Such shields may be flat, rectangular, or semi-cylindrical. Those shown in Figure 4.18 are hinged and particularly convenient to use.

Overflow

Once a laboratory distillation is set up and allowed to run overnight, the operation may prove so convenient and satisfactory that the scientist forgets his initial caution. He might, at this time, forget to empty the receiver at the proper time or might underestimate the distillation rate. A simple safeguard is to use a product receiver that is large enough to contain the entire charge to the distillation. If this is undesirable, one can use a smaller receiver, but connect it to a large overflow bottle. A five- or ten-liter bottle equipped with a two-hole rubber stopper and a drying tube in the vent line is a convenient safeguard.

Even after these precautions have been taken, it is worthwhile to mount the entire distillation apparatus in or over a metal tray large enough to contain all the liquid being distilled. A tray 8 cm (3.25 inch) deep is usually adequate and is not too cumbersome.

Monitoring by the Watchman

In large, well-staffed laboratories, the safety committee may insist that the night watchman, on his periodic trips through the laboratory, glance at any equipment that is running. Some scientists may feel that their "guard" has no feeling for scientific apparatus. Nevertheless, whenever apparatus is running at night, it is wise to have the apparatus so clearly labeled that an inexperienced, scientifically untrained guard could turn it off—preferably by turning no more than two or three conspicuously labeled switches and valves. Most universities with large numbers of buildings and hundreds of laboratories usually lock the doors to the laboratories when unoccupied to reduce the chance of theft. In most cases, the watchman does not enter laboratories unless specifically instructed to do so and few have the time to make many such inspections. This alternative is not recommended.

Conclusion

Many laboratories are quite cost conscious; other laboratories are hampered by shortages of personnel, while some are under great pressure to achieve results in a short period of time. Whatever the motivation, there are real advantages in being able to run laboratory operations, including distillations, in an unattended condition, provided this can be accomplished in a safe manner.

In safety programs it is often difficult to agree on the exact sources of potential trouble and their relative importance. Hence, it is difficult at times to agree on which precautions are necessary and which are simply advisable. Nevertheless, there are few distillations that cannot operate unattended—not only during the night, but during the day as well—if careful plans are worked out along the lines suggested here.

Figure 4.18. Modern acrylic shields protect apparatus against external blows, yet are easily opened for manipulation.

ACKNOWLEDGMENT

Reprinted from the *Journal of Chemical Education*, Vol. 43, A589 and A652, July and August 1966, with permission.

Note: The original article provided a list of specific devices, identified by brand name and catalog number, which has been omitted here. Because of the length of time since the original article was published, many of these items have been superseded by newer models and companies have changed hands or ceased operations, so that the reader is referred to catalogs of scientific supply houses to obtain equivalent equipment for the recommended devices.

REFERENCES

1. **Burford, H.C.,** Improved flow-sensitive switch, *J. Sci. Instrum.,* 37, 490, 1960.
2. **Cox, B.C.,** Protection of diffusion pumps against inadequate cooling, *J. Sci. Instrum.,* 33, 148, 1960.

3. **Houghton, G.,** A simple water failure guard for diffusion pumps and condensers, *J. Sci. Instrum.,* 33, 199, 1956.
4. **Mott, W.E. and Peters, C.J.,** Inexpensive flow switch for laboratory use, *Rev. Sci. Instrum.,* 32, 1150, 1961.
5. **Pike, E.R. and Price, D.A.,** Water flow controlled electrical switch, *Rev. Sci. Instrum.,* 30, 1057, 1959.
6. **Preston, J.,** Bellows operated water switch, *J. Sci. Instrum.,* 36, 98, 1959.
7. **Conlon, D.R.,** The safe operation of laboratory distillations overnight, in *CRC Handbook of Laboratory Safety*, 2nd ed., Steere, N.V., Ed., CRC Press, Boca Raton, FL, 1971, 227.

4.6.2.18. Control of Laboratory Processes*

4.6.2.18.1. Running Laboratory Reactions Under Safe Control

Introduction

Laboratory apparatus varies greatly in complexity—from the relatively simple thermostated ovens used for physical tests to quite complicated setups used for chemical operations and chemical reactions. Scientists and safety engineers differ greatly in their attitudes toward apparatus and in their willingness to have apparatus running unattended at any time. Therefore, in some laboratories, practically every piece of apparatus is shut down at the end of the working day. In others, numerous pieces of equipment and even chemical reactions run unattended for a few minutes, hours, or days.

The determining factors in deciding whether it is worth while to develop a set of conditions so that a laboratory apparatus can operate safely in the absence of the scientist (more likely a technician or graduate student at an academic institution) are

1. Is the operation repetitive and so demanding of close personal attention that it is quite tedious although not lengthy? (Exothermic reactions and vacuum-stripping operations are examples.)
2. Is it a lengthy experiment requiring many hours or days to run?
   a. If the experiment is lengthy, are there real disadvantages in interrupting the operation?
   b. If it were to be left running, could it be terminated automatically at some *desired* point?
   c. Could it be terminated at any point should some malfunction develop?

If the answer to these questions is yes, then one should examine the apparatus, determining which parts should be "watched" or monitored automatically. In such an examination, one usually finds there are very few units so complicated that they cannot safely be left running under automatic control; but also one finds there are few units so inherently safe that they do not need some safety monitoring.

This section considers both the simple and the complicated apparatus, both the endothermic and the exothermic reaction, both the short-term and the long-term operation, both "control" and "safety" monitoring.

---

*   Section 4.6.2.18 is a slightly edited version of an article that appeared in the *CRC Handbook of Laboratory Safety,* 2nd edition.[1]

Simple Endothermic Experiment

Figure 4.19 shows three apparatus types which from a safety point of view may be considered together: (A) the laboratory oven, (B) the constant temperature bath, and (C) the glass reaction flask. The assumptions made in grouping these together are: (1) only the variable of temperature need be "watched"; (2) should the temperature exceed a desired value, it can be detected by a monitor which automatically turns off the electric power; (3) since the operation is not exothermic, once the power is turned off, the apparatus will cool safely.

Apparatus such as ovens and constant temperature baths, (Figure 4.19, A and B) are not usually considered to be hazardous. Therein lies the danger; they are overlooked in safety inspections. This is particularly dangerous in the case of oil baths. If the oil bath thermostat fails, the oil will overheat and a fire may start. (Even if no fire results, valuable apparatus may be damaged and irreplaceable samples may be lost.)

Figure 4.19D represents schematically any experimental apparatus which is heated by an electric heater of either an external or an immersion type. Electrical power is supplied to the heater cyclically through a temperature controller connected to a sensing probe in the apparatus.

As shown in Figure 4.20, to detect thermal malfunction of the apparatus in Figure 4.19, one may use a monitor to watch:

1. The temperature of the apparatus
2. The temperature controller
3. The temperature sensing probe
4. The power supplied cyclically to the heater
5. The temperature of the heater

One might reason that since the experiment is primarily concerned with the temperature of the apparatus (item 1) this is what should be monitored. A good way of so monitoring is to use a secondary controller and probe that are independent of the primary temperature controller. If space is available, a relatively rugged bimetallic thermoregulator is ideal for the purpose. It should be set to operate 5° to 10° above the normal temperature of the apparatus and can be electrically connected to take over control of the temperature if the primary controller fails. In that event, it would hold the apparatus temperature at the 5° to 10° higher level. It may, on the other hand, be wired to a relay circuit (Figure 4.21) designed so that once the **THERMOSWITCH** opens, the relay will keep the power off until a push-button switch is reset. Figure 4.21 shows a commercially available "over-temperature guard" based on this design.

With smaller apparatus it may not be convenient to use a bulky bimetallic thermoregulator (although some are fairly small). If such is the case, one may use a thermocouple connected either to a pyrometer controller or to a meter relay. A thermistor plus a bridge-type controller may also be used.

If one prefers not to insert a secondary probe into the apparatus, then one should either use an external sensor or monitor items 2, 3, 4, or 5. In some cases (item 2) the primary controller may be partially able to monitor itself. Pyrometer controllers are often designed so that if certain internal components fail, the controller will cut off the heater power (this is known as *Fail Safe* design). A somewhat different approach is to make use of a pyrometer that has two independent control points. The second control point can be set at a higher level to sound an alarm, open a circuit, or take remedial action. Admittedly, the pyrometric

Figure 4.19. Examples of thermostated endothermic apparatus.

Figure 4.20. Five possible ways to monitor an apparatus for malfunction of the temperature controller.

OVER-TEMPERATURE CUT-OFF FOR OIL BATHS

Figure 4.21. Wiring diagram of an oven-temperature relay circuit. Reset switch must be pushed to re-energize circuit.

Figure 4.22. Compact relay unit based on circuit shown in Figure 4.21.

approaches are not 100% *Fail Safe*, but they do help to guard against trouble. This is also the case when one monitors item 3—the sensing probe. This, again, is common practice with pyrometer controllers; they are usually designed so that the controller will turn off the heat if there is a thermocouple break.

Item 4 calls for monitoring the ON and OFF power cycles to the heater. This can be accomplished with a so-called "reset" or "interval," timer or a "time delay" relay. Such units are designed to open or close a circuit after a period of seconds, minutes, or hours. When connected parallel with the heater, one of these can sound an alarm or turn off the heat if for any reason the electric power remains continuously ON. This approach has the advantage that it monitors the apparatus without physical contact with the apparatus.

Item 5 is sometimes the most convenient to monitor—particularly when one is using heating mantles into which the manufacturer has built a thermocouple. Again, one can attach the thermocouple either to a conventional pyrometer or to a meter relay.

Once the choice is made as to which approach (items 1 to 5) one prefers, one then proceeds to the next decision: What action should the monitor take if an excessive temperature occurs?

1. Should the monitor permanently shut down the apparatus until an operator pushes a reset button?
2. Should the monitor allow the apparatus to continue to operate but at some slightly higher value?
3. Should the monitor give the disturbance a short time to disappear and take no action if the disturbance is only momentary?
4. Should the monitor make any record that a disturbance occurred?

The above choices apply not only to the monitors described above but also to many of the monitors described in the following sections. For some situations, the monitor should shut down the apparatus rather than exert corrective action because normal operating conditions *cannot* possibly be restored. In other cases, however, the monitor should take corrective action because normal operating conditions *can* readily be restored.

Complex Apparatus

While the scientist is almost always concerned with temperature, the apparatus is often more complicated and involves still more variables than the examples shown in Figures 4.19 and 4.20. Figure 4.23 illustrates schematically a more complicated setup. The ten circles symbolize the components and/or variables the scientist would naturally watch if he were personally present. They represent, in some cases, malfunction of the apparatus and, in other cases, the progress of the reaction. They are as follows:

1. The rate of the stirring
2. The viscosity of the reaction mixture
3. The rate at which reflux is rising into the reflux condenser
4. The flow of cooling water to any water-cooled part
5. The rate of distillation of product or by-product out of the reaction flask
6. The volume of product or by-product in the distillate receivers
7. The pressure in the apparatus: either hydrostatic pressure, gaseous pressure, or vacuum, depending upon the nature of the apparatus
8. The rate of flow of gas (or liquid) to or from the reaction
9. Other indicators of the progress of the reaction, depending upon the specific reaction
10. The time

It should be remembered that with each variable one usually has the choice of several different approaches. To illustrate:

1. The rotation of the stirrer may be monitored by (a) a meter relay which measures the electric current to the stirrer motor, or (b) a tachometer with either a switch output or a meter relay output, or (c) an improvised tachometer such as shown in Figure 4.24.
2. The viscosity of the mixture is often more difficult to monitor. Three possibilities can be suggested: (a) Sometimes a commercially available viscometer can be adapted to the apparatus either by immersion into the reaction area or into a side stream. (b) If the rate of stirring is sensitive to the viscosity (and it should be if one selects an A.C. motor having "poor regulation") then one can monitor viscosity (1) as discussed above. (c) If the reactor is a continuous rather than batch-type reactor, the viscosity may be monitored by watching the back pressure as the reaction mixture is pumped.
3. The rate of reflux can be monitored quite readily by sensing the vapor temperature in the reflux condenser either in the region of the condensing zone or slightly above

Figure 4.23. Schematic illustration of a complex reaction apparatus.

Figure 4.24. An improvised stirrer monitor which uses Alnico magnets in an aluminum pulley to generate a rotating magnetic field. This magnetic field generates a current which is sensed by a meter relay connected to a pickup coil.

this zone. The temperature in these regions gives an indication of the reflux rate because an increase in the rate causes the condensing zone to move further up into the condenser. This results in a temperature increase at any given point in the condensation zone. Such an increase is easy to monitor using either a thermoregulator, a thermocouple plus pyrometer, or a thermometer with a capacitance-actuated controller attached (Figure 4.25). The monitor selected should be set 5° to 10° *below* the normal condensing temperature of the reflux vapors. The exact location of the sensing unit with respect to the condensing zone (and the manner in which the monitor is connected electrically) will depend on whether one wishes the monitor to turn off the heat cyclically whenever the rate of reflux gets too high, or to terminate the distillation should the rate drop too low or decrease too much.

4. The flow of cooling water to the condensers or to any other part of the apparatus can be monitored with either a commercially available *water flow controller* or the glass apparatus also described in the previous section (Figure 4.11A, section 4.6.2.16).

5. The rate of distillation is more difficult to monitor. One approach is discontinuous: to monitor the time required for successive samples of a given volume to accumulate. This could be done with either a relatively simple sample collector based on a syphon or one of the fixed-volume sample collecting and distributing devices which have been developed for chromatographic purposes. To monitor the time required for successive samples to accumulate, one would again use one of the time-delay relays or timers. The timer, in this case, could monitor the time elapsed per sample as being too long or too short, depending on how the timer was connected (to monitor both high and low rate would require two timers). The development of solid state timers, digitizing circuits, and digitally-based data acquisition and storage units since the original article was written provides enormous flexibility on the type of monitoring described here, as it is now very easy to compare and manipulate data from different channels.

6. The amount of product in a distillate receiver can be monitored either on a volume or weight basis as described in Section 4.6.2.16. It might also be monitored by counting the number of fixed volume fractions—if one were using a fraction collector as mentioned above.

7. The pressure within the apparatus can be monitored by use of either (a) commercially available pressure switches of the bourdon, bellows, or diaphragm type; or (b) mercury monitors with contact probes; or (c) mercury manometers to which have been clipped capacitance-actuated controllers which sense movement of the mercury levels. The two ways to use the mercury monitors are shown in Figure 4.26A.

8. The rate of gas or liquid flow either *to* the reaction or *from* the reaction can be a very important variable. The monitoring of a clean flow usually can be accomplished with a flow meter which develops a pressure differential across an orifice or capillary. Monitoring can be based on adjustable or fixed position contacts in a mercury U-tube connected across the capillary. It also can be monitored with a capacitance-actuated controller attached to one leg of the U-tube (Figure 4.26B).

9. This variable includes pH, color, presence of precipitate, etc. In general, it represents those variables that can be monitored only with rather sophisticated electronic devices. These may be either quite specialized units designed for a specific apparatus or general-purpose laboratory tools that the scientist uses quite frequently.

10. Time, the final variable on the list, can be monitored either (a) on the basis of the time of day or (b) elapsed time after a given stage of the reaction is attained. Scientists are quite accustomed to using electric clocks and timers to shut down apparatus at a predetermined time. They are less likely to be accustomed to the many possible ways of using the time-delay timers. Such a device, if interconnected to one of the other monitors listed above, can terminate the reaction n hours after the other monitor has indicated progress of the reaction. It is possible, for example, to

Figure 4.25. Reflux monitoring with a thermometer located in condensing zone of reflux condenser.

connect a time-delay relay to the primary temperature controller and have the timer triggered by the first OFF cycle of the temperature controller. This will enable it to control the time that the reaction was actually *at* temperature.

Coupling of Monitors

With a complex apparatus, it may be desirable to interconnect a number of monitoring devices.

In general, the monitoring units referred to above fall into several categories: mercury contact systems, mechanical-electrical devices (such as pressure switches), and electronic devices. These units generally have different electrical outputs.

Figure 4.26. Pressure and flow monitors based on manometers equipped either with contacts or capacitance-actuated controllers.

1. The mercury contact systems are either "normally open" or "normally closed," and maybe able to carry only a few milliamperes of current.
2. The electromechanicaldevices often have a microswitch with a single-pole, double-throw output that can be wired either "normally open" or "normally closed" and is able to carry at least 5 amperes of current. Microswitches, although simple and durable devices, are prone to eventual failure. No safety or critical control function should depend on a single micro-switch.
3. The electronic devices terminate in a variety of ways: a low voltage (or low current) control circuit, (b) an indicating meter that should be changed to a meter relay, (c) a 120-volt power circuit either normally ON or normally OFF, or (d) a single-pole, double-throw relay equivalent to the microswitch. Solid-state electronics usually depend on a voltage being "normally high," often around 5 volts or less, or "normally low," usually meaning 0 volts.

The output of any *single* monitor may be wired so that it interrupts the heat to the reaction (or causes some other type of corrective action to take place). When a number of monitors are being used, they can be wired together so that if any one monitor is actuated the heater can be turned OFF. A series circuit can be used if the output of every monitor is a "normally closed" relay or switch which opens upon malfunction (Figure 4.27). A parallel circuit would be used when every monitor has a "normally open" output which closes upon malfunction (Figure 4.28).

If one has a hybrid system of monitors—some "normally open," some "normally closed," some supplying a low-voltage output, and some supplying 120-volt output—then in order to combine them, one may need to add relays to at least some of the monitors or groups of monitors so that they can be connected in a series or parallel circuit. Again, solid state control equipment can be used to simplify the problem of working with a variety of inputs to make logical comparisons, i.e., "AND," "OR," "NAND," "NOR," etc., for control purposes.

Mention should be made of a more elaborate method of coupling monitors: the "sampling" technique in which the monitors are periodically rather than continuously checked. This may be accomplished with a stepping type switch wired to the monitors. As it steps, it connects each monitor in turn to the alarm and control circuits.

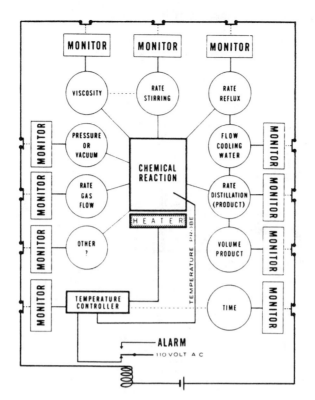

Figure 4.27. A number of normally closed circuit monitors may be readily connected in series.

Exothermic Reactions

Exothermic reactions, while usually not long-time reactions, at least so far as the exotherms are concerned, are particularly demanding of attention. When such reactions are running, the scientist may well feel he should not leave the laboratory for even a few minutes. This need not be, for many families of exothermic reactions are not dangerous and, once they are understood, can be entrusted to control by suitable monitors.

The monitors discussed under endothermic reactions may be used with exothermic reaction apparatus, both simple and complex, but with this difference: since the exothermic reaction evolves heat, the monitors may have to cause *positive* cooling to be applied to the apparatus. Thus, instead of wiring the monitor (or monitors) in such a way that it merely interrupts power to the heater, *it should be connected to turn on a cooling device.*

The nature of the cooling device will depend upon the nature of the exothermic reaction. Mildly exothermic reactions can be controlled by blowing air at the reaction flask. This is accomplished quite readily by having the temperature monitor turn OFF the heater and turn ON an electric fan or open a solenoid valve in an air tube connected to the laboratory air supply.

More active, but still "mild," exothermic reactions may require that the heater be removed from contact with the apparatus (so that any residual heat stored in the heater does not reach the reaction) and that simultaneously cooling air be turned ON. Still more active reactions require still more positive cooling. This can be accomplished if the cooling is applied in the form of an ice bath that can be raised automatically around the bottom of the reaction. (Heat to the reaction is applied from the top of the flask, by radiant heating.)

Figure 4.28. A number of normally open circuit monitors may be connected in parallel.

The steps described above can be applied automatically with a device known as a Jack-O-Matic that is now commercially available. When used in connection with a temperature-sensing system, Jack-O-Matic can either lower a heating bath from around a reaction flask and apply positive air-blast cooling, or turn off a radiant heater and raise an ice bath (Figure 4.29). It controls temperature by repeating these operations cyclically as long as cooling is required. Thus it is able to control automatically reactions that chemists have previously watched very closely. It is no longer necessary for the chemist to watch the thermometer, to lower the heating mantle, and to raise an ice bath around the reaction flask.

Other Factors

The foregoing notes discuss many of the points that often have to be considered in adding monitors and accessories to apparatus so that it can run unattended. The following are miscellaneous items that are less frequently encountered.

Programmed Control. Sometimes the reaction, instead of being carried out at constant temperature (or constant pressure, flow, etc.), should be carried out at increasing or decreasing levels. A suitable controller for this purpose usually can be designed with the assistance of the laboratory instrumentation group and a research machinist. For example, a programmed temperature control that step-wise increases the temperature can be built around a group of constant-temperature regulators, each accurate to a few hundredths of a degree. The regulators would be selected in sequence by a stepping switch controlled by a time-delay timer. This arrangement would make it possible to maintain an apparatus at each of a series of fixed temperatures for a predetermined time (this type of programming

Figure 4.29. Jack-O-Matic used to control exothermic reactions by raising and lowering ice baths.

is particularly useful in connection with physical test apparatus). The same result may be achieved with a computer interfaced with the monitors and control devices, to change or control a given parameter over virtually any desired range based on input from sensors. The computer can adjust a device, such as the heater voltage, to achieve the desired result. The major advantage of a computer-controlled system is that one can easily change the program by changing the software rather than hardware. A similar comment would be appropriate for each of the remaining topics in this particular section.

Programmed control in which the variable is allowed to increase or decrease gradually is sometimes desired. Programming controllers based on both linear and non-linear cams have been used for many years for plant-type process control. Although these units are usually rather large physically, they can be and are used with laboratory ovens and in-plant work. They may be also be used with laboratory reactions.

One can program-control other variables such as pressure or vacuum with plant-type cam controllers. One can also program-control these variables in an improvised manner by coupling a slow-speed motor to one of the commercially available pressure or vacuum regulators. The monitor will slowly drive the control point upscale. If a limiting value is desired, it can be achieved by adding a suitable cam which activates a microswitch and turns off the motor.

Sequential Control. Occasionally it may be desirable to program a sequence of operations. Thus, as an example, a reaction can be programmed so that after a predetermined number of hours a charge of reagent is added to the reaction flask, and after still another time period, still another charge is added. A different basis for sequential control is to monitor the reaction and use a given point in the development of the reaction, rather than time, to trigger a desired operation.

Foam-Level Control. In several rather different fields of laboratory work, excessive foaming in a laboratory operation can necessitate close personal attention and be a vexing problem. Vacuum stripping, although not a very long-time operation, is one example. As the lighter fractions are stripped from a mixture in a flask, foaming often occurs. If the stripping operation is not being watched quite carefully, part of the liquid in the still-pot may be swept over as foam into the condensing apparatus.

These boiling-type foams in glass flasks usually can be sensed either by probes coupled to capacitance-actuated controllers or by light beams using a light beam and a light beam photocell. In either case, the foam-sensing device can be connected to a solenoid valve which opens to admit a small amount of air or nitrogen to the apparatus (Figure 4.30). The resulting pressure pulse will cause the foam to subside almost immediately. As a result, vacuum stripping can be carried out much more rapidly and efficiently.

A second field for foam monitoring and control is in microbiological preparations such as fermentations. Such processes are relatively slow compared with the foaming referred to above. Moreover, the fermentation foams are not so clean as the boiling foams. Finally, the fermentation apparatus is often made of stainless steel and not readily adaptable to optical sensing. Figure 4.31 shows foam-sensing, using a Teflon-insulated probe and a capacitance-actuated control. The controller could either open a solenoid valve and admit a small volume of "antifoam" or could turn on a small pump and inject "antifoam."

Monitoring the Monitor

One can safely consider two different categories of monitoring—"control monitoring" and "safety monitoring." In designing apparatus that is to run unattended, one is concerned with both. The following notes illustrate the differences:

1. "Safety monitoring" is always an addition to "control monitoring." In some cases, the "safety monitors" will be *extra* monitors added to guard against failure of the "control monitors." In other cases, the "safety monitors" will be extra monitors added to watch variables that were not being controlled.
2. "Safety monitoring" systems usually need not be as precise nor as fast in response as "control monitoring" systems. For example, when "control monitoring" the temperature of a bath, one may use a sensitive mercury thermoregulator and hold the temperature constant to a few hundredths of a degree. When "safety monitoring" the same bath it would be quite satisfactory to add the relatively insensitive, slow responding, bimetallic thermoregulator.
3. Whenever possible, "safety monitors" should be unsophisticated devices with a mode of operation that is easily visible, easily understood by the scientist, and readily checked. The mercury manometer is almost ideal from these points of view. It can be made into an excellent monitor for pressure, or rate of pressure rise, rate of gas or liquid flow (or even flow of cooling water). The electronic viscosity-measuring devices are examples of the opposite extreme of complexity. Admittedly, one is grateful for any method of measuring viscosity, but one wishes for devices that would be simple and readily checked.

Figure 4.30. Photoelectric monitoring and control of stripping foams.

Figure 4.31. Monitoring and control of fermentation foams with a probe and a capacitance-actuated controller.

4. Although the "safety monitor" usually will be a device that is inherently simple and insensitive, there are some situations, particularly involving exothermic reactions, where the "safety monitor" must be a highly sensitive device with a fast response time. Under such conditions, one may well be willing to use as the "safety monitor" a device of greater complexity. In fact, it can be duplicate of the primary temperature controller.

5. One may be tempted, at times, to consider using the "safety monitor" as a "control monitor." This may even, at times, be quite reasonable, particularly when used to terminate a reaction. As an example, one can use a "safety monitor" attached to the stirrer motor to terminate a reaction when the reaction mixture has reached the desired viscosity. However, one must guard against the opposite approach: consider-

ing that a "control monitor" eliminates the need for a "safety monitor." The five possible monitors shown earlier in Figure 4.20 were really "safety monitors" added to guard against failure of the primary controller.

6. Finally, this philosophy of guarding against all possible malfunctions brings up the following question: If one always has to assume that the primary controllers can fail and therefore need monitoring, will the "safety monitor" sooner or later fail? Should one monitor the "safety monitor"? This is the age-old problem; just how far should one go in monitoring the monitor, in guarding the guard, in policing the policeman? Although not a complete answer to this question, a built-in or periodic check using an artificial signal to simulate a response of each "safety monitor" to an unsafe condition is highly recommended. This should be initiated as close to the original signal generation point as possible to test as much of the circuit as possible. Sometimes the solution to this problem is not difficult, since the variables are not independent. For example, if the temperature increases, other variables such as pressure or reflux rate will probably increase, and several different monitors will respond. In effect, the concept here is to supply sufficient redundancy to ensure safety. The system should be carefully analyzed to ensure that the failure of no single component defeats the safety systems. Sometimes the analysis is subtle and all failure modes are not fully considered. At least two competent persons should independently do a system analysis.

In considering monitors and their design, one should remember that the degree to which the monitors will be able to cope with both the "likely" and the "unlikely" malfunctions of the apparatus will depend upon the thought that has been given to their design and installation. In the final analysis, it is the good judgment of the scientist that is involved.

Other Safety Recommendations

The previous section (4.6.2.16) made a number of specific recommendations relative to safe operation of glass distillation equipment. Some of these same recommendations also apply to the simpler physical testing apparatus and some to the more complex reaction apparatus. To review the recommendations briefly:

1. If electrical power fluctuations would affect the apparatus, these effects would be minimized by use of constant-voltage transformers (or other voltage stabilizing devices).
2. Breakage of glass apparatus can be minimized by careful annealing, by inspecting the glass for strain, by proper support of the equipment, and by use of adequate safety shields.
3. Whenever liquid overflow of a liquid product is a possibility, oversized receivers should be used. It sometimes is advisable to place large trays under the apparatus.
4. The apparatus should be conspicuously labeled, so that an untrained person would have no trouble shutting it down in an emergency.

4.6.2.18.2. Conclusion

Although the suggestions made in this chapter are necessarily limited and will have to be modified to fit various local conditions, certainly there are great advantages in adapting laboratory apparatus so that it can run safely unattended or with minimal attention. The advantages are (a) apparatus that is run continuously produces the desired results much more rapidly than would otherwise be the case; (b) the quality, such as color, of the experimental samples produced when running without interruption is often better, and likewise, the quality of physical tests that are run continuously rather than interruptedly is better; (c) whenever the experiment can be run continuously, the scientist will have greater assurance that the run is valid and reproducible, in fact, when interrupted runs are made, the scientist may even have reservations about their validity; and (d) finally, whenever one is able to make use of the evening hours, one can greatly reduce the cost per experiment. There is one other benefit: provision of automatic controls eliminates or greatly reduces the opportunity for human error, and although there may be personnel in the laboratory the automatic controls will mean that they may not need to be in the immediate vicinity of the apparatus. Both of these factors should reduce the number of injuries.

The benefits of safe unattended operation of laboratory equipment are so great that creative efforts to achieve these are most worthwhile. In fact, such efforts by individual scientists and their supporting instrumentation personnel have been extremely profitable and are one of the most promising fields for the improvement of research.

ACKNOWLEDGMENT

Adapted from the *Journal of Chemical Education,* Vol. 43, A589 and A652, July and August 1966, with permission.

4.7. HAZARD AWARENESS (RIGHT-TO-KNOW)

In recent years, a movement has been growing to legally require that individuals working exposed to chemicals in the workplace are informed about the risks which these chemicals pose to their health. A number of states passed what are commonly referred to as "Right-to-Know" laws applicable to various groups of employees in their states. Eventually, after extended hearings, OSHA published a national version of this requirement as a hazard communication standard, 29 CFR 1910.1200. Comments were received from many sources, including industrial and trade associations, labor unions, academic groups, individuals, and others as to the content of the standard and which employees should be covered by the standard. As originally issued, the standard covered only the employers and employees in the Standard Industrial Classification (SIC) Codes 20 through 39. These classifications are given below:

20 Food & Kindred Products	30 Rubber & Plastic Products
21 Tobacco Manufacturers	31 Leather & Leather Products
22 Textile Mill Products	32 Stone, Clay & Glass Products
23 Apparel & Other Textile Products	33 Primary Metal Industries
24 Lumber & Wood Products	34 Fabricated Metal Products
25 Furniture & Fixtures	35 Machinery, Except Electrical
26 Paper & Allied Products	36 Electrical Equipment & Supplies
27 Printing & Publishing	37 Transportation Equipment
28 Chemicals & Allied Products	38 Instruments & Related Products
29 Petroleum & Coal Products	39 Miscellaneous Manufacturing Products

Laboratories in these industries were partially covered by some of the more critical portions of the standard. Laboratories in other industries and in academic institutions were not covered by the standard. Comments received during the hearings revealed marked differences of opinion on how the coverage should be applied to research laboratories. Some comments reflected an opinion that research laboratories, particularly academic laboratories, were among the most critical areas in which a definitive, explicit standard was needed while others felt that as knowledgeable, responsible individuals, research scientists needed at most only a voluntary standard.

Since many employees were omitted in the original version of the standard, court action followed almost immediately to force OSHA to cover all employees using or exposed to chemicals. After some delay in the court system, the legal issue was resolved in favor of OSHA being required to extend the coverage of the standard. The final rule was published in the *Federal Register* on August 24, 1987, to go into effect on May 23, 1988.

In a few areas the standard was applied to some research laboratories outside of the original list of industries. Virginia, for example, in adopting the federal standard added all public employees in the state to the list of those covered, which included laboratory personnel in the state's public universities and colleges as well as state-supported laboratories.

On May 1, 1990, 29 CFR 1910.1450, the OSHA standard for occupational exposure to hazardous chemicals in laboratories, went into effect. This standard preempted the general industry standards, including the hazard communication standard for laboratory scale use of chemicals, with a few exceptions. If the use of chemicals does not meet the definition of laboratory scale, even if the exposure takes place in a laboratory, the employer and employee must comply with the relevant part of 29 CFR 1910, Subpart Z. Even if the laboratory standard does preempt the hazard communication standard, in some key respects they are much alike. The definitions of hazardous chemical in the laboratory standard are specifically referenced to Appendices A and B of the hazard communication Standard for guidance. The OSHA PELs and action levels are still the same as those listed in Subpart Z. The MSDS requirements are basically the same. The training requirements differ very little. The requirements for informing the employees of the hazards associated with the materials are equivalent. The key differences are (1) the additional provision that a written chemical hygiene plan must be available for a laboratory, (2) standard operating procedures must be available, and (3) a chemical hygiene officer must be appointed. Since the use of chemicals in a laboratory is assumed to involve small quantities, the provisions in the hazard communication standard for notifying authorities of releases are not spelled out.

Although the hazard communication standard does not apply to employees involved with laboratory-scale use of chemicals, some of their activities may apply, even within the laboratory. Employees managing bulk chemical stores and individuals handling disposal of chemical wastes would not necessarily be engaging in laboratory-scale activities. Other support groups of a research facility would also be covered by the hazard communication standard in many of their activities. Therefore, the following sections will cover the basics of that standard. The standard as finally written is a performance standard, as is the laboratory safety standard. The required program is not spelled out except in very broad strokes, but the goals and objectives to be achieved are well defined. It is to be understood that the following material applies to the use of chemicals affecting nonlaboratory employees and to laboratory employees engaged in tasks not covered by the laboratory safety standard.

4.7.1. Basic Requirements*

Portions of the first few sections of 29 CFR 1910.1200, directly quoted below, spell out the basic concept of the standard.

(a) *PURPOSE*

(1)   The purpose of this section is to ensure that the hazards of all chemicals produced or imported are evaluated, and that information concerning their hazards is transmitted to employers and employees. This transmittal of information is to be accomplished by means of comprehensive hazard communication programs, which are to include container labeling and other forms of warning, material safety data sheets, and employee training.

(2)   This occupational safety and health standard is intended to address comprehensively the issue of evaluating and communicating the potential hazards of chemicals, and communicating information concerning hazards and appropriate protective measures to employees, and to preempt any legal requirements of a state, or political subdivision of a state pertaining to this subject....

(b) *SCOPE AND APPLICATION*

(1)   This section requires chemical manufacturers or importers to assess the hazards of chemicals which they produce or import, and all employers to provide information to their employees about the hazardous chemicals to which they are exposed, by means of a hazard communications program, labels and other forms of warning, material safety data sheets, and information and training. In addition, this section requires distributors to transmit the required information to employers.

(2)   This section applies to any chemical which is known to be present in the workplace in such a manner that employees may be exposed under normal conditions of use or in a foreseeable emergency.

(3)   This section applies to laboratories only as follows:

   (i)   Employers shall ensure that labels on *incoming* containers of hazardous chemicals are not removed or defaced;

   (ii)   Employers shall maintain any material safety data sheets that are received with *incoming* shipments of hazardous chemicals, and ensure that they are readily accessible to laboratory employees; and,

   (iii)   Employers shall ensure that laboratory employees are apprised of the hazards of the chemicals in their workplace in accordance with paragraph (h) of this section....

(h) *EMPLOYEE INFORMATION AND TRAINING*

Employers shall provide employees with information and training on hazardous chemicals in their work area at the time of their initial assignment, and whenever a new hazard is introduced into their work area.

---

*   The phrasing is not identical to the language of the laboratory safety standard, but these requirements are virtually identical to the corresponding ones in that standard.

(1) *INFORMATION.* Employees shall be informed of:

  (i)   The requirements of this section;
  (ii) Any operations in their work area where hazardous chemicals are present; and,
  (iii) The location and availability of the new written hazard communication program, including the required list(s) of hazardous chemicals, and material safety data sheets required by this section.

(2) *TRAINING.* Employee training shall include at least:

  (i)   Methods and observations that may be used to detect the presence or release of a hazardous chemical in the work area (such as monitoring conducted by the employer, continuous monitoring devices, visual appearance or odor of hazardous chemicals when being released, etc.);
  (ii) The physical and health hazards of the chemicals in the work area;
  (iii) The measures employees can take to protect themselves from these hazards, including specific procedures the employer has implemented to protect employees from exposures to hazardous chemicals, such as appropriate work practices, emergency procedures, and personal protective equipment to be used; and,
  (iv) The details of the hazardous communication program developed by the employer, including an explanation of the labeling system and the materials safety data sheet, and how employees can obtain and use the appropriate hazard information.

4.7.2. Written Hazard Communication Program

Based on the requirements outlined above, employers must develop a written program. The program must include provisions to ensure proper labeling of chemicals, to maintain a chemical inventory, to maintain a current and accessible MSDS file for all incoming chemicals, and to provide training and information to the employees in a number of relevant safety and health areas.

In order to develop an effective hazard communication program, it is desirable to have input from all of the groups that need to cooperate to make the program work. A straight-forward way of accomplishing this is to establish a hazards communications committee to help define the program and to monitor the performance of the program once it has become operational. There should be representatives on the committee from the administrative departments, including health and safety, personnel, purchasing, and physical plant, and from the research operations. In an academic setting, each college should be represented by a member from a major chemical-using department in the college, while in an industrial setting each major chemical-using division should be represented. One department should be assigned the leadership role in developing and implementing the program, probably the health and safety department, but it should be clear that all major constituencies share in the responsibility of formulating the program and also share the responsibility of making it work.

There are a number of things which must be in a written program:

  1. Although not explicitly required, all the employees that are exposed to chemicals in the organization during the normal course of their employment must be identified so that they can participate in the program. Note that this does not necessarily mean that the identified individuals use chemicals. An electronics technician might not use

chemicals himself, but could be considered to be exposed to those chemicals used by others working in the same room. An affected employee in this context would not be limited to salaried employees, but also to workers paid an hourly wage. On the other hand, a custodial worker would not be considered to have a significant exposure to a laboratory's chemicals if they only took out nonhazardous trash, but would have to be included if the cleaning materials which they use themselves contain hazardous chemicals. However, it would probably be desirable for a custodial employee to be informed of some of the general risks associated with chemicals in laboratories in their work area so they would understand the need to be careful while in the laboratory.

2. A list of hazardous chemicals in the workplace must be compiled. The provisions in the standard only require laboratories keep track of incoming chemicals since the effective date of the standard for employers, May 25, 1986. Existing inventories were in a sense "grandfathered." Eventually, as older stocks are disposed of or used, the list will come to reflect the actual holdings in a facility. The list should be kept as current as possible. If one person is assigned the responsibility to maintain the list, and the data kept in a personal computer data base, it is only necessary to keep track of additions and deletions in order to maintain a complete, current list. For the purposes of complying with this portion of the standard, the quantities of each chemical in the laboratory are not needed, although these data would be important for a sound management program and would be helpful in planning a safety program. The list of chemicals, in combination with the list of employees, will serve to help define the training program.

3. The written program must define how the employees are to be informed of the requirements of the standard. This will include: details of (1) how the employees are to be informed about the contents of the standard; (2) the contents of the written plan; (3) how they are to meet the labeling requirements; (4) how they are to learn of the methods available to them to warn them of exposures; (5) how to obtain and interpret a MSDS for a given chemical; (6) the hazards associated with the chemicals to which they are exposed; (7) how they are to be trained in procedures which will eliminate or reduce these chemical hazards; and (8) how they are to react in an emergency.

4. Many laboratory uses of chemicals involve repetitive tasks, while others do not. Employees must be made aware of the risks associated with the latter type of activities as well as those accompanying the more routine uses of chemicals, and the same basic type of information provided as in item 3.

5. Although pipes are not considered containers for the purpose of this standard and need not be labeled, the plan must include education of employees about the hazards associated with any unlabeled pipes containing chemicals in their work area and how to deal with these hazards.

6. There must be a procedure or statement in the plan as to how transient employees, such as persons working on contract, are to be informed of the chemical hazards to which they may be exposed, and for provision of information of protective measures for these transient employees. It is not specifically spelled out in the standard but there is a need for the converse as well. Contractors are often called in to do renovations, perform an asbestos abatement project, or to conduct a pest control program, as examples, and use hazardous chemicals in the process or expose personnel to airborne hazards. Provision should be made in the contracts for these groups for them to provide information to the occupants of the spaces where their work is being done.

4.7.2.1. Personnel Lists

This appears to be relatively straightforward, but in fact can be rather complicated. In a large academic institution, the actual duties associated with a given job classification often become blurred over a period of time. For example, a job title of laboratory technician might appear to logically relate to chemical exposure, but the duties of the individual may have changed so that the job may never bring the individual into contact with chemicals at all. It is not possible to simply have the personnel department list all persons in specific job classifications as well as professional staff or faculty in research areas.

As an initial step to determine which employees need to participate in a formal hazard communication program, a questionnaire can be sent to each department or other internal division asking them (1) to define those areas in which chemicals are used in their department, (2) to list each employee in those areas with their job title, and (3) for their appraisal of the involvement of these individuals with chemicals. This should be followed up with a second questionnaire to the managers of the individual areas, asking for the same information, and then the area should be visited to confirm the data provided. This sounds unnecessarily involved, but experience has shown that all three steps are necessary. In many cases, through oversights, individuals are not identified who should have been included and, occasionally someone is listed who has no exposure, usually because it was easier to list everyone rather than consider each individual case.

It is important to identify the position (most positions now have internal identification codes) so that when the position becomes vacant, a mechanism can be established to ensure that the new person filling the position receives a proper orientation program. Often, such a position might qualify for participation in a pre-employment medical screening examination as well so the effort to correlate positions with exposure to chemical hazards might be justified for more than one purpose.

All persons being considered for positions covered by the hazard communication program should receive a brief written statement concerning the program so that they may ask appropriate questions at the time of their job interview. If they are selected, a more extensive document should be provided so that they will be aware of the explicit requirements of the standard.

4.7.2.2. Chemical List

A list of all hazardous chemicals in the workplace is required as part of the standard. It may be difficult to convince many managers to take the time to go through their stocks of chemicals to prepare a list for their facilities. (The need for this information is implied in the laboratory safety standard, but not required.) This is one of the more burdensome tasks associated with the standard. However, it is very desirable that an effort be made. Not only does it provide information on which to base the training program, but it also is needed to prepare an MSDS file for the facility, although the MSDS file is only required for incoming chemicals. In practice, however, it is difficult to justify not having an MSDS for a hazardous chemical in use, based on a technicality. There is no real alternative for the initial survey as a basis for defining the scope of the program.

It would be desirable if the problem of maintaining the list of incoming chemicals could be centralized, perhaps as the purchase order is being processed. However, although surprisingly few chemicals are bought in quantity in most research institutions, even very large ones, there may be more than 1000 different substances bought during the course of a single year and several thousand purchased over a number of years. Many of these are bought under a number of synonyms, or as components of brand-name formulations. Commercial software is available by which a chemical can be identified by any number of synonyms, standard chemical name, trade names, or CAS numbers. Local information as to the purchaser and destination (building or facility) can be provided by the customer. Software is now available which can combine this information and more, e.g., date received and quantity, to generate a unique bar code that can be affixed to each container. These data can be used to maintain a continuing inventory for an entire organization and to track a chemical from the time of receipt to eventual full consumption or disposal. It is practical for individual laboratories to use microcomputers, even without these specialized programs to perform this task for themselves, using commercial database or spreadsheet programs. At most it would require a week or so by an employee to acquire the initial data and enter it. Maintenance of the data would involve only adding new containers and removing old ones.

4.7.2.3. Labeling

Current labels on original containers of chemicals as purchased from the distributor or manufacturer will almost certainly meet and exceed the requirements of the hazard communication standard (see Section 4.5.2). These requirements are

1. The identity of the hazardous chemical
2. Appropriate hazard warnings
3. The name and address of the chemical manufacturer, importer, or other responsible party

Item 2 is the only ambiguous requirement. Most commercially sold chemicals provide this sort of information on the label in a number of ways, such as:

1. A risk descriptor, i.e., *Danger, Warning, Caution*
2. The NFPA hazard diamond
3. A descriptive statement of the hazards
4. By use of stylized symbols, such as a radiation or biohazard symbol

Other useful safety-related information is normally provided as well, such as the flashpoint (if applicable), fire extinguisher type (if applicable), first aid and medical advice, a color code to aid in avoiding incompatible storage, and standard identifiers, such as a CAS number which can aid in referring to a MSDS data base and a UN number which is needed in disposing of the chemical as a hazardous waste.

Facilities are specifically enjoined by the terms of the standard from removing or defacing the labels on incoming containers of chemicals. However, it is relatively common to transfer a portion of the contents from the original container to a secondary container. If this material remains under the control of the individual responsible for the transfer, and is to be used during a single work session, then it is not necessary to label the secondary container. If it is not to be used under these conditions, then the secondary container must be marked with the identity of the chemical(s) in the container and with "appropriate" hazard warnings for the protection of the employee. These "appropriate" warnings need not be as comprehensive as the original label, but must provide adequate safety information.

The most likely occasions when secondary containers are used without proper labeling would be when chemicals are disbursed from a larger container into a smaller one at a central stockroom, and when containers are to be taken from the initial workplace into the field. Personnel must be sure to label the secondary containers in these cases and in any other comparable situation. If secondary containers are labeled properly, it also will help remedy one of the more troublesome problems associated with hazardous waste disposal, inadequately identified containers of chemicals.

The warning labels must be in English, although they may be provided in other languages *in addition*, if appropriate. In many academic institutions, in particular, graduate students who routinely use a language other than English as a primary language are becoming numerous, and, in some cases, consideration may be given to supplementing the commercial label with warnings in other languages. However, the majority of these graduate students can be expected to understand written English satisfactorily. Certain areas of the U.S. have changed demographically so that the use of languages other than English may have become predominant. All employees using chemicals must be instructed in how to interpret the hazard information on the labels.

A specific part of the written plan must address how the employees are to be made aware of the labeling requirements and how they are expected to comply with this standard. It would be highly desirable to develop a uniform program across an organization, particularly as to labeling of secondary containers, to avoid unnecessary confusion.

4.7.2.4. Material Safety Data Sheets

Since the receipt of MSDSs is tied so strongly to the purchase and receipt of chemicals, they were discussed in some detail in Section 4.3.3. The exact form of a MSDS is not mandated by the standard as long as the proper information is provided. Firms use a variety of formats to provide the required information. ANSI and the CMA have recommended a standard form which may be adopted.

There are two basic requirements associated with MSDSs in the hazard communication standard. Employees must be trained in how to use the information in them and the MSDSs must be readily available to the employees.

As discussed in Section 4.3.3, a major problem in a research institution, where the chemical users may operate virtually independently of each other and are likely to be in a number of different buildings, is to ensure that all users of a given chemical have ready access to a copy of the most recent version of the MSDS. The distributor is only required to send one copy of a MSDS to a purchaser and an updated version when a revision is necessary because of new information. Where several different components of an organization order independently, one purchaser may receive an update while the others do not, since the vendor technically has fulfilled its obligation by sending the MSDS to the first unit making the purchase. If a centralized mechanism for tracking chemical purchases has been established, then all MSDSs could be sent to a single location from which copies can be forwarded to all groups within the organization that need them. Another alternative, which does not provide as ready access to all users but does not require the tracking mechanism referred to above, would be to have all MSDSs received at one location and maintained in a master file, with copies placed in several secondary master files at locations reasonably convenient to the users. A third alternative, but still less accessible to users, would be for a single master file to exist, with copies of individual MSDSs provided upon request. This might not be considered to meet the accessibility requirement if the delay in receiving the MSDS is more than 1 or 2 working days. All of these mechanisms are at best cumbersome and manpower intensive.

Comprehensive generic MSDSs are now becoming commercially available on optical discs which can be processed by a computer and accessed at any time by the users. These typically are updated quarterly so that they can satisfy the need for the MSDS file to remain current. They are not inexpensive, but the cost is much less than for the amount of man-power needed to maintain an equivalent hard copy file, and they provide a comparable level of access. There also are firms which maintain computerized MSDS files available to subscribers as a database service. These are accessed by the users from their terminals using modems, but line and access charges are incurred.

No matter how a facility or organization sets up a MSDS file, a component of the training program for the employees must include an explanation of how an individual employee can obtain access to the file. It should be possible for an employee to obtain copies of a MSDS for a given chemical upon request. In addition to providing access to the MSDS file, part of the program training must also include instruction in how to interpret a MSDS to obtain appropriate hazard information.

The information presented in the various categories in a MSDS should pose no real difficulty to most technically trained persons. Some definitions of terms may need to be provided, such as LD_{50} (lethal dose, 50% of the time for the test species), if an individual is not accustomed to using such terms, but even these are straightforward. However, some persons will not be as scientifically sophisticated, and the training program for these individuals will need to be more thorough. In an instance at the editor's institution, grasping the distinction between a monomer and a polymer was a major problem for some clerical personnel who felt that they had been exposed to dangerous levels of a chemical due to some activities in the building where they worked. The health hazard data and the TLV values given in the MSDS for the chemical stated that they were for the monomer only and that the effects of exposure of the chemical would be serious at a few parts per billion in air. The employees' exposure was not to the monomer, but to very small quan-tities of the stable polymer, for which the health hazards were minimal. Extensive (and expensive) tests had to be run before the personnel were convinced (some perhaps still have doubts) that they had not been unduly exposed. During training, an effort needs to be made to ensure that understanding has been achieved.

Compliance with the training requirements for utilization of a MSDS as a source of hazard information can be readily achieved for technically trained personnel. For example, a written handout informing the employees how they can obtain access to a needed MSDS and a short videotape explaining the contents might be all that is needed. An individual capable of explaining any confusing points should administer such a program and be available to answer questions. A statement affirming that the training was received should be signed (and dated) by the employee after any questions were resolved. For less knowl-edgeable employees, a formal training session should be set up and an instructor-student format used. The handout and videotape mentioned above can still be part of the instruc-tion program, but the instructor should go over each of the categories in a MSDS and encourage the employee to ask questions. The employee also should be provided with a written version of the concepts covered, for later reference.

Some organizations document that an employee has not only been exposed to the infor-mation but also understands it by requiring that each employee take a very simple written quiz on the covered material, in place of signing a simple statement. This is not required to comply with the standard, but it does provide stronger documentation of an effective training program. Individuals should not resent imposition of such a requirement, but some professionals feel, rightly or wrongly, that they have demonstrated sufficient proficiency in their area by fulfilling the required educational and professional certifications. If a quiz is made part of a program, it should be expected that a number of individuals will object to taking it. Such a program, though, can be made to work with patience and support of the organization.

4.7.2.5. Employee Training and Information

Portions of the training program common to every employee—the basic concepts of the organization's program, how it is administered, the requirements of the standard, and how to read and understand information labels and MSDSs—have already been covered in the previous sections. These can be given by the lead department in the organization's hazard communication program, but a number of other areas will require a cooperative effort between the administrative department in overall charge of the program and the individual facilities and departments. In an organization which uses chemicals in a wide variety of activities, the local managers will need to be the primary parties responsible for providing much of the required training relating to operations specific to their program and facility.

The following topics need to be covered routinely in a training program to comply with the standard, in addition to those already discussed:

1. The physical and health effects of the hazardous chemicals which the employees may use or to which they may be exposed.
2. Means to detect the presence of toxic materials in the workplace. This should include means directly available to the employee, such as odor, presence of a respiratory irritant, and visual means or various symptoms such as dizziness, lassitude, etc. It also should include types of monitoring that can be done by laboratory personnel, by the organization's safety and health department, or by outside public and private agencies.
3. Means to reduce or eliminate the exposure of the employee to the risks associated with the hazardous chemicals in the workplace. This should include work practices that will reduce the exposures or the use of personal protective equipment.
4. Actions the organization has taken to minimize the exposure of employees to the chemical hazards. This can include the engineering controls which have been implemented such as ventilation or monitoring devices. It can also include policy positions which encourage or require that employees and their supervisors follow good safety practices at all times, and programs which provide incentives for them to do so or to penalize those who do not.
5. Emergency procedures to follow in the event of an accidental exposure to a hazardous material.
6. Procedures to warn nonorganizational personnel working in the area of potential exposures. Generally this will mean persons working under contract to the organization.
7. Measures to provide information as to the hazards and the protective measures which both the employer and employee can take to reduce or eliminate the hazards associated with a nonroutine task involving chemicals.
8. Measures to inform personnel of the hazards associated with unlabeled pipes carrying chemicals in their work area, and the safety precautions which should be taken.

The responsibility for training covering these topics should be shared between the local administrative unit, usually the individual facility or department, and the department assigned the lead in implementing the organization's program.

There are a number of generic chemical topics which would be essentially the same, no matter what type of chemical exposure is involved. Among these are

- flammable and combustible liquids
- corrosives, acids, and bases
- gases

- explosives
- toxic materials
- irritants

- carcinogens
- allergens
- pathogens

There also are a number of topics on protective measures to minimize exposures that could be made the topic of standardized presentations, such as:

- safe chemical working practices
- use of personal protective equipment
- fire safety
- safe working practices
- electrical safety
- emergency procedures

Standardized programs can be developed for these topics, as well as others, and videotapes made which can be used virtually anywhere. If the latter course is taken, much of the hazard communication standard training requirements could be met by requiring a new employee to view selected tapes, supplemented by an opportunity for the employee to ask questions. Standardized programs are especially useful where there is a continuing turnover, with customized training being required for single individuals on a sporadic basis.

A large number of companies offer hazard communication training programs, many of which do provide videotapes as discussed above. They need to be reviewed prior to purchase. Some, in order to demonstrate compliance with one aspect of a safety program, do not follow good safety practices in other areas. Cost is not necessarily a guide; some of the better tapes are among the less costly.

Although much of the training obligation can be met with standardized programs, the primary responsibility for training must be borne by local managerial personnel or persons delegated by them. The actual exposures vary from place to place. Generic programs can provide an excellent foundation for a hazard communication training program, but they must be interpreted and adapted to local environments and workplace practices. When training has been completed for a given area, or even for a single chemical and if this is all that is necessary when a new chemical is brought into the workplace, the employee should be asked to document that the training has been provided by signing a dated statement to that effect. It is not essential that the employee agree that all of the information has been understood, although it is certainly hoped that this is the case, but it is important to have a record documenting that the information has been provided. The employer may be called upon to provide this documentation during an OSHA inspection or in the event of litigation.

REFERENCE

1. Hazard Communication; Department of Labor, Occupational and Health Administration, 29 CFR Parts 1910 (Section 1200), 1915, 1917, 1918, 1926, and 1928, *Federal Register,* 52(163), 31852, August 24, 1987.

4.8. HEALTH EFFECTS

In recent years there has been increasing emphasis placed on the health effects of chemical exposures. However, as has been frequently noted, health effects are much more difficult to quantitatively characterize than most physical safety parameters. It is straightforward to define with reasonable accuracy a number of physical hazards, such as the upper and lower explosive limits of the vapors of a flammable material. However, the exposure levels (see Figure 4.31, taken from the *Federal Register,* Vol. 53, No. 109, June 7, 1988, p. 21342) which will cause a given physiological effect in humans are not nearly as precise, especially if the effect of interest is delayed or is due to prolonged exposure to low levels of a toxic material.

Even individual reactions to low levels of materials that cause serious immediate or acute effects at high doses are strongly dependent upon the inherent susceptibilities of individuals. Some will exhibit a reaction at extremely low levels, while others show no signs of responding at all to relatively high levels. Part of this is due to the natural range of sensitivity in a population, but part is also due to contributory effects. There are synergistic effects: for example, a heavy smoker may have developed emphysema due to the effects of inhaling smoke for an extended period. Such an individual would be affected by toxic materials which reduce pulmonary function before a person with healthy lungs. The sensitivity of individuals can change with time: an example is the common oleoresin allergen, poison ivy. The sensitivity of individuals is usually small to an initial exposure, but with successive exposures the sensitivity increases. Similarly, the sensitization of individuals to bee stings is well known. The effects of medication also can modify the sensitivity of individuals. The serious problems associated with simultaneously taking tranquilizers and drinking alcohol represent a well-known example of this phenomenon.

The natural differences in individuals due to genetic factors, age, sex, lifestyle, etc., make evaluation of laboratory tests quite subjective. On examination of the results of a typical blood panel, one notes that the patient's results are given for each parameter measured, as well as a *range* to be expected for a typical healthy person. An individual can have results somewhat outside the normal range due to hereditary or environmental conditions and still be perfectly healthy. Fasting and foregoing any medication prior to laboratory tests is an attempt to eliminate as many variables as possible which would affect the tests. Unless a baseline series of tests are available from a time prior to an exposure when, presumably, the quantities to be measured are "normal" for the individual, then it is often difficult to determine with certainty whether a given result is due to an exposure or not. Even then, the results could be distorted by extraneous factors occurring earlier.

Dependence on observed symptoms to indicate an exposure to a toxic material is also very subjective. Again, the wide variation in tolerance of individuals to concentrations of toxic agents causes a corresponding wide variation in the responses to an exposure. Many of the common symptoms associated with occupational exposures—shortness of breath, headaches, nausea, dizziness, etc.—often can also be the result of other problems, illnesses such as flu, lack of sleep, psychological problems such as stress due to personal problems or personality conflicts with a supervisor, overindulgence, etc. An individual needs not only to be aware of the symptoms which could result from an exposure, but also needs to try to distinguish when these are likely to be due to an occupational exposure and when they are likely not to be. If, for example, an individual normally enjoys good health, has not done anything which might result in any of the symptoms which he is experiencing, and there are no "bugs" going around, he might well suspect that he has suffered an exposure to some environmental hazard. In such a case, he should mention it to co-workers, report it to his supervisor, and leave the work area. Especially if others are experiencing similar symptoms, it is very likely that an exposure has occurred and appropriate steps should be taken to seek medical aid for the exposed personnel, limit the exposure of others, and correct the situation. If only one person is having difficulties, but the immediate work environment differs for each person, then an exposure may have occurred limited to the individual, and precautionary steps, such as leaving the area, lying down, and observation by a colleague, should take place. Often, prompt recognition of a problem is critical in minimizing the consequences, especially when the possible culprit is a material which provides no other warning signs. Whenever an exposure has occurred which has resulted in physical effects, an evaluation by a physician should be obtained promptly. Even if no exposure has occurred, an individual complaining of an illness should be taken seriously. There could be a medical disorder requiring care, intervention, or at least documentation.

Even if malingering is suspected, evaluation by a physician can help to confirm that the claim of illness is or is not valid.

Delayed effects due to prolonged exposures to relatively low levels of toxic materials or radiation rarely are reflected in immediate sensations of malaise to trigger concern about possible consequences of the exposures. If a material does not have any warning properties, then exposures may exist at unsafe levels indefinitely without the occupants of the area being aware of the exposure. The eventual consequences may be masked by naturally occurring illnesses of the same type. Lung cancer is, unfortunately, common and so the occurrence of lung cancer might not be recognized as due to an occupational exposure. Birth defects occur in about 3 to 6% of natural births (depending upon how birth defects are defined). What percentage might be due to an occupational exposure of the mother or father? Similarly, infertility is a problem for about 15% of married couples. What is the role of occupational exposures for the afflicted couple? Some neurotoxins cause deterioration of the central nervous system, but age and other illnesses may do the same. It is often difficult to establish a correlation between an occupational exposure and an illness, even statistically for a group, because it is difficult to isolate the effect from the interference of other variables or to define an equivalent control group. Anecdotal evidence citing an apparently unusual rate of a specific illness may be due to a statistically random occurrence. Often, unless the illness is rare, such as the angiosarcomas caused by exposure to vinyl chloride, it is impossible to definitely verify a causal relationship between an occupational exposure and a disease.

Not all delayed effects are due to low levels of exposure. The onset of cancer which may occur due to exposure to asbestos or radiation is often delayed for periods of 15 or more years, or, and this is a key point, they may not occur at all. By no means do all individuals exposed to even high levels of such hazards suffer the consequences.

There are basically three mechanisms by which health hazard data may be acquired: (1) epidemiological studies of groups of exposed individuals, (2) human experimentation, and (3) animal studies. There are problems associated with each of these three sources.

The major problem with epidemiological studies is that usually one does not have a controlled experiment; the data is either generated by an ongoing work situation or extracted from past medical records. In some instances, case reports are sufficiently unusual that they call attention to themselves, e.g., a reduction in fertility in a group of workers is so large that only a simple study to determine the cause-effect relationship between a common exposure factor and the resulting fertility depression is required, the effect being known; it only remains to determine what experience the workers have in common. Rarely are situations as simple as this, although they do occur.

In most instances, epidemiological studies to determine if an exposure to a substance results in a given effect take the form of cohort studies, in which two separate groups composed of exposed and unexposed individuals are studied. It is critical that the study be unbiased either by the way the participants are selected or by the manner in which the outcome is tested. Another critical factor is whether the two groups are in fact similar in all respects, which could affect the outcome of the study, or that the differences are such that they can be taken into account either in the design of the experiment or in the analysis of the data. In order to judge the validity of a study, all of the relevant factors must be completely documented and available for review.

Most of the epidemiological studies concerning exposure to toxic substances are from the industrial sector since only in such an environment is it likely that exposures would be limited to a single chemical or class of chemicals, and where the exposures would be relatively stable over a prolonged period of time. Most of the studies that are available tend to come from Scandinavia, where, for example, Finland maintains a computerized data base of the health records of all its citizens. Similar records do not exist in the U.S., although some categories of specialized health data are maintained. Many epidemiological studies of exposures in the U.S. depend upon records maintained by corporations, or equivalent public agencies such as the national laboratories, which are managed by industrial firms and have similar medical surveillance programs to them. The limited range of chemicals for which such work situations provide the basis for valid epidemiological studies limit the scope of this approach.

Human experimentation is limited by statute to studies in which there is no prospect of permanent harm to the volunteers participating in the study. This obviously limits the scope of the results obtained by this route, although it can be employed to determine the onset of early symptoms or to determine threshold levels for detection of odors or irritation as a potential warning mechanism. Any experiment of this type must be carefully reviewed by a human subject review committee of the institution or corporate research facility where the research is being contemplated. Any subject of such experimentation must be fully informed of any risks and normally must be given an opportunity to withdraw at any point. However, even with this restriction, many experiments using volunteers have been conducted and significant data have been obtained on symptoms initiated by modest levels of exposure. It should also be remembered that, though, "fully informing the volunteer of all known risks" can in and by itself skew the results. Therefore, published results on subjective symptoms using a small sample must be viewed as inconclusive or suspect.

There have been data obtained from direct human exposures due to accidental exposures to high levels of a number of hazardous chemicals. These, of course, are not controlled experiments and the dose levels must be inferred from the circumstances of the incident, but as direct evidence of the results of high exposures, they are extremely useful.

Since data from exposures directly to humans are limited, much of the available data on the toxic effects of chemicals is obtained from animal data. The easiest data to obtain are the median dose or the median concentration in air which is fatal to an animal under a standardized experimental protocol. The most common animals used for this purpose are strains of rats and mice because they can be obtained with uniform characteristics relatively inexpensively, and the cost of housing and feeding them is small compared to most other species. In addition to rats and mice, many other animals are used, such as primates (monkeys, chimpanzees), guinea pigs, rabbits, dogs, cats, and chickens, in efforts to obtain a model which would parallel the effect on humans. In the generic carcinogen standard, relevant animal studies are intended to specifically involve mammalian species.

The median lethal dose, written LD_{50}, is given in mg/kg and the species is given. The median lethal concentration, LC_{50}, may be given in mg/M^3 or ppm for a given species, and the exposure time interval is usually specified. Lesser amounts of data are given in the literature at other survival fractions, such as 25% or 75%. These data must be obtained under rigorously controlled conditions to be useful, and the experimental protocol must be totally documented. Among other things, enough animals must be used to provide statistical accuracy. Where the effect to be measured is less well defined than lethality, the number of animals needed to obtain the data may become quite large. Even after the data using animals has been obtained, the question remains in many cases of whether the animal model is sufficiently close to that of a human response to use it to determine the equivalent human response. Much of the controversy of using animal data to establish human exposure effects revolves around this question.

Another procedure which leads to varying interpretations on the health effects of tested materials is the practice of using large, nonlethal doses to reduce the number of animals required when studying other effects such as carcinogenicity. The premise is that a large dose given to a small number of animals is experimentally equivalent to small doses given to a large number of animals. A linear extrapolation hypothesis usually is used to estimate the effects at low exposures. This practice is not uniformly accepted and is often used as an argument to discredit the results which are obtained, but is the basis of much of the data on these nonacute effects. Other possibilities to estimate the effects of low doses would be to assume that the response will approach zero more or less rapidly than the dose. A linear extrapolation generally is considered conservative. Data from experiments such as this are frequently used in arriving at health standards, by pharmaceutical manufacturing companies and the media.

The use of animals has provided the greatest amount of health hazard data, but the practice has come under increasing attack by animal rights activists. Much of the public support for this movement originated from widely publicized instances in which animals were not well treated and undoubtedly suffered more than was necessary. As a result of public pressure and a concern on the part of many scientists, many new safeguards have been instituted to minimize the amount of pain and suffering experienced by laboratory animals. Animal care committees are now required to review experimental protocols and must approve the procedures in order to qualify the research for federal support. The number of animals involved in the research is limited to the number required to achieve meaningful results and the pain experienced by the animals must be no more than absolutely necessary. These committees must include persons not affiliated in any other way, directly or indirectly, with the institution, and who might be expected to be caring about the well-being of the animals.

Although conceding that improvements have been made in the care of the experimental animals, the animal rights activists' goal is to prevent the use of animals in any research which would adversely affect them. There have been instances in which animals have been "liberated" from facilities and instances where these "liberated" animals have been released into the environment. There are two practical problems with such actions, regardless of the ethics: (1) the animals usually are not accustomed to surviving in the wild and most often do not, and (2) there generally have been no efforts to ensure that the animals are healthy and, hence, a disease could be introduced into the environment.

The argument that there is no alternative to using animals to gain knowledge to prevent disease or to cure human illnesses, since experimentation on humans cannot be done, is rebutted in two ways by the animal activists, the first being a purely moral stance of "why do we assume that it is morally right to cause pain to animals to help humans?" This is an issue that each individual must answer for himself, unless a legal restriction is imposed. The second argument is that animal experimentation is no longer necessary. Computer modeling can provide equivalent information. Relatively few scientists accept this latter argument as a generalization, although it is agreed that computer modeling can be used in some cases.

There is merit on both sides, although the extremists of both groups are undoubtedly too extreme, and some middle position will eventually become acceptable practice. However, animal health hazard data may be less available in the future.

A newer but effective modality which can be statistically relevant is to use cell cultures. Direct effects can be seen and measured as to benefits or toxicity. It is more complicated and has seen limited use as yet.

A recent report by the EPA indicates that research now appears to indicate that some toxic substances are less dangerous than formally supposed, based on animal studies, specifically in regard to carcinogenicity. This finding, according to EPA spokespersons, is based on a better knowledge of how the metabolism of chemicals differs in various species, a better knowledge of how much of a chemical that has been taken into the body actually reaches an organ where it may do harm, and a better understanding of how the chemical influences the mechanisms that cause cancer. This is a controversial position since, in general, it leads to higher acceptable exposure levels of the chemicals under discussion, such as dioxin and arsenic. There are scientists that feel that the level of scientific knowledge does not as yet justify moving away from a very conservative approach. Both sides should avoid treating the issue as one that can be settled by politically biased discussion, and the data on which the findings are based should be evaluated, as should any other hypothesis, on the basis of their scientific merit.

As stated in the section on the hazard communication standard, 4.7.2, OSHA defines health effects, for the purposes of the standard, in Appendix A to 29 CFR 1910.1200. The definitions given below are from that appendix. The laboratory safety standard also specifically suggests using these same definitions for guidance in defining hazardous chemicals. The definitions originated in 21 CFR Part 191.* A few modifications have been made to number 7.

For the purposes of this section any chemicals which meet any of the following definitions, as determined by the following criteria, are health hazards:

Criteria:

1. *Carcinogenicity:* ...a determination by the NTP, the IARC, or OSHA that a chemical is a carcinogen or potential carcinogen will be considered conclusive evidence for purposes of this section.
2. *Human data:* Where available, epidemiological studies and case reports of adverse health effects shall be considered in the evaluation.
3. *Animal data:* Human evidence of health effects in exposed populations is generally not available for the majority of chemicals produced or used in the workplace. Therefore, the available results of toxicological testing in animal populations shall be used to predict the health effects that may be experienced by exposed workers. In particular, the definitions of certain acute hazards refer to specific animal testing results....
4. *Adequacy and reporting of data:* The results of any studies which are designed and conducted according to established scientific principles, and which report statistically significant conclusions regarding the health effects of a chemical, shall be a sufficient basis for a hazard determination....

Definitions:

1. *Carcinogen.* A chemical is considered to be a carcinogen if:
   a. It has been evaluated by the IARC, and found to be a carcinogen or potential carcinogen; or
   b. It is listed as a carcinogen or potential carcinogen in the *Annual Report on Carcinogens* published by the NTP (latest edition); or
   c. It is regulated by OSHA as a carcinogen.
2. *Corrosive.* A chemical that causes visible destruction of, or irreversible alterations in, living tissue by chemical action at the site of the contact. For example, a chemical is considered to be corrosive if, when tested on the intact skin of albino rabbits by the method described in the U.S. Department of Transportation in Appendix A to 49 CFR Part 173, it destroys or changes irreversibly the structure of the tissue at the

---

* Some of the material quoted is rearranged, and some nonessential wordage is deleted. The essential information is not changed.

site of contact following an exposure period of 4 hours. This term shall not refer to action on inanimate surfaces.

3. *Highly toxic.* A chemical falling within any of the following categories:
   a. A chemical that has a median lethal dose (LD_{50}) of 50 milligrams or less per kilogram of body weight when administered orally to albino rats weighing between 200 and 300 grams each.
   b. A chemical that has a median lethal dose (LD_{50}) of 200 milligrams or less per kilogram of body weight when administered by continuous contact for 24 hours (or less if death occurs within 24 hours), with the bare skin of albino rabbits weighing between 2 and 3 kilograms each.
   c. A chemical that has a median lethal concentration (LC_{50}) in air of 200 ppm by volume or less of gas or vapor, or 2 milligrams per liter or less of gas or vapor, or 2 milligrams per liter or less of mist, fume, or dust, when administered by continuous inhalation for 1 hour (or less if death occurs within 1 hour) to albino rats weighing between 200 and 300 grams each.

4. *Irritant.* A chemical which is not corrosive, but which causes a reversible inflammatory effect on living tissue by chemical action at the site of contact. A chemical is a skin irritant if, when tested on the intact skin of albino rabbits by the methods of 16 CFR 1500.41 for 4 hours' exposure or by other appropriate techniques, it results in an empirical score of five or more. A chemical is an eye irritant if so determined under the procedure listed in 16 CFR 1500.42 or other appropriate techniques.

5. *Sensitizer.* A chemical that causes a substantial proportion of exposed people or animals to develop an allergic reaction in normal tissue after repeated exposure to the chemical.

6. *Toxic.* A chemical falling within any of the following categories:
   a. A chemical that has a median lethal dose (LD_{50}) of more than 50 milligrams per kilogram but not more than 500 milligrams per kilogram of body weight when administered orally to albino rats weighing between 200 and 300 grams each.
   b. A chemical that has a median lethal dose (LD_{50}) of more than 200 milligrams per kilograms but not more than 1000 milligrams per kilograms per kilogram of body weight when administered by continuous contact for 24 hours (or less if death occurs within 24 hours), with the bare skin of albino rabbits weighing between 2 and 3 kilograms each.
   c. A chemical that has a median lethal concentration (LD_{50}) in air of more than 200 ppm by volume of gas or vapor, but not more than 2000 ppm by volume of gas or vapor, or more than 2 milligrams per liter but not more than 20 milligrams per liter of mist, fume or dust, when administered by continuous inhalation for 1 hour (or less if death occurs within 1 hour) to albino rats weighing between 200 and 300 grams each.

7. *Target organ effects.* The following is a target organ categorization of effects which may occur, including examples of signs and symptoms and chemicals which have been found to cause such effects. These examples are presented to illustrate the range and diversity of effects and hazards found in the workplace, and the broad scope employers must consider in this area, but are not intended to be all-inclusive.

   a. Hepatotoxins . . . . Chemicals which produce liver damage
      Signs and Symptoms . . . . Jaundice, liver enlargement
      Chemicals . . . . Solvents such as toluene, xylene, carbon tetrachloride, nitrosoamines
   b. Nephrotoxins . . . . Chemicals which produce kidney damage
      Signs and Symptoms . . . . Edema, proteinuria, hematuria, casts
      Chemicals . . . . Halogenated hydrocarbons, uranium

c. Neurotoxic . . . . Chemicals which produce their primary
effect on the nervous system
  Signs and Symptoms . . . . Narcosis, behavioral changes, coma,
decrease in motor functions
  Chemicals . . . . Mercury, carbon disulfide, lead

d. Agents which act on the blood . . . . Decrease hemoglobin function,
  or hematopoietic system   deprive the body tissue of oxygen.
  Signs and Symptoms . . . . Cyanosis, anemia, immune function
depression
  Chemicals . . . . Carbon monoxide, cyanide

e. Agents which damage the lung . . . . Chemicals which damage the
pulmonary function
  Signs and Symptoms . . . . Cough, tightness in chest, shortness
of breath
  Chemicals . . . . Silica, asbestos, organic fibers such as
cellulose-cotton

f. Reproductive toxins . . . . Chemicals which affect the reproductive
capabilities, including chromosomal
damage (mutations) and effects on
fetuses (teratogenesis).
  Signs and Symptoms . . . . Birth defects, sterility, functionality
  Chemicals . . . . Lead, DBCP, some blood pressure
medications

g. Cutaneous hazards . . . . Chemicals which affect the dermal layer
of the body
  Signs and Symptoms . . . . Defatting of the skin, rashes, irritation,
discoloration
  Chemicals . . . . Ketones, chlorinated compounds, soaps,
solvents

h. Eye hazards . . . . Chemicals which affect the eye or
visual capacity
  Signs and Symptoms . . . . Conjunctivitis, corneal damage,
blephaharitis
  Chemicals . . . . Organic solvents, acids, alkalis

4.8.1. Exposure Limits[*]

Consideration and attention to the concentrations of chemicals within any worker's environment are mandatory, such as chemicals in respired air, in water (used at work or for drinking), food contamination, contact with skin or eyes either directly or by vapor, and possibly radiation from specific chemicals used. The garnering of meaningful data has been slow, and changes in acceptable levels occur as new resources and studies are produced and policy formulations are agreed upon.[1]

The ACGIH has published TLVs for decades, and recent international meetings have formulated and categorized the values into what are called occupational exposure limits (OEL).[1,2] These limits refer to airborne values only. Where skin is involved an "s" may be appended. The latter values refer to time weighted averages (TWA) over a normal working day of 8 hours in a 40-hour week, extrapolated for the worker's lifetime. OSHA

---

*   This section was coauthored by Richard F. Desjardins and A. Keith Furr.

limits are defined as PELs, which are the legally enforceable limits for the materials for which they have been established.

The exposure limits should not be used to define a boundary, on one side of which working conditions are acceptable and on the other side, unacceptable. First, they are for a typical person, and individuals may not react in a typical way. Low levels for which a prolonged exposure could give rise to a delayed effect would not necessarily provide any warning for individuals who might be abnormally sensitive. Second, levels should be kept as low as possible, and preferably well below the exposure limits. Maintenance of air concentrations below the TLV, PEL, or OEL does not guarantee that a worker will or will not be affected at any given concentration. There is a very small percentage of workers who are so sensitive to particular chemicals that even minute quantities will precipitate severe reactions. It is not possible or feasible to try to make exposures acceptable for these individuals in most cases. Removing them from the environment and avoiding contact to the material causing the reaction is best for both the worker and the facility. The essential OSHA goal that "no worker will be affected" is unattainable in cases involving large numbers of persons. The variability of innate propensities, as well as such factors as daily diet variations, stress levels, and circadian rhythms, often relegates efforts to provide a "safe" environment to statistics and chance.

Dutch authorities have formulated a new means of evaluating exposures called the "maximal accepted concentration" (MAC), which helps assess the available data to derive a more meaningful OEL. Authorities state[3,4] that this formulation permits a better accommodation between the employee who does the work and the employer who orders or requires certain working conditions, procedures, and chemical usage. After deliberation by a working group of experts (WGE), a national Dutch MAC commission was founded (NMC) to promulgate the resultant report. This addresses the health-based OEL (HB-OEL) as recommended by WGE and the socioeconomic and technological constraints. The NMC then issues a recommendation for an "operational OEL" or MAC. This allows input from both employees and employers, and deliberations and reports are open to the public. The adoption by the Dutch has proven workable, but recommendations for international adoption have not occurred as yet. A very important concept must be stated—OELs are derived so as to maintain and preserve health, not just to prevent disease or death.[5-7]

One must differentiate between somatic effects and nuisance effects because:

1. Such a distinction is based on the nonexistent dichotomy between soma and psyche.
2. With adequate testing, a nuisance may be shown to actually be a somatic problem.
3. The nuisance may be the premonitory signs and symptoms of a slowly developing somatic problem.
4. The nuisance perception can be a statistically significant factor in diminished productivity.

Stokinger[7] appropriately considers nuisance perceptions and signs and symptoms as "co-inducers" of disease. Efforts to specifically categorize these areas might well be a fruitless endeavor. If they are considered mutually inclusive and are managed appropriately, the effect will be salubrious.

It would be valuable to keep in mind the effects of synergism in the workplace. In considering the many exposure/response relationships, extraneous factors must be considered as background variables which could skew results. Such factors include smoking, hypertension, heart disease, asthma or pulmonary fibrosis, diabetes, and many more. The preemployment examination made by the health facility has a responsibility through a good present and past health history, a work history with attention to types of exposures, a perceptive physical examination, and appropriate testings to determine whether that individual can healthfully be employed in a given area. This must be done at that time to: (1) avoid harmful exposures which would compromise the person's health, and (2) avoid even the slightest suspicion that a disease was evolved within the facility.

The Committee for the Working Conference on Principles of Protocols for Evaluating Chemicals in the Environment in the U.S.[9] defined adverse effects as "those changes in morphology, growth, development and life span that:

1. Result in impairment of functional capacity or in a decrement of the ability to compensate additional stress.
2. Are irreversible if such changes cause detectable decrements in the ability of the organism to maintain homeostasis.
3. Change the susceptibility of the organism to the deleterious effects of other environmental influences."

The "hygienic rule" states that exposure to chemicals at work should always be minimized as far as possible. Ultimately the goal is zero. Pragmatically, at this time this must be interpreted as at the lowest levels reasonably achievable. The airborne exposure limits, regardless of their source—ACGIH, NIOSH, MAC, or OSHA regulatory limits—are intended to represent TWA levels at which it should be acceptable for most workers to work normal 8-hour-day, 5-day work weeks without suffering any ill effects. As noted earlier, individuals vary in terms of susceptibility, and there may be contributory effects which would make these levels inappropriate for a specific person. The values are intended to be based on the best available data from the sources described earlier, as evaluated by committees composed of professionals with appropriate expertise. The OSHA values are an exception, to a degree, in that they are adopted as part of a legal standard, and changing them is often a long, involved process which frequently includes legal contests. Therefore, changes in OSHA values tend to lag somewhat behind those data such as the limits suggested by the ACGIH, which are reviewed and revised annually, if the data warrant a revision. However, even the OSHA values were originally based on the same evaluation procedure and were in most cases actually taken from the other sources. OSHA revised its air contamination standards in 1989, setting them in most cases to the ACGIH-recommended TLVs in effect at that time. Many of the values were lowered. However, in July 1992, a U.S. Court of Appeals struck down these new standards, and currently the federal standards have reverted to the pre-1989 values, except for specific substances for which individual revisions had been made.

Many of the newer legal limits established by OSHA include an action level concept, typically 50% of the OSHA PEL, at which an employer is to take action to reduce the level or to ensure that the 8-hour time-weighted PEL is not exceeded.

Although the 8-hour TWA, by definition, includes periods during which the level exceeds the exposure limits, the departure should not be gross. In some instances ceiling levels are recommended or mandated above the 8-hour TWA limits, but they should never be exceeded. In other instances, there are short-term exposure limits which are also set higher than the TWA limits, usually intended to be for durations of no more than 15 minutes, for a limited number of occasions (ACGIH uses four as a limit) and spaced widely throughout a work day. These are usually set for processes which may be sporadic or have occasions when higher levels are normal. Sometimes these higher numbers are based on toxicological data, and sometimes they are indirectly related to typical maximum excursions which might occur in an industrial environment. The TLV along with the TWAs establish guidelines for exposure control, both for the employee's health as well as for possible corporate or institutional responsibility.

The responsibility for maintaining levels of toxic vapors and gases below acceptable limits is usually assigned to management, preferably by engineering controls unless these can be shown to be impractical, and otherwise by procedural controls and the use of personal protective devices. However, it is equally important for the employees to properly use the equipment in the laboratories so as to not defeat engineering controls, to report promptly any safeguards needing repairs, to follow procedures which reduce safeguards, and to use and maintain personal protective equipment provided them. Neither a "macho" or a "scoff-law" attitude is appropriate for laboratory activities as regards compliance with acceptable exposure levels.

REFERENCES

1. **Zielhuis, R.L.,** Occupational exposure limits for chemical agents, in *Occupational Medicine*, Zenz, C. Ed., Chicago, 1988.
2. World Health Organization, Methods Used in Establishing Permissible Levels in Occupational Exposure to Harmful Agents, Tech. Report 601, WHO, Geneva, 1977.
3. World Health Organization, Principles and methods for evaluating the toxicity of chemicals, in *Environmental Health Criteria* 6, Part 1, WHO, Geneva, 1978.
4. **Zielhuis, R.L. and Notten, W.R.F.,** Permissible levels for occupational exposure: basic concepts, *Int. Arch. Occup. Environ. Health*, 42, 269, 1978–1979.
5. **Zielhuis, R.L.,** Standards setting for work conditions as risky behavior, in *Standard Setting*, Grandjean, P.H., Ed., Arbejosmiljofondet, Copenhagen, 1977, 15.
6. **Sherwin, R.P.,** What is an adverse health effect?, *Environ. Health Perspect.*, 52, 177, 1983.
7. **Stokinger, H.E.,** Modus operandi of Threshold Limits Committee of the ACGIH, *Am. Ind. Hyg. Assoc. J.*, 25, 589, 1964.
8. **Stokinger, H.E.,** Criteria and procedures for assessing the toxic responses to industrial chemicals, in *Permissible Levels of Toxic Substances in the Working Environment*, International Labor Organization, Geneva, 1970.
9. Principles for Evaluating Chemicals in the Environment: A Report of the Committee for the Working Conference on Principles of Protocols for Evaluating Chemicals in the Environment, National Academy of Science, Washington, D.C., 1975.

4.8.2. Environmental Monitoring*

Since many toxic materials provide no sensory warnings, and in many other cases, materials which do have an odor or cause irritation (the two most common warning properties) do not result in a recognizable warning at or near the recommended limits or even at the presence of levels of immediate danger, the establishment of limits is of no practical value unless the capability of monitoring exists. Standard monitoring programs, based on 8-hour TWA measurements, are normally not appropriate for the laboratory environment, where the duration of usage of a given material is rarely this long. However, average and peak levels for the various activities should be determined.

Monitoring programs are usually performed by an industrial hygienist instead of the laboratory employees, except where permanently installed fixed monitors are used. Ideally,

---

* This section was coauthored by A. Keith Furr and Richard F. Desjardins.

monitoring should measure the levels of chemical substances in the breathing zone of each employee. As will be discussed below, monitoring near the breathing zone is practical with small sampling systems which can be worn in a shirt pocket or attached to a lapel. Preferably, the employee should wear the monitoring device during an entire work day and, again preferably, over a period of several days to ensure that the results obtained will be truly representative, although in many research laboratories the variety of chemicals change so rapidly that there may not be a truly "typical" day. It would be good if this could indeed be done for every employee. The amount of time required would, in all likelihood, be prohibitive in most large research institutions to do this for every laboratory. However, for laboratories working with regulated carcinogens, as an example, or where other highly toxic materials are in use, it may be necessary or desirable to establish such an intensive and thorough program.

Where the activities of many employees are essentially similar, the exposures of a typical, representative person may serve to reflect the exposures of all the persons doing the same tasks. In some instances, the areas of exposure to chemical hazards are relatively small and well defined. Sampling devices may be placed at the location used to determine the average exposures per unit time of persons working in those areas. If these devices do not show concentrations above the acceptable limits, then individuals can be assured that their exposures do not as well.

In many instances the rate of release of toxic vapors and gases varies widely, and an extended sample taken for 8 hours may show a very low average level of exposure, but for brief intervals, the actual levels actually may be dangerously high. In such a case a monitoring program should include "instantaneous" or "grab" samples taken during these periods of higher-than-normal levels. All of the organizations which publish levels recognize that for some materials, even short-term high levels should not be permitted. This concept is embodied in the ceiling (C) or short-term exposure levels (STEL) incorporated in the standards for some materials. The monitoring program should be able to address the need to document compliance with these short-term excursions.

All data accumulated in the monitoring program should be maintained in a file, preferably for each employee but definitely for each facility. These records should be dated, signed by the individual responsible for the data, and documented as to the measurement parameters, i.e., instrument used, duration of sample collection, characteristics of the detection device, location or person wearing the device, and, if possible, supporting data on the operations being conducted at the time of data acquisition.

There are many different types of sampling instruments used, and it is not the authors' purpose to dwell on the characteristics of each. Some are direct reading while others require sophisticated follow-up procedures to analyze collected material. There is some legal merit in acquiring material for later analysis by accredited commercial laboratories, to provide additional credibility in case the results are contested at some later date. The delay in receiving results following this practice, which may vary from a day or so to weeks, may be unacceptable, and in-house analysis or alternative methods may be chosen to obtain results more rapidly.

Direct reading instruments, by definition, provide "instantaneous" or "grab" samples which may be entirely appropriate, as when the need is present to determine if an IDLH situation exists. A good example is the need to use an oxygen meter when entering any confined space. A combustible gas meter is often used at the same time to determine the presence of organic gases, and these two devices are often combined in the same instrument, using different sensors. Other instances are reentry to areas where the presence of a highly toxic gas may be suspected. Direct reading instruments also are useful in conducting surveys within a facility to determine the relative levels of pollutants in different parts of a space. There may be areas, for example, where air exchange is minimal and toxic levels of gases can accumulate. In such instances, the data can be used to help determine what remedial actions can be taken. If previous measurements have not been taken, as for example when an operation is to be performed for the first time, a direct reading instrument, perhaps incorporating an alarm, might well be used. In some cases, direct reading instruments may be incorporated into fixed monitoring devices to provide a warning and to automatically shut down an operation. If, for example, experimentation with explosive gases is being done, a sensor may be used to determine when the concentration of the gas reaches a predetermined level relative to the LEL of the gas and to automatically close a valve in the gas supply.

Some direct reading instruments are extremely sophisticated, and others are very simple. Some instruments include a sensor for a single substance, while others can serve to detect and provide quantitative information on a number of materials. Among the more sophisticated units which provide direct readings are compact, portable infrared spectrometers, atomic absorption units, and gas chromatographs. These may be set up to provide a high sensitivity for a single element, such as the familiar Bachrach atomic absorption unit for mercury detection, while the Miran infrared spectrometer can provide accurate readings for hundreds of organic solvents. Individuals using these instruments need to be properly trained to obtain accurate data. Interferences are a major problem, where two or more possible chemicals may contribute to the same "window" or peak area. If the potential presence of interfering chemicals is not recognized, erroneously high readings may be reported. If the problem is recognized, these same high readings still may be interpreted as an upper limit for the level of the chemical of concern, and if the combined contribution is still well below acceptable limits, then exposures may be no problem to the employees.

All of these instruments must be well maintained and kept in good calibration in order for the data to be meaningful. In some instances, as with radiation measuring instruments, the calibration may be done by a commercial facility to provide traceability to a National Bureau of Standards standard in order to be certified, while in other cases calibration can be done locally. Documentation of all maintenance and calibrations should be maintained in a log. Assurance of accuracy is especially important if the levels to be measured turn out to be near the acceptable limits. As a minimum, the accuracy should be such as to assure that the results are within ±25% of the actual value with a confidence level of 95%. It would, of course, be desirable to exceed this minimal goal.

A simple direct reading device is the familiar detector tube intended to provide a measurement of the airborne concentration of a specific chemical, in which a known volume of air is drawn through a tube containing a material with which the chemical of concern will react. The known volume of air is usually provided by a manually activated pump attached to one end of a glass tube, and air is drawn in through the other. The amount of air may be provided by a single cycle of the air pump, or several cycles may be required. The reaction of the chemical in the air and the material in the tube will begin at the end away from the pump, and the length of the stain will indicate the level of the airborne contaminant. There are now hundreds of tubes available for individual chemicals that provide a great deal of flexibility for a relatively modest cost. The units do have a limited shelf life so that it is generally not feasible to keep tubes on the shelf for every possible contaminant (newer versions have improved shelf lives for a number of common substances). However, they can be the basis for a monitoring program for a chemical which is in relatively steady use. They are, as noted, relatively inexpensive per tube, but if a large number of measurements are required and if a direct reading instrument is available, it may be a more economical long-term investment to buy a device which does not consume an expendable tube for each measurement. The detector tubes, although intended for a specific chemical, may in fact react with more than a single chemical, so that interferences may be a problem. These interferences generally will be identified in the literature accompanying each tube. A major use of detector tubes, other than being an alternative for a more expensive device, is for testing of atmospheres for entry into an area in which high levels of a specific contaminant are expected to exist.

The devices previously mentioned are not suitable for taking extended measurements since they provide only an instantaneous measurement, unless the instrument has an output which can be connected to a device to integrate the data over a period of time. There are a number of ways in which this can be done, from a relatively crude system based on a chart recorder in which the area under the trace can be translated into an integrated reading to a digital sampling device acting to transfer the data to a computer with an appropriate interface and software, or a dedicated multichannel scaler in which the data occurring in successive intervals are accumulated in successive channels. Newer versions of many of the instruments in common use have an integrating mode of operation or a data-logger feature to acquire these data over extended periods.

Most long-term measurements depend upon using a constant-volume air pump in which the pump pulls in a given volume of air over several hours, and the contaminating material in the air is collected within a collecting device. The type of contaminant will determine the choice of the collector. Many organic gases and vapors are adsorbed readily on the surface of materials such as activated charcoal, activated alumina, or silica gel from which they can be eluted within a laboratory or driven off by heat and the amounts adsorbed quantified using a gas chromatograph. In other instances, the materials are bubbled through a liquid into which they go into solution. The collection efficiency of such a unit can approach 100%. Particulates also may be absorbed into a liquid or, more frequently, collected by impact onto various types of filters, after which a number of techniques are available to quantify the amount of pollutant collected. The efficiency of the filter will depend upon the type of filter used.

In some cases a collecting device called a cascade impactor is used, in which the rapidly moving contaminated air passes through a number of stages. There are slits between each stage of decreasing size. At each stage some of the particles, because of their inertia, tend to continue in a straight line, impact on a collecting plate, and adhere to it. In the first stage the larger particles which have the most inertia are collected, while smaller ones are drawn through the exit slit into the next stage and are collected there. This continues until the smaller particles are collected at the last stage on a permeable membrane. Thus, not only are the contaminants collected but they are sorted into approximate sizes. Since the quantity of air drawn through the collector is known and the amount on the filter or collectors can be measured, it is a simple matter to compute the concentration of the particulates in terms of milligrams per cubic meter for each range of sizes.

In some instances in which there is only a single possible contaminant in the air, no further processing of the collected material is necessary, but in most cases, analysis of the collected material is necessary to separate and quantify the amounts present. NIOSH has published explicit analytical techniques to be used for a large number of airborne pollutants and for meaningful results, these procedures should be followed exactly. In many cases, the commercial laboratory will provide the appropriate collector, as well as explicit instructions on how they should be used, including, for example, the airflow rate and the length of time to be used.

Each time a pump is used it should be calibrated to ensure that the volume of air is accurately known. Most companies that sell sampling pumps also sell convenient calibration units, but if one is not available, the pumps should be calibrated using a graduated cylinder and determining the time a soap film across the cylinder moves a measured distance. If sampling pumps are to be mounted in a fixed location, they generally can be plugged in to ordinary electrical power outlets, but if they are to be worn by an individual, they will have to depend on battery packs. Battery packs are usually nickel-cadmium and these require some care to be able to run for long periods. Unless a nickel-cadmium battery is frequently taken through a major portion of its discharge cycle, it will lose capacity. It is said to "remember" the range over which it is used, and if it is typically drawn down only one third, for example, during each use, eventually this will be its capacity. Nickel-cadmium batteries are best suited for frequent, heavy use.

A simple sampling device, well suited for personnel monitoring, uses a permeable membrane over an activated charcoal collector in the form of a badge which may be worn during the period of interest. Normal movement by the wearer and the inherent motion of molecules in air are sufficiently characteristic to bring a definite quantity of the air bearing the contaminant into contact with the membrane, through which the contaminant passes and is adsorbed by the charcoal. From this point on, the adsorbed material is eluted as with any other charcoal collector and analyzed. Normally, the company which sells the badge offers the analytical service as well, but the badges also may be processed locally. If care is taken, several organic solvents (if present in the air) can be identified with a single badge. Badges are available at this time for a relatively small number of materials, but the convenience of wearing a badge instead of a pump, no matter how quiet and lightweight, make them desirable.

Sampling should be done as close to the breathing zone as practical. The results obtained as the worker moves around (and the air being sampled is representative of the individual's actual exposure) generally do not agree well with samples taken by fixed collectors. This raises some questions of the value of the grab samples obtained in surveys by portable instruments.

For various reasons, individuals asked to wear monitoring devices, such as a small sampling pump, are not always cooperative and do things which invalidate the data which are obtained. They may feel that they are being "checked on," perhaps for careless or sloppy work. They may simply feel that the pump is too noisy or that it gets in their way. They may not want to know the levels to which they are being exposed, for fear that they may be too high and that they could be in danger of losing their job. In any event, if the individual responsible for the monitoring program feels that the data obtained are not reliable, then steps should be taken to investigate and correct the problem. A direct person-to-person explanation of the reason for the monitoring program may be all that is required, persuasion may be effective, or in an extreme case of noncooperation in which there are legal issues at stake, it might be necessary to make it a disciplinary matter. It is hoped that the latter could be avoided since it would create the potential for a future adversarial relationship.

Any monitoring program only provides data (hoped to be valid) for the exposure levels present in the area where the program was carried out and under the circumstances as they existed at the time the program was conducted. If the circumstances change, then it may need to be repeated, extended to other materials, or modified because procedures have changed. Most laboratory organizations do not have sufficient manpower to monitor laboratories often enough to catch every variation. It is incumbent upon laboratory directors or managers or other employees to notify the industrial hygiene specialist of internal operational changes that would affect the airborne levels of any hazardous material.

It also should be remembered that the acceptable levels are those appropriate for an average person not to experience any adverse health effects. There are persons who are hypersensitive to a specific substance and the levels may not be suitable for them. It also is possible for persons not to be initially sensitive to a given material and to become more sensitive to it over a period of time. Sometimes this sensitization process may carry over to additional materials. We know or can infer a substantial amount about the acute effects of many chemicals at relatively high levels, primarily because of animal experimentation, but we do not know a great deal about the delayed effects of a single exposure to toxic material, the delayed effects of prolonged exposure to low levels of contaminants, or the synergistic effects of combinations of chemical exposures. It also is obvious that some individuals cannot tolerate particular environments because of heat, vibration, lighting conditions, ergonomics requirements, a phobic response such as claustrophobia, or because of the presence of chemicals or odors. However, most people can develop an equilibration with the environment, tolerating ambient conditions without any untoward effects. Occasionally a new element, such as a change in the process, additional or different chemicals, heat, vibration, etc., occurs, resulting in a hypersensitization wherein the previously acceptable pollutants are no longer tolerable. This is not a synergism, but a new condition. Precipitating concentrations of any pollutant may be below the TLV or TWA. Good housekeeping measures may be of some value, but when hypersensitization has occurred, there might be no alternative but to remove the employee from the locus. When asked for how long, only time will answer. Some individuals will gradually lose the sensitivity over 3 to 6 months, but some may retain it for life. Another condition can exist: the individual who has acquired the immune response or developed a lower threshold and is removed from the situation may become asymptomatic. The immune system, however, may remember and upon a reexposure the previous allergic response can occur in full force, just as it did at the time of primary acquisition. This is called the anamnestic response. Although it would be desirable to return an afflicted employee to the previous job, one must keep anamnesis in mind and proceed with caution. When reentry fails one must avoid the tendency to become angry, frustrated, or impatient with the employee. The reaction is not imaginary; it is more real and discomforting to the employee than it is to management. Therefore, the legal limits or the recommended levels should be considered as upper limits, not to be exceeded or approached if reasonable measures are available to reduce the actual levels in the workplace. There is no reason to excessively fear the laboratory environment, but it is foolish to scoff at or ignore reasonable measures to reduce levels of exposure.

4.8.3. Modes of Exposure

In order to better understand how exposure to hazardous materials in the laboratory enters into operational safety, a brief article from the 2nd edition of this handbook, pp. 314 to 316, by Herbert E. Stokinger, will be used to illustrate this point.

4.8.3.1. Means of Contact and Entry of Toxic Agents

Of the various means of body exposure to toxic agents, skin contact is first in the number of affections occupationally related. Intake by inhalation ranks second, while oral intake is generally of minor importance except as it becomes a part of the intake by inhalation or when an exceptionally toxic agent is involved. For some materials, as might be inferred, there are multiple routes of entry.

4.8.3.1.1. Skin Contact

Upon contact of an industrial agent with the skin, four actions are possible: (1) the skin and its associated film of lipid and sweat may act as an effective barrier which the agent cannot disturb, injure, or penetrate; (2) the agent may react with the skin surface and cause primary irritation; (3) the agent may penetrate the skin, conjugate with tissue protein, and effect skin sensitization; and (4) the agent may penetrate the skin through the folliculi-sebaceous route, enter the blood stream, and act as a systemic poison.

The skin however is normally an effective barrier for protection of underlying body tissues, and relatively few substances are absorbed through this barrier in dangerous amounts. Yet serious and even fatal poisonings can occur from short exposures of skin to strong concentrations of extremely toxic substances such as parathion and related organic phosphates, tetraethyl lead, aniline and hydrocyanic acid. Moreover, the skin as a means of contact may also be important when an extremely toxic agent penetrates body surfaces from flying objects or through skin lacerations or open wounds.

4.8.3.1.2. Inhalation

The respiratory tract is by far the most important means by which injurious substances enter the body. The great majority of occupational poisonings that affect the internal structures of the body result from breathing airborne substances. These substances lodging in the lungs or other parts of the respiratory tract may affect this system, or pass from the lungs to other organ systems by way of the blood, lymph, or phagocytic cells. The type and severity of the action of toxic substances depend on the nature of the substance, the amounts absorbed, the rate of absorption, individual susceptibility, and many other factors.

The relatively enormous lung-surface area (90 square meters total surface, 70 square meters alveolar surface), together with the capillary network surface (140 square meters) with its continuous blood flow, presents to toxic substances an extraordinary leaching action that makes for an extremely rapid rate of absorption of many substances from the lungs. Despite this action, there are several occupationally important substances that resist solubilization by the blood or phagocytic removal by combining firmly with the components of lung tissue. Such substances include beryllium, thorium, silica, and toluene-2,4-disocyanate. In instances of resistance to solubilization or removal, irritation, inflammation, fibrosis, malignant change, and allergenic sensitization may result.

Reference is made in the following material to various airborne substances, and to some of their biologic aspects.

Particulate Matter: Dust, Fume, Mist, and Fog

Dust is composed of solid particulates generated by grinding, crushing, impact, detonation, or other forms of energy resulting in attrition of organic or inorganic materials such as rock, metal, coal, wood, and grain. Dusts do not tend to flocculate except under electrostatic forces; if their particle diameter is greater than a few tenths micron, they do not diffuse in air but settle under the influence of gravity. Examples of dust are silica dust and coal dust.

Fume is composed of solid particles generated by condensation from the gaseous state, as from volatilization from molten metals, and often accompanied by oxidation. A fume tends to aggregate and coalesce into chains or clumps. The diameter of the individual particle is less than 1 μm. Examples of fumes are lead vapor on cooling in the atmosphere; and uranium hexafluoride (UF_6) which sublimes as a vapor, hydrolyses, and oxidizes to produce a fume or uranium oxyfluoride (UO_2F_2).

Mist is composed of suspended liquid droplets generated by condensation from the gaseous to the liquid state as by atomizing, foaming, or splashing. Examples of mists are oil mists, chromium trioxide mist, and sprayed paint.

Fog is composed of liquid particles of condensates whose particle size is greater than 10 μm. An example of fog is super-saturation of water vapor in air.

Gas and Vapor

A gas is a formless fluid which can be changed to the liquid or solid state by the combined effect of increased pressure and decreased temperature. Examples are carbon monoxide and hydrogen sulfide. An aerosol is a dispersion of a particulate in a gaseous medium while smoke is a gaseous product of combustion, rendered visible by the presence of particulate carbonaceous matter.

A vapor is the gaseous form of a substance which is normally in the liquid or solid state and which can be transformed to these states either by increasing the pressure or decreasing the temperature. Examples can include carbon disulfide, gasoline, naphthalene, and iodine.

Biologic Aspects of Particulate Matter

The size and surface area of particulate matter play important roles in occupational lung disease, especially the pneumoconioses. The particle diameter associated with the most injurious response is believed to be less than 1 μm; larger particles either do not remain suspended in the air sufficiently long to be inhaled or, if inhaled, cannot negotiate the tortuous passages of the upper respiratory tract. Smaller particles, moreover, tend to be more injurious than larger particles for other reasons. Upon inhalation, a larger percentage (perhaps as much as tenfold) of the exposure concentration is deposited in the lungs from small particles. This additional dosage and residence time act to increase the injurious effect of a particle.

The density of the particle also influences the amount of deposition and retention of particulate matter in the lungs upon inhalation. Particles of high density behave as larger particles of smaller density on passage down the respiratory tract by virtue of the fact that their greater mass and consequent inertia tend to impact them on the walls of the upper respiratory tract. Thus a uranium oxide particle of a density of 11 and 1 μm in diameter will behave in the respiratory tract as a particle of several microns in diameter, and thus its pulmonary deposition will be less than that of a low density particle of the same size.

Other factors affecting the toxicity of inhaled particulates are the rate and depth of breathing and the amount of physical activity occurring during breathing. Slow, deep respirations will tend to result in larger amounts of particulates deposited in the lungs. High physical activity will act in the same direction not only because of greater number and depth of respirations, but also because of the increased circulation rate, which transports the toxic amounts of certain hormones that act adversely on substances injurious to the lung. Environmental temperature also modifies the toxic response of inhaled materials. High temperatures in general tend to worsen the effect, as do temperatures below normal, but the magnitude of the effect is less for the latter.

Biologic Aspects of Gases and Vapors

The absorption and retention of inhaled gases and vapors by the body are governed by certain factors different from those that apply to particulates. Solubility of the gas in the aqueous environment of the respiratory tract governs the depth to which a gas will penetrate in the respiratory tract. Thus very little if any of the inhaled, highly soluble ammonia or sulfur dioxide will reach the pulmonary alveoli, depending on concentration, whereas relatively little of insoluble ozone and carbon disulfide will be absorbed in the upper respiratory tract.

Following inhalation of a gas or vapor, the amount that is absorbed into the blood stream depends not only on the nature of the substance but more particularly on the concentration in the inhaled air, and the rate of elimination from the body. For a given gas, a limiting concentration in the blood is attained that is never exceeded no matter how long it is inhaled, providing the concentration of the inhaled gas in the air remains constant. For example, 100 ppm of carbon monoxide inhaled from the air will reach an equilibrium concentration in the blood corresponding to about 13 percent of carboxyhemoglobin in 4 to 6 hours. No additional amount of breathing the same carbon monoxide concentration will increase the blood carbon monoxide level, but upon raising the concentration of carbon monoxide level in the air a new equilibrium level will eventually be reached.

4.8.3.1.3. Ingestion

Poisoning by ingestion in the workplace is far less common than by inhalation for the reason that the frequency and degree of contact with toxic agents from material on the hands, food, and cigarettes are far less than by inhalation. Because of this, only the most highly toxic substances are of concern by ingestion.

The ingestion route passively contributes to the intake of toxic substances by inhalation since that portion of the inhaled material that lodges in the upper respiratory tract is swept upwards within the tract by ciliary action and is subsequently swallowed, thereby contributing to the body intake.

The absorption of a toxic substance from the gastrointestinal tract into the blood is commonly far from complete, despite the fact that substances in passing through the stomach are subjected to relatively high acidity and on passing through the intestine are subjected to alkaline media.

On the other hand, favoring low absorption are observations such as the following: (1) food and liquid mixed with the toxic substance not only provide dilution but also reduce absorption because of the formation of insoluble material resulting from the combinatory action of substances commonly contained in such food and liquid; (2) there is a certain selectivity in absorption through the intestine that tends to prevent absorption of "unnatural" substances or to limit the amount absorbed; and (3) following absorption into the blood stream the toxic material goes directly to the liver, which metabolically alters, degrades, and detoxifies most substances.

ACKNOWLEDGMENT

Reprinted from *Occupational Diseases,* Publication No. 1097, U.S. Public Health Service, Washington, DC, 1964, pp. 7–12.

4.8.4. Health Assurance Program

Concern for the effect of the work environment on an employee's health should be the primary purpose of any medical monitoring program. The term "health assurance" not only implies that the health of the participant will be scrutinized, but carries the positive connotation that steps will be taken to eliminate or minimize any negative impact on the employee's health of work-related factors.

4.8.4.1. An Overview of Health Assurance Programs*

The laboratory environment, as well as that of many support activities in academic institutions and industrial organizations, has the potential of causing health problems for the employees. Since many laboratories are working on the forefront of knowledge, the persons working in these environs may be exposed to materials for which significant knowledge of the health effects are not yet known. In other cases, the exposures to known materials may be sufficient to cause an adverse effect on an individual. A very positive component of a health and safety program for any industry, but especially those industries for which the risks are incompletely known or where stress on an individual's health may be unusually high, is a health assurance program, i.e., a program in which the health of the individuals is monitored on a regular basis or as a result of any out-of-the ordinary work experience. Not only does such a program serve to alert the facility and the individual of possible health implications of their activities, but normally is viewed positively by the employees as an interest in their welfare.

It is incumbent on universities and/or industries to provide well thought out and implemented health assurance programs. The size and complexity of the programs obviously should be geared to the size, type, and activity of the facility. The recognition of the elements which contribute to poor environmental conditions or are harmful to employees is mandatory before control or corrective actions can be formulated and effected.

Progressive corporations and academic institutions are recognizing that a pleasant, safe, healthy environment and contented, healthy employees results in:

- Decreased employee turnover—less training for new employees
- Increased productivity—both quantity and quality
- Longevity—employees retire later
- Decreased lost time and insurance costs

The corporate health service can encompass many areas—biomechanics, psychology, ergonomics, health promotion, nutrition, CPR, first aid training, industrial hygiene, statistics, epidemiology, and occupational medicine. Related areas which could be in the program or closely tied to a health assurance program could be data processing, cost analysis, insurance evaluation, and workers' industrial compensation and laws. These areas might be addressed to whatever degree applicable by nurses, hygienists, technicians, or physicians. If the corporation is small, or if a large facility is near a specialized clinic, all of these parameters could be handled by that clinic without establishing an in-house health facility, although the organization might have more control over costs with an in-house program.

The interrelated services of a health assurance department must be correlated and administered by a highly motivated, knowledgeable manager. This may vary from a nurse in a small facility to an experienced health services professional in a large facility, such as

---

*   This section was written by Richard F. Desjardins, M.D.

a university with a multiplicity of situations. In a large factory, the gamut of exposures is relatively static, but in a technically oriented organization with multiple research programs covering many disciplines, there are myriad factors to be considered such as: (1) radiation—isotopes, reactors, X-ray machines, lasers, electron microscopes, etc., (2) diverse chemicals—solvents, pesticides, herbicides, carcinogens, metals, etc., (3) animal handling-research, veterinarian school, and agriculture, (4) genetic recombination-engineering, (5) employees working in power plants, steam tunnels, electric services, plumbing, carpentry, custodial work, etc., (6) air quality in buildings such as gymnasiums, laboratories, laboratory hoods, "energy efficient" buildings, dormitories, and classrooms, (7) emergencies—crowd control, fire control facilities, injuries whether in an athletic performance, fights, drug or alcohol problems, etc.

All of these areas must be dealt with, whether in a university or factory, as they arise in order to maintain a healthy, safe environment.

When environmental problems in the workplace arise, industry and university personnel are free to consult federal and state agencies for advice, to register complaints, or for help in controlling or solving certain conditions. Consultants who are specialists in nearly any given area can be called upon for help from such agencies as NIOSH and OSHA.

Qualifications of health physicists, safety analysts, technicians, and engineers can be obtained from the many texts relating to the organization, function, and running of a good health and safety department. There are many good references with details.[1-5,13]

4.8.4.1.1. Health Assurance Medical Departments

It is desirable that the health assurance program be a division of a health and safety department within an organization. Their missions are strongly interdependent. No nurse or physician has the specific knowledge, experience, or facilities to perform the tests, surveillance, or control as do the other members of the health and safety department, nor do the latter typically have the medical knowledge to operate independently. However, both groups should be able to obtain the help required from the other.

The size and complexity of the facility ultimately determines the size and complexity of the medical branch of the health and safety department. It is the responsibility of the employer to know when, where, and how to obtain medical personnel to provide medical services for the employees. The AMA (American Management Association) and the American Occupational Medical Association have plans and pamphlets containing much information to help devise and implement a medical facility.

In a small facility, the number of employees and the nature of the work dictate the extent of health monitoring. Clerical employees might be able to avail themselves of the medical facility, but may not need general physical or specific testing, unless, as is becoming more frequent, air quality problems arise stemming from the construction of energy efficient buildings in recent years. On the other hand, workers exposed to asbestos, toxic chemicals, or radiation would require more complete and frequent examinations, testing, and monitoring. In a small facility without undue usage of toxic chemicals or hazards, a trained nurse, paramedic, or physician's assistant might do all that is necessary and refer more complicated problems to a contract physician or hospital.

According to Dr. Marcus Bond,[5,7] a work force of 300 that includes employees who require periodic monitoring and medical examinations can justify an in-plant medical department with a full-time nurse and a part-time physician. For 100 to 300 employees, a part-time nurse, paramedic, or physician assistant might suffice; for 300 to 800 employees, one full-time nurse is usually sufficient. A part-time physician could be on call or spend a specified number of hours at the plant. A group of employees numbering 800 to 1500, including a substantial number who require physical and environmental monitoring, will justify a full-time physician and at least two industrial nurses.

This is an opportunity for a family practice physician[6] to participate in a part-time capacity in industry. It must be of some interest for the physician or else motivation would be missing in the proper medical and psychological care of employees. It is an opportunity to further involve themselves with patients and the functioning community. Statistics indicate that primary care physicians provide about 80% of employee medical care. Only 20% of plant physicians are occupational physicians. In the U.S., financially secure companies continue to have a full-time physician for approximately each 3000 employees, although some industries are known to have only one for about 15,000 employees. Indeed, there are a few substantial companies which have no full-time physician regardless of the number of employees. This might be because of low exposure or toxicity in the facility, or because a commercial medical clinic is contracted to care for their employees (HMO, PPO, emergency clinics).[7]

The acceptance of a health program by both management and by those actually manufacturing the product, whether it is paperwork, nuts, bolts, chemicals, or research, is by a slow accrual of satisfied patients. There is an inherent suspicion and skepticism by both areas. Management needs to know that they are complying with federal and state mandates and would like to see a positive productivity gain as a result of health expenditures. The workers find it hard to believe that management would venture a program other than for financial gain. It is gratifying to observe the progression of acceptance by both vital areas in programs that succeed.[8] It has been amply shown that it is cost effective to have a good health program with caring personnel. The employees are healthier, happier, and more productive; the turnover of workers decreases, lessening training expenditures; loss time for illness or injuries decreases both because of attitude and from instruction and training in "wellness"; workers who feel as though they are an integral, functional, and productive part of a facility are content to stay on the job to retirement. Environmental stress is diminished and managed.[14-17]

The physician who accepts the challenge of participation in a health program should obviously be able to perform the usual functions of a general physician, and, in addition, should be knowledgeable in the psychology of workers, the hazards and conditions in varying work sites, toxicology, and communicable diseases in the workplace.[11,12] Recognition of such problems as drug abuse, alcoholism, and the effects of smoking is mandatory. Although this is a very broad background, we must consider that patients/workers are male/female, young/middle-aged/elderly.

The occupational health examination[8,9] may be divided into three areas:

1. Pre-employment or pre-placement
   a. medical history
   b. occupational history
   c. physical exam
   d. laboratory and X-ray (if needed)
   e. multiphasic screening

2. Periodic examination (with interval history)
   a. annual physical examination or at desirable intervals
   b. executive physicals[10]
   c. toxic or hazardous exposure bioassays

3. Special examinations
   a. food handlers
   b. job transfers (if markedly different)
   c. return to work after serious illness or injury
   d. retirement examinations (document final condition and advise as to future health and wellness)
   e. fitness classification

These areas of assessment (1) determine the immediate health state, any change since previous examinations, and the suitability to work in any area; (2) will suggest advice and modalities to enhance or improve health; and (3) may indicate conditions of stress or unhealthy situations in the environment needing attention and change.[12]

The results of an examination should be kept confidential. A layperson should not be expected to interpret the results and make decisions as to employment on that data. The physician does not hire or advise that a person should be hired. That is a corporate decision. However, the physician can categorize the pre-employee or employee into several levels:

1. Fit for general work—physically and mentally.
2. Fit for work only in specific categories—physically or mentally.
3. Unfit for employment at this time—presence of medical condition requiring attention. When corrected may be eligible for employment.
4. Incapacitating condition—illness, injury (old or new), or mental illness of a chronic nature. These would prevent employment in either general category or specific category.[8]

The validity of the physician's assessment would depend on his knowledge of the required work conditions of the specific facility, and his ability in disability assessment.[11] It is also an introduction to the applicant or employee of a caring medical resource within the company or university. The perception of the physician as a "company doc" is really an uncomplimentary epithet. Fostering the perception of a caring physician really interested in the patient, who also happens to work for the establishment, will contribute to a more accurate assessment as well as help maintain healthiness at work.

Many applicants and employees are educationally deficient, but this must not be interpreted as low intellectual capacity. Very often, the appreciative patient will indicate that they had always wondered about a condition, but no doctor had ever taken the time to discuss it in understandable terms. The results of a small amount of expended time are very gratifying.

The initial assessment of an employee should be comprehensive. The elements of a specific company's products should dictate the specific areas to be assessed beyond the general. Obviously an individual who is to do heavy lifting of any kind should have musculoskeletal and neurological systems carefully examined. Whether X-rays are mandated is still a moot point; they really are of marginal value generally. If there is a suggestion of a problem which would make X-rays of value, either on the physical or from the medical history, X-rays must be taken. On the other hand, exposure to chemicals, gases, heat, cold, radiation, etc., would need more specific scrutiny in other systems. The initial medical history is completed by the applicant prior to seeing the physician, and carefully reviewed in the applicant's presence. Some areas can be amplified and clarified during this time. Encouraging questions by the patient at this time emphasizes caring. Following this, a careful physical examination is carried out.[8] It is probably a good idea to discuss findings on the physical examination as it proceeds. When the examination is completed, proceed to other testing. If specific findings during the physical mitigate against performing certain tests, the routine can be modified. Some tests may be done by the physician, nurse, or a technician. They might include color perception, visual acuity, audiometry, spirometry, glaucoma, and electrocardiography. Note that some tests, such as the one for glaucoma, are not normally occupationally related, but initial screening is relatively easy to do and detecting these things will enable the employee to seek additional medical help. There are some tests, such as a blood panel or urinalysis, which require laboratory services. All of these are mainly screening tests. Positive findings may, and probably should, be referred back to the patient's physician or to an appropriate specialist. The industrial physician must not be in competition with outside physicians. By the same token, if the patient signs an authorization, a complete copy of the examination can be forwarded to them or to the physician of their choice. Under ordinary conditions, it is not wise or proper to refer an employee to a specific physician unless that physician is the only one able to perform a given function. A list of qualified names can be provided for referral. This also contributes to harmonious relations with the local medical community.

Subsequent physical examinations of employees should be spaced appropriate to the nature of their jobs. The questionnaire may be abbreviated if they indicate the absence of changes. The physical, however, should be as careful and complete as at the beginning. This will naturally reveal any changing status, i.e., needs glasses or hearing aids, dermatitis, tumors, glaucoma, asbestosis, etc. Finding any deviation from normal early is a real bonus to treatment. By the time some symptoms are obvious to the patient, it may be too late. As an old medical professor once said, "There's a lot of pathology out there. All you have to do is find and recognize it." It is the responsibility of the physician to adhere to three dictums or duties:

1. Prevent disease.
2. Diagnose and treat to the best of your ability.
3. Help the patient's demise to be with as much dignity as possible. Good rapport will allow this.

The industrial physician may want to compose or purchase pamphlets appropriate to the specific facility and leaflets or booklets on general health-promoting ideas: smoking cessation, cholesterol control, back care, weight control, why and how to exercise, etc. How much good these actually do is not well documented, but employees do pick them up and carry them home. Perhaps even small dollops of advice absorbed will contribute to the enhancement of healthfulness. Recently, the author's institution added a wellness program available to all of the employees instead of just the occupationally challenged. An excellent response was obtained, which may indicate that there is a substantial level of interest in employer-sponsored health programs, and positive results may improve the overall health of the participants.

The highly motivated industrial physician should have the facilities to address most environmental, factory, and facility problems. If management recognizes and adequately funds a health assurance program, tangible evidence of the health improvement (reflected in decreased loss time) would be found to be cost effective. Intangible evidence is hard to accrue, but well employees are likely to feel more content with their work and remain with the company longer, and are more likely to accept healthful ways of work and pursue similar attitudes at home.

Health assurance is an important element of a well thought out and implemented health and safety program. It is cost effective, humane, and generally good administrative policy to provide such a program.

REFERENCES

1. **LaDou, J., Ed.,** *Introduction to Occupational Health and Safety,* National Safety Council, Chicago, 1986.
2. Council of Occupational Health, AMA: Scope, objectives and functions of occupational health programs, *JAMA,* 174, 533, 1960.
3. Council on Occupational Health, AMA: A management guide for occupational health problems, *Arch. Environ. Health,* 9, 408, 1964.
4. *Guide to Developing Small Plant Occupational Programs,* American Management Association, Chicago, 1983.
5. **Bond, M.B.,** Occupational health services for small businesses and other small employee groups, in *Occupational Medicine: Principles and Practical Applications,* Chicago, 89, 1988.
6. **Howe, H.F.,** Small industry: an opportunity for the family physician, *Gen. Pract.,* 26, 166, 1962.
7. **Knight, A.L. and Zenz, C.,** Organization and staffing, in *Occupational Medicine, Principles and Practical Applications,* Chicago, 1988.
8. **Collins, T.R.,** The occupational examination: a preventive medicine tool, *Continuing Med. Ed.,* 77, February 1982.
9. *Guiding Principles of Medical Examinations in Industry,* American Management Association, Chicago, 1973
10. **Thompson, C.E.,** The value of executive health examinations, *Occup. Health and Safety,* 49, 44, 1980.
11. Disability Evaluation Under Social Security. A Handbook for Physicians, Social Security Administration, Baltimore, August 1973.
12. Occupational Diseases. A Guide to Their Recognition, U.S. Dept. of Health. Public Health Ser., Pub. no. 1097, U.S. Government Printing Office, Washington, D.C., 1966.
13. **Felton, J.S.,** Organization and operation of an occupational health program, *J. Occup. Med.,* 6, 25, 1964.
14. **French, J.R., Caplan, R.D, and Von Harrison, R.,** *The Mechanism of Job Stress and Strain,* Wiley Series of Studies in Occupational Stress, John Wiley and Sons, New York, 1982.
15. **House, J.S., Wells, J.A., Landerman, L.R., et al.,** Occupational stress and health among factory workers, *J. Health Soc. Behavior,* 20, 139, 1979.
16. **Kahn, R.L.,** Conflict, ambiguity, and overload: three elements in job stress, *Occup. Ment. Health,* 3, 2, 1973.
17. **Selye, H.,** *The Stress of Life,* McGraw-Hill Book Co., New York, 1956.
18. Description and Evaluation of Medical Surveillance Programs in General Industry and Construction, Final Report, Office of Regulatory Analysis, Directorate of Policy, Occupational Safety and Health Administration, U.S. Department of Labor, Washington, D.C., 1993.
19. **Stokinger, H.E.,** Means of contact and entry of toxic agents, in *CRC Handbook of Laboratory Safety,* 2nd ed., Steere, N.V., Ed., CRC Press, Boca Raton, FL, 1971, 314.

4.8.4.2. A Health Assurance Program

A formal Health Assurance (HA) program should not be intended to replace an environmental monitoring program but to complement one. In an earlier section, some of the limitations of a monitoring program were mentioned, but the main point, which should be appreciated, is that we do not necessarily know what are "safe" exposure levels. Indeed, safe levels may not exist, only levels in which the negative effects attributable to an exposure disappear into the statistical noise caused by the presence of other parameters.

This is especially true for individuals, with their wide range of susceptibilities. In many instances, even this much is not known since the data are not available, simply because the studies have not been performed. For example, the ACGIH tables incorporate several hundred chemicals, but this is very small compared to the 50,000 to 100,000 (according to various estimates) commercial chemicals already in use, to which must be added hundreds of new ones developed yearly. Further, the number of studies establishing safe levels for synergistic interactions of combinations of materials is virtually nil, as often as not associating the effects of a material with an easily measurable personal habit, such as whether the exposed individual smokes. The prospect of conducting synergistic studies, with all the combinations involved, is obviously not bright.

Regulatory requirements for a number of materials now specifically mandate employee access to an employer-supported medical surveillance program designed to monitor problems associated with the specific chemical. Many authorities recommend participation in programs for anyone who works with toxic substances during the normal performance of their duties. However, there are few specific recommendation on what actually constitutes a significant involvement with toxic materials which should trigger participation in an HA program. This will be discussed further later in this section. The content of the examination will depend to a great extent on the duties associated with the job. However, there are some components of a HA program which are universally agreed upon as essential:

1. A medical history
2. A prior work history
3. A pre-placement examination
4. Periodic reexaminations
5. An end of employment examination

The entire program should, in addition to a number of standard components, be tailored to the anticipated types of exposures. As these exposures change during alterations in the research program, the periodic reexaminations can be modified to reflect the changing conditions.

Even before the HA program is initiated, there are a number of key ethical issues which must be addressed. If participation in a HA program is to be required as a condition of employment, the advertisement for position should so state. Further, the examination must be clearly intended to determine if the duties would be such as to make it unsafe for the employee to perform the work or aggravate an existing health problem. If reasonable adjustments can be made in the duties or responsibilities, then the examination cannot be used to discriminate against an otherwise qualified applicant. This issue is basically the premise of the Americans With Disabilities Act. The employee should have an assurance of confidentiality. There are factors which may be health related, but which have no bearing on the ability of an employee to do the work assigned, and which cannot harm those with whom the employee would come into contact. The employee has the right to expect that any such information remain confidential. Finally, the employee should have access to the results of the examination and any tests performed and should be able to authorize release of the data to others, such as his family physician, if so desired.

4.8.4.2.1. Participation

If the organizational approach is to provide an examination based on need, then the necessity arises to define criteria as to whom should be included and whom should not.

Individuals who do not work with chemical or biologic agents or where their duties are not unduly stressful can be justifiably excluded. Since the intent of a HA program, as applied to research personnel, is to monitor the impact of chemicals or pathological organisms on the health of the individual (either directly or indirectly, e.g., wearing a respirator can place a burden on a person with impaired pulmonary function), exposure to chemicals and biologically active (to humans) agents should be a major factor to be considered in the participation of an individual in a HA program. OSHA requires access to a medical program for persons working with regulated carcinogens or who are exposed to human blood, tissues, and other fluids, and those required to wear respiratory protection, among others (a list of the OSHA sections requiring medical surveillance programs is provided as an appendix to this section). However for other substances the toxicity of the material, the mechanism of exposure, the duration of the exposure, the intensity of the exposure, the safeguards available to prevent exposure, the current state of an individual's health, and prior exposures all play a part in the decision.

If an individual is working with an agent which is significantly infectious to humans, there appears to be little question that participation in a medical program is needed. Although the probability of contracting a disease increases with higher exposure rates, once contracted, the characteristics of the disease are not dependent upon continued exposure or the initial level of exposure.

If a major portion of an individual's time is spent working with a regulated carcinogen, other regulated materials such as lead or cotton dust, or other materials which meet the criteria for being highly toxic, corrosive, a sensitizer, or an irritant, then again it is usually desirable for the individual to be in a medical program. Even if facilities are available, such as totally enclosed glove boxes in which the work is done, it is arguable that un-planned exposures could occur, and the conservative approach would be to include rather than exclude the person. It could be argued also that if the exposure levels are maintained sufficiently low, then participation in a program is not needed. It is on this basis that OSHA defines exempt levels for meeting some of the regulatory requirements for some of the regulated carcinogens. However, documentation of the low levels would appear to be required to deny access of an individual to a medical surveillance program on this basis. Of course, participation in a HA program is clearly indicated if a monitoring program provides information indicating that the individual uses materials of concern or is in an area where others use them and is actually exposed to airborne concentrations which are typically a significant percentage of acceptable levels.

The OSHA standard for respiratory protection in 29 CFR 1910.134(b)(10) states:

"Persons should not be assigned to tasks requiring the use of respirators unless it has been determined that they are physically able to perform the work and use the equipment. The local physician shall determine what health and physical conditions are pertinent. The respirator user's medical status should be reviewed periodically (for instance, annually)."

Persons who do not have a continued exposure to chemicals, but periodically perform tasks requiring intense uses of chemicals for a brief period, such as in agricultural field experimentation, should probably be included in a HA program. Not only are many agricultural chemicals quite toxic, but the working conditions place severe physiological stress on the research personnel and their support staff. Respirators should be worn, as should clothing which will not be permeable to the chemical sprays. Respirators place stress on the pulmonary and cardiac system; protective clothing which is impermeable to fumes and vapors usually does not transpire either, and the body temperature will rapidly increase since the clothing prevents heat from being carried away from the body by evaporation, conduction, or convection of perspiration.

Persons who have known health problems that could be aggravated by the exposures involved in their job duties or who have had prior work histories where they could have had significant exposures to chemicals that could have sensitized them to chemicals in the workplace or that could have initiated delayed effects also would fall in a category which should be considered for participation in a HA program.

It is most difficult to determine whether persons for whom the exposures are marginal—i.e., where the portion of their duties in which they use chemicals is limited, but they do use materials with properties that could cause ill effects part of the time—should be included in a HA program. It would be easy to establish a criterion that any use whatsoever should qualify a person for participation. However, many activities of normal everyday life involve use of such items as gasoline and household products containing toluene, acetone, phenol, isopropyl alcohol, ethyl alcohol, hydrogen peroxide, acid, and caustic materials such as lye, which certainly are toxic materials. The "any use" criterion is undoubtedly too liberal, unless one simply admits that there are no selection criteria and includes every employee. A compromise which appears reasonable, but which has no other scientific justification, is to arbitrarily select a percentage (such as 10% of a typical work week) for actual use or exposure to a chemical or combination of chemicals of average health risk as a threshold. An employee approximating an exposure of this level could be asked to fill out a form listing the chemicals which are in use in their vicinity and an estimate of the average time each are used. This form should be reviewed by a physician (preferably with a background in environmental and occupational medicine, if one is available) and his recommendation should govern the question of participation or not. However, if an individual wishes to be included, but who might not be recommended, probably should be permitted to do so.

Although every person who has duties which could give rise to health problems should be a participant, it is especially critical to include permanent employees. Many of the tests which are run on the individual have a sufficiently wide "normal" range (except in extreme cases of acute exposure, where the individual will receive medical attention anyway) that a single examination may not be particularly informative. However, problems due to environmental work conditions, as will be discussed later, may be revealed by trends shown by comparison of successive examinations.

4.8.4.2.2. Medical and Work Histories

The medical history and prior work histories are key components of any health assurance program. There are any number of health-related factors for which a heredity predisposition exists, so that the medical history will normally include a segment concerning family members, particularly parents. Obviously, known prior medical conditions will be of importance. Emphysema, for example, would certainly be of concern if the employee were to have to wear a respirator frequently during the course of his duties. Hypertension and heart problems would clearly be of importance if the job involved significant physical stress, and of course there are chemicals which directly or indirectly affect the heart function. Medical history forms vary substantially in content but one used for a HA program should be comprehensive. It is part of a record which, along with the prior work history and the actual examination, including tests which may be run, will constitute the baseline against which changes in the employee's health will be compared to determine if occupational exposures are having a negative effect on the employee's health. Some prospective employees are inclined to conceal previous illnesses if it is likely to affect their chances of obtaining a desired position. This is unfortunate, but quite understandable. It is important, if these conditions later manifest themselves, that they have been detected, if possible, by appropriate questions or during the pre-placement examination. The need to do so stems from a desire (1) to avoid responsibility for the diseases and (2) to explain them on the basis of the occupational exposures experienced by the individual.

The prior work history serves essentially the same purpose as does the medical history. For example, a prospective employee who would be working in agricultural research programs might have had a previous period of employment working with pesticides and herbicides, which could have had a depressant effect on his cholinesterase enzyme levels. It would be important to include a test of this parameter in the pre-placement examination. Even some nonchemical activities, such as previous work in a heavily dust-laden atmosphere, might have caused a decrease of pulmonary function to an extent that an individual might find it impossible to wear a respirator to provide protection against solvent fumes. Previous exposures to some chemicals or substances may result in effects delayed for many years, such as the latency periods generally associated with carcinogens. Among other substances for which any history of prior exposures might be elicited would be asbestos, dusts, welding fumes, heavy metals, pesticides, herbicides, acids, alkalis, solvents, dyes, inks, paints, thinners, strippers, gases, radiation, etc. If there are any specific areas of concern because the work regimen will involve materials known to have a possible impact on a given physiological function or organ, then the physician should supplement the standard questionnaire for both the medical and work histories with questions designed to elicit as much relevant information as practicable.

There are any number of common diseases which are not necessarily related to an occupational exposure but which could be a risk in the work environment. Diabetes and hypertension are certainly not necessarily work related, but the individual, unless treated, could be a hazard to self and potentially to co-workers. Similarly, a disease of the eyes, such as glaucoma, could interfere with a person's ability to see properly, but many persons could be unaware of its onset as it is an insidious disease, primarily a problem to persons over 40. A relatively simple automatic instrument is available to detect pressure increases in the eyes, which is a sign of the disease, and which would permit the physician to refer persons to an ophthalmologist. Loss of hearing could be a problem if persons do not hear warnings and, again, many persons do not realize that this has become a problem or are reluctant to admit it, even to themselves, as a sign of increasing age. Although these problems should have been brought to the attention of the individuals by their family physician, a surprising number of persons do not have a family physician or do not see one frequently. At the author's institution, approximately 15 to 20% of the persons participating in the HA program for the first time had reasonably serious problems of which they were unaware, and which could have placed them at risk, or at best, reduced their efficiency and productivity. The scope of the examination should be sufficient to detect these conditions which may not be job related.

4.8.4.2.3. Pre-Placement Examination

It would be highly desirable if a pre-placement exam could be given prior to any work exposure to provide a true baseline for the individual. However, unless a medical examination has been an integral part of an organization's employment procedure since the inception of the company or institution, then instituting a HA program will always catch a number of current employees already in the midst of research programs involving exposure to hazardous materials. A medical examination at this time will still have significance in the sense that future examinations can still be compared to the earlier one to detect changes during the interval between examinations. However, the information gained in the exam, including any test results, will not necessarily reflect the normal conditions for the employee. If, for example, an individual has been working within the organization using agricultural chemicals and has a very low level of cholinesterase enzyme at the time of the initial examination, it may be suspected that the employee's work has caused the depressed level of the enzyme to occur, but it is uncertain. The individual may be one who has a naturally low level. If a person is tested at the time of initial employment, then the effect of the working environment on the parameters measured in the examination will be much more apparent, although the effect of work exposures on an individual may be confused if similar exposures are likely to occur outside the workplace. In the example just used, if the initial examination revealed a normal enzyme level and a later one showed a depressed value, perhaps after a suspected exposure, then the initial conclusion, barring any alternate exposure mechanism, would be that an exposure had occurred and remedial steps taken to prevent further exposures and to prevent future incidents of the same kind.

The other major purpose of a pre-placement examination would be to avoid placing a person in a position in which an existing condition would be aggravated or the individual could be injured by the work environment. A color-blind person, for example, should not be placed in a position in which the ability to distinguish colors is essential to being able to work safely. An individual with severely reduced pulmonary function should not be placed in a position requiring wearing a respirator much of the time. These restrictions may make it impossible for an applicant to be offered a position, and it should be clearly stated in the advertised job qualifications in such a case that passing a pre-placement examination is required as a condition of employment. A by-product of such a restriction is that the organization may be protected against acquiring a future liability. If, for example, a person with depressed pulmonary function is hired without an examination and placed in a situation in which exposures could cause the same result, it could be difficult to prove that the problem did not arise from a recent exposure. On the other hand, detection of the problem in the pre-placement examination might lead to a decision not to hire the individual because of the problem.

There are some pitfalls in using the pre-placement examination as an exclusionary device. This was alluded to in Section 4.8.3, in which a cautionary flag was raised against using it as a discriminatory device. This can occur with the best intentions in the world. An organization may decide to exclude women of fertile age from a position in which they may be exposed to a teratogen. Discrimination may be claimed if installation of engineering controls to reduce the exposures to well below the permissible limits are feasible, but not done. A woman may decide on her own not to work in an area where even low levels of an embryotoxin are present, but the decision should be clearly her own with no taint of coercion. In the recent revision to workplace rules involving radiation by the NRC, a provision has been added making it the employer's responsibility to limit the exposure to the unborn fetus, but before the employer can take the required steps, the woman has to declare her condition. If she should choose not to do so, no matter how obvious it is, the employer may not be permitted to take the necessary steps.

A fairly common practice in a pre-placement medical examination program, in addition to a thorough physical and a battery of tests, is to take a serum sample to be stored in an ultralow temperature freezer. These samples take up very little space and are valuable should a question later arise where a comparison between a current serum specimen and a baseline sample would be useful. It is also possible and feasible to lyophilize the serum for storage. This might be cost effective and space saving if a large number of specimens are to be kept.

4.8.4.2.4. Reexamination

Periodic reexaminations should be scheduled for all participants of a HA program, whether it is a part of a program mandated by a standard, as is becoming more common in newer regulations, or as a result of an internal decision based upon the level of usage. Note that the new laboratory safety standard does not call for periodic examinations, but only requires access to a medical surveillance program on an "as needed" basis should an overexposure occur or might have occurred. The frequency of the reexamination need not be any fixed interval, but should be based on the level of exposure. Returning to the use of pesticides, which could cause a depression in the cholinesterase enzyme level, as an example, it might be desirable to test for this one component prior to a period of active use, a second time at the height of the spraying season, and again at the end of the period of activity (assuming the material is significantly dangerous to humans). For less toxic pesticides, this amount of testing might be excessive, but for intensive use of an exceptionally dangerous material such as parathion, it might be desirable to test daily. However, normally there would be no reason to perform a complete examination at an accelerated schedule such as this.

An annual testing interval between scheduled examinations is probably the one most often used in HA programs for individuals with typical exposures in a representative laboratory. However, for persons only marginally meeting requirements for participation, the interval between examinations might be extended to 2, 3, or 5 years. Some programs use a 5-year interval for a complete physical, but recommend special tests more often. The National Institute of Health (NIH), in their program for their animal handlers, recommends taking a new serum sample every 5 years. The medical advisor or occupational physician should evaluate the requirements for each participant to establish the optimum period between examinations.

After each examination, the physician should compare the results of the current examination to the findings of previous examinations. Except in isolated instances, such as the depression of the cholinesterase enzyme which we have been using as an example, or unless there has been a severe exposure in which acute effects might be anticipated, the primary means of detecting problems will be the comparison between the results of successive examinations. Changes in various parameters which have been measured, such as pulmonary function, might vary slightly between two successive tests, but a persistent trend toward poorer performance would indicate damage to the respiratory system. Similarly, should a persistent trend develop for the other parameter measured, the examining physician should discuss the work environment and other possible contributing factors, such as leisure-time activities, with the employee. In at least one instance a spraying program to control insects at a cottage where weekends were spent was a major threat to an individual's health rather than any personal problem or exposure from any other source. The patient did not mention this factor to the physician because he did not recognize it as a potential problem. As a result, the examining physician had major problems identifying the cause of the individual's illness and was unable to treat it successfully. By the time the problem was recognized, the patient had been highly sensitized to any similar material and had some long-term health problems, that affected his capacity to perform many activities.

4.8.4.2.5. Utilization of Results

The primary purpose of the examination is to protect the employee, with a secondary purpose being to help protect the institution from the liability associated with unwittingly allowing an individual to become ill due to the work environment.

As noted earlier, a substantial number of persons involved in a HA program may have existing problems which are not job related or, as a normal course of events, develop health problems which are clearly not job related. These may be detected during the HA examination, as readily as in any other comparable comprehensive physical examination. Some organizations will assume direct responsibility for treating these illnesses, although most do not, leaving the burden of seeking treatment on the patient. Financially, there is only a moderate difference to the patient in many cases, due to the availability of group health insurance plans. If the patient has the responsibility of seeking out medical treatment, however, the condition may remain untreated, although the examining physician should certainly encourage the individual to seek assistance. In such cases, the employees should have the right to authorize the release of the medical records to their own physicians and to have this done promptly by the organization for whom they work.

Where the physical examination reveals a medical condition which may be job related or is aggravated by the duties of the person's job or the environment in which the individual works, steps should be taken to protect the employee's health. One of the first things to consider is to confirm that the condition exists or to obtain additional data to better understand the problem by seeking additional tests, obtaining a second opinion, or referring the patient to a specialist. These options should be discussed with the patient. In some cases, the situation is sufficiently straightforward so that these follow-up steps would not be necessary.

Whether one postpones gathering supportive data from additional examinations depends somewhat upon the seriousness of the problem which has been discovered and the work situation. The physician, in consultation with the individual, his supervisor, and usually a representative of the department with overall responsibility for the organization's health and safety program should meet to see what temporary steps can be taken to reduce the risk to the employee. In some cases, the head of the department in which the employee works may have to become involved if the supervisor does not have the authority or the flexibility to make changes.

Once all the data are available, consideration of the options that are available to protect the employee should be carefully reviewed. A number of these should be routinely considered.

1. If the condition can be treated, a temporary change in duties may be all that is needed.
2. It may be feasible to make engineering changes to modify the work environment.
3. Personal protective equipment or safety devices can be used to reduce an individual's exposure if engineering changes are not practicable.
4. It may be possible to change the job activity.
5. Job responsibilities may be distributed differently among personnel in the facility if the person has a unique problem and the duties causing the difficulties are not a problem to the others.
6. It may be possible to reschedule the activity causing the problem to another time or to modify the affected individual's schedule.
7. If there are no suitable options available within the individual laboratory, then relocation within the organization should be considered.

Some of these options are more easily applied in the industrial environment than in the typical academic laboratory, where each person may be supported on a grant and each laboratory is nearly autonomous. There may be very little flexibility available to the laboratory supervisor or laboratory director. This makes the task of treating the employee fairly much more difficult, since the work causing the problem usually must be done, and the laboratory supervisor does not have the funds to hire a new person and provide work to justify keeping the original employee as well. There may, in fact, be little flexibility of any kind if there are no positions available at the time for which the individual is qualified or is willing to accept. However, every avenue must be explored, because it is not permissible to maintain the individual in a situation in which their health may be endangered, even if the person wishes to do so. A waiver of responsibility for any future problems by the corporation or institution signed by the employee is not an acceptable alternative, nor is it legally defensible.

In extreme cases in which every option has been examined and none are feasible, individuals may have to cease to work in the organization either by resigning or being terminated for their own protection. An employee relations specialist in matching persons to jobs as well as an individual charged with seeing that employees are not discriminated against should be brought into the situation well before this drastic step is considered. In such a case a financial severance settlement, insurance such as workers' compensation, or disability retirement options may be available to the employee.

4.8.4.2.6. Physician Training

Any physician involved in a health assurance program will have had the usual training and exposure to a variety of medical experiences. It also is highly desirable for the individual to have specific training in industrial medicine. Since the actual conditions of employee exposure to hazardous materials will differ with each organization, the physician should be sufficiently familiar with the types of exposures represented by the job descriptions of the employees to be able to apply his own expertise and experience to the potential exposures. The more complex and diversified the research programs in an organization, the more difficult this task will be. It probably would be desirable for the physician to set aside some time to visit the various research areas, and visit with both the supervisors and individual employees.

Since OSHA requires that a medical surveillance program be made available to employees working with regulated carcinogens and a number of other materials, the physician should be provided with all current information related to these standards as well as appropriate technical information relating to these materials and other hazardous materials used by the employees. The physician, should, for example, have access to a set of all current MSDSs for the chemicals used by the employees. Subscriptions should be provided to some of the excellent services available to keep track of the rapidly changing regulatory and technical information, as well as the usual medical journals. The physician should have an opportunity to attend relevant workshops, seminars, and professional meetings to ensure that his background is maintained at a high standard.

Finally, the way in which the employees perceive the physician is an extremely important component of an organization's health and safety program. He should be perceived as professionally capable. It also is important that the employees do not perceive him as a "company" man. They must feel that their health is important to the physician and that if they are having a problem on the job, the physician is concerned about it for their sake, not because it will cause a problem for the organization. Certainly the physician should be concerned about the welfare of the organization, but this can be done by working to make sure that the health and safety of the organization's employees is protected. This is one reason, as noted earlier, why the name "health assurance program" is preferred over "medical surveillance program." The former has a much more positive sound than does the latter. Since a pre-placement medical examination is recommended for individuals exposed to hazardous materials, the physician has a superb opportunity to establish from the beginning that the company or institution is concerned about the employee's well-being.

4.8.4.2.7. Records

Because many materials are now known to have long-term effects and extended latency periods are known to exist for many carcinogens, it would be desirable to maintain all records pertaining to the medical examinations as well as those relating to exposures and monitoring for an extended period after the employee had left the organization. Many of the specific OSHA standards describe the records which must be maintained and the period for which the records must be kept. However, 29 CFR 1910.20 covers the topic of health and safety records in general. Some of the more critical portions of this section are given below. Some of these requirements are very detailed and demanding. The reader is referred to the OSHA Standards for General Industry for the complete version of the requirements for record maintenance.

Access to employee exposure and medical records

(a) *Purpose:* The purpose of this section is to provide employees and their designated representatives a right of access to relevant exposure and medical records, and to provide representatives of the Assistant Secretary a right of access to these records in order to fulfil responsibilities under the Occupational Safety and Health Act....

$$\vdots$$

(2) This section applies to all employee exposure and medical records, and analyses thereof, of employees exposed to toxic substances or harmful physical agents, whether or not the records are related to specific occupational safety and health standards.

$$\vdots$$

(4) "Employee" means a current employee, a former employee, or an employee being assigned or transferred to work where there will be exposure to toxic substances or harmful physical agents. In the case of a deceased or legally incapacitated employee, the employee's legal representative may directly exercise all the employee's rights under this section.

(5) "Employee exposure record" means a record containing any of the following kinds of information concerning employee exposure to toxic substances or harmful physical agents:

(i) environmental (workplace) monitoring or measuring, including personal, area, grab, wipe, or other form of sampling, as well as related collection and analytical methodologies, calculations, and other background data relevant to interpretation of the results obtained;

(ii) biological monitoring results which directly assess the absorption of a substance or agent by body systems (e.g., the level of a chemical in the blood, urine, breath, hair, fingernails, etc.) but not including tests which assess the biological effect of a substance or agent;

(iii) material safety data sheets; or

(iv) in the absence of the above, any other record which reveals the identity (e.g., chemical, common, or trade name) of a toxic substance or harmful physical agent.

(6)(i) "Employee medical record" means a record concerning the health status of an employee which is made or maintained by a physician, nurse, or other health care personnel, or technician, including:

(A) medical and employment questionnaires or histories (including job description and occupational exposures),

(B) the results of medical examinations (pre-employment, pre-assignment, periodic or episodic) and laboratory tests (including X-ray and all biological monitoring),

(C) medical opinions, diagnoses, progress notes, and recommendations,

(D) descriptions of treatments and prescriptions, and

(E) employee medical complaints.

(ii) "Employee medical records" does not include the following:

(A) physical specimens (e.g., blood or urine samples) which are routinely discarded as a part of normal medical practice, and are not required to be maintained by other legal requirements,

(B) records containing health insurance claims if maintained separately from the employer's medical program and its records, and not accessible to the employer by employee name or other direct personal identifier (e.g., social security number, payroll number, etc.), or

(C) records concerning voluntary employee assistance programs (alcohol, drug abuse, or personal counseling programs) if maintained separately from the employer's medical program and its records.

(7) "Employer" means a current employer, a former employer, or a successor employer.

(8) "Exposure" or "exposed" means that an employee is subjected to a toxic substance or harmful physical agent in the course of employment through any route of entry (inhalation, ingestion, skin contact or absorption, etc.) and includes past exposure and potential (e.g., accidental or possible) exposure, but does not include situations where the employer can demonstrate that the toxic substance or harmful agent is not used, handled, stored, generated, or present in the workplace in any manner different from typical non-occupational situations.

(9) "Record" means any item, collection, or grouping of information regardless of the form or process by which it is maintained (e.g., paper document, microfiche, microfilm, X-ray film, or automated data processing).

(d) *Preservation of records.* (1) Unless a specific occupational safety and health standard provides a different period of time, each employer shall assure the preservation and retention of records as follows:

(i) *Employee medical records.* Each employee medical record shall be preserved and maintained for at least the duration of employment plus thirty (30) years, except that health insurance claims records maintained separately from the employer's medical program and its records need not be retained for any specified period;

(ii) *Employee exposure records.* Each employee exposure record shall be preserved and maintained for at least thirty (30) years, except that:

(A) Background data to environmental (workplace) monitoring or measuring, such as laboratory reports and worksheets, need only be retained for one (1) year so long as the sampling results, the collection methodology (sampling plan), a description of the analytical and mathematical methods used, and a summary of other background data relevant to interpretation of the results obtained, are retained for at least thirty (30) years; and

(B) Material safety data sheets and paragraph (c)(5)(iv) records concerning the identity of the substance or agent need not be retained for any specified period as long as some record of the identity (chemical name if known) of the substance or agent, where it was used, and when it was used is retained for at least (30) years; and

(iii) *Analyses using exposure or medical records.* Each analysis using exposure or medical records shall be preserved and maintained for at least thirty (30) years.

(e) *Access to records.* (1) *General.* (i) Whenever an employee or designated representative requests access to a record, the employer shall assure that access is provided in a reasonable time, place, and manner, but in no event later than fifteen (15) days after the request for access is made.

(ii) Whenever an employee or designated representative requests a copy of a record, the employer shall, within the period of time previously specified, assure that either:

(A) a copy of the record is provided without cost to the employee or representative,
(B) the necessary mechanical copying facilities (e.g., photocopying) are made available without cost to the employee or representative for copying the record, or
(C) the record is loaned to the employee or representative for a reasonable time to enable a copy to be made.

⋮

Employers can charge reasonable direct expenses for additional copies of records, except that a certified collective bargaining agent for the employee can receive a copy without cost, and if new information is added to the record, this information is available to the employee without cost under the same conditions as the original record.

For certain medical records, there is protection for the privacy of the individual, in making records available:

⋮

(e)(2)(ii)(E) Nothing in this section precludes a physician, nurse, or other responsible health care personnel maintaining employee medical records from deleting from the requested medical records the identity of a family member, personal friend, or fellow employee who has provided confidential information concerning an employee's health status.

and, under *Analyses using exposure or medical records:*

(e)(2)(iii)(B) Whenever access is requested to an analysis which reports the contents of employee medical records by either direct identifier (name, address, social security number, payroll number, etc.) or by information which could reasonably be used under the circumstances indirectly to identify specific employees (exact age, height, weight, race, sex, dates of initial employment, job title etc.), the employer shall assure that personal identifiers are removed before access is provided. If the employer can demonstrate that removal of personal identifiers is not feasible, access to the personally identifiable portions of the analysis need not be provided.
New employees have certain rights concerning records from the beginning of their employment.

(g) *Employee information.* (1) Upon an employee's first entering into employment, and at least annually thereafter, each employer shall inform employees exposed to toxic substances or harmful physical agents of the following:

(i) The existence, location and availability of any records covered by this section;
(ii) the person responsible for maintaining and providing access to records; and
(iii) each employee's rights of access to these records.

⋮

Corporations are often bought out, merge with other firms, or cease to operate, and provision is made in the standard for the retention of records for the required periods by transfer of the records to the successor firm or under requirements of specific standards to NIOSH. This is rarely a problem for academic research institutions which seldom cease to operate, although semiautonomous components which retain their own records may cease to exist. In such a case their records should be subsumed into those of the parent institution.

4.8.4.2.8. CPR and First Aid Training

Subpart K—Medical and First Aid, 29 CHR 1910.151 of the OSHA Standards for General Industry describes the minimal medical care which must be available to employees, although there are references to first aid in several other sections of the standards. This short section is given below in its entirety:

Medical services and first aid

(a) The employer shall ensure the ready availability of medical personnel for advice and consultation on matters of plant health.

(b) In the absence of an infirmary, clinic, or hospital in near proximity to the workplace which is used for the treatment of all injured employees, a person or persons shall be adequately trained to render first aid. First aid supplies approved by the consulting physician shall be readily available.

(c) Where the eyes or body of any person may be exposed to injurious corrosive materials, suitable facilities for quick drenching of the eyes and body shall be provided within the work area for immediate emergency use.

The need for access to medical services in emergencies was discussed at some length in Chapter 1. However, prompt action can frequently save an individual's life or can significantly reduce the seriousness of injuries. Although not intended as an instruction manual, Chapter 1 presented some first aid procedures for accidents involving chemicals and CPR techniques. It would be highly desirable for individuals working in facilities where hazards are present to be trained in both of these subjects. By working carefully so as not to tempt fate too much, and with a great deal of luck, an individual may go through an entire working career without personally experiencing an accident or being present when someone else does, but this cannot be counted upon. Although you cannot perform CPR on yourself, and you may be incapacitated so that even simple first aid is beyond you, if enough personnel in a laboratory do make the effort to become trained, it is likely that someone will be available to start emergency aid while waiting for more skilled personnel to arrive. The training for basic first aid and single-person CPR is not difficult, and everyone should annually devote the few hours necessary to receive and maintain these skills. Many rescue squads, fire departments, hospitals, and other public service agencies offer the training at a minimal fee covering only the cost of the manuals and supplies.

4.8.4.2.9. Vaccinations

All of us as children probably received some vaccinations against a number of diseases. A number of common diseases afflicting children who were born in the first third of this century are now decreasing in frequency as a result of widespread vaccination programs. A recent controversy centered around whether the last smallpox virus in the world, being maintained in a laboratory, should be destroyed. Yet this used to be one of the world's great killers. Relatively recently, vaccinations for other diseases have been developed, and diseases such as polio and measles are relatively rare now in the United States, although, unfortunately, there has been some resurgence of these two illnesses. Tuberculosis is also on the rise as a consequence in the spread of AIDS. It would appear that with the obvious benefit, vaccination against a disease would be a matter of course, providing that a vaccine exists. This is not necessarily the case.

Several factors need to be considered in determining whether vaccination is desirable or not. The first clearly is: Does a safe, reliable vaccine exist? At one time, rabies vaccine using duck embryos was the best available. However, it did not always provide a reliable immunization, and a significant fraction of the persons on which it was used had reactions, some of which were neurologically very severe. Now, a much more reliable human diploid rabies vaccine is available which provides protection for a very high percentage of persons, and the incidence of untoward reactions is very low. It is probably desirable to mandate vaccination for all personnel who face a high risk of exposure to rabies, i.e., persons who work directly with animals that might be rabid, individuals who do necropsies on such animals, and technicians who work with untreated tissue from potentially rabid animals. The second question is: What is the risk-benefit to the individual if the disease is contracted? The disease may not be sufficiently serious as to warrant the risk of a possible reaction to a vaccination. On the other hand, if the disease is sufficiently life threatening, then the use of a vaccine would be indicated. Third: Is there a satisfactory postexposure treatment? This is really critical for life-threatening diseases. If there is not, and the exposure risk is significant, the use of even a less than totally satisfactory vaccine might well be considered. Other considerations would be the state of the individual's health. If the condition of the person is such that the possibility of an adverse reaction could have a strong negative impact on the individual, then one would question the desirability of using a vaccine, but one would also question placing such a person in an environment in which vaccination might be considered.

Booster injections are needed for some diseases to ensure an adequate protective level of antibodies. However, some patients may experience reactions to a booster. It is advisable to do a blood titer test prior to repeat injections. If the titer is adequate, no booster should be administered.

Although the laboratory supervisor should have considerable input in deciding whether a vaccine should be used or not, any decision to institute a mandatory vaccination program should be reviewed by a separate biosafety committee before implementation. Individuals must be fully informed of any possible risks.

APPENDIX

Materials that should be included in a medical surveillance program (note that not all of these necessarily involve laboratory usage of the material, some materials not normally found in the laboratory are included for completeness):

1. 29 CFR 1910.1014: Acetylaminofluorene, 2-
2. 29 CFR 1910.1045: Acrylonitrile
3. 29 CFR 1910.1011: Aminodiphenyl
4. 29 CFR 1910.1018: Arsenic, inorganic
5. 29 CFR 1910.1111 and 1101: Asbestos—nonlaboratory exposure
6. 29 CFR 1910.1010: Benzidine
7. 29 CFR 1910.1028: Benzene
8. 29 CFR 1910.1030: Bloodborne pathogens—postexposure
9. 29 CFR 1910.1027: Cadmium
10. 29 CFR 1910.1008: Chloromethyl ether, bis-
11. 29 CFR 1910.1029: Coke Oven emissions—exposure
12. 29 CFR 1910.1043: Cotton dust—exposure
13. 29 CFR 1910.1044: Dibromo-3-chloropropane, 1,2-
14. 29 CFR 1910.1007: Dichlorobenzidine, 3,3′ and its salts
15. 29 CFR 1910.1015: Dimethylaminoazobenzene, 4-
16. 29 CFR 1910.1012: Ethyleneimine
17. 29 CFR 1910.1047: Ethylene oxide
18. 29 CFR 1910.1048: Formaldehyde
19. 29 CFR 1910.120: Hazardous waste, emergency response personnel
20. 29 CFR 1910.1450: Laboratory chemicals, exposures above action, PEL levels, incidents, possible symptoms
21. 29 CFR 1910.1025: Lead
22. 29 CFR 1910.1006: Methyl chloromethyl ether
23. 29 CFR 1910.1009: Naphthylamine, beta
24. 29 CFR 1910.1003: Nitrobiphenyl, 4-
25. 29 CFR 1910.1016: Nitrosodimethylamine, N-
26. 29 CFR 1910.95: Noise, hearing conservation program required above action level
27. 29 CFR 1910.1013: Propiolactone, beta
28. 29 CFR 1910.1001 and 134: Respirator use, pulmonary function
29. 29 CFR 1910.1017: Vinyl chloride
30. 29 CFR 1910 Subpart Q: Exposures to welding fumes
31. 29 CFR 1928: Particulate respiratory exposures
32. 29 CFR 1928: Pesticide applicators

REFERENCES

1. **Hogan, J.C. and Bernaski, E.J.,** Developing job-related pre-placement medical examinations, *J. Occup. Med.,* 23(7), 469, 1981.
2. Health Monitoring for Laboratory Employees, Research and Development Fact Sheet, National Safety Council, Chicago, 1978.

4.8.5.   The OSHA Bloodborne Pathogen Standard: Infection from Work with Human Specimens

OSHA published 29 CFR 1910.1030, regulating exposures to bloodborne pathogens on December 6, 1991. The rule took effect on March 6, 1992. Under the standard, employers affected by the standard were to be in full compliance by July 6, 1992. An exposure control plan was to be in effect by May 5, 1992, and employee training made available by June 5, 1992. OSHA has taken a firm posture on implementation of the rule. OSHA can impose a fine of up to $70,000 per willful violation.

In the standard, OSHA defines *bloodborne pathogen* to mean "pathogenic micro-organisms that are present in human blood and can cause disease in humans." These pathogens include, but are not limited to, hepatitis B virus (HBV) and human immunodeficiency virus (HIV). The emphasis has been on these two diseases, although other diseases may be found in human blood and tissue and technically are covered. The standard also does not limit itself to human blood despite the name. The standard includes *other potentially infectious materials,* which are defined by the standard to mean (1) the following human body fluids: semen, vaginal secretions, cerebrospinal fluid, synovial fluid, pleural fluid, pericardial fluid, peritoneal fluid, amniotic fluid, saliva in dental procedures, any body fluid that is visibly contaminated with blood, and all body fluids in situations in which it is difficult or impossible to differentiate between body fluids; (2) any unfixed tissue or organ (other than intact skin) from a human (living or dead); and (3) HIV-containing cell or tissue cultures, organ cultures, HIV- or HBV-containing culture medium or other solutions, and blood, organs, or other tissues from experimental animals infected with HIV or HBV.

The CDC published guidelines for prevention of transmission of bloodborne diseases and identified certain substances in which the potential for transmittal of HBV and HIV was extremely low or nonexistent. These substances were feces, nasal secretions, sputum, sweat, tears, urine, and vomitus, unless they contain visible blood.

When the standard was published, it was predicted that the standard would prevent approximately 9000 infections of HBV and approximately 200 deaths per year from this disease. Far fewer persons are known to have acquired HIV from occupational activities—less than 100 at the time the standard went into effect. At the time of writing, there have been approximately 40 known cases of occupationally acquired HIV infection, although there are indications that the actual number could be twice as high.

4.8.5.1. Basic Provisions

There are several key components to a program to achieve compliance with the standard:

1. An exposure control plan
2. Exposure determination

The standard identifies several areas in which explicit guides are provided to assure compliance:

1. General
2. Engineering and work practice controls
3. Personal protective equipment
4. Housekeeping

Other requirements include:

1. HBV vaccination and postexposure evaluation and follow-up
2. Communication of hazards to employees
3. Record keeping

Each of the above areas will be discussed in the following sections.

4.8.5.2. Exposure Control Plan

The exposure control plan is the organization's written statement of how it plans to eliminate or minimize employee exposure. It must cover each of the broad areas listed in the previous section.

4.8.5.2.1. Exposure Determination

The employer must compile a list of job classifications in which (a) all employees in those jobs have occupational exposures and (b) some employees have occupational exposures. The employer must identify a list of tasks and procedures or groups of closely related tasks and procedures in which occupational exposure occurs and which the employees in the first two lists perform. The lists do not take into consideration any use of personal protective equipment to exclude personnel.

The first list in which all employees in those jobs do have occupational exposure might include for any research-oriented organization:

1. Medical doctor (organization's health service)
2. Nurse (organization's health service)
3. Research scientists
4. Technicians
5. Glassware cleaners
6. Laundry staff
7. Police/security
8. Cleaning staff
9. Maintenance staff
10. Athletic trainers (academic organizations)

The first two of these clearly would fit the requirement that all members of the group would be occupationally exposed, while the last eight could include some that would have occupational exposures while others do not.

The tasks and procedures associated with each of these groups is straightforward. A physician in the type of environment considered here, a research organization rather than a clinical hospital, would have to examine patients, perform pre-employment and periodic physical examinations, and treat minor injuries. The nurse would have to perform phlebotomies (draw blood), give shots, perform minor first aid, dress wounds, and collect contaminated clothing and dressings. The nurse would in the course of these actions handle syringes, blood vials, and possibly glass slides, and do pin pricks on fingers to obtain small blood samples. Technicians would handle vials containing blood and other fluids, pipettes, culture dishes, blenders, and sonicators, all of which could contain contaminated materials and result in an exposure used or handled improperly. Research scientists and their assistants can be exposed to virtually anything listed for a technician, but since they do not do as many things routinely, perhaps have a higher possibility for error.

Glassware cleaners, laundry staff, and cleaning staff are affected because of the operations of the professional staff. Glassware may have dried contaminated fluids or tissue still in or on the glassware. The laundry staff may be asked to wash gowns, sheets, and lab coats with dried blood on them. The cleaning staff may be exposed to broken glassware, lab instruments (e.g., syringes, scalpels, broken culture dishes), soiled bandages, and dressings improperly disposed of in the ordinary trash. The maintenance staff may be asked to repair equipment or change a HEPA filter in a biological safety cabinet.

Police and other emergency personnel, such as rescue squad members, are two groups with a high potential for exposure. The police and emergency rescue personnel are the first on the scene of an accident and are often called upon to administer emergency assistance such as CPR, treating wounds, or performing other actions in which their hands and clothing often are contaminated. The same sort of exposures occur for athletic trainers who, especially in the rougher contact sports, often have to bandage an injured player, remove bloody clothing, and collect potentially contaminated clothing after a game.

These sorts of tasks and procedures would need to be identified in the exposure plan.

4.8.5.2.2. Implementation of the Exposure Control Plan

General: The foundation of all the measures of compliance will be that universal precautions will be followed. In August 1987 the CDC published a document entitled "Recommendations for Prevention of HIV Transmission in Health-Care Settings." This document introduced the concept of universal precautions, which, simply recommended that steps be taken to prevent exposure of health care workers to the possibly contaminated blood, other body fluids, tissue from a human living or dead, HIV-containing cultures and other possibly contaminated items in the laboratory (see the first part of this section). The universal part of this concept comes from the assumption that all of these possible sources of infections are treated as if they *were* infected. This assumption extends to all personnel who may become infected by coming into contact with contaminated materials, from the physician or research scientist, to the laundry employee. The plan must make a commitment to adopt this policy and enforce it.

Engineering and work practice controls:

1. Engineering controls available could include:

   Biological safety cabinets, class I, II, or III
   Hand-washing sink in the laboratory
   Employees shall wash their hands (or any other body part) as soon as possible after removing gloves or other personal equipment or after contact of any part of the body with potentially contaminated materials
   Autoclave readily available
   Eyewash station, deluge shower available in the laboratory
   Plastic-backed absorbent paper or other work surface protective materials available
   Containers available to place under apparatus to catch spills
   Surfaces of doors, walls, floors, and equipment water-resistant for ease of decontamination.
   Joints, fixtures, and penetrations of the preceding items sealed to prevent contaminants from accumulating and to facilitate decontamination.
   Vacuum lines protected by disinfectant traps and HEPA filters

Ventilation inward into the facility, not recirculated but discharged directly to the outside

"Sharps" containers, as specified in the standard, readily available; in addition to syringes and needles, other "sharps" would include disposable pipettes, culture plates, capillary tubes, any broken or chipped glassware

Biological waste containers, as specified in the standard, available

2. Work practices could include:

Wear gloves whenever handling tissues or body fluids. Disposable gloves shall not be reused. Gloves shall be changed between contact with individual patients where such contacts are involved.

Wear appropriate personal protective equipment (protective equipment is appropriate only if prevents blood or other potentially contaminated equipment to pass through or reach the employee's work clothes, street clothes, undergarments, skin, eyes, mouth, or other mucous membranes, or normal usage of the equipment.

Smoking, eating, drinking, applying cosmetics, or handling contact lenses shall not be permitted in the laboratory.

No items intended for human consumption are to be stored in refrigerators, in cabinets, or on shelves or countertops with potentially contaminated materials.

Always use mechanical pipettes.

Procedures shall be used to avoid creation of aerosols, droplets, splashing; where these are unavoidable, containment shall be used.

Contaminated needles shall not be bent, recapped or removed, or sheared or broken except under conditions where no alternative is feasible.

All work surfaces will be decontaminated after a spill with a 1:10 solution of household bleach.

Clothing, lab coats, aprons, gowns, and other items of protective clothing worn in the facility will be removed prior to leaving the work area.

Avoid touching any item unnecessarily with contaminated gloves, including documents, pens, door knobs, telephones, etc. Any such items should be decontaminated after such contact.

The lists above of suggested environmental controls and work practices are not complete, since each facility will have some unique procedures, but clearly the principle behind all of these is to prevent employees from coming into contact with any contaminated materials and to prevent the spread of contamination to areas outside the work area or to other individuals.

Personal protective equipment (PPE): In the preceding section on engineering controls and work practices, numerous items of protective equipment were mentioned. There are several principles in the standard that govern the selection and use of PPE.

1. It must be appropriate. To be appropriate, it must not allow contaminated materials, including all of the materials which come under this classification, from reaching any part of the person wearing or using the PPE or any of their clothing. In selecting PPE the employer should consider the comfort and convenience of the wearer, as well as the equipment meeting the basic design standards for that item. Goggles, for example, need to be comfortable, not prone to fogging up, and permit wearing of prescription lenses, otherwise the employee may not wear them on every occasion when they need necessary.

2. Provision of PPE is the responsibility of the employer. It must be provided at no cost to the employee. Among items of PPE that are definitely included are goggles, respirators, masks, face shields, gloves, surgical caps, or hoods. Since some facilities

expect employees to provide their own uniforms, lab coats, and gowns, OSHA would consider these as general work clothes which the employer would not be required to provide them, but if any of these were to be considered as PPE, the responsibility would be that of the employee.

3. It is the responsibility of the employer to maintain, clean, and launder (if necessary) all items of PPE. Employees are not to be allowed to take clothing home for laundering.

4. OSHA requires the employer to "ensure" that the employee uses PPE when it is needed.

5. There are circumstances that make it acceptable to not wear PPE under conditions when it would normally be used. Section 1910.1030 (d)(11) states ..."unless the employer shows that the employee temporarily and briefly declined to use PPE when, under rare and extraordinary circumstances, it was the employee's judgement that in the specific instance its use would have prevented the delivery of health care or public safety services or would have posed an increased hazard to the safety of the worker or co-worker. When the employee makes this judgement, the circumstances shall be investigated and documented in order to determine whether changes can be instituted to prevent such occurrences in the future."

Housekeeping:

1. Employers shall ensure that the work site is maintained in a clean and sanitary condition. The employer must establish a written cleaning and decontamination schedule appropriate to the location of item to be cleaned within the facility, type of surface to be cleaned, type of soil present, and tasks or procedures being performed in the area.

2. All equipment and environmental and working surfaces shall be cleaned and decontaminated after contact with blood or other potentially contaminated infectious materials. The standard covers five specific situations:

   a. Work surfaces shall be decontaminated with an appropriate disinfectant (i) after completion of procedures, (ii) immediately or as soon thereafter as feasible when surfaces are overtly contaminated, (iii) after any spill of blood or potential infectious materials, or (iv) at the end of the work shift if the surface may have been contaminated since the last cleaning.

   b. Protective coverings of work surfaces, e.g., plastic-backed absorbent paper, shall be removed and replaced when they become overtly contaminated or at the end of a work shift if it may have become contaminated during the work shift.

   c. Bins, pails, cans, or similar receptacles intended for reuse that have a reasonable likelihood of becoming contaminated shall be inspected and decontaminated on a regular schedule and immediately cleaned and decontaminated as soon as feasible upon visible contamination.

   d. Broken, potentially contaminated glassware shall not be picked up with the hands. It shall be cleaned up using mechanical means (brush and dustpan, forceps, or tongs).

   e. Reusable "sharps" that are contaminated shall not be stored or processed in a manner that would require employees to reach into the containers where these "sharps" are placed by hand.

Regulated wastes: This section deals only with regulated medical or infectious waste. If waste materials have other hazardous characteristics, e.g., chemical or radioactive, handling of these mixed wastes will have to comply with the standards applicable to the other hazard characteristics as well.

1. Contaminated "sharps," discarding and containment: Contaminated "sharps" shall be discarded immediately or as soon as feasible into containers (Figure 4.32) that are closable, puncture-resistant, leakproof on sides and bottom, and labeled according to the requirements of the standard. The latter usually requires a biohazard symbol on the container and/or the container being red.

   a. The "sharps" containers shall be easily accessible to personnel, located as close as is feasible to where the "sharps" are used or can reasonably be expected to be found.
   b. Maintained upright throughout use.
   c. Replaced frequently and not be allowed to be overfilled.
   d. Closed immediately prior to being moved when being moved from the area of use to prevent spillage or protrusion of the contents.

   If any chance of leakage of the "sharps" primary containers exists, the container shall be placed in a secondary container which is closable, constructed to contain all the contents and prevent leakage during handling, storage, transport, or shipping, and labeled as was the original container.

2. Other regulated waste containment. The requirements for other regulated waste containers are essentially the same as for "sharps." If any contamination of the outside of the primary regulated waste container occurs, the container shall be placed in a secondary container.

3. Disposal of regulated waste. This topic will be covered in detail in a later section of this chapter, but basically it requires that the waste be rendered noninfectious, and the trend is to require it to be rendered unrecognizable as regulated medical wastes. Until recently the waste was either steam sterilized and subsequently discarded as ordinary trash or incinerated. Newer technologies are now being used in a number of locales, but steam sterilization and incineration remain the most common means of treatment of the wastes. The generator must be sure to follow the local regulations applicable to themselves. Otherwise, under the standard OSHA can issue a citation.

Laundry: Contaminated laundry shall be handled as little as possible, with a minimum of agitation. It shall be bagged or placed in an appropriate container at the location where it was used. Sorting or rinsing shall not take place at the location of use. The contaminated laundry shall be placed and transported in bags that are labeled or color coded as the standard requires unless the facility uses a laundry which utilizes universal precautions in handling all of the laundry received. If such is the case, alternative labeling or color coding is permitted if it permits the employees to recognize the laundry which requires compliance with universal precautions. The employer must ensure that employees who come into contact with contaminated laundry wear protective gloves and other appropriate personal protective equipment. If the contaminated laundry is wet and may soak or leak through the bag, the laundry must be placed in a secondary container to prevent this.

Hepatitis B vaccination:

1. The employer shall make available the HBV vaccine to all employees who have occupational exposures. The vaccinations shall be made available to the employee at no cost to the employee at a reasonable time and place. The vaccinations shall be performed by or under the supervision of a licensed physician or by or under the supervision of another licensed health care professional. The health care professional responsible for the vaccinations must be given a copy of the bloodborne pathogen standard. Any tests must be done by an accredited laboratory at no cost to the employee.

Figure 4.32. Typical sharps container.

2. The vaccinations must be made available after the employee has received training concerning information on the efficacy, safety, method of administration, benefits of vaccination, and the fact that the vaccine and vaccination will be free to the employee and within 10 working days of initial assignment to a position where occupational exposures are possible. If a titer reveals that the employee still has immunity from a prior vaccination or there are medical reasons why the employee should not receive the vaccination, the vaccination is not required.
3. The employer cannot make participation in a pre-screening program a prerequisite for receiving the vaccination.
4. The employee may decline to receive the vaccination, but the employer must provide the vaccination at a later time if the employee is still covered by the standard and decides to accept the vaccination.
5. If the employee declines to accept the vaccination, the employer must ensure that the employee signs the following statement:

"I understand that due to my occupational exposure to blood or other potentially infectious materials I may be at risk of acquiring hepatitis B virus (HBV) infection. I have been given the opportunity to be vaccinated with hepatitis B vaccine at no cost to myself. However, I decline hepatitis vaccination at this time. I understand that by declining this vaccination I continue to be at risk of acquiring hepatitis B, a serious disease. If in the future I continue to have occupational exposure to blood or other potentially infectious materials and I want to be vaccinated with hepatitis B vaccine, I can receive the vaccination series at no charge to me."

This statement should be signed and dated by the employee and kept as part of the records.
6. Any recommended future booster shots shall be made available under the same conditions.

Postexposure evaluation and follow-up: If an employee has an exposure incident, the employee should immediately notify the person in charge of the facility or another responsible person. Following the report, the employer shall immediately make available to the exposed employee a confidential medical evaluation and follow-up including the following elements:

1. Documentation of the incident, including the route of exposure and the circumstances under which the incident occurred.
2. Identification and documentation of the source individual, unless the employer can establish that identification is not feasible or prohibited by state or local law.
3. The source individual's blood shall be tested as soon as possible. If the source individual does not consent, the employer shall document that the legally required consent cannot be obtained. If consent is not required by law, the source individual's blood, if available, shall be tested and the results documented.
4. If the source individual is known to be infected with either HIV or HBV, testing is not required.
5. The exposed employee shall be informed of the results of the tests of the source individual. The employee shall also be informed of any laws or regulations affecting disclosure of the identity of the source individual.
6. The exposed employee's blood shall be collected and tested as soon as possible after consent is obtained from the exposed employee.
7. An employee may consent to have a blood sample collected but not consent to an evaluation for HIV. The sample can be retained for up to 90 days and tested as soon as possible for HIV if the employee elects to have it done during that interval.
8. Postexposure prophylaxis, when medically indicated as recommended by the U.S. Public Health Service, shall be made available, as well as counseling and evaluation of reported illnesses.
9. The employer must ensure that the health-care professional evaluating an employee after an exposure incident shall receive the following information: (a) a copy of the standard, (b) a description of the employee's duties relevant to the incident, (c) documentation on the route of exposure and the circumstances of the incident, (d) results of the source individual's blood tests, if available, (e) medical records relevant to the appropriate treatment, including vaccination status.
10. The employer shall obtain and provide a copy of the health care professional's written opinion within 15 days after completion of the evaluation.
11. The written opinion for HBV vaccination shall be limited to whether vaccination is indicated and if the employee has received the vaccination.
12. The health care professional's written opinion concerning postexposure evaluations and follow-up shall be limited to the following: (a) the employee has been told about the results of the evaluation and (b) any medical conditions resulting from exposure to blood or other infectious materials which require further treatment.

Hazard communication—Labels and signs:

1. Warning labels shall be affixed to containers of regulated waste, refrigerators and freezers containing potential contaminated materials, and other containers used to store, transport, or ship potentially infectious materials. The labels shall state BIO-HAZARD and be accompanied by the biohazard symbol. The labels shall be fluorescent orange or orange-red or predominantly so, with lettering or symbols in a contrasting color. The labels shall be firmly attached by adhesive, string, wire, or other methods that would prevent the loss of the label or unintentional removal. Red

bags or red containers can be substituted for labels. Labels required for contaminated equipment shall be the same as the above and must also state which portions of the equipment are contaminated.

2. Containers of blood, blood components, or blood products released for transfusion and labeled as to their contents are exempted from the other labeling requirements.
3. Individual containers placed in a properly labeled larger container need not be individually labeled.
4. Regulated wastes that have been decontaminated need not be labeled.
5. Work areas where work with potentially contaminated blood or other materials should be posted with the BIOHAZARD legend and symbol. This sign is required for HIV and HBV research laboratories and production facilities.

Information and Training: Employers must ensure that all employees with occupational exposures participate in a training program, at no cost to the employees and during normal working hours. The training shall be provided as follows:

1. At the time of initial assignment to tasks, where occupational exposure may take place.
2. All employees shall receive annual training within 1 year of their previous training.
3. Additional training is required for new exposure situations caused by new tasks or changes in procedures.
4. The training material will be appropriate in terms of content, vocabulary level, literacy level, and language for the training participants.

The content of the training program shall, at a minimum, contain the following elements:

1. An accessible copy of the standard and an explanation of its contents.
2. A general explanation of the epidemiology and symptoms of bloodborne diseases.
3. An explanation of the modes of transmission of bloodborne pathogens.
4. An explanation of the exposure control plan and how a written copy can be obtained.
5. An explanation of how to recognize tasks and other activities that may involve exposure to blood and other potentially infectious materials.
6. An explanation of how to prevent or reduce exposure by means of engineering controls, work practices, and use of personal protective equipment.
7. Information on the types, proper use, location, removal, handling, decontamination, and disposal of PPE.
8. An explanation of the basis for selecting PPE.
9. Information on the HBV vaccine as to its efficacy, benefits and method of administration and that the vaccine and vaccination will be free.
10. What actions to take and whom to contact in an emergency involving blood and other infectious materials.
11. What procedure to follow if an exposure incident occurs, including the method of reporting the incident and the medical follow-up that will be available.
12. Information on the postexposure evaluation and follow-up to be provided to the employee by the employer.
13. An explanation of the signs and labels and/or color coding required by the standard.

14. The employee will have opportunities to ask questions of the person providing the training. The person providing the training shall be knowledgeable about the subject matter in the context of the employee's workplace.

Record keeping—medical records: The employer shall establish and maintain records for each employee with occupational exposure in accordance with 29 CFR 1910.20 (see Section 4.8.2). The records are to be kept confidential and not disclosed or reported to anyone without the employee's written consent to any person within or outside the organization, except as required by the standard or by law. The records are to be maintained for at least the duration of employment plus 30 years. The content of the records shall consist of:

1. Name and Social Security number of the employee.
2. A copy of the employee's hepatitis B vaccination status, including the dates vaccination were received any medical records relevant to the employee's ability to receive vaccinations. If the employee had been involved in an exposure incident, the following additional information would be required:

   a. A record of all results of examinations, medical testing, and follow-up procedures for the employee.
   b. The employer's copy of the health care professional's written opinion.
   c. A copy of the information provided to the health care professional.

Record keeping—training records: The training records shall include the following information:

1. The dates of the training sessions.
2. The contents, or a summary thereof, of the training sessions.
3. The names and qualifications of the persons conducting the training.
4. The names and job titles of all persons attending the training sessions.

The training records shall be kept for a period of 3 years from the date the training occurred. The access to records shall be as provided in 29 CFR 1910.20. If the employer ceases to do business and there is no successor employee, the employer shall notify the OSHA Director at least 3 months prior to their disposal and transmit them to the Director if required by the Director to do so, within that 3-month period.

There are additional requirements for HIV or HBV research laboratories or production Facilities. If information on these requirements is needed, the reader is referred to the standard in such cases.

As a result of the increase in HIV infections, there has been an increase in the incidence of tuberculosis in the United States. Since 1985, the number of cases has increased by 18%. In 1990, there were 25,500 cases in the U.S. The disease is becoming more resistant to treatment. In New York City, the cure rate for the two normally most effective drugs dropped from 100% to 60%. As a result, in October 1993 OSHA issued guidelines applying specifically to health-care settings, correctional institutions, homeless shelters, long-term care facilities for the elderly, and drug treatment centers. OSHA, as they did originally for HIV and HBV before the current standard was enacted, intends to base citations on the General Duty Clause.

REFERENCES

1. Occupational Exposure to Bloodborne Pathogens, 29 CFR 1910.1030, FR 56, 235, December 6, 1991, 64175.
2. Recommendations for prevention of HIV transmission in health-care settings, in *Morbidity and Mortality Weekly Report,* Centers for Disease Control, U.S. Department of Health and Human Services, Public Health Service, 36, 2S, August 21, 1987.
3. **Domin, M.A. and Smith, C.E.,** OSHA's final rule on occupational exposure to bloodborne pathogens, *J. Health Care Material Manage.,* 10, 2, 1992.
4. Update: universal Precautions for prevention of transmission of human immunodeficiency virus, hepatitis B virus, and other bloodborne pathogens in health-care settings, in *Morbidity and Mortality Weekly Report,* Centers for Disease Control, U.S. Department of Health and Human Services, Public Health Service, 37, 24, June 24, 1988, 377–382, 387–388.
5. Guidelines for Handling Human Tissues and Body Fluids Used in Research, National Disease Research Interchange, Philadelphia, PA, 1987.

*4.8.5.3. Additional Information on HIV and HBV**

4.8.5.3.1. Human Immunodeficiency Virus

AIDS is the acronym for acquired immunodeficiency disease.[1] It is a fatal disease caused by a retrovirus presently called HIV (human immunodeficiency virus).[2] The disease has previously been called HTLV III and LAV. The incubation period is variable, but is considered to be from 6 months to 5 years or more.[3] It is a relatively new disease, having first been identified in 1981.

There is, as yet, no cure nor a vaccine to prevent the infection. The manifestations of AIDS are variable, but all are related to the compromising of the immune system. The result of destroying the essential cellular manager of the immune protection (the helper thymic lymphocyte) is to lose the capacity to exist healthfully in the presence of viruses, bacteria, fungi, and some forms of cancer.[4] As the disease progresses the patient loses weight, is very fatigued, develops large, swollen lymph nodes, develops infections such as herpes, moniliasis, pneumocystis carinii pneumonia, develops severe diarrhea, and most develop progressive neurological signs and symptoms. The usual cancer which develops used to be rare. It is not now. It is called Kaposi sarcoma and manifests itself as progressive purple to brown skin spots and splotches. Most patients die from massive infections not controlled or cured by any antibiotics. Persons exposed to HIV usually develop detectable antibodies against the virus within 6 to 12 weeks of becoming infected.

4.8.5.3.2. Viral Hepatitis

Hepatitis refers to inflammation of the liver. Hepatitis may be caused by trauma, toxic shock, hyperthermia, metabolic derangements, chemicals, radiation, bacteria, fungi, parasites, and a number of viruses. For the purposes of this section we refer only to type B (also called transfusion or serum hepatitis) hepatitis.

---

* The following section is a condensed excerpt of two sections which appeared in the *CRC Handbook of Laboratory Safety,* 3rd ed., CRC Press, Boca Raton, FL, 1971.

The usual onset of hepatitis symptoms occur 1 to 3 weeks prior to the obvious presence of the disease. These symptoms are nonspecific and may be seen with any of the types. There is the gradual development of vague aches and pains, fatigue, and anorexia. As this accelerates, the urine becomes dark, almost to a mahogany red, and the patient becomes jaundiced. This becomes evident when the blood bilirubin rises at least 2.5 to 3 times normal. Approximately 10% of patients have the triad of fever, arthralgia (bone and joint pain), and an urticarial rash. As the jaundice worsens, the patient usually feels better. The usual recovery period is 2 to 4 weeks. Differentiation of the types of hepatitis is helped by epidemiological considerations and by laboratory tests.[5] Following are some important statistical information for type B hepatitis:

Incubation period—45 to 160 days, average 60 to 120 days
Incidence of viral hepatitis—60%
Incidence of severe or fulminant hepatitis—60%
Viral carriers—0.5%
Chronicity (long-standing active hepatitis)—10%
Infectivity—Acquired by entry through the skin by cuts or abrasions, or by needle sticks when using fluids containing the virus. Positive seroconversion results in 19 to 24% of cases. A dilution of 1:1,000,000 of the virus is still infective.
Mortality rates—1 to 2%
Diagnosis[6,7]—There are five available tests for hepatitis B, which are used in various stages of the disease.
Prophylaxis[8]—There is a good hepatitis B vaccine that provides a good level of protection with little or no untoward reactions. This is called HEPTAVAX-B (made by Merck, Sharp and Dohme) and is rated as at least 96% effective after three doses. It works in vaccine responders against acute hepatitis B and asymptomatic infection and prevents the chronic state. If anti-hepatitis B develops, there is virtually 100% protection.

The liver is a beautifully productive factory within our bodies. It processes over 1500 known chemical reactions. It is estimated that we have about 80% more liver mass than we need, but when it is affected by such diseases as a viral hepatitis, 100% of it is involved. Fortunately, recovery is possible and probable in most cases. If recovery is not complete one could very well become a digestive cripple. Then too, the development of cirrhosis is devastating and ultimately fatal. A hepatoma may presumably be removed if discovered early, but often it is not.

REFERENCES

1.   **Koop, C.E.,** Surgeon General's Report on AIDS, U.S. Public Health Service, 1987.
2.   *Epidemiology Bull.,* Vol. 87, No. 12, December 1987.
3.   **Keeling, R.,** AIDS on the College Campus, ACHA Special Report, ACHA, 1986.
4.   On the front line, *Emerg. Med.,* January 15, 1986.
5.   **Halcrow, A.,** AIDS: The corporate response, *Personnel Journal,* August 1985.
6.   **Hoofnagle, J.,** The challenge of acute viral hepatitis, *Patient Care,* October 30, 1979.
7.   **Pribor and Duello,** Hepatitis, diagnostic profiles, *Laboratory Management,* 19(7), 8.
8.   **Fife and Cory,** Recent advances in the diagnosis of hepatitis, *Laboratory Medicine,* 11(10), 650.
9.   **Waldvogel, F.,** Viral hepatitis: Diagnose it fast and prevent its spread, *Modern Medicine,* October 1983.

10.   **Desjardins, R.F.,** *CRC Handbook of Laboratory Safety,* 2nd ed., Steere, N.V., Ed., CRC Press, Boca Raton, FL, 422, 1971.

4.8.5.4. Other Disease-Causing Agents[*]

4.8.5.4.1. Zoonotic Diseases

Zoonotic diseases are those transmitted from animals to humans, with a wide range of manifestations in humans from simple illness to death. Some of these are of special concern to pregnant women. Toxoplasmosis and listeriosis can cause, among other things, spontaneous abortions.

An understanding of the modes of transmission of these diseases and clinical signs observed in affected animals will help facility managers establish preventive measures to protect individuals who come into contact with the animals or their tissues.

Modes of Transmission

Disease agents can be transmitted either directly or indirectly. Bacterial, viral, fungal, and parasitic disease agents can be transmitted through direct contact with animal saliva, feces, urine, other body secretions, bites, scratches, aerosols, or excised body tissues. Humans can be protected through use of gloves, masks, gowns, and other protective clothing, and through the use of restraint techniques which minimize the possibility of bites and scratches.

One indirect means of disease agent transmission involves fomites, inanimate objects (boots, brooms, cages, instruments, etc.) that can transport the agent following contact with animals, secretions, or wastes. Disease agents may be short lived when outside the body, or persistent for years on fomites if the object is not cleaned or disinfected. The Orf virus, from sheep and goats, has remained viable for 15 years in dried scabs. Rooms, cages, and equipment should be adequately disinfected with virucidal disinfectant agents.

A second indirect means of transmission involves vectors, living organisms (insects) which can extract or carry the disease agent from one animal to other animals or humans. A mechanical vector extracts and carries the disease agent without any change occurring in the agent. In a biological vector, the disease agent undergoes changes in one or more stages of its life cycle before becoming an infective form.

An effective vermin and insect control program is needed to eliminate these indirect means of transmission.

Routes of Exposure

Barkley and Richardson[1] listed the four primary routes of exposure or entry of a disease agent:

1. Ingestion (i.e., placing contaminated fomites in one's mouth, or contaminated hand contact with food)

---

[*]   This section was written by David M. Moore, D.V.M.

2. Inhalation (i.e., aerosolized material—urine, feces, saliva, or other bodily secretions; these materials may also become aerosolized when using high pressure water hoses to clean rooms or cages)
3. Contact with mucous membranes (i.e., contact with nose, mouth, eyes through spills, contaminated hands, or aerosolized material)
4. Direct parenteral injection (i.e., bites, cuts, scratches, accidental needle sticks)

Each should be handled accordingly by prohibiting food consumption in animal holding areas, requiring the practice of good hygiene, altering sanitation procedures to lessen aerosol production, providing protective garments and safety items (safety goggles, respirators, gloves, and masks), and establishing safety awareness training programs to advise employees as to risks and preventive measures.

REFERENCES

1. **Barkley, E.W. and Richardson, J.H.,** Control of biohazards associated with the use of experimental animals, in *Laboratory Animal Medicine*, Fox, J.G. et al., Eds., Academic Press, Orlando, FL, 1984, 595.
2. **Fox, J.G., Newcomer, C.E., and Rozmiarek, H.,** Selected zoonoses and other health hazards, in *Laboratory Animal Medicine,* Fox, J.G. et al., Eds., Academic Press, Orlando, FL, 1984, 613.
3. Biological Hazards in the Nonhuman Primate Laboratory, Office of Biohazard Safety, NCI, Bethesda, MD, 1979.
4. **Richardson, J.H. and Barkley, E.W.,** *Biosafety in Microbiological and Biomedical Laboratories,* PHS/NIH, HHS Publication No. (CDC) 84-8395, 37, 1984.
5. **Hellman, A., Oxman, M.N., and Pollack, R.,** *Biohazards in Biological Research,* Cold Spring Harbor Laboratory, New York, 1973.
6. *Am. Ind. Hyg. Assoc. J.,* 54(3), 113, 1993.

4.8.5.4.2. Allergies

Some investigators and animal care technicians who have prolonged contact with laboratory animals may develop allergies to animal dander, hair, urine, tissues, or secretions. Reactions to skin contact or inhalation of these materials vary from a wheal and flare phenomenon (a firm, red raised area at the site of skin contact which develops within several minutes) to life-threatening anaphylactic shock.

Olfert[2] lists the species most commonly associated with allergic reactions in a laboratory setting: rat, rabbit, guinea pig, and mouse. When transfer of personnel to a nonanimal area is not a viable option, other measures should be taken to avoid exposure to specific allergens. Lutsky et al.[1] suggests the use of gloves, masks, protective outer garments, and filtered cages as methods to reduce exposure. Additionally, eliminating recirculation of room air will decrease the levels of allergens, as will more frequent cage cleaning.

REFERENCES

1. **Lutsky, I.T., Kalbfleisch, J.H., and Fink, J.N.,** Occupational allergy to laboratory animals: employer practices, *J. Occup. Med.*, 25(5), 272, 1983.

2. **Olfert, E. D.,** Allergy to laboratory animals - an occupational disease, *Lab Animal,* 15(5), 24, 1986.
3. **Krueger, B.,** Lab animal allergies: a manager's perspective, *LAMA Lines,* 2(6), 16, 1987.

4.8.5.4.3. Waste Collection and Storage*

Shearing of hypodermic needles following injection of infectious or toxic agents or following routine clinical use in animals should be avoided. Aerosolization of the contents of the needle can occur during shearing, posing a hazard to humans or other animals in the room.[1]

4.8.5.4.4. Bedding

Bedding from cages housing animals treated with biohazardous microbial or chemical agents should be considered contaminated and disposed of appropriately.[2] If an incinerator is not on site or available for direct dumping of bedding, then bedding should be double bagged and tagged as hazardous material prior to transport to the incinerator, to avoid contamination of personnel or of the work environment. The incineration of carcinogen-contaminated bedding requires an incinerator capable of operating at a temperature range of 1800° to 1900°F with a retention time of 2 seconds.[3]

REFERENCES

1. **Barkley, W.E. and Richardson, J.H.,** Control of biohazards associated with the use of experimental animals, in *Laboratory Animal Medicine,* Fox, J.G. et al., Eds., Academic Press, Orlando, FL, 1984, 595.
2. **Wedum, A.G.,** Biohazard control, in *Handbook of Laboratory Animal Science,* Melby, E.C. and Altman, N.H., Eds., CRC Press, Cleveland, 1974, 196.
3. Chemical Carcinogen Hazards in Animal Research Facilities, Office of Biohazard Safety, NCI, Bethesda, MD, 15, March 1979.
4. **Dimmick, R.L., Vogl, W.F., and Chatigny, M.A.,** Potential for accidental microbial aerosol transmission in the biological laboratory, in *Biohazards in Biological Research,* Hellman, A. et al., Eds., Cold Spring Harbor Laboratory, Cold Spring Harbor, NY, 246, 1973.

4.8.6. Pregnancy in the Laboratory**

Pregnancy is not a disease. It is the ultimate expression of a couple's heritage, the desirable as well as the less-than-desirable elements. During the 9-month development period of the fetus, there is a dynamically changing physical and chemical status of both

---

*   In addition to this and the following section, the reader should refer to the later material in the handbook on handling regulated medical wastes (infectious waste) for a more complete treatment of handling not only animal waste contaminated by pathogens contagious to humans, but also other medical wastes potentially contaminated with contagious organisms.

** This section was written by Richard F. Desjardins, M.D.

the mother and child. It might be stated that the pregnant state imbues the fetus with a metabolic priority to obtain oxygen, chemicals, and nutrition from the only source available, the mother. It would therefore behoove the mother to sustain her physicochemical status at an optimum, both to provide requirements for development of the fetus and to prevent depletion of her own requirements for a healthful state.

By the same token, chemical exposure by the mother will result in fetal absorption of sometimes deleterious concentrations of either the parent elements or compounds and their metabolites. The noxious potential may be more or less in the parent element or compound or of the metabolites. Estrogens are less stimulatory to the liver, where most of the elements or compounds are metabolized, than the male testosterone. Therefore, if a parent element or compound is more toxic than the metabolite, it would have a more injurious effect on a female than it would on a male. This is at least theoretically valid, but there have been so few relevant studies that it must be held as unsubstantiated at this time. In 1946, Baetjer[1] made a critical review of the literature and data which stated that women were more susceptible than men to chemical exposures. She found that the conclusions were based on an inadequate number of poorly constructed studies as well as failures to include both chemical quantity and duration of exposures. She found no significant basis for conclusions that nonpregnant women were more liable to injury by chemicals than were males.

As a result of legislation passed in these last two decades by OSHA (1970), Title VII of the Civil Rights Act of 1964, and guidelines adopted by Equal Employment Opportunity Commission (EEOC), more women are entering the workforce than ever (presently the workforce is over 40% women), and there can be no discrimination in the area of work as to gender. Basic to the provisions of the federal regulations is the inherent assumption that tolerance to industrial stresses, susceptibilities to toxic substances, and performance capabilities are similar in all workers. Furthermore the "preservation of our human resources" implies not only the protection of the health of our working men and women, but also the preservation of their reproductive capacities and products of such capacity.

With a few exceptions, current standards for occupational exposures to chemicals and physical agents in the work environment do not consider the reproductive effects, including mutagenesis, teratogenesis, transplacental carcinogenesis, or infections.[*] The latest available data[4] indicates that of 3,034,000 women who had a live birth during a 12-month period, an estimated 1,260,000 women (41.5%) had worked during their pregnancy. This, however, represents only about 8.8% of the estimated 14,357,000 married women of reproductive age in the workforce at that time. A significant number of the women work in laboratories and in the health services.

It should be noted that consideration of the effects of toxicants on the paternal elements is mandatory, if seldom addressed. The long-term effects of vinyl chloride and dibromo-chloropropane are well documented in this respect, but other male reproductive toxicants are being identified, albeit slowly (see the most recent "California" list).

The American Management Association[5] has recognized reproductive hazardous materials in the workplace and prepared a report in 1985, reviewing 120 chemicals. They pointed out general principles, clinical applications, and aids to recognizing human teratogens. Also presented are reviews and opinions for three representative chemicals, i.e., acrylonitrile, inorganic arsenic, and carbon disulfide. Table 4.19 considers the dynamics of reproductive toxicology.

---

[*]   The situation is changing to some extent. Note the number of reproductive toxicants in the most recent "California" list earlier in this chapter.

Table 4.19. Dynamics of Reproductive Toxicology

Stage of Development	Stage of Pregnancy		
	Preconception	Intrauterine	Perinatal
	Gametes (sperm, ova) organogenesis	1st trimester organogenesis, 2nd, 3rd trimester fetus	Infant
Vulnerable areas	Spermatogenesis Oogenesis Fertilization		Lactation
Major developmental effects	Mutagenesis	Teratogenesis	CNS—late transplacental carcinogenesis
Adverse manifestations	Sterility, decreasing fertility, chromosomal aberrations	Implantation defects, spontaneous abortions	Stillbirth; structural, behavioral, or functional alternations
Parental source of problem	Maternal and paternal	Maternal (3rd trimester)	Maternal (lactation)

Chemicals or physical agents can act directly on germ cells so that fertilization or implantation may not occur. The result could be spontaneous abortion. Anomalies are about 60 to 100 times more frequent in spontaneous abortions than for live births. Birth defects may occur in up to 8% of births, being revealed up to the age of 10, and 6% of these are due to known environmental agents. Of the rest of the birth defects, the causes are not known at this time. It is believed some will be of occupational environmental etiology.

Specific data and tables can be found in the section on "Reproductive Toxicology and Occupational Exposure" found in Chapter 54 of the excellent book by Zenz.[3]

Rarely recorded or discussed are the acquired infections of the mother and developing fetus. AIDS and hepatitis are known to cross the placental barrier, and other viruses and infectious agents might also do so. Laboratory precautions are particularly essential during pregnancy due to the increased vulnerability at that time.

In general, it is prudent for any pregnant woman to be assiduous in her precautions at work. Considering the dearth of relevant data on the effects of environmental toxicants while pregnant, the best advice should be to minimize exposures to any chemical which can be inhaled, swallowed, or absorbed cutaneously. We must consider the chemical type, quantity, means of exposure, the time of exposure (both the length of exposure, and at what stage of pregnancy), and individual susceptibility. What is known is that some chemicals can cause maternal and fetal problems, including death. The liability of these problems is real and must be addressed by both knowledge of the possible effects of the specific chemical being used and by exerting proper laboratory precautions.

A beautiful newborn is happiness. A defective child evokes guilt and unhappiness. It is then too late to fantasize as to the causative etiology, if any can be known. A few precautions taken appropriately will at least reassure the parents that they did nothing wrong in the formative period of their child.

REFERENCES

1. **Baetjer, A.,** *Women in Industry, Their Health and Efficiency,* W.B. Saunders Co., Philadelphia, 1946, 145.
2. **Henschel, A.,** Women in industry—the difference, Trans. of 33rd Mtg. of the Amer. Conf. of Gov. Ind. Hyg., Toronto, Canada, 73, 1971.
3. **Messite, J. and Bond, M.B.,** Reproductive toxicology and occupational exposure, in *Occupational Medicine,* Zenz, C., Ed., Year Book Pub., Inc., Chicago, 1988.
4. National Center for Health Statistics, Statistics needed for determining the effects of the environment on health, U.S. Dept. of HEW, *Health Resources,* 77, 1457, 1977.
5. **Council on Scientific Affairs,** Effects of toxic chemicals on the reproductive system, *JAMA,* No. 23, 253, 1985.
6. Assessment of reproductive and teratogenic hazards, in *Advances in Modern Environmental Toxicology,* Vol. 3, Christen, M.S., Galbraith, W.W., Voytek, P., and Mehlman, M.A., Eds., Princeton Scientific Publ., Princeton, NJ, 1983.

4.8.7. Regulated and Potential Carcinogens

An individual planning to work with carcinogenic material must be prepared to ensure that the laboratory hygiene plan developed to meet the performance standards implicit in 29 CFR 1910.1450, the laboratory safety standard, meets all of the criteria set forth in the standard in terms of quality of the facility, training and information made available to the employee, operational procedures, and availability of personal protective equipment. These considerations should arise at least by the time of purchase of the material or before. Therefore, the list of chemicals considered to be probable carcinogens was placed in the section on purchasing earlier in this chapter.

4.8.7.1. Carcinogens (Ethylene Oxide)

Ethylene oxide is used here as an example of a carcinogenic compound and is one for which a specific standard was adopted by OSHA relatively recently (August 24, 1984). The standard, in its appendices, provides an unusually complete guide to the safe use of this material. The laboratory safety standard may modify some of the specific requirements, but adhering to them would form the basis of a sound management strategy.

Ethylene oxide (C_2H_4O; CAS No. 75-21-8) is a gas at normal temperatures (boiling point = 10.7°C, 51.3°F). The specific gravity of the gas with respect to the density of air is 1.49. It dissolves readily in water. It has an ether-like odor when concentrations are well above the OSHA PEL, so that it cannot be considered to warn adequately of its presence by its odor. It is a significant fire hazard in addition to being a health hazard. The lower and upper explosion limits are, respectively, 3% and 100%. It will burn without the presence of air or other oxidizers, with a flash point below 0°F (-18°C) and may decompose violently at temperatures above 800°F (444°C). It will polymerize violently when contaminated with aqueous alkalis, amines, mineral acids, and metal chlorides and oxides. It would be classified as a class B fire hazard for purposes of compliance with 29 CFR 1910.155. Locations defined as hazardous due to its use would be class I locations for purposes of compliance with 29 CFR 1910.307.

Although dangerous because of its physical properties, the primary reasons for regulating the material by a separate standard were related to health effects—specifically its identification as a human carcinogen, adverse reproductive effects, and ability to cause chromosome damage. Because of the latter two problems, women who suspect or know that they are pregnant should take special care to avoid exposures above the acceptable limits. There are a number of other adverse health effects in addition to these relatively newly identified problems.

Acute effects from inhalation include respiratory irritation and lung damage, headache, nausea, vomiting, diarrhea, shortness of breath, and cyanosis. Ingestion can cause gastric irritation and liver damage. It is irritating on contact to the eye and skin and can cause injury to the cornea and skin blistering on extended contact. Contact with pressurized, expanding vapor can cause frostbite. Individuals using this material should not wear contact lenses. It has also been associated with mutagenic, neurotoxicity, and sensitization effects.

Safe work practices for ethylene oxide fall into two areas: (1) the normal practices associated with the use of a flammable gas and (2) health practices needed to reduce exposures to the vapors. For the former set of problems and the more common health problems, the procedures are relatively straightforward, the same as for other chemicals with similar properties: keep ignition sources and reactive materials away from the material; do not smoke, eat, or drink in the area; and wear personal protective equipment (goggles, gloves, respiratory protection, and protective clothing) as needed to prevent exposure.

The means of preventing exposure to the gas to protect against the carcinogenic and reproductive hazards are spelled out in considerable detail in the appendices to the OSHA standard (29 CFR 1910.1047) as well as the specific measures that would meet the needs of the standard. A major use of the material is as a sterilizing agent in medical care operations, so much of the material in the standard is concerned with means of avoiding release of the gas or to capture gas that has been released. Other portions of the appendices deal with achieving compliance with other parts of the standard. The basic features of the standard are briefly given below as representative of those of the other regulated carcinogens. Many details are omitted for which the reader is referred to the complete current standard, available from any local OSHA office.

A key provision of the laboratory safety standard is that procedures be developed to ensure that it would be unlikely that any individual on a worst-case basis will exceed a 0.5 ppm action level for an 8-hour TWA exposure. These levels are measured without taking into account any respiratory protection provided by personal protective equipment. When any circumstance changes in such a way that the levels of exposure may increase, it will probably be necessary to demonstrate anew that the levels are lower than the action levels. Records of these data must be kept and available for examination. If the action levels are exceeded, steps must be taken to reduce the levels, preferably by means of engineering controls. Medical surveillance provisions of the laboratory safety standard would be invoked by persistent levels above the OSHA action levels.

Probably the most critical requirement of the standard is to ensure that no employee is allowed to be exposed to an airborne concentration level of EtO in excess of 1 ppm as an 8-hour TWA. This must be demonstrated for each employee, although, where the exposure conditions are sufficiently similar, measurements need only be made for representative employees. The method used to make the measurements must be capable of providing an accuracy of ±25% in the range of the PEL of 1 ppm with 95% confidence limits and ±35% in the range of the action limit of 0.5 ppm. It should be noted that a STEL of 5 ppm has recently been adopted.

If the measured levels are between 0.5 and 1 ppm, monitoring is required to be repeated at least every 6 months, while if the levels are in excess of 1 ppm, monitoring is required to be done at least every 3 months. The results of the monitoring program must be made available to the employees as well as the outcome of any corrective actions taken to reduce the levels.

A regulated area must be established wherever the airborne concentrations may exceed 1 ppm. Access to this area must be limited to authorized personnel and the number of these persons must be kept at a minimum. The entrances to this area must be clearly marked with the following sign:

DANGER
ETHYLENE OXIDE
CANCER HAZARD AND REPRODUCTIVE HAZARD
AUTHORIZED PERSONNEL ONLY
RESPIRATORS AND PROTECTIVE CLOTHING MAY BE
REQUIRED TO BE WORN IN THIS AREA

Any containers of EtO with the potential for causing an exposure at or above the action level must be labeled with the legend:

DANGER
CONTAINS ETHYLENE OXIDE
CANCER HAZARD AND REPRODUCTIVE HAZARD

The label also must warn against breathing EtO. If EtO is to be used as a pesticide, the container labeling requirements of the Federal Insecticide, Fungicide, and Rodenticide Act (FIFRA) preempt the OSHA requirements.

If the PEL of 1 ppm is exceeded, the employer must establish and implement a written program to reduce actual employee exposures to below this level. Preferably, this protective program should be based on engineering controls and work practices, but also may include the use of approved respiratory protection where alternate measures are not feasible. Approved respiratory protection means those respiratory devices specifically approved for protection for EtO exposure by either the Mine Safety and Health Administration (MSHA) or by the NIOSH, under the provisions of 30 CFR, Part 11. Employee rotation is not an acceptable means of achieving compliance. The laboratory safety plan also must include means of leak detection, and an emergency plan. The plan must be reviewed and revised as needed at least annually. The written emergency plan must provide for equipping employees with respiratory protection. It must include those elements required under 29 CFR 1910.38. Provision must be made for alerting the employees of an emergency and for evacuation of employees from the danger area.

A medical surveillance and consultation program must be available for any employee who may be exposed to EtO at or above the action level of 0.5 ppm for 30 or more days a year, or in an emergency situation. There must be a pre-employment examination, a medical and work history, an annual examination for each year the 30-day criterion is met, and a post-employment examination. In addition to these requirements, examinations may be indicated for employees exposed in an emergency, and as soon as possible for any employees who believe that they are exhibiting symptoms of exposure to ethylene oxide. Employees may also request and be given medical advice about the effects of their exposures to EtO on their ability to produce a child. The physician may recommend other examinations. For example, an employee may wish to obtain fertility and pregnancy tests, and they are to be given these tests if the physician considers the tests appropriate under the circumstances.

The surveillance program must include a medical and work history, with emphasis placed on the pulmonary, hematologic, neurologic, and reproductive systems, the eyes, and the skin. The physical examination must emphasize the same areas. A complete blood count is to be part of the examination as well as any other appropriate tests designated by the physician.

The results of the tests must be made available to the employer, employee, and others (such as the employee's physician) upon the employee's written authorization. The physician must provide a written opinion including the results of the examination, whether the examination revealed any conditions that the employee's occupational exposure would aggravate, and whether there should be any restrictions placed on the employee or modification to the employee's duties to reduce exposure. The physician must also state that he has discussed the results of the examinations with the employee and any follow-up actions that should ensue. If there are any extraneous medical factors not pertinent to the work-related activities of the employee discovered in the course of the examination, the employee has the right to expect complete confidentiality of this information. The medical records must be maintained, according to the provisions of 29 CFR 1910.20, by the employer for the duration of the employee's employment plus 30 years. Much of the supporting data records, such as exposure information, must be kept for a similar period.

In addition to the labeling and signs that have already been discussed, a written laboratory safety plan must be established by the employer. Before an employee is assigned duties which could result in an exposure above the action level, they must be provided training and information about EtO. The information and training program must be repeated at least annually, and must be made current as needed during the course of the year. The program must include information about the requirements of the OSHA standard including: where a copy of it can be obtained, operational procedures, the medical surveillance program, methods available to detect EtO, measures taken by the employer to ensure compliance with the standard, measures the employees can take to protect themselves, the emergency plan, the hazards of EtO, where the current MSDS can be found and how to interpret the information on it, and the details of the laboratory and corporate or institutional industrial hygiene plan.

REFERENCES

1. Occupational Safety and Health Administration, General Industry Standards, Subpart Z - Occupational Health and Environmental Controls, Ethylene Oxide, 29 CFR 1910.1047, Washington, D.C.
2. Occupational Safety and Health Administration, Occupational Exposure to Hazardous Chemicals in Laboratories, 29 CFR 1910.1450, Washington, D.C.
3. Occupational Safety and Health Administration, Employee Emergency Plans and Fire Prevention Plans, 29 CFR 1910.38, Washington, D.C.

4.8.8. Neurological Hazards of Solvents

The Environmental Protection Agency (EPA) in August 1993 published a document in the *Federal Register* entitled *Draft Report: Principles of Neurotoxicity Risk Assessment,* and called for comments. As the document stated, there are data indicating that exposure to neurotoxic agents may constitute a significant health problem. Assuming that this is so, a need exists to develop guidelines to assess the hazards. The document covered in some detail current knowledge of the characteristics of neurotoxicity, methods of assessing human neurotoxicity, and the use and efficacy of animal studies in neurotoxicity investigations. The goal is to combine the three key components of a hazard assessment program—hazard identification, dose-response relationship, and exposure assessment—to characterize the risk that neurotoxins may pose to exposed individuals. In addition, under the Toxic Substances Control Act, the EPA (also in August of 1993) required companies that make, import, or produce any of ten high-volume organic solvents to perform a series of four tests on rats for the ten solvents. The tests are functional observational battery (FOB), motor activity, neuropathology, and schedule-controlled operant behavior (SCOB). The ten chemicals (numbers in parentheses are the Chemical Abstract Service identifiers) are Acetone (67-64-1), 1-butanol (71-36-3), diethyl ether (60-29-7), 2-ethoxyethanol (110-80-5), ethyl acetate (108-10-1), methyl isobutyl ketone (108-10-1), N-amyl acetate (628-63-7), N-butyl acetate (123-86-4), isobutyl alcohol (78-83-1), and tetrahydrofuran (109-99-9). Several major companies objected to testing the first six of these and the use of the SCOB test, and in October of 1993 filed a lawsuit to express their objections.

Many solvents have been recognized or suspected of being carcinogenic or to pose reproductive problems (see Section 4.3.4.4). However, many of the recognized health effects due to solvents are neurotoxic.* In commonly available references, such as Sax or Merck, much of the descriptive material on health symptoms due to exposure are based on neurotoxic actions. As with most other health effects, neurotoxic problems may be divided into acute or immediate effects, or those due to chronic exposures which lead to delayed and possibly persistent health changes.

Much of the epidemiological data on health effects has been due to exposures in an industrial setting rather than the laboratory, since the exposures are liable to be at relatively stable levels and involve a relatively small number of solvents, rather than the extremely complicated and rapidly changing laboratory environment.

The primary modes of uptake of solvents by the body are inhalation and by absorption through the skin. The rate of absorption by inhalation is affected by a number of factors, some due to the properties of the material and the interaction of the solvent vapors with the lungs, and some due to other factors such as the concentration of the solvent fumes in air, the duration of the exposure, and the level of exertion at the time of exposure by the exposed individual. The rate of intake is significantly increased by elevated levels of physical activity.

The rate of intake through the skin is dependent upon the duration of the contact, skin thickness, degree of hydration of the skin, and possible breaks in the integrity of the skin, i.e., injuries or skin disorders. An example of the comparative rates of intake by the two routes is that an immersion of both hands in xylene for 15 minutes gives about the same levels in the blood as an exposure to an airborne concentration of 100 ppm for the same period. However, this cannot be assumed to be an accurate reflection of the comparative rates for other solvents.

After exposure, the material is often transformed by the liver into less toxic water-soluble compounds or in some cases into more toxic intermediate metabolites. In other cases, the solvents are lipophilic, i.e., may be taken up and accumulate in lipid-rich tissues such as the nervous system.

---

* The information in the remainder of this section is derived primarily from NIOSH Current Intelligence Bulletin 48, Organic Solvent Neurotoxicity, DHHS (NIOSH) Publication No. 87-104, National Institutes for Occupational Safety and Health, U.S. Department of Health, Education and Welfare, Washington, D.C., 1987.

The acute, short-term neurotoxic effects result from action of the solvent on the CNS. The effects may range from symptoms resembling intoxication to CNS depression, psychomotor impairment, narcosis, and death from respiratory failure. At intermediate levels of exposure, common effects are drowsiness, headache, dizziness, dyspepsia, and nausea. Short-term effects may cause mood changes as well, as reflected by increasing feelings of physical and mental tiredness during an exposure period corresponding to a typical work-shift, for example.

The effects of extended or chronic exposures have been divided into three categories of varying severity at two relatively recent international workshops, minimal, moderate and pronounced, although the nomenclature differed slightly at the two conferences. The last category, which so far has not been observed in an occupational situation but has in persons deliberately exposing themselves to solvent fumes, would be reflected in serious deterioration of the nervous system, including mental capacity and function. The effects of such extreme exposures would be only partially reversible at best.

The least severe category would be characterized by deterioration in memory function and ability to concentrate, physical fatigue, and irritability, while the second level would involve sustained mood and personality changes, as well as further deterioration of intellectual functions, including learning capacity. At least some of the effects of chronic effects appear to persist well beyond the termination of the exposure, and may be permanent.

Acute effects appear to be caused by the solvent itself, while the chronic effects may be associated with the intermediate metabolic reaction products. The effects of short-term exposures appear to be reversible, while the effects of prolonged exposures leading to changes in nerve tissue may be irreversible. Although the data on chronic exposure toxicity are not as abundant or definitive as desired, the data that are available definitely appear to support a conservative approach to personnel exposure. Levels to which an individual is exposed, especially over an extended period, should emulate the NRC "as low as reasonably achievable" (ALARA) philosophy. The various levels established as regulatory by OSHA (PEL), recommended exposure levels (REL) by NIOSH, or the TLVs recommended by the ACGIH should be adopted as a maximum permitted occupational level, using the lowest of the three values as a guide. An action level of 50% of the appropriate level is recommended as a trip point for initiating remedial steps.

Adequate general ventilation, provision of appropriate containment equipment (fume hoods, safety cabinets, or localized exhaust systems), procedural controls, or use of personal protective equipment, respiratory protection, and protective gloves, clothing, and eye protection are all part of a possible program to reduce exposures.

It is difficult in the laboratory to monitor the exposure levels as recommended above, but monitoring should be performed wherever possible. Even partial data or data taken under nonstandard conditions are better than none at all. As a supplement to a monitoring and exposure limitation program, those individuals who actively use solvents as a routine part of their job a significant portion of their time should participate in a medical surveillance program. Individuals should receive a pre-employment examination, which should include a prior work history and medical history. The examination should emphasize the nervous, cardiovascular, respiratory, and reproductive systems as well as the liver, kidneys, blood, gastrointestinal tract, eyes, and skin. A comprehensive blood panel should be run as well as a complete blood count and a urinary test. Some special tests might be suggested by the physician, such as a cholinesterase enzyme test for information on the CNS. A blood serum sample might be stored for later comparison. Periodic reexaminations should be given, annually if the exposure is heavy, but no less often than every 5 years.

Table 4.20. OSHA PELs, NIOSH RELs and ACGIH TLVs for Some Organic Solvents[a]

Compound (CAS No.)	OSHA PEL (Ceiling)		NIOSH REL (Ceiling)		ACGIH TLV (STEL, 15 min)	
Alkanes (C5-C8)						
Pentane (109-66-0)	1000		120	(610/15 min)	600	(750)
Hexane (110-54-3)	500		100	(510/15 min)	50	
Heptane (142-82-5) (pns)	500		85	(440/15 min)	400	(500)
Octane (111-65-9)	500		75	(385/15 min)	300	(375)
Allyl chloride (107-5-1)	1		1	(2/15 min)	1	(2)
Benzene (71-43-2) (cns) (carcinogen)	1		0.1	(1/15 min)	10	
Benzyl chloride (100-44-7)	1			(1/15 min ceiling)	1	
Carbon disulfide (75-15-0) (cns, pns)	20	(30 ceiling; 100/30 min)	1	(10/15 min)	10	
Carbon tetrachloride (56-23-5) (carcinogen) (cns)	10	(25 ceiling; 200/5 min in 4 hours)	2	(1 hour)	5	
Chloroethane (75-00-3)	1000		Handle with care		1000	
Chloroform (67-66-3) (carcinogen)		(50 ceiling)	2	(1 hour)	10	
Chloroprene (126-99-8) (carcinogen)	25			(1/15 min ceiling)	10	
Cresol (1319-77-3) (cns)	5	mg/m^3	2.3		5	
Di-2-ethyl-hexylphthalate (117-81-7) (carcinogen)	5	mg/m^3	5	(10) mg/m^3	5	(10) mg/m^3
Dioxane (123-91-1) (carcinogen)	100			(1/30 min ceiling)	25	
Epichlorohydrin (106-89-8) (carcinogen)	5		Minimize occupational exposure		2	
Ethylene dibromide (106-93-4) (carcinogen)	20	(50/5 min)	0.045	(0.13/15 min)	Suspect skin carcinogen	

Table 4.20.   OSHA PELs, NIOSH RELs and ACGIH TLVs for Some Organic Solvents[a] (continued)

Compound (CAS No.)	OSHA PEL (Ceiling)	NIOSH REL (Ceiling)	ACGIH TLV (STEL, 15 min)
Ethylene dichloride (107-06-2) (carcinogen) (cns)	50 (200/5 min in 3 hours)	1 (2/15 min)	10
Furfuryl alcohol (98-00-0)	50	10 (15)	10 (15)
Glycol ethers			
2-Methoxyethanol (cns) (110-86-4)	25	Reduce exposure to lowest feasible level	5 (skin)
2-Ethoxyethanol (110-80-5)	200	Reduce exposure to lowest feasible level	5 (skin)
Isopropyl alcohol (67-63-0)	400	400 (500)	400 (500)
Ketones			
Acetone (67-64-1)	1000	250	750 (1000)
Methyl ethyl ketone (78-93-3)	200	200	200 (300)
Methyl n-propyl ketone (107-87-9)	200	150	200 (2500)
Methyl n-butyl ketone (591-78-6)	100	1	5
Methyl n-amyl ketone (110-43-0)	100	100	50
Methyl isobutyl ketone (108-10-1)	100	50	50 (75)
Diisobutyl ketone (108-83-8)	50	25	25
Cyclohexanone (108-94-1) (cns)	50	25	25 (skin)
Mesityl oxide (141-79-7) (cns)	25	10	15 (25)
Diacetone alcohol (123-42-2)	50	50	50
Isophorone (78-59-1)	25	4	5 (ceiling)
Mercaptans			
Butanethiol (Butyl mercaptan) (109-79-5) (cns)	10	(0.5/15 min)	0.5
Ethanethiol (Ethyl mercaptan) (75-08-1)	(10 ceiling)	(0.5/15 min)	0.5

Table 4.20. OSHA PELs, NIOSH RELs and ACGIH TLVs for Some Organic Solvents[a] (continued)

Compound (CAS No.)	OSHA PEL (Ceiling)		NIOSH REL (Ceiling)		ACGIH TLV (STEL, 15 min)	
Methyl alcohol (67-56-1)	200		200	(250)	200	(250)
Methyl bromide (74-83-9) (carcinogen)		(20 ceiling)	Reduce exposure to lowest feasible level		5	(skin)
Methyl chloride (74-87-3) (carcinogen) (cns)	100	(200 ceiling; 300/5 min in 3 hours)	Reduce exposure to lowest feasible level		50	(100)
Methylene chloride (75-09-2) (carcinogen) (cns)	500	(1000 ceiling; 2000/5 min in 2 hours)	Reduce exposure to lowest feasible level		50	
Methyl iodide (74-88-4) (carcinogen) (cns)	5	(skin)	Reduce exposure to lowest feasible level		2	(skin)
Nitriles						
Acetonitrile (75-05-8) (cns)	40	(60)	20		40	(60)
Tetramethyl succinonitrile (3333-52-6) (cns)	0.5	(skin)	0.5	(skin)	0.5	(skin)
2-Nitropropane (79-46-9) (carcinogen) (cns)	25		25		10	
Styrene, monomer (100-42-5)	100	(800 ceiling; 600/5 min in 3 hours)	50	(100 ceiling)	50	(100)
1,1,2,2-Tetrachloroethane (79-34-5) (carcinogen) (cns)	5	(skin)	1	(skin)	1	(skin)
Tetrachloroethylene (127-18-4) (carcinogen) (cns)	200	(200 ceiling; 300/5 min in 3 hours)	Minimize workplace exposure levels; limit no. of workers		25	(100)
Toluene (108-88-3) (cns)	200	(300 ceiling; 500/10 min)	100	(150)	100	(skin)

Table 4.20.   OSHA PELs, NIOSH RELs and ACGIH TLVs for Some Organic Solvents[a] (continued)

Compound (CAS No.)	OSHA PEL (Ceiling)		NIOSH REL (Ceiling)		ACGIH TLV (STEL, 15 min)	
1,1,1-Trichloroethane (71-55-6) (cns)	350	(450)		(350 ceiling/15 min)	350	(450)
1,1,2-Trichloroethane (79-00-5) (carcinogen) (cns)	10	(skin)	10	(skin)	10	(skin)
Trichloroethylene (79-01-6) (carcinogen) (cns)	100	(200 ceiling; 300/5 min in 2 hours)	25		50	(100)
Xylene (1330-20-7) (cns)	100		100	(200/10 min)	100	(150)

[a] All values in parts per million except where noted.

Table 4.20 provides, for a number of common solvents, current recommended PELs, RELs, and TLVs for 8-hour TWAs. The OSHA PEL values were lowered for a time to, in most cases, the ACGIHs TLV. In the context of this table, those solvents that affect the CNS are so noted with (cns) while those that affect the peripheral nervous system are indicated with (pns).

As can be seen from Table 4.20, the values from the three sources frequently disagree. Usually, the ACGIH values are the most current, while the OSHA levels often go back to the original adoption of the OSHA act. The OSHA values are binding legally, however, while those of the other two are guidelines only. The NIOSH values, where available, are often the most conservative.

REFERENCES*

1. *TLVs©, Threshold Limit Values and Biological Exposure Indices for 1993–94*, American Conference of Governmental Industrial Hygienists, Cincinnati, OH, 1993.

2. **Astrand, I.,** Uptake of solvents in the blood and tissues of man, *Scand. J. Work Environ. Health,* 199, 1975.

3. **Baker, E.L. and Seppalainen, A.M.,** Session 3: human aspects of solvent neurological effects, report on the workshop session on clinical and epidemiological medicine, *Arch. Ind. Health,* 13, 581, 1986.

4. **Baker, E.L. and Vyskocil, J.,** The neurotoxicity of industrial solvents: a review of the literature, *Am J. Ind. Med.*, 8, 207, 1956.

5. **Bird, M.,** Industrial solvents: some factors affecting their passage into and through the skin, *Ann Occup. Hyg.,* 24, 235, 1981.

6. **Browning, E.,** *Toxicity and Metabolism of Industrial Solvents,* Elsevier Publishing Company, Amsterdam, the Netherlands, 1965.

7. **Cherry, N., Venables, H., and Waldron, H.,** The acute behavioral effects of solvent exposure, *J. Soc. Occup. Med.,* 33, 13, 1983.

8. **Gamberale, F.,** Behavioral effects of exposure to solvents, experimental and field studies, in *Adverse Effects of Environmental Chemicals and Psychotropic Drugs,* Vol. 2, Horvath, M., Ed., Elsevier Publishing Company, Amsterdam, the Netherlands, 1976, 111.

9. **King, M., Day, R., Oliver, J., Lush, M., and Watson, J.,** Solvent Encephalopathy, *Br. Med. J.,* 283, 663, 1981.

10. **Knave, B., Anselm-Olson, B., Elofsson, S., Gamberale, F., Isaksson, A., Mindus, P., Persson, H.E., Struwe, G., Mennberg, A., and Westerholm, P.,** Long-term exposure to jet fuel. II. A cross-sectional epidemiologic investigation on occupationally exposed industrial workers with special reference to the nervous system, *Scand J. Work. Environ. Health,* 4, 19, 1978.

11. **Lindstrom, K.,** Changes in psychological performances of solvent-poisoned and solvent-exposed workers, *Am. J. Ind. Med.,* 1, 69, 1980.

12. Criteria for a Recommended Standard: Occupational Exposure to Refined Petroleum Solvents, U.S. Department of Health, Education and Welfare, Public Health Service, Centers for Disease Control, National Institute for Occupational Safety and Health, DHEW (NIOSH) Pub. No. 77-192, Cincinnati, 1977.

13. **Orbaek, P., Risberg, J., Rosen, I., Haeger-Aronson, B., Hagstadius, S., Hjortsberg, U., Regnell, G., Rehnstrom, S., Svensson, K., and Welinder, H.,** Effects of long-term

---

* The references are the more general and comprehensive ones from the list of references in the original article with some more recent ones.

exposure to solvents in the paint industry, *Scan. J. Work Environ. Health*, 11 (Suppl. 2), 1, 1985.

14.   **Politis, M., Schaumburg, H., and Spencer, P.,** Neurotoxicity of selected chemicals, in *Experimental and Clinical Neurotoxicology*, Spencer, P. and Schaumburg, H., Eds., Williams and Wilkins, Baltimore, 1980, 613.

15.   **Seppalainen, A.M.,** Neurophysiological aspects of the toxicity of organic solvents, *Scand. J. Work. Environ. Health,* 11 (Suppl. 1), 61, 1985.

16.   **Seppalainen, A.M. and Antti-Poika, M.,** Time course of electrophysiological findings for patients with solvent poisoning, *Scand. J. Work. Environ. Health.,* 9, 15, 1983.

17.   **Seppalainen, A. M., Husman, K., and Martenson, C.,** Neurophysiological effects of long-term exposure to a mixture of organic solvents, *Scand. J. Work. Environ. Health,* 4, 304, 1978.

18.   **Seppalainen, A.M., Lindstrom, K., and Martelin, T.,** Neurophysiological and psychological picture of solvent poisoning, *Am. J. Ind. Med.,* 1, 37, 1980.

19.   **Spencer, P. and Schaumburg, H.,** Organic solvent neurotoxicity, facts and research needs, *Scand. J. Work. Environ. Health,* 11 (Suppl. 1), 53, 1980.

20.   **Toftgard, R., and Gustafsson, J.,** Biotransformation of organic solvents, a review, *Scand. J. Work. Environ. Health,* 6, 1, 1980.

21.   **Valciukas, J., Lilis, R., Singer, H., Glickman, L., and Nicholson, W.,** Neurobehavioral changes among shipyard painters exposed to solvents, *Arch. Environ. Health,* 40, 47, 1985.

22.   **Waldon, H.A.,** Solvents and the brain, *Br. J. Ind. Med.,* 43, 73, 1986.

23.   Organic Solvents and the Central Nervous System, EH5, World Health Organization and Nordic Council of Ministers, Copenhagen, 1985, 1.

24.   EPA Draft Report: Request for Comments on Principles of Neurotoxicity Risk Assessment, 58 *Fed. Reg.,* 41558, Aug. 4, 1993.

25.   **Parkinson, D.K. et al.,** *Am. J. Indust. Med.,* 17, 661, 1993.

26.   **National Research Council,** *Environmental Neurotoxicology,* National Academy Press, Washington, D.C., 1992.

27.   **Tilson, H. and Mitchell, C., Eds.,** *Neurotoxicology,* Raven Press, New York, 1992.

4.9. SPILLS AND EMERGENCIES

A chemical spill is probably the most common type of laboratory accident and potentially one of the most serious if the material gives rise to hazardous vapors, interacts with the laboratory environment in a violent physical fashion, or is toxic or corrosive upon contact with a person's body. Most accidents involving chemical spills do not have such dramatic consequences, but they must all be handled correctly. The response to an emergency spill may invoke meeting the requirements of 29 CFR 1910.120—Hazardous Waste Operations and Emergency Response. In most facilities using chemicals there will need to be a plan to comply with the requirements of this standard.

4.9.1. Small- to Moderate-Scale Spills

Chemical spills generally involve only small quantities of materials, such as would be contained in a single reagent container or reaction flask. Unless the chemical does have unusually hazardous properties, procedures to correct such minor spills are relatively straightforward. If the material is a solid, it is simply swept up into a container with which it will not react for disposal. However, relatively few materials may then be disposed of as ordinary trash. Custodians should not, under normal circumstances, be expected to remove trash containing chemical materials from laboratories. Chemicals known to be harmless if placed in a municipal landfill and which are to be disposed of in this way should be placed in a separate container, labeled with the identity of the material, and marked "**safe for disposal in regular trash.**" Some materials which are soluble in water or, if liquid, miscible in water may also be disposed of into the sanitary system, while others may be rendered harmless by reaction with other chemicals in the laboratory. However, before any of these things are done, it should be confirmed that the material does in fact not fall under the provisions of the RCRA which prohibits such disposal procedures. If restrictions exist, the materials must be treated as hazardous waste.

Individuals handling even small quantities of hazardous waste such as might result from the remediation of a small spill should be provided with basic training on what to do with the materials, how to identify and label them, suitable containers for various classes of waste materials, and, of course, basic safety measures needed to protect themselves and others. This level of training should be provided to every new employee in an area or to each new graduate student. The material in this section will apply to handling spills that do not involve a release of a hazardous material into the environment, and that can be handled by in-house personnel, meaning either laboratory employees or safety employees who have received appropriate training prescribed by the standard. Unless the training is provided, an employee or student should not be allowed to be involved with cleaning up a spill or handling hazardous waste. It is possible that the organization could be cited under 1910.120 otherwise. Basic training such as this may be provided using the employer's own staff.

4.9.1.1. Spill Response Procedures

Spilled liquids often may be diluted with water and simply mopped up or, in some cases, eliminated by spreading an absorbent material, such as vermiculite or a clay absorbent (such as calcium bentonite) on the spilled material or, a bit more neatly but much more expensively, by placing pillows or pads containing an absorbent material on the liquid, after which the absorbent material is collected into containers for later disposal.

Spills of larger quantities of solids are usually no different from smaller spills except in scale. The spilled solids simply remain as they fell. As long as the solids do not react with materials with which they may have come into contact or the dust is not a breathing hazard, the clean up is simply mechanically larger, with more containers being set aside for evaluation as to the mode of disposal, i.e., to determine if they are to be treated as hazardous waste or not.

Spills of liquids, even of relatively innocuous materials, are messier and more likely to come into contact with other materials with which they may interact. Absorbent materials (such as loose absorbent, spill control pillows, and absorbent pads) immediately available in the laboratory quickly confine the area of the spill, the immediate problems are minimized, and the clean-up is considerably easier.

Relatively few laboratory spills involve totally innocuous materials. However, there are degrees of hazards associated with spilled materials, from the mildly troublesome to IDLH materials. The risk also depends upon the exposure mode or the character of an ensuing physical hazard. A material which emits deadly fumes is infinitely more dangerous to most laboratory personnel than one which is highly corrosive to tissue. Unless an individual is directly injured by contact with the material, the latter hazard may be avoided by removing oneself from the immediate area, and protective clothing is often enough to protect those engaged in the ensuing clean up. However, generation of deadly vapors or gases will usually mandate an evacuation of at least the laboratory and perhaps the entire building. Individuals correcting the situation will require air-supplied breathing apparatus and often wear garments protecting their entire body from contact with the airborne materials. Similar and often more cumbersome protective gear will be required if a fire occurs after a spill of flammable materials. Remedying the situation involving a corrosive spill is very likely to be within the capacity of laboratory employees, but incidents involving toxic gases or fire would almost certainly mandate the participation of trained emergency personnel.

The following material will address handling spills and emergencies which are confined to the laboratory. Those emergencies not confined to at least the building in which they originate will be treated in a later section in the context of the specific requirements of the OSHA standard.

There should be some basic emergency equipment and supplies readily available to every laboratory using hazardous chemicals. It is essential that some items be available within the laboratory itself. For example, in any laboratory in which flammables are stored, OSHA standards (1910.106(d)(7)(b)) require that at least one 12-B portable fire extinguisher be located not less than 10 feet nor more than 25 feet from a flammable material storage area within a building. If other types of chemicals are in use, other types of fire extinguishers should be available, such as class D units if reactive metals are stored or in use or pressurized water units if there are substantial amounts of ordinary combustible materials present.

Spill control materials to absorb spilled materials are available commercially. Some materials are universal, and some are intended to not only absorb materials, but also to neutralize specific materials such as acids, hydrofluoric acid, caustics, and mercury. Some of the products for coping with acids and caustics contain color indicators to determine when the spill is neutralized. If bought in convenient small packages suitable for cleaning up modest spills of about half a liter, the cost per unit of absorbent is high for these commercial products. However, they can usually be bought in small drums at a substantial saving. If bought in this way, the materials can be repackaged into convenient smaller sizes for use in individual laboratories. For many purposes, calcium bentonite, bought in bulk, serves very nearly as well as most commercial absorbents and is much less costly.

Whether commercial kits of absorbent materials or cheaper substitutes are used, some capability to quickly soak up spilled materials needs to be readily available. Bulk quantities can be kept in a central location for replenishing supplies in individual laboratories or for use in atypical larger spills, but a sufficient supply should be kept within an individual laboratory to use on a spill of up to a gallon or two.

In addition to absorbent material, the following items should be maintained in an individual laboratory (Chapter 1 also provides a more comprehensive list of items and equipment that should be available within the building, although not necessarily within the laboratory):

- Bags, large, 6-mil polyethylene
- Brooms
- Brushes, hand
- Bucket, plastic (polyethylene)
- Containers, plastic (5-gallon)
- Coveralls, chemically resistant treated, lightweight
- Dustpan
- Gloves, chemically resistant
- Goggles, chemical splash
- Mops
- Paper, plastic-backed, absorbent roll
- Paper towels, regular
- Respirators, organic, acid, dust, caustics
- Scoops, shovel
- Shoe covers, high-topped, chemically resistant
- Soap, detergent
- Tape, duct

For typical laboratory spills, many of these items will not be needed, but all of them could have a use, depending upon the type of spill. For example, an acid spill could quickly destroy a person's shoes during the clean up, and drops of acid on a person's clothes would ruin them. Since those doing this clean up work are likely to be laboratory personnel, not only should they be protected against injury, but they also should not be expected to incur any economic loss.

Large quantities of each item should not be needed in a typical spill kit. One or two persons actually working on the clean up, with one person bringing supplies and taking waste away, is probably about optimum, unless the spill is unusually large. Aisle widths in the typical laboratory would preclude easy access of more than a few persons to the spill at a time, so that even if relatively complete protective clothing and equipment were needed, no more than three sets would be needed at a time.

None of the equipment listed above requires an extraordinary level of training to use properly. It should be possible for most technically trained personnel to clean up fairly substantial spills safely. There are some straightforward guidelines to aid the user in preventing any material from getting on their persons. If any substantial amount of the equipment were needed, it would be desirable for the work to be done under the supervision of a person trained as required by 1910.120.

Chemically resistant clothing generally snaps, buttons, or zips up the front, and although the two edges overlap, material can still enter the front seam. Overlapping layers of duct tape will seal this opening. The sleeves are usually loosely fitting and should be folded over around the outside of the glove, and again duct taped, with the tape in contact with both the sleeve and the glove. The trouser cuffs should be brought down over the top of the shoe covers, and folded tightly around them. Duct tape is then used to seal this opening by wrapping it around the ankle so that the adhesive is in contact with both the shoe cover and the coverall. With this done, the entire body below the neck is protected from incidental contact with materials. With the hands and feet attached to the body of the coverall, movement will tend to strain the garment unless there is some slack in the fit. It would be desirable if the coverall or suit were somewhat oversized to provide the needed freedom of motion. The head can be covered with a hood of the same material as the coverall and the face protected with a full-face respirator, if this is required. Otherwise, a half-face respirator and goggles will provide almost as complete protection.

There are two notes of caution about the apparel described above. Lightweight coated garments are usually not rated for chemical protection, although for the level of contact which should be experienced if one works carefully, the wearer should be reasonably well protected against light exposures for the time needed to clean up a minor spill. If substantial contact with the chemical or its vapors is anticipated, a heavier coverall designed to provide protection against chemicals should be substituted. Another problem with the clothing is that it does not "breathe," and after a relatively short period—30 minutes to 1 hour depending upon the level of exertion—it is necessary to cease work for a period, open the coveralls, and "cool-off." Heat exhaustion can occur, and even heat stroke. The latter can be fatal. However, it should be possible to clean up most small to moderate laboratory spills in less than 1 hour.

If the outside of the coverall is contaminated, there is a definite procedure to be used to avoid the outside surface from coming into contact with skin and clothes. If a hood and face mask are worn, these should be removed first. The duct tape sealing the hood to the coverall is peeled off and then the hood is grasped at about the level of the cheekbones and peeled backward so that it tends to turn inside out. The hood is put in a plastic bag. The respirator (and goggles, if worn separately) is then removed and placed in a plastic bag, since it is likely to be worn again, whereas the coveralls may not be reused. The duct tape down the front of the coverall is then removed and the front undone. The two sides are then separated and pulled back off the shoulders so that the inside is exposed, the outside doubling back over itself. The duct tape is removed from one sleeve, and the arm of the coverall on that side is pulled down over the hand, so that the sleeve is turned inside out. The operation is then repeated on the other side, to free both arms. The duct tape is removed from the cuffs of the pants and the leg of the coverall pulled down over the foot so that the leg is turned inside out. The operation is then repeated on the other leg. The coverall is now off, with the exposed surface being the clean, interior surface. The coverall is put in the plastic bag with the hood. The shoe covers can then be removed carefully (so that the outside does not come into contact with the skin and clothing) and placed in the plastic bag with the clothing. Finally, the gloves are removed in such a way that they also are turned inside out and placed into the contaminated clothing bag. If carefully done, there should be no opportunity for the contaminated outer surface of any of the protective gear from ever coming into contact with the wearer.

The clean up also has an optimum procedure. If two persons are working directly on the spill, then a third person should be available to bring fresh supplies and to take waste away. Individuals working directly on the clean up should not leave the immediate vicinity to avoid contaminating a wider area. It normally is desirable for spills of substantial sizes and involving materials of significant hazard to establish a formal entrance and exit control point, one side of which is to be considered dirty and the other maintained clean. The clean up should start at the perimeter of the contaminated area and move inward so as to steadily reduce the area involved. In principle, the workers should clean in front of themselves so that they create a clean area in which to work. A previously cleaned area should not be readily recontaminated if this procedure is followed.

Any waste from the dirty area should be placed in plastic pails and passed to the person outside the dirty area. If the pails are contaminated they should be placed in double, clean, heavy-duty (preferably 6 mil or better) plastic bags before being set on a floor area covered with plastic-backed absorbent paper.

Once the spilled chemical has been removed from the surface, it is always good practice to thoroughly scrub or mop the previously contaminated area. Occasionally, especially if the floor is permeable as a wooden or untreated concrete floor could be, the contamination may have penetrated the surface and a layer might have to be removed.

These procedures seem very formal on first reading if one is not familiar with decontamination work. However, it takes very little additional time to do the job properly. The procedures ensure that no one is likely to be injured or that no clothes are damaged in the process, and it is unlikely that the clean up will have to be repeated.

For large spills, especially as the hazards associated with the material become more serious, it would be highly desirable for the work to be done under the guidance and assistance of an experienced safety professional. At some higher level of risk, the danger to inexperienced personnel, as most laboratory personnel usually are in the context of a serious chemical emergency, should cause one to consider not using laboratory personnel at all, other than as an information resource. These are judgment calls to be made by the organization's emergency coordinator, who should be called to the scene of any major chemical emergency, especially ones in which a building has been evacuated. Really large emergencies may be beyond even the most experienced local personnel, and hazard material response teams should be called in from outside agencies as quickly as possible. One of the more important factors in reducing the scope of most emergencies is prompt, effective action. If local capacity to deal with an emergency is questionable, outside aid should be called for immediately while local efforts continue to protect human life and confine the scope of the incident. Most emergency groups would rather be called for unnecessarily, rather than to arrive at a scene where the situation has deteriorated to the point of being out of control.

4.9.2. Large-Scale Releases of Chemicals

In the last few years there have been some noteworthy releases of chemicals from facilities using chemicals, which have caused substantial numbers of injuries and deaths in nearby communities. There already had been a substantial movement in many states to require operators of facilities using hazardous chemicals to provide information on these chemicals to the nearby communities, and to work with the communities on emergency planning. Many responsible chemical firms had individually begun to provide such data. The accidents which occurred in 1984 in Bhopal, India, and in 1985 in Institute, West Virginia, undoubtedly provided additional motivation for federal action in this area. On October 17, 1986, the Emergency Preparedness and Community Right-To-Know Act was signed into law. This law is more commonly known as Title III of the SUPERFUND Amendments and Reauthorization Act (SARA, Title III), and this appellation will be used henceforth. This act extended and revised the authorities established under the original SUPERFUND Act (the Comprehensive Environmental Response, Compensation, and Liability Act of 1980, CERCLA). The extended act specifically establishes new authorities for emergency planning and preparedness, community right-to-know reporting, and toxic chemical release reporting.

The OSHA standard covering hazardous waste operations and emergency response, 29 CFR 1910.120, is clearly intended primarily for hazardous waste site operations, but the standard covers all operations authorized under RCRA (Resource Conservation and Recovery Act) and required to have a permit or interim status under EPA, pursuant to 40 CFR 270.1 or from a state agency pursuant to RCRA. If an organization or institution is a large generator of hazardous waste, then they are covered if they have personnel who collect and manage the hazardous waste prior to disposal and if they respond to chemical emergencies. A very large number of organizations fall under this definition. Training to comply with this standard is known commonly as HAZWOPER training as a convenient term for "hazardous waste operations."

Hazardous Waste Operations and Emergency Response and SARA Title III requirements will be covered in the following two sections. In many ways, they are two different aspects of the same effort, to ensure that hazardous materials are handled in such a way as to minimize risks to employers, the public, and the environment.

4.9.2.1. Hazardous Waste Operations and Emergency Response

Under 29 CFR 1910.120, Hazardous Waste Operations and Emergency Response, an employer who has employees engaged in managing hazardous wastes and responsible for responding to emergencies involving the wastes and other chemical incidents is required to have a written safety and health program. Programs already in place to satisfy other state and federal regulations are acceptable if they already cover or can be modified to cover hazardous waste operations.

In either case, they must contain the following elements:

An organizational structure
A comprehensive work plan

A site-specific safety and health plan
A safety and health training program
A medical surveillance program
The employer's standard operating procedures for safety and health
Any necessary interface between the general program and site-specific activities

Relatively few commercial research organizations or academic institutions operate their own treatment, storage, disposal, and recycling facilities, although a few do. These larger operations should be familiar with all of the rules and regulations affecting their personnel, while this book is intended for the laboratory facility. The program as described in this section will assume that the organization is a large-scale generator of hazardous waste, i.e., more than 1000 kg per month but uses a commercial hazardous waste disposal firm. For a large-scale generator, there will be typically at least one full-time employee and frequently several who will be covered by the standard.

4.9.2.2. Safety and Health Program

4.9.2.2.1. Organizational Structure

The "site" for most research facilities, and especially for academic facilities can be considered as encompassing all of the buildings in which chemicals are used, since hazardous waste operations are conducted in these buildings, such as collection of the wastes, or more strictly as the area where hazardous waste is stored temporarily while awaiting being picked up. There can be a primary storage site and several satellite storage sites. Hazardous wastes are not only stored in these areas for up to 90 days, but preparation of some of the wastes for disposal, such as bulking of compatible liquids and segregation into various classes, is also done. Some organizations pack their own wastes and are simply spot-checked by the disposal firm. These storage areas most closely match the intended scope of the standard. Since this situation is generally much simpler than the operations intended to be covered by the standard, some aspects of the standard will be passed over. However, the hazardous waste operational group should go through the entire standard in preparing their own plan.

The employees who perform these tasks are usually members of the environmental health and safety group who manage the waste program but also provide the required safety oversight. The organizational structure for this type of operation might identify the hazardous waste manager as the general supervisor and safety and health official. If an alternate is needed, the individual to whom the safety and health manager reports can be identified as the ultimate authority for safety and health. The organizational structure should also identify all other personnel associated with the operations and their general functions and responsibilities. The organizational structure would include an organizational chart identifying the lines of authority, responsibility, and lines of communication.

4.9.2.2.2. Comprehensive Workplan

The comprehensive workplan must include the following components:

1. Normal operating procedures and activities required for incident response and clean up
2. Definition of the work tasks and objectives and methods used to accomplish them

3. Personnel requirements
4. Provisions for implementation of the required training and information programs
5. Provisions for the implementation of the medical surveillance program

4.9.2.2.3. Site-Specific Safety and Health Plan

The safety and health plan shall, at a minimum, include:

1. A hazard analysis for each task and operation involved in managing the hazard waste and emergencies
2. Employee training assignments to ensure compliance
3. Personal protective equipment (PPE) to be used by the employees for the different tasks and operations
4. Medical surveillance requirement
5. Description of the air monitoring programs, including methods of calibration and maintenance of the monitoring equipment
6. Site control measures
7. Decontamination procedures
8. An emergency response plan, including necessary PPE and other equipment
9. Confined space entry procedures, if applicable
10. Spill containment program

4.9.2.2.4. Hazard Identification and Relevant Information

The areas where hazardous chemicals are stored prior to shipment and where any ancillary operations are conducted should be carefully evaluated for hazards. Access and evacuation routes should be clearly marked and checked to ensure freedom from obstruction. Any very hazardous materials, especially any IDLH materials, should be in specific, well-marked areas. Means of annunciating an alarm should be readily available. Assistance agencies, e.g., fire departments, rescue squads, police, and haz-mat teams, should be identified.

4.9.2.2.5. Personal Protective Equipment

Personal protective equipment should be available for evacuation, such as 5-minute escape air packs, normal operations, and reentry into an area in which a chemical emergency has occurred. Among items needed would be positive-pressure supplied-air units, protective clothing, goggles, gloves, head covers, and shoe covers, appropriate to the conditions.

4.9.2.2.6. Monitoring

Monitoring equipment appropriate to the hazards present shall be available. This may include organic vapor detectors of various types, combustible gas detectors, oxygen detectors, detector tubes and pumps, and specialized detectors such as mercury detectors.

4.9.2.3. Hazard Communication Program

A program embodying the requirements of the OSHA 29 CFR 1910.1200 hazard communication program shall be in effect.

4.9.2.3.1. Training

The standard has specific initial training requirements that have become known as HAZWOPER training. Hazardous waste personnel are required to receive either 40 hours of training prior to working on a site, followed by 3 days of field training under the supervision of a knowledgeable individual, or 24 hours of training followed by 1 day of supervised field experience. Academic training and work relevant experience can be substituted for some or all of these requirements if the alternatives can be shown by the employer to be equivalent. However, it is recommended that training programs offered by professional groups or individuals be offered to hazardous waste employees, if possible. This would eliminate any question of conflict of interest and could reduce the liability to the employer should an accident occur.

The 40-hour HAZWOPER course is recommended for senior staff and those immediately involved in activities where there is a real potential for exposure to hazardous chemicals. The 24-hour course will probably be sufficient for most of the hazardous waste employees who are covered by the standard due to the organization being a large generator of hazardous waste. Much of the time, these employees' time is not spent in directly working with hazardous waste that is likely to cause exposures. Even during collection of materials from individual generating sites, the wastes are in sealed containers so that the risk of exposure is minimal. Neither the 40-hour nor 24-hour programs qualify an individual to manage a large-scale clean up effort, although both would enable the employee to understand the need for various aspects of the clean up and be able to assist under the supervision of professionals experienced in handling emergency situations. Note that in many jurisdictions the responsibility for remediation of a hazardous material release that extends beyond the boundaries of a facility is established by law as being that of a haz-mat team or a local fire department.

Following is a syllabus for a 24-hour HAZWOPER training course that would meet the requirements of the standard:

1. A review of the requirements needed to comply with the 29 CFR 1910.120 regulation
2. A review of the employer's work and safety program and comprehensive safety plan
3. General nonchemical safety hazards, electric-powered equipment, walking surfaces, hot and cold temperature-related hazards
4. Health and safety effects of hazardous chemicals
5. Employer's medical surveillance program, recognition of signs and symptoms of overexposure
6. Means of detection of hazardous vapors and gases, monitoring technology, procedures, and strategies
7. Engineering and procedural control of environmental releases and exposures
8. Personal protection equipment, selection and use, respiratory protection, clothing, etc.
9. Contents of an effective health and safety plan
10. Identification and recognition of hazards, site safety planning
11. Emergency response planning, emergency procedures
12. Decontamination procedures
13. Sources of information, assistance, emergency response agencies

14. Employee and employer rights and responsibilities
15. Field exercises, table-top exercises
16. Final exam

The employee who successfully completes a HAZWOPER training program must receive a written certification to that effect. Unless employees have been certified or could be shown by the employer to have had the equivalent training by other means, they shall be prohibited from working in hazardous waste operations. The employees are to receive 8 hours of refresher training annually.

In academic institutions, students are often used to supplement permanent employees working with hazardous waste. Usually this participation involves little risk. However, even if the involvement is limited to transporting small containers from a laboratory to a satellite collection site within the same building, they should receive training about the potential risks and measures to minimize the risks to themselves and others.

4.9.2.4. Medical Surveillance Program

The employer shall provide a medical surveillance program for employees handling hazardous materials. This program is very similar to medical programs required by other standards, but does have a few specific additional requirements. Employees are covered if (1) they may be exposed above the PELs or, if none exist, other published exposure levels, for 30 days or more per year; (2) if they wear a respirator more than 30 days per year; or (3) they are injured or develop signs or symptoms of possible overexposure to hazardous substances during an emergency response or from hazardous waste activities. The examinations are to be at no cost to the employee, without loss of pay and at convenient times and places.

The employee shall receive an examination (1) prior to assignment of work involving hazardous waste, (2) annually or at least biannually, at the discretion of the examining physician, (3) at termination of employment or reassignment unless the employee had an examination within the previous 6 months, (4) as soon as possible after an incident or overexposure, or upon developing signs and symptoms of an over exposure, and (5) more frequently if the examining physician deems it medically necessary.

The examinations are to include a medical history and work history, and are to emphasize symptoms associated with exposure to hazardous chemicals, and health hazards and the employee's fitness for duties requiring the wearing of PPE under conditions that might be expected during the employee's work assignments. The examinations are to be done by or under the supervision of a licensed physician, preferably one with a background in occupational medicine.

The employer is required to provide the physician with a copy of the standard, including appendices, and (1) the employee's duties relevant to chemical exposure, (2) the exposure levels or anticipated exposure levels, (3) PPE used or expected to be used, (4) information from prior examinations, and (5) any relevant information required by 29 CFR 1910.134.

As with other OSHA required exams, the physician is required to provide a written opinion to the employer who shall provide a copy to the employee. The opinion shall include any medical conditions which would place the employee at a significant health hazard from work with hazardous waste operations or use of a respirator. This last requirement would imply that a pulmonary function be given. The opinion would also include the physician's recommendations on any limitations on the employee's work. The physician's opinion must include the statement that the employee had been informed of the results of any medical findings, especially those that would require further evaluation. Any findings not related to the employee's occupational exposures are not to be included in the report. The medical records are to be retained as defined by 29 CFR 1910.20.

4.9.2.5. Engineering Controls, Work Practices, and Personal Protective Equipment for Employee Protection

As with other OSHA standards, the preferred hierarchy of protection is (1) engineering controls, (2) work practices and procedures, and (3) use of PPE. In hazardous waste operations all three are likely to be needed and used. The standard contains very specific recommendations based on rather rigorous exposure scenarios.

The latter half of the standard goes in depth into emergency planning and emergency response. Since the hazardous waste RCRA regulations require comparable contingency plans and emergency response, the discussion of hazardous material emergency response will be deferred until Section 4.10.

4.9.2.6. SARA Title III, Community-Right-to-Know

The SARA provisions apply to those industries covered under the OSHA Hazard Communication Act which has been extended to all chemical users in addition to the originally covered industry groups in the Standard Industrial Codes 20–39 and which handle, use, or store certain chemicals of extremely hazardous chemicals at levels in excess of limits established by EPA under the Act. The limits are such that many facilities do not have quantities in excess of the limits.

There are currently a number of important exemptions under SARA, one of which is specifically applicable to the research facility. In Title 40 CFR 370.2(5) a "hazardous chemical" means any hazardous chemical as defined under 29 CFR 1910.1200(c), except that such a term does not include the following substances:

1. Any food, food additive, color additive, drug, or cosmetic regulated by the Food and Drug Administration
2. Any substance present as a solid in any manufactured item to the extent exposure to the substance does not occur under normal conditions of use
3. Any substance to the extent it is used for personal, family, or household purposes, or is present in the same form and concentration as a product packaged for distribution and use by the general public
4. Any substance to the extent that it is used in a research laboratory or a hospital or other medical facility under the direct supervision of a technically qualified individual
5. Any substance to the extent it is used in routine agricultural operations or is a fertilizer held for sale by a retailer to the ultimate customer

The fourth exemption is obviously the one of importance in the context of this book. Note that it does not exempt research facilities; it simply redefines the definition of a hazardous chemical to exclude those used in small quantities in laboratories under the direct supervision of a technically knowledgeable person. It does not exclude the same chemicals stored in a warehouse or stores area, nor does it exclude chemicals used in maintenance or support operations, except as they fall under the other exemptions.

There are a number of critical requirements under SARA for a facility which falls under the provisions of the Act. A facility is subject to the provisions of the Act if it has any (nonexempt quantities) of the 360 extremely hazardous chemicals currently on the list which EPA originally issued (with 406 chemicals) on April 22, 1987, in excess of the threshold planning quantity established by EPA for that substance.

Individual facilities were sent the list of 406 extremely hazardous chemicals by their state commissions in the spring of 1987 and required to inventory their holdings and make a report by May 17, 1987, of those which met the threshold planning quantity. It was also necessary to list the chemicals which the facility had in excess of higher thresholds for emergency planning. Eligible facilities were then required by September 17, 1987, to notify the local planning committee established under the act of their facility emergency planning coordinator. The local committee was required to complete their local emergency plan by October 17, 1988. Large facilities may be represented directly on the planning committee since membership is required to be drawn from elected state and local officials, law enforcement, civil defense, firefighting, first aid, health, local environmental, hospital, transportation personnel, broadcast and print media, community groups, and facility owners and operators subject to SARA.

The first request for information was originally designed to identify agencies which would be likely to fall under the provisions of SARA. Organizations subject to SARA are required to prepare certain inventory forms and make them available to: (a) the appropriate local emergency planning committee, (b) the state emergency response commission, and (c) the fire department with jurisdiction over the facility.

There are basically two inventory forms containing certain levels of information, defined as Tier I and Tier II. The inventory form with Tier I information is required to be submitted by March 1, 1988, and annually thereafter. If Tier II forms are requested by the groups listed in the preceding paragraphs for a given year, then Tier I forms are not required for that year.

Tier I information required on the inventory form is

1. An estimate (in ranges) of the maximum amount of hazardous chemicals in each category of health and physical hazards, as set forth by the Occupational Health and Safety Act, and regulations published under that act, at the facility during the preceding calendar year
2. An estimate (in ranges) of the average daily amount of hazardous chemicals in each category present at the facility during the preceding calendar year
3. The general location of hazardous chemicals in each category

Tier II information is required to be provided only upon request to the same three groups. They in turn can make the information which they have been provided available to others, such as other state and local officials or the general public under certain conditions.

Tier II information applies to each hazardous chemical at the facility. The following information is required:

1. The chemical name or the common name of the chemical as provided on the MSDS for the chemical
2. An estimate (in ranges) of the maximum amount of the hazardous chemical present at the facility at any time during the preceding calendar year
3. A brief description of the manner of storage of the hazardous chemical
4. The location at the facility of the hazardous chemical

5. An indication of whether the owner elects to withhold location information of a specific hazardous chemical from disclosure to the public under Section 324 (which allows the owner to do so upon request)

Leaks, spills, and other releases of specified chemicals into the environment require emergency notification under both SARA Title III and section 103 of CERCLA. Under CERCLA, those in charge of a facility must report any spill or release of a chemical on a list of hazardous chemicals included in that act in excess of a reportable quantity (RQ) that is substance specific. The report must be made immediately to the National Response Center (1988 telephone numbers 800-424-8802 or 202-267-2675). Under SARA Title III, the notification process now includes all the 360 extremely hazardous chemicals. In addition to the National Response Center, the state commission and the local committee must be notified. Note that if the release is confined totally within a facility and does not enter the "environment", so that it affects only the employees within the building, it does not have to be reported, unless it falls under an OSHA regulation and would have to be reported to OSHA.

Generally, the following information must be reported:

1. The name of the chemical(s) (trade name protection of the chemical is not permitted)
2. Identification of whether the chemical is on the extremely hazardous chemical list
3. The quantity released or an estimate of the quantity released
4. The location, time, and duration of the release
5. Weather conditions, wind speed, and direction
6. Medium (air, water, soil) into which the chemical(s) was released
7. Known acute or chronic risks, and any available helpful medical data
8. Precautions to take, including evacuation, if necessary
9. The names and telephone numbers of persons to be contacted for further information

Obviously if a facility has an emergency contingency plan covering releases of materials from its facilities to the environment, these should be coordinated with emergency plans for the local region developed by the district emergency planning committee. Under the current standard, it appears that most laboratories are, in effect, exempt from the act due to the provision that chemicals in research laboratories shall not be considered hazardous if they are used under the direct supervision of a qualified individual, but it should be remembered that the facilities themselves are not exempt and there may be chemicals elsewhere on site which trigger coverage by SARA Title III. Also, these provisions may well be modified at a later date. Finally, if an emergency does occur within a laboratory such that a substantial portion of the stock becomes involved in an incident so that there is a significant release to the environment, a responsible position may be to have actively participated in the emergency response process in the local district.

REFERENCES

1. Resource Conservation and Recovery Act (RCRA) of 1986, SARA, Title III, Sections 300 to 330, Washington, D.C., 1986.
2. Extremely Hazardous Substance List, Sections 302 to 304, *Fed. Reg.*, 13378, 1987.
3. Emergency Planning and Hazardous Chemical Forms and Community Right to Know Reporting; Final Rule, Title III, Sections 311-312, *Fed. Reg.*, 52, 38344, 1986.
4. Toxic Chemical Release Reporting; Community Right-to-Know; Final Rule Title III, Section 313, *Fed. Reg.*, 53, 4500, 1986.

5. Occupational Safety and Health Administration, Hazardous Waste Operations and Employee Right-to-Know, 29 CFR 1910.120, Washington, D.C.
6. Occupational Safety and Health Administration, Employee Emergency Plans and Fire Emergency Plans, 29 CFR 1910.38, Washington, D.C.

4.10. CHEMICAL WASTES

Until a relatively few years ago, disposal of many chemical wastes from laboratories was down the nearest convenient drain. However, this practice is now illegal as a general procedure, although to a limited degree it is still permissible for certain wastes. The change has been due to the growing concern about the impact of chemicals on the environment. Virtually all the chemical wastes in this country are due to industrial sources, with less than 1% being due to laboratory wastes. However, regulations established to govern the disposition of chemical wastes include the wastes generated by laboratories, so that organizations of which the laboratories are a part must have a hazardous waste program. Locally, a major research university or corporate research laboratory may be one of the largest, if not the largest, generators of hazardous waste.

The following sections will provide the basics of a hazardous waste program, and an overview of the regulations that apply to the program. It would be impossible to provide here a complete guideline covering every contingency. Such a document would be comparable in size to the entire handbook. Operators of hazardous waste programs should receive specialized training from any of the many commercial training programs and acquire a complete set of copies of the regulations and appropriate reference materials.

4.10.1. Resource Conservation and Recovery Act

Many individuals who are now responsible for the management of laboratory facilities received their initial training when there were no regulations governing the disposal of laboratory wastes, and they may not fully appreciate the need for the newer procedures. However, especially in the academic area, the training of students should be in the context of compliance with regulations so that when these students embark on their own careers, they will be trained to comply with legally applicable standards. Senior personnel need to adapt, if they have not already done so, and manage their operations by current standards.

Organizations involved with excess hazardous chemical materials must conform to the provisions of the Environmental Protection Agency Resource Conservation and Recovery Act (RCRA) 40 CFR Parts 260 to 265, as most recently amended. In 1984, the Hazardous and Solid Waste Amendments (HSWA) of 1984 required that the EPA ban the land disposal of over 400 wastestreams unless the wastes are treated or unless it could be demonstrated that there would be no migration of the waste while the waste remains hazardous. The portion of these rules covering dioxin and solvents went into effect on November 8, 1986. Although there was a partial extension for some wastes, the extension did not include solvents. The rationale for the extension was the lack of alternate capacity for the large amounts of soil contaminated by dioxin or solvents. The second portion covering the "California" list, including several heavy metals, went into effect on July 8, 1987. The remainder of the restrictions went into effect on three different dates: August 8, 1988, June 8, 1989, and July 8, 1990. These amendments have already begun to significantly change the options available to laboratories for disposing of their waste chemicals.

Laboratory facilities or their parent organizations have important decisions to make concerning the scope of their activities that will be subject to the provisions of the RCRA regulations. Unless they produce less than 100 kg per month, at which level they are defined as a small generator, they are subject to portions of 40 CFR Part 262, which applies to generators. If they produce more than 1000 kg per month, they are "large" generators and they must comply with all of the requirements applicable to generators.

Other portions of the act cover the operations of treatment, storage, disposal, and recycling (TSDR) facilities, which require additional permitting. Many large industrial organizations, in which laboratory operations represent only a small part of their activities involving hazardous chemicals, may choose or have chosen to go through the permitting process. Some also transport their own hazardous waste and are subject to the regulations on transporters besides the regulations applying to generators. However, only a few academic institutions have the capacity or interest required to meet the stringent regulations covering operations other than those of the generators of the waste chemicals, and the same is true of many small industrial operations and their affiliated laboratories. Those organizations which intend to treat, store, dispose, and recycle their hazardous waste normally will be sufficiently sophisticated and knowledgeable about the procedures to not need additional information, so this section will concern itself only with waste management programs for generators.

Before a program of waste management appropriate to a research organization is discussed, some additional background information will be helpful. What constitutes a hazardous waste needs to be defined and the essential portions of 40 CFR Part 262 need to be addressed. It is virtually a full-time job to keep up with the changing requirements in the context of laboratory operations, even for relatively small organizations. It is desirable to transfer at least the transportation of the wastes to an outside contractor, but substantial savings in hazardous management costs can be realized by performing much of the waste management operations internally.

4.10.1.1. Definition of a Hazardous Waste

There are relatively frequent changes in detail as to what constitutes a hazardous waste according to the EPA, but the main features of the definitions have been consistent. Frequently, there are evaluations and rulings pertaining to the hazardous nature of various wastestreams, for example, reported in the *Federal Register,* but these are based on certain guides in 40 CFR 261 Subpart D. There have already been some changes in some definitions, specifically that of toxicity. The announced requirements for treatment before land disposal are examples of the more stringent requirements. There have also been significant changes in information that the generator is required to provide the transporter.

The criteria are (a) it is a waste listed in Part 261 (currently there are about 600 commercial chemicals listed in 261.31 to 261.33), and (b) it meets certain criteria of reactivity, ignitability, corrosiveness, and toxicity.

Wastes which meet certain hazardous criteria may be placed on the lists under three different categories:

Hazardous wastes from nonspecific sources—Examples of laboratory wastes that would fall in this category would be spent solvents, the residue resulting from distillation recovery of used solvents, materials left over from silk screening and electroplating procedures in electronic laboratories, and other sources of used chemicals.

Hazardous wastes from specific sources—Unless the laboratory is a pilot operation simulating an industrial process, it is unlikely that most research laboratories would fall within this category. Note however, that hazardous chemicals used in a pilot plant operation normally would not be exempt from the regulatory provisions of SARA Title III, discussed in Section 4.9.2.

Discarded commercial chemical products, off-specification species, containers, and spill residues thereof—There are two levels in this category (a) those materials which are acutely hazardous, and (b) those which are less so. There are more stringent limitations on the former.

The following excerpts are from 40 CFR 261.11: *Criteria for listing hazardous waste.*

(a) The Administrator shall list a solid* waste as a hazardous waste only upon determining that the solid waste meets one of the following criteria:

(1)   It exhibits any of the characteristics of hazardous waste identified in Subpart C: ignitability, corrosivity, reactivity, and toxicity.

(2)   It has been found to be fatal to humans in low doses or, in the absence of data on human toxicity, it has been shown in studies to have an oral LD_{50} toxicity (rat) of less than 50 mg/kg, an inhalation LC_{50} toxicity (rat) of less than 2 mg/L, or a dermal LD_{50} toxicity (rabbit) of less than 200 mg/kg, or is otherwise capable of causing or significantly contributing to an increase in serious irreversible, or incapacitating reversible, illness. (Waste listed in accordance with these criteria will be designated Acute Hazardous Waste.)

(3)   It contains any of the toxic constituents listed in Appendix VIII (in 40 CFR 261), unless after considering any of the following factors, the Administrator concludes that the waste is not capable of posing a substantial present or potential hazard to human health or the environment when improperly treated, stored, transported, or disposed of or otherwise managed:

    i.   The nature of the toxicity presented by the constituent

    ii.   The concentration of the constituent in the waste

    iii.   The potential of the constituent or any toxic degradation product of the constituent to migrate from the waste into the environment under the types of improper management considered in paragraph (a)(3)(vii) of this section of the standard

    iv.   The persistence of the constituent or any toxic degradation product of the constituent

    v.   The potential for the constituent or any toxic degradation product of the constituent to degrade into non-harmful constituents and the rate of degradation

    vi.   The degree to which the constituent or any degradation product bioaccumulates in ecosystems

    vii.   The plausible types of improper management to which the waste could be subjected

---

*   "Solid" waste need not be physically solid, as will be noted from some of the definitions in this section.

viii.   The quantities of the waste generated at individual generation sites or on a regional or national basis

ix.   The nature and severity of the human health and environmental damage that have occurred because of the improper management of wastes containing the constituent

x.   Action taken by other governmental agencies or regulatory programs based on the health or environmental hazard posed by the waste or waste constituent

xi.   Such other factors as may be appropriate

Substances will be listed in Appendix VIII only if they have been shown in scientific studies to have toxic, carcinogenic, mutagenic, or teratogenic effects on humans or other life forms.

⋮

If the material is not a listed waste, it is a hazardous waste if it meets any of the criteria in Subpart C, Characteristics of Hazardous Waste. Following are the essential sections of this part of the regulations.

261.21 Characteristic of Ignitability

(a)   A solid waste exhibits the characteristics of ignitability if a representative sample has any of the following properties:

(1)   It is a liquid other than an aqueous solution containing less than 24% alcohol by volume and has a flash point less than 60°C (140°F)....

(2)   It is not a liquid and is capable, under standard temperature and pressure, of causing fire through friction, absorption of moisture, or spontaneous chemical changes and, when ignited, burns so vigorously and persistently that it creates a hazard.

(3)   It is an ignitable compressed gas as defined in 49 CFR 173.300 and as determined by the test methods described in that regulation or equivalent test methods approved by the Administrator under paragraphs 260.20 and 260.21....

(4)   It is an oxidizer as defined in 49 CFR 173.51.

(b)   A solid waste that exhibits the characteristic of ignitability has the EPA Hazardous Waste Number of D001.

261.22 Characteristic of Corrosivity*

(a)   A solid waste exhibits the characteristics of corrosivity if a representative sample of the waste has either of the following properties:

(1)   It is aqueous and has a pH less than or equal to 2 or greater than or equal to 12.5....

(2)   It is a liquid and corrodes steel (SAE 1020) at a rate greater than 6.35 mm (0.250 inches) per year at a test temperature of 55°C (130°F)....

(b)   A solid waste that exhibits the characteristic of corrosivity has the EPA Hazardous Waste Number of D002.

---

*   Note that the earlier OSHA definition of a corrosive material in terms of a health hazard was explicitly limited to action of a material on tissue.

261.23 Characteristic of Reactivity

(a)   A solid waste exhibits the characteristic of reactivity if a representative sample of the waste has *any* of the following properties:

    (1)   It is normally unstable and readily undergoes violent change without detonating.

    (2)   It reacts violently with water.

    (3)   It forms potentially explosive mixtures with water.

    (4)   When mixed with water, it generates toxic gases, vapors, or fumes in a quantity sufficient to present a danger to human health or the environment.

    (5)   It is a cyanide or sulfide bearing waste which, when exposed to pH conditions between 2 and 12.5, can generate toxic gases, vapors or fumes in a quantity sufficient to present a danger to human health or the environment.

    (6)   It is capable of detonation or an explosive reaction if it is subjected to a strong initiating source or if heated under confinement.

    (7)   It is readily capable of detonation or explosive decomposition or reaction at standard temperature and pressure.

    (8)   It is a forbidden explosive as defined in 49 CFR 173.51, or a Class A explosive as defined in 49 CFR 173.53, or a Class B explosive as defined in 49 CFR 173.88.

(b)   A solid waste that exhibits the characteristic of reactivity has the EPA Hazardous Waste Number of D003.

261.24 Characteristic of Toxicity

(a) A solid waste exhibits the characteristic of toxicity if, using the test methods described in Appendix II [of the regulation] or equivalent methods approved by the Administrator under the procedures set forth in paragraphs 260.20 and 260.21, the extract from a representative sample of the waste contains any of the contaminants listed in Table 1 [of the regulation] at the concentration equal to or greater than the respective value given in that Table. Where the waste contains less than 0.5 percent filterable solids, the waste itself, after filtering using the methodology outlined in Appendix II, is considered to be the extract for the purposes of this section.

(b) A solid waste that exhibits the characteristic of toxicity has the EPA Hazardous Waste Number specified in Table 4.21 which corresponds to the toxic contaminant causing it to be hazardous.

Since the appendices and lists referred to in the above material change as materials are added to and deleted from the list, laboratory personnel should subscribe to an information service which will provide information to permit maintenance of a current valid list.

A generator's hazardous waste must be evaluated to decide if it may be exempt from being considered a hazardous waste or, if not, if it is a listed waste, as defined above, or meets any of the criteria for a hazardous waste. Most laboratory wastes are likely to fall under one of these provisions and must be treated as a hazardous waste. A major problem with many chemicals produced by laboratory activities, as opposed to chemicals that are purchased, is that they are insufficiently identified. The identity of the chemical in a bottle, for example, labeled "solution A" may have been known perfectly well by the individual who labeled it at the time the label was affixed to the bottle, but even this person may not recall the contents several months later. As often happens, the chemical is an "orphan" left behind by a departed graduate student or employee, in which case the identity of the contents may be even more uncertain. Commercial waste disposal firms will normally not

Table 4.21. Maximum Concentration of Contaminants for the Toxicity Characteristic

EPAHW Number	Contaminant	CAS Number	Regulatory Level (mg/L)
D004	Arsenic	7440-38-2	5.0
D005	Barium	7440-39-3	100.0
D018	Benzene	71-43-2	0.5
D006	Cadmium	7440-43-9	1.0
D019	Carbon tetrachloride	56-23-5	0.5
D020	Chlordane	57-74-9	0.03
D021	Chlorobenzene	108-90-7	100.0
D022	Chloroform	67-66-3	6.0
D007	Chromium	7440-47-3	5.0
0023	o-Cresol	95-48-7	+200.0
D024	m-Cresol	108-39-4	+200.0
D025	p-Cresol	106-44-5	+200.0
D026	Cresol (total)**	----------	+200.0
D016	2,4-D	94-75-7	10.0
D027	1,4-Dichlorobenzene	106-46-7	7.5
D028	1,2-Dichloroethane	107-06-2	0.5
D029	1,1-Dichloroethylene	75-35-4	0.7
D030	2,4-Dinitrotoluene	121-14-2	0.13*
0012	Endrin	72-20-8	0.02
D031	Heptachlor (and its epoxide)	76-44-8	0.008
0032	Hexachlorobenzene	118-74-1	0.13*
D033	Hexachlorobutadiene	87-68-3	0.5
D034	Hexachloroethane	67-72-1	3.0
D008	Lead	7439-92-1	5.0
D013	Lindane	58-89-9	0.4
D009	Mercury	7439-97-8	0.2
D014	Methoxychlor	72-43-5	10.0
D035	Methyl ethyl ketone	78-93-3	200.0
D036	Nitrobenzene	98-95-3	2.0
D037	Pentachlorophenol	87-86-5	100.0
D038	Pyridine	110-86-1	5.0*
D010	Selenium	7782-49-2	1.0
D011	Silver	7440-22-4	5.0
D039	Tetrachloroethylene	127-18-4	0.7
D015	Toxaphene	8001-35-2	0.5
0040	Trichloroethylene	79-01-6	0.5
D041	2,4,5-Trichlorophenol	95-95-4	400.0
D042	2,4,6-Trichlorophenol	88-06-2	2.0
D017	2,4,5-TP (Silvex)	93-72-1	1.0
D043	Vinyl chloride	75-01-4	0.2

*   Quantitation limit is greater than the calculated regulatory level. The quantitation limit therefore becomes the regulatory level.
**   If o-, m-, and p-cresol concentrations cannot be differentiated, the total cresol (D026) concentration is used.

accept these unknown containers until they are characterized. Other examples of unknown chemicals which also must be identified are older containers that have lost their labels or whose labels may have become damaged so that the contents cannot be determined. In all cases of unknowns, there is a certain amount of risk associated with opening an unlabeled container for any purpose, including that of identifying the contents. The container may contain, for example, degraded chemicals that have become explosive, such as ether with a high content of peroxides or dry picric acid. Procedures should be adopted that require all personnel to properly identify the contents of any container which will not be used in its entirety by the maker or the person transferring a quantity from an original container to a secondary container. The latter requirement is written into the OSHA hazard communication standard, and the former should be adopted as a matter of good laboratory practice in developing a laboratory chemical hygiene plan.

Most generators of laboratory waste use commercial firms as the means of removing hazardous waste from their facility rather than doing it themselves. Sometimes, the commercial firms do virtually a turnkey job, i.e., they go to the individual laboratories and perform all the tasks necessary to prepare and dispose of the waste properly. This is appropriate for smaller facilities that do not have the resources to have their own staff to do many of these tasks. Larger operations often find it economical to do most of the work themselves and use the commercial firm only to transport the material to an ultimate disposal site. Single laboratories rarely have the resources to use the latter method, and in most comprehensive universities and research organizations the process of handling hazardous waste has been assigned to specialists within the organization, usually as an adjunct of the safety and health organization.

4.10.1.2. Requirements for Generators of Hazardous Waste

Title 40 CFR Part 262 defines the requirements for generators of hazardous waste. As in other parts of this section, it will be assumed that the generators referred to will be those who are not owners or operators of treatment, storage, disposal, or recycling facilities and who dispose of most of their waste by shipping it away from the generating facility, except for the modest amount of processing permitted under the law. As will be noted, the generator assumes substantial responsibilities in initiating a shipment of hazardous waste from the facility. The information in this section is taken directly from the standard. In some cases the language has been simplified or made more concise for brevity and clarity, while in other cases the exact wording of the standard has been followed where it did not lend itself to modification.

1. The generator must first decide if the waste meets the criteria for hazardous waste. The previous section describes the basic criteria which define a hazardous waste, and the waste must be evaluated in terms of these criteria. Figure 4.33 shows this process graphically. In addition, it may be decided that a given material is a hazardous waste based on the materials or processes used. This last interpretation might apply to much of the materials synthesized or generated in a research laboratory, where the detailed characteristics of the waste might be difficult to determine, but the procedures make it likely that it is a hazardous material. The material classification can be assigned on the basis of this information and simple tests for at least the ignitability, corrosivity, and reactivity characteristics.

2. If the generator intends to initiate a shipment of hazardous waste, an EPA identification number must be assigned by the EPA Administrator.

3. The generator must be sure that anyone engaged to handle or receive his hazardous waste has a valid EPA identification number. This step is required but it is not sufficient to assure that waste generated at a facility will actually be handled legally and environmentally soundly. Although it may consume additional time and re-

HAZARDOUS WASTE
FEDERAL LAW PROHIBITS IMPROPER DISPOSAL

IF FOUND CONTACT THE NEAREST POLICE OR
PUBLIC SAFETY AUTHORITY OR THE
U.S. ENVIRONMENTAL PROTECTION AGENCY

PROPER D.O.T.
SHIPPING NAME _____   UN OR NA NO. _____

GENERATOR INFORMATION

NAME _____

ADDRESS _____   PHONE _____

CITY _____   STATE _____ ZIP _____

EPA
ID. NO. _____   EPA
                                WASTE NO. _____

ACCUMULATION   MANIFEST
START DATE _____   DOCUMENT NO. _____

CONTAINS HAZARDOUS OR TOXIC WASTES

PROPER SHIPPING NAME _____

TECHNICAL CHEMICAL NAME _____

UN OR NA ID NO. _____

D.O.T. PROPER NAME AND AND UN OR NA NO. WITH PREFIX

HANDLE WITH CARE !

Figure 4.33. Hazardous waste label meeting EPA and DOT regulations.

sources, visits to a waste-brokers' storage facility and/or to the eventual site in which the material will be placed in a landfill, incinerated, or otherwise processed is highly recommended. All of the organizations willing to bid on laboratory waste disposal services purport to offer virtually the same degree or quality of service. Many major, reputable firms have had to close facilities, pay fines, or substantially modify their procedures because they did not comply with acceptable operating practices. In the worst possible case, where the operator of a facility does not have resources available to correct a problem which has been identified by the EPA, the waste generators can be required to share the cost of cleaning up a substandard facility.

On December 21, 1990, The Department of Transportation (DOT) published new "Performance-Oriented Packaging Standards." These standards essentially brought the United States into conformance with the United Nations Dangerous Goods regulations. The result is that the manner in which a generator (and other groups) package and document hazardous waste for shipment has changed in almost every detail. The new procedures are complex, and generators who use commercial disposal firms to dispose of their hazardous waste may prefer that the disposal firm help them prepare the needed information for the documentation to accompany the waste offered for transportation, and to ensure that the waste is properly packaged. The new regulations are being phased in over a period of several years. Many of them are already in effect or will be in effect by October 1, 1994. All of the regulations will be in effect by October 1, 1996. Most of the critical ones affecting generators are already in effect. The following section covers the basics of the new requirements under Department of Transportation, HM-181.

4.10.1.2.1. Basics of Compliance with HM-181

The new DOT hazard identification system has been changed to a numerical one. Table 4.22 shows the new and old classifications and some new terms, "Packaging Groups and Hazard Zones." This information is part of that required for preparation of shipping papers.

The definitions of packing groups are

I—Great Danger
II—Medium Danger
III—Minor Danger

Packing groups, where applicable, are provided for all of the materials listed in 49 CFR 172.101.

Hazard zones for gases that are poisonous by inhalation are

A—$LC_{50} \leq 200$ ppm
B—$LC_{50} > 200$ ppm to ≤ 1000 ppm
C—$LC_{50} > 1000$ ppm to ≤ 3000 ppm
D—$LC_{50} > 3000$ ppm to ≤ 5000 ppm

Hazard zones for peroxides are assigned from a generic system consisting of the seven letters, A to G. The organic peroxide table in 49 CFR 173.225 contains all of the peroxides assigned to a generic type. If not listed, the material cannot be transported without written approval from the associate administrator for Hazardous Materials Safety.

The hazard zones for poisonous materials that are toxic because of inhalation of vapors from the material are (for a single toxic substance)

A—vapor concentration ≥ 500 LC_{50} where $LC_{50} \leq 200$ mL/m^3
B—vapor concentration ≥ 10 LC_{50} to < 500 LC_{50}
    where $LC_{50} > 200$ mL/m^3 to ≤ 1000 mL/m^3

Both of these hazard zones would correspond to Packing group I. There are no hazard zones for Packing groups II and III. There are four hazard zones A to D for mixtures containing more than one poisonous substance.

Following are a few comments about some rationales for the various divisions in Table 4.22. Many of the divisions are self-explanatory. To adequately perform the required procedures, it is essential to have a copy of the relevant hazardous material data tables identified in HM-181.

Class 2.1 gases are gaseous at standard atmospheric temperature and pressure, or have boiling points at standard pressure less than standard temperature (20°C or 68°F) and can be ignited at air mixtures of less than 13%.

A flammable liquid is defined differently in this context from that of an OSHA flammable liquid. It is any liquid with a flash point between 141° and 200°F (OSHA definition is a liquid with a flash point of 100°F).

Infectious materials are those that can cause disease in humans or animals. Note that this is a broader definition than that described earlier in the section on bloodborne pathogens.

The definition of corrosive has been extended beyond that in 40 CFR Part 261 to include materials that cause visible destruction or irreversible alterations of the skin tissue at the site of contact when tested on the intact skin of an animal for ≤ 3 minutes besides its effect on steel.

Table 4.22. DOT Hazard Identification Scheme

Name of Class/Division	Class Number	Class Division	Previous DOT Hazard Class	Packing Groups	Other
Flammable gas	2	2.1	Flammable gas	None	
Nonflammable gas	2	2.2	Nonflammable gas	None	
Poisonous gas	2	2.3	Poison A	None	Hazard zones, A to D
Flammable liquid	3	None	Flammable liquid	I, II, III	
Combustible liquid	None	None	Combustible liquid	III	
Flammable solid	4	4.1	Flammable solid	II, III	
Spontaneously combustible solid	4	4.2	ORM-C	I, II, III	
Dangerous when wet solid	4	4.3		I, II, III	
Oxidizer	5	5.1	Oxidizer	I, II, III	
Organic peroxide	5	5.2	Organic peroxide	II	Types A to G
Poisonous material	6	6.1	Poison B, irritating material, etiological agent, ORM-A	I, II, III	Hazard zones A to D, PIH vapors only, liquids/mixtures
Infectious substance	6	6.2		None	
Corrosive material	8	None	Corrosive material, ORM-A	I, II, III	
Miscellaneous hazardous materials	9	None	ORM-A, B, C, E	II, III	
ORM-D (consumer commodities)	None	None	ORM-D	None	

The primary source of information needed to prepare the proper shipping descriptions and manifests is 40 CFR 172.101 Subpart B, Table of Hazardous Materials, and the appendix to the table. Table 101 is divided into ten columns.

Column 1—Symbols. There are only five symbols and most of the materials in the table do not have anything in this column. The symbols are "+" which *fixes* the proper shipping name regardless of whether the material meets the criteria under the hazard class or packing group column; A or W is used to designate a material which is hazardous only when transported by air or water, unless it is a hazardous material or hazardous waste; D indicates that the proper shipping name is only correct for domestic shipments; and I indicates the case is for international shipments.

Column 2—Proper shipping name. The only proper shipping name is the one that appears in this column. No other name is acceptable, including chemical formulas.

Column 3—Hazard Class number, see Table 4.22. A "+" in the symbol column supersedes the hazard class/division normally associated with the material.

Column 4—A number that specifically corresponds to the proper shipping name of the material and provides identification of the material in case of an accident. If the number is preceded by "UN" it is a United Nations number that is recognized both internationally and domestically. If preceded by NA, the numbers are only valid in the United States and Canada.

Column 5—Packing Group, see Table 4.22.

Column 6—Labels required on a package filled with a material (materials if there is a secondary material, in which case the label for the secondary materials must not indicate the Class Number/Division. Only the label for the primary material may show this and it must do so. The labels would be a 4 inch × 4 inch diamond shape. There are exceptions when no label is required or, in some cases, allowed.

Column 7—Special provisions. A letter is followed by a number; these are defined in 40 CFR 172.102.

Column 8—Packing Authorizations. Column 8 is divided into three subcolumns; the first identifies exceptions, the second nonbulk requirements, and the third, bulk requirements. The numbers identify paragraphs in Part 173.304.

Column 9—Quantity Limitations. The maximum amount allowed on a passenger-carrying aircraft or rail car.

Column 10—Quantity Limitation. The maximum value allowed when shipped by cargo aircraft.

The appendix to 49 CFR 172.101 lists the reportable quantities (RQ) for many "hazardous substances." For DOT purposes, a hazardous substance must be listed in this appendix to be considered a hazardous substance and also there must be an amount equal to or greater than the RQ in one package or, if in a mixture or solution, the amount present is in concentrations by weight equal to or greater than is given in the table below from 49 CFR 172.8.

RQ (lb)	Concentration by weight (%)
5000	10
1000	2
100	0.2
10	0.02
1	0.002

Table 172.203 is comparable to 172.101 materials shipped as generic, not otherwise specified (n.o.s.).

49 CFR 172.200 describes the following information that must be provided on the shipping papers to describe a hazardous material offered for transportation. The required information must be in English.

1. A hazardous material identified by the letter X in a column labeled HM placed before the proper shipping name (if appropriate, X may be replaced by RQ but RQ cannot be replaced by X)
2. Proper shipping name (this may be one of the generic descriptions in 172.203)
3. Hazard class or division
4. The identification number (UN or NA) from 172.101 or 172.203
5. The Packing Group, from 172.101 or 172.203, if any
6. Quantity

The above information must be given in that order with no other information interspersed, except that technical or chemical group names may be entered in parentheses between the proper shipping name and hazard class or following the basic description. Two other items must be on the shipping paper:

1. An emergency response telephone number
2. Certification—"This is to certify that the above-named materials are classified, described, packaged, marked, and labeled and are in proper condition for transportation according to the applicable Department of Transportation." An acceptable alternate statement is similar, but also identifies the transporter.

Other information may be on the shipping papers as long as it is placed after the basic description and is not inconsistent with it. Normally, the shipping paper will contain the name and address of the shipper and consignee. For hazardous waste shipments over highways, as most of the waste covered in this section will be, the shipping papers must be within the cab with the driver, or in the absence of the driver, on the seat or in a holder fixed to the inside of the door on the driver's side. This is to ensure that they are available during an inspection or in case of an accident are in a standard location so that they could be found, if necessary. The requirements for labels, placards, manifests all have this underlying purpose, to provide information in case of an emergency so that emergency response personnel will have vital information to guide their response.

A manifest must be prepared according to 40 CFR Part 262.20. The original copy of the manifest must be dated by and bear the handwritten signature of the person representing (1) the generator of the waste at the time it is offered for transportation, and (2) the initial carrier accepting the waste for transportation. The manifest prepared according to 40 CFR Part 262 can also serve as the shipping paper as long as it contains all of the information required by 49 CFR 172.200-205.

Copies of the manifest must be dated and bear the handwritten signature of the person representing each subsequent carrier of the waste and the designated facility receiving the waste, upon receipt.

Copies of the manifest bearing all of the required dates and signatures must be

1. Given to the person representing each carrier involved in the transportation of the waste. The copies must be carried during transportation.
2. Given to the person representing the designated facility receiving the waste.
3. Retained by the generator and each carrier for 3 years from the date the waste was accepted by the initial carrier. The copies must bear all of the required signatures and dates up to and including those entered by the next person receiving the waste.

4.10.1.2.2. Container Labels

The containers, normally drums, offered to the transporter or used by the commercial disposal firm must be marked with an appropriate label as defined in the tables in Parts 1-72.101 and 172.203. No container can be marked with a label for a material not in the container. Besides these DOT labels, EPA/RCRA regulations require each container to have a label stating prominently that the container contains a hazardous waste. The EPA/RCRA label contains much of the same information found on the shipping papers or manifests: generator name and address, accumulation start date, manifest number, proper shipping name, UN or NA number. Additional information may be provided by the generator. An example of a hazardous waste label is shown in Figure 4.33. If the container contains a hazardous chemical regulated by OSHA in a substance-specific health standard, a label identifying the material and the hazard must be on the container. Labels are to be placed within 6 inches of each other and must be on the side of the container, not the top.

The vehicle carrying the hazardous waste must be properly placarded, as required by DOT regulations and given in the tables in 49 CFR 172.101 and 172.203, or as required under any special provisions. It is the responsibility of the generator to ensure that this is done, if necessary by providing the appropriate placards to the transporter. It is unlikely that this will be required since transporters routinely provide the correct placards.

All shipments of hazardous materials shall be transported without unnecessary delay, from and including the time of commencement of the loading of the cargo until its final discharge at its destination. The generator must receive a completed copy of the manifest signed by a representative of the TSDR facility within 35 days or must initiate action to track the wastes. If the waste is not located within 45 days, an exceptions report must be sent to the EPA or the equivalent state agency. An intermediate destination at a suitable storage facility is permitted, in which waste materials from different sources can be combined for delivery to a point of final disposition of the material. If the materials on the vehicle will not be treated identically, then the broker or processor can break up a shipment, as long as complete track of each container is maintained and records kept.

4.10.1.2.3. Local Waste Management

Unless the generator has been granted an extension by the EPA, the generator must either have a permit as a storage facility or must not collect and store hazardous waste for more than 90 days. During this storage interval, the materials must be kept in appropriate containers, the beginning of the accumulation period must be shown on each container, and they must be labeled as hazardous waste. The label shown in Figure 4.33 would be affixed when the container was put into storage.

It may be appropriate to store the material within the individual laboratories until shortly before a waste disposal firm is scheduled to pick up an organization's waste. Under appropriate conditions, as much as 55 gallons of hazardous waste (or 1 quart of acutely hazardous waste) may be accumulated at or near the point of generation, i.e., the laboratory, beyond 90 days without a permit as a storage facility. The containers used must be in good condition so that they do not leak, and they must be constructed so that the hazardous waste does not react with the container to impair the ability of the container to store the material. Except for periods during which materials are to be added to the container, the container must remain closed. The container must be treated to keep it from leaking. As a minimum, the containers must be labeled with the words "Hazardous Waste," but the label should provide a comprehensive description of the contents.

4.10.1.3. Record Keeping Required of the Generator

Copies of the manifests signed by the transporter must be maintained for at least 3 years. Retention substantially beyond this date would probably be advisable. Similarly, copies of required biennial reports (see following paragraph) and exception reports must be kept for at least 3 years beyond the date on which they were due. Any test data relevant to the shipment must also be kept for 3 years beyond the date of the shipment.

Biennial reports must be prepared by each generator initiating shipments of hazardous waste away from the facility. These must be prepared on EPA Form 8700-13A and submitted by March 1 of each even-numbered year and must cover the previous calendar year. The report must provide: (a) the EPA identification number, name and address of each facility to which waste was shipped during the year, (b) the name and EPA identification number of each transporter used during the year, (c) a description, EPA identification waste number and amount of each hazardous waste shipped off-site. This information must be identified further with the identification number for each site to which it was shipped. The report must be signed by the generator or an authorized representative to certify that the report is correct.

4.10.1.4. Personnel Training

A record must be maintained at the facility engaged in hazardous waste activities of each position related to management of the waste. The name of the current person filling each of these positions must be recorded. A written job description must be available for each position, describing the skills, education, and other qualifications needed for each job and the actual duties involved for each position.

The training requirements for haz-mat personnel managing waste under 49 CFR, Subpart H, are similar to the requirements for haz-mat personnel covered under 29 CFR 1910.120 and 40 CFR Subpart 265. Where they overlap, the training requirements for each of the three standards can be used rather than being duplicated for an employee. The owners and operators of a facility engaged in a hazardous waste management program must ensure that the personnel identified as working in the facility are properly trained so that the program will be conducted safely and with minimal risk to the environment and the general public. This can be done by the haz-mat employers providing classroom instruction and effective on-the-job training themselves or by providing access to public or commercial training programs. The person or persons responsible for the training must themselves be knowledgeable concerning hazardous waste management procedures. The training should be tailored to the specific duties of each person working in the facility. The employees must receive training in the facility's contingency plan, including initiation of an alarm and evacuation procedures, and appropriate responses to a fire, explosion, or an environmental contamination incident. They should know how to safely secure the facility in an emergency.

A written description of the initial and maintenance training programs for each person or position must be available. Persons assigned to duties in a hazardous waste management program must successfully complete the training programs before they are allowed to work unsupervised and within 90 days after employment or assignment to a new task involving haz-mat duties. Each employee shall take part in a review of the initial training appropriate to their duties every 2 years. Records of the training for each continuing employee must be kept until closure of the facility, and for employees who leave the facility, records must be kept for at least 3 years after the employee has left.

4.10.1.5. Preparedness and Prevention

Every facility involved with hazardous waste must be operated in such a manner as to minimize the chances of any incident, such as a fire, explosion, and release of a toxic substance into the environment or which could endanger human life. However, planning for such emergencies must take place even though it is intended that there will never be a need for an emergency response.

Communications are a vital part of any emergency system. Provision must be made for an emergency alarm in the facility. Ready access must be available to either a telephone or two-way radio system to summon emergency help from appropriate agencies, e.g., fire department, police, rescue squads, and emergency response teams. Whenever hazardous material is being handled, access to communication equipment is especially important and must be available. Another occasion when communication capability is especially critical is when an employee is alone in a facility while operations are being conducted. Such a situation is undesirable and should be avoided. Local hospitals should be made aware of the potential for exposures of personnel to hazardous materials and the types of injuries that could result from explosions, spills, or airborne releases of hazardous or toxic materials.

Other emergency equipment that must be present are portable fire extinguishers (of different fire suppression classes, if needed), spill control equipment, and decontamination equipment. Other desirable emergency equipment would be positive air-supplied breathing apparatus, fire and chemical protective clothing, and turnout clothing. Access to any area of the facility must be available in an emergency, so adequate aisle space must be provided. All emergency equipment must be tested and maintained as needed to ensure that it is operational in case of an emergency.

Participation with a local area emergency plan, such as required under Title III SARA, is essential. Arrangements should be made with local fire departments and police departments to familiarize them with the facility, its operations, and access to the facilities for emergency vehicles. Agreements should be established with state and local emergency response agencies, as well as commercial firms that might be needed to assist in an emergency. Other internal groups, such as physical plant or building maintenance groups, and internal security personnel should be similarly involved. Working agreements with nearby medical facilities should be established, and they should be made aware of the type of exposures or medical injuries which they might be asked to handle because of operations within the facility. Periodic drills would be highly desirable.

It is the responsibility of the owners and operators of an organization engaging in hazardous waste operations to make these emergency arrangements. If an agency refuses to accommodate the facility in establishing an appropriate arrangement, this refusal must be documented.

4.10.1.6. Contingency Plan

The previous section described some emergency provisions that must be made. However, a requirement of every hazardous waste facility is that a written contingency plan must be adopted. Copies of this plan must be maintained at the facility and provided to any group that may be called upon to assist in an emergency incident, including police, fire department, emergency response teams, and hospitals. This plan must be designed to reduce the adverse effects of any emergency incident due to the operations of the facility. The plan must be immediately activated upon any such emergency. There are a number of essential components of the plan.

1. The contingency plan must describe the actions facility personnel will take in case of emergencies such as fires, explosions, or releases of hazardous substances into the environment.
2. It must describe the agreements or arrangements made with the emergency response groups to coordinate the activities of these agencies in case of an emergency.
3. A person must be identified as the primary emergency coordinator. Others who are qualified to act as emergency coordinator are to be listed in the order they would be expected to act as alternates. The addresses and home and business phone numbers of these individuals must be included on the list. At all times, at least one of these persons must be at the facility or on call in case of an emergency.
4. A current list of all the emergency equipment at the facility, and the specific location of each item, must be included. Each item must be identified by a brief physical description and its capabilities.
5. The plan must include an evacuation plan for facility personnel, identifying the signal(s) used to initiate the evacuation and primary and alternate escape routes.

4.10.1.7. Emergency Procedures

If an emergency appears likely or actually occurs, the individual acting as the emergency coordinator must immediately initiate an emergency response, including the following actions:

1. Activate the alarm system to alert facility personnel.
2. If necessary, notify emergency response groups to secure their aid.
3. If there has been a release of hazardous substances, a fire, or explosion, the emergency coordinator must identify the nature of the material released, the exact source of the release, and the actual extent of the release.
4. The emergency coordinator must also immediately assess the possible impact of the release on human health or the environment away from the facility, from both direct causes and secondary effects, and include in the latter any adverse effects such as water runoff from efforts to control the incident.
5. If it is determined that adverse effects on humans or the environment could occur and evacuation of areas near the facility is needed, the emergency coordinator must immediately notify appropriate local authorities and remain available to help in making decisions as to which areas require evacuation.
6. In the event of an emergency of possible harm to humans and the environment away from the facility, the emergency coordinator must also report the incident to the government coordinator at the scene or directly to the National Response Center (24-hour telephone number at time of writing, 1-800-424-8802). The report must include:

   a. Name, address and telephone number of the owner or operator of the facility
   b. Name, address, and telephone number of the facility
   c. Date, time, and character of incident (fire, explosion, spill, etc.)
   d. Name and quantity of materials involved, to the extent known
   e. Extent of injuries, if any
   f. Assessment of potential hazards to humans or the environment outside the facility, where applicable

7. The emergency coordinator must take all reasonable steps to contain the scope of the incident. If operations have not already ceased, shutting down operations must be

considered. Waste containers and other hazardous materials not already involved in the incident should be moved to safer locations. The shutdown of operations must be done to ensure there will not be any untoward equipment failures that could cause additional problems.

8. After the emergency, all recovered waste and contaminated soil, runoff water, etc., must be properly handled according to the provisions applicable to generators of hazardous waste.
9. All emergency equipment listed in the contingency plan must be cleaned and fit for its intended use before operations are resumed.
10. The operator or owner must notify the regional administrator and appropriate state and local authorities that the facility is again in compliance with the required standards before operations are resumed.
11. The operating records of the facility must reflect the time, date, and cogent details of any incident that requires implementation of the contingency plan. A report must be made to the regional administrator within 15 days after the incident. Besides the information listed under item 6 above, the report must include:

   a. An assessment of the actual or potential damages to human health or the environment because of the incident, where applicable
   b. The estimated quantity and disposition of recovered material that resulted from the incident

4.10.2. Practical Hazardous Waste Management Program

Sections 4.10.1.1–7 briefly covered basic portions of the RCRA, OSHA regulations, and the DOT regulations affecting hazardous materials and hazardous waste. Hazardous waste management programs for generators must comply with applicable portions of these acts, including requirements that were not provided directly or paraphrased in those sections. It is incumbent upon generator waste management personnel to become familiar with all of the requirements, including changes as they occur.

4.10.2.1. Internal Waste Management Organization

In most organizations, it would be impractical for each individual laboratory or even department to establish a separate program, even if the regulatory organization in the state in which the facility was located was willing to deal separately with hundreds of different generator facilities at a single institution or research establishment. Thus, in most academic institutions and at many corporate research facilities, the responsibility for the general management of the hazardous waste program has been assigned to a centralized unit. This does not mean that the individual laboratory and laboratory personnel do not play a key role in an effective hazardous waste management program. Since they are the generators of the wastes, they can control, to a major degree, the amount of hazardous waste generated, and they are the primary source for identifying waste generated in their facility, especially that which is not in the original containers.

The hazardous waste organization thus starts with the laboratory itself. It is at this level that the waste is generated, collected, identified, and sometimes processed for reuse or rendered harmless. Each individual in the laboratory need not know all of the requirements under RCRA, but each person must be trained in the procedures by which the waste generated in the laboratory is to be handled before removal. Each worker must know what precautions to take to ensure that the waste will be acceptable for disposal, what information is required for each waste generated, and what records must accompany waste offered for disposal.

Unless the facility is so small that outside waste disposal firms come directly to the individual laboratories to package and remove the material, a waste management unit normally will be established within the organization to pick up the waste, classify it (according to the RCRA and DOT criteria), store it temporarily (up to 90 days), prepare appropriate documentation for each container, package it (in some cases), and arrange for the material classified as excess hazardous waste to be taken from the facility, according to the RCRA and DOT requirements.

It will be the responsibility of this internal agency to keep up with the current regulations, make all of the required reports, maintain all of the required records, and handle the waste itself. It will also probably be called upon to help develop and publish procedures for the institution's researchers to use in their part of the waste management program.

It is also likely that the same internal agency is the one assigned the responsibility of preparing and carrying out the required emergency contingency plan. This program must be integrated with the disaster emergency planning for the entire organization. As a technically trained group, some emergency responsibilities outside the waste program may also be assigned to waste management personnel.

4.10.2.2. Reduction of Hazardous Waste Volume

An effective hazardous waste management program cannot concern itself solely with disposal of unneeded materials. This is far too expensive and contradicts the RCRA concept. Chemicals are a resource that should be conserved, and where practical and legal it is desirable for surplus chemicals to be put to a beneficial use rather than buried in a landfill, burned, or otherwise rendered unusable. However, as will be discussed later, there are limitations on how much effort an organization can devote to salvaging or treating a chemical waste without becoming a treatment and disposal facility.

One of the current requirements on generators is that they must document that they are making an effort to reduce the amount of waste that they generate. There are several effective ways in which laboratory operators themselves can aid in this effort. Among the more successful are

1. *Planning of experiments.* Anticipation of waste generation should be a part of each operation. The research should be designed to reduce or eliminate the amount of dangerous substances generated as a by-product of the research to the fullest extent possible, either as an ultimate or intermediary waste. Alternative reagents which would result in waste that is less dangerous or would reduce the volume of hazardous waste should be evaluated. The experiment should be planned to include rendering harmless any excess material generated in the research as a part of and logical continuation of the work.
2. *Purchasing in smaller quantities.* It has been noted earlier that a substantial premium per unit chemical purchased is paid by buying chemicals in smaller containers rather than in larger ones. The cost of eight 1-pint bottles of a given material, purchased singly, is virtually certain to be much greater than the same material in a 1-gallon container, especially if one adds in the surcharge for buying the material in glass bottles covered with a protective plastic safety film. However, the total cost to the organization, if the cost of disposal of unused material is included, might favor the smaller container. If individual containers are packed in a drum properly with an

absorbent material, one typically can get only the equivalent of about 15 gallons of waste in the drum. If the original cost of the material is paid for by the laboratory facility, while waste costs are paid for out of the organization's total budget, it is difficult to persuade the laboratory to pay the additional costs. It may require an explicit organizational policy to mandate purchasing of all chemicals in smaller sizes, unless the laboratory can document that their rate of use is sufficient to ensure the use of all the material in a short period, so that the material will not become degraded or become suspected of possible contamination and therefore unusable. If all materials in an institution or organization are purchased and distributed centrally, this becomes an easier policy to enforce. This centralization also allows bulk purchasing of cases of smaller containers, perhaps making up for the original differential in cost. Proposals to charge for waste are usually self-defeating. Either the costs of collecting the charges in time and money exceed the funds collected, or some individuals attempt to evade the charges by disposing of their waste improperly down the sink.

3. *Using smaller quantities in experiments.* New technology has made it possible to do much chemical research with smaller amounts of materials than was possible only a few years ago. More sensitive analytical instruments, balances, instruments operating on new principles, or more versatile and reconfigured instruments based on older principles often make it possible to perform an experiment with very small amounts of the reagents involved. For example, interfacing a small computer to an experiment may allow the simultaneous recording of many variables, with automatic control of the experiment to continuously modify the experiment, so that one run may replace a dozen or more. Not only will this reduce the volume of waste generated, but the experiments should become inherently safer.

4. *Redistribution of used chemicals.* Excess chemicals are not necessarily waste chemicals. A specific part of any chemical waste program should be a classification and evaluation step, before which a chemical is only "surplus," not "waste." Clearly, many materials are obviously clearly unusable and must be classified as waste and, if they meet any of the RCRA hazardous waste criteria, classified as hazardous waste or acutely hazardous waste if they fall in that category. However, many surplus chemicals are still usable for research or, if not, often are adequate for instructional laboratories. Unopened containers are the most likely to be attractive to potential users, but even partially used containers of reasonably fresh materials, which do not normally degrade by contact with air or moisture, are often desirable to someone. A list of these materials should be maintained and circulated frequently to potential "customers" within the same organizations. Sometimes these materials are offered at a lower price than fresh material, but in most cases, charging will reduce the amount of surplus redistributed and may wind up increasing the cost to the organization because of the cost of disposing of the surplus material. There are "waste exchanges" to which lists of unwanted materials may be submitted which then circulate the information to organizations that might be interested.

Another category of materials that can often be reused are contaminated solvents. Many materials, such as xylene, methanol, and ethanol are heavily used materials and frequently are available in the waste in sufficient volume to make it practical to reclaim by distillation. An automatic spinning band still can easily handle 20 to 30 liters per day. There will still be residues that must be treated as waste, but the volume will be far less than the original amount. The wastes must be treated carefully; to ensure that two materials that form azeotropic mixtures are not mixed; otherwise simple distillation will not produce pure solvents. Recent prohibitions on disposal of liquids in landfills make this method of waste reduction even more attractive.

Other materials such as mercury and silver from photographic processes are also readily recovered, and it is possible for the value of the reclaimed silver, for example, to more than pay for the effort involved.

5. *Management by tracking materials that degrade and become dangerous with time.* There are many chemicals that deteriorate over time, usually as they react with air or moisture. Ethers are a well-known example of this class of materials. The contents of partially empty containers of ethers kept in storage are especially likely to form peroxides which are heat, shock, and friction sensitive so that an attempt to use them could result in a fatal explosion. Chemical waste vendors will not collect these potentially dangerous containers and remove them from the facility as part of a normal waste shipment. There are firms that specialize in disposing of such dangerous materials, but their services are extremely expensive. An effective program should be established for materials falling in this category for tracking them, beginning by dating each container when it is acquired, setting a target date for beginning checking the containers for dangerous degradation products, and establishing a date when they must be removed from the facility. Other materials that should be included in a tracking program should be pyrophoric materials, highly water-reactive materials or the converse, materials which become dangerous as they dry out, such as perchloric acid, and picric acid. Computer software now available makes it possible to identify with a bar code every container received so that every container can be tracked. This software, coupled with a central receiving point for chemicals, should make it possible to virtually eliminate outdated containers as a problem.

6. *Elimination of unknowns.* There are two different ways in which unknowns are generated in a laboratory. One way is for information on the original labels on commercial containers to become lost either by becoming defaced or by disappearing. The other is when inadequate information is placed on labels to begin with, such as when a laboratory worker labels a container "Solution A" and then leaves the facility, so that no useful records are available. Unknown chemicals are among the most troublesome of all materials to handle as waste. There is a certain amount of risk in opening an unknown container, although the type of operations conducted in the laboratory in which it is found may provide enough information to reduce or eliminate the concern about shock or friction causing an explosion, or the possibility of a reaction with air or moisture in the air when the container is opened. Even after opening the container, the characteristics of the material may be sufficiently ambiguous to make identification very difficult, especially if it is not originally a commercial product but has been generated in the laboratory.

The solution to the degraded or destroyed label problem is relatively straightforward. If the label on a container is damaged, then it should be laboratory policy for a new label to be affixed immediately by the first individual to be aware of the problem. With the coming of the new OSHA hazard communications standard, and its secondary container requirements, every laboratory should have an ample supply of generic labels available for use. These generic labels should have space on them for the common name of the chemical, its CAS number, and basic hazard data, such as the NFPA 704 system color-coded numbering system or diamond.

The use of computer software for tracking chemical containers mentioned in item 5 would make it possible to eliminate truly old containers from accumulating if the organization were to adopt a policy that all chemical containers greater than 5 years old (with some exceptions) were to be discarded. The failure to properly label containers in the first place is a more difficult management problem, since there is typically only one person who knows what the container contents really are. In academic laboratories the turnover in graduate students and postdoctoral research associates (and to some extent, faculty as well) is high. A policy should be adopted and enforced to have every person leaving the facility identify the contents of each container for which the individual has been responsible. Withholding graduation or requiring a financial deposit have been suggested as means of enforcing such a policy. Often, however, there would be few difficulties if the laboratory director or supervisor simply met with the individual shortly before the scheduled departure and went over the questionable containers. Unfortunately, some laboratory directors do not wish to take the time. There also have been cases where the individual in charge has compounded the problem by having the unknowns surreptitiously carried to an ordinary trash container outside the laboratory. Where this occurs and those responsible can be identified, they should be subject to severe punishment by the organization's authorities, because, in addition to the inappropriateness of such behavior, the organization can be held responsible for the consequences of such actions.

7. *Placing severe restrictions on "free" research materials or chemicals accepted by the research personnel.* It is frequently simpler, for example, for a supplier of agricultural chemicals to provide a researcher with a drum of one of their products (which may be a "numbered" experimental material) than to weigh out the amount that the person actually needs. A policy should be introduced that a supplier of a "free" material must either take back any excess material or guarantee to pay for the disposal cost of the excess. Institution of such a policy at the author's institution substantially reduced the amount of hazardous agricultural waste chemicals that required disposal. Other disciplines depend less on free materials, but the problem does exist elsewhere and should be restricted in the same manner.

8. *Miscellaneous.* There are a number of waste reduction techniques that are more appropriate to the laboratory than the internal waste disposal agency, which could result in the institution being considered as a treatment and disposal facility, if the procedures were employed after collection from the laboratory. For example, a container of ether, if there is only a small quantity remaining in the container, should be placed in a suitable fume hood and allowed to evaporate. This eliminates the material as a solvent waste, as well as any chance of dangerous peroxides forming. Small quantities of other solvents can be dealt with in a similar fashion. Small quantities of acids and bases can be used in elementary neutralization processes. There are other purification, reclamation, and neutralization techniques suggested in the literature for specific materials.

Since a nearly empty bottle requires as much space as a full one when packed in a drum for disposal, containers should be filled in the laboratory with either the same waste material or with compatible materials before being transferred from the laboratory to the waste collection group. In the latter case, a complete, accurate record of the contents of the container must be maintained and placed on the waste label affixed to the container when the transfer takes place.

A modest amount of material can be safely disposed of into the sanitary system. However, there may be local restrictions that preclude this type of disposal, and if encouraged, some materials may be disposed of in this fashion for which the procedure is not appropriate. Unless done carefully, under the direct supervision of a qualified laboratory employee, preferably a senior person who normally exercises managerial authority, disposal of materials into the sanitary system should be discouraged. The amount of money saved over several years of doing so safely would probably be minor compared to the cost of a single incident that subsequently causes a problem.

An excellent reference for means to render chemicals in the laboratory less hazardous is *Prudent Practices for Disposal of Chemicals from Laboratories,* published by the National Academy Press in 1983. Although the regulatory material included in the book is outdated, the chemistry remains valid.

4.10.2.3. Waste Collection

The RCRA regulations allow temporary storage of hazardous waste at or near the point of generation, which in the present document, means in or near a laboratory. The amount is limited to 55 gallons of hazardous waste material or 1 quart of acutely hazardous material. If the procedure is to pick up the waste directly from the laboratory, then the only requirement will be to provide the proper documentation to the collection firm. Normally, the storage in the laboratory will only be a stage in the internal handling of the waste, with, typically, the waste being picked up and transferred to a central location, where most of the waste management activities take place.

There are advantages if the determination of whether the surplus chemical meets the RCRA criteria for hazardous waste is delayed until after the material is taken to the central facility. If the excess chemical is defined at the laboratory as hazardous waste, transportation of the material may be a problem since one of the restrictions on transportation of hazardous waste, without meeting the requirements of a hazardous waste transporter, is that hazardous waste cannot be transported *along* a public street, although the waste can be moved *across* a public street. This problem does not arise for many corporate facilities since internal roads at corporate sites often are not accessible by the public. Most academic campuses, however, are open to the public and few streets in the vicinity of academic research buildings are restricted to the public. Sometimes, the strict interpretation of this requirement may be avoided by using small vehicles that can move on sidewalks and across limited access parking lots and other open spaces.

The delay of classification until the material is taken to a central location is itself a dubious circumvention of the standard if there is no real effort to classify the material brought to the facility as usable materials, harmless waste and hazardous waste at the facility. Only modest amounts of material should be transported at any time across a crowded campus to limit the consequences of an accident. Except in unusual circumstances, it would be desirable to limit the size of individual containers to 5-gallon pails for liquids and the equivalent for solids, which would be approximately 50 to 75 pounds, again as a safety measure. Containers of these sizes can be handled by hand, while 55-gallon drums would require equipment to lift in the event they were spilled in an accident. No chemicals should be moved in damaged containers or ones not sealed tightly. The containers should be securely loaded on the vehicle being used for transportation, so that they will not shift while the vehicle is in motion. Larger amounts in 55-gallon drums can be accumulated in secondary storage areas to avoid moving them across a facility. Waste management agencies must be notified of the locations of these satellite accumulation areas and the facility emergency contingency plan must apply to these areas.

Two separate trips should be made if two or more highly reactive materials or incompatible materials are to be transported. DOT regulations list which materials cannot be transported in the same vehicle or which can be transported simultaneously only under certain restrictions to protect against interactions of the materials.

It would be a good idea to have emergency supplies on the vehicle used for transportation of the chemicals, including: respiratory protection, gloves, chemical splash goggles, chemically protective coveralls, two class ABC fire extinguishers, a bag or two of an absorbent material, a package or two of Plug-n-Dike™ for plugging holes in a leaking container, a coil of rope and at least three lightweight standards (for establishing a temporary barrier), and a two-way radio. All of this emergency gear should be where it is likely to be accessible in case of an accident. The individuals who pick up the material must be trained in how to respond in the event of an emergency.

4.10.2.4. Packaging

An upper limit to the size of the container packages was recommended in the previous section. Most surplus chemicals are not in 5-gallon metal cans, but are typically in smaller glass bottles of various English or metric sizes. The caps on containers should be required to be screw type, which are neither cracked nor corroded. Bottles with other types of stoppers, such as rubber, cork, or glass, may come open during transportation, allowing the contents to spill and cause a dangerous incident. The outside of any container offered for transportation should be **clean**. Small vials containing liquids may be placed in a larger container for transportation.

Container sizes for flammable solvents, if the material has been transferred from the original container, cannot exceed the permissible sizes allowed by OSHA for flammable liquids of the appropriate class. The quality of the containers must be comparable to those used for the commercial versions of the chemicals. Chemical wastes should not be placed in polyethylene milk jugs or the equivalent. These containers are not sufficiently sturdy.

Some solid materials normally come in bags. The original bags should be placed within one or two additional bags. One may suffice if the original bag is intact, while two should be used if the original bag is damaged and leaking. The external bag(s) preferably should be clear so the contents can be seen, especially if the original bag bears a legend identifying the contents. The thickness of the outer bag(s) should be no less than 6 mils. As noted earlier, compatible chemicals can be combined in a single container.

4.10.2.5. Characterization of the Waste

A typical waste label that might be used at any facility to identify and document waste internally is shown in Figure 4.34. A form such as this should be attached to each container of surplus chemical before it is removed from the laboratory and should remain with the container until final disposition of the material. A copy should be retained by the waste facility as part of their records. There are two sections to this form, one identifying the source of the material and the other characterizing the material itself. The latter information is to be used by the personnel responsible for classifying the material and if necessary to complete the preparations for its disposal as a hazardous waste. Note that the material is not classified as a hazardous waste, allowing it to be evaluated, perhaps as a usable or recyclable material.

As can be seen, the information pertinent to the source is sufficiently complete to trace the material to its origin. A very important piece of information is the name of the person originating the waste, both as a printed name and as a full signature. The signature is important since it represents an acknowledgment by the individual signing the form that the remainder of the information is correct and eliminates confusion that might arise from a set of initials. If problems develop with the material, the person signing the form can be looked to for further information.

Providing information identifying the contents is the responsibility of the generator, not the waste management personnel. They can provide a reference or telephone numbers, but it is not their responsibility to perform this task for the generator. They are justified in refusing to take containers in which the contents have not been properly identified. All constituents of the material in the container must be listed by their proper chemical names. Abbreviations and formulas may cause confusion and should not be used as the sole identification. Similarly, trade names may be helpful, but again should not be used as a substitute for the correct chemical name. One useful identifier would be a CAS number, which,

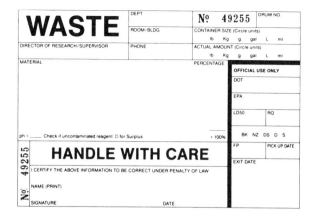

Figure 4.34. Waste label for internal use. Information from labels such as this are used to provide the data for labels meeting DOT requirements.

if not conveniently available to the researcher elsewhere, can be found for most common chemicals in virtually any chemical catalog. The term "inert ingredients" should not be used. All materials should be completely identified on the label.

Waste liquids containing acidic or basic materials must show the approximate pH of the liquid. If the liquid is not homogeneous but is stratified into layers, the pH of each layer must be provided.

Any material that is equivalent to a commercial product and is uncontaminated should be specifically marked as such as a candidate for the redistribution program.

The approximate percentage of each constituent should be provided, and the sum of these percentages must add up to 100%. If the container is one to which materials have been added from time to time, the log of these additions can be used to estimate the percentages of the various materials in the container.

It is advisable for waste management personnel to at least spot-check materials that they have collected from the various laboratories. A simple gas chromatograph analysis can be done on most organic solvents in about 15 minutes if the machine has the proper column and has been calibrated with appropriate standards. The pH of aqueous solutions can also be easily checked. If there is any doubt about the identity of most solid materials, the physical characteristics may be enough to resolve the question and at least a rough qualitative identification can be made for most commonly used materials. Unknowns, of course, are a different story, and the analysis of these materials should be the responsibility of the originating laboratory, if possible. There are times when this is impossible, but these should be rare. The cost of analysis of unknowns by commercial disposal firms is high. In an academic institution with a graduate program in chemistry, unknowns can be characterized much less expensively by supporting a graduate student to do this task.

4.10.2.6. Packing of Waste for Shipment

The manner of packing the waste in preparation for transportation to a disposal facility depends upon the eventual means of disposal. The normal means of packaging bulk liquids is in 55-gallon steel drums, either DOT 1A1 closed-head or DOT 1A2 open-head steel drums, which are used for packing smaller containers. These are intended to be used only once, i.e., single-trip containers (STC), and are identified as such by stamping them with this legend. Normally containers are placed in the drum with enough filler material to completely fill the empty space. If the waste is a liquid, the filler material should be sufficient to absorb the liquid should the containers break. The space occupied by the awkward shape of most chemical containers, plus the need for enough inexpensive filler to completely absorb the liquid, limits the total volume of the bottles which can be placed in a 55-gallon drum to between 12 and 17 gallons. Smaller containers of incompatible chemicals, i.e., ones which can react vigorously together, cannot be placed in the same container. Smaller drums or pails also can be used if they meet DOT specifications for the materials placed in them. Occasionally there is a need to place a drum inside a larger one if there is a potential for risk of a leak. This is called an overpack and is relatively expensive. Very few liquids now are destined to be placed in a landfill, but these types of containers are used for transporting liquids to other types of disposal facilities or for recycling.

Some firms that accept materials for incineration will accept waste in metal drums, if they remove the waste containers from the drums at the disposal site. However, most firms accepting waste for incineration prefer the waste to be placed in combustible 1G2 fiber or 1H1 or 1H2 plastic drums that can be placed directly in the incinerator. These combustible fiber drums cannot be placed in landfills.

The drum types mentioned in the material above are the types meeting the new international requirements. Older drums can be used until October 1, 1996.

4.10.2.7. Restrictions on Wastes

Some materials are forbidden to be transported as waste by DOT restrictions. Among these are

1. Reactive wastes that can explode or release toxic vapors, gases, or fumes at standard temperature or pressure
2. Reactive materials that, when mixed with water or are exposed to moisture, can explode or generate toxic vapors, fumes, or gases
3. Materials that are shock sensitive
4. Materials that will explode, etc., if heated while confined
5. Class A or class B explosives

These restrictions are imposed to ensure that the containers will not explode or evolve toxic gases, vapors, or fumes during transportation while loaded on a waste carrier and endanger the public. If a container of material were to explode on a truck loaded with other wastes, a major catastrophe could result.

The Hazardous and Solid Waste Amendments prohibit land disposal of some wastes unless the wastes are treated or it can be shown that there will be "no migration as long as the waste remains hazardous." Dioxin- and solvent-contaminated soils, and also dilute wastewaters contaminated with solvents, were originally to be excluded, but there was not sufficient capacity to handle these otherwise; therefore, an extension was granted to continue placing these in landfills.

A more important restriction affecting laboratories went into effect on November 8, 1986, when it became illegal to place in a landfill wastes containing more than 1% of many solvents. These are listed in 40 CFR 268.30 and 268.31:

F001—The following spent halogenated solvents used in degreasing: tetrachloroethylene, trichloroethylene, methylene chloride, 1,1,1-trichloromethane, carbon tetrachloride, and chlorinated fluorocarbons, all spent solvent mixtures/blends used in decreasing containing before use, a total of 10% or more (by volume) of one or more of the above halogenated solvents or those solvents listed in F002, F004, and F005, and still bottoms from the recovery of these spent solvent mixtures.

F002—The following spent halogenated solvents: tetrachloroethylene, methylene chloride, trichloroethylene, 1,1,1-trichloromethane, chlorobenzene, 1,1,2-trichloro-1,2,2-trifluoroethane, ortho-dichlorobenzene, and trichlorofluoromethane; ortho-dichlorobenzene, and trichlorofluoromethane; all spent solvent mixtures/blends containing, before use, a total of 10% or more (by volume) of one or more of the above halogenated solvents or those solvents listed in F001, F004, F005, and still bottoms from the recovery of these spent solvents and spent solvent mixtures.

F003—The following spent nonhalogenated solvents: xylene, acetone, ethyl acetate, ethyl benzene, ethyl ether, methyl isobutyl ketone, n-butyl alcohol, cyclohexanone, and methanol; all spent solvent mixtures/blends containing, solely the above spent nonhalogenated solvents; and all spent solvent mixture/blends containing before use, one or more of the above nonhalogenated solvents, and a total of 10% or more (by volume) of one or more of those solvents listed in F001, F002, F004, and F005; and still bottoms from the recovery of these spent solvents and spent solvent mixtures.

F004—The following spent nonhalogenated solvents; creosols and cresylic acids and nitrobenzene: all spent solvent mixtures/blends containing, before use, a total of 10% or more (by volume) of one or more of the above nonhalogenated solvents or those solvents listed in F001, F002, and F005; and still bottoms from the recovery of these spent solvents and spent solvent mixtures.

F005—The following spent nonhalogenated solvents; toluene, methyl ethyl ketone, carbon disulfide, isobutanol, and pyridine; all spent solvent mixture/blends containing, before use, a total of 10% or more (by volume) of one or more of the above nonhalogenated solvents or those solvent listed in F001, F002, F004; and still bottoms from the recovery of spent solvents and solvent mixtures.

F020—Wastes (except wastewater and spent carbon from hydrogen chloride purification) from the production or manufacturing use (as a reactant, chemical intermediate, or component in a formulating process) of tri- or tetrachlorophenol, or of intermediates used to produce their pesticide derivatives. (This listing does not include wastes from the production of hexachlorophene from highly purified 2,4,5-trichlorophenol.)

F021—Wastes (except wastewater and spent carbon from hydrogen chloride purification) from the manufacturing use (as a reactant, chemical intermediate, or component in a formulating process) of pentachlorophenol, or of intermediates used to produce its derivatives.

F022—Wastes (except wastewater and spent carbon from hydrogen chloride purification) from the manufacturing use (as a reactant, chemical intermediate, or component in a formulating process) of tetra-, penta-, or hexachlorobenzenes under alkaline conditions.

F023—Wastes (except wastewater and spent carbon from hydrogen chloride purification) from the production of materials on equipment previously used for the production or manufacturing use (as a reactant, chemical intermediate, or component in a formulating process) of tri- and tetrachlorophenols. (This listing does not include wastes from equipment used only for the production or use of hexachlorophene from highly purified 2,4,5-trichlorophenol.)

F026—Wastes (except wastewater and spent carbon from hydrogen chloride purification) from the production of materials on equipment previously used for the manufacturing use (as a reactant, chemical intermediate, or component in a formulating process) of tetra-, penta-, or hexachlorophene under alkaline conditions.

F027—Discarded, unused formulations containing tri-, tetra-, or pentachlorophenol or discarded, unused formulations containing compounds derived from these chlorophenols. (This listing does not include formulations containing hexachlorophene synthesized from prepurified 2,4,5-trichlorophenol as the sole component.)

Many other materials have been placed on the restricted lists since this first list. These are listed in 40 CFR 268.32 through 40 CFR 268.35. Among the more critical in the context of laboratory wastes are (1) liquid hazardous wastes having a pH ≤2, (2) liquid hazardous wastes containing PCBs at concentrations ≥50 ppm, (3) liquid hazardous wastes that are primarily water and halogenated compounds at concentrations ≥1000 mg/L and ≤10,000 mg/L, (4) mixed radioactive/hazardous wastes, and (5) RCRA hazardous wastes that contain naturally occurring radioactive materials.

As time passes, it is becoming clearer that there will be more limitations on the materials to be placed in landfills. Clearly, there will be additional restrictions on the construction and management of landfills, which will make them more expensive to use.* These costs will necessarily be passed on for payment to the generator. Obviously, it is going to become more important to the academic institution or corporation to restrain the amount of waste being produced. Disposal costs already are not a negligible item for a facility that is actively pursuing a waste program fully complying with the standards.

4.10.2.8. Shipping Waste

Once the waste has been collected and classified as hazardous, it must be disposed of by shipping it off-site for all those who are only classified as generators and do not have a permit for treatment, storage, disposal, or recycling facilities. A substantial amount of money can be saved if packing and preparation of the manifest are done by the internal waste management personnel rather than the disposal firm. This requires the management personnel to be thoroughly familiar with the current requirements for generators initiating shipments of hazardous waste equivalent to those given in Section 4.10.1.2. Some hazardous waste disposal firms will not allow this, feeling that they cannot be sure that be waste is packaged properly. Others will accept waste packaged by the generator but will spot-check individual drums or will check the pH, for example, of some individual containers to confirm that the materials are packed properly, that the drums contain materials that agree with the accompanying lists, and that the chemicals are identified properly. It is essential that the generators subscribe to a reference service or have access to an on-line computer resource that will provide them with a current set of regulations for both the EPA requirements and DOT. The latter, especially, change too frequently for the preparers of the shipments to be sure otherwise that they are in compliance, and, in any event, they need the references for the chemical identification numbers, proper shipping names, packing groups, labeling and placarding requires, and any restrictions on shipping.

They will also need to conform to the specific requirements of the firm taking the waste. Some firms are essentially pure brokers, having no facilities of their own for the ultimate disposal of the waste. These firms sometimes collect wastes from several small generators at a central facility, group those materials that will be landfilled, incinerated, or used in a

---

* All landfill costs are increasing. Disposal of radioactive waste will rise by at least a factor of 4 during the next 2 years in the author's area, while disposal of ordinary solid wastes has gone up by a factor of 5. Due to landfill regulations, the cost of constructing and managing landfills has dramatically increased. As a result, many landfills are being closed. Similar increases in costs of landfill burials of hazardous waste or use of alternative technology are to be expected.

fuel-blending cement kiln facility, and then take the materials to firms that provide these services. If the size of the shipment is large enough, they may take it directly from the generator to the appropriate disposal site. In a case such as this, the generator will in effect have to conform to both the broker's requirements and those of the second firm, although it may make little difference to the generator.

Some firms offer a complete range of services for disposing of prepared waste. They offer transportation and either landfill disposal or incineration facilities. In principle at least, these firms should provide disposal at a lower cost per unit volume since they do not have to share the profits with an intermediary firm.

As noted earlier, there are materials that transporters are forbidden to carry because of the risk of an accident. There are firms that specialize in handling these dangerous materials, which may have special carriers that allow them to be moved safely or offer on-site destruction of materials such as shock-sensitive chemicals, explosives, highly reactive chemicals, gas cylinders that cannot be returned to the manufacturer, or those that pose other special hazards. Some firms bring special equipment to a facility that allows them to take the materials, such as very old ethers containing large quantities of peroxides and other comparably dangerous materials, directly from the laboratory in cases of extreme risk. Local destruction will normally require permits and an isolated area in which to perform the task. Usually these will be the responsibility of the generator, and sometimes both may be difficult to obtain. Disposal of these items will be expensive.

The generator must accept one last responsibility that is not explicitly stated by the standards, and that is that the disposal firm must be capable of handling the waste properly. It would be assumed that the possession of an EPA permit and an Identification Number and being financially secure and able to provide a certificate of insurance would suffice to guarantee proper handling of the waste. However, there have been numerous fines, and closures of facilities operated by even major firms. It could be cost effective for a knowledgeable representative of the waste management group, or a colleague in the area whose judgment is known to be reliable, to visit a facility before selecting a firm. At the author's institution, such a practice has resulted in negative decisions in several cases. The facilities involved eventually were required by regulatory agencies to take corrective actions and in some cases ceased operations. Other protective steps a generator may take are (1) to require the disposal firm to provide a list of all citations for the past 5 years, (2) an insurance certificate or bond, and (3) very thorough checking of references. There are many cases in which the costs of remedying problems at disposal sites eventually have had to be borne by the generators that shipped waste to the sites.

4.10.2.9. Landfill Disposal

Until recently landfill disposal has been the favored method of disposing of laboratory wastes. Because of potential problems such as damage to the incinerators and control of toxic emissions and hazardous constituents in the residual ash due to the variety of materials in mixed laboratory wastes, most incineration facilities did not wish to accept this category of materials, although some would accept properly segregated materials. In addition, or partially because of the operational problems, incineration was somewhat more expensive. However, because of the growing number of restrictions on materials that can be put into landfills (see Section 4.10.2.7) and the increasing costs of constructing and managing landfills, incineration is becoming more popular and more cost effective.

Landfills are essentially elongated rectangular trenches cut into the earth in which the drums of hazardous waste are placed. Landfills are preferably sited in areas in which the water table is low compared to the depth of the trench and where the underlying earth consists of deep beds of clay with very low permeation properties. Clay layers underlying some of the more heavily used existing hazardous waste landfills are several hundred feet thick. Any material leaking (the current restrictions on liquid wastes make this eventuality unlikely) into these landfills from the waste containers should be contained much as water in a bathtub. The containment is enhanced by current requirements to also use manufactured liners for the trench. Although steel drums would not deteriorate as rapidly as a combustible fiber drum, they still have a short lifetime compared to that of many materials contained within the drums. Orderly placement of the drums and having the drums completely filled is essential to keep the surface from sagging as the drums deteriorate. To prevent leakage from these sites, they should also be located in areas in which surface flooding is highly unlikely. All other things being equal, it would be desirable to have them in areas of low precipitation, but if the surface is properly configured, capped properly, and maintained, reasonable amounts of precipitation should not be a problem.

Low population density, the absence of commercial deposits of valuable minerals, and no historic or scenic attractions in the immediate vicinity also are desirable. The availability of good roads for bringing waste to the site is essential.

The typical size of these facilities is not too large, ranging from a few tens to a few hundreds of acres, and if managed properly they are not an environmental problem to the surrounding area. Many existing landfills were not designed or managed properly and are among the SUPERFUND sites that must be cleaned up. Permitted landfills today must be lined with synthetic materials and have a leachate collection system and a groundwater monitoring system. Because there are relatively few landfills that meet all of the essential criteria, waste disposal firms often transport waste for long distances to them.

A fundamental concern about the landfilling of hazardous waste is the fact that many materials retain their hazardous properties for very long periods. Although currently these sites can be managed properly, it is possible to imagine future circumstances that could arise in which good management would no longer be possible and even where records would be lost identifying the material buried at the site. Individuals raising these issues admit that these scenarios are remote, but the possibility of a negative impact on future generations has been raised as an issue weighing against the use of landfills for disposal of hazardous chemicals. The history of mismanagement by some collectors of hazardous waste has also contributed to an emotional reaction against landfills, so that siting additional landfills (or other hazardous waste operations) has become increasingly more difficult because of public concerns. Instead of being the method of choice, use of landfills is now often the last resort.

4.10.2.10. Incineration of Hazardous Waste

The two primary options of waste disposal off-site which have been available to generators of hazardous chemical waste have been landfills and incineration. It is unusual for a laboratory facility or even a facility with many laboratories at the same site to generate a sufficient wastestream that is consistent enough to be usefully reprocessed or recycled by a waste processor, although occasionally some operations may generate enough clean, nonhalogenated organic solvents to be of interest as fuel. Landfilling has been the predominant disposal mechanism in the past, but incineration is now the preferred method by regulatory agencies.

Incineration has many major advantages. Many materials that are currently prohibited from being placed in landfills can, in principle, be destroyed by incineration. Regulatory requirements for most materials are that 99.99% be destroyed, although for some materials such as dioxin, the required destruction efficiency is 99.9999%. It is possible for most organic material to meet the RCRA requirements. The result is a much smaller volume of solid waste that may or may not have to be handled as a hazardous waste depending upon the materials in the original wastestream. Therefore, the size of the long-term management problem, if not totally eliminated, is at least greatly diminished. As a result, the residual liability for the generator for waste that has been incinerated may be eliminated.

There are some disadvantages. The emission of toxic materials, such as NO_x, SO_x, HCl, and some particulates containing metals from the stack must be controlled within stringent standards established by the EPA. This can be expensive for some wastestreams. There are products of incomplete combustion (PICs), which may or may not be a problem. If an incinerator is managed properly, the PICs do not appear to be a problem in the large incinerators currently being used for waste disposal. The ash residue may contain concentrated quantities of heavy metals that must be disposed of as hazardous waste in a landfill. The combustion chamber itself may suffer physical damage or corrosion, and maintenance may be a problem. On the whole, however, especially as EPA standards on alternatives become more restrictive, the advantages outweigh the disadvantages as long as the facilities are operated properly.

There are two types of incinerators used predominantly for incineration of hazardous waste. The first type is the rotary kiln, which operates at temperatures of approximately 2300°F. More hazardous wastes are burned in this type of kiln than any other and it is the one for which most tests on destruction efficiencies have been done. A second type is the cement kiln, which is being used more often. The temperature of the cement kiln reaches 3000°F. The hazardous waste is blended with the fuel and used in the making of cement. By modifying the fuel blend, it is possible to obtain efficient destruction, even for chlorinated solvents.

The article by Lawrence G. Doucet in this chapter on alternative medical waste disposal techniques provides a review of the factors that must be considered in establishing a local incineration program for an organization's hazardous waste.

REFERENCES

1. Environmental Protection Agency, Hazardous Waste Regulations, 40 CFR Parts 260 - 268, Washington, D.C.
2. Environmental Protection Agency, Designation, Reportable Quantities ("RQs"), and Notification, 40 CFR Part 302, Washington, D.C.
3. Environmental Protection Agency, Worker Protection in Hazardous Waste Operations, 40 CFR Part 311, Washington, D.C.
4. Environmental Protection Agency, Emergency Planning and Notification, 40 CFR Part 355, Washington, D.C.
5. Environmental Protection Agency, Hazardous Chemical Reporting: Community Right-to-Know, 40 CFR Part 370, Washington, D.C.
6. Occupational Safety and Health Administration, Employee Emergency Plans and Fire Prevention Plans, 29 CFR 1910.38, Washington, D.C.
7. Occupational Safety and Health Administration, Hazardous Waste Operations and Emergency Response, 29 CFR 1910.120, Washington, D.C.
8. Occupational Safety and Health Administration, Hazard Communication, 29 CFR 1910.1200, Washington, D.C.
9. Department of Transportation, Hazardous Material Regulations 49 CFR Parts 171–173, 177, Washington, D.C.
10. Department of Transportation/United Nations Performance-Orientated Packaging Standards HM-81, 49 CFR Parts 172, 173, 177, 178., Washington, D.C.

Table 4.23.   Listed Regulated Medical Wastes

A. *Cultures and stock of microorganisms and biologicals*—Discarded cultures, stocks, specimens, vaccines, and associated items likely to have been contaminated by them are regulated medical wastes if they are likely to contain organisms likely to be pathogenic to healthy humans. Discarded etiologic agents are regulated medical waste. Wastes from the production of biologicals and antibiotics likely to have been contaminated by organisms likely to be pathogenic to healthy humans are regulated medical wastes.

B. *Cultures and stock of microorganisms and biologicals*—Discarded cultures, stocks, specimens, vaccines, and associated items likely to have been contaminated by them are regulated medical wastes if they are likely to contain organisms likely to be pathogenic to healthy humans. Discarded etiologic agents are regulated medical waste. Wastes from the production of biologicals and antibiotics likely to have been contaminated by organisms likely to be pathogenic at healthy humans are regulated medical wastes.

C. *Pathological wastes*—All pathological wastes and all wastes that are human tissues, organs, body parts, or body fluids are regulated medical waste.

D. *Sharps*—Sharps likely to be contaminated with organisms that are pathogenic to healthy humans, and all sharps used in patient care or veterinary practice are regulated medical wastes.

E. *Animal carcasses, body parts, bedding, and related wastes*—When animals are intentionally infected with organisms likely to be pathogenic to healthy humans for the purposes of research, *in vivo* testing, production of biological materials, or any other reason, the animal carcasses, body parts, bedding material and all other wastes likely to have been contaminated are regulated medical wastes when discarded, disposed of, or placed in accumulated storage.

F. *Contaminated spill materials*—Any residue or contaminated soil, water, or other debris resulting from the clean up of a spill of any regulated medical waste is regulated medical waste.

G. *Other*—Any solid waste contaminated by or mixed with regulated medical waste is regulated medical waste.

4.10.2.11. Infectious Waste

The EPA published 40 CFR Part 259—Standards for the Tracking and Management of Medical Wastes March 24, 1989. On July 2, 1990, the EPA issued a clarification to this rule. This was intended to be a demonstration rule to be in effect for a limited time. It only became effective in New York, New Jersey, Connecticut, Rhode Island, and Puerto Rico. It has not been extended to the remainder of the country. However, since then at least 14 states have passed their own chain-of-custody tracking regulations that are more stringent than the federal standards, and 45 states have passed regulations defining medical wastes and medical waste management requirements. Generally, these have followed definitions similar to those given in Table 4.23. The EPA position now is that there is no fundamental reason why medical wastes should not go into a normal solid waste landfill and so they have not taken action after the 1990 clarification. As noted, at least 14 states have done so, however, and others are likely to do so. The OSHA bloodborne pathogen standard, if for no other reason, will dictate safe disposal of contaminated materials as defined by that standard.

The characteristics of a regulated medical waste are: Any solid waste is a regulated medical waste if it is suspected by the health care professional in charge of being capable of producing an infectious disease in humans. A solid waste shall be considered to be capable of producing an infectious disease if it has been or is likely to have been contaminated by an organism likely to be pathogenic to healthy humans, the organism is not routinely and freely available in the community, and the organism has a significant probability of being present in sufficient quantities and with sufficient virulence to transmit disease. If the exact cause of a patient's illness is unknown, but the health care professional in charge suspects a contagious disease is the cause, the likelihood of pathogen transmission should be assessed on the basis of the pathogen suspected of being the cause of the illness.

Until the late 1980s in most areas the only acceptable means of rendering infectious "waste safe" was to autoclave the materials, which was only suitable for relatively small quantities, or incineration used for larger amounts. Most medical facilities and many research facilities had incinerators used for burning pathological materials. These older incinerators rarely meet current requirements for complete destruction of infectious wastes. There are also operational problems when sharps (plastics, glass, and metal objects) are included in the medical wastes, as these materials will damage the refractory lining over time. Destruction, or at least rendering them unrecognizable, of syringes and other recognizable medical wastes has become an important factor due to the concerns of the public when they see these items in solid wastes. Although it is suitable to take autoclaved wastes at least to a landfill, the operators will frequently refuse the materials.

A major concern with the use of incineration is its cost. It is likely to be 10 to 20 times more expensive to dispose of a solid waste by incineration as compared to ordinary burial in a landfill. This cost has led to a strong interest in developing efficient alternatives which will kill the pathogenic organisms to no more than 10^{-6} of the original amount, render the waste unrecognizable, and leave an end product that can be disposed of in either a sanitary sewer or an ordinary landfill (or if not, result in a large decrease in the volume that would have to be treated as hazardous waste).

There are now many, in theory, suitable alternatives to incineration of infectious wastes. These are described in the next section. This section provides an in-depth discussion of incinerators used to treat medical wastes, by the same author and was in the third edition of the *CRC Handbook of Laboratory Safety,* CRC Press, Boca Raton, FL.

4.10.2.12. Alternative Technologies for Disposal of Regulated Medical Waste

4.10.2.12.1.   The Latest Developments in Alternative and Emerging Medical Waste Treatment Technologies[*]

Introduction

Alternative medical waste treatment technologies have undergone a great amount of research and development over the last five years. New processes and technologies are being introduced and marketed at a rapid rate, and it seems that new systems are being reported in technical journals just about every month. Simultaneously, there has been a high rate of system vendors going out of business or ceasing operations. Such rapidly changing conditions make it difficult to stay current of developments with existing systems, let alone to keep apprised of the features of those newly introduced.

---

[*]   This article was written by Lawrence G. Doucet, P.E., D.E.E.

In 1990, the author published a technical paper through the American Society for Healthcare Environmental Services and the American Society for Hospital Engineering entitled *Update of Alternative and Emerging Medical Waste Treatment Technologies*. That paper summarized and described the basic features of all of the alternative medical waste treatment processes and vendors which had been reported or documented at that time. However, not too long after the paper was published, a substantial amount of information relating to specific vendors and their processes were out of date.

The purpose of this paper is to present and discuss the current status of developments of alternative medical waste treatment processes and technologies. The data and descriptions of the various vendors and their processes which are included in this paper are based upon extensive correspondence, numerous telephone discussions, and vendor responses to a specially prepared 13-page questionnaire form which was completed by each. Data are also based on information developed on numerous, related consulting and engineering projects for various clients. In addition, the data and descriptions included in this paper were sent to each vendor for review and approval as to accuracy. Accordingly, data and descriptions in this paper are believed to be current and accurate as of March, 1994.

Historical Perspective

Medical waste treatment and disposal was not a major problem or issue for most hospitals throughout the United States prior to 1987. Medical waste incinerators were relatively easy to permit, most landfill owners and operators did not find hospital waste objectionable, and off-site disposal costs were reasonable and competitive in most parts of the country. Also, medical waste was not considered a potentially hazardous or dangerous material, there were no particular reasons for the general public to be concerned with it, and regulations covering medical waste management and disposal were minimal and not very stringent.

In the summers of 1987 and 1988, medical waste dominated the media and public attention. There seemed to be daily accounts of medical debris washing up on public beaches and broken containers of medical waste showing up in trash dumpsters and in alleyways. Those events served to spark promulgation of stringent federal and state regulations, a trend which has continued through today. The U.S. EPA enacted the *Medical Waste Tracking Act of 1988*, the Centers for Disease Control issued *Universal Precautions,* and the Occupational Safety and Health Administration enacted *Bloodborne Pathogen* regulations and standards dealing with medical waste management, treatment, and disposal.

On a national level, the net immediate effect of the flurry of regulatory activities was a rapid and substantial increase in medical waste volumes coupled with a simultaneous, substantial decrease of viable, cost-effective treatment and disposal options. This situation, in turn, encouraged research and development efforts to find better, more cost-effective alternatives and methods for treating and disposing of medical waste.

4.10.2.12.2. Process and Technological Overview

Criteria and Requirements

Medical waste treatment systems are designed and operated to achieve the basic requirements of disfigurement and decontamination. Most systems achieve these requirements as a function of their internal processes, while others require add-on components or additional processing steps.

Disfigurement relates to the recognizability of the treated medical waste. Disfigurement is necessary to alter the appearance of the waste such that it is no longer perceived to be dangerous, and it is also necessary to render "sharps" in the waste stream nonhazardous from a puncture standpoint. Disfigurement is achieved through shredding, combustion, and sometimes through reactions with strong chemicals. Several states have specific regulations and criteria covering medical waste disfigurement and recognition.

Decontamination relates to a requirement for rendering the medical waste biologically safe for disposal. State requirements for decontamination range from total sterilization to a high degree of disinfection. Typically, a minimum of 6 logs reduction are required for viruses, parasites, fungi, vegetative bacteria, and mycobacteria. Many states have specific requirements for monitoring decontamination performance, including biological indicators, frequency of sampling and required methodologies. Decontamination is achieved through the use of chemical disinfectants, thermal incineration, combustion, irradiation, and biochemical reactions.

Waste Treatment Process Categories

There are five broad categories of medical waste treatment technologies as follows:

1. Mechanical Processes
2. Thermal Processes
3. Chemical Processes
4. Irradiation Processes
5. Biological Processes

The following sections generally discuss each of these process categories.

Mechanical Processes

Mechanical processes are used to change the physical form or characteristics of the waste either to facilitate waste handling or to process the waste in conjunction with other treatment steps. The two primary processes are compaction and shredding. Compaction involves compressing the waste into containers to reduce its volume. Shredding, which also includes granulation, grinding, pulping, and the like, is used to break the waste into smaller pieces.

Compaction and shredding are not considered acceptable alternative medical waste treatment systems by themselves. In fact, many state regulations, as well as good practice, prohibit compacting or shredding untreated medical waste because of concerns of aerosoling or spilling microorganisms.

Shredding is the first step of various alternative treatment processes. In those states where shredding is prohibited prior to treatment, such systems require special permitting variations. However, there are typically no restrictions for shredding wastes after it has been decontaminated to make it unrecognizable.

Thermal Processes

Thermal processes use heat to decontaminate or destroy medical waste. Most microorganisms are rapidly destroyed at temperatures ranging from 120°F to 195°F, and most living organisms are killed at 212°F.

There are two categories of thermal processes, namely, low heat systems and high heat systems. Low heat systems use steam, hot water, or electromagnetic radiation to heat and decontaminate the waste. They typically operate at temperatures of less than 300°F which is insufficient to combust or destroy the materials.

High heat systems employ combustion, pyrolysis, and high temperature plasmas to decontaminate and destroy the waste. These systems operate at temperatures ranging from as low as 1,100°F to more than 10,000°F.

The basic thermal treatment processes are as follows.

- **Autoclaving:** Autoclaving is a low heat thermal process which uses steam to disinfect the waste. There are several different types and designs of autoclave systems which basically vary in their ability to improve steam contact efficiencies such that processing times can be reduced and processing capacities can be increased. Gravity-type autoclaves rely on the steam pressures alone to evacuate air from the autoclave chamber; pre-vacuum type systems utilize vacuum pumps to evacuate air from the autoclave chamber and thereby increase steam penetration and contact efficiencies; and retort-type autoclave systems use large chambers designed to operate at higher temperatures and pressures than other type systems. Retort-type autoclaves are typically large capacity units, and they have been used on applications having capacities of up to 60 tons per day per unit. Until recently, conventional steam autoclaving was not used for treating large volumes of medical waste. This is primarily because autoclaving does not routinely provide sterilization and because a post-shredding step is required to provide disfigurement. There currently about ten different vendors or manufacturers of autoclave systems for treating medical waste. These include those offering variations to standard autoclave designs, such as systems using high temperature hot water, very high steam temperatures and pressures, and those including integral shredding processes.

- **Incineration:** Incineration systems use high temperature combustion to destroy medical waste. Modern incineration systems are well-engineered, high technology processes which are designed to maximize combustion efficiencies and minimize emissions. Modern incineration systems require state-of-the-art controls, high-efficiency air pollution control systems, the application of extensive and complex continuous emissions monitoring systems, and highly skilled and trained operators. There are approximately a dozen different vendors or manufacturers of conventional, or standard, incineration systems and equipment. Approximately one-half dozen vendors offer non-conventional or innovative incineration systems which have either been transferred from other applications or which reflect new inventions.

- **Pyrolysis:** Pyrolysis is a process category involving high temperature combustion in the absence of oxygen. Pyrolysis has long been used for such applications as coal gasification and petroleum processing. Recently, several vendors have developed or are developing systems featuring pyrolysis for treating medical waste. These systems typically operate with temperatures ranging from as low as 1000°F to as high as 2000°F. Pyrolysis is very similar to incineration in many respects, and most states require pyrolysis systems to meet the same requirements as incinerator systems.

- **Plasma and Plasma Pyrolysis Systems:** These systems utilize a plasma-torch or burner for heating the waste to super high temperatures. some plasma-fired systems are reportedly designed to operate with furnace temperatures as high as 10,000°C. Several vendors have been developing plasma systems for medical waste treatment for several years. Plasma and plasma-pyrolysis systems are also comparable to incinerator systems in many respects, except that conventional combustion does not take place. Accordingly, most states require these systems to meet the same requirements as incineration systems.

Excluding the categories of autoclaving and incineration, a total of 18 different thermal treatment processes and systems are currently being developed and marketed by different vendors. Exhibits 1-1 through 1-3 list the different thermal process vendors, briefly describe their different processes, and include an overview of the processes and technical parameters.

Chemical Processes

Chemical treatment is synonymous with chemical disinfection or simply disinfection. Most chemical waste treatment systems use a disinfectant solution in combination with shredding to provide decontamination and disfigurement. However, several systems feature strong chemical reactions which destroy or disintegrate the waste, and a few use chemical polymers to encapsulate the waste.

Twelve different chemical treatment processes are being developed or marketed by different vendors. Exhibits 2-1 through 2-3 list the different chemical process vendors, briefly describe their different processes, and include an overview of the processes and technical parameters.

Irradiation Processes

Irradiation is synonymous with electromagnetic or ionizing irradiation. Currently, one vendor is developing a process utilizing Cobalt-60, and two other vendors are developing systems utilizing an electron beam accelerator unit, or electron beam gun, for irradiating and sterilizing the medical waste. These systems require post-shredding to render the waste unrecognizable.

Exhibits 3-1 through 3-3 list the different irradiation process vendors, briefly describe their different processes, and include an overview of the processes and technical parameters.

Biological Treatment Processes

One manufacturer is developing a system that uses biological enzymes for treating medical waste. It is claimed that biological reactions will not only decontaminate the waste but also cause the destruction of all the organic constituents such that only plastics, glass, and other inerts will remain in the residues. Specifics of the vendor and parameters of the system are included in Exhibits 3-1 through 3-3.

4.10.2.12.3. Evaluations, Comparisons, and Ratings

There are significant comparative advantages and disadvantages between each of the various alternative treatment technologies. However, because of rapidly changing developments and a sparsity of data, it is typically difficult to make a comparative assessment and select the best technology for a specific application. This difficulty is compounded by the uncertain and changing regulatory requirements of many states.

Because most of the technologies are new or under development and because most have no full-scale operational systems, reliable and substantive technical, performance, and economic data are meager or non-existent. To compound the problem, some vendors have promotional literature and issue press releases containing misleading and exaggerated claims about their systems. In such cases, system capacities and costing are distorted and overly optimistic.

There are seven key criteria which are recommended for consideration and analysis when evaluating medical waste treatment technologies for a particular application. These are as follows:

1. Demonstrated Performance
2. Technical and Performance Criteria
3. Vendor Qualifications
4. Environmental and Permitting Issues
5. Occupational Health and Safety Issues
6. Facility and Infrastructure Requirements
7. Economics

In assessing each particular technology with respect to the above, a comparative rating value can be established. For example, for each criteria the "lowest rating" could be given a point value of zero and the "highest rating" could be given a value of 5 or 10. Also, depending upon specific preferences and individual judgements, each of the criteria could be given a different weight. For example, "Demonstrated Performance" could be given a weight value which is much higher than "Facility and Infrastructure Requirements." Multiplying the point values by the weight values would give a total value which could serve as a ranking of the different technologies. This process would also identify critical flaws which may eliminate particular processes.

The seven criteria and their sub-categories are described below along with recommended "rating" comparisons for each.

Demonstrated Performance

Stage of Development

What is the status of the system in terms of the commercial viability and proven application, i.e., conceptual, under development, prototype testing, full-scale testing, commercially available, etc.?

- **Lowest Rating:** Conceptual and/or under development and no operational systems
- **Highest Rating:** Commercially available and proven successful performance of full-scale applications

Number of Operational Systems

How many full-scale systems have been installed and are operational?

- **Lowest Rating:** No operational systems
- **Highest Rating:** The more the better, but *at least* 5 full-scale operational systems are recommended

Years of Successful Operations

How many years of successful, full-scale operations have been achieved, both elapsed years of experience and combined years of all systems?

- **Lowest Rating:** No experience with operational systems
- **Highest Rating:** The more the better, but *at least* 3 years of continuous operational experience with a single system and 10 years of combined experience are recommended

Technical and Performance Criteria

Process Capacity

How much waste can the system process on an hourly or daily basis? Also, does process capacity match required treatment/disposal rates or are several units required?

- **Lowest Rating:** Process capacity cannot match required treatment/disposal rates unless many units are required
- **Highest Rating:** Process capacity readily matches required treatment/disposal rates within an acceptable time period and with only a single unit

Waste Exclusions or Limitations

Are there waste types, components, or characteristics which must be excluded or limited because of adverse effects on system performance?

- **Lowest Rating:** Extensive exclusions and/or limitations that substantially reduce the volumes of waste can be processed
- **Highest Rating:** No or insignificant exclusions and/or limitations

Waste Size Limitations

Are there any limitations on individual waste container sizes or waste loading cart volumes that will affect overall waste handling and/or packaging requirements?

- **Lowest Rating:** Limitations require waste repackaging and/or excessive extra waste handling
- **Highest Rating:** Waste packaging and handling practices are not affected

Weight Change

What is the reduction or increase in the weight of the treated residue?

- **Lowest Rating:** Substantial weight gain (a 100% weight increase is not uncommon)
- **Highest Rating:** Substantial weight reduction (a 20-to-1 weight reduction is achievable)

Volume Change

What is the reduction or increase in the volume of the treated residue?

- **Lowest Rating:** Overall increase in volume
- **Highest Rating:** Substantial volume reduction (a greater than 20-to-1 volume is achievable)

Recognizability/Disfigurement

What is the physical appearance and consistency of the treated residue?

- **Lowest Rating:** Fully recognizable, poses sharps hazards, or is not acceptable per state criteria/regulations
- **Highest Rating:** Unrecognizable, contains no sharps hazards, and meets state criteria/regulations

Decontamination

What is the degree of decontamination of the treated residue?

- **Lowest Rating:** Routinely has low disinfection or log reduction rates and is not acceptable per state criteria/regulations
- **Highest Rating:** Routinely sterile or consistently has high log reduction rates and meets state criteria/regulations

Performance Data

What data are available to demonstrate that the system can achieve a required, claimed or specified degree of decontamination?

- **Lowest Rating:** Either no data or data from unrepresentative testing procedures
- **Highest Rating:** Data readily available from full-scale systems operating under representative conditions

Process Complexity

How standardized or complex is the basic process and associated components and operational features?

- **Lowest Rating:** Highly complex system comprising exotic components and using unconventional and numerous processing steps
- **Highest Rating:** Simple system using standard, proven components and materials and comprising only a few well-established processing steps

Operator Training

How much operator training is required?

- **Lowest Rating:** Extensive operator training and certification required due to complex and specialized O&M procedures
- **Highest Rating:** Minimum operator training required due to process simplicity and ease of Operation and Maintenance (O&M)

Vendor Qualifications

Number of Vendors

How many different vendors offer a similar or comparable process?

- **Lowest Rating:** A single vendor having no competition
- **Highest Rating:** Several vendors offering comparable systems on a competitive basis

Vendor Resources

Is the vendor sound, stable, and proven?

- **Lowest Rating:** Single-person enterprise with no organization, questionable or minimal financial resources, and no track-record of successful endeavors
- **Highest Rating:** Financially sound, long-established, and multi-faceted corporation with proven track record of successful endeavors

Vendor Diversity and Services

What are the ranges and quality of services available directly from the vendor?

- **Lowest Rating:** Only developer or seller of the system with no other capabilities
- **Highest Rating:** Complete range of services, including installation, start-up, testing, and maintenance and repair services over the life of the system

Environmental and Permitting Issues

Air Emissions

What air pollutants are emitted into the atmosphere from stacks, vents, and other discharge areas; how well are they controlled or abated; and what are their potential environmental impacts?

- **Lowest Rating:** High, uncontrolled emission rates for an array of potentially hazardous or toxic pollutants are discharged at low elevations and poorly discharged such that ambient concentrations can become excessive

- **Highest Rating:** System has no emissions, no potentially hazardous or toxic pollutant emissions, and/or an abatement system to remove or eliminate all potential hazardous or toxic emissions. Also, emissions are properly discharged such that ambient concentrations are always well below safe levels

Liquid Effluents

What pollutants are in the liquid effluents from the system?

- **Lowest Rating:** Large flow of liquid effluents having unacceptably high concentrations of regulated, toxic, or hazardous pollutants or chemicals
- **Highest Rating:** System either has no liquid effluent discharges, no potentially hazardous or toxic pollutants or chemicals in the effluents, or an abatement system to remove potentially hazardous or toxic pollutants or chemicals

Treated Residues

Are the solid residues from the treatment process an environmental concern?

- **Lowest Rating:** Treatment residues contain high concentrations of potentially hazardous or toxic constituents or chemicals, or they pose potential health or environmental threats due to their characteristics
- **Highest Rating:** Treated residues are benign and inert and contain no significant concentrations of toxic constituents or chemicals

Permitability

How difficult is it to permit the system?

- **Lowest Rating:** Either state regulations prohibit installation of the system or the permitting process is difficult, complex, lengthy, and costly with variances and public hearings involved
- **Highest Rating:** State regulations do not restrict the installation of the system and the permitting process is simple and routine with no obstacles

Public Perception

How will the general public perceive the environmental impacts of the system and what is the likelihood of public opposition?

- **Lowest Rating:** Substantial public opposition to perceived environmental problems
- **Highest Rating:** No public interest or concerns of environmental problems

Occupational Health and Safety Issues

Routine Exposures

What are the potential, occupational (worker) exposure hazards or dangers during normal or routine operations?

- **Lowest Rating:** High risks of potentially hazardous or injurious emissions or conditions, i.e., chemical concentrations, heat, radioactivity, pathogens, sharps, noise, smoke, etc.
- **Highest Rating:** No or insignificant exposure hazards or dangers

Maintenance and Repair Exposures

What are the potential occupational (worker) exposure hazards or dangers during unexpected maintenance and repair (M&R) procedures?

- **Lowest Rating:** Difficult M&R procedures and high risk potentials from exposed, untreated infectious waste items or chemicals
- **Highest Rating:** Simple M&R procedures with minimal or no risks of exposure to potentially hazardous or dangerous conditions

Facility and Infrastructure Requirements

Space Requirements

How much overall space or room area is required for installing a system with allowances for proper operations and ease of maintenance?

- **Lowest Rating:** Very large space requirements, room height extending beyond a typical floor, multiple levels or pits, and unusual shape or odd configuration
- **Highest Rating:** Small or moderate space requirements, suitable for location within typical room, and single level and with no pits

General Construction Requirements

Is the system room or enclosure required to have special construction?

- **Lowest Rating:** Special and costly construction such as vibration isolation, noise containment, chemical corrosion resistance, mechanical abrasion resistance, electromagnetic radiation containment, and extensive fire protection
- **Highest Rating:** No special construction requirements

Utility Requirements

What are the utility demands on the system?

- **Lowest Rating:** High peak and operational demands for such utilities as electrical power, natural gas, fuel oil, water, and steam
- **Highest Rating:** Low peak and operational demands for all utilities

Space/Facility HVAC Requirements

What are the heating, ventilation, and air conditioning (HVAC) requirements for the system?

- **Lowest Rating:** Special HVAC system required due to such conditions as odors, chemical fumes, heat, smoke, steam, dust, etc.
- **Highest Rating:** No special HVAC requirements

Economics

Capital Costs

What is the *total* capital cost for an installed system, including associated support facilities, permitting, and compliance testing?

- **Lowest Rating:** Highest capital cost—particularly in comparison to other alternatives
- **Highest Rating:** Lowest capital costs

Annual Costs

What is the total annual cost for operating labor, utilities, and treated residue disposal, as well as the disposal of waste items excluded from the system?

- **Lowest Rating:** Highest annual costs—particularly in comparison to other alternatives
- **Highest Rating:** Lowest annual cost

Life-Cycle Costs

What are the life-cycle costs for the system over its expected, useful economic life, i.e., present worth value, annualized owning and operating costs, and unit cost ($/lb) inclusive of escalation and interest rates?

- **Lowest Rating:** Highest life-cycle costs—particularly in comparison to alternatives and with no basis to predict an estimated long-term system life
- **Highest Rating:** Lowest life-cycle cost with reliable data to predict or estimate long-term system life

4.10.2.12.4. Process Statistics and Status Summary

Number of Vendors

In 1988, only a single alternative treatment process was commercially available and being used for treating medical waste in the United States other than incineration and autoclaving. This was a chemical treatment process located in a hospital in Indiana. By 1991, a total of 18 different alternative treatment processes were reportedly being developed or commercially available other than incineration and autoclaving. However, by 1993, ten of the 1991 vendors were out of business or had ceased developing or offering systems. Today a total of more than 55 different vendors reportedly offer or are developing alternative treatment systems. These comprise 34 different non-incineration and non-autoclave processes. These are illustrated in Exhibit 4.

The changing number of vendors over the last 5 years is indicative of several key points. First, it is obvious that there continues to be a lot of activity in the alternative treatment system marketplace. Second, the trend in the number of vendors illustrates a general instability in the marketplace and the fact that most vendors have very little, if any, long-term experience. A total of 25 of the 34 vendors in existence today have been in business only two years or less, and 32 have been in business in the United States for less than five years. This instability may create uncertainties and difficulties when it comes to selecting a particular vendor or technology for a specific application.

Number of Systems

Exhibit 5 tabulates the approximate total number of the different alternative medical waste treatment systems which have been installed or are under construction in the United States having capabilities greater than 50 pounds per hour. Of the total, only about 60 systems are nonincineration and nonautoclave systems, and of the 60 about half are either still under construction, have yet to obtain permitted full-scale operations, or are not suitable for on-site applications. This serves to illustrate an overall lack of long-tern experience of alternative treatment processes applications.

4.10.2.12.5. Summary

Developments in alternative medical treatment technologies have been ongoing at a rapid pace over the last three to five years, and there are signs that they will continue for at least another several years. It is likely that many new technologies have yet to be introduced and at least several of the current technologies will soon be out of existence. Regulatory developments and concerns of emissions and environmental impacts for some alternatives will impact their viability. However, over time several alternatives will likely prove to offer viable, long-term, cost-effective solutions for hospitals and other health care facilities.

When considering an alternative technology for a selected application, it is very important to obtain reliable data, and extra efforts in evaluating the various alternatives on the basis of key criteria will be well-justified.

Exhibit 1-1: Alternative Thermal Treatment Processes
General Process Descriptions

ABB Sanitec, Inc.

Carted waste is dumped into a hopper where it is shredded and moistened with steam. An auger conveyer transports the shredded waste from the hopper past a series of magnetron units, or microwave generators, which heat it. The heated residue is deposited into a retention chamber from which an auger conveyer transports it either to a dumpster or through a secondary shredder and then into a dumpster.

Biomedical Waste Systems, Inc.

A conveyer transports waste to a hopper which feeds it into a shredder. The shredded waste enters a chamber where it is indirectly heated by a heated oil system. The heated residue is then conveyed to an insulated, heated retention chamber, and then it is conveyed to a dumpster. Treated plastic residues, such as "sharps," can be conveyed to a separate system for recovery.

Eshland Environmental Technologies

A conveyer transports containerized waste to a loading device that feeds into an electrically heated pyrolysis chamber operating at temperatures between 200°F and 600°F. Resultant pyrolytic gases are combusted in an electrically heated secondary chamber operating at 2000°F. Inorganic residues are collected and disposed of as ash.

Medical Waste Tech, Inc.

Waste is loaded into a processing vessel where it is shredded and mixed with super-heated steam/water. The residue is cooled within the vessel, and solids are then strained from the liquids for disposal.

Medifor-X Corp.

Waste is loaded into a treatment chamber and heated by a combination of infrared radiation and forced hot-air convection. The treated residue is automatically transferred to a cooling chamber where it is compacted.

MedMark International

Waste is loaded into a treatment chamber where it is heated by infrared radiation to temperatures at about 500°F. The treated residue is removed from the unit for disposal.

Partnerships Limited, Inc.

Waste is put into a canister which is loaded into a chamber for combusting under controlled, pyrolytic conditions. Inorganic residues are collected and disposed of as ash.

Penberthy Electromelt International, Inc.

Waste is loaded into an electrically heated glass furnace where it is combusted in the presence of air and water vapor. During the process, hydrogen chloride is converted to calcium chloride and inorganics are melted into glass.

Plasma Energy Applied Technology

An automatic ram feeder loads waste into a reactor chamber fired by a plasma torch burner. Organics are combusted at very high temperatures and inorganics are melted into a slag. Off-gases pass through a coke bed filter where they are combusted with air and then discharged through a stack.

Risk Management Instruments, Inc.

A single stainless steel needle is placed into a hand-held processing unit where it is electrically oxidized to the basic components of chromium oxide, nickel oxide, and ferris oxide.

Roatan Medical Services Corp.

Waste is collected in specially designed collection containers, and full containers are sealed and loaded into a processing chamber. Water is injected into each container and microwave generators heat the water to produce superheated steam. The containers of treated waste are then removed from the chambers and waste is loaded into a shredder.

San-I-Pak, Inc.

Waste is loaded into a conventional pre-vacuum type steam autoclave chamber for treatment. Afterwards, it is automatically ejected into a shredding unit or directly into a compactor/container.

SAS Systems, Inc.

Waste is loaded into a multi-stage, pre-vacuum type autoclave where it is subjected to high steam temperature and pressures. During the process, waste is shredded into an ejection chamber from which it is discharged into a container for disposal.

Spintech, Inc.

"Sharps" or laboratory waste items are put into a plastic cylinder. A filled cylinder is loaded into a metal canister which is then placed into a table-top processing unit. Electrical heating melts the plastic cylinder to a liquid which flows through the waste contained within it. The molten plastic solidifies into a cylinder block which is removed from the unit for disposal.

Stericycle, Inc.

A conveyer transports waste into the hopper of an enclosed shredding system. Waste is pulled through the shredders by a vacuum system and then deposited and compressed into large tubs within an enclosed area. The loaded tubs are automatically conveyed to a dielectric heating unit where low frequency radio waves (macrowaves) heat the waste to 200°F. The tubs are then conveyed out of the heating unit and the release is disposed. Some plastics are separated for recycling.

Tempico, Inc.

A conveyer transports waste to an autoclave-type vessel where it is subjected to very high steam temperatures and pressures. Waste within the vessel is mechanically agitated to enhance steam contact. Treated waste is discharged from the unit and conveyed to a shredder.

Thermokill, Inc.

Waste is loaded into a hopper which discharges to two grinders which, in turn, discharge into a thermal screw device where it is indirectly heated through a jacketed vessel up to 360°F. The treated residue is then discharged through an air-lock to a cooling screw and then through an auger conveyer to a dumpster.

Vance IDS

Waste is loaded into a hopper from which an auger transports it into a pressurized chamber. the waste is exposed to incandescent heat which is electrically generated in an argon plasma cloud at temperatures ranging from 15,000°F to 20,000°F. Inorganic residues are automatically discharged from the unit, and off-gases are cleaned in a wet scrubbing system.

Exhibit 1-2: Alternative Thermal Treatment Processes Process Overview

Vendors	Status	First U.S. Installation	No. of U.S. Installations	Process Capacity Range	No. of States Permitted
1. ABB Sanitec, Inc.	CA	1990	25	220–290 lb/h	40
2. BioMedical Waste Systems, Inc	D	1993	1	3000 lb/h	6
3. Eshland Environmental Technologies	CA	1993	2	40–125 lb/h	1
4. Medical Waste Tech, Inc.	D	Expected 1994	0	14 g/cycle to 100 32-g	49
5. Medifor X-Corp.	CA	1993	2	48 lb/h	20
6. MedMark International	CA	1993	0	5 cf/cycle	23
7. Partnerships Limited, Inc.	D	Prototype 1991	1	34 g/cycle	2
8. Penberthy Electromelt International, Inc.	CA	Prototype 1991	0	NR	NR
9. Plasma Energy Applied Technology	D	Prototype 1991	0	3–250 ton/module	NR
10. Roatan Medical Services Corp.	CA	NR	NR	30 cf/cycle	NR
11. Risk Management Instruments, Inc.	CA	NR	>2000	1 needle/cycle	50
12. San-I-Pak, Inc.	CA	1979	204	30–600 lb/h	49
13. SAS Systems, Inc.	CA	1992	2	300–900 lb/h	24
14. Spintech, Inc.	CA	NR	NR	1 qt–2 g/cycle	32
15. Stericycle, Inc.	CA	1990	5	4000–9000 lb/h	12
16. Tempico, Inc.	CA	1992	5	100–1200 lb/h	47
17. Thermokill, Inc.	CA	1993	1	To 1000 lb/h	10
18. Vance IDS	D	Expected 1994	0	100–6000 lb/h	NR

CA—Commercially available; D—under development; NR—not reported.

Exhibit 1-3: Alternative Thermal Treatment Processes Technical Overview

| Vendors | Waste Load Restrictions | | Reported Degree of Decontamination | Reported Volume Reduction | Reported Weight Reduction |
	Maximum Weight	Maximum Load Rate			
1. ABB Sanitec, Inc.	None	Varied	6 log reduction	0.8	To 5% gain
2. BioMedical Waste Systems, Inc.	400 lb	3000 lb/h	Disinfection	0.8	15–30%
3. Eshland Environmental Technologies	NR	125 lb/h	Sterilization	0.99	0.94
4. Medical Waste Tech, Inc.	Small scale	1–2 load/h	Sterilization	88%	NR
5. Medifor-X Corp.	20 lb	2.4 load/h	Sterilization	0.9	None
6. MedMark International	40 lb	2/3 cycle/h	Sterilization	To 95%	To 50%
7. Partnerships Limited, Inc.	NR	>1/4 load/h	Sterilization	0.99	0.98
8. Penberthy Electromelt International, Inc.	NR	NR	Sterilization	NR	NR
9. Plasma Energy Applied Technology	NR	NR	Sterilization	NR	NR
10. Roatan Medical Services Corp.	NR	10-3 cf	NR	80%	None
11. Risk Management Instruments, Inc.	NA	1 needle/load	Sterilization	0.99	0.99
12. San-I-Pak, Inc.	600 lb	1 load/h	Disinfection	83% (compacted)	To 5% gain
13. SAS Systems, Inc.	250 lb	4 load/h	Sterilization	0.8	None
14. Spintech, Inc.	Small scale	NR	Sterilization	0.8	None
15. Stericycle, Inc.	NA	NA	6 log reduction	0.85	Minimal
16. Tempico, Inc.	1000 lb	1 load/h	Disinfection	0.85	None
17. Thermokill, Inc.	200 lb	15 load/h	Disinfection	0.8	None
18. Vance IDS	1600 lb	4 load/h	Sterilization	0.97	0.95

NA—Not applicable; NR—not reported.

Exhibit 2-1: Alternative Chemical Processes
General Process Descriptions

Ecomed, Inc.

"Sharps" or laboratory wastes are either put directly into a processing chamber or into a separate container which fits into a processing chamber. The processing chamber is then loaded into a dishwasher-sized processing unit in which a disinfectant solution has been added. Wastes are ground within the processing unit and then discharged.

The Futures Group

Batch waste loads are fed into a reactor vessel where they are subjected to a liquid chemical oxidant heated to temperatures ranging between about 570°F and 950°F. Organics are converted to carbon dioxide, water vapor, and a salt. Inorganics are captured in the liquid reaction and separated by precipitation.

Isolyzer Company, Inc.

Loaded sharps containers are filled with a solution containing a phenolic disinfectant. Two catalysts are added to the container to create an exothermic reaction which solidifies to encapsulate and disinfect the sharps.

Medical SafeTec, Inc.

A conveyer transports waste to a hopper where it is sprayed with a sodium hypochlorite solution while it passes through a high-torque shredder followed up by a high-speed shredder. The resultant slurry enters a rotary separator/dryer which separates the solids from the liquids. Solid residues are deposited in a cart for disposal and liquids are recirculated.

Mediclean Technology, Inc.

A conveyer transports waste to an enclosed granulator chamber and it is shredded to a small particle size which passes through a classifier screen into an auger bin. Chlorine dioxide is sprayed through ports above and below the granulator. Processed residue is transported from the auger bin to a dewater device prior to disposal.

Medzam, Division of SafeTec of America, Inc.

A powder is applied to or mixed with liquid waste at the point of generation to disinfect it and to convert it to a solid.

Premier Medical Technology, Inc.

Waste is loaded into a processing unit where it is shredded and mixed with a proprietary dry chemical disinfectant. The treated residue is conveyed to a dumpster.

Sexton Environmental Systems, Inc.

Waste is loaded into a primary shredder and mixed with ozone enriched water. It is then passed through two additional shredding stages. The resultant slurry then undergoes conventional waste water treatment steps with ozone being a sterilant. The treated slurry may be dewatered and disposed or its components may be recovered for recycling.

SPS Medical Equipment Corp.

Sharps are put into a table-top processing unit and shredded. The residue is then evacuated into a disposable container filled with a liquid disinfectant. The container can hold up to 70 processed syringes at a time.

Steris Corp./Ecomed, Inc.

"Sharps" or laboratory wastes are either put directly into a processing chamber or a separate container which fits into a processing chamber. The processing chamber is then loaded into a dishwasher-sized processing unit in which a disinfectant solution has been added. Wastes are ground within the processing unit and then discharged.

Ster-O-Lizer, Inc.

Waste is loaded into a 100-gallon processing unit where a sodium chloride solution is added. Electrodes within the processing unit liberate chloride, ozone, and hydroxyl radicals which sterilize the water.

Winfield International

Waste is loaded through a hopper feeding into a shredding chamber where it is sprayed with a chlorine dioxide solution. It is then further ground in a secondary granulating chamber before falling through a screen and into a heated tank of chlorine dioxide solution. An auger conveyer transports the slurry through a tube and dewatering screen, and a conveyer then transports the solids to a dumpster.

Exhibit 2-2: Alternative Chemical Treatment Processes Process Overview

Vendors	Status	First U.S. Installation	No. of U.S. Installations	Process Capacity Range	No. of States Permitted
1. Ecomed, Inc.	CA	1992	28	12–20 lb/h	36
2. The Futures Group	D	None	0	NR	0
3. Isolyzer Company, Inc.	CA	NA	NR	NA	25
4. Medical SafeTec, Inc.	CA	1985	80	200–1200 lb/h	39
5. Mediclean Technology, Inc.	CA	1991	2	600–800 lb/h	40
6. Premier Medical Technology, Inc.	CA	1993	2	1000 lb/h	28
7. Sexton Environmental Systems, Inc.	CA	1993	1	To 100 ton/d	1
8. SPS Medical Equipment Corp.	CA	1993	NR	60–250 sharps/h	43
9. Steris Corp./Ecomed, Inc.	D	None	0	12–20 lb/h	0
10. Ster-O-Lizer, Inc.	CA	NR	NR	100–120 lb/h	NR
11. Winfield Environmental	CA	1992	7	600 lb/h	42

CA—Commercially available; D—under development; NR—not reported.

Exhibit 2-3: Technical Overview

| Vendors | Waste Load Restrictions | | Reported Degree of Decontamination | Reported Volume Reduction | Reported Weight Reduction |
	Maximum Weight	Maximum Load Rate			
1. Ecomed, Inc	5 lb	5 load/h	6 + log reduction	70–80%	None
2. The Futures Group	NR	NR	NR	NR	NR
3. Isolyzer Company, Inc.	NR	NA	Disinfection	NR	NA
4. Medical SafeTec, Inc.	None	1200 lb/h	6 + log reduction	To 90%	To 10% Gain
5. Mediclean Technology, Inc.	None	600-800 lb/h	Disinfection	80–90%	Varies
6. MedZam, Division of SafeTec of America, Inc.	NA	NA	Disinfection	NA	NA
7. Premier Medical Technology, Inc.	100 lb	750 lb/h	6 = log reduction	To 90%	To 7% Gain
8. Sexton Environmental Systems, Inc.	None	By design	6-9 log reduction	80–90%	To 20% and Gain
9. SPS Medical Equipment Corp.	Small Scale	NR	Disinfection	80%	None
10. Steris Corp./Ecomed, Inc.	5 lb	5 load/h	6 log spore reduction	70–80%	None
11. Ster-O-Lizer, Inc.	200 lb	6 load/h	Sterilization	None	Some Gain
12. Winfield Environmental	75 lb	600 lb/h	Disinfection	70–80%	None

NA—Not Applicable; NR—Not Reported

Exhibit 3-1: Alternative Irradiation and Biological Treatment Processes, General Process Descriptions

Irradiation Treatment Processes

Iotron Industries Canada, Inc.: Batched waste volumes are conveyed past an electron beam accelerator for sterilization.

Nordion International, Inc.: Waste is placed in a metal box and a conveyer transports it past a Cobalt-60 irradiation sterilizer. Treated waste is removed from the box and transported to a shredder and baler unit.

Nutek Corp.: A conveyer transports waste past an electron beam accelerator unit for sterilization. Treated waste is then conveyed to a shredder/compactor prior to disposal.

Biological Treatment Processes

SciMed Technologies, Inc.: A conveyer transports waste to a shredder. The shredded residue is mixed with water and pumped as a slurry to a tank containing enzymes where biological reactions take place. Treated residue is pumped from the tank through a screw press to separate solids. Solid residues are compacted and collected in a bin for disposal.

Exhibit 3-2: Irradiation and Biological Treatment Processes Process Overview

	Vendors	Status	First U.S. Installation	No. of U.S. Installations	Process Capacity Range	No. of States Permitted
Irradiation	1. Iotron Industries, Canada, Inc.	CA	1993	0	3500 lb/h	NR
	2. Nordion International, Inc.	CA	NR	0	5–25 ton/d	NR
	3. Nutex Corp.	CA	Expected 1994	0	800 lb/h	2
Biological	1. SciMed Technologies, Inc.	CA	1993	1	To 5000 lb/8 h	NR

Exhibit 3-3: Irradiation and Biological Treatment Processes Technical Overview

	Vendors	Waste Load Restrictions		Reported Degree of Decontamination	Reported Volume Reduction	Reported Weight Reduction
		Maximum Weight	Maximum Load Rate			
Irradiation	1. Iotron Industries Canada, Inc.	NR	NR	Sterilization	None	None
	2. Nordion International	NR	255 ton/d	Sterilization	None	None
	3. Nutek Corp.	NR	800 lb/h	Sterilization	To 80%	None
Biological	1. SciMed Technologies, Inc.	150 lb	NA	6 + log reduction	80–90%	Varies

NA—Not applicable; NR—not reported.

Exhibit 4-1: Total Number of Alternative Treatment System Vendors

	1988	1991	1991 Vendors Still in Existence	1993
Thermal process vendors	0	11	3	18
Chemical process vendors	1	6	4	12
Irradiation process vendors	0	1	1	3
Biological process vendors	0	0	0	1

Exhibit 5-1: Total Number of Alternative Medical Waste Treatment Systems Installed Nationwide as of January 1994

Incinerators	6,000+	includes 2,440 hospitals
Autoclaves	500+	
Other thermal processes	30	
Chemical processes	28	
Irradiation processes	1	
Biological processes	1	

Note: Table includes those of more than 50 lb/hour capacity which are either installed or under construction.

4.10.3. State-of-the-Art Hospital and Institutional Waste Incineration, Selection, Procurement, and Operations[*]

4.10.3.1. Introduction

On-site incineration is becoming an increasingly important alternative for the treatment and disposal of institutional waste. Incineration reduces the weight and volume of most institutional solid waste by upwards of 90 to 95%, sterilizes pathogenic waste, detoxifies chemical waste, converts obnoxious waste such as animal carcasses into innocuous ash, and also provides heat recovery benefits. At most institutions, these factors provide a substantial reduction in off-site disposal costs such that on-site incineration is highly cost effective. Many systems have payback periods of less than 1 year. In addition, on-site incineration reduces potential exposures and liabilities associated with illegal or improper waste disposal activities.

Clearly, the most important factor currently affecting the importance of on-site incineration for health-care organizations and research institutions across the country relates to infectious waste management and disposal. First of all, recent legislation and guidelines have dramatically increased the quantities of institutional waste to be disposed as "potentially infectious." For many institutions, particularly hospitals, incineration is the only viable technology for processing the increased, voluminous quantities of waste. Second, about half of the states and several major cities currently mandate that infectious waste be treated on-site, restrict its off-site transport, and/or prohibit it from being landfilled. Many

---

[*]   This article was written by Lawrence G. Doucet, P.E., D.E.E.

additional states are planning similar, restrictive legislative measures within the next few years (see the preceding article by Doucet, written 4 years later, and the earlier section on infectious waste regulations).

Off-site disposal difficulties and limitations probably contribute the greatest incentives for many health care and other institutions to consider or select on-site incineration as the preferred infectious waste treatment method. It has become increasingly difficult, if not impossible, to locate reliable, dependable infectious waste disposal service contractors. Many institutions able to obtain such services are literally required to transport their infectious waste across the country to disposal facilities. Furthermore, such services are typically very costly, if not prohibitive. Off-site disposal contractors are typically charging from about $0.30 to about $0.80 per pound of infectious waste, and some are charging as much as $1.50 per pound. For many hospitals and other institutions, this equates to hundreds of thousands of dollars per year, and several are paying in excess of $1 million per year.

4.10.3.2. Incineration Technologies

Before the early 1960s institutional incineration systems were almost exclusively multiple-chamber types, designed and constructed according to Incinerator Institute of America (IIA) *Incinerator Standards*. Since these systems operated with high excess air levels, most required scrubbers in order to comply with air pollution control standards. Multiple-chamber type systems are occasionally installed at modern facilities because they represent proven technology. However, the most widely and extensively used incineration technology over the past 20 years is "controlled-air" incineration. This has also been called "starved-air" incineration, "two-stage" incineration, "modular" combustion, and "pyrolytic" combustion. More than 7000 controlled-air type systems have been installed by approximately 2 dozen manufacturers over the past 2 decades.

Controlled-air incineration is generally the least costly solid waste incineration technology—a factor that has undoubtedly influenced its popularity. Most systems are offered as low cost, "pre-engineered" and prefabricated units. Costly air pollution control equipment is seldom required, except for compliance with some of the more current, highly stringent emission regulations, and overall operating and maintenance costs are usually less than for other comparable incineration technologies.

The first controlled air incinerators were installed in the late 1950s, and the first U.S. controlled-air incinerator company was formed in 1964. The controlled-air incineration industry grew very slowly at first. The technology received little recognition because it was considered unproven and radically different from the established and widely accepted IIA *Incinerator Standards*.

Approximately every 5 years the controlled-air incineration industry has gone through periods of rapid growth. In the late 1960s this was attributable to the Clean Air Acts, in the early and late 1970s to the Arab oil embargoes, in the early 1980s to the enactment of hazardous waste regulations, and, recently, to the enactment of infectious waste disposal regulations. Dozens of "new" vendors and equipment suppliers appeared on the scene during each of these growth periods. However, increased competition and rapid changes in the technology and market structure forced most of the smaller and less progressive companies to close. Generally, the controlled-air incineration industry has been in a state of almost constant development and change, with frequent turnovers, mergers, and company failures.

Today there are approximately a dozen listed "manufacturers" of controlled air incinerators. However, only about half of these have established successful track records with demonstrated capabilities and qualifications for providing first-quality installations. In fact, some of the "manufacturers" listed in the catalogs have yet to install their first system, and a few are no more than brokers who buy and install incinerator equipment manufactured by other firms.

Controlled-air incineration is basically a two-stage combustion process. Waste is fed into the first stage, or primary chamber, and burned with less than stoichiometric air. Primary chamber combustion reactions and turbulent velocities are maintained at very low levels to minimize particulate entrainment and carryover. This starved air burning condition destroys most of the volatiles in the waste materials through partial pyrolysis. Resultant smoke and pyrolytic products, along with products of combustion, pass to the second stage, or secondary chamber. Here, additional air is injected to complete combustion, which can occur either spontaneously or through the addition of auxiliary fuel. Primary and secondary combustion air systems are usually automatically regulated, or controlled, to maintain optimum burning conditions despite varying waste loading rates, composition, and characteristics.

Rotary kiln incineration systems have been widely promoted within the past few years. A rotary kiln basically features a cylindrical, refractory-lined combustion chamber which rotates slowly on a slightly inclined, horizontal axis. Kiln rotation provides excellent mixing, or turbulence, of the solid waste fed at one end—with high-quality ashes discharged at the opposite end. However, in general, rotary kiln systems have relatively high costs and maintenance requirements, and they usually require size reduction, or shredding, in most institutional waste applications. There are only a handful of rotary kiln applications in hospitals and other institutions in the U.S. and Canada.

"Innovative" incineration technologies also frequently appear on the scene. Some such systems are no more than reincarnations of older "failures," and others feature unusual applications and combinations of ideas and equipment. Probably the best advice when evaluating or considering an innovative system is to first investigate whether or not any similar successful installations have been operating for a reasonable period of time. Remember, so-called innovative systems should still be designed and constructed consistent with sound, proven principles and criteria.

4.10.3.3. Sizing and Rating

Classifications systems have been developed for commonly encountered waste compositions. These systems identify "average" characteristics of waste mixtures, including such properties as ash content, moisture, and heating value. The classification system published in the IIA *Incinerator Standards* is the most widely recognized and is almost always used by the incinerator manufacturers to rate their equipment. In this system, shown in Table 4.24, wastes are classified into seven types. Types 0 through 4 are mixtures of typical, general waste materials, and Types 5 and 6 (not listed in table) are industrial wastes requiring special analysis.

4.10.3.4. Primary Combustion Chambers

4.10.3.4.1. Heat Release Rates

Incinerator capacities are commonly rated as pounds of specific waste types, usually Types 0 through 4, that can be burned per hour. Incinerators generally have a different

Table 4.24. Classification of Wastes

Classification of Wastes			Approximate Composition (% by Weight)	Moisture Content (%)	Incombustible Solids (%)	Value/lb of Refuse as Fired
Type	Description	Principal Components				
0	Trash	Highly combustible waste, paper, wood, cardboard cartons, treated papers, plastic or rubber scraps; commercial and industrial sources	Trash 100%	10%	5%	8500
1	Rubbish	Combustible waste, paper, cartons, rags, wood scraps, combustible floor sweepings; domestic, commercial and industrial sources	Rubbish 80% Garbage 20%	25%	10%	6500
2	Refuse	Rubbish and garbage; residential sources	Rubbish 50% Garbage 50%	50%	7%	4300
3	Garbage	Animal and vegetable wastes, restaurants, hotels, markets; institutional, commercial and club sources	Garbage 65% Rubbish 35%	70%	5%	2500
4	Animal solids and organic wastes	Carcasses, organs, solid organic wastes, hospital, laboratory, abattoirs, animal pounds, and similar sources	100% animal and human tissue	85%	5%	1000

capacity rating for each type. For example, an incinerator rated for 1000 pounds per hour of Type 1 waste may only be rated for about 750 pounds per hour of Type 0 waste or about 500 pounds per hour of Type 4 waste. Such rating variations exist because primary chamber volumes are sized on the basis of internal heat release rates, or heat concentrations. Typical design heat release values range from about 15,000 to 25,000 BTU per cubic foot of volume per hour. In order to maintain design heat release rates, waste burning capacities vary inversely with the waste heating values (BTU/pound). As heating values increase less waste can be loaded.

Since Type 3 waste, food scraps, and Type 4 (pathological) waste have heat contents of only 3500 and 1000 Btu per pound, respectively, it might be assumed that even higher capacity ratings could be obtained for these waste types. However, this is not the case. The auxiliary fuel inputs required to vaporize and superheat the high moisture contents of Types 3 and 4 wastes limit effective incineration capacities.

In essence, primary chamber heat release criterion establishes primary chamber volume for a specific waste type and charging rate. Heat release values are simply determined by multiplying burning rate (pound per hour) by heating value (BTU per pound) and dividing by primary volume (cubic foot).

4.10.3.4.2. Burning Rates

The primary chamber burning rate generally establishes burning surface, or hearth area, in the primary chamber. It indicates the maximum pounds of waste that should be burned per square foot of projected surface area per hour (pounds per square foot per hour). Recommended maximum burning rates for various waste types are based upon empirical data, and are published in the IIA *Incinerator Standards*. Table 4.25 tabulates these criteria.

The figures in the table are calculated as follows: maximum burning rate (pounds per square foot per hour, or lbs/ft^2/hr) for Types 1, 2, and 3 wastes using factors as noted in the formula:

B_R = Factor for type waste × log of capacity/hr
B_R = Maximum burning rate lbs/ft^2/hr

For example, for an incinerator capacity of 100 lbs/hr, for Type 1 waste:

B_R = 13 (factor for Type 1 waste) × log 100 (capacity/hr)
    = 13 × 2 = 26 lbs/ft^2/hr

4.10.3.4.3. Secondary Combustion Chambers

Secondary chambers are generally sized and designed to provide sufficient time, temperature, and turbulence for complete destruction of combustibles in the flue gases from the primary chamber. Unless specified otherwise, secondary chamber design parameters are usually manufacturer specific. Typical parameters include:

- Flue gas retention times ranging from 0.25 seconds to at least 2.0 seconds
- Combustion temperatures ranging from 1400°F to as high as 2200°F

Table 4.25.   Maximum Burning Rate (lbs/sq. ft/hr) of Various Type Wastes

Capacity (lbs/h)	Logarithm	Type of Waste			
		1 (Factor 13)	2 (Factor 10)	3 (Factor 8)	4 (No factor)
100	2.00	26	20	16	10
200	2.30	30	23	18	12[a]
300	2.48	32	25	20	14[a]
400	2.60	34	26	21	15[a]
500	2.70	35	27	22	16[a]
600	2.78	36	28	22	17[a]
700	2.85	37	28	23	18[a]
800	2.90	38	29	23	18[a]
900	2.95	38	30	24	18[a]
1000	3.00	39	30	24	18[a]

[a] The maximum burning rate in $lb/ft^2/hr$ for Type 4 waste depends to a great extent on the size of the largest animal to be incinerated. Therefore, whenever the largest animal to be incinerated exceeds 1/3 the hourly capacity of the incinerator, use a rating of 10 $lb/ft^2/hr$ for the design of the incinerator.

- Turbulent mixing of flue gases and secondary combustion air through the use of high velocity, tangential air injectors, internal air injectors, abrupt changes in gas flow directions, or refractory orifices, baffles, internal injectors, and checker work in the gas flow passages

Retention times, temperatures, and turbulence are interdependent. For example, secondary chambers that are specially designed for maximum turbulence but that have relatively short retention times may perform as well as other designs with longer retention times but less effective turbulence. On many applications, increased operating temperatures may allow for decreased retention times, or vice versa, without significantly affecting performance. Regulatory standards and guidelines often dictate secondary chamber retention time and temperature requirements.

Flue gas retention time(s) is determined by dividing secondary chamber volume (cubic foot) by the volumetric flue gas flow rate (ft^3/s) quantities and operating temperatures. They can be calculated or measured. However, during normal incinerator operations, flue gas flow rates vary widely and frequently.

4.10.3.4.4. Shapes and Configurations

Primary and secondary chamber shapes and configurations are generally not critical as long as heat release rates, retention time, and air distribution requirements are satisfactory. Chamber geometry is most affected by the fabrication and transport considerations of the equipment manufacturers. Although some primary and secondary chambers are rectangular or box-like, most are cylindrical.

Controlled air incinerators with a capacity of less than about 500 pounds per hour are usually vertically oriented, with primary and secondary chambers integral, or combined, within a single casing. Larger capacity controlled air incinerators are usually horizontally oriented and have nonintegral, or separated, primary and secondary chambers. A few controlled air incinerator manufacturers offer systems with a third stage, or tertiary chamber, following the second stage. One manufacturer offers a fourth stage, termed a "reburn tunnel," which is primarily used to condition flue gases upstream of a heat recovery boiler.

Most manufacturer "variations" are attempts to improve efficiency and performance. However, some of these may be no more than "gimmicks" that offer no advantages or improvements over standard, conventional systems. Adherence to proper design fundamentals, coupled with good operations, are the overall keys to the success of any system. Acceptance of unproven variations or design deviations is usually risky.

4.10.3.5. Selection and Design Factors

Highly accurate waste characterization and quantification data are not always required for selecting and designing incineration systems; however, vague or incomplete data can be very misleading and result in serious problems.

Waste characterization involves identification of individual waste constituents, relevant physical and chemical properties, and presence of any hazardous materials. A number of terms commonly used to characterize waste can be very misleading when used in specifications. As examples, vague terms such as "general waste," "trash," "biological waste," "infectious waste," and "solid waste" provide little information about the waste materials. An incineration system designed for waste simply specified as "general" waste would probably be inadequate if waste contained high concentrations of plastics. Likewise, the term "pathological" waste is frequently, but incorrectly, used to include an assortment of materials, including not only animal carcasses but also cage waste, laboratory vials, and biomedical waste items of all types. "Pathological" incinerators are usually specifically designed for burning animal carcasses, tissues, and similar types of organic wastes. Unless the presence of other materials is clearly specified, resultant burning capacities may be inadequate for wastestreams to be incinerated.

Waste characterization can range from simple approximations to complex and costly sampling and analytical programs. As discussed, the most frequently used approximation method is to categorize "average" waste mixtures into the five IIA classes, Types 0 through 4. The popularity of this waste classification system is enhanced by the fact that most of the incinerator manufacturers rate their equipment in terms of these waste types. However, it should be noted that actual "average" waste mixtures rarely have the exact characteristics delineated for any of these indicated waste "types."

The other end of the characterization spectrum involves sampling and analysis of specific "representative" waste samples or items in order to determine "exact" heating values, moisture content, ash content, and the like. This approach is not generally recommended because it is too costly and provides no significant benefit over other acceptable approximation methods.

Virtually all components found in typical institutional-type solid waste have been sufficiently well characterized in various engineering textbooks, handbooks, and other technical publications. An example is presented in Table 4.26. In many cases a reasonably accurate compositional analysis of the wastestream, in conjunction with such published data and information, could provide reliable and useful characterization data.

Table 4.26.   Waste Data Chart

Material	BTU Value per lb as Fired	Weight (lb/ft³) (Loose)	Weight (lb/ft³)	Content by Weight (%)	
				Ash	Moisture
Type 0 waste	8,500	8–10		5	10
Type 1 waste	6,500	8–10		10	25
Type 2 waste	4,300	15–20		7	50
Type 3 waste	2,500	30–35		5	70
Type 4 waste	1,000	45–55		5	85
Acetic acid	6,280		65.8	0.5	0
Animal fats	17,000	50–60		0	0
Benzene	18,210		55	0.5	0
Brown paper	7,250	7		1	6
Butyl sole composition	10,900	25		30	1
Carbon	14,093		138	0	0
Citrus rinds	1,700	40		0.75	75
Coated milk cartons	11,330	5		1	3.5
Coffee grounds	10,000	25–30		2	20
Corn cobs	8,000	10–15		3	5
Corrugated paper	7,040	7		5	5
Cotton seed hulls	8,600	25–30		2	20
Ethyl alcohol	13,325		49.3	0	0
Hydrogen	61,000		0.0053	0	0
Kerosene	18,900		50	0.5	0
Latex	10,000	45	45	0	0
Linoleum scrap	11,000	70–100		20–30	1
Magazines	5,250	35–50		22.5	5
Methyl alcohol	10,250		49.6	0	0
Naphtha	15,000		41.6	0	0
Newspaper	7,975	7		1.5	6
Plastic-coated paper	7,340	7		2.6	5
Polyethylene	20,000	40–60	60	0	0
Polyurethane (foamed)	13,000	2	2	0	0
Rags (linen or cotton)	7,200	10–15		2	5
Rags (silk or wool)	8,400–8,900	10–15		2	5
Rubber waste	9,000–11,000	62–125		20–30	0
Shoe leather	7,240	20		21	7.5
Tar or asphalt	17,000	60		1	0
Tar paper 1/3 tar, 2/3 paper	11,000	10–20		2	1
Toluene	18,440		52	0.5	0
Turpentine	17,000		53.6	0	0
1/3 wax, 2/3 paper	11,500	7–10		3	1
Wax paraffin	18,621		54–57	0	0

Table 4.26.   Waste Data Chart (continued)

Material	BTU Value per lb as Fired	Weight (lb/ft^3) (Loose)	Weight (lb/ft^3)	Content by Weight (%) Ash	Moisture
Wood bark	8,000–9,000	2–20		3	10
Wood bark (fir)	9,500	12–20		3	10
Wood sawdust	7,800–8,500	10–12		3	10
Wood sawdust (pine)	9,600	10–12		3	10

Table 4.25 shows the various BTU values of materials commonly encountered in incinerator designs. The values given are approximate and may vary based on their exact characteristic or moisture content.

A key factor is that incineration systems must be designed to handle the entire range of the wastestream properties and characteristics, not just the "averages." System capacity and performance may be inadequate if the waste data does not indicate such ranges.

4.10.3.5.1. Capacity Determination

One of the primary criteria for selecting incineration system capacity is the quantity of waste to be incinerated. Such data should include not only average waste generation rates, but also peak rates and fluctuation cycles. The most accurate method of determining such data is a comprehensive weighing program over a period of about 2 weeks. However, the most common procedure has been to estimate waste quantities from the number and volume of waste containers hauled off-site to land disposal. Large errors have resulted from such estimates because of failures to account for container compaction densities or from faulty assumptions that the waste containers were always fully loaded.

Three major variables affect the selection of incineration system capacity, or hourly burning rate: waste generation rates; waste types, forms, and sizes; and operating hours.

The quantity of waste to be incinerated is usually the primary basis for selecting system capacities. When waste generation rates are grossly underestimated, incineration capacity may be inadequate for the planned, or available, periods of operation. In such cases the tendency is to overload the system, and operational problems ensue. On the other hand, incineration systems must be operated near their rated or design capacities for good performance; an oversized system must be operated less hours per day than may have been anticipated.

Such reduced operating hours could cause difficult problems with waste handling operations, particularly if waste storage areas are marginal. Furthermore, if waste heat recovery is necessary to justify system economics, insufficient waste quantities could be a serious problem.

Since incinerators are primarily sized according to heat release rates, waste heating value is a fundamental determinant of capacity. However, the physical form or consistency of waste may have a more significant impact on burning capacities. For example, densely packed papers, books, catalogs, and the like may have an effective incinerability factor of only about 20% compared to burning loosely packed paper. Likewise, animal bedding or cage waste typically has high ash-formation tendencies that may reduce burning rates by as much as 50%. Furthermore, highly volatile wastes, such as plastics and containers of flammable solvents, may require burning rate reductions of as much as 65% to prevent smoking problems.

The physical size of individual waste items and containers is also an important factor in the selection of incineration capacity. One rule-of-thumb is that an average incinerator waste load, or largest item, should weigh approximately 10% of rated, hourly system capacity. On this basis, a minimum 300 pounds per hour incinerator would be required for, say, Type 1 waste packaged in up to about 30-pound containers or bags. This capacity would be required regardless of the total daily quantity of Type 1 waste requiring incineration.

A typical daily operating cycle for a controlled air incinerator without automatic ash removal is as follows:

Operating Steps	Typical Durations
Clean-out	15–30 minutes
Preheat	15–60 minutes
Waste loading	12–14 hours
Burn-down	2–4 hours
Cool-down	5–8 hours

It is important to note that waste *loading* for systems with manual clean-out is typically limited to a maximum of 12 to 14 hours per operating day.

4.10.3.5.2. Burning Rate vs. Charging Rate

When evaluating incineration equipment, it is important to distinguish between the terms "burning or combustion rate" and "charging or loading rate." Manufacturers may rate their equipment or submit proposals using either term. "Burning rate" refers to the amount of waste that can be burned or consumed per hour, while "charging rate" is the amount of waste that can be loaded into the incinerator per hour. For systems operating less than 24 hours per day, "charging rates" typically exceed "burning rates" by as much as 20%. Obviously, failure to recognize this difference could lead to selecting a system of inadequate capacity.

4.10.3.6. Incinerator System Auxiliaries

The incinerator proper is only a single component in a typical incineration "system." Other components or subsystems which require equal attention in the design and procurement process, include:

- Waste handling and loading systems
- Burners and blowers
- Residue or ash removal and handling systems
- Waste heat recovery boiler systems
- Emissions control systems
- Breeching, stacks and dampers
- Controls and instrumentation

Features of the major subsystems follow.

4.10.3.6.1. Waste Handling and Loading Systems

Incinerators with capacities less than about 200 pounds per hour are usually available only with manual loading capabilities. Manual loading entails charging waste directly into the primary chamber without the aid of a mechanical system. Units with capacities in the 200 to 500 pounds per hour range are usually available with mechanical loaders as a special option. Mechanical loaders are standardly available for most incinerators with capacities of more than about 500 pounds per hour.

The primary advantage of mechanical loaders is that they provide personnel and fire safety by preventing heat, flames, and combustion products from escaping the incinerator. In addition, mechanical loading systems prevent or limit ambient air infiltration into the incinerator. In most cases air infiltration affects combustion conditions and, if excessive, substantially lowers furnace temperatures and causes smoking at the stack and into charging room areas. Infiltration also increases auxiliary fuel usage and usually accelerates refractory deterioration. Several states have recently enacted regulations requiring mechanical loading on *all* institutional waste incinerators.

Mechanical loaders enable incinerators to be charged with relatively small batches of waste at regulated time intervals. Such charging is desirable because it provides relatively stabilized combustion conditions and approximates steady-state operations. Limiting waste batch sizes and loading cycles also helps protect against overcharging and resultant operating problems.

The development of safe, reliable mechanical loaders has been a major step toward modernizing institutional waste incineration technology. The earliest incinerators were restricted to manual charging, which limited their capacities and applications. Of the loader designs currently available, most manufacturers use the hopper/ram system. With this system, waste is loaded into a charging hopper, a hopper cover closes, a primary chamber fire door opens and a charging ram then pushes the waste into the primary chamber. Hopper/ram systems are available with charging hopper volumes ranging from several cubic feet to nearly 10 cubic yards. The selection of proper hopper volume is a function of waste type, waste container size, method of loading the hopper, and incinerator capacity. An undersized hopper could result in spillage during waste loadings, an inability to handle bulky waste items, such as empty boxes, or the inability to charge the incinerator at rated capacity. On the other hand, an oversized hopper could result in frequent incinerator overcharging and associated operational problems.

A few manufacturers have recently developed mechanical loader systems capable of accepting as much as an hour's worth of waste loading at one time. These systems use internal rams to charge the primary chamber at intervals, as well as to prevent hopper bridging. Although these systems have had reportedly good success, they are still generally in the developmental stage.

One particular rotary kiln manufacturer uses an integral shredder at the bottom of the waste feed hopper. This system is termed an "auger feeder." It basically serves to process waste into a size that is compatible with the kiln dimensions.

With small capacity incinerators, less than about 500 pounds per hour, waste is usually loaded manually, bag by bag, into the charging hopper. Larger capacity systems frequently employ waste handling devices such as conveyors, cart dumpers, and sometimes skid-steer tractors to charge waste into the hopper. Pneumatic waste transport systems have been used to feed incinerator loading hoppers at a few institutions, but these have had limited success.

A cart-dumper loader basically combines a standard hopper/ram system with a device for lifting and dumping waste carts into the loading hopper. Several manufacturers offer these as integrated units. Cart dumpers also can be procured separately from several suppliers and adapted or retrofitted to almost any hopper/ram system. Cart-dumper loader systems have become increasingly popular because using standard, conventional waste carts for incinerator loading reduces extra waste handling efforts and often eliminates the need for intermediate storage containers and additional waste handling equipment.

Most modern hopper/ram assemblies are equipped with a water system to quench the face of the charging ram face after each loading cycle. This prevents the ram face from overheating due to constant, direct exposures to high furnace temperatures during waste injection. Without such cooling, plastic waste bags or similar materials could melt and adhere to the hot ram face. If these items did not drop from the ram during its stoking cycle, they could ignite and be carried back into the charging hopper, where they could ignite other waste remaining in the hopper or new waste loaded into the hopper. For additional protection against such possible occurrences, loading systems can also be equipped with hopper flame scanners and alarms, hopper fire spray systems, and/or an emergency switch to override the normal charging cycle timers and cause immediate injection of hopper contents into the incinerator.

4.10.3.6.2. Residue Removal and Handling Systems

Residue, or ash, removal has always been a particular problem for institutional type incineration systems. Most small capacity incinerators (less than about 500 pounds per hour) and most controlled-air units designed and installed before the mid-1970s must be cleaned manually. Operators must rake and shovel ashes from the primary chamber into outside containers. Small capacity units can be cleaned from the outside, but large capacity units often require operators to enter the primary chamber to clean ashes. The practice of manual clean-out is especially objectionable from many aspects, including:

- Difficult labor requirements
- Hazards to operating personnel because of exposures to hot furnace walls, pockets of glowing ashes, flaming materials, airborne dusts, and noxious gases
- Daily cool-down and start-up cycling requirements, which substantially increase auxiliary fuel usage and reduce available charging time
- Detrimental effects of thermal cycling on furnace refractories
- Severe aesthetic, environmental, and fire safety problems when handling hot, unquenched ashes outside the incinerator
- Possible regulatory restrictions

In multiple-chamber incinerators, automatic ash removal systems usually feature mechanical grates or stokers. In rotary kiln systems, ash removal is accomplished via the kiln rotation. However, automatic, continuous ash removal has historically been difficult to achieve in controlled air systems which have conventionally featured stationary, or fixed, hearths.

Early attempts at automatic ash removal in controlled air incinerators employed a "bomb bay" door concept. With these systems, the bottom of the primary chamber would swing open to drop ashes into a container or vehicle located below. Serious operating problems led to the discontinuance of these systems. More recent automatic ash removal systems use rams or plungers to "push" a mass of residue through the primary chamber and out a discharge door on a batch basis. Most of these systems have had only limited success.

Controlled-air incinerator automatic ash removal systems that have shown the most promise use the waste charging ram of the hopper/ram system to force waste and ash residues through the primary chamber to an internal discharge or drop chute for removal.

Although charging rams usually extend no more than about 12 to 18 inches into the furnace during loading, this is sufficient to move materials across the primary chamber via the repetitive, positive-displacement actions of the rams. With proper design and operations, the waste should be fully reduced to ash by the time it reaches the drop chute. For incinerators with capacities greater than about 800 to 1000 pounds per hour, internal transfer rams are usually provided to help convey ashes through the furnace to the drop chute. Transfer rams are necessary because the ash displacement capabilities of charging rams are typically limited to a maximum length of about 8 feet. Primary chambers longer than about 16 feet usually have two or more sets of internal transfer rams.

The most innovative residue removal system uses a "pulse hearth" to transfer ashes through the incinerator. The entire floor of the primary chamber is suspended on cables and pulses intermittently via sets of end-mounted air cushions. The pulsations cause ash movement across the chamber and toward the drop chute.

After the ashes drop from the primary chamber through the discharge chute, there are two basic methods, other than manual, for collecting and transporting them from the incinerator. The first is a semiautomatic system using ash collection carts positioned within an air-sealed enclosure beneath the drop chute. A door or seal gate at the bottom of the chute opens cyclically to drop ashes into ash carts. Falling ashes are sprayed with water for dust suppression and a minor quenching. Because of weight considerations, ash cart volumes are usually limited to about 1 cubic yard.

Loaded ash carts are manually removed from the ash drop enclosure and replaced with empty carts. After the removed carts are stored on-site long enough for the hot ashes to cool, they are either emptied into a larger container for off-site disposal or are brought directly to the landfill and dumped. Adequate design and proper care are needed when dumping ashes into larger on-site containers to avoid severe dusting problems. In addition, some ashes could still be hot and may tend to ignite when exposed to ambient air during the dumping operations.

The second method of ash removal is a fully automatic system using a water quench trough and ash conveyor that continuously and automatically transports wet ashes from the quench trough to a container or vehicle. With these systems, the discharge chute terminates below water level in a quench trough in order to maintain a constant air seal on the primary chamber. Most manufacturers use drag, or flight, type conveyors, but a few offer "back-hoe" or "scoop" type designs to batch grab ashes from the quench trough. The important factor is that the selected ash conveyor system be of proven design and of heavy-duty construction for the severe services of ash handling.

4.10.3.6.3. Waste Heat Recovery

In most incineration systems, heat recovery is accomplished by drawing the flue gases through a waste heat boiler to generate steam or hot water. Most manufacturers use conventional firetube type boilers for reasons of simplicity and low costs. Both single and multipass firetube boilers have been used successfully at many installations. Several facilities incorporate supplemental fuel-fired waste heat boilers so that steam can be generated when the incinerator is not operating. Also, automatic soot blowing systems are being installed on an increasing number of firetube boilers, in order to increase on-line time and recovery efficiencies.

One manufacturer uses single-drum, watertube type waste heat boilers on incineration systems. Watertube boilers are also used by other manufacturers on installations where high steam pressures and flow rates are required. Another manufacturer offers heat recovery systems with waterwall or radiant sections in the primary chamber. These waterwall sections, which are usually installed in series with a convective type waste heat boiler, can increase overall heat recovery efficiencies by as much as 10 to 15%.

Many incinerator manufacturers typically "claim" system heat recovery efficiencies for their equipment ranging from 60 to as high as 80%. However, studies and EPA-sponsored testing programs have shown that realistic heat recovery efficiencies are typically on the order of 50 to 60%. The amount of energy or steam that can be recovered is basically a function of flue gas mass flow rates and inlet and outlet temperatures. Depending on boiler type and design, gas inlet temperatures are usually limited to a maximum of 2200°F. Outlet temperatures are limited to the dewpoint temperature of the flue gases in order to prevent condensation and corrosion of heat exchanger surfaces. Depending upon flue gas constituents, incinerator dewpoint temperatures are usually on the order of 400°F.

For estimating purposes, about 3 to 4 pounds of steam can be recovered for each pound of typical institutional type solid waste incinerated. However, the economic feasibilities of providing a waste heat recovery system usually depend upon the ability to use the recovered energy. If only half of recovered steam can be used because of low seasonal steam demands, heat recovery may not be cost effective.

Some controlled air incinerator manufacturers offer air preheating, or "economizer packages," with their units. These primarily consist of metal jacketing, or shrouds, around sections of the primary or secondary chambers. Combustion air is heated by as much as several hundred degrees when pulled through the shrouds by combustion air blowers. This preheating can reduce auxiliary fuel usage by as much as 10 to 15%. In addition, the shrouding on some systems also helps limit incinerator skin temperatures to within OSHA limits.

For safety and normal plant shutdown, waste heat boilers are equipped with systems to divert flue gases away from the boiler and directly to a stack. One such system comprises an abort, or dump, stack upstream of the incinerator and stack. Another system includes a bypass breeching connection between the incinerator and stack. Modern, well-designed bypass systems are equipped with isolation dampers either in the dump stack or in the bypass breeching section. In systems without isolation dampers, either hot flue gases can bypass the boiler or ambient air can dilute gases to the boiler. Because of these factors, boiler isolation dampers may improve overall heat recovery efficiencies by at least 5%.

4.10.3.6.4. Chemical Waste Incineration

An increasing number of institutions are disposing of chemical waste in their incineration systems. Incinerated chemicals are usually flammable waste solvents that are burned as fuels with solid waste. A simple method of firing solvents has been to inject them through an atomizer nozzle into the flame of an auxiliary fuel burner. Larger capacity and better designed systems use special, packaged burners to fire waste solvents. Such burners are either dedicated exclusively for waste solvent firing or have capabilities for switching to fuel oil firing when waste solvents are not available. Waste solvent firing is usually limited to the primary chamber in order to assist in the burning of solid wastes and to maximize retention time by fully utilizing secondary chamber volumes. Injectors and burners must be located and positioned so as not to impinge on furnace walls or other burners. Such impingement results in poor combustion and often causes emission problems.

Chemical waste incineration systems must also include properly designed chemical waste handling systems. These include a receiving and unloading station, a storage tank, a pump set to feed the injector or burner, appropriate diking and spill protection, and monitoring and safety protection devices. Most of these components must be enclosed within a separate, fire-rated room that is specially ventilated and equipped with explosion-proof electrical fixtures.

When transporting, storing, and burning chemical waste, local, state and federal hazardous waste regulations must be followed. If the incinerated waste is regulated as a "hazardous waste," very costly trial burn testing, (Part B) permitting, and monitoring equipment are required. In addition, obtaining the permits could delay starting a new facility by as much as 12 to 18 months. Incinerators burning chemical solvents which are only hazardous due to "ignitability" are not likely to be considered hazardous waste incinerators, and the costly and lengthy hazardous waste incinerator permitting process is avoided. However, the storage and handling of these solvents will likely require a hazardous waste (Part B) permit.

At many institutions, bottles and vials of chemical wastes are often mixed with solid waste for incineration. If the quantities or concentrations of such containers and chemicals are very small with respect to the solid waste, incinerator operations may be unaffected. However, whenever solid waste loads are mixed with excessive concentrations of chemical containers, serious operating problems are likely, including rapid, uncontrolled combustion and volatilization resulting in heavy smoke emissions and potentially damaging temperature excursions. In addition, glass vials and containers tend to melt and form slag that can damage refractory materials and plug air supply ports.

4.10.3.6.5. Emission Control Systems

In general, only controlled air incinerators are capable of meeting the stringent emission standard of 0.08 grains of particulate per dry standard cubic foot of flue gas (gr/DSCF), corrected to 12% carbon dioxide, without emission control equipment. However, no incineration systems can meet the emission limits being recently enacted by many states which require compliance with best available control technology (BACT) levels. The BACT particulate level identified by many of the states is 0.015 gr/DSCF, corrected to 12% carbon dioxide. However, this is a controversial level which is being challenged by some in that it is only applicable to municipal waste incineration technology. Compliance with a 0.015 level will likely require a very high pressure drop, energy intensive, venturi scrubber system. Although "dry scrubbers," which comprise alkaline injection into the flue gas stream upstream of a baghouse filter, may also achieve a 0.015 level, as of this writing this technology has yet to be demonstrated on an institutional waste incineration system.

Most institutional solid wastestreams, particularly hospitals, include significant concentrations of polyvinyl chloride (PVC) plastics. Upon combustion, PVC plastics break down and form hydrogen chloride (HCl) gas. The condensation of HCl gases results in the formation of highly corrosive hydrochloric acid. Therefore, flue gas handling systems, and particularly waste heat boilers, must be designed and operated above the dewpoint of the flue gases. Protection of scrubbing systems typically includes the provision of an acid neutralization system on the scrubber water circuitry and the use of acid-resistant components and materials.

Some states have identified BACT for HCl emissions as either 90% removal efficiency or 30 to 50 ppm, by volume, in the exhaust gases. For most well-designed wet scrubbers, 99% removal efficiencies are readily achievable. With respect to minimizing emissions of PICs, such as carbon monoxide and even dioxins and furans, the keys are proper furnace sizing, good combustion controls designed to accommodate varying waste compositions and charging rates, good operations and proper care, and adjustment of system components. Inadequacies in any of these could result in objectionable emissions.

4.10.3.7. Incineration Performance and Procurement

4.10.3.7.1. Success Rates

Incineration is considered proven technology in that a great many systems readily comply with stringent environmental regulations and performance requirements. Properly designed and operated incineration systems provide "good" performance if they satisfy specific user objectives in terms of burning capacity; throughput, burnout, or destruction; environmental integrity; and on-line reliability. However, many incineration stems of both newer and older designs perform poorly. Performance problems range from minor nuisances to major disabilities, and needed corrective measures range from simple adjustments to major modifications or even total abandonment. Furthermore, performance problems occur as frequently and as extensively in small, dedicated systems as in large, complex facilities. The most common incineration system performance problems are shown in Table 4.27.

It has been estimated that roughly 25% of incineration systems installed within the last 10 years either do not operate properly or do not satisfy user performance objectives. A 1981 University of Maryland survey of medical and academic institutions incinerating low-level radioactive wastes indicated that only about 50% of the institutions surveyed (23 total) "reported no problems," and about 47% of the institutions (20 total) reported problems ranging from mechanical difficulties to combustion difficulties. A survey conducted by the U.S. Army Corps of Engineers Research Laboratory in 1985 at 52 incineration facilities reported that 17% of the users were "very pleased with their systems," 71% were "generally satisfied with the performance of their systems" (but indicated that minor changes were needed to reduce maintenance and improve efficiency), and 12% were "not happy with their systems" (reporting severe problems). Results of this Army survey are summarized in Table 4.28.

4.10.3.7.2. Fundamental Reasons for Poor Performance

Underlying causes or reasons for poor incineration system performance are not always obvious. When performance difficulties are encountered, a typical reaction is often to "blame" the incinerator contractor for furnishing "inferior" equipment. While this may be the case on some installations, there are other possible reasons which are more common and sometimes more serious. Generally, incineration system performance problems can be related to deficiencies or inadequacies in any or all of three areas:

1. Selection and/or design—*before* procurement
2. Fabrication and/or installation—*during* installation
3. Operation and/or maintenance—*after* acceptance

Examples of deficiencies in these three areas are as follows.

Table 4.27.   Incineration System Performance Problems

Major Performance Difficulties	Examples
Objectionable stack emissions	Out of compliance with air pollution regulations Visible emissions Odors Hydrochloric acid gas (HCl) deposition and deterioration Entrapment of stack emissions into building air intakes
Inadequate capacity	Cannot accept "standard" size waste containers Low hourly charging rates Low daily burning rates (throughput)
Poor burnout	Low waste volume reduction Recognizable waste items in ash residue High ash residue carbon content (combustibles)
Excessive repairs and downtime	Frequent breakdowns and component failures High maintenance and repair costs Low system reliability
Unacceptable working environment	High dusting conditions and fugitive emissions Excessive waste spillage Excessive heat radiation and exposed hot surfaces Blowback of smoke and combustion products from the incinerator
System inefficiencies	Excessive auxiliary fuel usage Low steam recovery rates Excessive operating labor costs

System Selection and/or Design Deficiencies

Deficiencies in this area are usually the result of basing incineration system selection and design decisions on incorrect or inadequate waste data, as well as failures to address specific, unique facility requirements. The resultant consequences are that system performance objectives and design criteria are also inadequate. An example of this is the procurement of an incineration system of inadequate capacity because of underestimated waste generation rates. Not so obvious examples include the relationships between operating problems and inadequate waste characterization data. Since incinerators are designed and controlled to process specific average waste compositions, vague identification of waste types or wide variances between actual waste parameters and "selected" design parameters often result in poor system performance. Significant deviations in parameters such as heating values, moisture, volatility, density, and physical form could necessitate a capacity reduction of as much as two thirds in order to avoid objectionable stack emissions, unacceptable ash quality, and other related problems. Table 4.29 indicates examples of improper waste characterization affecting incineration capacity. The establishment of good performance objectives based upon sound data and evaluations is only the initial step toward procuring a successful installation. The next step would be to assure that system design criteria and associated contract documents are adequate to satisfy the performance objectives. A prime example of design inadequacies is the failure to relate incinerator furnace volumes to any specific criteria such as acceptable heat release rates. Another

Table 4.28.   Twenty Common Problems Found in Small Waste-To-Energy Plants[a]

Problems	Installations Reporting (%)
Castable refractory	71
Underfire air ports	35
Tipping floor	29
Warping	29
Charging ram	25
Fire tubes	25
Air pollution	23
Ash conveyer	23
Not on-line	21
Controls	19
Inadequate waste supply	19
Water tubes	17
Internal ram	15
Low steam demand	13
Induced draft fans	12
Feed hopper	10
High pH quench water	8
Stack damper	4
Charging grates	2
Front-end loaders	2
Consensus:	
17% very pleased	
71% generally satisfied—minor improvements needed	
12% not happy	

[a] Results of 1983 survey of 52 heat recovery incineration systems (5-50 TPD) conducted by U.S. Army Construction Engineering Research Laboratory.

example is the specification of auxiliary components, such as waste loaders and ash removal systems, that are not suitable for the required operating schedules or rigors.

Fabrication and/or Installation Deficiencies

Deficiencies in this area relate to inferior workmanship and/or materials in either the fabrication or installation of the system. The extent and severity of such deficiencies are largely dependent upon the qualifications and experience of the incinerator contractor. Unqualified incineration system contractors may be incapable or disinterested in providing a system in compliance with specified criteria. This could be either because of general inexperience in the field of incineration or because of a disregard of criteria that is different from their "standard way of doing business or furnishing equipment.

Table 4.29.   Waste Characterization Data Deficiencies Necessitating System Capacity Reductions[a]

Actual Waste Characterizations (Deviations from Selected "Design" Values)	Typical Examples	Basic Reasons for Reduced Capacities
Heating values (BTU/lb) excessive	Greater concentrations of paper and plastic components (or less moisture) than originally identified and specified	Incinerator volumetric heat release rates (BTU/ft³/h) exceed design limits[b]
Moisture concentrations excessive	Greater concentrations of high water content wastes, such as animal carcasses or food scraps (garbage), than originally identified and specified	Increased auxiliary fuel firing rates and additional time required for water evaporation and superheating
Volatiles excessive	Greater concentrations of plastic (such as polyethylene and polystyrene) or flammable solvents than originally identified and specified	Rapid (nearly instantaneous) releases of combustibles (volatiles) in large quantities along with excessively high temperature surges
Densities excessive	Computer printout, compacted waste, books, pamphlets, and blocks of paper	Difficulties in heat and flames penetrating and burning through dense layers of waste
High ash formation tendencies	Animal bedding or cage wastes—wood chips, shavings, or sawdust	Ash layer formation on surface of waste pile insulates bulk of waste from heat, flames, and combustion air

[a] Failure to reduce capacities, or hourly waste loading rates, to accommodate indicated deviations would likely result in other more serious operational problems.

[b] Based upon accepted, empirical values, primary chamber heat release rates should be in the range of 15,000 to 20,000 BTU/ft³/h.

It is typical for even the most experienced and qualified incineration system contractors to deviate to some extent from design documents or criteria. This is largely because there are no such things as "standard" or "universal" incineration systems or "typical" applications or facilities. Unless design documents are exclusively and entirely based upon and awarded to a specific, preselected incinerator manufacturer, different manufacturers usually propose various substitutions and alternate methodologies when bidding a project. The key to evaluating such proposed variations is to assess whether they comply with fundamental design and construction criteria and whether they reflect proven design and application. On the other hand, allowing such variations without proper assessment could have unfortunate consequences.

The number and severity of fabrication and installation deficiencies are also directly related to quality control efforts during construction phases of a project. For example, a review of contractor submittals, or shop drawings, usually helps assure compliance with contract documents *before* equipment is delivered to the job site. Site inspections during installation work may detect deficiencies in design or workmanship before they lead to operational problems and performance difficulties. In addition, specific operating and performance testing as a prerequisite to final acceptance is a key element in assuring that a system is installed properly.

Table 4.30 lists some of the most common reasons for deficiencies in the fabrication and installation of incineration systems.

Operational and/or Maintenance Deficiencies

Deficiencies in this area are basically "self-inflicted" in that they usually result from owner or user omissions or negligence, and related problems occur *after* a system has been successfully tested and officially accepted.

Successful performance of even the best designed, most sophisticated, and highest quality incineration systems is ultimately contingent upon the abilities, training, and dedication of the operators. The employment of unqualified, uncaring, poorly trained, and unsupervised operators is one of the most positive ways of debilitating system performance in the shortest time.

Incineration systems are normally subject to severe operating conditions, and they require frequent adjustments and routine preventive maintenance in order to maintain good performance. Failures to budget for and provide such adjustments and maintenance on a regular basis lead to increasingly bad performance and accelerated equipment deterioration. Also operating incineration equipment until it "breaks down" usually results in extensive, costly repair work and substantially reduced reliability.

Table 4.31 lists some of the most common operational and maintenance deficiencies which could result in poor incineration system performance.

The above problems are usually interrelated, and they usually occur in combination. They occur as frequently and as extensively in small, dedicated facilities as in large, complex facilities. They may range in severity from objectionable nuisances to major disabilities. Also, required corrective measures may range from minor adjustments to major modifications or even total abandonment.

Selection and design deficiencies are probably the most common as well as the most serious causes of problem incineration systems. Reputable incinerator contractors usually make every effort to satisfy specified design and construction criteria and meet their contractual obligations. Operating and maintenance deficiencies can usually be corrected. However, once a system has been installed and started, very little can be done to compen-

Table 4.30.   Common Reasons for Fabrication and Installation

Incineration equipment vendor (manufacturer) unqualified

Equipment installation contractor (GC) unqualified

Inadequate instructions (and supervision) from the manufacturer for system installation by the GC

No clear lines of system performance responsibility between the manufacturer and the GC

Failure to review manufacturer's shop drawings, catalog cuts, and materials and construction data to assure compliance with contract (design) documents

Inadequate quality control during and following construction to assure compliance with design (contract) documents

Payment schedules inadequately related to system performance milestones

Final acceptance testing not required for demonstrating system performance in accordance with contract requirements

---

Table 4.31.   Common Reasons for Operational and Maintenance Deficiencies

Unqualified operators

Negligent, irresponsible, and/or uncaring operators

Inadequate operator training programs

Inadequate operating and maintenance manuals

No record keeping or operating logs to monitor and verify performance

Inadequate operator supervision

Lack of periodic inspections, adjustments, and preventive maintenance

Extending equipment usage when repairs and maintenance work are needed

---

sate for fundamental design inadequacies. Major, costly modifications and revisions to performance objectives are usually required.

The relatively frequent occurrence of design deficient systems may largely be attributable to a general misconception of the incineration industry as a whole. Incinerators are often promoted as standard, off-the-shelf equipment that can be ordered directly from catalogs, shipped to almost any job site and literally "plugged in." This impression has been enhanced by many of the incinerator vendors in a highly competitive market. Exaggerations, half-truths, and sometimes false claims are widespread relative to equipment performance capabilities. In addition, attractive, impressive brochures often suggest that implementation of an incineration system is simpler than it really is.

Incineration systems are normally subject to extremely severe operating conditions. These include very high and widely fluctuating temperatures, thermal shock from wet materials, slagging residues which clinker and spall furnace materials, explosions from items such as aerosol cans, corrosive attacks from acid gases and chemicals, and mechanical abrasion from the movement of waste materials and from operating tools. These conditions are compounded by the complexity of the incineration process. Combustion processes are complicated in themselves, but in incineration this complexity is magnified by frequent, unpredictable, and often tremendous variations in waste composition and feed rates. To properly manage such severe and complex operating conditions, incineration systems require well-trained, dedicated operating personnel, frequent and thorough inspections, maintenance and repair, and administrative and supervisory personnel attuned to these requirements.

At many facilities, the practice is to operate the incineration system continuously until it breaks down because of equipment failures. This type of operation accelerates both bad performance and equipment deterioration rates. Repairs done after such breakdowns are usually far more extensive and costly than those performed during routine, preventive maintenance procedures. Also, items which are typically capable of lasting many years can fail in a fraction of that time if interrelated components are permitted to fail completely.

4.10.3.8. Key Step

A first step in procuring a good incineration system is to view the incineration "industry" in a proper perspective. There are four basic principles to bear in mind:

1. Incineration technology is not an "exact" science—it is still more of an art than a science, and there are no shortcuts, simplistic methods, or textbook formulas for success.
2. There is no "universal" incinerator—no design is universally suited for all applications. Incinerators must be specifically selected, designed, and built to meet the needs of each facility on an individual basis. Manufacturers' catalogs identify typical models and sizes, but these are rarely adequate for most facilities without special provisions or modifications.
3. There is no "typical" incinerator application—even institutions of similar type, size, and activities have wide differences in waste types and quantities, waste management practices, disposal costs, space availability, and regulatory requirements. Each application has unique incineration system requirements that must be identified and accommodated on an individual basis.
4. Incinerator manufacturers are not "equal"—there are wide differences in the capabilities and qualifications of the incinerator equipment manufacturers. Likewise, there are wide differences in the various systems and equipment which are offered by different manufacturers.

4.10.3.9. Recommended Procurement Steps

Table 4.32 outlines six steps, recommended for implementing an incineration system project. Each is considered equally important towards minimizing or eliminating the deficiencies discussed above and for increasing the likelihood of obtaining a successful installation.

Performance difficulties on *most* problem incineration systems can usually be traced to a disregard or lack of attention to details in the first two steps; namely, (1) evaluations and selections and (2) design documents. For example, many facilities have been procured strictly on the basis of "purchase orders" containing generalized requirements such as:

> "Furnish an incineration system to burn
> ____ lb/hr of institutional waste
> in compliance with applicable regulations."

Obviously, the chances for success are marginal for any incineration system procured on the basis of such specifications.

On many projects, incinerator contractor evaluation and selection under step 3 involve no more than a solicitation of prices from a random listing of vendors with the award of

Table 4.32.   Recommended Incineration System Implementation Steps

Evaluation and selections
    Collect and consolidate waste, facility, cost, and regulatory data
    Identify and evaluate options and alternatives
    Select system and components
Design (contract) documents
    Define wastes to be incinerated—avoid generalities and ambiguous terms
    Specify performance requirements
    Specify *full* work scope
    Specify minimum design and construction criteria
Contractor selection
    Solicit bids on quality and completeness—not strictly least cost
    Evaluate and negotiate proposed substitutions and deviations
    Negotiate payment terms
    Consider performance bonding
Construction and equipment installation
    Establish lines of responsibility
    Require shop drawing approvals
    Provide inspections during construction and installation
Startup and final acceptance
    Require "punch-out" system for contract compliance
    Require comprehensive testing: system operation, compliance with performance
      requirements and emissions
    Obtain operator training
After final acceptance
    Employ qualified and trained operators
    Maintain operator supervision
    Monitor and record system operations
    Provide regular inspections and adjustments
    Implement preventive maintenance and prompt repair; consider service contract

a contract to that firm proposing a system for the "least cost." There are two basic problems with this approach. First, the selected incinerator contractors are assumed to have equivalent capabilities and qualifications. Second, "least cost" acceptance assumes that the equipment offered by each of the contractors is equivalent or identical. A comparative "value" assessment of proposals usually results in the procurement of a superior quality system for a negligible price difference. It is not uncommon to see cost proposals "low" by no more than 10%, but the equipment offered of only half the quality of the competition.

Again, although incineration is considered a proven technology, in many ways it is still more of an art than a science. There are no textbook formulas or shortcut methods for selecting and implementing a successful system, and there are no guarantees that a system will not have difficulties and problems. However, the probabilities of procuring a successful, cost-effective system increase proportionally with attention to details and utilization of proven techniques, methodologies, and experience.

REFERENCES

1.   **Bleckman, J., O'Reilly, L., and Welty C.,** *Incineration for Heat Recovery and Hazardous Waste Management,* American Hospital Association, Chicago, 1983.

2. **Boegly, W.J., Jr.**, *Solid Waste Utilization—Incineration with Heat Recovery*, prepared for Argonne National Laboratory under Contract W-31-109-Eng-38 with the U.S. DOE, Publication No. ANL/CES/TE 78-3, Washington, D.C., 1978.

3. **Cooley, L.R., McCampbell, M.R., and Thompson, J.D.**, *Current Practice of Incineration of Low-Level Institutional Radioactive Waste,* prepared for U.S. DOE, Publication No. EGG-2076, Washington, D.C., 1981.

4. **Doucet, L.G.**, Waste handling systems and equipment, *NFPA Fire Protection Handbook,* Chapter 14, Section 12, National Fire Protection Association, Quincy, MA, 1985.

5. **Doucet, L.G.**, Incineration: State-of-the-Art Design, Procurement and Operational Considerations, *Technical Document No. 055872*, American Society for Hospital Engineering, Environmental Management File, Chicago, 1988.

6. **Doucet, L.G. and Knoll, W.G., Jr.**, The craft of specifying solid waste systems, *Actual Specif. Eng.*, 107, May 1974.

7. **Doucet, L.G.**, Institutional waste incineration problems and solutions, *Proceedings of Incineration of Low Level and Mixed Wastes Conference*, St. Charles, IL, April 1987.

8. **Ducey, R.A., Joncich, D.M., Griggs, K.L., and Sias, S.R.**, Heat Recovery Incineration: A Summary of Operational Experience, Technical Report No. CERL SRE-85/06, prepared for U.S. Army Construction Engineering Research Laboratory, Champaign, IL, 1985.

9. **English, II, J.A.**, Design aspects of a low emission, two-stage incinerator, *Proceedings 1974 National Incinerator Conference,* ASME, New York, 1974.

10. Environmental Protection Agency, Small Modular Incinerator Systems with Heat Recovery: A Technical, Environmental and Economic Evaluation, prepared by Systems Technology Corporation, Environmental Protection Agency, Publ. SW-177c, Washington, D.C., 1979.

11. **Hathaway, S.A.**, Application of the Packaged Controlled Air-Heat Recovery Incinerator of Army Fixed Facilities and Installations, Technical Report No. CERL-TR-E-151, prepared for U.S. Army Construction Engineering Research Laboratory, Champaign, IL, 1979.

12. *Incinerator Standards,* Incinerator Institute of America, New York, 1968.

13. **Martin, A.E.**, Small-Scale Resource Recovery Systems, Noyes Data Corporation, Park Ridge, MN, 1982.

14. **McColgan, I.J.**, Air Pollution Emissions and Control Technology: Packaged Incinerators, Economic and Technical Review, EPS-3-AP-77-3, Canadian Environmental Protection Service, Burlington, Ontario, 1977.

15. **McRee, R.E.**, Controlled-air incinerators for hazardous waste application theory and practice, *Proc. APCD Int. Workshop Ser. on Hazardous Waste,* New York, April 1, 1985.

16. **McRee, R.E.**, Waste heat recovery from packaged incinerators, *Proc. ASME Incinerator Division Conf.,* Arlington, VA, January 25, 1985.

17. **Theoclitus, G., Liu, H., and Dervay, J.R., II**, Concepts and behavior of the controlled air incinerator, *Proc. 1972 National Incinerator Conf.,* ASME, New York, 1972.

4.11. LABORATORY CLOSEOUT PROCEDURES

Occasionally laboratories cease operations and totally close down, due to the laboratory director retiring, changing jobs or locations, or for other reasons. In these circumstances, there are almost always large quantities of surplus chemicals for which disposal must be arranged. Prior planning and a cooperative effort involving laboratory personnel and the waste management group will ensure that there will not be substantial quantities of chemicals left behind which cannot be identified. It should be mandatory, wherever possible, that at least 30 days notice, and preferably more, should be given to the waste group by the persons responsible for the laboratory, to allow the identification procedure to be done carefully and thoroughly.

Most of the chemicals in the laboratory should be in the original containers and should pose no difficulty in identification. Many of these will still be useful and should be either directly distributed by the laboratory personnel to others who might use them, or should be transferred to the organization's redistribution program. Unfortunately, some will not be useful and must be placed in the category of material which will require disposal. The worst situation, for those materials which are readily identified, will be those that have become unstable and require special measures to move and handle safely. Often stuck at the back of cabinets and long forgotten will be ancient bottles of ether, perchloric acid, picric acid, etc. However, even in this difficult situation, at least the material is known and is subject to known procedures.

It will require the assistance of the laboratory personnel to identify those materials which do not have labels or labels which have no significance to those who have not been working in the laboratory. However, with the aid of the individuals currently working in the laboratory, at least, it should be possible to identify the contents of the containers for which they are directly responsible. If the labels bear initials so that it can be determined who prepared the material, then from laboratory notebooks, reports, theses or dissertations, or even some familiarity with what the individual had been working on, it may be possible to determine what the contents are likely to be, which will greatly facilitate identification, if tests are required. In some cases, it may be possible to make a reasonable estimate from other containers in the vicinity, if they are obviously grouped by category, for example.

There will almost always be some containers left over which cannot be identified. Some may be dangerous while others may not. If materials are routinely used in the laboratory which degrade to dangerous materials, then these unknown containers must be treated with care. In some cases it may be necessary to call in a firm which specializes in handling dangerous materials, rather than take the risk of handling the material at all. When the materials normally generated in the laboratory are not shock, heat, or friction sensitive and procedures have not changed in character for several years, it is probably permissible to handle these unknown materials with only normal care. If the laboratory personnel were no longer available, it might not be possible to make this distinction.

Even if there are still unknowns left over after the best efforts of the laboratory personnel and waste group working together, there will certainly be far fewer than if the latter group had to do all the work themselves, or if the evaluation of the containers had to be done hastily because of inadequate warning. Most individuals will be surprised at the effort and time required to close out an average laboratory. It will rarely be done in a single day. Providing adequate warning will allow waste personnel to schedule their own duties to take maximum advantage of their own time and those of the laboratory personnel. The reduction in the amount of waste generated and the reduction in the number of containers requiring analysis will amply repay the effort.

CHAPTER 5

Nonchemical Laboratories

5.1. INTRODUCTION

The emphasis in the previous chapters has been on laboratories in which the primary concerns were due to the use of chemicals, although in order to not completely avoid a topic unnecessarily, some of the problems arising in other types of operations have been addressed. For example, in the latter part of the last chapter some serious issues involving problems with specific contagious diseases were discussed due to the increasing importance these diseases have in our society and in our laboratories, as a continuation of the topic of health effects. The problems of dealing with infectious waste were considered at some length as well as chemical wastes as part of the larger problem of dealing with hazardous wastes. In this chapter laboratory operations which involve special problems will be presented in greater detail. However, in responding to these special problems, one should be careful not to neglect the safety measures associated with more common hazards.

5.2. RADIOISOTOPE LABORATORIES

Exposure of individuals to ionizing radiation is a major concern in laboratories using radiation as a research tool or in which radiation is a by-product of the research. Although there are many types of research facilities in which ionizing radiation is generated by the equipment, e.g., accelerator laboratories, X-ray facilities, and laboratories using electron microscopes, the most common research application in which ionizing radiation is a matter of concern is the use of unstable forms of the common elements which emit radiation. A given element must have a fixed number of protons in the nucleus, but can differ in the number of neutrons, the different forms being called isotopes. It is the property of the unstable forms, or radioisotopes, to emit radiation which makes them useful, since their chemical properties are essentially identical to the stable form of the element (where a stable form exists; for elements with atomic numbers greater than that of bismuth, there are no completely stable forms). The radiation which the radioisotopes emit allows them to be distinguished from the stable forms of the element in an experiment.

The radiation which makes radioisotopes useful also makes their use a matter of concern to the users and the general public. Exposure to high levels of radiation is known to cause health problems; at very high levels, rapid death can follow. At lower, but still substantial levels, other health effects are known to occur, some of which, including cancer, can be delayed for many years. At very low levels, knowledge of the potential health effects is much more uncertain. The generally accepted practice currently is to extrapolate statistically known effects on individuals exposed to higher levels to large groups of persons exposed to low levels of radiation in a linear fashion. The concept is similar to the use of higher concentrations of chemicals using a limited number of animals in health studies of chemical effects, instead of more normal concentrations in a very large number of test animals. There are some who question the validity of this assumption in both cases, but it is a conservative assumption and, in the absence of confirmed data, is a generally safe course of action to follow. However, the practice may have led to a misleading impression of the risks of many materials. When a scientist makes the statement that he does not know whether a given material is harmful or not, he is often simply stating in a very honest way that the data do not clearly show whether, at low levels of use or exposure, a harmful effect will result. It does not necessarily imply, as many assume, that there is a lack of research in discovering possible harmful effects. In many cases, major efforts have been made to unambiguously resolve the issue, as in the case of radiation, and the data do not support a definite answer. In the case of radiation, there is even a substantial body of experimental data (to which proponents of a concept called "hormesis" call attention) that supports possibly positive effects of radiation at very low levels. This position is, of course, very controversial. However, in chemical areas there are many examples of chemicals essential to health in our diets in trace amounts that are poisonous at higher levels. It is not the intent of this section to attempt to resolve the issue of the effects of low-level radiation, but to emphasize that there are concerns by many employees and the general public. It may well be that, by being very careful to not go beyond known information, scientists have actually contributed to these concerns. Another way of looking at the issue, and certainly a more comforting way, is that many unsuccessful attempts have been made to demonstrate negative effects at low levels. Radiation levels which normally accompany the use of radioisotopes are deliberately kept low, and the perception of risk by untrained individuals may be overstated. However, a linear dose-effect relation is the accepted basis for regulatory requirements at this time, and until better data are available scientists using radioactive substances must conform to the standards. Users owe it to themselves and the public to use the materials in ways known to be safe. However, as a general concept, it would be wise for scientists, when speaking to persons not trained in their field, to be sure that when they say they do not know of possible harmful effects of any material that this statement is understood to be an informed uncertainty, where this is the case, as opposed to being based on a lack of effort.

It is unfortunate that there is so much concern about radiation since there are many beneficial effects, but because of the dramatization of the concerns, many individuals fear radiation out of all proportion to any known risks. In an opinion poll in which members of the general public were asked to rank the relative risks of each of a number of hazards, nuclear radiation was ranked highest but in reality, the *least* dangerous of all the other risks based on known data was much more likely to cause death or injury than radiation. Individuals who have badly needed X-rays, the application of diagnostic use of radioactive materials, or radiation therapy have declined to have them because of this heightened fear. Used properly, radiation is an extremely valuable research tool and has many beneficial aspects. Used improperly, it can be dangerous, but so can many other things.

5.2.1. Natural Radioactivity

A common misconception is that radiation is an artificial phenomenon. Many of the most commonly used radioisotopes have been created artificially, but there are abundant sources of natural radiation. Elements such as potassium and carbon, which are major constituents of our body, have radioactive isotopes. Many other elements, such as the rare earths, have radioactive versions. Every isotope of elements with atomic numbers (i.e., the number of protons in the nucleus of the element, or the number of electrons around the nucleus in a neutral atom) above 83 is unstable and these elements are common in the soils and rocks which make up the outer crust of the earth. There are areas in the world in which the natural levels of radiation could exceed that permitted for the general public resulting from the operation of any licensed facility using radioisotopes. Radiation constantly bombards us from space due to cosmic rays. Persons who frequently take long airplane flights receive a significantly increased amount of radiation over a period of time compared to persons who fly rarely or not at all. Arguments that these natural forms of radiation are acceptable because they are natural has absolutely no basis in fact. As will be discussed later, there are only a modest number of varieties of radiation, and these are produced by both natural and artificially produced radioactive materials. Similarly, there are only a few ways in which radiation may interact with matter, and they also are the same for all sources of radiation.

The acknowledgment of natural sources of radiation is not intended to minimize concerns about radiation, even the natural forms, but to point out that if there are concerns about low levels of radiation, then these natural levels must be considered as well as the artificial sources. One of the naturally occurring radioactive materials, radon, has been receiving much attention and may be a significant hazard, perhaps contributing to an increase of 1 to 5% of the number of lung cancer deaths each year. This estimate, as in most cases dealing with attribution of specific effects of low levels of radiation, is supported by some and disputed by others. Note, however, that even in this case at least 95 to 99% of the lung cancer deaths are attributable to other causes. Radon as an issue will be discussed in a separate section later in this chapter. An isotope of potassium, an essential element nutritionally and present in substantial amounts in citrus fruits and bananas, for example, emits very penetrating radiation.

There are various estimates of the average source of radiation exposure for most individuals. An article by Komarov,[1] who is associated with the World Health Organization, provides the following data about sources of radiation: 37% from cosmic rays and the terrestrial environment, 28% from building materials in the home, 16% from food and water, 12% from medical usage (primarily X-rays), perhaps 4% from daily color television viewing, 2% from long-distance airplane flights, and 0.6% (under normal operating conditions) from living near a nuclear power plant. The Komarov article was written before the Chernobyl incident, but even this outstanding example of poor management is not sufficient to change the general picture. Unlike the Chernobyl reactor, commercial nuclear power plants in the United States are protected by very strong confinement enclosures to prevent unscheduled releases. In the case of the Three-Mile Island incident, in which the reactor core melted down, the confinement enclosure performed as designed and minimal amounts of radioactive material were released. As the news media reported some time after the initial furor, "the biggest danger from Three-Mile Island was psychological fear," to which the media contributed significantly.

In summary, radiation is a valuable research tool. In order to prevent raising public concerns and perhaps lead to further restrictions on its use, scientists need to be scrupulously careful to conform to accepted standards governing releases or overexposures. Fortunately, for common uses of radiation in research laboratories, this goal is easily achieved with reasonable care.

5.2.2. Basic Concepts

Each scientific discipline has its own special terms and basic concepts on which it is founded. This section is, of course, not necessary for most scientists who work with radiation, but it may be useful for establishing a framework within which to define some needed terms. As scientists work with accelerators of higher and higher energies, the concept of matter is at once growing more complex and simpler; more complex in that more entities are known to make up matter, but simpler in that theorists working with the data generated by these gigantic machines are developing a coherent concept unifying all of the information. For the purposes of this discussion, a relatively simple picture of the atom will suffice.

5.2.2.1. The Atom and Types of Decay

In the simple model of the atom employed here, the atom consists of a very small dense nucleus, containing positively charged particles called protons and neutral particles called neutrons, surrounded by a cloud of negatively charged electrons. The number of protons and the number of electrons are equal for a neutral atom, but the number of neutrons can vary substantially, resulting in different forms of an element called, as already noted, isotopes of the element. Some elements have only one stable isotope, although tin has ten. There are unstable isotopes, logically called radioisotopes, in which, over a statistically consistent time, a transition of some type occurs within the nucleus. Different types of transitions lead to different types of emitted radiation. Hydrogen, for example, has two stable forms and one unstable one, in which a transition occurs to allow an electron to be generated and emitted from the nucleus, producing a stable isotope of helium. Prior to the transition, the electron did not exist independently in the nucleus. A neutron is converted to a proton in the process, and the electron is created by a transformation of energy into matter. This process is called beta decay. No element with more than 83 protons in the nucleus has a completely stable nucleus, although some undergo transitions (including by processes other than beta decay) extremely slowly.

In some cases, the energy of the nucleus favors emission of a positive electron (positron) instead of a normal electron which has a negative charge. This is called positive beta decay or positron decay. Here a proton is converted into a neutron. A competitive process to positive beta decay is electron capture (ε) in which an electron from the electron cloud around the nucleus is captured by the nucleus, a proton being converted into an neutron in the process. In the latter process, X-rays are emitted as the electrons rearrange themselves to fill the vacancy. However, following positron emission, the positive electron eventually interacts with a normal electron in the surrounding medium, and the two vanish or annihilate each other in a flash of energy. The amount of energy is equal to the energy of conversion of the two electron masses according to $E = mc^2$. This amounts to, in electron volts, 1.02 million electron volts, or 1.02 MeV. In order to conserve momentum, two photons or gamma rays of 0.511 MeV each are emitted 180° apart in the process.

In many cases, the internal transitions accompanying adjustments in the nucleus result in emission of electromagnetic energy, or gamma rays. These can be in the original or parent nucleus, in which case they are called internal transitions, and the states are called metastable states. More often, the gamma-emitting transitions occur in the daughter nucleus after another type of decay, such as beta decay (metastable states can exist in the daughter nucleus also). The gamma emission distribution can be very complex. In some instances, the internal transition energy is directly transferred to one of the electrons close to the nucleus in a process called internal conversion, and the electron is emitted from the atom. In this last case, energy from transitions in the orbital electron cloud is also emitted as X-rays.

Finally, the most massive entity normally emitted as radiation is the alpha (α) particle which consists of a bare (no electrons), small nucleus having two protons and two neutrons.

The nucleons making up an alpha particle are very strongly bound together, and unlike electrons the alpha particle appears to exist in the parent nucleus as a unit prior to the decay in our simple model.

The processes briefly described above are the key ones in terms of safety in the use of radioisotopes. There is another very important aspect of the decay processes, and that is the energy of the emitted radiation. The electrons emitted in beta decay can have energies ranging from a few eV to between 3 and 4 MeV. There is an unusual feature of the beta decay process in that the betas are not emitted monoenergetically from the nucleus as might be expected and as does occur for alpha and gamma decay. The reason is that, in addition to a beta being emitted, another particle, called a neutrino, of either zero mass or very close to it, is emitted simultaneously and shares the transitional energy. This particle reacts minimally with matter and, except for modifying the energy of the electrons, does not play a role in radiation safety, although its existence is very important for many other reasons. Gammas can have a similar range of energies to that of electrons, but the energies of the gammas are discrete instead of a distribution.

Alpha particles have a relatively high energy, normally ranging from 4 to 6 MeV. The decay of alphas with lower energies is so slow that it occurs very rarely, while with an energy just a little higher, the nucleus decays very rapidly. The high energy, accompanying the high mass and the double positive charge, make the alpha particle a particularly dangerous type of radiation, if it is emitted in the proximity of tissue which can be injured. This is an important qualification as will be seen later.

Table 5.1 summarizes the properties of the types of radiation.

Graphically, the decay process can be depicted as shown in Figure 5.1, where A = the atomic mass number and Z = the nuclear charge. The box with A,Z is the parent nucleus and the others are the possible daughters for the processes shown.

5.2.2.2. The Fission Process

A major omission from the above discussion is the process of fission, which normally will not be of direct concern in most laboratories using radioisotopes. However, without this process many of the commonly used radioisotopes would not be available, since they are obtained from reprocessing spent fuel and recovery of the remnants left over after the fission process. The process of fission describes the process by which a few very heavy atoms decay by splitting into two major components and usually a few neutrons, accompanied by the release of large amounts of energy, ≈ 200 MeV. The process can be spontaneous, but also can be initiated by exposing specific heavy nuclei to neutrons. There are no radioisotopes that normally emit neutrons, but there are several interactions in which a neutron is generated. Among these are several reactions in which a gamma ray interacts with beryllium to yield neutrons, so that a portable source of neutrons can be created. There are many other ways to generate neutrons but there is no need to describe these in this book. However, if a source of neutrons, n, is available and is used to bombard an isotope of uranium, ^{235}U, the following reaction can occur.

$$n + {}^{235}U \rightarrow X + Y + \approx 2.5n + \text{energy} \tag{1}$$

Here, X and Y are two major atomic fragments or isotopes resulting from the fission process. The process is enhanced if the initiating neutrons are slowed down until they are in thermal equilibrium with their surroundings. X and Y themselves will typically decay after the original fission event, a few by emitting additional neutrons, as well as betas and gammas. As noted earlier, about 200 MeV of energy are released in the process, much of

Table 5.1.   Properties of Radioactive Emissions

Type	Mass (amu)	Charge (Electron Units)	Range of Energy
Alpha (α)	4	+2	$\approx$ 4–6 MeV
Beta (ß)	1/1840	±1	$\approx$ eVs–4 MeV
Gamma	0	0	$\approx$ eVs–4 MeV
X-Rays	0	0	$\approx$ eVs–100 KeV

it as kinetic energy shared by the particles. Some of these fission fragments are long lived, and can be chemically separated to provide radioisotopes of use in the laboratory. These are the major source of the by-product radioisotopes regulated by the Nuclear Regulatory Commission (NRC). The fission reaction can, under appropriate circumstances, be self-sustaining in a chain reaction. In some configurations the chain reaction is extremely rapid, and an atomic bomb is the result. However, by using the neutrons emitted by the fission fragments (called delayed neutrons), the process can be controlled safely in a reactor. Over a period of time, the fission products build up in the uranium fuel and eventually can be recovered when the fuel element is reprocessed.

Other by-product materials or radioisotopes are made by the following reaction:

$$n + {}^{A}X \rightarrow {}^{(A+1)}Y^{*} + a \qquad (2)$$

The asterisk indicates that the product nucleus, Y, may be unstable and will undergo one (or more) of the modes of decay discussed previously. The a indicates that there may be a particle directly resulting from the reaction. In many cases, the source of neutrons for radioisotopes created by this reaction is a nuclear reactor, so these also are "by-product materials."

Plutonium is made in nuclear reactors by the above reaction where ${}^{238}U$ is the target nucleus. Although there are other reactions using different combinations of particles in the reaction, in most cases these require energetic bombarding particles generated in accelerators. Also, since there are no common radioisotopes that generate neutrons, there is essentially no probability that other materials in laboratories will be made radioactive by exposure to radiation from by-product materials.

Materials which will undergo fission and can be used to sustain a chain reaction are, in the nomenclature of the NRC, "special" nuclear materials. These include the isotopes of uranium with mass numbers 233 and 235 materials enriched in these isotopes, or the artificially made element plutonium. Materials which have uranium or thorium in them to the extent of 0.05% are called source materials.

5.2.2.3. Radioactive Decay

An important relationship concerning the actual decay of a given nucleus is that it is purely statistical, dependent only upon the decay constant for a given material, i.e., the activity, A, is directly proportional to the number, N, of unstable atoms present:

$$\text{Activity} = A = dN/dt \equiv C\,N \qquad (3)$$

This can be reformulated to give the number of radioactive atoms N at a time t in terms of the number originally present.

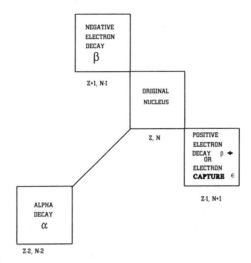

Figure 5.1. Schematic representation of radioactive decay processes.

$$N(t) = N_0 e^{-\lambda t} \tag{4}$$

where
$$\lambda = \ln 2 / \tau$$

Equation 4 shows that during any interval, τ, theoretically half of the unstable nuclei at the beginning of the interval will decay. In practice, *approximately* half will decay in a half-life, τ. This is illustrated in Table 5.2.

The data in this table illustrate clearly that when small numbers are involved, the statistical variations cause the decrease to fluctuate around a decay of about one half of the remaining atoms during each successive half-life, but obviously between 3 and 4 half-lives in this table, it would have been impossible to go down by precisely half. The table also illustrates a fairly often used rule-of-thumb: after radioactive waste has been allowed to decay by 10 half-lives, the activity has often decayed sufficiently to allow safe disposal. This, of course, depends upon the initial activity.

The daughter nucleus can also decay, as can the second daughter, and so forth. However, eventually a nucleus will be reached which will be stable. This is, in fact, what occurs starting with the most massive natural elements, uranium and thorium. All of their isotopes are unstable, and each of their daughters decays until eventually stable isotopes of lead are reached. The existence of all of the elements above atomic number 83 owe their existence to the most massive members of these chains that have half-lives that are comparable to the age of the earth so a significant fraction remains of that initially present.

5.2.2.4. Units of Activity

The units of activity are dimensionally the number of decays or nuclear disintegrations per unit time. Until recently, the standard unit to measure practical amounts of activity was the curie (Ci), which was defined to be 3.7×10^{10} disintegrations per second (dps). Other units derived from this were the millicurie (mCi) or 3.7×10^7 dps, the microcurie (μCi) or 3.7×10^4 dps, the nanocurie (nCi) or 37 dps and the picocurie (pCi) or 0.037 dps. Many health physicists prefer to use disintegrations per minute (dpm), and the NRC also prefers the data logged in laboratory surveys to be expressed in dpm. The curie was originally

Table 5.2.   Typical Decay of a Group of Radioactive Atoms

Number	Time (τ)	Number	Time (τ)
1000	0	14	6
502	1	7	7
249	2	4	8
125	3	1	9
63	4	1	10
31	5	0	11

supposed to equal the amount of activity of 1 g of radium. This unit, and the derivative units, are still the ones most widely used in this country; however, an international system of units, or SI system, has been established. In this system, 1 disintegrations per second is defined as a becquerel (Bq). Larger units, which are multiples of 10^3, 10^6, 10^9, and 10^{12}, are indicated by the prefixes kilo, mega, giga, and tera, respectively. In most laboratories that use radioisotopes as tracers, the quantities used are typically about 10^4 to 10^8 dps. There are other uses of radioisotopes (e.g., therapeutic use of radiation) which use much larger amounts.

5.2.2.5. Interaction of Radiation with Matter

5.2.2.5.1. Alphas

As an alpha particle passes through matter, its electric field interacts primarily with the electrons surrounding the atoms. Because it is a massive particle, it moves comparatively slowly and spends a significant amount of time passing each atom. Therefore, the alpha particle has a good opportunity to transfer energy to the electrons by either removing them from the atom (ionizing them) or raising them to higher energy states. Because it is so much more massive than the electrons around the atoms, it moves in short, straight tracks through matter and causes a substantial amount of ionization per unit distance. An alpha particle is said to have a high linear energy transfer (LET). A typical alpha particle has a range of only about 0.04 mm in tissue or about 3 cm in air. The thickness of the skin is about 0.07 mm so that a typical alpha particle will not penetrate the skin. However, if a material that emits alphas is ingested, inhaled, or, in an accident, becomes imbedded in an open skin wound, so that it lodges in a sensitive area or organ, the alpha radiation can cause severe local damage. Since many heavier radioactive materials emit alpha radiation, this often makes them more dangerous than materials that emit other types of radiation, especially if they are chemically likely to simulate an element retained by the body in a sensitive organ. If they are not near a sensitive area, they may cause local damage to nearby tissue, but this may not cause appreciable damage to the organism as a whole.

5.2.2.5.2. Betas

Beta particles are electrons, they have a single negative or positive charge and are the same mass as the electrons around the atoms in the material through which they are moving. Normally, they also are considerably less energetic than an alpha particle. They typically move about two orders of magnitude more rapidly than alpha particles. They still interact with matter by ionization and excitation of the electrons in matter, but the rate of interaction per unit distance traveled in matter is much less. Typically, beta radiation (≈ 1 MeV) can penetrate perhaps 0.5 cm deep into tissue, or about 4 meters of air, although this is strongly dependent upon the energy of the beta. Low-energy betas, such as from ^{14}C, would penetrate only about 0.02 cm in tissue or about 16 cm in air. Therefore, only those organs lying close to the surface of the body can be injured by external beta irradiation and then only by the more energetic beta emitters. Radioactive materials emitting betas taken into the body can affect tissues further away than those that emit alphas, but the LET is much less.

There is a secondary source of radiation from beta emitters. As the electrons pass through matter, they cause electromagnetic radiation called "bremsstrahlung," or braking radiation to be emitted as their paths bend. The energy that appears as bremsstrahlung is approximately ZE/3000 (where Z is the atomic charge number of the absorbing medium and E is the ß energy in MeV. This is not a problem for alpha particles since their paths through matter are essentially straight. Bremsstrahlung radiation can have important implications for certain energetic beta emitters such as ^{32}P. Protective shielding for energetic beta emitters should be made of plastic or other low-Z material instead of a high-Z material such as lead. Because of the silicon, even keeping ^{32}P in a glass container can substantially increase the radiation dose to the hands while handling the material in the container as compared to the exposure that would result were bremsstrahlung not a factor.

5.2.2.5.3. Gammas

Since gamma rays are electromagnetic waves, they are not charged and do not have any mass, they interact differently with matter than do alpha and beta particles. They can, however, interact with the electrons in matter by three different mechanisms.

Photoelectric Effect

The gamma ray interacts with electrons around an atom. Electrons with which they interact normally are completely removed from the atom, i.e., the atom is ionized. In a photoelectric effect process, all the energy of the gamma ray is transferred to the electron and the gamma ray no longer exists. The photoelectric effect mechanism is dependent upon the gamma energy as shown in Figure 5.2. For low energy photons, the photoelectric effect depends upon the charge of the absorber, approximately as Z^4. As energies increase, the importance of the atomic number of the absorber decreases.

Compton Effect

The gamma ray can also scatter from an electron, transferring part of its energy to the electron and thus becoming scattered as a lower energy gamma. There is an upper limit to the amount of energy that can be transferred to the electron by this mechanism, so that in every scattering event, a gamma ray remains after the interaction. Dependent upon the energy transferred, the residual gamma can be scattered in any direction, relative to the original direction, up to 180°. This has important implications on shielding, since gammas

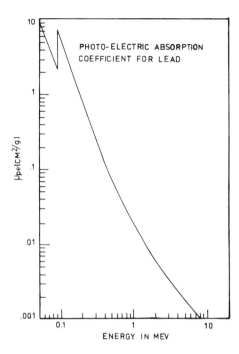

Figure 5.2. Photoelectric absorption for lead as a function of energy.

Figure 5.3. Scattering coefficient for lead as a function of energy.

can be scattered by the shielding itself, or by other nearby materials into areas shielded from a direct beam. Equation 5 gives the energy of the scattered gamma as a function of the angle of scattering. The interaction with matter is considerably less strongly dependent upon the energy of the gamma. This is also shown in Figure 5.3. Compton scattering is the primary mechanism of interaction for low atomic number elements, and decreases as the atomic number increases.

$$E_s = \frac{E}{1 + (E/m_e C^2)(1 - \cos\theta)} \tag{5}$$

Pair Production

If the energy of the gamma is greater than the energy needed to create an electron-positron pair, 1.02 MeV, then the gamma can interact with the absorbing medium to create the pair of electrons, an electron and a positron. The probability of this process increases as the energy increases. The excess energy over 1.02 MeV is shared by the two particles. The energy dependence of this process is also shown in Figure 5.4. The probability of the reaction increases with the atomic number of the absorber, approximately proportionally to $Z^2 + Z$.

Gamma rays can penetrate deeply into matter, in theory infinitely, since unless the gamma interacts with an electron it will go on unimpeded, just as in theory a rifle bullet fired into a forest can continue indefinitely unless it hits a tree (assuming no loss of energy for the bullet due to air friction). The intensity I of the original radiation at a depth x in an absorbing medium compared to the intensity of the radiation at the surface I_0 is

$$I = I_0 e^{-\mu x} \tag{6}$$

This equation is literally true if only gammas of the original energy are considered. If Compton scattering and the pair-production process are included, the decrease in the total number of gammas is less than that given by Equation 6, because of the scattered gammas from the Compton process, and the contribution of the annihilation gammas as the positron eventually is destroyed by interacting with a normal electron. The actual increase in the radiation levels is dependent on the gamma energy and the geometry of the scattering material.

If the total effect of all three mechanisms is considered at low and high energies, higher Z absorbers interact with gammas more strongly. However, between about 1 and 3 MeV, there is a relatively minor difference in the total absorption coefficient as a function of atomic number.

As can be noted, all of the mechanisms by which a gamma interacts with matter (except the very small number of instances in which the gamma ray interacts with a nucleus) result in the energy being transferred to an electron, so a gamma is considered to have the same low LET characteristics as do betas. At very low energies the linear energy transfer characteristics of electrons increase some. However, unless a beta emitter is taken into the body, most internal organs will not be affected by beta radiation, while gammas can penetrate deeply into the body and injure very sensitive organs such as the blood-forming tissues.

Neutrons

As noted earlier, neutron radiation is rarely encountered in most laboratories that use radioisotopes in research programs. However, it is useful to understand the difference in the mechanisms by which a neutron interacts with matter compared to those involving other

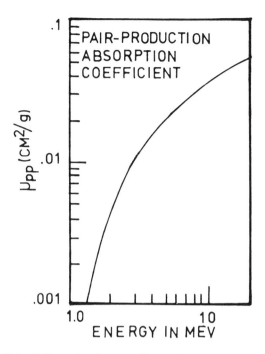

Figure 5.4. Pair-production coefficient for lead as a function of energy.

types of radiation since neutron radiation *may* make the matter with which it interacts radioactive. The neutron has no charge, but it does have about one fourth of the mass of an alpha particle, so that it does have an appreciable mass compared to the atoms with which it interacts.

An equation similar to Equation 6 gives the number of neutrons N, with an initial energy E, of an original number N_0 penetrating to a depth x in matter. Note that in both Equations 6 and 7, the units of x are usually converted into mg/cm^2 for the commonly tabulated values of μ and σ.

$$N = N_0 e^{-\sigma x} \qquad (7)$$

A neutron does not, as do electrons, alpha particles, and gammas, interact with the orbital electrons, but instead interacts directly with the nucleus. The neutron is not repelled by the positive charge on the nucleus as is the alpha because it has no charge. It is either scattered (elastically or inelastically) or captured by the nucleus. In a typical capture process usually several "capture" gammas with a total energy of about 8 MeV are emitted (it is somewhat less for some lighter nuclei). Thus, by this mechanism alone, the neutron could be considered more harmful than other radiations. Further, the nucleus in which it is captured may have been made radioactive, and the charge on the nucleus could change, so that the atom would no longer be chemically equivalent to its original form. In any event, the energy transferred to the participants in the interaction normally would be sufficient to break the chemical bonds.

Scattering events also typically would transfer enough energy to break the chemical bonds as long as the initial energy of the neutron is sufficiently high. As with Compton-scattered gammas, the scattered neutrons can be scattered into virtually any direction, so that the equivalent of Equation 7 for neutrons of all energies would, as for gammas, have to be modified to include a buildup factor.

No figure showing the systematics of the reaction mechanisms will be given here because the relationships are extremely complex, varying widely not only between elements, but between isotopes of the same element. In addition, the interaction probabilities can vary extremely rapidly as a function of energy, becoming very high at certain "resonant" energies and far less only a few electronvolts above or below the resonances. However, a few generalizations are possible. The probability of the capture process, excluding resonance effects, typically increases as the energy of the neutrons become lower, and for specific isotopes of certain elements, such as cadmium, gadolinium, samarium, and xenon, is extremely high at energies equivalent to thermal equilibrium (about 0.025 eV for room temperature matter). Energy can be lost rapidly by neutrons in scattering with low-Z materials, such as hydrogen, deuterium ($^2_1$H), helium, and carbon. Interposing a layer of water, paraffin, or graphite only a few inches thick, backed up by a thin layer ($\approx 1/32$ inch) of cadmium, in a beam of fast neutrons makes an effective shield for a beam of neutrons. Paraffin wax, in which boric acid has been mixed also makes an effective and cheap neutron shield (^{10}B has quite a respectable capture cross-section at thermal neutron energies).

Overall, the estimate of the danger of neutrons interacting with matter is estimated to be about ten times that of a gamma or electron, although this varies depending upon the energy of the neutrons, thermal neutrons are about two times as effective in causing atoms in tissue to be ionized, for example, as are betas and gammas while neutrons of 1 to 2 MeV energy are about 11 times more damaging.

5.2.2.6. Units of Exposure and Dose

There are two important concepts in measuring the relative impact of radiation on matter: one is the intensity of the radiation field, which represents a potential exposure problem, and the other is the actual energy deposited in matter, or the dose. Further, as far as human safety is concerned, the amount of energy absorbed in human tissue is more important than that absorbed in other types of matter. Each of these quantities have been assigned specific units in which they are measured.

The original unit of measuring radiation intensity was the roentgen, defined as the amount of X-ray radiation that would cause an ionization of 2.58×10^{-4} coulombs per kilogram of dry air at standard temperature and pressure. As noted, the dose or energy deposited in matter is more important, so another unit was defined, the rad, which was defined as the deposition of 0.01 joules per kilogram of matter. An exposure to 1 roentgen would result in an absorbed dose of 0.87 rads in air. A third unit, the rem, was subsequently defined which measured the equivalent dose, allowing for the relative effectiveness of the various types of radiation in causing biological damage. This originally was allowed for by multiplying the absorbed dose in rads by a relative biological effectiveness factor (RBE), to obtain a dose equivalent for tissue for the different varieties of radiation. Later, it was decided to restrict the term RBE to research applications and an equivalent multiplier called the quality factor, Q, was substituted. For practical purposes, RBE and Q factors are equivalent, although the latter is the one commonly used.

The terms rads and rems are still used by most American health physicists, and the current NRC regulations use these terms, as they do the curie and its derivative units. However, there are SI units for dose and also for activity. The equivalent units are.

1 Gray (Gy) ≡ 1 joule/kilogram = 100 rad = absorbed dose
1 Sievert (Sv) ≡ 1 Gray × Q × N = dose equivalent (N is a possible modifying factor, assigned a value of 1 at this time)
1 Sievert = 100 rem = dose equivalent

The quality factors for the various types of radiation are listed in Table 5.3.

This concludes this very brief discussion of some basic terms and concepts in radiation physics that will be employed in the next few sections. Many important points and significant features have been omitted that would be of importance primarily to professional health physicists, but are of less importance to those individuals that use radiation as a research tool to serve their more direct interests.

5.2.3. Licensing

This section will be restricted to a discussion of licensing of radioisotopes or by-product materials, rather than other types of applications such as a research reactor. It has been some time since any new applications for construction of a nuclear power plant in the United States has been approved, and the number of operating nongovernmental research reactors has been diminishing. Several of these research units are either in the process of terminating their license or going into an inactive status. At least some research reactors have closed rather than renew their license, as they must do periodically, because of excessive costs needed to meet the concerns of the public. The other major type of facility involved with radiation—laboratories using X-ray units—are usually regulated by state agencies, although the federal Food and Drug Administration sets standards for the construction of the machines and their applications. X-ray facilities will be discussed in a separate section.

Radioactive materials fall into two classes as far as regulation is concerned. Radioactive materials "yielded in or made radioactive by exposure to the radiation incident to the process of producing or utilizing special nuclear material" are regulated by the NRC, or by equivalent regulations in states with whom the NRC has entered into an agreement allowing for the states to act as the regulatory agency within their borders. Radioactive materials that are naturally radioactive or produced by means such as a cyclotron are regulated by the states in most cases.

The licensing of by-product material is regulated under 10 CFR Part 30 or 33. Licenses are issued to "persons," a term which may refer to an individual but may also mean organizations groups of persons, associations, etc. It is possible for individuals within an organization to have separate licenses, although it is more likely that instead of several individuals having separate licenses, an institution will apply for and be granted a license covering the entire organization, if they can show that they can ensure that the individual users will conform to the terms of the license and regulations governing the use of radioactive materials.

For most types of licenses of interest to research laboratories, the NRC has delegated licensing authority to five regional offices in Pennsylvania, Georgia, Illinois, Texas, and California. The current addresses of these regional offices can be obtained by writing to:

Director
Office of Nuclear Material Safety and Safeguards
U.S. Nuclear Regulatory Commission
Washington, DC 20555

Table 5.3.   Quality Factors

Type of Radiation	Q
Betas, gammas, X-rays	1
Alphas	20
Thermal neutrons	2
Fast neutrons ($\approx$ 1 MeV)	20
Neutrons (unspecified energy)	10

Not all uses of radioisotopes require securing a license. There are many commercial products, such as watch dials and other self-luminous applications and some types of smoke detectors, that contain very small quantities of radioactive materials which the owner obviously does not require a license to own. However, the most significant class of exemptions is the "exempt quantities" listed in Schedule B, 10 CFR Section 30.71. Schedule A lists exempt concentrations in gas, liquids, and solids. In Table 5.4, the units are in microcuries. To convert to becquerels, multiply the number given in microcuries by 37,000.

Any by-product material that is not listed in Table 5.4, other than alpha-emitting by-product material, has an exempt quantity of 0.1 µCi or 3700 Bq.

Most users of radioisotopes would find it necessary to use more than the exempt quantities in Table 5.4 and should apply for a license. This is done through NRC Form 113, which can be obtained from the NRC office in the local region. If the activity planned has the potential for affecting the quality of the environment, the NRC will weigh the benefits against the potential environmental effects in deciding whether to issue the license. For most research-related uses of radioisotopes, environmental considerations will not usually apply, although where the isotopes will be used in the field, outside of a typical laboratory, the conditions and restrictions on their use to ensure that there will be no meaningful release into the environment will need to be fully included in the application.

There are three basic conditions that the NRC expects the applicant to meet in their application. In this context, "applicant" is used in the same sense as the word "person," which can be an individual or an organization, as noted earlier.

1. The purpose of the application is for a use authorized by the Act. Legitimate basic and applied research programs in the physical and life sciences, medicine, and engineering are acceptable programs.
2. The applicant's proposed equipment and facilities are satisfactory in terms of protecting the health of the employees and the general public, and being able to minimize the risk of danger to persons and property. The laboratories in which the radioisotopes are to be used need to be in good repair and contain equipment suitable for use with radioisotopes. Depending upon the level of radioactivity to be used and the scale of the work program, this may mandate the availability of hoods designed for radioisotope use. It could require specific areas designated and restricted for isotope use only, or the level of use and the amounts of activity may make it feasible to perform the research on an open bench in a laboratory. In any event, it must be shown in the application that the level of facilities and equipment must be adequate for the proposed uses of radiation.
3. The applicant must be suitably trained and experienced so as to be qualified to use the material for the purpose requested in a way that will protect the health of individuals and minimize danger to life and property. The experience and training must be documented in the application.

Table 5.4.   Exempt Quantities

By-Product Material	μCi	By-Product Material	μCi
Antimony 124	10	Indium 113m	100
Antimony 125	10	Indium 114m	10
Arsenic 73	100	Indium 115m	100
Arsenic 74	10	Indium 115	10
Arsenic 76	10	Iodine 125	1
Arsenic 77	100	Iodine 126	1
Barium 131	10	Iodine 129	0.1
Barium 133	10	Iodine 131	1
Barium 140	10	Iodine 132	10
Bismuth 210	1	Iodine 133	1
Bromine 82	10	Iodine 134	10
Cadmium 109	10	Iodine 135	10
Cadmium 115m	10	Iridium 192	10
Cadmium 115	100	Iridium 194	100
Calcium 45	10	Iron 55	100
Calcium 47	10	Iron 59	10
Carbon 14	100	Krypton 85	100
Cerium 141	100	Krypton 87	10
Cerium 143	100	Lanthanum 140	10
Cerium 144	1	Lutecium 177	100
Cesium 131	1,000	Manganese 52	10
Cesium 134m	100	Manganese 54	10
Cesium 134	1	Manganese 56	10
Cesium 135	10	Mercury 197m	100
Cesium 136	10	Mercury 197	100
Cesium 137	10	Mercury 203	10
Chlorine 36	10	Molybdenum 99	100
Chlorine 38	10	Neodymium 147	100
Chromium 51	1,000	Neodymium 149	100
Cobalt 58m	10	Nickel 59	100
Cobalt 58	10	Nickel 63	10
Cobalt 60	1	Nickel 65	100
Copper 64	100	Niobium 93m	10
Dysprosium 165	10	Niobium 95	10
Dysprosium 166	100	Niobium 97	10
Erbium 169	100	Osmium 185	10
Erbium 171	100	Osmium 191m	100
Europium 152 (9.2 hr)	100	Osmium 191	100
Europium 152 (13 yr)	1	Osmium 193	100
Europium 154	1	Palladium 103	100
Europium 155	10	Palladium 109	100
Fluorine 18	1000	Phosphorus 32	10
Gadolinium 153	10	Platinum 191	100
Gadolinium 159	100	Platinum 193m	100
Gallium 72	10	Platinum 193	100
Germanium 71	100	Platinum 197m	100
Gold 198	100	Platinum 197	100
Gold 199	100	Polonium 210	0.1
Hafnium 181	10	Potassium 42	10
Holmium 166	100	Praseodymium 142	100
Hydrogen 3	1000	Praseodymium 143	100

Table 5.4.   Exempt Quantities (continued)

By-Product Material	μCi	By-Product Material	μCi
Promethium 147	10	Tellurium 125m	10
Promethium 149	10	Tellurium 127m	10
Rhenium 186	100	Tellurium 127	100
Rhenium 188	100	Tellurium 129m	10
Rhodium 103m	100	Tellurium 129	100
Rhodium 105	100	Tellurium 131m	10
Rubidium 86	10	Tellurium 132	10
Rubidium 87	10	Terbium 160	10
Ruthenium 97	100	Thallium 200	100
Ruthenium 103	10	Thallium 201	100
Ruthenium 105	10	Thallium 202	100
Ruthenium 106	1	Thallium 204	10
Samarium 151	10	Thulium 170	10
Samarium 153	100	Thulium 171	10
Scandium 46	10	Tin 113	10
Scandium 47	100	Tin 125	10
Scandium 48	10	Tungsten 181	10
Selenium 75	10	Tungsten 185	10
Silicon 31	100	Tungsten 187	100
Silver 105	10	Vanadium 48	10
Silver 110m	1	Xenon 131m	1000
Silver 111	100	Xenon 133	100
Sodium 24	10	Xenon 135	100
Strontium 85	10	Ytterbium 175	100
Strontium 89	1	Yttrium 90	10
Strontium 90	0.1	Yttrium 91	10
Strontium 91	10	Yttrium 92	100
Strontium 92	10	Yttrium 93	100
Sulfur 35	100	Zinc 65	10
Tantalum 182	10	Zinc 69m	100
Technetium 96	10	Zinc 69	1000
Technetium 97m	100	Zirconium 93	10
Technetium 97	100	Zirconium 95	10
Technetium 99m	100	Zirconium 97	10
Technetium 99	10		

Before granting the license, the NRC may require additional information, or may require the application to be amended. The license is issued to a specific licensee and cannot be transferred without specific written approval of the NRC. The radioisotopes identified in the license can be used only for the purposes authorized under the license, at the locations specified in the license. If the licensee wishes to change the isotopes permitted to be used, to significantly modify the program in which they are used, or to change the locations where they are to be used, the license must be amended. This typically takes a substantial length of time, 1 to 3 months or even more not being unusual. Consequently, most substantial users of radioisotopes often apply for a "broad" license under 10 CFR Part 33.

Under the terms of a broad license, the application usually covers a request to use radioisotopes with atomic numbers from 3 to 83, with individual limits on the quantities held of specific isotopes, and an overall limit of the total quantity of all isotopes held at once.

In addition, there should be specific identification of sealed sources held separately by the applicant on the license. The purposes still must be explained and the locations still must be identified. However, under the terms of a broad license, an internal procedure can be established to review individual uses, personnel, and spaces in which the materials can be used and to monitor the operations. This will require the formation of a radiation safety committee and the naming of a radiation safety officer. This infrastructure must be described in the license application, and the credentials of the individual members of the committee and the safety officer included as part of the application. To obtain a license of broad scope, applicants must prove that they have the personnel and administrative structure to make sure that the radioisotopes will be used safely and in compliance with regulations.

The license will be granted for a specific period, and the ending date will be written into the license. If the licensee wishes to renew the license as the end of the license period approaches, the applicant must be sure to submit a renewal request at least 30 days before the expiration date of the license. If this deadline is met, the original license will remain in force until the NRC acts on the request.

While the license is in effect, the NRC has the right to make inspections of the facility, the by-product material, and the areas where the by-product material is in use or stored. These inspections have to be at reasonable hours, but they normally are unannounced. The inspector also will normally ask to see records of such items as surveys, personnel exposure records, transfers and receipts of radioactive materials, waste disposal records, instrument calibrations, radiation safety committee minutes, documentation of any committee actions, and any other records relevant to compliance with the terms of the license and compliance with other parts of 10 CFR, such as 19 and 20. Failure to be in compliance can result in citations of various levels or of financial penalties. Enforcement will be discussed further later. The NRC can require tests to be done to show that the facility is being operated properly, such as asking for tests of the instruments used in monitoring the radiation levels, or ask the licensee to show that security of radioactive material in the laboratory areas is not compromised.

Under Section 30.51, records of all transfers, receipts, and disposal of radioactive materials normally must be kept for at least 2 years after transfer or disposal of a radioactive material, or in some cases until the NRC authorizes the termination of the need to keep the records. There are other record keeping requirements in other parts of Title 10.

If for any reason there is a desire to terminate the license on or before the expiration date, under 10 CFR 30.36 there are procedures that must be followed.

1. Terminate the use of by-product material.
2. Remove radioactive contamination to the extent practicable (normally to background level, see 5 below).
3. Properly dispose of by-product material.
4. Submit a completed Form NRC-314.
5. Submit a radiation survey documenting the absence of radioactive contamination, or the levels of residual contamination. In the latter case, an effort will be required to eliminate the contamination.
   a. The instruments used for the survey must be specified and then certified to be properly calibrated and tested.
   b. The radiation levels in the survey must be reported as follows:
      (1) Beta and gamma levels in microrads per hour at 1 cm from the surface, and gamma levels at 1 meter from the surface
      (2) Levels of activity in microcuries per 100 cm^2 of fixed and removable surface contamination
      (3) Microcuries per mL in any water
      (4) Picocuries per gram in contaminated solids and soils

6. If the facility is found to be uncontaminated, the licensee shall certify that no detectable radioactive contamination has been found. If the information provided is found to be sufficient, the NRC will notify the licensee that the license is terminated.

7. If the facility is contaminated, the license will continue after the normal termination date. However, the use of by-product materials will be restricted to the decontamination program and related activities. The licensee must submit a decontamination plan for the facility. They must continue to control entry into restricted areas until they are suitable for unrestricted use, and the licensee is notified in writing that the license is terminated.

In principle, the NRC has the right to modify, suspend, or revoke a license for a facility that is being operated improperly or if the facility were to submit false information to the NRC. If the failure to comply with the requirements of the license and other requirements for safely operating a facility can be shown to be willful or if the public interest, health, or safety can be shown to demand it, the modification, suspension, or revocation can be done without institution of proceedings, allowing the licensee an opportunity to demonstrate or achieve compliance.

Normally, an inspection will be followed up with a written report by the inspector in which any compliance problems will be identified. These may be minimal, serious (which would require immediate abatement), or between these two extremes. The facility can (1) appeal the findings and attempt to show that they were complying with the regulations or that the violation was less serious than the citation described or (2) accept the findings. Unless the facility can show compliance, it must show how they will bring the facility into compliance within a reasonable period.

In recent years, there have been increasing numbers of occasions when the NRC has imposed substantial financial penalties on research facilities, including academic institutions, as they are entitled to do under Section 30.63 for violations that are sufficiently severe. Further, a few years ago, one city filed 179 criminal charges against a major university and several of its faculty members for failure to comply with radiation safety standards. Many individual violations were relatively minor, but apparently the city attorney thought he had a substantial case for a pattern of failure to comply with the terms of the license and the regulations.

The use of radioisotopes in research is continuing to increase, while the public concern about the safety of radiation continues unabated. It behooves all licensees to follow all regulations scrupulously, not only to ensure safety, but also to avoid aggravating the concerns of the public unnecessarily.

5.2.3.1. Radiation Safety Committees

The primary function of the radiation safety committee (RSC), which is required under 10 CFR 33, is to monitor the performance of the users of ionizing radiation in a facility. It is, in effect, a local surrogate of the NRC or the equivalent state agency in an agreement state. Usually, it is the ultimate local authority in radiation matters. In this one area at least, it is assigned more responsibility than the usual senior administrative officials. It is an operational committee, charged with an important managerial role in the use of ionizing radiation within the organization. Due to this power, the NRC holds the committee responsible for compliance and will cite the committee for failure to provide appropriate oversight if the radiation users or radiation safety personnel under its supervision fail to ensure compliance with the regulations.

In addition to the responsibility of the RSC to ensure compliance with the provisions of the by-product license and the other regulatory requirements of Title 10 CFR, it also must establish internal policies and procedures to guide those wishing to use radiation and to provide the internal operational structure in which this is done. The committee has other duties as well, which will be discussed after the makeup of the committee is considered.

The membership of an RSC should be carefully selected. It would be highly desirable to select much of the membership from among the active users of radiation within the organization and across the major areas or disciplines represented among the users. Each prospective member should be scrutinized very carefully. A RSC must enforce regulations set by one of the stronger regulatory agencies, and it must be willing to accept the delegated authority. Individuals on the committee must be willing, if necessary, to establish policies that many users may feel are too restrictive. As active users themselves, they have a better chance of achieving compliance, if the other users realize that the members of the RSC have accepted imposition of these same policies on their own activities. The members of the committee should have a reputation for objectivity, fairness, and professional credibility. A prima donna has no place on such a committee.

As professional scientists in their own right, they also will understand the impact of a given procedure or policy on laboratory operations and can often find legitimate ways to develop effective policies and procedures that are less burdensome on the users to carry out than would otherwise be the case.

The radiation safety officer (RSO) of the organization must be part of the RSC and is a person who must maintain a current awareness of the rules and regulations required by the NRC and of radiation safety principles. This individual will serve to carry out the policies of the committee, and should be the individual to do the direct day-to-day monitoring of the operations of the laboratories using radioactive material. The decisions of the committee can be burdensome not only on the users, but, without the ideas of the RSO, can be equally burdensome for this individual to carry out.

The working relationship between the RSO and the RSC is extremely important. No committee can effectively administer a program of any size on a daily basis. It must delegate some of its authority to a person, such as the RSO, or to an alternate agency such as a safety and health department charged with the daily administration of the area of responsibility assigned to an operational committee. However, especially when the RSO is a dynamic, effective person there is a tendency to defer to this person and to abrogate some of the committee's oversight responsibility. Both the RSC and the RSO should guard against this possibility. The RSO should have a voice and an influential one in the committee's deliberations, but should not be allowed to dictate policies independently.

The membership need not be limited to the persons already defined. A relatively new NRC requirement is that a senior management representative must be an *ex officio* member of the committee and no official meeting can be held should the senior management representative not be present. This individual cannot veto the actions of the committee, but by being present guarantees that higher management is aware of the actions of the committee. The head of the health and safety department, if different from the radiation safety officer, may be a member because this individual would bring in a wider perspective than would the RSO alone on the implications of some issues brought before the committee. Some large organizations may wish to have a representative of the organization's legal department as a member. Some may wish to have a representative of the public relations area as a member, especially if the facility is in an area in which there has been vigorous public opposition to the use of ionizing radiation. Some may wish to include a layperson, if not as a voting member, then perhaps as an observer. The membership should not become too large, however, so that it will be practical to set up meetings without too much concern for having a quorum. Committees that are too large also tend to be less efficient, because of the time required for all the members to participate in discussions. On the other hand, each major scientific discipline using radiation should be represented. A reasonable size might be between 9 and 15 members, with a quorum established at between 5 and 8 members.

It is essential for the chair of the committee to be someone with prior experience with radiation, but it is also highly desirable if the chair is an individual with administrative credentials. Such a person will normally ensure that committee meetings will be conducted efficiently, but if the administrative experience is at a level carrying budgetary and personnel responsibilities, the chair will bring still another dimension to the committee. Some actions of the committee may carry cost or manpower implications; an individual with managerial experience will recognize and perhaps have a feel for the feasibility of accommodating these requirements.

Besides the monitoring of existing programs, establishing policies, and providing guidance to radiation safety personnel, there are at least four other important functions that the committee must perform. The first of these is to perform the same function as the NRC in authorizing new participants to use radiation or radioactive materials. Basically, the same information that the NRC requires for new applicants for a license should be required when a new facility is involved. The adequacy of the facility, the purpose of the program for which the use of radiation is involved, and the qualifications of the users should all be reviewed. At academic institutions especially, there is a considerable turnover in users, represented by graduate students, postdoctoral research associates, and even faculty. Often individuals come from other facilities where internal practices may differ from local practices. To ensure that all users are familiar with not only the basic principles of radiation safety, but also with local internal procedures, a simple written test, administered as part of the authorization procedure, is an effective and efficient means of documenting that the prospective users have familiarized themselves with the information. To avoid setting standards on who should take the test, it should be administered universally. Some faculty may object, but it serves an important legal point. A passed quiz shows unequivocally that the individual is familiar with the risks and the requirements associated with the use of radiation at the facility.

An internal authorization should be issued to an individual. Others may be added to the authorization, but one person should be designated as the ultimately responsible person, responsible for compliance with applicable safety and legal standards related to the use of radiation under the authorization. In laboratories that involve multiple users, it may be necessary to formally identify a senior authorized user to provide the additional authority to this individual.

The second additional function is to carefully review research or "new experiments," substantially different in the application of radioactive materials or radiation envisioned from work previously performed under the license. This role is easy to play when it is part of a new request for an authorization, but when an ongoing operation initiates a new direction in their program, it will be necessary for the committee to make it clear that the user must address the question to himself, "Is this application covered under the scope of work previously reviewed by the committee in my application?" If the answer is "no," or "possibly not" then the responsible individual should ask for a review by the committee. The need to do this must be explicitly included in the internal policies administered by the committee. The RSC then must consider the proposed program in the same context as the institution's application to the NRC. Is the purpose of the work an approved purpose? This question must be answered positively in the context of the NRC facility license. Are the facilities adequate to allow the work to be done safely? Are the persons qualified because of training or experience to carry out the proposed research program safely? Incidentally, it is *not* within the purview of the committee's responsibility to judge the validity or worth of the research program, but only if the proposed research can be done safely according to radiation safety and health standards. Of course, frivolous research is unacceptable.

In the past the use of proven research technology was sufficient to approve most research and routine experiments did not receive the scrutiny that new experiments did. In the last few years, the NRC has required the investigators to formally review even standard procedures for possible hazards, to establish procedures to prevent these potential hazards from occurring, and to develop a response protocol. Worst case failures must be reviewed. It is enlightening to see the results of these analyses. There is far more potential for failure than most would anticipate. The committee must review and approve of these hazard analyses.

The fourth function not previously discussed is the role of the RSC as a disciplinary body. Occasions will arise when individual users will be found to not be in full compliance with acceptable standards. Often this will be done by the RSO in his periodic inspections, but many will be reported by the users themselves. The NRC will expect these situations to be evaluated and appropriate actions taken, which can include disciplinary measures. Not all violations are equally serious. Categories of violations should be established by the RSC to guide the RSO and the users. Faulty record keeping is not as serious as poor control over by-product material usage. Allowing material to be lost or radioactive material to escape into the environment is a serious violation. If the loss or discharge is due to an unforeseen accident, it is less serious than if the cause is negligence. However, a continuing pattern of minor violations may show carelessness or lack of concern about compliance with the standards that could eventually lead to a more serious incident. Such a pattern should be considered as a serious violation. Possible penalties should be listed in the internal policies and guidelines issued by the RSC.

Every case in which noncompliance is discovered or reported should be carefully investigated by the RSO and a report made to the RSC. The person responsible for the noncompliance and the responsible individual from the facility (if not the same person) should be invited to meet with the committee and present their sides of the issue if they wish. Few will normally contest minor citations, but most will contest a citation that could affect their ability to use radioactive materials. After hearing both sides, the committee should take the appropriate measures. Issues are rarely black or white, and the penalties or corrective actions should be adapted to the circumstances. An initial minor violation, for example, might elicit no more than a cautionary letter from the committee. However, a series of minor violations within a short interval probably should result in a mandatory cessation of usage of radioisotopes until the user can show a willingness and capability to comply with acceptable practice. A proven serious accidental violation should result in an immediate cessation of operations until procedures can be adopted to prevent future recurrence of the problem. A serious violation due to willful noncompliance should result in a mandatory cessation of the use of radioisotopes for a substantial length of time or even permanently. The elimination of the right to use radioisotopes is a very serious penalty, since the user's research program may depend upon this capability. In an academic institution, even a relatively short hiatus in a research program could result in the loss of a research grant or failure to get tenure. As a result, a permanent or extended loss of the right to use radioactive material should not be imposed lightly, but if the user shows by action and attitude that future violations are likely to occur, the committee may have no practical alternative except to do so to protect the rest of the organization's users from the loss of the institution's license or the imposition of a substantial fine by the NRC. They must be willing to accept the responsibility, unpleasant though it may be. If the users believe that the RSC is willing to be firm and fair, they will be more likely to comply with the required procedures.

The role of the RSO could be construed as the enforcement arm of the RSC and the radiation safety office (which may be part of a larger organization). If this were the case, the RSO would be the equivalent of an NRC inspector. However, as with many other persons working in safety and health programs who have enforcement duties, their primary function is service to the users. In later sections the other duties will make this clear.

5.2.4. Radiation Protection, Discussion, and Definitions

Many basic terms were defined in some detail in the several subsections of Section 5.1. However, several additional concepts will be introduced in the following sections, and some additional terms need to be defined.

The original definitions of dose units primarily were employed by users of radioactive materials and other applications of ionizing radiation for external exposure. Concerns relating to internal exposure generally were covered by establishing maximum permissible concentrations of radionuclides in air and water, in terms of the workplace and in terms of the public, in Appendix B to 10 CFR 20, Tables I and II, respectively. Protection was provided by considering the amount of radiation given to the most critical organ by the intake of specific radioisotopes and their physical or chemical form. The exposure limits to the whole body were established by the organs assigned the lowest dose limits, the bone marrow, gonads, and lens of the eye. The reasons for establishing these organs as the most sensitive to radiation were concerns about leukemia, hereditary effects, and cataracts, respectively.

The amounts of a radionuclide in organs, or the "burden," were calculated based on a constant exposure rate, maintained for a period sufficiently long so that an equilibrium would be established between the intake of the material and the effective elimination rate. The effective elimination rate, or effective half-life, is a combination of the radioactive half life and the biological half-life based on the rate at which the material would be eliminated from the body. The relationship is given by the following equation:

$$\frac{1}{T_{eff}} = \frac{1}{T_r} + \frac{1}{T_b} \tag{8}$$

where T_{eff} = Effective half-life
T_r = Radiological half-life
T_b = Biological half-life

The maximum permissible concentrations (MPC) were those that correspond to an organ burden that would cause the annual dose limits to be attained. Control measures, therefore, were designed to maintain the concentrations below the MPCs.

The current 10 CFR Part 20 that went into effect January 1, 1994, sets standards for protection for users of radioisotopes and requires combining internal and external exposures. The revision contains several other changes in Part 20 based on many of the recommendations contained in the International Commission on Radiological Protection (ICRP) Publications 26, 30, and 32. The remainder of this section and some succeeding sections will discuss the new regulations and will draw attention to significant changes.

5.2.4.1. Selected Definitions

In order to discuss the current Part 20 and the proposed Part 20, some terms introduced into the latter from ICRP 26 and 30 are needed.

The first six main definitions are related to dose terms:

1. Dose equivalent means the product of absorbed dose, quality factor, and all other necessary modifying factors at the location of interest in tissue.
2. External dose means that portion of the dose received from radiation sources from outside the body.
   a. Deep dose equivalent (H_d) applies to the external whole-body exposure and is taken as the dose equivalent at a tissue depth of 1 cm.
   b. Eye dose equivalent (H_e) applies to the external exposure of the lens of the eye and is taken as the dose equivalent at a tissue depth of 0.3 cm.
   c. Shallow dose equivalent (H_s) applies to the external exposure of the skin or an extremity and is taken as the dose equivalent at a tissue depth of 0.007 cm.
3. Internal dose is that portion of the dose equivalent received from radioactive material taken into the body.
   a. Committed dose equivalent ($H_{c,T}$) means the dose equivalent to organs or tissues of reference (T) that will be received from an intake of radioactive material by an individual during the 50-year period following the intake ($H_{50,T}$).
   b. Effective dose equivalent (H_E) is the sum of the products of the dose equivalent (H_T) to the organ or tissue (T) and the weighting factors (w_T) applicable to each of the body organs or tissues that are irradiated.

$$H_E = \Sigma W_T H_T \tag{9}$$

   c. Committed effective dose equivalent is the sum of the products of the weighting factors applicable to each of the body organs or tissues that are irradiated and the committed dose equivalent.
   d. Collective effective dose equivalent is the sum of the products of the individual weighting dose equivalents received by a specified population from exposure to the given source of radiation.
4. W_T is the weighting factor assigned to an organ that is proportional to the estimate of risk to that organ relative to the estimate of risk per unit of dose for a uniform whole body exposure. In Table 5.5, except for the gonads for which the risk is based on serious hereditary effects in the first two generations of the exposed person, the weighting factors are based on the risk of inducing cancer in the organ.
5. Occupational dose means the dose received by an individual in a restricted area or during employment in which the individual's assigned duties involve exposure to radiation and to radioactive material from licensed and unlicensed sources of radiation, whether in the possession of the licensee or other person. This dose does not include exposure from natural background, medical exposure as a patient, participation in a research program, or as a member of the public.
6. Public dose is an exposure of a member of the public to radiation or to the release of radioactive material, or to another source, either in a licensee's controlled area or in unrestricted areas. This does not include medically related exposures.

The next six definitions relate to dose control factors:

7. ALARA is an acronym that stands for "as low as reasonably achievable." It is a policy that involves making every reasonable effort to reducing the exposures of

Table 5.5.   Definitions: Weighting Risk Coefficient

Organ or Tissue	Weighting Factor (w_T)	Risk Coefficient (per rem)	Probability (per rem)
Gonads	0.25	4×10^{-5}	1 in 25,000
Breast	0.15	2.5×10^{-5}	1 in 40,000
Red bone marrow	0.12	2×10^{-5}	1 in 50,000
Lung	0.12	2×10^{-5}	1 in 50,000
Thyroid	0.03	5×10^{-6}	1 in 200,000
Bone surfaces	0.03	5×10^{-6}	1 in 200,000
Any remaining organs or tissues receiving the highest dose at a relative sensitivity of 0.06 each	0.30	5×10^{-5}	1 in 20,000
Total	1.0	1.65×10^{-4}	1 in 6,000

personnel below the dose limits, by considering the purpose of the utilization of radiation, technological feasibility, economics, benefits, etc. It does not mean taking extreme measures, where the benefit would not be cost effective.

8. Annual limit of intake (ALI) means the derived limit for the amount of radioactive material taken into the body of an adult worker by inhalation or ingestion in 1 year. It is the smaller of (1) the value of the intake of a given radionuclide in 1 year by a reference man that would result in a committed dose equivalent of 5 rems (0.05 Sv); (2) a committed dose equivalent of 50 rem (0.5 Sv) to an organ or tissue, i.e., the quantity of radioactive material such that,

$$0.05 \text{ Sv} = 5 \text{ rem} = \Sigma \, H_T W_T \tag{10}$$

or
$$0.5 \text{ Sv} = H_{c,T} \tag{10a}$$

9. Derived air concentration (DAC) means the concentration of a given radionuclide in air which, if breathed by a reference man for a working year of 2000 hours under conditions of light activity (corresponding to an inhalation rate of 1.2 m^3 air per hour), results in an inhalation of 1 ALI. These are comparable to the MPCs in the older Part 20.

10. Dose limits means the permissible upper bounds of radiation doses. These are usually set for a calendar year. They apply to the dose equivalent received during the set interval, the committed effective dose equivalent resulting from the intake of radioactive material during the interval or the effective dose equivalent received in 1 year. The external dose and the internal dose must be combined so as not to exceed the permissible limits. The following equation can be used to compute the relative amounts of each, for the annual intake I_j of nuclide j:

$$\frac{H}{5} + \frac{\Sigma \, _j I_j}{(ALI)_j} \leq 1 \tag{11}$$

Two terms are used to describe two different classes of effects of radiation.

11. Stochastic effects refers to health effects that occur randomly, so that the probability (generally assumed to be linear, without a threshold) of an effect occurring, such as the induction of cancer, is a function of the dose rather than the severity of the effect.
12. Nonstochastic effects are health effects for which the severity depends upon the dose, and for which there is probably a threshold.

With these definitions in mind, the following section presents selected parts of the current 10 CFR Part 20, in effect as of January 1, 1994. Holders of NRC licenses must have formal programs to ensure that all programs covered by the regulations and all individuals working with radioactive materials comply with the revised standard.

5.2.4.2. Selected Radiation Protection Standards from 10 CFR Part 20

The following section contains the sense of the revised Part 20 regulations, paraphrased in some instances for clarity and brevity. A few comments are added in some sections, where they were felt to be helpful. Most of the areas covered by the standard will be gone into in more detail in later sections describing practical implementation of a program complying with the standard.

5.2.4.2.1. Occupational Limits for Adult Employees

The dose limits are the levels not to be exceeded, but they should not be considered as a goal. The revised standard requires the licensee to use to the extent practicable procedures and engineering controls based upon sound radiation protection principles to achieve occupational doses and doses to members of the public that are ALARA.

Whole body, head, trunk, arm above elbow, and leg above knee

The more limiting of:

- 5 rems per year (0.05 Sv per year)—includes summation of (external) deep dose equivalent and (internal) committed effective dose equivalent, or
- The sum of the deep-dose equivalent (the dose equivalent at a depth of 1 cm in tissue) and the committed dose equivalent to any individual organ or tissue other than the lens of the eye being equal to 50 rems (0.5 Sv).

Without any intake of radioactive materials, the limits above correspond to the external exposure limits of the previous standard.

Eyes, Skin, Arms Below Elbows, Legs Below Knees

- 15 rems per year (0.15 Sv year) to lens of eye
- A shallow dose equivalent (at a tissue depth of 0.007 cm over a 1 cm^2 area) of 50 rems (0.5 Sv) to the lower arms, hands, lower legs, and feet.

The assigned deep-dose equivalent and shallow-dose equivalent must be for the part of the body receiving the highest exposures. These data can be inferred from surveys or other measurements if direct data are not available.

The DAC and ALI values are given in Table 1, Appendix B, of Part 20 and can be used in the determination of an individual's internal dose and to demonstrate compliance with required occupational dose limits.

The limits specified are for an individual. The total exposures from work for different employers cannot exceed these limits.

Significant changes are that the previous standard did not require addition of external and internal doses; the basic interval over which radiation is measured has been extended from a quarter to a year; the 5(N-18) cumulative dose limit has been deleted; the higher limits to eye exposure and the extremities reflects more information on the sensitivity to radiation for these areas. In addition, the revised standard contains a new provision for planned special exposures for limited higher doses to an individual in exceptional cases when other alternatives are unavailable or impractical. The licensee must specifically authorize the planned special exposure in writing before the event. The individual who will receive the exposure must be informed in advance of (1) the purpose of the operation, (2) the estimated doses and potential risks involved in the operation, and (3) measures to be taken to comply with the principles of ALARA, taking into account other risks that might be involved. The licensee must also determine in advance the cumulative lifetime dose of the individual participating in the exercise. With these steps completed, the participating individual must not be caused to receive a dose from all planned special exposures and all other doses in excess of

- The numerical values of the allowable dose limits in any year, and
- Five times the annual dose limits during the individual's lifetime.

The licensee must retain records of all planned special exposures. In addition, the licensee must provide exposed individuals with a report of the dose received and submit a report to the regional director of the NRC within 30 days of the event.

It is not necessary in every case to sum the external and internal doses if the licensee can show that the internal dose does not contribute significantly. If, for example, the only intake of radionuclides is by inhalation, the total effective dose equivalent is not exceeded if the deep-dose equivalent divided by the total effective dose equivalent, plus an estimate of the internal dose as determined by one of three procedures stipulated in the regulation does not exceed 1, the internal dose need not be added to the external dose. Similarly, unless the amount of radioactivity ingested is more than 10% of the applicable ALI, it need not be included in the total dose equivalent. Most laboratories using radioactive materials at reasonable levels under normal conditions will find that they need only consider external exposures as they have been doing.

Only individuals likely to receive within 1 year more than 10% of the allowable dose limits are required to be monitored by the licensee. However, unless the dose is monitored, it is difficult to establish with certainty that an active user of radioisotopes may not have exceeded the 10% limit. Many licensees do monitor most users of radioactive materials by providing personnel dosimeters to measure external exposures, excluding those who only work with weak beta emitters. In order to monitor internal exposures, the licensee can perform measurements of (1) concentrations of radioactive materials in the air in the workplace, (2) quantities of radionuclides in the body, (3) quantities of radionuclides excreted from the body, or (4) combinations of these measurements.

The personnel monitors normally used for individual monitoring must be processed by a dosimetry processor holding personnel dosimetry accreditation from the National Voluntary Laboratory Accreditation Program (NVLAP) of the National Institute of Standards and Technology for the type of radiation or radiations included in the NVLAP program that most closely approximates the exposure characteristics for the individual wearing the dosimeter.

5.2.4.2.2. Occupational Limits for Minors (Under 18) and to an Embryo/Fetus

The annual occupational dose limits for minors are 10% of the limits for adult workers.

The limits to a fetus are based on an exposure over the entire 9 months of the pregnancy of a declared pregnant woman. If the woman chooses not to declare her pregnancy to her employer, the licensee may be in a legally difficult position. Antidiscrimination laws prevent an employer from discriminating against a woman, but the normal work regimen may not be compatible with the NRC regulation. If the woman were to make an issue of a job change based on an obvious but undeclared pregnancy, there could be problems. It will be assumed here that nearly all women in the workplace would wish to limit the exposure to their unborn child as soon as possible and declare their pregnancy when they are sure. This could still cause up to an approximately 2-month period of higher exposure than desirable due to uncertainty in the early stages of pregnancy.

The responsibility of the licensee is to ensure that the total dose on the fetus during the entire pregnancy does not exceed 0.5 rem (5 mSv), and to make efforts to avoid substantial variations in a uniform monthly exposure rate. The dose to the fetus shall be taken as the sum of the deep-dose equivalent to the pregnant woman and the dose to the fetus from radionuclides in the fetus and in the declared pregnant woman.

If the dose to the embryo is found to have exceeded the 0.5 rem limit or is within 0.05 rem of it by the time the woman declares her pregnancy, the licensee will be considered to be in compliance if the additional exposure during the remainder of the pregnancy does not exceed 0.05 rem (0.5 mSv) during the remainder of the pregnancy.

5.2.4.2.3. Dose Limits for Individual Members of the Public

The licensee must conduct operations in such a way as to ensure (1) that the total effective dose to individual members of the public from the licensed operation does not exceed 100 m rem in 1 year, exclusive of the dose contribution from the licensee's disposal of radioactive materials into sanitary sewerage, and (2) the dose in any unrestricted area from external sources does not exceed 2 mrem in any 1 hour. The licensee must be able to demonstrate that if an individual were continuously present in an unrestricted area, they would receive a dose of 2 mrem in 1 hour and 50 mrem in 1 year. The licensee must make surveys, measurements, and/or calculations to prove that the limits would not be exceeded for an individual likely to receive the highest dose from the licensed operations.

There are also EPA regulations governing releases of radioactive materials into the environment that will be treated separately.

5.2.4.2.4. Surveys and Monitoring

Monitoring requirements have been covered in the section on occupational exposures for adult employees. In addition to personnel monitoring, surveys are required. The surveys can be done with portable instruments or samples can be obtained and tested in laboratory instruments. These surveys are used to evaluate (1) the extent of radiation levels, (2) concentrations or quantities of radioactive materials, and (3) the potential radiological hazards that could be present. The equipment used for the surveys must be calibrated periodically for the radiation measured. Normally this is done at least annually. A description of personnel monitoring and survey devices will be found later in this chapter, and recommended survey procedures.

Controlled Areas, Restricted Areas, Radiation Areas, High Radiation Areas, and Very High Radiation Areas

Most laboratories using radioisotopes do not use sufficient amounts of radioactive materials so that they evoke the NRC requirements for areas defined by these labels. Access to a laboratory may be controlled for a number of other reasons—security, use of biological pathogens, toxic chemicals, or explosives—but the level of radiation and use of radioactive materials in most research laboratories is normally quite modest. Exceptions would be reactor facilities, medical labs for therapeutic radiation, fuel fabrication facilities, etc. These facilities normally will have their own radiation safety specialist who would be thoroughly familiar with these requirements, so they will not be covered further here.

Storage and Control of Licensed Material

A matter of serious concern to the NRC is security of licensed materials. Although not explicitly stated in the regulations, there have been interpretations by individuals in the NRC that since the regulation specifies licensed materials, exempt quantities of materials may be construed as exempt from security requirements, but this interpretation is not necessarily firm. It would be advisable to treat exempt amounts of licensed materials as not exempt.

- A licensee must secure from unauthorized removal or access licensed materials stored in controlled or unrestricted areas.
- A licensee must control and maintain *constant* surveillance of licensed materials that are in a controlled or unrestricted area and that are not in storage.

The word "constant" was italicized in the last requirement for emphasis. It is not permissible to leave material unsecured for brief visits to the soft drink machine, to step outside to smoke, or even to go to the bathroom.

Posting of Areas or Rooms in which Licensed Material is Used or Stored

The licensee must post each area or room in which an amount of licensed material more than 10 times the quantity specified in Appendix C to 10 CFR Part 20 is used or stored with a conspicuous sign or signs bearing the radiation symbol and the words **CAUTION, RADIOACTIVE MATERIALS** or **DANGER, RADIOACTIVE MATERIALS.**

There are some exceptions to these posting requirements, but it is normally appropriate to post. There are additional posting requirements specified in 10 CFR Part 19 that will be discussed later.

Labeling Containers

Every container of licensed material must bear a clear, durable label bearing either **CAUTION** or **DANGER, RADIOACTIVE MATERIAL**. The label must also provide additional information, such as the radionuclide in the container, the amount present, the date on which the activity was estimated, radiation levels, and the kinds of materials in order for individuals to determine the level of precautions needed to minimize exposures. The labels on empty, uncontaminated containers intended for disposal must be removed or defaced before disposal. There are exceptions to these labeling requirements, but most containers in use in a facility should be labeled.

Procedures for Ordering, Receiving, and Opening Packages

A supplier of radioactive materials must ensure that the facility ordering the material has a valid radioactive materials license before filling an order for radioactive materials. This is accomplished by the vendor being provided with a copy of the license. Although not required, it is desirable if all radioactive materials are ordered through the radiation safety office and delivery specified to that office. With all materials being ordered through one point of sale and receipt, it is sometimes possible to arrange discounts for bulk purchases, even if individual items are ordered separately.

Licensees expecting to receive packages containing more than a type A quantity (defined in 10 CFR Part 71 and Appendix A to Part 71) must make arrangements to receive the package upon delivery or receive notification of its arrival by a carrier, and arrange to take possession of it expeditiously. In most facilities the responsibility for receiving radioactive deliveries for the entire organization is delegated to the radiation safety office. The package must be monitored for surface contamination and external radiation levels. The limits for loose contamination are 22,000 dpm, 200 mrem (2 mSv) at the surface, or 10 mrem (0.1 mSv) 3 feet away. Levels in excess of these limits would require immediate notification of the NRC. If the levels are higher than anticipated or the package is damaged, the package should be carefully opened, as a minimum safety precaution, in a fume hood designated for use with radioactive materials. A glove box would be better.

Many carriers deliver packages outside normal working hours, and some organizations will receive them at portions of their facility that function 24 hours per day, such as a security office. If the packages are received during normal working hours, required monitoring must be done no later than 3 hours after receipt. If packages are received outside normal working hours, monitoring must be done no later than 3 hours from the beginning of the next working day.

Disposal of Radioactive Waste

The activity in radioactive waste from radioisotopes is included in the possession limits for a licensed facility until disposed of in an authorized manner. Therefore, records of materials transferred into waste must be maintained as with any other transfer by the individual user. At some point, the waste from individual laboratories will be combined and disposed of by one of the means approved by the NRC. The disposal options available to a generator of radioactive waste are

- Transfer to an authorized recipient according to the regulations in 10 CFR Parts 20, 30, 40, 60, 61, 70, or 72
- Decay in storage
- By release of effluents into the sanitary system subject to specific limitations
- By other approved technologies by persons authorized to use the technologies

A person can be specifically licensed to receive radioactive waste for disposal from a facility for:

- Treatment prior to enclosure (this often means solidification of liquid wastes, or compaction of dry wastes)
- Treatment or disposal by incineration
- Decay in storage
- Disposal at a land disposal facility licensed under 10 CFR Part 61
- Disposal at a geologic facility under 10 CFR Part 60

The limitations on release into the sanitary system are

- The material must be readily soluble in water or is biological material readily dispersible in water.
- No more of a specific radioisotope can be released into the sanitary system in 1 month than an amount divided by the total volume of water released into the system in 1 month that does not exceed the concentration listed in Table 3 of Appendix B to Part 20.
- The total quantity of licensed and unlicensed material released into the sanitary system in 1 year does not exceed 5 Ci (185 GBq) of ^{3}H, 1 Ci (37 GBq) of ^{14}C, and 1 Ci of all other radioactive materials combined.
- Excreta containing radioactive material from persons undergoing medical diagnosis or therapy is not subject to the disposal limits.

The licensee can dispose of the following licensed materials as if they were not radioactive:

- 0.05 μCi (1.85 kBq) or less of ^{3}H or ^{14}C per gram of medium used for liquid scintillation counting. (Note that some liquid scintillation fluids are EPA-regulated RCRA wastes and must be treated as hazardous wastes; alternative liquid scintillation fluids are available.)
- 0.05 μCi (1.85 kBq) or less of ^{3}H or ^{14}C per gram of animal tissue averaged over the weight of the entire animal.

Animal tissue containing radioactive material in the amounts specified cannot be disposed of in a way that would allow its use as either animal or human food. The licensee may use incineration to dispose of radioactive materials subject to the limitations of this section.

Records must be maintained by the licensee of disposal of radioactive wastes. Material transferred to a disposal firm is subject to requirements very similar to those covering chemical wastes. The manifest has specific requirements for the description of the physical and chemical form of the radioactive materials and the radiation characteristics of the wastes. There are specialists in transporting radioactive waste who normally serve as the intermediary or broker for the generator and disposal facility.

Records of Survey, Calibration, and Personnel Monitoring Data

Records of facility surveys, package monitoring data, and equipment calibrations must be retained for 3 years after they are recorded. Where required to sum internal and external doses, data required to determine by measurement or calculation the doses to individuals or radioactive effluents released to the environment must be kept until the termination of the license requiring the record.

Personnel dose records, including internal and external doses where required, for all personnel for whom monitoring is required must be kept until the NRC terminates the license for which the record was required.

Loss or Theft of Material

Losses of significant amounts of radioactive materials must be reported to the NRC both by telephone and by a written report. If the amount that is not accounted for is equal to or greater than 1000 times the quantity specified in Appendix C to 10 CFR Part 20, and it appears that an exposure could affect persons in unrestricted areas, the loss shall be reported by telephone as soon it is known to the licensee. Within 30 days after lost, stolen, or missing material is discovered, the NRC must be notified by telephone of any amount greater than 10 times the quantity specified in Appendix C that is still missing at that time. Most facilities will make their report to the NRC Operations Center.

After making the telephone report, the licensee must within 30 days make a written report to the NRC providing details of the loss, including (1) the amount, kind, chemical, and physical form of the material, (2) the circumstances of the loss, (3) the disposition or probable disposition of the material, (4) exposures to individuals, how the exposures occurred, and the probable dose equivalent to persons in unrestricted areas, (5) steps taken or to be taken to recover the material, and (6) procedures taken or to be taken to ensure against a recurrence of the loss. If at a later time the licensee obtains additional information, it must be reported to the NRC within 30 days. Names of individuals who may have received exposure due to the lost material are to be stated in a separate and detachable portion of the report.

5.2.4.2.5. Notification of Incidents

Immediate Notification—Any event due to licensed materials that would cause any of the following conditions: (1) an individual to receive a total effective dose of 25 rem (0.25 Sv) or more, (2) an eye dose of 75 rems (0.75 Sv) or more, (3) a shallow dose equivalent to the skin or extremities of 250 rads (2.5 Gy) or more, (4) a release of material into the environment such that an individual, if present for 24 hours, could have received an intake of five times the occupational annual limit, (5) a loss of 1 working week or more of the operation of any of the facilities affected, or (6) a loss to property in excess of $200,000.

24-Hour Notification—Within 24 hours of the discovery of an event involving loss of control of licensed radioactive material that may have caused or may cause any of the following conditions: (1) an individual to receive within a 24-hour period an effective dose exceeding the annual limits for an effective dose equivalent, and eye dose equivalent, a shallow dose equivalent to the skin or extremities, or an intake dose, (2) a loss of 1 day or more of the operation of any facility affected, or (3) damage to property in excess of $2000.

The times given are calendar days not working days. For example, a facility shut down for 2 days during which time remediation efforts were conducted only during normal working hours would have triggered the 24-hour notification requirement.

For most facilities, the reports must be made by telephone to the NRC Operations Center *and* to the administrator of their regional office by telegram, mailgram, or facsimile.

Reports of Exposures, Radiation Levels, and Concentrations of Radioactive Materials Exceeding the Limits

A written report must be submitted within 30 days after learning of the event for any of the following conditions: (1) an event requiring immediate or 24-hour notification, (2) doses in excess of the occupational dose limits for adults, the occupational dose limits for a minor, the limits for a fetus, the limits for an individual member of the public, (3) any applicable limit in the license, or (4) levels of radiation or concentrations of radioactive materials in a restricted area in excess of any applicable license limits or more than 10 times any applicable license limits in an unrestricted area. An individual need not be exposed in the last two conditions. The reports must contain the following information: (1) the extent of exposure of individuals, including estimates of each exposed individual's dose, (2) the levels of radiation and concentrations of radioactive materials involved, (3) the cause of the incident, and (4) corrective steps taken or planned to ensure against a recurrence, including the schedule for achieving conformance with applicable regulations.

The reports must include in a separate, detachable part the exposed individual's name, Social Security Number, and birthdate of individual involved. The reports are to be sent to the U.S. NRC Document Control Desk, Washington, D.C., 20555, with a copy to the regional administrator.

5.2.4.3. EPA, National Emission Standards for Hazardous Air Pollutants, Radionuclides

The EPA enacted a rule limiting emissions of radionuclides into the atmosphere for several classes of source facilities, among them facilities licensed by the NRC. The rule became effective on December 15, 1989. However, shortly after that the rule was stayed for NRC facilities and federal facilities not owned by the Department of Energy. On January 28, 1994, NRC license holders were informed that the rule was now in effect, and unless licensees could show that they were exempt, a report would have to be filed by March 31, 1994, and annually thereafter. The primary concerns were with the potential for increased risk of cancer due to inhalation of radionuclides. For laboratory facilities using radioactive materials, fume hood exhausts normally would be the source of emissions. The groups of concern would be the receptors (homes, offices, schools, resident facilities, businesses) nearest the emission source. Another source of concern would be farms, including meat, vegetable, and dairy farms where the radionuclides could enter the food chain. The EPA provides several alternative means of determining whether a facility is exempt, in compliance, or needs to come into compliance. In their compliance guide, they state that if a facility uses more than about six nuclides and has multiple release points, the licensee would be best served by using a computer program called COMPLY, available from the EPA. The results depend upon many input parameters, but once these have been determined, the program quickly will compute the results. The basis for compliance is that the effective-dose equivalent is less than 10 mrem per year for all radionuclides and less than 3 mrem per year for radioiodine. If the effective-dose equivalent from all radionuclides is less than 1 mrem per year and from radioiodine, 0.3 mrem, for the entire facility the facility including any new construction or modification, the licensee is exempt from reporting. One requirement that would be especially restrictive would be if the facility would have to request approval for new construction, a possibility under the standard. This would not be required if the new construction would not cause an effective-dose equivalent of 0.1 mrem per year.

5.2.5. Radioisotope Facilities and Practices

The following sections describe facilities and practices that comply with the regulations outlined in Sections 5.2.4 to 5.2.4.3.

5.2.5.1. Radiation Working Areas

The areas in which radiation is used should meet good laboratory standards for design, construction, equipment, and ventilation, as described in Chapter 3. The International Atomic Energy Agency (IAEA) has defined three classes of laboratories suitable to work with radionuclides. Their class II facility is essentially equivalent to a good quality level 2 facility as described in that chapter, while their class I facility would be similar to the level 3 or 4 laboratory, depending upon the degree of risk, especially equipped to handle even high levels of radioactive materials safely.

One additional feature which may need to be included in the design of a laboratory using radioactive materials is provision for using a HEPA (high efficiency particulate aerosol) filter or an activated charcoal filter for nonparticulate materials on the exhaust of any fume hood in which substantial levels of radioactive materials are used.* A problem with this requirement is the possibility of the filter becoming rapidly "loaded up" by the chemicals used in the hood. A velocity monitor should be mandatory on any fume hood, but especially one equipped with a HEPA filter, to provide a warning should the face velocity fall below 100 fpm. The definition of "substantial" will depend upon the type of radioactive material used. The levels of activity from an unfiltered hood exhaust should not exceed the levels permitted in Appendix B, Table II (10 CFR Part 20) under the worst possible circumstances, such as a spill of an entire container of a radioisotope in the hood. The volume of material, the physical properties of the material, and the rate at which air is pulled through the system should allow the maximum concentration in the exhausted air to be computed.

Most laboratories using radionuclides are not required to be limited access facilities. However, any area in which radioactive material is used or kept should be identified with a standard radiation sign such as shown in Figure 5.5. Unless the amount of material or exposure levels in the facility trigger a more explicit warning, the legend should say only **CAUTION, RADIOACTIVE MATERIAL**. Individuals working with the radioactive materials must be trained in safe procedures in working with radiation under the rules and

---

*   If appropriate filters are not available or are too costly, the maximum levels of activity exhausted by the hood must be restricted to levels which will not allow members of the public to be exposed to levels that exceed those permitted by Part 20 for unrestricted areas or results in the facility violating the EPA 40 CFR Part 61 air emissions standard.

Figure 5.5. Standard radiation symbol.

regulations described in Section 5.2.4.2. Individuals working in the same area who do not use radioactive materials need not be as thoroughly trained, but should be sufficiently informed so that they understand the reasons for the care with which the others using the facility handle the radioactive materials, so they will not inadvertently become exposed to radiation levels in excess of those permitted for members of the public, and so that their actions or inactions do not cause an incident involving radiation.

If licensed materials are stored in an unrestricted area, the materials must be securely locked to prevent their removal from the area. Many materials used in the life sciences must be kept either in refrigerators or freezers, which should be purchased with locks or equipped with padlocks afterwards. If the radioactive materials in an unrestricted area are not in storage, they must be under the constant surveillance and immediate control of the licensee. As noted in Section 5.2.4.2, this precludes leaving them unattended for any reason for even brief periods.

Some laboratory facilities, or portions of them, may need to be made into "restricted areas" because of the type of activities conducted within them or to ensure that members of the public will not be exposed to radiation in excess of that allowed by Part 20 for occupational exposures and members of the public. Access to a restricted area is not prohibited to a member of the public, but it must be controlled to provide the proper assurance of protection. A sign should be posted at the entrance to a restricted area stating: **RESTRICTED AREA, ACCESS LIMITED TO AUTHORIZED PERSONNEL ONLY.** The entrance always should be locked when the area is unoccupied and at such times that unauthorized individuals could enter the area under such circumstances that the occupants would be unaware of or unable to control their entrance.

Within a restricted area, there may be specifically defined areas where the levels of radiation may be significantly above the levels that would be acceptable for personnel to work on a normal 40 hours per week schedule. Depending upon the level, these areas would be designated as Radiation Areas, High Radiation Areas, and Very High Radiation Areas. Additional restrictions would be imposed on entering these areas.

5.2.5.2. External Radiation Exposure Areas

A Radiation Area within a restricted area is an area accessible to personnel where there exists radiation (according to the legal definition) arising in whole or in part within licensed material at such levels that a major portion of the body could receive a dose of 5 mrem (0.05 mSv) in 1 hour, or in excess of 100 mrem (1 mSv) in 5 consecutive days. A Radiation Area must be conspicuously posted with one or more signs bearing the radiation symbol and the words:

CAUTION
(or DANGER)

RADIATION AREA

Similarly, a High Radiation Area is an area within a restricted area accessible to personnel where there exists radiation, arising in whole or in part within licensed material at such levels that a major portion of the body could receive a dose of 100 mrem (1 mSv) in 1 hour. A High Radiation Area must be posted with one or more signs carrying the radiation symbol and the legend:

CAUTION
(or DANGER)

HIGH RADIATION AREA

In addition to the warning signs, additional measures must be taken to prevent individuals from accidentally entering a High Radiation Area. These measures would include one or more of the following.

1. An automatic device must reduce the level to the 100 mrem per hour level upon entry of a person into the area.
2. The area must be equipped with an automatic visual or audible alarm to warn the individual and the licensee, or a supervisor, of the entry into the area.
3. The area must be kept locked except at such time that entry into the area is required and positive control is maintained over entry to the area at such times.
4. In place of the three alternatives just listed, the licensee provides continuous direct or electronic surveillance that is capable of preventing unauthorized entry.

None of the control measures that might be adopted can be configured to restrict individuals from leaving the high radiation area.

The revised Part 20 defines a Very High Radiation area, where even higher levels of activity may be encountered where levels in excess of 500 rads per hour (5 grays per hour) could be received at 1 meter from a sealed source that is used for irradiation of materials.

A Very High Radiation area requires very stringent control measures to either prevent entry into the area or to automatically reduce the radiation levels to those for which an individual would receive less than 100 mrem in 1 hour. The requirements address the possibility that the control measures may fail and require additional control devices to effect the same reduction in the radiation levels and to provide a visual and audible alarm of the failure of the safety barriers to the person trying to enter the area, the licensee, and at least one other knowledgeable person, who can render or secure assistance. Similar protective devices must be provided to reduce the exposure level and provide alarms should the shielding (other than the shielding used in storage of the irradiation device) fail or be removed.

The irradiation facility must be equipped with devices to automatically provide an adequate, timely warning allowing those in the area to leave the irradiation area prior to such an irradiation device being operated. A clearly identified backup control must be provided in the irradiation area so that anyone in the area can prevent activation of the irradiation source. Administrative controls must also be in place to ensure that no one might have entered the area since the previous use of the irradiation source and is still in the area, before the irradiation source device is put into operation. No access control measures may be established which would prevent a person from leaving the area. The entry control devices must be checked for proper functioning each time before the initial operation on any given day, or after any occasion when operations were unintentionally interrupted. A periodic test schedule of the entry control devices is required.

Direct radiation measurements must be made prior to the entry of the first person into an area after a use of the source to determine that the level is below the 100 mrem per hour level. A fixed radiation monitoring system or a portable instrument with suitable response characteristics (will not saturate in high radiation fields) and ranges may be used.

Any other means of access to the area, such as for delivery of equipment or for removal of processed materials, other than the ones normally used for facility personnel entering or leaving the irradiation area, must be controlled to prevent inadvertent entry of anyone through these additional entrances.

5.2.5.3. Areas with Possible Internal Exposures

The previous limits within a restricted area were based on external exposures to a major portion of the whole body. The most probable means of radioactive materials entering the body to cause an internal exposure is through inhalation.* Therefore, Section 20.1202-1204 establishes requirements for spaces in which airborne radioactivity is present. Any area defined as an airborne radiation area is one in which airborne radioactive material in excess of the derived air concentrations (DACs) listed in Appendix B to Part 20 or, which averaged over the number of hours present in the area in a week a person is in the area without respiratory protection, the person could exceed could exceed 0.6% of the annual limit on intake (ALI) or 12 DAC-hours. Each area meeting or exceeding the required limits must be conspicuously posted with one or more signs with the radiation symbol and with the legend:

CAUTION
(or DANGER)

AIRBORNE RADIOACTIVITY AREA

Normally, individuals without respiratory protection should not work in such areas. If feasible, engineering practices should be used to eliminate the need for individual respiratory protective devices. The NRC in 10 CFR Part Subpart H, section 1703-1704, defines the procedures to be followed should a respiratory protection program be required. A discussion on the usage of respiratory protection will also be found in Chapter 6, Section 6.1.

In most research laboratory situations, it would be unusual to find an airborne radioactivity area on other than a short-term basis. If there is any possibility of an approach to the limits while working on an open bench, the use of radioactive materials should be restricted

---

* Radiation can enter the body by ingestion, by absorption through the skin, or through a break in the skin, such as a cut. However, the last of these is more likely to be due to unusual circumstances rather than due to the presence of a continuing source.

to a hood. In a facility with a broad license, it is recommended that at least some individuals using radioisotopes should be included in a bioassay program for the same reason that others wear a personal monitoring device, to ensure and document that no one is receiving an internal dose over the limits.

5.2.6. Material Control Procedures

The amount of materials that a licensee may possess at any time is established by the terms of the license. As long as the total holdings are maintained within those limits licensees may order radioactive materials, maintain them in storage, use them, transfer them to other licensees, and dispose of waste amounts following regulated procedures. Accurate records must be maintained for individual laboratories and for a facility as a whole where the facility has a broad license. For short-lived isotopes, there will be a continuing reduction of the amount on hand of a given material due to decay alone, while the distribution of the remainder of the material in use, storage, or transferred to waste changes continuously. Written records should be maintained of the amounts involved in each of these processes, and it should be possible to account for nearly all of the material from the time it is received until disposed of as waste. Some uncertainty is certain to be introduced during actual experimentation, especially if there is a gaseous metabolic or combustion product released through the exhaust from a hood. There also will be small quantities that will be retained on the interior of vessels containing radioactive liquids that will escape into the sanitary system when the container is washed. However, it should be possible to estimate these types of losses with reasonable accuracy.

Facility records of receipt, transfer, use, and disposal should be maintained at each individual laboratory authorized to use radioisotopes. This is the responsibility of the senior individual in charge of the facility to whom has been assigned the internal equivalence of a radioisotope license. A technician may perform the actual record keeping, but the ultimate responsibility for the radioactive material in the facility belongs to the principal authorized user. Except for the removal of material from the storage container for use within the facility, the organization's radiation safety officer (RSO) should be involved in each of these transactions and should be able to detect any anomalies or disparities. It is the RSOs responsibility to maintain records for the overall inventory of the radioisotopes within the organization, and these records can be used to compare with or audit the records for each facility. Laboratory managers may be assured that an inspector will check the records of some laboratories at a facility during each unannounced inspection, so records must be able to pass these inspections at any time.

It has already been noted that materials in unrestricted areas must be kept in secure storage when the user is not present. This requirement for security is applicable whenever the facility is left vacant and open, even for short periods for going to the restroom, the stockroom, checking mail, etc. If multiple keys are available to a facility, with some being held by nonlicensed personnel, it is not an acceptable alternative to lock the doors to the room while absent, if unsecured radioactive materials are accessible. In restricted areas, which always should be locked when they are not occupied or when access of unauthorized persons can be controlled, radioactive material can, in principle, be kept in storage areas that are not necessarily locked. However, if substantial amounts of radioactive materials are to be kept on hand, an effective key-control program must be in place or other security measures taken to protect against loss of the material.

Whenever radioactive material is removed from or returned to storage, the type of material transferred, the amount, the time, and the person doing the transfer should be entered into a log.

Perhaps the most common type of loss is due to waste material being disposed of as ordinary trash. Up to this stage, the radioactive material is normally maintained in containers labeled as containing radioactive materials as required by Part 20, but contaminated trash may appear the same as ordinary waste to a custodian. This can be due to laboratory workers inadvertently putting the material in, with, or near ordinary trash. Under such circumstances, a custodian is liable to take the material away due to failure to recognize the material as being different from other waste, even if the radioactive waste containers may have "radioactive waste" written on them, may be marked with the radiation symbol, or may be in distinctive colors. Unfortunately, an increasing percentage of the population is becoming functionally illiterate (a recent estimate is 30%), individuals may be color-blind, and if the symbol were to be concealed the custodian may very conscientiously remove the material as ordinary trash unless additional precautions are taken. These errors can be reduced if the internal locations of radioactive waste (and also broken glass, waste, or surplus chemicals) and ordinary solid waste are well separated, and radioactive waste is placed in distinctively shaped containers, used for no other purpose. These additional measures have proven helpful, if all personnel have been fully informed and cooperate. It would be desirable for new custodial personnel to receive special training, not only to reduce the possibility of loss of material, but also to ensure that they are not unduly concerned about servicing a laboratory in which radioactive materials are used.

If it is suspected or known that radioactive material has been lost from a laboratory, either by direct knowledge of the disappearance or as inferred by an examination of the records, the organization's RSO must be notified immediately. If after investigation the loss or possible theft is confirmed, it is always desirable to notify the NRC. It is required that this be done for significant losses, according to the requirements of Section 20.2201 as described in Section 5.1.4.2. The situation should be thoroughly investigated, independent of the size of the loss, to attempt to recover the material or to find out its actual fate. If the loss appears to be deliberate, it may be necessary to solicit police assistance. Although the actual monetary loss may be small, individuals who are not knowledgeable about the potential hazards have suffered severe injuries from the possession of radioactive materials. If the loss is a strong-sealed source used for irradiation, it is very important, as will be discussed in a later section, that the material be recovered quickly.

It is rare for radioactive material to be taken deliberately; one of the most common loss mechanisms is loss as trash, as already discussed. If it is a reportable incident, the NRC must be informed immediately by telephone or by a written report filed within 30 days of the specifics as to what was lost and the circumstances concerning the disappearance of the material. They will wish to know, to the best of the licensee's knowledge, what happened to the material and the possible risks to individuals in unrestricted areas. The steps taken to recover the material will need to be in the report and perhaps most important, the steps the facility intends to take to ensure that a similar incident will not recur.

Even if the amounts lost are minimal and pose no danger to the public (typically the amounts at any given time in the waste in a laboratory are in this category), the RSO and the person in charge of the laboratory in which the loss occurred should be required to make a full report to the organization's RSC. The RSC should consider whether disciplinary measures are needed. It should be determined if the loss was due to a failure by the laboratory to follow established policies or if the internal policies are sufficient to prevent future incidents of the same type. If the policies need to be modified, the committee should develop new requirements and inform all users within the organization. A mechanism that works well to reduce such occurrences is to send a safety note to all authorized users, informing them of the incident (with the identity of the laboratory involved not identified) and require that all users acknowledge reading the safety note in a log.

In a well-run organization with sound policies guiding the use of radiation, incidents involving loss of radioactive material should occur very rarely. In any active research institution, especially in academic ones where graduate students and even faculty change frequently, mistakes will occur. A fairly common defense in instances in which radioactive material is lost is that the amounts lost were small and no one will be harmed. The RSC, however, needs to determine if internal procedures or enforcement of procedures is the basic problem, not the quantity lost. Individuals who have made a mistake should not be treated harshly if they recognize that a mistake has been made and learn from it. On the other hand, the RSC should take very firm steps to correct the situation if a facility shows a continuing pattern of laxity in following sound procedures or, worse, deliberate neglect of safety and regulatory policies.

5.2.6.1. Ordering and Receipt of Materials

A very sound procedure to follow in ordering and receiving radioactive materials is for all orders and receipt of radioactive materials to be handled by the radiation safety office. This has several advantages to the organization in terms of record maintenance, as discussed in the previous section, and to the individual facility since it provides a parallel set of records. Some suppliers of radioactive materials will provide a discount for volume purchases if all the orders go through a common ordering center, even if the materials are for different users. Virtually the only problems involve the occasional user who wishes to have a custom compound prepared with the radioisotope in a specific location within the compound. They may wish to discuss their requirements by telephone with the vendor. Even such telephone orders can be handled through the radiation safety office satisfactorily if the vendor, the user, and the radiation safety office responsible for ordering work together. Since many biological materials deteriorate with time and at normal temperatures, they are sent packed in dry ice to keep them cool. If the radiation safety office is aware of the anticipated delivery date, the staff can ensure that it will be delivered promptly.

A major safety function of the radiation safety office is to receive all packages of radio-isotopes and process them according to the requirements of Section 20.1906. Unfortunately, there are occasional errors in shipping and packages can be damaged. On at least two occasions at the author's institution, packages have been received which contained substantially larger amounts of material, with consequently higher radiation levels, than should have been the case. Most packages of radioactive materials are shipped in what are called type A packages, which are only required to withstand normal transportation conditions without a loss of their contents. The amounts of the various nuclides that can be shipped in a type A package are given in 10 CFR 71, Appendix A. It is the responsibility of the shipper to conform to these limits.

The receiver of packages of radioactive materials must be prepared to receive them when delivered, or pick them up at the carrier's location at the time of arrival or promptly upon notification of the arrival of the material. Most packages are probably delivered to the purchaser's location by the many freight and parcel delivery services that are available. Some of these deliver 24 hours per day; except in the largest organizations that conduct research on a 24-hour basis, however, most radiation safety groups work a normal daytime schedule. At a university, for example, the only operation likely to be functioning on a 24-hour basis is the security or police unit, who would not normally be knowledgeable about how to do the required checks on packages received. It is desirable for packages to be placed in a secure area until they can be inspected, so one alternative would be to place a few permanently mounted, lockable boxes in the security area in which packages could be placed until the following morning.

Packages are to be checked for external radiation levels, including loose surface contamination and direct radiation, within 3 hours of receipt during the day or within 3 hours after the beginning of the next working day if received after normal working hours. Levels that would require immediate notification of the NRC would be 200 mrem per hour (2 mSv per hour) on the surface or 10 mrem per hour (0.1 mSv per hour) 3 feet from the surface of the package. A limit of 22,000 dpm is set as well for loose contamination on the surface.

The radiation safety officer normally should not open individual packages (although some do so) unless there is reason to suspect that the material contained within the package is not consistent with the packing list, but rather should deliver the material to the user or have the users pick up their packages. The users should exercise due care in opening packages to reduce exposures to the persons performing this task. The precautions that need to be taken will be dependent upon the nature of the material, the anticipated radiation levels, or the potential for loose material to become airborne should the package not be tightly sealed. If there is any risk of personnel becoming contaminated while opening a package, it should be done in a radioisotope hood, with the package placed in a pan large enough to retain any spilled material, and with the pan sitting on a layer of plastic-backed absorbent paper. The employee should be double gloved and should be wearing a lab coat. For radioactive materials with high specific activity or with a small ALI, it could be desirable to open suspect packages in a glove box providing complete containment. If loose contamination were found on the surface of the package, a HEPA filtered respirator would be recommended for persons handling it besides the other protective clothing. The need for shielding would depend upon the radiation characteristics of the material.

Any package found to be damaged such that more than the allowable limits of contamination is released must be reported at once to the NRC by telephone and telegraph mailgram (or facsimile) and to the final delivering carrier. It is recommended to report any significant problems with the package, whatever the amounts.

5.2.7. Operations

"Operations" is an all-inclusive term encompassing the program in which radioactive materials are used and the support programs necessary to allow the materials to be used safely and in compliance with all regulations. Many of the latter requirements have been discussed in some detail, including, in some instances, the means by which compliance or safety can be enhanced, in preceding sections.

Although a direct relationship between low level exposures and adverse biological effects has not been shown conclusively, it is assumed, as a conservative premise, that many effects that are known to be caused by high doses will occur on a statistical basis to some of a large population exposed to lower levels. Stochastic effects that are based on probabilities, such as induction of cancer or genetic damage, are assumed to have no threshold, while nonstochastic effects, such as the causing of cataracts by radiation, are assumed to be related to the intensity of the exposure and are assumed to have a threshold. Often, the stochastic effects are the ones that limit the permissible exposures in using radioisotopes.

There is a wide variation in the susceptibility of individuals to radiation as with most other agents that can cause harm to humans. Some individuals are much more resistant than others, while there are individuals that are hypersensitive. There are many factors that can affect the damage done to an individual. Some of these may be sex linked and some dependent upon the age of the individual. Some may depend on exposure to other agents (synergistic effects), and others will depend upon the health of the individual at the time of an exposure. This section will not attempt to pursue these modifying factors but will assume, as did the sections in which chemical agents were discussed, that the workers are normal, healthy individuals in their normal working years, unless otherwise specified.

The following sections will be concerned with means to reduce exposures while using radioisotopes.

5.2.7.1. Reduction of Exposures, ALARA Program

The goal of an ALARA program is to make personnel exposures to radiation as low as reasonably achievable in the context of reasonable cost, technical practicality, and cost-benefit considerations. In such a program, each aspect of an operation is evaluated to decide if there are alternatives or means by which the procedure can be modified to reduce exposures.

One of the first considerations is to select the radionuclide that is to be used in the research. The half-life, the type of radiation emitted, the energies of the emissions, and the chemical properties of the radioisotopes are factors to be evaluated.

Control of the external hazard is based on manipulation of three primary variables, time, distance, and shielding.

Control of internal hazards is more complex. The properties of the materials must be considered, but the experimental techniques to be used, the design and construction of the facility, the way it is equipped, and the use of personal protective equipment are all factors that must be considered in establishing an effective ALARA program.

5.2.7.1.1. Selection of Radioisotopes

To select the appropriate radioisotope to use in a given research program, several factors must be considered. If the choice is strictly based on relative safety involving nuclear properties, a beta emitter generally would be preferable to a gamma emitter, since gammas typically are much more penetrating than betas. Although alphas are much less penetrating than betas, an alpha emitter normally would not be selected over a beta emitter because, if ingested, among other reasons, the damage caused by an alpha as it moves through matter is so much greater than for a beta. However, the primary reason is that alpha emitters are generally not found among the useful elements used in biological research. For the same type of emitter, the isotope with the lower energy emission should normally be chosen. When half-lives are considered, a half-life as short as practicable would be preferable, so that the problem of disposal of the nuclear waste could be solved by simply allowing the radiation in the waste products to decay rather than having to ship them away for burial. A second consideration is that if the materials may be ingested by humans or used as a diagnostic aid, the total dose to an individual could be much less if the isotope used has a short half-life.

The desired chemical properties and costs may dictate the isotope chosen. Some isotopes are much more expensive than others, so for economic reasons the less expensive choice might be preferable. A custom-prepared, labeled compound is usually much more expensive than a commercially available labeled compound, so that unless the safer isotope is available in a previously prepared compound, economic reasons would probably indicate using what is readily available.

Chemical properties may dictate the choice of materials because of the risk to personnel ingesting them. An examination of Appendix B, 10 CFR Part 20, shows wide variation in the ALIs or DACs, even among those that have similar chemical properties. The isotope with the greatest ALI or DAC should be selected.

Carbon-14 and tritium are two of the most frequently used radioisotopes since they can readily be incorporated into organic compounds. Unfortunately, ^{14}C has a very long half-life, 5730 years, while tritium has a much shorter one, 12.3 years. Even the latter is long enough that it represents a disposal problem. Otherwise, these two isotopes have excellent radiological properties. The beta emitted by carbon has an energy of only 0.156 MeV, so that its range in tissue is less than 0.3 mm and about 10 inches in air. The beta emitted by tritium is much weaker (0.0186 MeV) so that its range in tissue is only about 0.006 mm, and about one fifth of an inch in air. Clearly, both isotopes would pose little risk as far as external exposures are concerned, although both can represent an internal problem. Tritium, for example, can easily become part of a water molecule that would be treated biologically by the body virtually the same as would any other water molecule. Other popular beta-emitting isotopes are ^{35}S, which emits a 0.167 MeV beta and has an 87.2-day half-life, and ^{32}P, which emits a 1.71 MeV beta and has a 14.3-day half-life. As noted in an earlier section, the high-energy beta from ^{32}P can cause a substantial radiation exposure problem due to emission of bremsstrahlung radiation, if the betas are allowed to interact with materials of significant atomic numbers. Another radioisotope of phosphorus can be used, ^{33}P, which has a longer half-life, 25.2 days, but a much lower energy beta, 0.248 MeV.

Some of the other more commonly used radioactive isotopes used in research are listed in Table 5.6.

5.2.7.1.2. Shielding

Shielding is usually the first protective measure that comes to mind when considering radiation protection. In reviewing properties of radioactive particles, the ranges or penetrating capacity of the various types were briefly discussed to provide some measure of understanding of their characteristics. These properties will now be considered to decide how best to provide shielding. The discussion will be limited to shielding against beta and gamma radiation.

Betas

The range of betas in any material can be calculated by the following two equations:

$$\text{For energies} \quad 0.01 \leq \text{Energy (MeV)} \leq 2.5$$
$$R = 412E^{(1.265-0.0954\ln E)} \tag{12}$$

$$\text{and for energies} \quad E \geq 2.5 \text{ MeV}$$
$$R = 530E - 106 \tag{13}$$

The range, R, in these equations is given in mg/cm^2. This can be converted into centimeters in a given material by dividing the ranges given by these equations by the density of the material in mg/cm^3. The range R can be considered as the area density of a material, in contrast to the usual volume density.

Table 5.6.   Properties of Some Selected Nuclides

Nuclide	Half-Life	Beta(s) (MeV)	Gamma(s) (MeV)
^{45}Ca	163 days	0.257	0.0124
^{109}Cd	453 days	ϵ	0.088
^{36}Cl	3.0×10^5 years	0.709	—
^{60}Co	5.27 years	0.318	1.173, 1.332
^{51}Cr	27.7 days	ϵ	0.320
^{137}Cs	30.2 years	0.512, 1.173	0.6616
^{59}Fe	44.6 days	0.467, 0.273	1.099, 1.292
^{203}Hg	46.6 days	0.212	0.279
^{125}I	59.7 days	ϵ	0.0355
^{131}I	8.04 days	0.606, others	0.364, others
^{54}Mn	312.5 days	ϵ	0.835
^{22}Na	2.6 years	ϵ, ß$^+$ 0.545	1.27, 0.511 (annihilation gammas)
^{63}Ni	100 years	0.0659	—
^{65}Zn	243.8 days	ϵ, ß$^+$ 0.325	1.116, 0.511 (annihilation gammas)

The range of the 1.71 MeV beta from ^{32}P in mg/cm^2 calculated from Equation 12 is 785 mg/cm^2. Tissue and many plastic materials have a density near 1, so that this beta would have a range of about 0.8 cm, or about 0.3 inches in such materials. In lead the range would be 11.34 times less or 0.7 mm (0.027 inches). However, it was mentioned earlier that bremsstrahlung would be a major problem for such an energetic beta emitter if a high atomic number absorber were used. Therefore, a low Z material should be used for shielding of pure beta emitters. A piece of plastic 3/8 of an inch thick would shield against nearly all common betas used as radioisotopes in research, since there are few commercially available beta emitters that emit more energetic betas (quite a few neutron-activated nuclides emit higher energy betas, but these are not often used as radioisotope materials). Research personnel who work with ^{32}P can reduce hand exposures from bremsstrahlung by slipping thick-walled plastic tubing over test tubes and other containers.

Gammas

Equation 6, repeated below as Equation 14, can be used to calculate the attenuation of a monoenergetic narrow beam of gamma rays by a shielding material.

$$I = I_0 e^{-\mu x} \tag{14}$$

This equation, as noted earlier, will not take into account gammas scattered into the shielded area by Compton interactions by the shield, and by the walls (and floor and ceiling, if the gammas are not collimated so as to avoid these surfaces). Figure 5.6 shows an optimal shielding geometry in which a shield just subtends a sufficient solid angle to

Figure 5.6. Optimum shielding to minimize scattered radiation from the shield.

Figure 5.7. Typical shielding configuration for hood usage of radioactive materials.

completely block the direct rays from the source. This geometry reduces inscatter. In Figure 5.7, a more typical geometry is shown where a worker is using a source in a fume hood. In this geometry the shielding would not be quite as effective.

For work with gamma emitters and low energy betas, lead is a good shield. It is dense, 11.4 g/cm³, and has a high atomic number, 82. It is relatively cheap, and is more effective than any other material that is comparably priced. The mass attenuation coefficient for lead is shown in Figure 5.8. The coefficients read from the graph can be converted to linear coefficients by multiplying them by the density expressed in g/cm³. If the absorption coefficients for lead at 0.5, 1.0, and 1.5 MeV are used to compute the thickness of lead required to attenuate a beam of lead by factors of 2 and 10, the half-value thicknesses would be 0.4, 1.1, and 1.5 cm, respectively, and for the tenth-value thicknesses, 1.25, 3.5, and 5.0 cm for the three energies. Thus, a lead brick 2 inches thick would reduce the intensity of the gammas from most of the common radioisotopes used in research by a factor of 10 or more, which is often enough for the average source strengths used in research. Shields

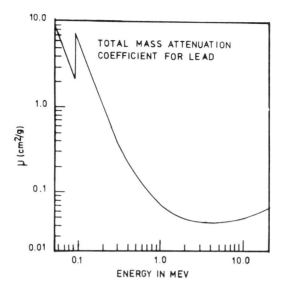

Figure 5.8. Total mass absorption coefficient for lead as a function of energy.

should, however, be built of at least two layers of bricks so that the bricks can overlap to prevent radiation streaming in a direct path through a seam.

Occasionally, individuals will set up a shield to protect themselves and forget that radiation from an open source extends over 4π steradians. Levels through the top, bottom, sides, and rear of the work area, and perhaps in the next room, may be excessive.

Extremely high-level gamma radiation operations are conducted in a special shielded enclosure called a "hot cell." Usually the shielding in such units is provided by making the enclosure of high density cast concrete. The thickness needed to reduce the radiation to levels acceptable for occupational exposures would depend upon the strength of the source. Often, the top of a hot cell is covered as well as the sides to avoid "sky-shine" radiation, i.e., radiation scattered from the air and ceiling due to radiation from the source within the cell. The researcher would perform the operations required in the cell using mechanically or electrically coupled manipulators. Vision typically would be through a thick leaded glass window, often doped with a material such as cerium to reduce the tendency of glass to discolor upon exposure to radiation, or closed-circuit television.

5.2.7.1.3. Distance

Distance is an effective means of reducing the exposure to radiation. The radiation level from a point source decreases proportionally to $1/R^2$, while from a point near an extended source, such as a wall, the level falls off in a more complicated manner, since the distance from each point on the wall to the measuring location will vary and the contribution of each point must be considered. There are many excellent references in which specific geometrical sources are discussed, but unfortunately, most "real world" situations often do not lend themselves to simple mathematical treatment. If the distance to the source is large compared to the size of the source, then the approximation of a point source is reasonably accurate.

Many radionuclides emit more than one gamma, and for many the gamma spectrum is extremely complex. For each original nuclear disintegration, which defines the number of curies or becquerels represented by the source, a specific gamma will occur a certain fraction, f_i, of the time. If a gamma is emitted for every original nuclear decay, then f_i for that gamma would be 1. With this concept in mind, there are two simple expressions that can be used to calculate the dose at a distance R from a point source. The difference in the two expressions is simply a matter of units.

To compute the dose, D, in μSv per hour, the expression is

$$D = \frac{\Sigma_i M f_i E_i}{6R^2} \tag{15}$$

Here, M is the source strength in MBq, E_i is the energy of the ith gamma in MeV, and R is the distance in meters.

The same expression, expressed in the more traditional units, is

$$D = \frac{6\Sigma_i C f_i E_i}{R^2} \tag{16}$$

Here, C is the source strength in Curies, E_i is the energy in MeV, R is in feet, and D is given in rads per hour.

It should always be kept in mind that even from a point source, both expressions give only the dose due to the *direct* radiation from the source. They do not include the radiation scattered from objects in the room, the walls, floors, or ceiling or the workbench on which the source is sitting. It also should be kept in mind that the radiation dose can be reduced further by placing a shield in the path of the radiation. If this is done, the levels calculated by either Equations 15 or 16 would be reduced by the factor $e^{-\mu x}$, where μ would be the linear attenuation coefficient for the shield material at the energy, E_i, of the gamma. The equation would have to be calculated for each gamma and the contribution of all the gammas would have to be summed. The result would be the *minimum* dose that would be received from the source. The actual dose would be higher.

Because the exposure rate does go down rapidly with distance, sources should always be handled with tongs rather than directly. Whenever a source is to be transferred, only an individual trained to handle the equipment and the source should do the transfer with perhaps an assistant nearby, and the transfer should be done as expeditiously as possible.

When thinking of distance as a protective measure, the mental image is of a source with the user standing back from it. However, one way of thinking of contamination on the skin is of there being virtually a zero distance separating the irradiating material and the tissue, so that even small amounts of radioactive material could eventually cause a significant local exposure, even if the material did not penetrate or permeate through the skin. To prevent contamination of the skin, gloves and protective covering such as a laboratory coat, should always be worn when working with material that is not a single solid piece, such as a liquid or a powder that could be spilled or otherwise contaminate the worker and the experimental area.

5.2.7.1.4. Time

Time is perhaps the most easily achieved means of reducing the dose received. If the radiation level remains constant, reducing the exposure time reduces the dose received by the same proportion that the time is shortened. If the radioactivity of the material in use decays during the interval, the reduction in exposure would be even greater. This should be kept in mind when developing experimental procedures. Personnel should leave sources within their containers as much of the time as possible while working.

Another way in which time can be used effectively to at least reduce the level of exposure to a single individual is to share the exposure time among more individuals. Statistically, for those events that depend on the probability of an event being caused by exposure to individuals, i.e., stochastic events, little has been gained if the exposure to the population of the entire group of persons is considered, but it does reduce the probability of a problem developing for a specific individual and reduces the nonstochastic effects that depend upon the dose received to the person.

5.2.7.1.5. Quantity

Modern techniques often allow work with much smaller amounts of material than was possible only a few years ago. When working with radioactive materials, full advantage should be taken of any efficient and effective procedure for using less of the radioactive substance, not only at any given time, but over the course of the research. A carefully constructed written research plan and hazard analysis, in which the amounts of materials needed for each procedure are calculated, will help the minimization process.

5.2.7.1.6. Example of Time and Distance

Small sealed sources are frequently used in laboratories. If a source is dropped from a storage container, an employee may pick it up in his hands to return it to the storage container, handling it in direct contact with the hands for perhaps 1 minute. Even this short interval with weak sources can produce significant doses to the hands. Data have been published showing that at or very near the surface of a steel-clad sealed gamma source, electrons from the source cladding caused by interactions of the gammas with the material can contribute between 25 and 45% of the dose to the hands. The primary exposure to the hands still is due to direct radiation from the source, which is high because of the small distance between the source and the tissue. The tissue of the hands is estimated to absorb about 5 to 10% of the gamma energy per cm. With these facts in mind, if a 1 Ci source of ^{60}Co is 1/4 of an inch in diameter, and is clad with 1/32 of an inch of 304 stainless steel, the approximate surface dose rate per minute for the source would be 2075 rads per minute. In less than 2 1/4 seconds, the hand of a person picking up the source would receive, locally, the entire year's permissible exposure of 75 rads. The rate falls off rapidly with depth in the tissue; at 1 cm deep, the rate is estimated to be 114 rads per minute and at 3 cm, 16 rads per minute. A reduction of any one of the three contributing effects, the quantity of radioactive material in the source, the time it is handled, or increasing the distance from the source will diminish the harm done to the individual.

Although this may be dismissed as unlikely, there have been a number of cases in which persons unfamiliar with what they were handling have picked up sources and placed them in their pockets. Often, the persons have been custodians or other support personnel. In some cases, many persons outside laboratory facilities have been exposed to high levels of radiation, because pellets of material from irradiation sources that were no longer being used were disposed of improperly and wound up in trash. Deaths have resulted from such instances. Not all high exposure incidents have been among the untrained and poorly educated. There have been instances of laboratory personnel who have had substantial exposures because of failure to exercise sufficient care in working with radioactive materials.

In several cases, the workers were exposed to substantial radiation levels due to exposed sources of which they were unaware because an interlock had failed. Safety features which depend upon a single microswitch do not provide adequate protection. Some persons have died and a number have been seriously injured. It should be noted that the exposures in these cases have been in the several hundreds of rads (or rems) range or greater, and sometimes localized exposures have been in the many tens of thousands of rad range, with the person surviving, but with extremely severe damage over the parts of the body exposed to these extraordinarily high levels.

In summary, to minimize external exposure:

1. Work with the safest isotope appropriate to achieve the desired results.
2. Use the smallest amount of radioactive material possible, consistent with the requirements of the experimental program.
3. Reduce the time exposed to the radiation.
4. Use shielding wherever possible.
5. Take advantage of distance as much as possible. Use extension tools to avoid direct handling of the material.
6. Make sure that any protective devices are fully functional. A checklist that must be followed before each use is a very inexpensive safety device.

5.2.7.1.7. Internal Dose Limiting

There are, as with chemicals, only four basic means for radiation to enter the body: inhalation, ingestion, absorption through the skin, or through a break in the skin. The basic means of control of internal exposure is to restrict the entry into the body by any of these routes. However, other factors will affect the consequences of radioactive material that has succeeded in bypassing the defenses to prevent entry into the body. The form of the material is important. Is it in a soluble or insoluble form? Insoluble particles carried into the lungs following inhalation may remain there for long intervals. Soluble materials may be absorbed into body fluids. If the material is incorporated in a form in which it is likely to be metabolized, it is more likely to reach a critical organ if ingested than otherwise. The size of the particle is critical if it is an airborne contaminant in determining in what part of the respiratory system the material will be deposited. Part of the material that is not exhaled may be swallowed.

The chemical properties of the material are important in determining what organs are likely to be involved. In some cases, such as tritium, the tissue of the entire body is likely to be involved, since often the chemical form of released tritium will be as HTO, and the behavior within the body will be the same as that of ordinary water. Other materials, such as strontium, are bone seekers if in soluble form, while iodine in soluble form will most likely go to the thyroid.

Entry Through the Skin

Entry through the skin, either percutaneously or through a break, is normally the least likely to occur and is the most easily prevented. Usually, entry would follow an inadvertent spill in which material reached the skin by the worker handling contaminated objects without using proper equipment, or by failure of protective items of apparel. However, in an exposure of an otherwise unprotected worker to water vapor containing an appreciable amount of HTO, a significant fraction of the total exposure could be through skin absorption (normally assumed to be one third of the total, the remainder being due to inhalation).

Prevention of contamination of items likely to be handled is the first step in prevention of skin exposure. For example, the work surface should be covered with a layer of plastic-backed absorbent paper to ensure that the permanent work surface is not contaminated by spilled materials, by aerosols generated in many different laboratory procedures, or even by vapors that might diffuse from a container, since vapors of most chemicals are heavier than air. Aerosols are perhaps the most likely source of local contamination as the droplets settle. Aerosols can come from opening a centrifuge in which a tube has been broken, use of a blender or sonicator, the last drop falling from the end of a pipette, or even from dragging a wire across the surface of an uneven agar surface in a petri dish. The most obvious solution to most aerosol problems is doing such procedures in a fume hood, but some procedures still would contaminate the hands and arms. Frequent surveys for contamination of work surface and hands should help avoid contamination from aerosols. Setting up the apparatus in a shallow container, such as an inexpensive plastic or aluminum pan that can be obtained from a department store, large enough to contain all the spilled liquid in use at the time, is a simple way to confine the consequences of an accident to items that are disposable rather than ones that would have to be decontaminated. Covering some items of equipment with the plastic-backed absorbent paper, or with aluminum foil for warm surfaces, may prove profitable for some levels of work using materials of high toxicity or amounts of radioactive materials in excess of the typical laboratory application, or in procedures in which the possibility of contamination is higher than normal. Frequent checks of the apparatus in use and the immediate work area near the apparatus with survey instruments or smear or wipe tests of work surfaces and equipment being handled are good means of limiting contamination before it becomes widespread. Checks of the survey instrument itself and such areas as faucet handles on sinks used in the work are often overlooked in contamination evaluations, but if the hands are contaminated, any items that may have been handled need to be checked.

Protective gloves and laboratory coats or coveralls are the first level of protective gear that should be adopted in laboratory operations involving the use of radioisotopes. Wraparound lab coats may be a better choice than those that button up the front to avoid material passing through the front opening. The choice of gloves should be chosen to be resistant to the chemicals of the materials in use. The dexterity allowed by the gloves also is a factor. Where the possibility of damage to the gloves is significant or the properties of the radioactive materials are especially dangerous if contact occurs, wearing two pairs of gloves should be considered. Many facilities make this a standard requirement instead of optional. Gloves used for protection while working with radioactive material should be discarded as radioactive waste after use. The cuffs on gloves should be long enough to allow sealing (with duct tape or the equivalent) to the sleeves of a garment worn to provide body protection if needed.

Any cuts or abrasions on the hands, or other exposed skin areas on the forearms, should be covered with a waterproof bandage while actively working with radioactive materials.

Cotton laboratory coats provide reasonable protection for the body of the worker in most laboratory uses of radioisotopes. However, if they do become contaminated, they must be washed to decontaminate them. Laundering facilities appropriate for cleaning of radioactive materials may not be available, and so consideration of disposable protective clothing made of materials such as Tyvek™ may be desirable. These garments are not launderable, but if a thorough survey shows that they are free of contamination and if they are handled with reasonable care, they can be worn several times before being discarded. However, they are inexpensive enough to allow disposal whenever necessary if they do become contaminated. Disposable garments should have both welded and sewn seams to provide assurance that the seams will not split during use (wearing a garment a size or even two larger than is actually needed also helps avoid this). Body contamination surveys should be made of the outer protective garment before removal and of the street clothes afterwards. Checking the outerwear is important to find out if the procedures used generate contamination and to avoid contamination from a discarded garment. If the protective clothing is found to be contaminated while following routine procedures, the procedures need to be revised to eliminate the problem. The clothing is a final barrier, not the first line of defense.

Hands should be carefully washed at the conclusion of any operation in which radioactive materials are handled in procedures which offer any opportunity for contamination. If the skin has become contaminated, the affected area should be washed with tepid water and soap. The use of a soft brush aids in the removal of material on the surface. Harsh or abrasive soap should not be used. After washing for a few minutes, the contaminated area should be dried and checked for contamination. If the area is not free of radioactivity, rewashing can be tried. A mild detergent also can be tried, but repeated or prolonged application of detergents to an area may damage the skin and increase the likelihood that surface contamination will penetrate it. Organic solvents, acid or alkaline solutions should not be used since they will increase the chances of skin penetration. Difficult-to-clean areas should be checked with extra care.

Any contamination that cannot be removed by the above procedures should be reviewed by a radiation safety specialist. Further measures that could cause abrasion or injury to the skin should only be done under the advice and supervision of a physician.

Ingestion

The inadvertent ingestion of radioactive material is often the consequence of transferral of radioactive material from the hands to the mouth while eating or drinking. No ingesting of food or drink, chewing gum, smoking, using smokeless tobacco products, or application of cosmetics should be allowed in the active work area of a laboratory in which radioactive materials are used. Before leaving the laboratory after working with such materials, as noted in the previous section, the hands should always be carefully washed and surveyed to ensure that no material is on the hands. If gloves and protective outerwear have been worn, the gloves should be discarded and the outerwear left at the entrance to the work area after ensuring that it is not contaminated. Washing the hands should take place after handling the clothing. It should be possible to effectively prevent ingestion of radioactive materials using reasonable care to prevent transfer of any activity to food and drink, or any other item that might be put in the mouth such as a cigarette or gum.

Inhalation of Radioactive Materials

Airborne radioactivity is the most likely means by which radioactive materials can enter the body, resulting in an internal exposure. Most mechanisms by which either skin contamination or ingestion occurs involve contact of the user with the active material in a fixed location. However, airborne materials are not constrained to a given space, and unless the work done is confined to a glove box, they can fill the entire volume of the laboratory, surround the occupants of the room, and contaminate the air they breathe. The radioactive material also can be discharged from the laboratory by the building exhaust system or through the fume hoods. It is this last possibility that is the concern addressed by the EPA restrictions on air emissions.

The size of particles deposited in the deep respiratory tract reaches a maximum at 1 to 2 μ. Unfortunately, particles in this size range tend to remain suspended in air for relatively long periods of time. The gravitational settling rate for a 1-μ particle with a density of one is about 0.0035 cm per second. It would require about 12 hours for such a particle to settle about 5 feet under the influence of gravity, if allowed to do so without being disturbed. Thus, the particles of airborne radioactive contamination which have the greatest potential for deposit in the deep respiratory tract are among those that would remain airborne long enough for a worker to have an opportunity to inhale them.

It has been estimated that about 25% of all soluble particles inhaled are exhaled, about the same fraction are dissolved and absorbed into the body fluids. The remaining 50% are estimated to be deposited in the upper respiratory tract and swallowed within 24 h after intake. For insoluble particles, about seven eighths are exhaled or deposited in either the upper or deep respiratory tract, but swallowed within 24 hours after intake. However, the remaining one eighth is assumed (lacking specific biological data appropriate for the material in question) to be deposited and retained in the deep respiratory tract for 120 days. After being swallowed, the radioactive particles may either descend through the gastrointestinal system and be excreted, or it may pass into the body. Varying portions may be excreted through the kidneys.

Engineering controls are the preferred means of keeping the DACs below the limits in 10 CFR Part 20, Appendix B. Areas in which radioactive materials are used should be ventilated with nonrecirculated air. If the workplace is designed properly so that air movement is away from the workers' breathing zone, work in which airborne releases are possible are done in an effective fume hood, and the quantities of activity are typical of most research programs that do not require access control. An air exchange of 6 to 10 air changes per hour should be sufficient to provide the needed environmental control for normal operations. The preferred hood type would be a good quality standard hood rather than an add-air type, since the latter type, with two airstreams and a more complicated design, have been found on occasion to be more prone to spillage if not used and maintained properly. An airflow velocity through the face of the hood of approximately 125 fpm (38 m/min) would be a good design target. The hood should be equipped with an alarm to warn if the airflow through the face were to fall below 100 fpm (30.5 m/min). Making sure that any source of radioactive gaseous or vapor effluent is at least 8 inches (20 cm) from the entrance to the hood and keeping the sash down approximately halfway will further significantly improve the ability of a hood to capture and confine airborne hazardous materials.

If for a valid reason full control of the levels of airborne contaminants cannot be maintained within the legal limits, it is permissible to use respiratory protection under some conditions. These are found in the current Part 20 in Section 20.1703 and in Appendix A to Part 20. A key provision in 20.1703(a)(3) is for the licensee to have a formal written respiratory protection program.

There are several different types of respirators that provide various levels of protection. The various types and their characteristics will be discussed briefly below and more extensively in Chapter 6. However, their function is to reduce the level of radioactive material in the air being breathed by the user as far below the acceptable levels as is reasonably achievable.

Respirators fall into two primary classes, those in which air is supplied and those in which the air is purified by being passed through filters. The most common type of non-air-supplied units used for radioactive materials are HEPA filter-equipped respirators since most of the problem materials involve particulates. Supplied-air units are used where greater protection factors are needed. Supplied-air respirators themselves fall into several different classes. Units in which air is supplied on demand, which do not provide a positive pressure inside the face mask with respect to the ambient, air are the least effective because contaminants can enter should the seal between the mask and the face fail. These are rated to provide a protection factor of 5. Self-contained units in which air is supplied continuously or in which the air is always at positive pressure are rated highest and provide protection factors up to 10,000. There are two other types, one in which the air is recirculated internally and purified chemically but is always at a positive pressure with respect to the outside, and another in which air is supplied to a hood that does not fit snugly to the face, but has a skirt that comes down over the neck and shoulders. These latter two types provide actual protection factors of 5000 and up to 1000, respectively, but are rated more conservatively. Self-contained units have a severe limitation in that they provide air for only a limited period. The most popular SCUBA type, with a 30-minute air tank, may provide air for only 15 minutes if the wearer is under heavy physical stress. Larger versions of the air-recirculating and air-purifying type can provide air for up to 4 hours, but even these are not suitable for continuous wearing.

Respirators that purify the air by filtering require a snug fit to the face to ensure that they are effective. Unfortunately, as a person works, the contact may be temporarily broken and the protection provided by the mask diminished. Facial hair will prevent a good seal, so that bearded workers cannot be allowed to use respirators requiring a good facial seal for protection. Some facial types are very hard to fit successfully, although there are increasingly more models available which allow different types of features to be fitted. More models are also becoming available with a mask made of hypoallergenic materials that permit more persons to wear them without suffering irritation.

A major requirement is that the individual be physically able to wear a respirator. It takes some effort to breathe through a cartridge respirator, and a person with emphysema or some other breathing impairment would not be able to wear such a unit for extended periods. Passing a physical examination that includes a pulmonary function test is required of employees who would be expected to wear a respirator in doing some tasks assigned them. The employer is required to have a comprehensive respirator program if advantage is taken of this method of achieving compliance with the permissible air limits of airborne concentrations of radioactive materials for the employees.

Although respiratory protection does offer an alternative to engineering controls to provide suitable breathing air quality, the use of respiratory protection is more commonly used for providing suitable air for temporary or emergency work situations.

In summary, to reduce internal exposure:

1. Do a hazard analysis on all procedures, including routine ones, for failure modes where releases of materials, contamination, or exposures could occur.
2. Use the isotope that would be the least dangerous if taken into the body.
3. Use the smallest amount of radioactive material possible consistent with the requirements of the experimental program.
4. Reduce the time in which an airborne exposure could occur.
5. Do the hands-on work in a radiological fume hood for most radioisotope research, or in a glove box if the possibility of airborne exposure mandates it because of the type, the amount, or the form of the material.
6. Take measures to reduce contamination with which it would be possible to come into contact, e.g., using absorbent plastic-backed paper to line the work surface, procedures to reduce the production of aerosols, and a tray underneath the experimental apparatus to catch any spills.

7. Wear gloves (two pairs when working with material in solvents that could cause a glove material to soften or weaken, or where a single pair might be breached due to abrasion, or puncture).
8. Use protective outerwear that can be discarded if contaminated.
9. Do not eat, drink, chew gum, apply cosmetics, or smoke in the area in which radioactive materials are used.
10. Wash your hands after any use of radioactive material, before leaving the work area, and before eating, drinking, smoking, or any other activity in which radioactive material could be ingested.
11. Supplement the protection provided by the hood with the use of respiratory protection, if needed, to achieve compliance with airborne concentration limits.
12. Cover any cut with a waterproof bandage. If a minor cut occurs while working, wash the wound out, allowing it to bleed freely for a while. Then check for contamination. If uncontaminated, cover with a waterproof bandage. If still contaminated, immediately contact radiation safety personnel for assistance.

Limitation of Dose to the Fetus

The revised Part 20.1208 places the responsibility of reducing the exposure to a fetus/embryo on both the employer and the women carrying the fetus/embryo. The woman's formal responsibility is met by the woman *declaring*, i.e., notifying her employee, that she is pregnant. It is then the responsibility of the employer to limit the total effective dose (including external and internal exposure, if applicable) to the fetus/embryo to 0.5 mrem (5 mSv) during the pregnancy.

It is generally accepted that rapidly dividing cells are unusually radiosensitive, which, of course, is the situation that obtains as the embryo develops. The early stages of fetal development, especially between the 10th and the 17th week, appear to be a critical time during which the fetus is unusually susceptible to radiation damage. Unfortunately, during these first few critical months of a pregnancy, some women might not realize that they are pregnant and thus could begin to take precautions to limit the exposure to the developing fetus. However, to avoid any unusually high exposures at potentially critical stages of development, the recommendation is that the exposure be spread uniformly throughout the prenatal period, or an average of 0.054 rem (0.54 mSv) per month.

Because of the uncertainty in the early critical stages of pregnancy of whether an individual is pregnant or not, a restriction on all fertile women to an annual dose of 0.5 rem (5 mSv) might be considered desirable. Since only a limited number of women of fertile age are pregnant at any given time, this could cause discrimination against hiring women and unnecessarily affect their employment opportunities. Radiation workers, on average, receive a dose equivalent of less than 0.5 rem per year, so that it does not appear justifiable to impose such a limit on fertile women as a class. It would appear reasonable for women to continue to exercise reasonable care in reducing the radiation exposure of any unborn child that they might be unknowingly carrying.

Part 20, Section 20.1208 sets limits to the occupational exposure of a woman who has voluntarily *declared* or informed her employer that she is pregnant. These limits include both external and internal exposures. Many materials do pass from the women to the unborn child through the placenta. Sometimes the fetus receives a disproportionate share. In addition, the fetus may be exposed to radiation from materials in the organs of the woman's body. The proposed rule requires that the licensee ensure that the embryo/fetus in such a case not receive an effective dose more than 0.5 rem (0.5 mSv) during the entire pregnancy. The effective dose is the sum of the deep dose equivalent to the declared pregnant woman and the dose to the fetus from radionuclides in the fetus and radionuclides in the declared pregnant woman. As noted in Section 5.2.4.2, because of the uncertainty in establishing the actuality of the pregnancy, some women may not inform the licensee that they are pregnant until the fetus has already received a dose of 0.5 rem (0.005 Sv) or more. In such a case the licensee is enjoined from allowing the fetus to receive more than 0.05 rem (0.5 mSv) during the remainder of the pregnancy.

Compliance with the limitations on exposure to the embryo/fetus should not involve an economic penalty on the woman, nor should it jeopardize the woman's job security or employment opportunities.

Personnel Monitoring

Determination of whether or not individuals receive doses over the permissible levels requires that a personnel monitoring program be in effect, in which the doses to individuals are measured. Until the recent revision to Part 20, programs generally measured and maintained records for external doses only, since the exposures to airborne concentrations are limited to a relatively small number of specialized facilities. Part 20.1204 specifies the conditions under which it is necessary to include the internal doses in the total committed effective dose for an individual. Most programs will still find it necessary to monitor only external exposures.

Personnel monitoring requirements are given in Section 20.1501. The licensee must provide personal monitoring devices and require their use by employees who may receive an occupational dose in excess of 10% of the limits in 1 year specified in Section 20.1201, or of any individual who may enter a high radiation area. Personnel monitoring records must be kept according to Section 20.2106 and shall be retained until the NRC terminates the license requiring the records.

Part 19 provides in Section 19.13 that workers or former workers can have access to their radiation monitoring records. Individual workers may request an annual report of their radiation exposure. Former workers may request their exposure history from an employer, and this must be provided within 30 days, or if the records have not been completed for the current quarter by the time the request is made, within 30 days after the data are available. An employee who is terminating employment can also request and receive his monitoring records. These data must provide the dates and locations at which the exposures occurred. There also are additional technical requirements for the content and form of the data provided.

Maintenance of these records by each licensee is important since each employer must attempt to determine a worker's prior exposure record. Under Section 20.2104, for records that are not available, the current employee must assume a prior exposure of 1.25 rem (12.5 mSv) per calendar quarter for each quarter the employee could have received an occupational exposure. The individual would not be available for planned special exposures if prior records were not available. These restrictions could restrict the employee's usefulness to the current employer.

5.2.7.1.8. Methods of Monitoring Personnel Exposures

There are several types of devices, or dosimeters, which can be used to implement a personnel radiation monitoring program. These devices are worn by the worker so they represent the actual radiation received by the wearer at the location on the body where they are worn. Since the most radiologically sensitive organs are the blood-forming tissues, the genitals, and, to a lesser extent, the eyes, the most logical place to wear a dosimeter usually is on the outside of a shirt pocket. Working at a laboratory bench or in front of a hood in which the radioactive material is located, the distance to the source will probably be least in this location, and so the reading will most likely be higher than average over the body. This is a conservative procedure. The most common monitoring devices are film badges, thermoluminescent dosimeter badges, and direct or indirect reading pocket ionization chambers (the latter are becoming much less common). Except for the latter type of dosimeters and dosimeters used to measure the dose to the extremities, Section 20.202(c) requires that personnel dosimeters used by licensees to comply with the monitoring requirements must be processed and evaluated by a dosimetry processor:

1. Holding current personnel dosimetry accreditation from NVLAP of the National Bureau of Standards and Technology.
2. Approved in this accreditation process for the type of radiation or radiations included in the NVLAP program that most closely approximates the type of radiations for which the individual wearing the dosimeter is monitored.

To accurately reflect the dose received by the body, a dosimeter ideally should have an energy response that would be the same as the tissue of the body. None of the dosimeters mentioned above fully meet this requirement, although the use of various filters in badges allows correction factors to be applied to the results. Each of the three types of units mentioned above will be discussed in the paragraphs below.

There are other dosimetry materials that are used in certain circumstances, for example, at very high exposure levels. There also are devices that electronically integrate the output of an electronic radiation detector and provide a digital reading of the accumulated dose, serving the same purpose as a simpler pocket ionization chamber, but at a considerably higher cost.

Versions of each of the following devices can be used to measure exposure to neutrons. However, since the use of neutrons outside reactor facilities or some types of accelerator facilities is uncommon, these will not be discussed here.

Film Badges: The sensitivity of film to radiation was the means by which Roentgen discovered the existence of penetrating radiation, and it still serves as one of the more commonly used materials to measure radiation exposures. The radiation level to which the film has been exposed is proportional to the blackening of the film. For gammas below about 200 KeV, however, the energy dependence of the film depends very strongly on the energy, increasing rapidly as the gamma (or X-ray) energy decreases. In a typical case, the sensitivity of the film peaks at about 40 KeV, falling off below this energy because of the attenuation of the soft gammas by the cover on the film to protect it from light. The sensitivity curve can be modified by using an appropriate filter, as shown in Figure 5.9. With filters, the sensitivity of a dosimeter using film can be made to be uniform to within about ±20% over an energy range of about 0.12 to 10 MeV. For intermediate energy gammas, film is usable (with care) as a dosimeter for exposures of about 10 mrad (0.1 mSv) to about 1800 rad (18 Sv). Film covered only by a light-tight cover also can be used to measure the dose from a beta emitter if the energy of the beta is above about 0.4 MeV, for exposure levels of about 50 mrad to about 1000 rad (0.5 mSv to 10 Sv). There is no point in providing badges to individuals who work with low-energy pure beta emitters such as tritium, ^{14}C, and ^{35}S since those betas would not pass through the light-tight cover of the film.

Figure 5.9.   Film badge for personal radiation monitoring. The various areas shown on the front of the badge provide for information to be obtained about the radiation reaching the film.

Film badges can be fairly sophisticated in their design. Even the simplest will have at least one film packet (or part of a larger piece of film) covered only by its light cover, to allow detection of betas and soft gammas or X-rays.

Another area of the film covered only by the plastic of which the badge holder is typically made will help distinguish between the betas and less energetic gammas. Usually one or more metal filters of different atomic numbers will be used that will change the response characteristic of the film, as well as help define the energies of the incident radiation. Usually the film has two emulsion layers, one of which is "faster," i.e., more sensitive than the other, to provide a wider response to different levels of radiation. Additional filters can be placed within the badge to allow it to detect and distinguish between thermal and fast neutrons. A diffuse source of radiation (such as scattered radiation) can be distinguished from a point source since the former will give diffuse edges to the images while the latter causes the edges of the filter images to be more sharply defined.

There are factors other than the varying energy response of the film that cause the reproducibility of their readings to vary. Each batch of film is a little different, the processing is subject to variation, and the temperature in the environment in which it is worn will affect the results. The average accuracy of film badge dosimeters to doses of 100 mrad (1 mSv) or more is about ±25%. Values below about 50 mrad (0.5 mSv) are often relatively meaningless. The requirement that dosimeter evaluation be done by processors belonging to NVLAP is a measure taken to control as many of the variable factors as possible. Additional variations are introduced by the way they are handled and worn. They should be worn on the body of the worker, on the outside of the clothes, since clothes would modify the radiation received. The wearer should be careful to orient the badge properly since the various sections of the badge depend upon radiation being incident upon the front of the badge. Obviously, the badges should never be placed in the employee's trouser pocket. Not only will the orientation be incorrect, but the badge will be shielded by coins, keys, and any other objects that might be present.

Although film badges have some disadvantage—specifically relatively poor accuracy and a nonuniform energy response for low energy gammas—they have a major advantage in that they represent a long-lasting record. If the original developed film is stored properly, it can be reevaluated later if there is reason to suspect the original record is incorrect.

Thermoluminescent Dosimeters (TLDs): The two most serious problems with a film badge—nonuniform energy response and reproducibility —, are eliminated by a TLD dosimeter. Since a Lithium Fluoride (LiF) TLD with an effective atomic number of 8.1 is similar to tissue which has an effective atomic number of 7.4, the LiF energy response is about the same as tissue. The response of a LiF TLD is almost energy independent over an energy range of 0.1 to about 3 MeV. The accuracy of a TLD can be better than 10% over an exposure range of 10 mrad to 1000 rad (0.1 mSv to 10 Sv). The upper limit for a quantitative response is on the order of 100 times as great.

TLD materials are crystals in which radiation excites the atoms of the material, leaving them in long-lived metastable energy levels, locking the excitation energy into the crystal. Heating the crystals releases the energy in the form of light. If the heating process takes place at a uniform, controlled rate, the crystal will release the trapped energy in a reproducible "glow" or light curve as the temperature of the crystal rises. The area under the light curve generated in this manner is proportional to the quantity of the original radiation. The TLD material is often in the form of a small chip, although one vendor of TLD badges uses Teflon impregnated with the LiF instead. Obviously, the light seen by the reader from the detector will be affected by the condition of their surfaces, but if handled with care, a TLD dosimeter can be recycled many times.

As noted above, the energy response and reproducibility of a LiF TLD dosimeter is much better than that of a film badge. Badges can be configured as with film badges to provide dose information on betas, X-rays, soft gammas, and more energetic gammas. However, unlike film, the process of reading the dosimeter to obtain the exposure information will destroy the information stored in the crystal. They also cannot readily distinguish between a point and diffuse source. TLD dosimeters have gained a large share of the market in recent years because of their overall advantages, despite the deficiencies noted.

Ionization Dosimeters: Both of the two previous types of dosimeters require processing in order to be read. The two most common types of ionization dosimeters can be read immediately, although the indirect type requires an auxiliary reading device to do so. Ionization dosimeters to be used as personnel dosimeters are typically about the size of a pen and are usually called pocket dosimeters.

The indirect reading type is simply a high-quality cylindrical capacitor in which the outer cylinder, made of an electrically conducting plastic or one having a conducting surface, forms one side or plate of the capacitor, and a wire along the axis of the cylinder forms the other side. The capacitor is charged to a predetermined voltage, and radiation impinging on the walls and the air in the cavity acts to discharge it, lowering the potential difference between the two electrodes. The potential difference is then read by an electrostatic voltmeter, calibrated so that the scale reads the amount of radiation to which the dosimeter and the wearer were exposed. The process of reading this type of ionization chamber discharges them so they must be recharged to be useful again.

The direct reading type also is a simple device. A fine, gold-plated quartz fiber, acting as a gold-leaf electroscope, is charged to a predetermined potential. As the air surrounding the fiber is exposed to radiation, the "electroscope" discharges, and the quartz fiber moves toward the discharged position. The motion of the fiber, superimposed on a scale, is viewed through a simple microscope incorporated in the device. The scale is calibrated so that the change in position can be read as the radiation to which the dosimeter has been exposed.

Both of these two types of dosimeters are primarily intended to read gammas, and provide a response which is within about 15% of the actual exposure over gamma energies of about 0.04 to 2 MeV. Both can be configured to read over a wide range of exposures. Most of those used in the laboratories are usually set up to read from 0 to 200 mrem. The readings of the units are conservative since charge leakage, or discharging due to factors other than radiation, will cause the readings to be too high.

5.2.7.1.9. Bioassays

The dosimeters discussed in the previous section measure the external fields of radiation to which the wearer is exposed. They do not measure the intake of any materials that will contribute to internal exposures. Methods of measuring airborne concentrations of radionuclides will be discussed in the following section, as part of the general subject of area radiation surveying. Measurements of the actual uptake of radioactive materials by the body is done by doing bioassays on individuals. There are routine laboratory uses of radioactive materials that are biologically active, such as the use of ^{125}I in certain laboratory analytical tests, which call for tests of specific organs to see if any of the material has been taken into the body. On other occasions, bioassays are used to determine if there has been any intake of material after a spill or other types of accidents in which there has been an unplanned release.

There are two distinct aspects to doing a bioassay: (1) measurement of an activity related to the original intake, and (2) inferring the exposure from the measured activity. Of the two, measurement of an activity associated with the intake is the easier.

The basic assumption on which many bioassay procedures are based is that the activity in the excreta, e.g., the urine or feces, will be related to the amount of activity taken into the body. Urine is the usual choice for monitoring, especially if the original material is soluble, while for particulates that are less soluble both the feces and urine might be used. The radioactivity from ^{40}K (0.012% isotopic abundance, 1.28×10^9 half-life, 1.31 MeV ß, 1.461 MeV γ) in the potassium in urine typically results in about 45 dps/L as an interfering activity when trying to measure the activity due to an intake of another radioactive material. Chemical removal of the potassium activity is possible and usually desirable to ease the measurement of the intruding isotope. Where the activity to be measured is due to a pure beta emitter and bremsstrahlung radiation does not accompany the beta emission, determining the activity in the urine and feces is almost the only way to obtain a measurement.

The interpretation of the data obtained from urine and fecal measurements in terms of the activity in the body and the original amount taken in is complicated. Soluble materials containing radioisotopes that are distributed throughout the body lend themselves to the easiest analysis. Soluble materials which concentrate themselves in an organ represent a more complicated situation since they first must build up in the organ from the body fluids and then leave the organ to enter the excreta. Those that are absorbed in the bone have very long clearance times. Unfortunately, there are many factors that affect the metabolic processing of the materials and affect the interpretation of the results in terms of the actual internal exposure.

Urine analysis can be done for gamma emitters as well, but in cases in which the isotope is strongly concentrated in a single organ, a direct measurement of the activity in the organ using a scintillation counter (or perhaps a Germanium detector) may be feasible. It also is possible to obtain a relatively accurate measurement of the amount in the organ by using a known amount of the isotope (or an isotope with a comparable gamma spectrum) in a model (or phantom) which simulates the physical characteristics of a person, as far as absorption and scattering of radiation from the organ are concerned. This is commonly done for iodine uptake measurements in the thyroid. The use of radioactive iodine is significant, so this technique is widely used.

An extension of the concept of measuring the activity in a single organ is to use a "whole-body-counter" in a scan of the activity throughout the body, which provides a measurement of the entire body burden of gamma-emitting radioisotopes. These are large, expensive devices and are not often found in a typical radiation safety operation, although there are a number of locations where access to one is available in emergencies.

If it is suspected that a radioactive material has been inhaled, measuring the activity from nose swipes or in a tissue after the nose is blown can be used to confirm the inhalation. Measurement of radon concentrations in the breath is a technique that can be used if materials are inhaled which have radon as a decay product. Carbon-14 can be detected in exhaled breath as well.

5.2.7.1.10. Radiation Surveys

Measurements of the radiation levels in an area are a vital aspect of radiation safety. Every laboratory using radioisotopes should have ready access to the means to measure the ambient activity in the facility. In areas where there are fixed sources of substantial radiation such as a ^{60}Co irradiator, some types of accelerators, or a nuclear reactor, fixed radiation monitoring systems usually are installed. These permanently installed systems need to be supplemented by portable instruments to be used to detect radiation levels in localized areas that might not be "seen" by the fixed system, or arising from movable sources.

There are several types of instruments that can be used for radiation monitoring, each with its own characteristics, advantages, and limitations. There have been serious radiation exposures when individuals used an instrument that was not appropriate for the radiation field involved. This section will be limited to the type of radiation detection instruments that are commonly used for radiation surveys rather than those used for laboratory research. The instruments will primarily be those intended to measure external exposure rates.

A steady exposure to radiation that would result in a dose of approximately 100 mrem (1 mSv) per week or 2.5 mrem per hour (25 μSv per hour) for a 40-hour work week would allow a worker to meet the annual limits of occupational exposure for the whole body of 5 rem. The hands are often placed in radiation fields that would be at much higher levels than this, although they would rarely remain exposed to higher levels for an entire work week. The current limit of 75 rem per year would translate to about 1.5 rem (15 mSv) per week or approximately 37 mrem per hour (0.37 mSv per hour). A survey instrument for a laboratory using by-product materials should be able to read levels that are a small percentage of the average whole-body exposure rate that would remain within the limits (10 CFR Section 35.72 stipulates that a survey instrument should be able to read a minimum of 0.1 mrem per hour), and should, at a minimum, be able to read levels at or above the equivalent rate for extremities. The high end of the instrument's range probably should be selected to be somewhat greater than this guide would suggest, especially for spills in which an amount of material larger than that typically used in a single procedure may be involved. The maximum amounts of radioactive material that could be in the laboratory at a given time could be used to determine the high range of a survey instrument that might be selected. Survey instruments are required by 10 CFR Section 35.51 to be calibrated on all scales up to 1000 mrem per hour if they have that range.

Most laboratory survey instruments use gas-filled chambers as their detectors. In this type of instrument, as discussed earlier, the passage of radioactive emissions (α, β, or γ)

through matter creates a large number of ion pairs that can be translated into an electrical signal to detect the radiation. There are several factors that affect the ability of an instrument to detect an interaction with the gas in the chamber as a discrete event or not. The combination of these factors is called the resolving time, with a short resolving time allowing more events per unit time to be detected individually than a longer resolving time. Other factors would affect the sensitivity of the detector and its ability to distinguish between emissions of different energies or types. Usually, survey instruments for daily use in a laboratory are needed to simply detect the presence or absence of radiation, without regard to the energy or type. Most, however, do have the ability, by mechanical means, to distinguish between types of radiation. (Taking readings with and without a cap over the detector will allow the instrument to distinguish between betas and gammas.) Since the radioisotope in use is usually known, and laboratories often use only one type of material, or at most a limited number, the person doing the survey rarely will need an instrument that can measure energies or varieties of radiation for simple surveys. When an instrument with more sophisticated capabilities is needed, the radiation safety office normally can be counted upon to provide one. For more information on the properties of gas filled counters, the reader is referred to any number of references on nuclear instrumentation.

Although there are many commercial varieties of instruments available with various advanced and useful features, they usually are only elaborations, with more versatile and sophisticated circuitry, of the simple Geiger counters and ionization chambers that have been available for decades. The basic instruments are adequate to illustrate the safety features that are most critical.

All survey instruments kept in the laboratory should be calibrated at least once per year on all scales. If facilities are not available to do this locally, there are many commercial firms that will provide the service.

Geiger-Müller Counter

The G-M counter, or Geiger counter as it is usually called, is the least sophisticated type of instrument that might be employed as a laboratory survey instrument. Its simplicity allows it to be built inexpensively as a rugged and, if used properly, relatively foolproof device. However, many knowledgeable and competent scientists have managed to use one incorrectly.

A simple Geiger counter in its basic form often has a cylindrical probe approximately 1 inch in diameter and about 6 inches long. The outer cylinder forms the cathode of the detector (the electrode toward which the positive ions move from the ion pairs created by the passage of a particle or ray). A center wire forms the anode toward which the negative ions, normally electrons, move from the ion pairs. The center wire is connected to the external circuitry. One end of the cylinder usually has a thin "window" with an aerial density of about 1 to 2 mg/cm^2, through which betas of approximately 30 KeV and greater can pass. In order to protect the fragile window, it usually is covered with a removable cap. Alpha particles also can be detected with a thin window counter such as this, if the counter is held very close to the source of the alphas. A 4.5 MeV alpha has a range in air of about 3 cm, but it loses energy as it passes through the air, so that at a distance of a little less than 2 cm the alpha particle would not retain sufficient energy to penetrate even a 1 mg/cm^2 window. The geometry just described is not ideal for surveying a surface contaminated with emitters of alphas and betas where the detector has to be held very close to the contamination, but many thousands of such instruments have been constructed and used.

Since gammas can penetrate the cylindrical walls of the counter as well as the thin end window, the geometry is not as disadvantageous for them.

When a charged particle, gamma or X-ray enters the active volume of a Geiger counter, not only does it create ion pairs, but the potential between the cathode and anode is set sufficiently high so that these ion pairs create an "avalanche" of charge along the entire length of the central anode so that a substantial positive space charge is developed around the center wire. The negative ions are electrons and move very swiftly to the anode while the positive ions, being atoms, move far more slowly toward the cathode, represented by the outer wall of the tube. While the positive ions are moving, the detector tube is not able to initiate another pulse, or at least not one as large as the first. The minimum time that must elapse before another event can be detected is the resolving time. For most Geiger counters, the resolving time is about 100 to 200 microseconds. This seems short, but it places a severe limitation on the number of events per second that can be detected. For example, let us calculate what a counter with a 200 μs resolving time would measure if it is operated so that it should have a true count rate of 10,000 counts per second, if there were no problems with resolving time.

The equation that relates the true count rate C, the observed count rate c, and the resolving time t, is given by Equation 17.

$$C = \frac{c}{(1 - c\tau)} \tag{17}$$

If the assumed numbers are inserted in this equation, the observed count rate is 3333 cps, or a discrepancy of two thirds. Obviously, a Geiger counter cannot be used at a high interaction rate to obtain an accurate exposure rate. If the number of particles or rays becomes too high, the counter will saturate so that it will appear that no pulses are occurring and the apparent count rate will become zero. Under these conditions, the instrument would be worse than useless since a zero reading might be construed by the user as the true one, and on that basis enter a dangerously high radiation field. This has happened, and serious radiation injuries have occurred as a result.

Because of the inability to function properly at high count rates, simple Geiger counters are not usable in high fields. A Geiger counter can be calibrated against a known source of a known specific activity in terms of mrad (mGy) or mrem (mSv). A basic Geiger counter usually has a maximum scale of 20 to 50 mrem per hour (0.2 to 0.5 mSv per hour). If the activity being used is small enough, this may be adequate, although instruments that can read levels 10 to 20 times as high, but still have a scale on which the maximum reading is 1 to 2 mrem per hour (10 to 20 μSv) would be a more versatile general purpose instrument. Ionization chambers are available that meet this criterion, as discussed in the next section.

Ionization Chambers

The basic geometry is similar to that of the Geiger tube except that the diameter of the chamber usually is substantially larger so that the volume of the gas in which the radiation may interact is much larger. An ion chamber is operated at a high enough potential between the electrodes so that the ion pairs created by the impinging radiation do not recombine, but not so high that the moving ions create any significant number of secondary ions. The ions are collected on the anode and cathode and can be detected as a weak current. A more intense (stronger) source will create more ion pairs and would produce a stronger current. Actually the current is not measured, but a large resistor (10^{10} ohms or larger) is connected from the anode to ground so that a voltage is developed across the resistor. A sensitive and stable electrometer circuit is used to measure the voltage. The larger the resistor, the more sensitive the counter. However, using larger resistors results in a penalty due to the time it requires for the current through the resistor to reach equilibrium (the time constant of the response is determined by the load resistor and the capacitance of the chamber). On a very sensitive scale, it could take 1 minute or more for a steady-state voltage to be attained. An ion chamber requires more care than does a Geiger counter. Accuracy requires that the resistance values not change and dirt and moisture can cause the effective resistance to decrease.

The advantage of the ion chamber is not that it can read very low fields, but that it can be designed, if the necessity arises or exists routinely, to read higher levels that represent a danger to the workers. At the higher levels, the slow response time observed at lower levels should pose no problems since the resistor and, therefore the time constant, can be substantially lower and still provide an adequate voltage that can be measured.

As with the simple Geiger counter, the end window of the counter chamber can be made thin so that the unit can be capable of counting alphas, betas, and gammas. If only one survey instrument can be afforded by a facility, an ion chamber with an upper range of 10 R per hour (0.1 Sv per hour) would be a good choice.

Other Concepts

There are units designed to operate in what is called the proportional region of the confined gas, in which there is some multiplication of the charge deposited by the impinging radiation, but which permit much shorter resolving times so that count rate losses are much less severe. There are other arrangements of the electrodes, so that the detector presents a large sensitive area to the surface being surveyed and a shallow depth for the sensitive volume, instead of a long cylindrical geometry. This usually will result in increased sensitivity, especially for alphas and betas that have short ranges in matter. It also allows "seeing" a larger surface area and facilitates a survey of an extended area, such as a work surface. If energy discrimination and sensitivity are needed, portable scintillation counters are available, incorporating NaI(Tl) crystals. If there is a possibility of personnel contamination, counters can be set up at the points of egress which can be used to check automatically the hands and feet of the employees as they leave the facility. Friskers, which use a pancake-shaped detector, are available to rapidly scan a person's body to detect contamination on areas other than the hands and feet. There are specialized counters for virtually any application.

General Surveys

Maps of the laboratory area should be prepared and, periodically, a general contamination survey of the area should be made using a general purpose survey meter, especially if the isotope in use emits gammas or strong beta emitters such as ^{32}P. The frequency of these general surveys should depend upon the level of use of material in the facility. If the use is very heavy, a daily survey of the immediate work area and a general survey of the entire laboratory at an interval of once a week might be necessary, while very light use might make once a quarter adequate. The records of these surveys should be maintained in a permanent log and audited periodically by the institution's radiation safety officer. They should also be retained for the NRC, should they be needed.

More detailed surveys of the immediate work area and certain critical areas should be carried out as part of contamination control, as discussed in the following sections.

5.2.7.1.11. Measurement of Airborne Activities

The collection of samples for testing airborne radioactive materials follows essentially the same procedures as described in Section 4.8.2 for airborne monitoring of chemical contaminants. The basic procedure is to use a pump to circulate a known quantity of air through a collection device with known characteristics for capturing contaminants. If there are individuals working in an area where it is possible that airborne radioactivity may be a problem, it is recommended that routine periodic sampling of the air for activity be done and that the activities to which individuals may be exposed be directly and specifically monitored. Sampling can be done for an entire work day or sometimes during times of potential peak exposures by using small, lightweight pumps and collectors worn near the breathing zone. If significant levels are found, a program to control the airborne contamination should be promptly initiated.

If the suspected contaminants are particulates, membrane filters are frequently used which have collection efficiencies of 100% for particles most likely to enter the deep respiratory tract (>0.3 to around 2 μ). A membrane filter traps particles on the surface, which is desirable for beta and alpha emitters, since energy losses in the filter material could be important if the particles were carried into the filter. The concentration of a specific airborne contaminant in air is derived by measuring the activity of the collected sample, correcting for counting efficiencies, and dividing by the volume of air circulated through the filter.

Natural airborne activity will contribute to most airborne measurements of activity. This arises primarily from the decay products of the various radon isotopes. Radon is a gaseous radioactive element, arising from the decay of thorium and radium. The two principal radon isotopes of concern are ^{220}Rn (55 second half-life) and ^{222}Rn (3.82 day half-life). The first of these has such a short half-life that it will tend to decay before the gas diffuses very widely; therefore, the daughter products, which are not gaseous, will remain in the same vicinity as the ^{224}Ra, the parent nuclide for ^{220}Rn. Radon-222, with its much longer half-life, will diffuse away from the original location of its parent. The short-lived daughter products of ^{222}Rn, with half-lives ranging from 164 μs for ^{214}Po to 26.8 minutes for ^{214}Pb, tend to become ionized and attach themselves to dust particles.

The activity in the particulates in the air is typically at an approximate state of equilibrium, since additional radon gas is diffusing into the air constantly. However, the activity due to the daughter products of ^{222}Rn on the particulates that have been collected on the filter is not being replenished and will decay until a much longer half-life daughter is formed, ^{210}Pb (half-life 22.3 years), which will have a much lower specific activity. A 4- or 5-hour wait will usually see most of the activity from the daughter products of ^{222}Rn gone. Although not as significant originally, the activity from the daughter products of ^{220}Rn, such as that of ^{212}Pb and its daughter products, may dominate the natural activity contribution after 4 or 5 hours. The half-life of ^{212}Pb is longer (10.6 hours) than that of any of the following daughters (or of the preceding ones) until a completely stable isotope is reached, so the activity in the collected sample due to the daughters of ^{220}Rn will have an equilibrium half-life of 10.6 hours. The activity of most radioisotopes used in the laboratory is typically much longer than this, so the activity of the airborne research isotope can readily be mathematically separated from the background natural activity by counting the collected sample several times over the next 2 or 3 days. If t_c is the half-life of the contaminant and t_n is the half-life of the natural background radiation in the air sample, the activity due to the contaminant in the collected sample can be obtained using any two of the delayed counts and by using the expression:

$$C = \frac{C_2 - C_1 e^{-0.693 \ t/t_n}}{e^{-0.693 \ t/t_c} - e^{-0.693 \ t/t_n}} \tag{18}$$

Here, t is the time interval between the two counts employed in the calculation.

If several counts are taken, Equation 18 can be used for several pairs to make sure that only two half-lives are contributing to the activity. Within statistical counting variations, the results should be the same for each pair. If they change systematically, as the time from the beginning of the initial count changes, then additional components are contributing to the activity.

If the airborne contaminant is contained in a solvent aerosol or is gaseous instead of being a particulate, the basic procedures for sampling for nonparticulate materials in Section 4.8.2 can be used. The counting procedures may differ. Instead of using a standard counter, such as a proportional counter (for betas and gammas) or a solid scintillation detector (for gammas), a liquid scintillation detector might be used instead. The latter type of counter is particularly effective for betas, as its efficiency can approach 100%. If the contaminant is a gamma emitter, a more sophisticated counting system using either a NaI(Tl) or a germanium detector and a dedicated multichannel analyzer or computer can differentiate the energies of the gammas. These latter types of counting systems, with either a "hardwired" or software-based analytical program, cannot only positively identify the isotopes contributing to the energy, but can automatically determine the activity in the sample. These sophisticated analytical instruments are expensive compared to the type of instruments previously discussed, so that they normally are found only in laboratories heavily engaged in work with gamma emitters or in the radiation safety facility.

5.2.7.1.12. Fixed and Loose Surface Contamination

Contamination is the presence of undesirable radioactivity. It may be either fixed or loose. Usually the acceptable limits for these in laboratories are set by the organization, except for certain items such as packages received in the laboratory for which Part 20 establishes acceptable standards, at levels that should be sufficient to assure the occupational safety of personnel and that of members of the public. Although acceptable levels are generally set by the licensee, 10 CFR Section 35.70 stipulates that in medical facilities involving humans, the licensee must have instruments capable of measuring radiation levels as low as 0.1 mrem per hour or detect contamination of 2000 dpm with a wipe test. Although not explicitly stated, it could be inferred that these would be acceptable limits for surface contamination by the NRC in these facilities.

Fixed Contamination

Fixed contamination, by definition, is unlikely to either become airborne so that it may be inhaled or adhere to the hands upon being touched, so that it is also not likely to become ingested. Material that has soaked into a porous surface would be a typical fixed source of radioactive contamination. The hazard is an external one. Under the circumstances, an acceptable level could be defined by the acceptable levels for external exposures. In an unrestricted area a level of 0.25 mrem per hour (2.5 μSv per hour) would barely meet this criterion. A level substantially below this level would be preferable, such as 50 μrem per hour (0.5 μSv per hour). The NRC Regulatory Guide No. 1.86, for the purposes of releasing an area to unrestricted use, uses a value of 5 μR per hour (0.05 μSv per hour) above background at a distance of 1 meter from the source. This is roughly one third of the average background in most parts of the country. Such a level would be for a permanent cessation of operations, and is probably lower than is necessary while the facility is licensed. While normal operations are being conducted and radiation safety survey and monitoring programs are actively done, a level five times as high would correspond to a conservative value of 10% of the permissible level for unrestricted areas. For betas, a value of 50 μR per hour (0.5 μSv per hour) for a reading close to the surface would not appear unreasonable for an unrestricted area. An alpha level that is detectable at all in an unrestricted area would suggest a possible dispersal of some of the more toxic radionuclides in the area, which should not be allowed to occur, because of a research program. A nondetectable level of alpha contamination for fixed material in an unrestricted area would be desirable. However, levels equivalent to about 0.3 nanoCi or 10 Bq of loose contamination per 100 cm² are considered acceptable by some agencies.

For a restricted area, a level of fixed contamination well below the occupational level of 5 rem per year (50 mSv per year) or about 2.5 mrem per hour (25 μSv per hour) would be appropriate. Some organizations have set the level at 1 mrem per hour (10 μSv per hour). Another option might be to set it below a level that would require personnel monitoring, currently the equivalent of about 0.25 mrem per hour (2.5 μSv per hour).

Loose Surface Contamination

The limits for loose surface contamination should be related to the amount of airborne contamination that would result if the material were to be disturbed so it could enter the body by inhalation or indirectly by ingestion through handling objects on which the material rested. The limiting factor is internal exposure. An examination of 10 CFR Part 20 readily reveals that the permissible DACs for the various isotopes vary by more than seven orders of magnitude, although the majority cluster around intermediate levels between the extremes. For the isotopes most commonly used in the laboratory, the maximum permissible concentrations still differ by almost three orders of magnitude. This disparity makes it difficult to establish a single limit for loose surface contamination.

As noted above, a level of removable alpha contamination between nondetectable and about 0.3 nanoCi (10 Bq) per 100 cm² is considered acceptable by many different groups for an unrestricted area. In restricted areas some organizations still do not accept any stray alpha contamination, except in the immediate work area, such as the inside of a hood. However, a level of about 0.8 nanoCi or 30 Bq per 100 cm² is considered acceptable by others.

Loose surface contamination is usually measured by rubbing or swiping an area of about 100 cm² with a filter paper, and then counting the activity that adheres to the paper. An estimate must be made of the fraction of the loose material that is removed from the surface and remains on the paper, to make a judgment of the amount of surface contamination.

For beta and gamma contamination, otherwise unspecified, levels of 0.1 to 1 nanoCi per 100 cm² are considered acceptable by various organizations for unrestricted areas. For restricted areas, these organizations usually make the acceptable levels ten times higher, i.e., 1.0 to 10 nanoCi per 100 cm². The United Kingdom and some others attempted to roughly take into account the differences in the MPCs and DACs by grouping the nuclides into toxicity classifications and recommending acceptable levels for each group. For the most toxic radionuclides (which includes the alpha emitters), an acceptable level of loose contamination in restricted areas is set at 0.8 nanoCi (30 Bq) per 100 cm^2. The levels for medium toxicity radioisotopes are increased by a factor of 10, while the levels for lesser toxicity materials are increased by another factor of 10 to 100. This range of four orders of magnitude does not span the entire range of DACs, but it is more reasonable than a single number for all species of radioactive materials.

All the numbers are for general area contamination. For personal clothing they should be at least ten times less and for the skin of the body, the levels should be either zero or as close as can be obtained by decontamination efforts that do not cause damage to the tissue.

Frequency of Surveys for Contamination

In laboratories where unsealed radioactivity is in use, a contamination survey of the work areas directly involved in the operations should be made at the end of each workday, or at the conclusion of that day's operations involving radioactive materials. Surveys of the workers' hands and clothing also should be checked at the end of each day's operations, or more frequently if they leave the work area or take a break to get something to eat or drink. In any week that radioactive materials have been used, a contamination survey of the entire facility should be made. Particular attention should be paid to the work areas, items that might have been handled during the day, including controls of equipment, sink faucets, and the survey meter, radioactive material storage area, radioactive waste storage area, and the points of egress from the facility. The latter data are for proving that no radioactive materials have been transported from the room on the soles of the employees' shoes.

Accidents and Decontamination

Contamination may arise from poor laboratory practices, in which case the organization's radiation safety program must be prepared to take whatever steps are necessary to ensure that these practices are corrected. No such situation should be allowed to persist for long in a well-managed organization. In a properly run laboratory accidents should be the major source of contamination of the facility. Even accidents should be infrequent in a well-run facility.

The scale of the accident will determine the response to the incident. If, for example, the accident is a spill in a hood, if the base of the hood has been lined with absorbent paper, and the apparatus set up in a pan or tray that is sufficient to contain all the spilled fluids, there may be little or no contamination or dispersal of airborne activity within the room because the vapors will be discharged in the hood exhaust. The individual working at the hood, if not personally contaminated, should notify the laboratory supervisor and the material removed from the hood and placed in a suitable container. The container can be sealed and placed in the area reserved for radioactive waste. Any additional protection required for cleaning up, such as the use of two pairs of gloves, respirators, and coveralls, should be decided upon before beginning the remedial work. Every accidental spill should be reported to the radiation safety office. The RSO may wish to help, even for a minor spill, to ensure that all personnel are protected. After the clean up is completed, the entire area involved in the incident should be surveyed as well as all personnel involved in the original incident and the clean up. Controls of instruments that may have been handled should not be overlooked. The problem of personal contamination is covered below.

Larger accidents pose more risk that individuals may have become directly contaminated or may have inhaled or ingested radioactive materials. Other individuals in the laboratory other than the ones directly using the material may have been exposed as well. A contingency plan should have been developed that should be put into effect when a radioactive emergency occurs. The following two scenarios illustrate some of the more common problems that should be considered in preplanning for an emergency:

A Spill Directly upon a Person

The individual should stay where he is or, at most, move a few feet away to avoid standing in spilled material and immediately remove all garments that are or may be contaminated by the spilled material. The spread of contamination should be minimized. The garments should be placed in a pail or other liquid-tight container, preferably one that can be sealed. The pail can be placed in a double garbage bag and the bags closed with a twist if a covered container is not available. After removing all contaminated clothing, the individual involved in the spill should wash all contaminated areas of the body, being especially careful to wash areas where tissue creases or folds. If the spilled material is volatile, the contaminated individual and those assisting him should don appropriate respirators as soon as possible. If it is volatile, the most likely respirator needed would be one equipped with an organic material cartridge, unless the activity were in a different type of liquid. Individuals who assist the contaminated person should also thoroughly wash their hands. Potentially contaminated persons should be checked for contamination. Persons not immediately involved in the incident and not needed should immediately leave the area until the potential for aerosol dispersal is determined. However, those leaving the area should be checked as they leave to confirm that they are not carrying contamination with them.

Radiation safety personnel and laboratory employees should jointly decide how to clean up the spill after the immediate contamination to personnel problem has been resolved, and determine a schedule that will allow operations to resume. An investigation should take place for any incident of this size or greater, and the RSC should review the circumstances, the immediate actions taken, and decontamination program to ensure that all actions taken were consistent with maximum protection of personnel and compliance with regulatory standards.

A Large Incident in Which Significant Airborne Activity is Generated

In any incident in which there is a likelihood of airborne contamination being generated, the most conservative approach is to immediately evacuate all personnel. This is based on the assumption that the situation will not rapidly worsen if it is left alone. An example of a situation that could worsen would be a small fire that involves small quantities of radioactive material and that could spread to involve not only larger quantities of radioactive materials, but other dangerous materials as well. If the fire appears controllable with portable fire extinguishers (which should be conveniently available), one or two persons might choose to remain to attempt to rapidly put out the fire, but all other occupants should immediately evacuate. The individuals staying behind should leave as soon as possible. Note that this approach will depend upon factors such as the toxicity of the materials, the possibility of inhalation of airborne material, and the amounts involved, as well as the possibility of controlling the fire. If staying to fight the fire only to reduce loss of property would endanger a person's life, the recommendation is to evacuate all personnel immediately.

In a situation where the only problem is the generation of airborne radioactivity, individuals should leave the space in which the airborne activity is present, closing the doors behind them and turning off any ventilation in the area, if the controls are in the room. If the ventilation controls are in a switch room or breaker room elsewhere, they can be turned off from this location. If the accident occurs in or near a fume hood that would vent the material from the laboratory, it may be appropriate to allow the hood to remain on to reduce the airborne concentration of activity within the room; however, this would depend upon the radioactive toxicity of the material involved. If it is one of the more toxic materials, it would usually be preferable to avoid dispersal into a public area even if the levels are low, because of the possible exposure to persons in these areas. The laboratory represents a controlled space that can be decontaminated and, in general, the best approach is to restrict the contaminated area as much as possible.

Individuals within the laboratory should, as in any other emergency, evacuate the area following routes that would reduce their risks. Any location within the facility should have two evacuation routes. The storage location of portable radiation survey instruments should be near the entrance to the facility, or along the evacuation route deemed most likely to be safe in a radiation-related emergency.

Once the occupants have left the immediate area of danger and it has been isolated, the next steps to take in the emergency should be considered. The first order of business is for an individual (preferably one with the least potential of having experienced personal exposure in the incident) to notify the radiation safety department and other responsible authorities to request assistance. The police or security staff may be needed to control access to the facility. Concurrently, others should begin to evaluate any personal radiation exposure that may have taken place, by using the laboratory's portable instrumentation to check each individual. Assuming it is safe to remain, no one should leave the immediate area until they have been thoroughly checked for contamination, for their own safety and to prevent spread of contamination. In the case of airborne contamination, it would be essential to check the accessible portions of the respiratory tract for traces of inhalation. Nasal swipes should be taken. Personnel might blow up balloons, if available, for later checking. All individuals who are knowledgeable about the incident should remain to provide information to those persons arriving to take charge of the emergency response.

Once radiation safety personnel, security staff, and other persons such as the building authority have arrived and have been made thoroughly aware of the circumstances of the accident, a plan can be developed to respond to the situation. Other temporary steps to isolate the problem can be taken, such as taping around the edges of the doors and defining a secure area with barricades to prevent persons from entering or approaching the area and interfering with the response program.

Groups involved in planning the response should include the participation of laboratory personnel, radiation safety specialists, managerial personnel, media relations staff, and other support groups as dictated by the nature of the incident.

An operations center should be established outside the immediate area of the incident, and all operations and information releases should be managed from this center to ensure that all actions to control and remedy the situation are properly coordinated. Every incident is different and the correctional plan also would be different. However, in general, the approach is to establish a control point at the boundaries of the affected area through which all personnel, supplies, and waste would enter and leave the area. All persons and material leaving this area must be checked for contamination and all individuals entering would be checked to see that they are suitably protected. The normal program is to start at the edge of the contaminated area and work to reduce the area involved until all contamination has been removed or reduced to an acceptable level.

Assistance should be sought immediately for individuals who might have received internal and external contamination. Contact with a regional radioactive incident response center is advisable, since few local medical centers have sufficient training, experience, or facilities to respond to an internal radiation exposure incident. If the telephone number of the appropriate center is not immediately available, the regional NRC office, which must have been notified according to Part 20 2202, can provide it. The regional center may recommend transfer of the affected personnel to a facility in their area. Any information that could possibly aid medical personnel in their evaluation and treatment is invaluable. The response center may be able to suggest short-term measures to reduce the amount of radioactive material taken up by the body.

Decontamination Techniques

Whether contamination has been caused by routine operations or by a major or a minor spill, the problem of decontamination arises. The cost and feasibility of decontamination of individual pieces of equipment and the added exposure risk to personnel must be weighed against the cost of replacement equipment and the cost of disposal of any material considered as radioactive waste. The latter costs are not negligible.

The facility itself, including major items of equipment such as hoods, work benches, and valuable specialized items of laboratory apparatus, are generally worth substantial decontamination efforts since they are expensive to replace. At current average building costs of $150 to $300 per square foot for laboratory space, a facility is not lightly abandoned or left idle for extended periods. If the facility has been built originally to simplify decontamination of spills of all types, and the major items of equipment specified for ease of maintenance, the decontamination process may be straightforward. Personnel, dressed in coated Tyvek™ coveralls, shoe covers, and head covers and wearing gloves and respirators, can clean most such areas by washing the surfaces with detergent and water since the surfaces should have been selected to resist absorption of water and chemicals. The equipment in the room should have few seams or cracks in which contamination can become lodged.

Unfortunately, many laboratory facilities using radiation have been located in older buildings with vinyl tile, unsealed concrete, or even wooden floors in which radioactive contamination can become fixed, and equipment has many seams and cracks that lend themselves to becoming contaminated. If surfaces are porous or if water would collect in the cracks, the use of detergent and water may not be desirable. In addition, materials on surfaces tend to adhere more strongly to them as time passes, due to moisture, oils, etc., in the air, so if it is not desirable to use a cleansing solution to remove the material, it would be best to begin decontamination as soon as practicable while material is still relatively loose. If a very limited area is involved, loose particulates often may be picked up by pressing the sticky side of masking or duct tape against the contaminated surface. If a larger area is contaminated, much of the loose material can be picked up by using a HEPA filter-equipped vacuum cleaner. The filter in a vacuum cleaner of this type is capable of removing 99.97% of all particles of 0.3 μm or greater from the air moving through it. Brushing lightly with a soft brush can break loose additional material that can be vacuumed with the special vacuum. To keep airborne material from spreading, the entrance to the area being cleaned can be isolated with a temporary "airlock" made of 6-mil polyethylene plastic that will block any air moving to and from the isolated area, but will allow passage of workers in and out of the room. If needed, the plastic enclosure can include a change area or even a temporary shower. Airborne material also can be continuously removed from the area by placing a movable, HEPA filter-equipped, air circulator within the space being cleaned. The circulator should be sized to pass the air in the space through the filter frequently. A 2000-cfm unit would pass the air (if thoroughly mixed) in one of the standard laboratory modules described in Chapter 3 through the filter about every 2 minutes.

If contamination has permeated deeply into the work and floor surfaces and into seams and cracks in the floor, it may be simpler to remove some of the surface and to remove and replace floor tile than to attempt cleaning. If the floor tile contains asbestos, the removal would have to be done according to OSHA and EPA restrictions on asbestos removal.

Every area that has been decontaminated must be thoroughly surveyed by the organization's radiation safety office and certified as meeting the limits for surface contamination established by the organization before release for use.

Pieces of electronic equipment are perhaps the most difficult items to clean since they frequently are equipped with fans that draw the air through them; besides normal lubricants in them, over time they all accumulate at least some dust and grime to which the contamination is likely to adhere, even in rooms in which the air is carefully filtered. Normally, washing the interior components is impractical because of the potential damage to the components. It may be possible to use a very fine nozzle on a HEPA vacuum to clean the bulk of the removable dust from inside the instrument, and the remainder may be loosened and removed with the careful use of small swabs and solvent, followed by another vacuuming. The decision to make the effort will depend on the cost of the equipment, and the difficulty of the decontamination effort. A factor that might influence the decision to make the effort would be the possibility of a successful damage claim on the organization's insurer. The disposal cost of the equipment should be considered as well as the replacement cost.

Small tools and glassware with hard, nonporous surfaces may be cleaned by standard techniques. Washing with a detergent and water may be sufficient. More resistant contaminating material in glassware can be cleaned with any of the standard solutions used in laboratories for cleaning glassware, such as a chromic acid solution. Contaminated metal tools that resist cleaning with detergent and water, accompanied by brushing, can be washed with a dilute solution of nitric acid. Sulfuric acid can be used on stainless steel tools. Again, the choice must be made among the relative cost of the tool, the cost of disposal, and the effort required to decontaminate it. Tools that have grease in and on them and hidden crevices in which radioactivity may become embedded may be too difficult to decontaminate to make the effort cost effective.

If disposable protective clothing is used throughout the decontamination process the problem of laundering contaminated garments can be minimized. However, any items that need to be cleaned should be checked by the radiation safety office before sending them to a commercial laundry. If the quantity is not large, manual washing of some items might be possible, taking care not to get skin contaminated. Where more clothes are involved, purchasing a washing machine for the job might be cost effective. A simple, inexpensive model would normally be sufficient and probably can eliminate or reduce the contamination to an acceptable level unless the clothes are heavily contaminated in a form physically difficult to remove. Any wastewater should be captured and retained until certified by the radiation safety officer as suitable to be disposed of in the sanitary system.

5.2.7.1.13. Radioactive Waste Disposal

In recent years almost all off-site low-level* radioactive waste disposal has been sent to three commercially operated waste disposal burial grounds located in South Carolina, Nevada, and the state of Washington. The Nevada site no longer is accepting waste and the Barnwell facility is planning to shut down in 1996, to be replaced by a facility in North Carolina. There have been other commercial facilities closed for various environmental problems. The choice of the location to which generators of radioactive waste sent their waste was dictated by a number of operating restrictions for the three facilities that remained operating. As a result, many institutions found themselves sending much of their laboratory-type radioactive waste from the east coast and elsewhere to the Washington facility.

Because it was deemed inappropriate by Congress for one region to be responsible for handling another region's wastes, they dictated that a system of regional compacts was to be established, and each region would be required, when the system became fully operational, to take care of its own radioactive waste. Some regional compacts are fully operational while others are not. At this time, it is still possible for a generator in some regions to send waste to another area, but at some point in the future, this will no longer be allowed. The date at which this will occur is not wholly certain because it may be modified by Congress to allow additional time for regions who have not set up facilities to do so.

Although the nation's research and medical facilities do not produce the volume of waste generated by the electric utilities, they are much more vulnerable to the lack of a disposal facility for their waste because they do not have adequate facilities in which to store low-level waste on a long-term basis. It is essential for many research applications to use certain isotopes, which, because of their radiation characteristics, must go to a long-term disposal facility. The inability to use these isotopes would seriously hamper many research programs. These research applications often are in the life sciences and involve such critical areas as cancer, the search for cures to other diseases, recombinant DNA, and other comparably important areas of basic and applied research.

The actual amount of waste shipped off-site at many research facilities has decreased or stabilized as more facilities use waste reduction techniques to reduce the amount of radioactive waste that must be shipped away.

Decay-in-storage is the simplest method available for the reduction in the amount of waste that must be disposed of in a burial facility. Many of the most commonly used radioisotopes have sufficiently short-half lives so that the radioactive content will decay to a "safe" level within a reasonably short time. For example, all of the following have half-lives of less than 65 days, so that any material in the waste will go through at least 10 half-lives in less than 2 years, which is generally accepted as a reasonable storage period: ^{131}I, ^{32}P, ^{47}Ca, ^{33}P, ^{51}Cr, ^{59}Fe, ^{203}Hg, and ^{125}I. Also the combination ^{99}Mo $\rightarrow$ ^{99m}Tc, which has important nuclear medicine applications, has a short effective half-life so that any waste from its use also can be allowed to decay in storage. The initial activity of any radioactive material when placed in storage will determine if ten half-lives is sufficient to reduce the level to a point at which an unshielded reading with a survey meter on its most sensitive scale shows no reading over background. If this condition is not met, then additional decay time must be allowed before the material is disposed of as ordinary waste.

Another significant means of reducing the amount of waste is to take advantage of the provisions of Section 20.303 to dispose of some of the radioactive waste into the sanitary

---

* Low-level as opposed to the high-level waste engendered by the use of uranium as a fuel.

system. However, Section 20.2003 prohibits the disposal of licensed material into the sanitary system unless it is readily soluble or dispersible in water. Some recent concern has been expressed about the dispersible materials and this may be modified. Many fluids used in liquid scintillation fluids do not meet these criteria, but there are now commercially available alternative scintillation liquids that do. The use of these newer scintillation fluids allows licensed or unlicensed material to be put into the sanitary system as long as the amounts are within the limits specified in 20.2003(a). Another restriction limits the yearly amounts to 5 Curies of ^{3}H or 1 Curie of ^{14}C and a total of 1 Curie of all the other radioisotopes combined. Since ^{14}C and ^{3}H, two of the most commonly used radioactive research materials, have long half-lives and advantage cannot be taken of the decay-in-storage technique, the capability of using the sanitary system is of significant help in reducing the waste volume requiring commercial burial. Records of the amounts disposed of in this method must be maintained, and also the amounts disposed of by other means.

Very low specific activity (0.05 µCi or less per gram) ^{3}H or ^{14}C in scintillation fluid or in animal tissue (averaged over the weight of the entire animal) can be disposed of without consideration of the radioactivity but is normally incinerated. The scintillation fluids can be incinerated only if they do not meet the EPA ignitability criterion. Some generators have incinerators licensed for disposal of radioactive waste that are used for this purpose.

Some dry solid wastes can be compacted to reduce their volume. Some generators have developed centralized local facilities to do this, while others place these materials into separate containers for the waste disposal firm to carry out this waste volume reduction technique.

These methods, plus encouraging the users to choose a material that can be allowed to decay in storage, can be employed to substantially reduce the amount and expense of radioactive waste disposal.

To take advantage of these waste and cost-reduction measures, users must cooperate to segregate the waste into appropriate categories: (1) aqueous, (2) nonaqueous, (3) aqueous containing ^{3}H and ^{14}C from other aqueous, (4) vials of nonmiscible scintillation fluids containing less than 0.05 µCi/g of ^{3}H and ^{14}C from other scintillation fluids and EPA compliant scintillation fluids from others, (5) dry solid wastes, and (6) animal carcasses separated into those containing more than 0.05 µCi/d per g of tissue of ^{3}H and ^{14}C from those that contain less than that amount. The generators of the waste must label each container of waste with the amount of each radioactive isotope contained in it. This information is used by either the radiation safety office to prepare the waste for shipping and to prepare the required manifests and forms as specified in Section 20.2006, or by a waste disposal firm who may do this service as well as transporting the waste to a burial ground. Each container or package of waste shipped must be classified and labeled according to 10 CFR Part 61, Section 60.55–57.

Almost all waste is originally packed into 55-gallon steel drums that meet DOT specifications, segregated by types. Generally, landfills will not take liquid wastes, even if they have been absorbed in an appropriate absorber. They must be solidified before being offered for burial. Some organizations have a commercial service do this procedure. The choice in most cases of whether to do a given operation locally or have it done by an outside contractor is usually based on economic factors. Nonaqueous waste is usually treated as RCRA waste as well.

Radioactive waste allowed to decay in storage may be disposed of according to the following schedule:

- Dry solids are taken to an ordinary landfill. Any labels on any container showing that the contents are radioactive must be removed or defaced.

- Nonhazardous water, miscible waste is put into the sanitary system.
- Chemically hazardous waste is disposed of through the hazardous waste disposal program. Labels on a container showing that the contents are radioactive must be removed or defaced.
- Carcasses are normally incinerated but new, primarily state, regulations may affect this method.

Proper disposal of all radioactive waste is a major responsibility of the organization's radiation safety office. Proper records must be maintained of all transfers of radioactivity from the laboratory into the wastestream.

5.2.7.1.14. Individual Rights and Responsibilities

Individuals who are employed in licensed programs in which they work with by-product materials (or those employed in several other activities regulated under Title 10 Chapter I, of the Code of Federal Regulations) have their rights and responsibilities spelled out in the regulations, primarily in Parts 19 and 21. Part 21 requires that the responsible parties in any licensed activity report to the NRC any safety deficiency, improper operations or defective equipment that could pose a "substantial safety hazard...to the extent that there could be a major reduction in the degree of protection provided to public health and safety..." However, any individual can make such a report. If a responsible party fails to do so, substantial civil penalties can be imposed.

Part 19 is the section of the regulations that defines the rights of the employees and those of the NRC, and also the responsibilities of the licensee to make these rights available. The major provisions of Part 19 are summarized below.

Information Requirements

The licensee must post several items of information relating to the operations under the license or post information describing the documents and where they may be examined. The latter choice is the one most often exercised. The documents involved are

- The regulations in 10 CFR Part 19
- The regulations in 10 CFR Part 20
- The license, license conditions, and documents incorporated into the license by reference
- The operating procedures applicable to the licensed activities
- Form NRC-3, "Notice to Employees"

The following items must be posted, without the option of simply informing the employees of their existence and where they may be examined

- Any violations of radiological working conditions
- Proposed civil penalties or orders
- Any response of the licensee

A critical section of Part 19 is 19.12, "Instructions to Workers," given in its entirety below. It is brief, but is functionally equivalent to the OSHA Hazard Communication Standard for employees engaged in activities regulated by the NRC.

Section 19.12 Instructions to Workers

"All individuals working in or frequenting any portion of a restricted area shall be kept informed of the storage, transfer, or use of radioactive materials or of radiation in such portions of the restricted areas; shall be instructed in the health protection problems associated with exposure to such radioactive materials or radiation, in precautions or procedures to reduce exposure, and in the purposes and functions of protective devices employed; shall be instructed in, and instructed to observe, to the extent within the worker's control, the applicable provisions of Commission regulations and licenses for the protection of personnel from exposure to radiation or radioactive materials occurring in such areas; shall be instructed of their responsibility to report promptly to the licensee any condition which may lead to or cause a violation of Commission regulations and licenses or unnecessary exposure to radiation or to radioactive material; shall be instructed in the appropriate response to warnings made in case of any unusual occurrence or malfunction that may involve exposure to radiation or radioactive material; and shall be advised as to the radioactive exposure reports that workers may request pursuant to § 19.13. The extent of these instructions shall be commensurate with potential radiological health protection problems in the restricted areas."

Implicit in the above instructions is that the employees must comply with the regulations and with procedures adopted to protect not only themselves but others, including the public outside the areas in which the licensed activities take place. However, it is not sufficient for an employer to place the responsibility on the employee. The employer must see that the employees comply with all applicable standards. There have been several instances in which heavy fines have been imposed on corporations and academic institutions that have failed to enforce compliance.

Monitoring Data

Individual workers have the right to information concerning their radiation exposures, and directly related supporting data, under several provisions of Section 19.13:

- The worker can request an annual report.
- A former worker can request a report of his exposure records. The report must be provided within a 30-day interval or within 30 days after the exposure of the individual has been determined, whichever is later.
- An employee terminating work involving exposure to radiation either for the licensee or while working for another person can request the information. An estimate, clearly labeled as such, may be provided if the final data are not yet available at the time of the request.
- If the licensee is required to make a report to the NRC of an individual's exposure data, a report must be made to the individual no later than the report made to the NRC.

Inspection Rights

The NRC has the right to make unannounced inspections of a licensed facility, including the "...materials, activities, premises, and records...." During an inspection, the inspectors have the right to consult privately with the workers. However, at other times, the licensee or a representative of the licensee can accompany the inspector.

The workers may have an official representative who may accompany the inspector during the inspection of physical working conditions. Normally, this is an individual who is also a worker engaged in licensed activities under the control of the licensee, and therefore could be expected to be familiar with the instructions provided to the employees by the employer. However, if agreeable to the employer and to the employees, an outside person can be allowed to accompany the inspector during the inspection of the physical working conditions. Different employee representatives can accompany the inspector during different parts of the inspection, but no more than one at any given time. If a person deliberately interferes with an inspector while conducting a reasonable inspection, he can refuse to let that person continue to accompany him. Certain areas, such as classified areas, or areas where proprietary information is involved, would require that only individuals normally having access to these areas or information can accompany the inspector.

During an inspection, an employee can bring privately to the attention of the inspector any matter relating to radiological safety pertaining to the licensee's operations that he wishes, either orally or in writing.

An employee or an employee representative can report a violation of the terms of the license or regulations covering the radiological activities engaged in by the employee. This should be done in writing to the commission or to the director of the commission's regional office and should provide pertinent details of the alleged violations. The identity of the employee making the report will not appear in any public report except under demonstrated good cause.

If the director of the regional office feels that there is valid cause to believe that a problem exists as described by the complainant, he will initiate an inspection as soon as practical. If an inspection does take place, the inspector will not necessarily confine himself to the original problem. The director may decide that an inspection is not justified. However, a complainant may request an informal conference to press his case or a licensee also may request one, although in the latter case the complainant has the right to not allow his identity to be made known.

The licensee is specifically prohibited against any act of discrimination against an employee who makes a complaint or asks for an inspection.

Penalties

Violations of Part 19, as with other regulatory sections, can be classed as *de minimis* or higher. If the violations are sufficiently serious, the commission can obtain a court order prohibiting a violation or one can be obtained to impose a fine or a license may be revoked. If a violation is willful, in principle the person responsible can be considered guilty of a crime and be subject to a fine and/or imprisonment. Of course, none of the penalties can be invoked without the licensee having ample opportunity to respond to the charges and to propose a plan of correction.

REFERENCES

1.   **Komarov, E.I.,** Radiation in daily life, *Int. Civ. Defense*, No. 295, 3.
2.   New a-bomb studies alter radiation estimates, *Science*, Vol. 212, May 22, 1981.

3. Ionizing Radiation: Sources and Biological Effects, United Nations Scientific Committee on the Effects of Atomic Radiation (UNSCEAR), 1982 Report to the General Assembly, with Annexes, United Nations, NY, 1982.

4. **Fabrikant, J.I.,** Carcinogenic effects of low-level radiation, *Health Phys. Soc. Newsl.,* X(10), October, 1982.

5. Radiation Exposure of the Population of the United States, NCRP Report No. 93, National Council on Radiation Protection, Bethesda, MD, 1987.

6. **Luckey, T.D.,** Physiological benefits from low levels of ionizing radiation, *Health Phys.,* 43(6), 771, 1982.

7. Special issue on radiation hormesis, *Health Phys.,* 52(5), May, 1987.

8. **Cohen, B.L.,** Alternatives to the BEIR relative risk model for explaining atomic-bomb survivor cancer mortality, *Health Phys.,* 52(1), 55, 1987.

9. *Location and Design Criteria for Area Radiation Monitoring Systems for Light Water Nuclear Reactors,* ANSI/ANS-HPSSC-6.8.1, American National Standards Institute, New York, 1981.

10. *Performance Requirements for Pocket-Sized Alarm Dosimeters and Alarm Rate Meters,* ANSI N13.27, American National Standards Institute, New York, 1981.

11. **Hodges, H.D., Gibbs, W.D., Morris, A.C., Jr., and Coffey, W.C., II,** An improved high-level whole-body counter, *J. Nucl. Med.,* 15(7), 610, 1974.

12. **Graham, C.L.,** A survey-instrument design for accurate ß dosimetry, *Health Phys.,* 52(4), 485, 1987.

13. **Kathren, R.L.,** *Radiation Protection, Medical Physics Handbook 16,* Adam Hilger, International Publishers Services, Accord, MA, 1985.

14. Standards for Protection Against Radiation, 10 CFR Part 20, May 31, 1983.

15. Standards for Protection Against Radiation, 10 CFR Part 20, *Fed. Reg.,* 56(97), 1991.

16. Rules of General Applicability to Domestic Licensing of Byproduct Material, 10 CFR Part 30, *Fed. Reg.,* March 31, 1987.

17. Licensing Requirements for Land Disposal of Radioactive Waste, 10 CFR Part 61, *Fed. Reg.,* 47(248), 57446, 1982.

18. Decontamination Limits, Regulatory Guide 1.86, Nuclear Regulatory Commission, Washington, D.C., 1974.

19. *Decontamination Limits,* ANSI/ANS-N13.12, American National Standards Institute, New York.

20. Instruction Concerning Prenatal Radiation Exposure, Regulatory Guide 8.13, Nuclear Regulatory Commission, 1975.

21. Information Relevant to Ensuring that Occupational Exposures at Medical Institutions Will Be As Low As Reasonably Achievable, Regulatory Guide 8.18, Nuclear Regulatory Commission, 1982.

22. Instructions Concerning Risks from Occupational Radiation Exposure, Regulatory Guide 8.29, Nuclear Regulatory Commission, 1981.

23. National Emission Standards for Hazardous Air Pollutants: Radionuclides, 40 CFR Part 61, *Fed. Reg.,* 54(240), 51654, 1989

24. A Guide for Determining Compliance with the Clean Air Act Standards for Radionuclide Emissions from NRC-Licensed and Non-DOE Federal Facilities, Revision 2, EPA 520/1-89-002, Environmental Protection Agency, Washington, D.C., 1989.

25. User's Guide for the Comply Code, 520/1-89-003, Revision 2, Environmental Protection Agency, Washington, D.C., 1989.

26. **Steere, N.V., Ed.,** Safe handling of radioisotopes, in *CRC Handbook of Laboratory Safety,* CRC Press, Boca Raton, FL, 1971, 427.

27. **Strom, D.J.,** The four principles of external radiation protection: time, distance, shielding, and decay, *Health Phys.,* 54(3), 353, 1988.

28.   **Saenger, E.L.,** Acute local irradiation injury, in *Proc. REACTS Int. Conf. Medical Basis for Radiation Accident Preparedness,* Hübner, K.F., and Fry, S.A., Eds., Elsevier/North-Holland, 1980.

29.   **Martin, A. and Harbison, S.A.,** *An Introduction to Radiation Protection,* 3rd. ed., Chapman and Hall, New York, 1986.

30.   **Morgan, K.Z. and Turner, J.E.,** *Principles of Radiation Protection,* John Wiley & Sons, Inc., New York, 1967.

31.   **Pelá, C.A., Ghilardi, A.J.P., Netto, and Ghilardi, T.,** Long-term stability of electret dosimeters, *Health Phys.,* 54(6), 669, 1988.

32.   **Cember, H.,** *Introduction to Health Physics,* Pergamon Press, Oxford, 1969.

33.   *Radiological Health Handbook,* U.S. Department of Health, Education and Welfare, Washington, D.C., 1970.

5.2.8. Radon

The problem of radon is not necessarily connected with laboratory operations involving radiation other than its being a contributor to background for counters used in these facilities and representing a problem when taking an air sample. However, many laboratories are built of concrete blocks and have concrete floors, drains that open to storm sewers, and other means for radon to enter a facility. The sand and aggregate used for the concrete in both the blocks and cement slabs may contain substantial amounts of natural radioactivity. Currently, a considerable amount of concern is being raised about the possibility of radon-induced lung cancer from this source. Since a typical research worker may spend an average of 8 or more hours per day in a research facility, it may be useful to provide a brief section on radon. A discussion of the physics of radon activity was provided in Section 5.2.7.1.10. Figure 5.10 provides some additional information on the decay products of ^{222}Rn, the major environmental problem.

A study by the National Research Council has resulted in a new estimate of the risk of lung cancer that is intermediate between those of earlier estimates. The report also suggests that the risk to smokers is ten or more times greater than that to nonsmokers. Although compiled by an eminent group of scientists using a more sophisticated set of assumptions than in the past, other well-known scientists have questioned some of the assumptions made and some features of the model. The results are given below, but before presenting them, a brief discussion of some of the terms is needed.

Risk estimates are made in terms of working-level-months (WLM). A WLM is defined as "an exposure to a concentration of any combination of the short-lived radon daughters in 1 L of air that results in the ultimate release of 1.3×10^5 MeV of potential alpha energy for a working month of 170 hours." This amount of radiation is approximately equivalent to the radiation of a radon daughter in equilibrium with 100 pCi of radon. A level of 4 pCi/L of radon is equal to about 0.02 working levels. Over an entire working year, equal to 2000 hours, the equivalent exposure would be equal to about 0.25 WLM.

The BEIR IV report estimates that the excess lifetime risk of death from lung cancer from this source is 350 deaths per million person WLM. Based on this model and the assumed levels of exposures throughout the U.S., this is equivalent to an estimated 5000 to 20,000 lung cancer deaths due to radon per year in the U.S.

At the time of this writing, there are reputable individuals who both support and object to these findings. However, because there is some concern about the potential of radon as a cause of lung cancer, many commercial services have been established which will provide devices that can be placed in the home for a period of time, and then can be sent to their laboratories for analysis of the ambient radon levels. These firms place advertisements in

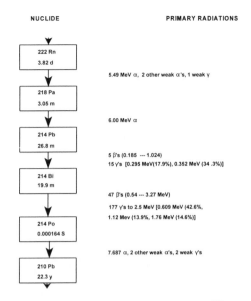

NUCLIDE   PRIMARY RADIATIONS

222 Rn
3.82 d

5.49 MeV α, 2 other weak α's, 1 weak γ

218 Pa
3.05 m

6.00 MeV α

214 Pb
26.8 m

5 β's (0.185 --- 1.024)
15 γ's [0.295 MeV(17.9%), 0.352 MeV (34 .3%)]

214 Bi
19.9 m

47 β's (0.54 --- 3.27 MeV)
177 γ's to 2.5 MeV [0.609 MeV (42.6%,
1.12 Mev (13.9%), 1.76 MeV (14.6%)]

214 Po
0.000164 S

7.687 α, 2 other weak α's, 2 weak γ's

210 Pb
22.3 y

Figure 5.10. Primary decay path for Rn^{222}.

the media so that anyone wishing the analysis can readily have it done. In some states, state agencies will make a survey upon request.

REFERENCES

1. *Health Risks of Radon and Other Internally Deposited Alpha-Emitters*; BEIR IV, National Academy Press, Washington, D.C., 1988.
2. **Cohen, B.L. and Gromicko, N.,** Adequacy of time averaging with diffusion barrier charcoal adsorption collectors for ^{222}Rn measurements in homes, *Health Phys.*, 54, 195, February, 1988.
3. Radon in Buildings, NBS Special Publication 581, Collé, R. and McNall Jr., P.E., Eds., National Bureau of Standards, Gaithersburg, MD, 1980.
4. **Puskin, J.S. and Yang, Y.A.,** A retrospective look at Rn-induced lung cancer mortality from the viewpoint of a relative risk model, *Health Phys.*, 54(6), 635, 1988.
5. **Cohen, B.L.,** Relationship between exposure to radon and various types of cancer, *Health Phys.*, 65(5), 529, 1993.
6. **Cole, L.A.,** *Element of Risk: The Politics of Radon,* AAAS Press, Washington, D.C., 1993.

5.2.9. Acute Radiation Syndrome

It is improbable that an individual working with by-product materials will ever be exposed to radiation levels that would cause him to be concerned about the immediate acute effects of radiation, leading to serious injury or death. However, there have been a small number of incidents in which massive exposures have occurred. These have normally involved large sources used for radiography, radiation therapy, sterilization of insects, machines producing intense beams of radiation, X-ray units, and critical reactor assemblies. Individuals have survived very high localized radiation doses, in the range of 10,000 rems or more, if the areas affected were relatively small and did not expose the more sensitive organs to a substantial dose. Following is a brief discussion of the immediate effects of a short-term exposure to high levels of radiation. It should be noted that the levels that cause immediate harm to individuals are hundreds of times greater than that allowed individuals over a period of months.

There is significant variation in the responses of individual persons to exposures of the whole body. The least amount of radiation for which clinical evidence is discernible in a "typical" person is approximately 25 rad (0.25 Gy). The effect will be noted as changes in the blood system. Above about 50 rad (0.5 Gy), most persons will show some effect in their blood system. There may be temporary sterility in men for exposures to the gonads at these levels. However, it is unlikely that an individual will personally feel any immediate discomfort at these relatively low levels.

As the exposure levels increase to approximately 100 rad (1 Gy), a small percentage of sensitive persons will begin to experience some of the symptoms of discomfort associated with acute exposures to substantial levels of radiation, nausea, fatigue, loss of appetite, sweating, and a general feeling of malaise. As the exposure level increases, the percentage of individuals exhibiting symptoms increases, until at the point a person experiences an exposure of 200 rad (2 Gy) most will be affected to some degree. At about this same level, some individuals who do not receive treatment may not survive. At an exposure between 300 and 450 rad (3 to 4.5 Gy), the survival rate without aid decreases to 50% by 30 days after the exposure. With proper treatment, survival is a distinct possibility up to about 500 rad (5 Gy,) above which, even with optimal treatment, survival chances decrease rapidly, until at about 800 rad (8 Gy) survival is highly unlikely.

Many individuals who have a finite chance of survival will begin to feel better after a day or two, but after a few weeks, their situation appears to worsen. They may have fevers, infections, loss of hair, severe lethargy, hemorrhaging, and other problems with their cardiovascular systems. During this 2- to 3-week interval, blood tests will reveal that significant changes are occurring in the blood system which reflect the damage to the blood-producing bone marrow. At about 200 rad (2 Gy) depression of the bone marrow function is apparent. At higher doses, complete ablation of the bone marrow occurs. Up to some exposure level, if the patient can be kept alive, the bone marrow will regenerate, but at a level in the area of 400 to 600 rad (4 to 6 Gy), this will become unlikely, and bone marrow transplants will be necessary for survival. There are problems with this technique, including the need for a close genetic match for the donor material and the possibility of infection. However, if the blood-forming function can be restored, and death does not occur due to secondary infections because of the lack of white blood cells, survival for 60 days or more is a promising sign of eventual survival.

As the level of radiation increases, the survival period decreases. At about 1000 rad (10 Gy), there will be serious damage to the gastrointestinal tract and survival may be a few weeks. At exposures of several thousand rad, the central nervous system will be strongly affected and survival will be measured in terms of hours.

REFERENCES

1. **Upton, A.C. and Kimball, R.F.,** Radiation biology, acute radiation syndrome, in *Principles of Radiation Protection,* Morgan, K.Z. and Turner, J.E., Eds., John Wiley & Sons, New York, 1968, 427.
2. **Wald, N.,** Effects of radiation over-exposure in man, in *Principles of Radiation Protection,* Morgan, K.Z. and Turner, J.E., Eds., John Wiley & Sons, New York, 1968, 457.

3. **Andrews, G.A., Sitterson, B.W., Kretchner, A.L., and Brucer, M.,** Criticality accident at the Y-12 Plant, Proc. Scientific Meeting on the Diagnosis and Treatment of Acute Radiation Injury, World Health Organization, Geneva, 17 October, 1960.

4. **Andrews, G.A., Hübner, K.F., and Fry, S.A.,** Report of 21-year medical follow-up of survivors of the Oak Ridge Y-12 accident, in *The Medical Basis for Radiation Accident Preparedness,* Hübner, K.F. and Fry, S.A., Eds., Elsevier/North-Holland, New York, 1980, 59.

5. Radiation accident grips Goiania, *Science,* 238, 1028, 1987.

6. **Schauer, D.A., Coursey, B.M., Dick, C.E., MacLaughlin, W.L., Puhl, J.M., Desrosiers, M.F., and Jacobson, A.D.,** A radiation accident at an industrial accelerator facility, *Health Phys.,* 65(2), 1993.

7. **Levin, S.G., Young, R.W., and Stohler, R.L.,** Estimation of median human lethal radiation dose computed from data of reinforced concrete structures in Nagasaki, Japan, *Health Phys.,* 63(5), 522, 1992.

5.3. X-RAY FACILITIES

There are many radiation safety issues associated with the use of X-rays in the laboratory. Much of the health physics information in standard texts, as applied to X-ray systems, discusses radiation safety in terms of a determination of the amount of shielding necessary in medical X-ray installations to protect persons in adjacent areas from the effects of radiation. The radiation sources are the primary beams scattering from the patient or target and leakage from the radiation source. Except in specialized texts, there is little discussion of operational personnel protection. Although there are many research applications in which X-ray machines are used in a manner equivalent to medical usage, even here there are many applications in which there are significant differences in safety considerations. For example, in veterinary medicine, X-rays of large animals are not only likely to cover a larger area and involve more scattering mass, but it is far more likely that a holder (or more than one) will be needed than for most human patients. Besides briefly covering shielding requirements, the sections that follow will emphasize the exposures that the researcher receives, the doses to the patient, and the means by which these exposures can be reduced. Shielding design cannot be neglected, but will often be decided by the architect and his consultant before the research scientist being involved. Often, the research scientist will have relatively little control over his physical environment, especially those who begin working in existing facilities. Further, most will seek out a consultant to assist them should a need arise to create an X-ray facility in an existing structure. Most X-ray users' expertise and interests are primarily in applications, not facility design other than those features that enhance its usefulness to them. However, the scientist will exercise essentially complete control over the operations of the X-ray facility and should be held responsible for establishing operating procedures that maximize radiation protection for everyone.

Many individuals associate anything to do with radiation with the NRC, but it does not regulate the use of X-ray machines nor establish exposure levels for individuals working in these facilities. The Food and Drug Administration in Title 21 of CFR provides standards for X-ray devices, but for the most part the responsibility for regulating the use of equipment that emits X-rays is left up to the states. The degree of regulation provided varies among the various states. The National Committee on Radiation Protection and Measurements (NCRP) has recommended limits of exposure that are similar to those required by the NRC, basically 100 mrem per week (1 mSv per week) as an occupational limit and 10 mrem per week (0.1 mSv per week) for the public.

5.3.1. Generation of X-Rays

The general form of a device in which X-rays are generated is one in which electrons emitted from a cathode are accelerated through a voltage, V, and focused on a target that has a high atomic number. The primary mechanism by which the X-rays are produced is bremstrahlung, as the electrons slow within the target. A substantial amount of heat is generated in the process, since only about 1% or less of the energy of the electron beam is converted to X-rays. In order to have a high atomic number for the maximally efficient conversion of energy into X-rays, the target is typically made of tungsten, which also has a high melting point. The target is usually embedded in an efficient heat conductor, usually copper, serving as the anode. Some designs add fins to conduct heat away from the anode, some cause the anode to rotate so the beam will not be continuously focused on the same spot, and other designs circulate cooling liquids through the anode.

The maximum value of the energy of the X-rays generated in an X-ray tube is determined by the accelerating potential, usually denoted by kVp. However, the energy spectrum of the X-rays is a continuum extending downward from this maximum energy. Thus, in any X-ray spectrum there are many "soft" X-rays that are often not of any significant value, although if the normal operating voltage is low, part of the wall of the tube may be made of a thin piece of beryllium to allow more of the lower energy components to escape from the tube. Usually, these unwanted X-rays are eliminated by interposing aluminum filters (or for higher energy machines, aluminum and copper are used to extend the energy of the filtered component to higher levels) in the beam. These energy-modifying filters are typically 0.5 to 3 mm thick, depending upon the kVp of the X-ray machine.

5.3.2. Types of Machines

As noted above, X-ray machines for medical applications represent only one type of device that emits X-rays used in the laboratory. There are open and closed beam analytical instruments, used for X-ray diffraction work, cabinet systems, electron microscopes, and microprobes, among others. Each of these devices can, under appropriate circumstances, pose radiation exposure problems. Besides installing these in properly designed facilities, operating procedures must be established to limit occupational exposures, as well as those of the patients (for medical machines) and members of the public. Users of these machines should have access to a copy of the publication, *A Guide to Radiation Protection in the Use of X-Ray Optics Equipment*,[2] which provides much practical guidance to assist X-ray equipment users to do so safely.

5.3.2.1. Diagnostic Machines

There are three sources of radiation for a diagnostic X-ray machine (and for an open-beam analytical machine): direct radiation from the primary beam, scattered radiation from the target, and leakage from the protective tube housing. If the orientation of the machine can change, it may be necessary to design all the walls of the structure to provide adequate shielding for the primary beam. If the machine has a permanent fixed orientation, one wall can be designed to protect against the primary beam, but the other walls, and possibly the floors and ceiling, will only have to be designed to reduce the intensity of the scattered and leakage radiation to an acceptable value. An acceptable value for the radiation level in a given space will depend upon whether the area is a controlled access area or not, and whether the exposures of the personnel working within the space are monitored. Maximum permissible exposure levels, P, are 100 mrem per week (1 mSv per week) in a controlled area, and 10 mrem per week (0.1 mSv per week) in other areas outside the controlled area. The primary shielding barrier thickness required can be obtained from the characteristics of the machine, the manner in which it is used, and the occupancy level of the spaces on the other side of the barrier.

5.3.2.1.1. Primary Beam Shielding

The information required to design an adequate shield for the primary beam is (1) the peak voltage (kVp) at which the machine may be operated, (2) the maximum current, I, of the electron beam used to excite the X-rays, usually expressed in milliamperes (mA), (3) the amount of time per week the unit is in use, usually expressed in minutes, and the fraction of time the machine is in use that it is aimed at the wall for which shielding is being calculated, U, and (4) a factor, T, which gives the occupancy level of the adjacent space. The required thicknesses of shielding can be computed by using this information and experimental data for the effectiveness of shielding of different materials for broad beam X-ray irradiation at different kVps. The compiled data is given in terms of a factor K.

For the primary beam, K can be calculated from:

$$K = \frac{Pd^2}{ItUT} \tag{19}$$

Here, d is the distance from the beam target in the X-ray tube to the desired point on the other side of the protective barrier. Usually the product, It, is replaced by the single factor W. All the factors on the right-hand side of the equation are normally known, so that K can be calculated. All that remains to obtain the required thickness is to find K, select the given kVp curve from a family of published curves for a given shielding material, and read the shielding thickness from the abscissa.

The factor T ranges from 1 for full occupancy (space occupied by full-time employees, children play areas, living quarters, employee restrooms, occupied space in immediately adjacent buildings, etc.) to 0.25 for areas with pedestrian traffic but unlikely to have long-term occupancy (corridors too narrow for desks, utility rooms, public parking lots, public rest rooms) to 1/16 for areas occupied occasionally (stairways, automatic elevators, closets, outside areas used by pedestrians or vehicular traffic).

5.3.2.1.2. Scattered Radiation

The corresponding equation for radiation scattered from the object being X-rayed is

$$K_{ux} = \frac{P \times (d_{sca})^2 \times (d_{sec})^2 \times 400}{a \times W \times T \times F \times f} \tag{20}$$

Here: K_{ux} =   scattered radiation rate/workload in rads per milliampere per minute
                    at 1 meter
     a   =   scattered-to-incident ratio
    d_{sec} =   distance from a point to the scattering object in the X-ray beam in meters
    F   =   area of beam impinging on scatterer (cm^2)
    d_{sca} =   distance from X-ray source to scatterer
    f   =   factor that depends on the kVp and adjusts Equation 2 for the enhanced
                    production of X-rays as the energy increases; below 500 kVp, f is taken
                    to be 1

5.3.2.1.3. Leakage Radiation

A properly shielded diagnostic X-ray tube is limited to a maximum leakage of 100 mrem per hour (1 mSv per hour) at a distance of 1 meter. In order not to exceed the maximum weekly exposure at any distance, an attenuation of the leakage radiation is required as given by Equation 21:

$$B_{1x} = \frac{P \times d^2 \times 600 \ I}{WT} \tag{21}$$

The thickness of shielding required to provide this degree of attenuation for a broad beam can be found by computing the number of half-value layers (thickness required to reduce the intensity of an incident beam by one half) that would be needed. To attenuate both the scattered and leakage radiation, the shielding required for each component is computed using Equations 20 and 21. If one of the answers is more than a tenth-value layer larger than the other, the thicker shield can be considered adequate. If not, then the thicker shield should be increased by one half-value layer.

Equation 21 also can be used for a therapeutic X-ray machine if the factor 600 in the numerator is changed to 60, reflecting the factor of 10 higher leakage allowed for this type of machine. Tables 5.7 and 5.8, adapted from NCRP 49, provide half- and tenth-value layers for lead and concrete for several peak voltages, and the scattered-to-incident ratio, a.

It will be noted in Table 5.8 that the minimum for the scattered ratio a in the energy range covered by Table 5.8 comes at an angle ranging from 45° to 90° with respect to the incident beam. At lower energies, the minimum is closer to 45°, while at the upper end of the voltage range, the minimum shifts toward 90°. Since the range of energies in Table 5.8 covers a large portion of the diagnostic voltages used, a person acting as a holder should position themselves accordingly, and at as great a distance as feasible from the scatterer and the X-ray tube.

A recent article concluded that the procedure described gives a value that will be too thick, and consequently more expensive than necessary, and provides an alternate procedure which, however, is somewhat more complicated. Since the method in NCRP 49, and briefly outlined here, will result in a conservative shield, the additional thickness and cost may be justified since it does provide some margin for error and provides for lower exposures.

5.3.2.1.4. Exposure to Users

Many people are abnormally afraid of radiation because of a lack of knowledge or because of misinformation that they may have. A questionnaire given to second-year

Table 5.7.   Half-Value (HVL) and Tenth-Value (TVL) Layers

Voltage (kVp)	Lead		Concrete	
	HVL (mm)	TVL (mm)	HVL (cm)	TVL (cm)
50	0.06	0.17	0.43	1.5
70	0.17	0.52	0.84	2.8
100	0.27	0.88	1.6	5.3
125	0.28	0.93	2.0	6.6
150	0.30	0.99	2.24	7.4
200	0.52	1.7	2.5	8.4
250	0.88	2.9	2.8	9.4
300	1.47	4.8	3.1	10.4

Table 5.8.   Scattered to Incident Exposure Ratio

Voltage (kV)	Scattering Angle (Relative to Incident Beam)					
	30	45	60	90	120	135
50	0.0005	0.0002	0.00025	0.00035	0.0008	0.0010
70	0.00065	0.00035	0.00035	0.0005	0.0010	0.0013
100	0.0015	0.0012	0.0012	0.0013	0.0020	0.0022
125	0.0018	0.0015	0.0015	0.0015	0.0023	0.0025
150	0.0020	0.0016	0.0016	0.0016	0.0024	0.0026
200	0.0024	0.0020	0.0019	0.0019	0.0027	0.0028
250	0.0025	0.0021	0.0019	0.0019	0.0027	0.0028
300	0.0026	0.0022	0.0020	0.0019	0.0026	0.0028

medical students before they had a course in biophysics or radiation applications in medical school, but after they had completed an undergraduate, presumably scientifically-oriented program, revealed a striking and troubling level of confusion about X-rays. If this is the situation that obtains for a well-educated group (the entrance requirements for medical schools are notoriously high), the lack of knowledge of the public is likely to be far worse. Four of the questions and the test results are given below.

1. Following the completion of an X-ray radiographic examination, objects within the room
   a. emit a large amount of radiation   29.3%
   b. emit a small amount of radiation   43.9%
   c. do not emit radiation   26.8%
   d. don't know   0.0
2. Intravenous contrast materials used in angiograms and intravenous pyelograms are radioactive.
   a. true   37.0%
   b. false   61.0%
   c. don't know   2.0%

3. Gamma rays are more hazardous than X-rays.
   a. true   58.5%
   b. false   22.0%
   c. don't know   19.5%
4. Nuclear materials used in nuclear medicine are potentially explosive.
   a. true   24.4%
   b. false   68.3%
   c. don't know   7.3%

The normal practice in taking medical X-rays is for the patient to be situated with respect to the source and film cassette by the technician and, where possible, the patient is asked to "hold still" while the technician retreats to a shielded control room and activates the X-ray unit. The technician, under such circumstances, does not receive a dose other than that allowed by the scattered and leakage radiation levels behind the shield. Since the permissible limits are 100 mrem per week (1 mSv per week), the X-ray technician must be provided with a dosimeter capable of reading X-rays accurately. Both TLD and film badges are commercially available which are satisfactory for the purpose. The reproducibility and inherent radiological similarity to tissue of commonly used TLD dosimeters are desirable features for this type of dosimeter. The ability of film to distinguish between a diffuse extended source, as would be the case for the scattered radiation and that of a localized source, is an advantage of film dosimeters. The patient is exposed to not only the direct beam, but also the scattered and leakage radiation. However, the exposure of the patient will usually be a single episode, or at least a limited number of exposures, while the occupational exposures of the technician will continue. When X-ray personnel are not wearing the dosimeter badges, they should be stored or kept in an area at background radiation levels, so that the dosimeter readings will reflect true occupational exposures.

The control room booth should have walls and a viewing window with sufficient shielding capability to ensure that the operator's exposure does not exceed the permissible limits for a controlled area. The dimensions of the booth must be of adequate height and width to ensure complete shielding while the beam is on, and the distance between the activation switch for the beam and the entrance to the door to the X-ray area should be sufficient to guarantee that an X-ray technician cannot operate the machine in an unshielded location. The entrances to an X-ray area should be posted with a sign bearing the words (or equivalent) **CAUTION: X-RAY EQUIPMENT.** It would also be desirable if a visual or audible warning were provided should anyone attempt to enter the beam area while the beam is on. Another precaution that can be taken is to have the doors leading into the X-ray area other than from the control area interlocked so that they can only be opened from the outside when the beam is off or so that the X-ray machine will be turned off automatically upon the doors being opened.

Not all X-rays are taken under ideal circumstances. The patient may be in pain and find it difficult to stay in position. Children may be frightened or simply be too young to follow instructions. In certain fluoroscopic examinations, even healthy and cooperative patients may need assistance. Passive restraints should be used wherever possible, but on occasion patients must be held. This must not always be the responsibility of a single individual or a limited group of persons. Excessive personal doses are likely unless this duty is shared among several persons. Since the X-ray beam is not on except for well-defined and determinable intervals, there should be no increase in the collective dose over that of the single individual by following this practice. No one who is not essential to the procedure should remain in the room while the X-ray is being taken, although relatives could restrain patients, especially children, who might need a familiar person present to avoid being frightened.

Holders should remain as far from the primary beam as possible. As noted earlier, the intensity of the scattered radiation is a minimum of 45° to 90° with respect to the incident beam for most commonly used diagnostic X-ray voltages. Either a lead apron or a movable shield can substantially reduce the whole-body exposure. As noted in Table 5.7, less than 1 mm of lead will reduce the intensity of a broad beam of 150 kVp X-rays by a factor of 10. This amount of lead might be too heavy in an apron for comfort for some individuals, but even half a mm of lead will substantially reduce the level of radiation to the torso, where the critical organs are.

If a lead apron is not worn, the dosimeter badge should be worn at the waist or above, so that the measured dose will be representative of a whole-body exposure. If it is worn on the upper part of the body, the dosimeter will not be partially shielded by the X-ray table in a vertical beam exposure when holding a patient lying down.

When a lead apron is worn, the dosimeter should be worn above the apron on the shirt collar or on the lapel of a laboratory coat so that it provides a reading of the dose to the exposed area above the apron. Especially in the early stages of her pregnancy, a pregnant woman should not act as a holder, to avoid an exposure to the fetus. Should it be necessary for a pregnant woman to act as a holder, she should wear a second dosimeter under the lead apron near the fetus. If an employee is aware that she is pregnant, it is recommended that she "declare" (as under 10 CFR Part 20, although the NRC regulations do not apply to X-ray radiation) or inform her supervisor of her pregnancy. The supervisor, with the cooperation of the pregnant employee, should see that she limits the fetal exposure to 0.5 rem (5 mSv) over the entire pregnancy. To confirm that this is so, a more frequent monitoring schedule should be established for this individual, unless work assignments can be made which preclude her exceeding the recommended levels.

Persons who assist with fluoroscopic procedures should wear ring badges and also body dosimeters. The beam stays on for longer periods than for a normal X-ray film exposure (sometimes longer than necessary when the radiologist is not conscientious about turning off the beam as promptly as possible at intervals during the procedure), and consequently the potential for exposure of the assistant is higher. Additional shielding to protect the viewer and any aides from scattered radiation should be provided as needed. If a direct viewing screen is employed, the entire primary beam must be intercepted by the screen, or the equipment should be interlocked so that it will not work. Units are now available so that the image is viewed on a high-resolution television screen instead of directly, which substantially reduces the amount of radiation exposure to the radiological physician.

Film and TLD dosimeters are normally read on a monthly or quarterly basis, instead of immediately, although the latter can be read in a few minutes if a calibrated and certified TLD dosimeter reader is available. If there is concern about potential exposures and an immediate reading is needed or desired, pocket dosimeters designed to be properly responsive to X-ray fields should be used to supplement the standard dosimeters, to provide an immediate exposure reading.

Many devices that decrease the dose to the patient, such as faster film or screen intensifiers, also serve to reduce the exposure to persons required to be in the room. Beam filters, usually of aluminum, are used to eliminate the soft X-rays from the X-ray beam. This reduces the skin exposure to the patient and the exposure to scattered radiation from this source to the worker. The recommended filter thicknesses are 0.5 mm aluminum (or equivalent) for machines operating at up to 50 kVp, 1.5 mm for operations between 50 and 70 kVp, 2 mm for 70 to 125 kVp, and 3 mm for 125 to 300 kVp. Collimators, which are used to define and limit the incident beam to only the area of clinical interest, also reduce the amount of scattering mass within the beam. The irradiated area is required by current standards to be defined by a light that is accurately coincident with the area exposed to the

X-ray beam. The visible target will allow any required assistant to take a position that will be as far from the primary beam as possible and which will reduce the exposure to scattered radiation.

The genital areas of patients must be protected with a shield if the patient is of reproductive age, unless the shield would interfere with the diagnostic procedure. It would be desirable to know if a woman is pregnant before an X-ray of the pelvic area, and it is recommended that the physician suggest a pregnancy test to fertile women if they are sexually active. This is obviously a very sensitive topic and must be done extremely tactfully. However, failure to discuss the risk and warn the patient of possible problems with the development of the fetus could result in allegations of liability by the physician. A sign prominently posted in the X-ray facility such as **IF YOU ARE PREGNANT, IT IS IMPORTANT THAT YOU TELL THE TECHNICIAN** could be helpful. In an emergency this concern should not delay taking an X-ray. Since the patient's body usually is exposed to scattered and leakage radiation from the areas being X-rayed, shields for other sensitive parts of the patient's body should be considered if a substantial series of "shots" are to be taken.

No more X-rays than necessary should be taken. No X-rays of humans should be taken for training, demonstration, or other purposes not directly related to the treatment of the patient.

One of the more disturbing problems of diagnostic X-rays is the frequent failure of either the physician or the radiologist to be responsive to the question by a patient of "how much radiation will be received." Too often the attitude appears to be that "it has been decided that the patient needs the X-ray and that is all that matters." There is substantial concern about the effects of radiation, and a substantial amount of misconceptions. This fear and confusion may inhibit the patient from having a needed X-ray. Not only should a reasonable answer be given but the significance of the answer should be explained, since it is unlikely that most persons will be knowledgeable about radiation safety terminology. The output from a typical X-ray machine is generally quite substantial. At 50 kVp, with no filtration and a thin beryllium window, the output of an X-ray tube is about 10 rad/min/mA at 1 meter or 100 mGy/min/mA at 1 meter. For a 100 kVp machine, with a 3 mm external aluminum filter, a typical exposure rate is 3 rad/min/mA at 1 meter or 30 mGy/min/mA at 1 meter. These rates, or preferably, the actual rates for the machine being used, can be used to provide an estimate of the doses to the patient. The next section will describe the process of inspection and calibration of X-ray machines that should be followed. Since one of the results of the evaluations will be an exposure chart representative of various types of X-ray applications, the information is readily available if these evaluations are done. Techniques to reduce the exposures to the patient should be used. A normal "good" level of exposure to the patient for a chest X-ray is 10 mrad (1 mGy) or less.

The performance of a diagnostic X-ray machine should be checked at least annually with appropriate instruments by a qualified person. If the capability to do this is not available in-house, a qualified consultant should be hired to do the task. If any maintenance is done or if the machine is relocated, a survey should be undertaken for leakage radiation from the source. If the unit is moved to another facility, the exposure levels in the adjacent areas should be tested to ensure that the exposure levels are within the permissible limits for controlled and uncontrolled areas. Records of all maintenance, surveys, leakage checks, calibration, personnel monitoring, etc. should be maintained at the facility and at the radiation safety office. Because of the long latency period for cancer developing from radiation exposures, it would not be unreasonable to maintain personnel exposure records for up to 40 years.

5.3.2.2. X-Ray Quality Control in Medical X-Ray Laboratories*

This section is restricted to X-ray facilities, although another major use of radiation in larger nuclear facilities is the use of radioactive materials. The radioisotope usage is regulated by the NRC in 10 CFR Part 35, and by a medical industry group, the Joint Council on Accreditation of Healthcare Organizations. In terms of employees and the public, the same basic NRC restrictions on exposure generally must be followed. However, all hospitals use X-ray machines of various types.

Performance standards for X-ray equipment are developed by the Food and Drug Administration under 21 CFR Part 1020, Performance Standards for Ionizing Radiation Emitting Products. Regulation of the use of X-ray equipment is essentially left to the individual states and to the same health care industry group which monitors hospital quality control in many areas other than use of X-rays and radioisotopes, the Joint Council on Accreditation of Healthcare Organizations.

Inspections and tests of individual X-ray units and evaluation of general compliance with radiation safety rules and operating procedures is normally done by independent consultants, hired by the hospitals. The regulations of each state may vary, but the accrediting organization is a national organization, so their standards provide a baseline for guidance on the standards to be met during an inspection. Generally, X-ray units are checked at least annually, but may be checked more often if needed. Each type of X-ray unit used for different applications has slightly different features that are checked. Fluoroscopic systems generally have much higher exposure rates than those used for general X-ray procedures, so they include more checks on scattered radiation. Machines used for soft tissue evaluation, e.g., mammography, use lower accelerating voltages and a "softer" X-ray beam extending to lower energies, so the filters to remove the lowest energy X-rays are considerably thinner than for general application units. A typical state inspection form is shown in Table 5.9. The forms provide the range of acceptable values except for the filter thicknesses. For higher voltage exposures the half-value filter thickness (HVL) may range from 2.3 to 3.2 mm of aluminum, while for low voltages, the HVL is much less, about 0.27 to 0.33 mm of aluminum.

The entrance skin exposures are the critical parameters in terms of safety to the patient. Generally, the higher kVp exposures deposit less energy in the patient and reach the film, providing less patient exposure for the same film darkening. However, the contrast in the image for soft tissue is better for lower kVp exposures. A balance is determined for a given procedure to provide the best image with the least exposure. The values in Table 5.10 are for average individuals with thickness through the chest of 23 cm. Adjustments would be required for very obese patients and for other nontypical physical types.

The consultant provides data to the facility on the results of the measurements made. Tests are run at a variety of tube voltages and using different filters in the beam. The test equipment measures the kVp, time, and exposure simultaneously. The patient exposure is computed from the measured data by a computer program for a representative.

Table 5.10 provides exposure/dose guides for average patients for a variety of diagnostic procedures. The information represents data from facilities of all sizes, ranging from private practices to large hospitals. The group that compiled the data, the Committee on Quality Assurance in Diagnostic Radiology (CQADR), did so to provide goals to help reduce excessive exposures to the patients. As they point out in the introduction to their

---

*   This section was coauthored by John Cure and A. Keith Furr.

Table 5.9.　General Radiographic Systems

TUBE REGISTRATION # _____

Machine

Registrant Name _____　Make _____ Purpose _____

Surveyor Name _____　Model _____ Serial # _____

Surveyor Signature _____　Max. kVp _____ Max. mA _____

Survey Date _____　Room # _____ Phase _____ 1 _____ 3

OK　N/A　N/S　S

___ ___ ___ ___ 1. Warning label not present. **NON-SERIOUS**

___ ___ ___ ___ 2. HVL at ____ kVp is ____ mm Al. Minimum is ____ mm Al. Deficiency of ≤ 0.2 mm, **NON-SERIOUS**, deficiency ≥ 0.2 mm, **SERIOUS**.

___ ___ ___ ___ 3. The length _____ width _____ misalignment between the X-ray and visual fields is ____ %. >2% to < 5% **NON-SERIOUS**, ≥ 5%, **SERIOUS**.

___ ___ ___ ___ 4. Misalignment between center of X-ray field and center of image receptor is ____% of the SID. >2% to < 5% **NON-SERIOUS**, ≥ 5%, **SERIOUS**.

___ ___ ___ ___ 5. SID not indicated to within 2% of SID. **NON-SERIOUS**.

___ ___ ___ ___ 6. The length ____ width ____ dimensions of the X-ray field not indicated to within 2% of the SID. **NON-SERIOUS**.

___ ___ ___ ___ 7. The length ____ of the X-ray field exceeds that of the image receptor by ____ % of the SID (special purpose systems only). >2% to < 5% **NON-SERIOUS**, ≥ 5% **SERIOUS**.

___ ___ ___ ___ 8. Radiographic control does not require constant operator pressure or does not terminate the exposure properly. **SERIOUS**.

___ ___ ___ ___ 9. Radiographic control switch not permanently located in an appropriate protected area or on a stretch cord of sufficient length as required. **SERIOUS**.

___ ___ ___ ___ 10. X-ray control not equipped with both visual and audible indication of X-ray production. **NON-SERIOUS**.

___ ___ ___ ___ 11. Timer reproducibility, coefficient of variation is ____ % at a technique setting of _____. > 10% to < 15%, **NON-SERIOUS**, ≥ 15% **SERIOUS**.

___ ___ ___ ___ 12. Exposure reproducibility: Coefficient of variation is ____ % at a technique setting of _____. > 10% to < 15%, **NON-SERIOUS**, ≥ 15% **SERIOUS**.

___ ___ ___ ___ 13. Timer accuracy is ± ____ % of the indicated time at ____ sec. > 10% to < 15%, **NON-SERIOUS**, ≥ 15% **SERIOUS**.

___ ___ ___ ___ 14. Standby radiation exposure is more than 2 mR/h Capacitor discharge systems only) **SERIOUS**.

___ ___ ___ ___ 15. mA Linearity: mR/mAs values at ____ Ma and ____ mA differ by more than 10% of their sum. **NON-SERIOUS**.

___ ___ ___ ___ 16. Positive beam limitation, if present, does not operate properly. **NON-SERIOUS**.

___ ___ ___ ___ 17. Light field illuminance is ____ ft. candles. Should be no less than 10 ft. candles at 100 cm. or at the max. SID whichever is less. <10 to 9 ft. candles: **NON-SERIOUS**, 9 ft. candles, **SERIOUS**.

___ ___ ___ ___ 18. kVp accuracy is ± _____ of indicated kVp at _____ kVp. > 10% to < 15%, **NON-SERIOUS**, ≥ 15% **SERIOUS**.

___ ___ ___ ___ 19. Exposure data: projection: _____ technique factors: _____ kVp ____ mA ____ MS ____ inches SID. Exposure results _____ mR.

___ ___ ___ ___ 20. Other/Remarks: _____

SID = source to image distance

Table 5.10.   Entrance Skin Exposures

Projection	Patient Thickness (cm)	Grid	SID (in.)	ESE (mr) Film Speeds	
				300	400
Abdomen	23	yes	40	490	300
Lumbar spine	23	yes	40	570	330
Full spine	23	yes	72	260	145
Cervical spine	13	yes	40	135	95
Skull	15	yes	40	145	70
Chest	23	no	72	15	10
	23	yes	72	25	15

MAMMOGRAPHY (CRANIOCAUDAL VIEW)
Mean Glandular

Imaging System	Dose (mrad)
Screen film	
w/out grid	70
with grid	150
Xerox	
Positive mode	400
Negative mode	300

Computerized Tomography (CT Scans) Doses

Typical 3rd generation unit (tube and detector rotates)	4.4 Rads
Typical 4th generation unit (tube rotates, detectors stationary)	5.4 Rads

Fluoroscopic Exposure Rates

Examination	Dose Rate (Roentgens/min)
Routine typical patient	4.5
With barium	7.8
Maximum rate	8.1

SID = source to image distance.

report, the largest contributor to total population exposure to radiation from manmade sources is from diagnostic (dental and medical radiography). Clearly, X-rays need to be taken only when needed and under circumstances to reduce exposures to the patient, the medical staff, and others involved.

The fluoroscopic portion of Table 5.10 shows that the actual exposure depends upon the time the beam is on. If the beam is left on while adjustments are made to the patient's position, as occurs during the barium enema used in the table, the exposures can be much higher than necessary. Note that the units are not mrem or mrads but are in effect rads, a thousand times larger, since there is only a modest difference in the energy deposited in tissue by rads and roentgens.

Although not mandated by a hospital accrediting group and possibly not regulated by a state agency, it is recommended that a qualified X-ray consultant periodically inspect and evaluate the X-ray machines described in the following sections.

5.3.2.3. Open-Beam Analytical Machines

Open-beam analytical X-ray machines have many research applications that involve frequent manipulations of the equipment and samples. There are many opportunities for excessive occupational exposures. In crystallography studies, for example, a very high intensity beam, although confined to small areas, is involved. If an individual were to expose the hands or any other portion of the body to the direct beam, very high localized exposures could occur in a few seconds. The occupational radiation levels considered acceptable for general whole body exposure are the same as for diagnostic machines, 100 mrem per week (1 mSv per week) in controlled areas. The maximum hourly levels are the levels that would be equivalent to this rate for a 40-hour week, 2.5 mr per hour (25 µSv per hour). Acceptable levels for hand exposures are 15 times greater than those for whole body exposures, 37.5 mrem per hour (37.5 µSv per hour). For exposures in uncontrolled areas, the levels should not exceed 2 mrem per hour (20 µSv per hour) or 100 mrem per week (1 mSv) on a short-term basis or 500 mrem (5 mSv) averaged over 1 year. Note that these levels are comparable to the levels recommended by the NRC for exposure to radioactive materials in the 10 CFR Part 20 before January 1, 1994.

Laboratories that use open-beam analytical machines should possess a survey instrument capable of measuring the scattered radiation from the machine and associated apparatus around the experimental area. The scattered radiation from a crystal is nonuniform, which can cause narrowly defined "hot" spots if the radiation is allowed to escape from the apparatus into the laboratory area. A careful survey needs to be made whenever maintenance is done on any part of the system that could allow radiation to escape from the apparatus or when the system is reconfigured. If an alignment or maintenance procedure must be done with the beam on, surveys must be performed while this work is in progress, preferably by two persons, one of whom does the actual work while the other one monitors the radiation levels. Periodic surveys should also be made to ensure that no changes have been made or have occurred without everyone's knowledge which could affect the radiation field near the apparatus. A quarterly schedule is recommended while work is actively in progress. A survey should also be made any time anyone suspects that an abnormal condition could exist that might increase occupational exposures.

Personal dosimeters should be worn by individuals in the laboratory whenever the machine is operated, and finger badges should be worn whenever working directly on the apparatus, making adjustments, or doing any action that could cause the fingers or hand to receive an abnormal exposure. All badges, when not being worn by the individual, should be stored in a background level radiation area so that they will reflect the actual occupational exposure of the person to whom they have been assigned.

A procedure manual must be prepared for systems as potentially dangerous as open beam analytical systems, and all persons must be made familiar with the procedures applicable to safe use of the equipment. The person in charge of the facility is responsible for ensuring that all personnel are informed of safe procedures and that it is required that all personnel must comply with the written safety policies. This should be done not only by instruction but by the laboratory director or supervisor setting an appropriate example as well. Every individual working in the facility should be trained in the basic principles of radiation safety, the permissible exposure levels for controlled and uncontrolled areas, what dosimeters to wear and when, what surveys are required and what records need to be maintained. All employees should also be informed of their rights to limit their exposures and of their rights to their exposure records.

Laboratories in which analytical X-ray machines are located should be posted with signs at all entrances, bearing the legend **CAUTION—APPARATUS CAPABLE OF PRO-**

DUCING HIGH-INTENSITY X-RAYS or equivalent. Any time the machine is on, it should be attended or the doors to the experimental area should be locked. Keys should be restricted to those persons who work within the facility. It is especially critical to prevent access, or not to leave the apparatus unattended if the equipment is on and partially opened for alignment.

Measures must be taken to warn persons unfamiliar with the risks associated with the equipment in case they enter the facility while no one is present, even if the equipment is not on. The X-ray source housing of analytical X-ray equipment should have a conspicuous sign attached to it bearing the words **CAUTION—HIGH INTENSITY X-RAYS**, and a similar sign placed near any switch that could energize the beam, stating **CAUTION—THIS APPARATUS PRODUCES HIGH-INTENSITY X-RAYS WHEN THE SWITCH IS ENERGIZED**. In addition, a lighted sign stating **X-RAY SYSTEM ON**, or the equivalent, should be near the switch. The light should be incorporated in the activation circuit so that if the light fails, it will be impossible to activate the system. A key-operated switch would be an added safety feature.

It is highly desirable that an interlock system be incorporated into the system that will prevent any part of the body from being directly exposed to the primary beam, either by making it physically impossible to do so or by entry causing the beam to be shut off. Where, for some compelling reason, this is not feasible, alternatives need to be in place to reduce the chance of an accidental overexposure. Conspicuous warning devices to alert persons working on or near the machine should be in place if the status of the apparatus is such that a danger could exist. These may take the form of lighted signs, for example, stating **CAUTION—SYSTEM IS ON** or **DANGER—SHUTTER OPEN** if a beam port is open. These warning devices should be fail safe. If they are not functional, the system should not be able to operate. Shutters on any unused ports should be firmly secured in the closed position. It would be desirable if the system could not operate with a port open, unless deliberately bypassed, in which case a fail safe warning sign would have to be used. Similarly, it should be impossible to open a shutter unless a collimator or an experimental device were connected to the port. This feature must be incorporated in commercial systems constructed after January 1, 1980. Beam catchers should be in place to intercept any beam beyond the point at which the beam has served its purpose.

Interlock systems should not depend solely upon the operation of a device such as a microswitch. These devices can operate for years without failing and then fail without warning. A more basic device, such as incorporating a male and female plug assembly into the shutter mechanism to complete a circuit, is virtually foolproof. A light in the circuit confirms that the circuit is complete. The completed circuit should always provide a warning and preferably be a condition of a safe system, i.e., if the light is on, the system is safe. If it is off, the activation circuit is interrupted by an unsafe condition or the light is burned out, so the system will not operate in either case. In some cases, as with the open shutter above, a warning system needs to operate to warn of an unsafe condition. Here, the failure of the system to operate if the light is not on simply guarantees that the warning sign is on during the less safe condition.

Repair and alignment activities provide some of the best opportunities for accidental overexposures. Every precaution should be taken to prevent these incidents. Several measures are listed below. Most of these should be made mandatory by the organization's written radiation safety policies.

1. Alignment procedures recommended by the manufacturer should be used. Departures from these procedures should require prior approval by radiation safety personnel.

2. Alignment procedures must be in writing, and the personnel doing them must have specific training on doing the procedures.
3. The radiation from an X-ray tube shall be confined by a suitable housing during alignment and maintenance activities.
4. The main switch, instead of safety interlocks, must be used to turn the equipment off.
5. If the interlocks are bypassed to allow alignment or other work to be done, a sign must be placed on the equipment stating **DANGER—SAFETY INTERLOCKS BYPASSED**.
6. Personnel dosimeters and finger badges must be worn at all times the beam is on, by the person doing the work and the assistant (see item 11 below).
7. The smallest practical voltage and beam current that allow the alignment to be done shall be used.
8. Temporary shielding should be used to reduce the exposure to scattered radiation.
9. Long-handled tools and extension devices should be used to reduce the risk of direct exposure to the hands.
10. If the alignment and maintenance operations are such as to cause the radiation in an uncontrolled area to exceed the permissible levels for such an area, the area should be secured against entry by locks, surveillance, or both.
11. Alignment and repair operations with the beam on should not be done while working alone. A second person should be present as a safety observer, equipped with a survey meter to check radiation and prepared to immediately shut off the beam in case of a direct exposure of any part of the person working on the equipment.
12. The system should be checked to ensure all interlocks have been reconnected and to ensure that leakage radiation is at an acceptable level.
13. A record of all repairs and the results of surveys taken subsequently to the work done should be recorded in the permanent log.

5.3.2.4. Closed-Beam Analytical Systems

Closed-beam diffraction cameras are used for purpose similar to those of open-beam systems. However, they are designed so that the X-ray tube, sample, detector, and diffracting crystal are enclosed in a chamber that should prevent entry into the system by any part of the body. However, some older systems allowed the hands to be put into the sample chamber without turning off the beam, although this was not supposed to be done. Instances of exposures of 50,000 to 100,000 rems (500 to 1000 Sv) to the hands are known to have occurred. The ports on such units must now be equipped with safety interlocks by the manufacturer, incorporating a warning light in series with the interlock, which guarantees that if the light is off, the beam is off. Any older units not so equipped should be retrofitted with this safety feature. Because there should be little or no leakage requirements with a closed beam analytical system, personnel need not be required to wear dosimeters.

5.3.2.5. Cabinet X-Ray Systems

Cabinet X-ray systems are enclosed systems that not only enclose the X-ray system but also the object to be irradiated. They are normally designed with a key-activated control so that X-rays cannot be generated when the key is removed. A conspicuous sign with the legend **CAUTION: X-RAYS PRODUCED WHEN ENERGIZED** must be placed near the controls used to activate the generation of X-rays.

The doors to the cabinet are required to have two or more safety interlocks and each access panel is required to have at least one. If X-ray generation is interrupted through the functioning of a safety interlock, resetting the interlock shall not be sufficient to resume generation of X-rays. A separate control must be provided to reinitiate X-ray production.

The generation of X-rays must be indicated by two different indicators, one being a lighted warning light labeled **X-RAYS ON**, and the other can be the meter that reads the X-ray tube current.

The radiation at 5 cm from the surface of the cabinet must be no more than 0.5 mrem per hour (5 μSv per hour). The radiation limits in uncontrolled area due to the operation of the machine are 2 mrem per hour (20 μSv per hour) or 100 mrem (1 mSv) in a 7-day period or 500 mrem per year (5 mSv per year). At least an initial radiation survey and an annual survey thereafter should be made to ensure conformity with these limits. The interlock systems should be tested periodically if the mode of operations does not automatically cause their function to be tested.

Personnel using a cabinet X-ray unit should be provided with a dosimeter that they should wear them whenever the machine is in operation.

5.3.2.6. Miscellaneous Systems

There are any number of systems that generate X-rays, some as dangerous as or more so than the ones described above. Precautions appropriate to the system, including personnel dosimetry, surveys, safety interlocks, etc., must be taken to ensure the safety of all workers and those persons in uncontrolled areas nearby. An appropriate safety program must be worked out with the radiation safety department within the organization to ensure that the program not only meets all safety requirements recommended by the manufacturer, but also is in conformity with regulatory requirements, including those policies established within the organization.

REFERENCES

1. **Marlin, E.B.M.,** *Guide to Safe Use of X-Ray Diffraction and Spectrometry Equipment,* Science Reviews, Leeds, England, 1983.
2. *A Guide to Radiation Protection in the Use of X-Ray Optics Equipment,* Science Reviews, Leeds, England, 1986.
3. **Simpkin, D.J.,** A general solution to the shielding of medical X and gamma rays by the NRCP report no. 49 methods, *Health Phys.,* 52, 431, 1987.
4. *Structural Shielding Design and Evaluation for Medical Use of X-Rays and Gamma Rays of Energies Up to 10 MeV,* NCRP Report No. 49, National Council on Radiation Protection and Measurements, Washington, D.C., 1976.
5. *X-Ray Equipment,* UL-187, Underwriters Laboratories, Chicago, IL, 1974.
6. Radiological Safety Standard for the Design of Radiographic and Fluoroscopic Industrial X-Ray Equipment, NBS Handbook 123, National Bureau of Standards. Washington, D.C., 1976.
7. *Radiation Safety for Diffraction and Fluorescence Analysis Equipment,* ANSI N43.2, American National Standards Institute, New York, 1971.
8. Performance Standards for Ionizing Radiation Emitting Products, 21 CFR Chap I, 1020.30, 1988.
9. **Kaczmarek, R., Bednarek, D., and Wong, R.,** Misconceptions of medical students about radiological physics, *Health Phys.,* 52, 106, 1987.

10.   **Fleming, M.F. and Archer, V.E.,** Ionizing radiation, health hazards of medical use, *Consultant*, p. 167, January 1984.
11.   **Lubenau, J.O., David, J.S., McDonald, D.J., and Gerusky, T.M.,** Analytical X-ray hazards: a continuing problem, *Health Phys.*, 16, 739, 1969.
12.   **Weigensberg, I.J., Asbury, C.W., and Feldman, A.,** Injury due to accidental exposure to X-rays from an X-ray fluorescence spectrometer, *Health Phys.*, 39, 237, 1980.
13.   **Cember, H.,** *Introduction to Health Physics*, Pergamon Press, New York, 1969.
14.   **Shapiro, J.,** *Radiation Protection, A Guide for Scientists and Physicians,* 3rd ed., Harvard Press, Cambridge MA, 1990.
15.   Average Patient Exposure Guides, CRCPD Pub. No. 92.4, Conf. of Radiation Control Program Directors, Frankfort, 1992.
16.   Nationwide Evaluation of X-Ray Trends (NEXT), Conf. of Radiation Control Program Directors, 1974–89.
17.   Accreditation Manual for Hospitals, Joint Council on Accreditation of Healthcare Organizations, Chicago, 1982.

5.4. NONIONIZING RADIATION

The International Non-Ionizing Radiation Committee of the International Radiation Protection Association (IRPA/INIRC) has developed a revised version of the 1984 Interim guidelines on limits of exposure to radiofrequency electromagnetic fields in the frequency range from 100 kHz to 300 GHz. The revised guidelines were approved by the IRPA Executive Council on June 2, 1987. The following material is an abbreviated version of the guidelines prepared under the auspices of a committee chaired by Jammet.[1] Only the basic material and the recommended exposure numbers are to be found below. The interested reader should refer to the complete document and to a special section on nonionizing radiation found in *Health Physics Journal*.[2]

The current federal standards are given in 29 CFR 1910.97. From 10 MHz to 100 GHz, the radiation protection guide provides that the incident radiation not exceed 10 mW/cm², as averaged over any possible 0.1-hour period. This refers to the power density. The guide provides that the energy density be no more than 1 mW/cm² for any 6-minute period.

5.4.1.   Guidelines on Limits of Exposure to Radiofrequency Electromagnetic Fields in the Frequency Range From 100 kHz to 300 GHz

The exposure limits can be expressed in terms of either the energy absorbed by the presence of a body in the area or the power density in an area in the absence of a physical body to absorb the energy. The basic limits of exposure are given by the specific absorption rate (SAR) and expressed in units of power absorbed per unit mass or (W/kg). Obviously, the SAR represents the body-present situation. The absorption can be averaged over the entire body mass of a person present, a defined time interval, a modulation period, or a single pulse of the radiation.

In the absence of a body being present, the electromagnetic power density can be calculated if the electric and magnetic field strengths are known. In the far-field region, well away from the source of radiation, the relationship between the electric and magnetic field strengths is straightforward, since the field is essentially a plane wave. Here, the electromagnetic power density is given by:

$$S = E^2/120\pi = 120\pi/H^2 \ \text{W/m}^2$$

In the near-field region, and where there are multiple paths, the relationship between the electric and magnetic field components is not straightforward, and both must be measured to compute the "equivalent plane wave power density." This is generally the situation that obtains in cases below 100 MHz.

Below 10 MHz, the occupational exposure to radiofrequency radiation "should not exceed the levels of the unperturbed RMS electric and magnetic field strengths given in Table 5.11, "when the squares of the electric and magnetic field strengths are averaged over any 6-minute period during the working day, provided that the body to ground current does not exceed 200 mA and that any hazards of RF burns are eliminated." Currents at the point of contact above 50 mA pose the risk of burns.

Above 10 MHz, the occupational exposures should not exceed 0.4 W/kg averaged over the whole body for a 6-minute period, provided that 2 W per 0.1 kg will not be exceeded in the hands, wrists, arms, and legs and that 1 W per 0.1 kg will not be exceeded in any part of the body. The values for the field strengths in Table 5.11 are the working limits derived from the 0.4 W/kg SAR limit. The guidelines also recommend that as long as the occupational limits over a 6-minute interval are not exceeded, the equivalent plane wave power density averaged over the pulse width for a pulsed field does not exceed 1000 times the P_{eq} limits for the frequency involved.

The limits for exposure to radiofrequency radiation for the public are given in Table 5.12. Below 10 MHz, the limits are defined in the same way as for an occupational exposure, but are lower. Above 10 MHz, the limits for the various frequencies are derived from an SAR of 0.08 W/kg averaged over the entire body over any 6-minute period.

REFERENCES

1. **Jammet, H.P. et al.,** Guidelines on limits to radio frequency electromagnetic fields in the frequency range from 100 kHz to 300 GHz, *Health Phys.,* 54, 115, 1988.
2. Special section: non-ionizing radiation, *Health Phys.,* 53, 567, December, 1987.
3. Nonionizing Radiation, 29 CFR 1910.97, Occupational Health and Safety Administration, Washington, D.C., 1988.
4. Safety Level of Electromagnetic Radiation with Respect to Personnel, ANSI C-95.1, American National Standards Institute, New York, 1982.

5.5. LASER LABORATORIES*

The risk to personnel from the use of a laser in the laboratory depends upon several factors. The first consideration is the classification of the laser itself, which is based on several parameters. Among these are (1) the frequency or frequencies of radiation emitted; (2) for a pulsed system, the pulse repetition frequency (PRF), the duration of each pulse, the maximum or peak power P in watts or maximum energy Q in joules per pulse, the average power output, and the emergent beam radiant exposure; (3) for a continuous wave (CW) machine, the average power output; and (4) for an extended-source laser, the radiance of the laser and the maximum viewing angle subtended by the laser. Relatively few lasers used in the laboratory fall in this latter class, and the following discussion will be limited to lasers that are not considered extended source units.

---

* The material in this section for laser operations complies with Title 21 CFR Part 1040.1 for laser products, as well as ANSI Z-136 and the Threshold Limit Values (TLV, a copyrighted trademark) of the American Conference of Governmental Industrial Hygienists).

Table 5.11.   Occupational Exposure Limits to Radiofrequency Electromagnetic Fields

Frequency (MHz)	Unperturbed RMS Field Strength		Equivalent Plane Wave Power Density P_{eq} (W/m^2)
	Electric E (V/m)	Magnetic H (A/m)	
0.1–1	614	1.6/f	—
>1–10	614/f	1.6/f	—
>10–400	61	0.16	10
>400–2000	$3f^{1/2}$	$0.008f^{1/2}$	f/40
>2000–300,000	137	0.36	50

Table 5.12.   General Public Exposure Limits to Radiofrequency Electromagnetic Fields

Frequency (MHz)	Unperturbed RMS Field Strength		Equivalent Plane Wave Power Density P_{eq} (W/m^2)
	Electric E (V/m)	Magnetic H (A/m)	
0.1–1	87	$0.23/f^{1/2}$	—
>1–10	$87/f^{1/2}$	$0.23/f^{1/2}$	—
>10–400	27.5	0.073	2
>400–2000	$1.375f^{1/2}$	$0.0037f^{1/2}$	f/200
>2000–300,000	61	0.16	10

All commercial units currently being built must have the classification identified on the unit. The classification must be according to that given in ANSI-Z 136.1 and in 21 CFR 1040. However, facilities that work with experimental units should determine the class in which their lasers fall. The specifications of noncommercial units should be compared with Tables 5.13 and 5.14 to determine the classification.

The least powerful class of laser is class I. In this class the power and energy of the unit are such that the TLV$^{©*}$ for direct viewing of the laser beam, if the entire beam passes through the limiting aperture of the eye, cannot be exceeded for the classification duration (the maximum duration of the exposure inherent in the design of the laser). In the spectral region of 400 to 1400 nanometers (1 nm = 10^{-9} meters), the limiting aperture of the eye is taken to be 7 mm. Class I lasers must emit levels below the accessible exposure limits (AEL) which are given in the first part of Table 5.13. Many of the operating procedures are based on not exceeding the AELs for class I lasers.

These low power lasers require no control measures to prevent eye damage. Some class I lasers incorporate more powerful lasers, but are designated Class I because they are in an enclosure. If the more powerful enclosed laser is accessible, control measures appropriate to the higher class are required. Class II lasers are not considered hazardous for momentary (0.25 second) and unintentional direct viewing of the laser beam, but should not be deliberately aimed at anyone's eyes. A class IIa laser must have a label affixed to the unit which bears the words

---

* A copyrighted trademark of the American Conference of Governmental Industrial Hygienists.

Table 5.13A.   Class I-Accessible Emission Limits for Laser Radiation

Wavelength (nanometers)	Emission Duration (seconds)	Class I-Accessible Emission Limits (power [W] or energy [J])
180–400	$t \geq 3 \times 10^4$	$2.4 \times 10^{-5} k_1 k_2{}^a$ J
180–400	$t \geq 3 \times 10^4$	$8.0 \times 10^{-10} k_1 k_2$ W $>400 -$
>400–1400	$t > 1 \times 10^{-9} - 2 \times 10^{-5}$	$2.0 \times 10^{-7} k_1 k_2$ J
>400–1400	$t > 2 \times 10^{-5} - 1 \times 10^1$	$7.0 \times 10^{-4} k_1 k_2 t^{0.75}$ J
>400–1400	$t > 1 \times 10^1 - 1 \times 10^4$	$3.9 \times 10^{-3} k_1 k_2$ J
>400–1400	$t > 1 \times 10^4$	$3.9 \times 10^{-7} k_1 k_2$ W
>1400–2500	$t > 1 \times 10^{-9} - 1 \times 10^{-7}$	$7.9 \times 10^{-5} k_1 k_2$ J
>1400–2500	$t > 1 \times 10^{-7} - 1 \times 10^1$	$4.4 \times 10^{-3} k_1 k_2 t^{1/4}$ J
>1400–2500	$t > 1 \times 101$	$7.9 \times 10^{-4} k_1 k_2$ W
>2500–1×10^6	$t > 1 \times 10^{-9}$ to 1×10^{-7}	$1.0 \times 10^{-2} k_1 k_2$ J/cm^2
>2500–1×10^6	$t > 1 \times 10^{-7}$ to 1×10^1	$5.6 \times 10^{-1} k_1 k_2 t^{1/4}$ J/cm^2
>2500–1×10^6	$t > 1 \times 10^1$	$1 \times 10^{-1} k_1 k_2 t$ J/cm^2

[a]   Class I accessible emission limits for wavelengths ≥ 180 nm to 400 nm shall not exceed the Class I accessible limits for the wavelengths ≥ 1400 nm to 1×10^6 nm with a k_1 and a k_2 of 1.0 for comparable sampling intervals, t.

Table 5.13B.   Class II- and Class III-Accessible Emission Limits for Laser Radiation.

Wavelength (nanometers)	Emission Duration (seconds)	Accessible Emission Limits (Power [W] or Energy [J])
Class IIa Accessible Emission Limits for Laser Radiation		
400–710	$t > 1 \times 10^3$	3.9×10^{-6} W
Class II Accessible Emission Limits for Laser Radiation		
400–710	$t > 2.5 \times 10^{-1}$	1×10^{-3} W
Class IIIa Accessible Emission Limits for Laser Radiation		
400–710	$t > 3.8 \times 10^{-4}$	5×10^{-3} W
Class IIIb Accessible Emission Limits for Laser Radiation		
180–400	$t \leq 2.5 \times 10^{-1}$	$3.8 \times 10^{-4} k_1 k_2$ J
180–400	$t > 2.5 \times 10^{-1}$	$1.5 \times 10^{-3} k_1 k_2$ W
>400–1400	$t > 1 \times 10^{-9} - 2.5 \times 10^{-}$	$10 \, k_1 k_2 t^{1/3}$ J/cm^2 to a maximum value of 10 J/cm^2
>400–1400	$t > 2.5 \times 10^{-1}$	5.0×10^{-1} W
>1400–1×10^6	$t > 1 \times 10^{-9} - 1 \times 10^1$	10 J/cm$^2 t$
>1400–1×10^6	$t > 1 \times 10^1$	5.0×10^{-1} W

Note: Class IIa, II, III, and IIIb limits are the same as for class I except within the these wavelength and emission durations.

Table 5.14.   Values of Wavelength-Dependent Correction Factors K_1 and k_2

Wavelength	k_1	k_2		
180–302.4	1	1		
>302.4–315	$10^{\left(\frac{\lambda - 302.4}{5}\right)}$	1		
>315–400	330	1		
>400–700	1	1		
>700–800	$10^{\left(\frac{\lambda - 700}{515}\right)}$	if: $t \le \dfrac{10100}{\lambda - 699}$ then $k_2 = 1$	if: $\dfrac{10100}{\lambda - 699} < t \le 10^4$ then $k_2 = \dfrac{t(\lambda - 699)}{10100}$	if: $t > 10^4$ then $k_2 = \dfrac{\lambda - 699}{1.01}$
>800–1060	$10^{\left(\frac{\lambda - 700}{515}\right)}$	if: $t \le 100$ then $k_2 = 1$	if: $100 < t \le 10^4$ then $k_2 = \dfrac{t}{100}$	if: $t > 10^4$ then $k_2 = 100$
>1060–1400	5			
>1400–1535	1	1		
>1535–1545	$t \le 10^{-7}$ $k_1 = 100$ $t > 10^{-7}$ $k_1 = 1$	1		
>1545–1×10^6	1	1		

CLASS IIA LASER PRODUCT—AVOID LONG-TERM VIEWING OF DIRECT LASER RADIATION. A class II laser must have a label as shown in Figure 5.11 affixed to the unit in a conspicuous location, bearing the words **LASER RADIATION—DO NOT STARE INTO BEAM** in position 1 on the label. The words **CLASS II LASER PRODUCT** must appear in position 3. The labels of all class II, III, and IV units must give in position 2, in appropriate units, the maximum output of laser radiation, the pulse duration (when appropriate), and the laser medium or emitted wavelength(s).

Class III laser devices are hazardous to the eyes if the direct beam is looked at directly, or at specular reflections of the beam. Class IIIa units must have a label, as in Figure 5.11, affixed to the unit with the words **LASER RADIATION—DO NOT STARE INTO BEAM OR VIEW DIRECTLY WITH OPTICAL INSTRUMENTS** in position 1 and **CLASS IIIa LASER PRODUCT** in position 3. A Class IIIb unit would use a label, as in Figure 5.12, and have the words: **LASER RADIATION—AVOID DIRECT EXPOSURE TO THE BEAM** in position 1 and the words **CLASS IIIb LASER PRODUCT** in position 3. Mirror-like, smooth surfaces of any material that would reflect the beam as a beam should be eliminated as much as possible from the area in which the laser is situated. At a minimum, the user should take every precaution to avoid aiming the laser at such surfaces. It would be desirable to have the beam terminate on a surface that would only provide a diffuse reflection. It should, however, reflect well enough for a well-defined beam spot to be observable so that it is possible to visibly determine the point of contact. The laser should be set up in an area to which access can be controlled. When no responsible person is in the room, the room should be locked to

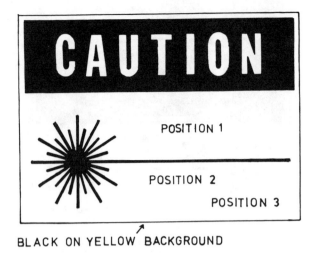

BLACK ON YELLOW BACKGROUND

Figure 5.11. Caution sign for low-to-moderate power lasers.

can be controlled. When no responsible person is in the room, the room should be locked to prevent others from entering the room and changing the physical configuration, or accidentally exposing their eyes to the laser beam.

If there is any way in which it would be possible for either the direct or specularly reflected beam to enter the eye, appropriate eye protection must be worn in an area where a class IIIa laser is operated. Operation of class IIIb lasers generally should follow the practices for class IV lasers.

Class IV lasers must use a label as in Figure 5.12 and bear the words **LASER RADIATION—AVOID EYE OR SKIN EXPOSURE TO DIRECT OR SCATTERED RADIATION** in position 1 and the words **CLASS IV LASER PRODUCT** in position 3. Class IV lasers require many precautions to be used safely. The radiation from pulsed or CW units that lie in the visible and near-infrared can be focused by the eye and can cause damage to the retina from either direct and specular beams or diffuse reflections. The skin can also be injured. Pulsed units and CW units operating in the infrared and ultraviolet are a danger to the skin and the external portions of the eye and can provide enough energy to cause a fire in combustible materials.

Lasers of any class, except medical units and class II lasers that do not exceed the accessible exposure limits of class I lasers for any exposure duration of 1000 seconds or less, must have a label affixed near any aperture through which laser radiation is emitted in excess of the limits for class I lasers bearing the words **AVOID EXPOSURE—LASER RADIATION EMITTED THROUGH THIS APERTURE.**

Table 5.14 provides the means to calculate the values of the wave-dependent factors k_1 and k_2. Table 5.15 provides the TLV©s for direct ocular exposure from a laser beam.

5.5.1. Protective Procedures for Class IIIb and Class IV Lasers

The primary means of protection is to physically prevent exposure. For laboratory workers, baffles may be used to physically intercept or terminate the primary beam and any reflected or secondary beams. Any windows in the facility should be covered during operations. Safety glasses are to be worn by workers while within the facility during operations. Interlocks of various sorts are another avenue of protection. Entrance to the facility by unauthorized

Figure 5.12. Danger sign for higher-powered lasers.

personnel or unexpected entry by laboratory employees should be prevented by safety interlocks while the laser is in an operating condition, i.e., when it is on and in a condition to emit radiation. A warning light should be placed at the entrance. The interlocks should be capable of being bypassed to allow individuals to pass in and out of the controlled area in an emergency or to allow controlled access as needed to allow operations to be conducted. In the latter case, which might be considered a routine bypassing of the interlocks, the activation of the bypass should be limited to the person in charge of the operations at the time, who is aware of the condition of the system. In an emergency, there should be one or more readily available rapid shutoff switch to immediately disconnect power to the laser. One of these can be the required remote control unit stipulated in 21 CFR 1040.1. The access points to the controlled area should be so situated that there should is little or no likelihood that there would be dangerous levels of emitted radiation to persons entering or leaving through the portals or in areas beyond the entrances. Guests or visitors should be allowed in the controlled area during operations only under carefully controlled conditions, and with everyone in the area aware of their presence. Special care must be taken to provide such persons with protective eyewear or other protective gear as needed to ensure their safety. The applicable TLV© levels should not be exceeded for either employees or visitors.

It should be made impossible, by designing the firing circuit with sufficient fail safe safeguards, for a class IV pulsed laser to be discharged accidentally. A warning mechanism should be incorporated into the design to ensure that all persons in the room are aware that the discharge cycle has been initiated. If the laser unit and its power supply are more than 2 meters apart, both the laser and the power supply must have separate emission warning devices. Both the fail safe system and the warning system should be designed so that no single component failure or a shorted or open circuit can disable the protective and warning features of the system. The system should not be capable of being operated if the redundancy of these circuits or of the emergency shut-off system has been compromised.

Very high-power infrared CW lasers, such as CO_2 units, represent not only a danger to operating personnel, they also constitute a fire hazard if their beams come into contact with combustible material. It would be desirable to have these units operated remotely or have them totally within an enclosure that affords good fire stopping capability. Asbestos should be avoided because it is a regulated carcinogen under current OSHA regulations, and under intense radiation could become friable.

Table 5.15. Threshold Limit Values for Direct Ocular Exposures (Intrabeam Viewing) from a Laser Beam

Spectral Region	Wavelength (nm)	Exposure Time (t)(seconds)	TLV©	
UVC	200–280	$10^{-9} - 3 \times 10^4$	3 mJ/cm	
UVB	280–302	$10^{-9} - 3 \times 10^4$	3 mJ/cm^2	
	303		4 mJ/cm^2	
	304		6 mJ/cm^2	
	305		10 mJ/cm^2	Not
	306		16 mJ/cm^2	to
	307		25 mJ/cm^2	exceed
	308		40 mJ/cm^2	$0.56t^{1/4}$
	309		63 mJ/cm^2	J/cm^2
	310		100 mJ/cm^2	$\leq$ 10 s
	311		160 mJ/cm^2	
	312		250 mJ/cm^2	
	313		400 mJ/cm^2	
	314		630 mJ/cm^2	
UVA	315–400	$10^{-9} - 10$	$0.56t^{1/4}$ J/cm^2	
		$10 - 10^3$	1.0 J/cm^2	
		$10^3 - 3 \times 10^4$	1.0 m W/cm^2	
Light	400–700	$10^{-9} - 1.8 - 10^{-5}$	5×10^{-7} J/cm^2	
	400–700	$1.8 \times 10^{-5} - 10$	$1.8(1/t^{0.75})$ mJ/cm^2	
	400–549	$10 - 10^4$	10 mJ/cm^2	
	550–700	$10 - T_1$	$1.8(1/t^{0.75})$mJ/cm^2	
	550–700	$T_1 - 10^4$	$10\ C_B$ mJ/cm^2	
	400–700	$10^4 - 3 \times 10^4$	C_B μW/cm^2	
IR-A	700–1049	$10^{-9} - 1.8 \times 10^{-5}$	$5\ C_A \times 10^{-7}$ J/cm^2	
	700–1049	$1.8 \times 10^{-5} - 10^3$	$1.8\ C_A(1/t^{0.75})$ mJ/cm^2	
	1050–1400	$10^{-9} - 10^{-4}$	5×10^{-6} J/cm^2	
	1050–1400	$10^3 - 3 \times 10^4$	$320\ C_A$ μW/cm^2	
IR-B and C	1.4 μm $- 103$ μm	$10^{-9} - 10^{-7}$	10^{-2} J/cm^2	
		$10^{-7} - 10$	$0.56\ t^{0.75}$ J/cm^2	
		$10 - 3 \times 10^4$	0.1 W/cm^2	

Note: For C_A see Figure 5.13; $C_B = 1$ for $\lambda = 400$ to 549 nm; $C_B = 10^{[0.015(\lambda-550)]}$ for $\lambda = 550$ to 700 nm; $T_1 = 10$ s for $\lambda = 400$ to 549 nm; $T_1 = 10 \times 10^{[0.02(\lambda-550)]}$ for $\lambda = 550$ to 700 nm.

Commercial lasers incorporate many safety features. Following is a brief list of system safety features, besides the ones already mentioned, that either should be available in the laser as manufactured or should be incorporated in the system when set up for use. The items refer primarily to class IIIb and class IV systems, although they also apply to class II units, except

Figure 5.13. Correction factor for TLV©s.

primarily to class IIIb and class IV systems, although they also apply to class II units, except those in which the acceptable emission limits of class I are not exceeded for any emission duration up to 1000 seconds. This list complies with the provisions of 21 CFR 1040 and ANSI Z136. When possible, systems should be upgraded to the requirements of the latest editions of these two documents as they are revised.

1. All lasers are required to be in a protective housing, but safety interlocks should be provided on any portion that could be removed when the unit is operating if the exposure limits for the class would be exceeded. Normally the enclosures should limit the radiation to no more than the limits for a class I laser. Some powerful laser systems should be required to be within enclosures, including the target or irradiation area, to protect personnel. Any portion of the housing that is not interlocked and could emit radiation over the accessible emission limits for each class must have a label attached to it bearing the words previously mentioned for position 1 for the basic identifying label for each class.
2. Interlocks that are designed to prevent firing of a pulsed laser by turning off the power supply or interrupting the beam must not automatically allow the power supply to be reactivated when they have been reset after serving their protective function.
3. An audible or visible warning device is needed if a required interlock is bypassed or defeated. The warning system should be fail safe, i.e., if it became inoperative the unit should be inoperable.
4. Class IV lasers must be key interlocked or activated, and it is recommended that class IIIb lasers have the same feature. The key must be removable and the unit must not be operable without the key in place and in the on position.
5. A portal, viewing window, or an attached optical device must be designed to prevent any exposure above the permissible TLV©.
6. Any class II (except units emitting less than class I limits for no more than 1000 seconds), III, or IV laser should be equipped with an appropriate safety device, such as a beam stop or optical attenuator, which will prevent emission of radiation in excess of those for a Class I unit in the controlled area.
7. The control units for laser units should be located in areas where the accessible emission limits for a Class I device are not exceeded.
8. Signs bearing essentially the same information as on the labels on the lasers in use within a controlled area should be posted at each entrance to the laser laboratory.

9. There is a possibility of radiation being emitted by other parts of the laser system, such as the power supply. This radiation associated with the operation of the laser system is called collateral radiation. The limits on this collateral radiation are (from Table 4, 21 CFR 1040.1):

   a. *Accessible emission limits* for collateral radiation having wavelengths greater than 180 nanometers but less than or equal to 1.0×10^6 nanometers are identical to the accessible emission limits of Class I radiation...

      (1). In the wavelength range of less than or equal to 400 nm, for all emission durations;

      (2). In the wavelength range of greater than 400 nm, for all emission durations less than or equal to 1×10^3 seconds and, when applicable...[see f(8) of 21 CFR 1040.1], for all emission durations.

   b. *Accessible emission limits* for collateral radiation within the X-ray range of wavelengths are 0.5 m roentgen in an hour, averaged over a cross-section parallel to the external surface of the product, having an area of 10 square centimeters with no dimension greater than 5 centimeters.

5.5.2. Eye Protection

The shorter wavelength ultraviolet radiations, UV-C (100 to 280 nm) and UV-B (280 to 315 nm) are primarily absorbed within the conjunctiva and corneal portions of the eyes, causing corneal inflammation. The ultraviolet frequencies just below the visible range, UV-A nm, (315 to 400) are absorbed largely within the lens of the eye. Although little UV-B radiation reaches the lens, UV-B is much more effective in causing cataracts to form than is UV-A. Radiations in the range of 400 to 1400 nm, which includes visible light and the near-infrared, are focused by the lens of the eye on the retina. The focusing properties of the lens may increase the energy per unit area for a point source by as much as 100,000 times. The energy is primarily absorbed by the retinal pigmented epithelium and the choroid. At wavelengths longer than 1400 nm, the water in the eye tissues becomes opaque so that most of the energy is absorbed in the corneal region. The major mechanism for damage is thermal absorption, although there is some damage due to photochemical reactions.

Protective eyewear appropriate to the laser system in use should be worn if there is any eye hazard. The filter or filters in the protective goggles should be matched to the wavelength of the emissions of the laser. Since some lasers emit radiation at more than 1 wavelength, it may be necessary to have filters to cover each range of frequencies. It would be desirable to have the filters in the protective eyewear attenuate only narrow wavelength regions spanning those emitted by the laser, thereby allowing as much visible light through as possible to facilitate seeing by the wearer.

The term optical density is a convenient way to define the attenuation of incident radiation by a filter. A difference of one unit in the optical density of two filters corresponds to a transmission difference of a factor of ten. Thus, a filter that attenuates the incident radiation by a factor of 10 would have an optical density of 1, attenuation by a factor of 100 would mean an optical density of 2, and so forth. Since this is a logarithmic scale (to the base 10) the optical density of two filters stacked together is the sum of the optical densities of the two filters. Note, however, that this is true only if the two filters are for the same wavelength. If they are for different wavelengths and transmit essentially all of the radiation at the other's wavelength, the filtration of the stacked filter would be the same as that of the two filters considered individually.

When working with lasers of very high power or beam intensities, the absorption of energy in the filter can cause damage to the filters. For pulsed lasers, the threshold for damage to glass filters is approximately 10 to 100 joules/cm^2 and for plastic and dielectric coatings, also between 10 and 100 joules/cm^2. A continuous-wave laser operating at 10 watts or more can cause glass filters to fracture and can burn through plastic filters. If these numbers appear small, it should be considered that the power per square centimeter of a heating element on a range is approximately 15 to 20 watts. The filters should be inspected routinely to be sure that they are not damaged.

In Table 5.16 there are two columns for each type of laser, one labeled "Maximum Output Power or Energy," which should be used for a focused beam, and the other labeled "Beam Radiant Exposure," which should be used for an unfocused beam that is larger than the pupil of the eye. Note that the pulsed laser outputs are defined in terms of energy (joules and joules/cm^2) while the CW laser outputs are defined in power (watts or watts/cm^2).

5.5.3. Medical Surveillance

All persons who routinely work with class III or class IV lasers should have a pre-employment medical examination. Others who occasionally work with lasers or in tasks for which it is conceivable that they could experience an eye exposure to the laser radiation should at least have an eye examination. The examination for those that normally work with lasers should include a medical history, stressing the visual and dermatologic systems. The examination should measure visual acuity, and special attention should be paid to those tissues most likely to be affected by the wavelengths emitted by the lasers that the employee will use. If employees suffer injuries or suspect that they have had a potential injury to the eye or skin, an examination of the potentially affected tissues should be done by a qualified physician and the employees should be provided treatment as determined to be appropriate by the physician.

Although periodic examinations are not required by current guidelines, it would not be unreasonable to have a thorough eye examination on a 3- to 5-year schedule to determine if there are subtle changes occurring in the various systems of the eye.

REFERENCES

1. Performance Standards for Light-Emitting Products-Laser Products, 21 CFR Chap. I., Part 1040.1, Food and Drug Administration, 1987.
2. *A Guide for the Control of Laser Hazards,* American Conference of Governmental Industrial Hygienists, Cincinnati, OH, 1981.
3. American National Standard for the Safe Use of Lasers, ANSI Z-136, American National Standards Institute, New York, NY, 1980.
4. **Goldman, L., Rockwell, J.R., Jr., and Hornby, P.,** Laser design and personnel protection from high energy lasers, in *CRC Handbook of Laboratory Safety,* 2nd ed., Steere, N.V., Ed., CRC Press, Inc., Boca Raton, FL, 1971.
5. *Threshold Limit Values,* American Conference of Governmental Industrial Hygienists, Cincinnati, OH, 1993.

Table 5.16. Selection Guide for Eye Protection for Direct Viewing of Laser Beams Emitting Radiation Between 400 and 1400 Nanometer Wavelengths

| | Q-Switched (1 ns–0.1 ms) | | Non-Q-Switched (0.4–10 ms) | | Continuous | | | | Attenuation | |
| | | | | | Momentary 0.25–10 s | | Long-Term Staring ≥3 h | | | |
Maximum Output Energy (J)	Maximum Beam Radiant (J/cm²)	Maximum Output Energy (J)	Maximum Power Radiant (J/cm²)	Maximum Beam Output (W)	Maximum Power Irradiance (W/cm²)	Maximum Beam Output (W)	Maximum Attenuation Irradiance (W/cm²)	Factor	OD
10	20	100	200	NR	NR	NR	NR	10^8	8
1	2	10	20	NR	NR	NR	NR	10^7	7
10^{-1}	2×10^{-1}	1	2	NR	NR	1	2	10^6	6
10^{-2}	2×10^{-2}	10^{-1}	2×10^{-1}	NR	NR	10^{-1}	2×10^{-1}	10^5	5
10^{-3}	2×10^{-3}	10^{-2}	2×10^{-2}	10	20	10^{-2}	2×10^{-2}	10^4	4
10^{-4}	2×10^{-4}	10^{-3}	2×10^{-3}	1	2	10^{-3}	2×10^{-3}	10^3	3
10^{-5}	2×10^{-5}	10^{-4}	2×10^{-4}	10^{-1}	2×10^{-1}	10^{-4}	2×10^{-4}	100	2
10^{-6}	2×10^{-6}	10^{-5}	2×10^{-5}	10^{-2}	2×10^{-2}	10^{-5}	2×10^{-5}	10	1

NR = Not recommended.

5.6. MICROBIOLOGICAL AND BIOMEDICAL LABORATORIES

5.6.1. Introduction

Individuals who work in microbiological laboratories face a number of special problems when they work with organisms that are infectious to humans. There is evidence to show that biological laboratory personnel in such circumstances do have higher rates of incidence for a number of diseases associated with selected types of organisms than do comparable personnel working elsewhere, but there is no comprehensive system of data collection that defines the extent of the problem. However, virtually none of the primary laboratory infections which have occurred appear to have led to secondary infections for families, friends, or members of the general public. Only about 20% of the laboratory infections that have been reported have been attributed to specific incidents. The remainder have been assumed to be related to work practices within the laboratories, especially those which generate aerosols. In recent years the fear which many laboratory workers feel about contracting the human immunodeficiency virus (HIV) virus, with the consequences which that implies, has emphasized the need for providing guidelines for safety in laboratories working with all types of infectious organisms. In 1988 the Centers for Disease Control (CDC) published its "Universal Precautions" due to the concern for transmittal of the HIV virus, but it has become the basis for handling blood and other body fluids that could contain contagious organisms. In 1991, the OSHA incorporated the Universal Precautions in their standard for Occupational Exposure to Bloodborne Pathogens, covered in Chapter 4 of this book. The CDC has published comprehensive guidelines for Biosafety in Microbiological and Biomedical Laboratories, covering many other pathogenic organisms in addition to the HIV and hepatitis B (HBV) viruses which were emphasized in the OSHA standard. This section is based on the third edition of these guidelines, published in 1993.

Persons who work in microbiological laboratories in which the research involves organisms infectious to humans are usually well aware of the risks to themselves posed by the infectious agents. However, not everyone recognizes the many different operations and procedures which may result in significant quantities of aerosols being generated. Some of this awareness can be taught, some can be dictated by firm rules of laboratory practice, but some must be gained through experience. On the other hand, familiarity with procedures can occasionally lead to a casual attitude toward the risks associated with the research activities. Whether due to inexperience, too relaxed an attitude, or poor work performance, it only requires one mistake to cause a problem for everyone. Unlike radiation, there is no simple means to detect contamination should it occur. There are methods for surveying for biological contamination, but they are more complicated than those for radiation, so that it is possible for an unsafe condition to persist for some time without the workers being aware of the situation. Although some diseases require a substantial exposure to the infectious organism, some can be transmitted with only a minor exposure so that it does not require a major incident to create a problem. In addition, one can become as ill once the disease has been contracted whether it was due to an exposure to a few organisms or to a multitude.

5.6.2. Laboratory Safety and Hazard Communication Standards

Neither the OSHA Laboratory Safety nor the OSHA Hazard Communication standards explicitly cover exposures to contagious organisms. The OSHA bloodborne pathogen standard, by extension to other sources of infection than humans, comes closest to being an appropriate standard, but as it is written, the emphasis is on human blood, tissues, and other bodily fluids. However, all of these standards contain the basic requirement that all employees must be made aware, by documented training programs, of the hazards associated with materials with which they work. The laboratory standard does not cover hazards associated with infectious organisms, but many of the procedures in the microbiological laboratory involve chemicals covered by the standard. Individuals already working in the laboratory and all new employees, at the time of employment, must be instructed about the dangers associated with the specific chemical agents involved in their work and the safety procedures which have been adopted. Instruction on new materials brought into the laboratory must take place before employees begin using the material. As noted, the legal standard only applies to chemicals, but the concept is entirely appropriate to any microbiological hazards to which the workers may be exposed. If an employee is exposed to a known pathogenic organism due to a lack of information on the risks, means of preventing the exposure, and treatment options, it would be possible for citations to be issued under the general duty clause of the OSHA General Industrial standards. This is the approach by which the initial concerns about HIV and HBV were addressed, and has been considered for a response to the problems of exposure by health care and emergency personnel to resistant strains of tuberculosis.

5.6.3. Laboratory Director

The key person in any laboratory, but especially so in a microbiological laboratory, is the person in charge of the facility, the laboratory director. As noted in the previous section there are guidelines and recommendations provided by many agencies, but there are relatively few mandatory governmental requirements for this class of laboratory. In the absence of a substantial body of formal regulations other than the internal policies which may have been set within the organization of which the laboratory is a part, the laboratory director is usually the individual that has been assigned the responsibility and the discretionary authority to set work practices. The attitude of this person will be reflected by others working in the facility. It is this individual who must see that laboratory workers are aware of the risks associated with their environment and that they are capable of performing all procedures safely and effectively. The laboratory authority must also determine if the available facilities are adequate for the research to be done, or to determine if special safety practices are needed.

5.6.4. Miscellaneous Safety Practices

The microbiological laboratory shares many of the same safety needs as a standard chemical laboratory. Some of these are briefly summarized below. For a more detailed discussion of good general laboratory safety practices, refer to the comparable sections in the first four chapters.

5.6.4.1. Laboratory Line of Authority

The telephone numbers for the laboratory authorities and emergency phone numbers should be posted within the laboratory near the telephone and near the primary entrance, outside the laboratory. In addition, a list of the significant hazards in the laboratory should be posted at the entrances to the area in the event that emergency personnel cannot reach knowledgeable persons, or if the time scale of the emergency requires immediate entrance.

Emergency workers always should be alerted to any risk to which they might be exposed. They are much less likely to have received training in infectious disease problems than in chemical emergencies. Exposure of emergency personnel to diseases has recently been addressed in the Ryan-White Act for Emergency Response Workers. This act *does* address potential infectious diseases other than HIV and HBV.

5.6.4.2. Spills and Emergencies Involving Chemically Dangerous Materials

Minor spills should be cleaned up immediately by laboratory personnel, providing that the material is not immediately dangerous to life and health (IDLH) and that equipment and supplies to do so are available. A basic emergency spill kit should always be maintained within a facility.

Moderate spills of ordinarily dangerous materials may require technical guidance, supplies in excess of those normally kept on hand, or assistance of the organizations safety or emergency response personnel.

For moderate to large spills of IDLH-level materials or for large spills of ordinarily dangerous materials, e.g., acids, etc.:

1. Evacuate the area and initiate an evacuation of the building either personally or with the assistance of the building authority. The persons leaving the building should gather at a designated point upwind from the building for further instructions.
2. Call the local emergency number (911, if available) and report the incident. The type of emergency must be clearly described so that the dispatcher can send the appropriate emergency responders. Do this from a location away from the affected area.
3. Those individuals directly involved in the incident are to remain available outside the building to assist the emergency group unless they require immediate medical attention, which has first priority. After the initial notification of emergency personnel, the laboratory authority and the department head are to be notified.

The building should be evacuated and the fire department called for any fire other than a very small one, where there is confidence that it can be put out without risk of spreading or danger to anyone.

If there is an emergency involving an injury, be sure to alert the dispatcher of the circumstances of the injuries. If chemicals are involved, specific members of the local rescue squad having special training may need to respond. If there is exposure to the eye or skin, assist the injured person to an eyewash station, a deluge shower, or to a combination unit.

5.6.4.3. Emergency Equipment for Nonbiohazardous Spills

Everyone should be familiar with the location and use of all equipment in their laboratory area. This includes means to initiate an evacuation (fire alarm pull stations, etc.), fire extinguishers, fire blankets, eyewash stations, deluge showers, first aid kits, spill kit materials, respiratory protective devices, and any other materials normally kept in the area for emergency response.

5.6.4.4. Evacuation

Everyone should be familiar with the primary and secondary evacuation routes from their area to the nearest exit or an alternative one if the primary exit is blocked. Everyone should be told what method is used to signal a building evacuation, where to go, to check in with a responsible person, and not to reenter the building until an *official* clearance to do so is given.

5.6.5. Accidents and Spills of Biohazardous Materials

Most public emergency agencies are not specifically equipped to handle biohazardous incidents so that laboratories handling organisms that are infectious to humans or organisms that could harm the environment should plan for emergencies involving these materials. Concern for damage to the environment has caused an increasing number of emergency groups to obtain equipment that provides total personnel protection from airborne contaminants. The availability of these resources could make it possible to handle situations in which airborne dispersal of contagious organisms could be involved. The emergency respondents would depend greatly on the information available from laboratory personnel in determining the scope of their response. They could feel that the risks are unacceptable to them unless a person's life was in danger.

Each facility should perform a hazard analysis of their operations to determine the worst-case scenarios involving all of their procedures, and to develop a plan to prevent these worst-case situations or to cope with them. The maximum credible event should be specifically addressed. A simple transfer of a container of active agents from one area of the room could result in dropping the container and releasing the organism to the room in the splashed material and in the form of aerosols. Some of the latter, if small enough, could remain airborne for hours.

A minor spill would be one that remains contained within a biological safety cabinet which provides personnel protection so no potentially infectious aerosols escape. It is assumed that no one is contaminated by direct contact with the spilled materials. Essentially, the procedures to be used in such a case should take advantage of the protection afforded by the cabinet, i.e., the cabinet should continue to operate. Eventual decontamination of the cabinet would be the measure to be taken to remedy the situation. In some minor cases, this can be done by using the routine procedures normally used for surface contamination. However, this will not decontaminate the fan, filters, and the airflow plenums which normally are not accessible. When there has been a substantial spill within the cabinet, the cabinet should be decontaminated by agents that could reach these concealed spaces. It could be necessary to use a strong agent such as formaldehyde gas. There are standard procedures for doing this available from a number of government agencies. Individuals should be trained in how to perform this procedure safely and may wish to use it on a number of occasions, such as before filter changes, before maintenance work, before relocating a cabinet, and upon instituting a different program in a biosafety cabinet.

A major spill would be one that is not contained within a biological safety cabinet. The response would depend upon the nature of the organism involved and the size of the spill, i.e., the probability of individuals being infected. In a later section, the design, special equipment, and practices will be described for laboratories for biosafety levels of 1 through 4. At the upper end of this scale, biosafety levels 3 and 4, there is real concern about the possibility of infections being carried out of the laboratory, so that the response would be different for these facilities than for the areas designed and operated to work with less dangerous organisms.

Following a spill of an infectious agent, one should immediately evacuate the room, breathing as little as possible of any aerosols. The door to the contaminated area should be closed to avoid airborne transfer of the material from the area in which the spill occurred. Any outer garments that were contaminated, including laboratory coats, shoes, trousers, gloves, etc., should be removed and placed in sealed containers. These can be pails with covers, autoclavable bags, or, if no other alternatives are available, double plastic garbage bags. Everyone in the room should at least thoroughly wash their face and hands with a disinfectant soap and if possible, shower thoroughly. Once the area is isolated and the possibility of retaining infectious material on individuals minimized, the situation should be reviewed and a plan of correction determined by the responsible parties. Because of the possibility of liability, institutional and corporate management personnel as well as safety specialists should participate in the discussion. Any persons potentially infected should be referred to a physician as soon as possible. They should not participate in the clean up efforts until it is determined whether they were infected.

The potentially contaminated materials should be autoclaved as soon as possible, or, depending on the value of the materials and the risk of infections to someone trying to sort the material, it may be decided that disposal as a hazardous waste is the proper action. Most organizations carry insurance to cover losses due to accidents.

If the material is not highly infectious and the quantity is not great, it may be sufficient to have someone reenter the area after aerosols have had an adequate time to settle, wearing a coverall covering the entire body, gloves, shoe covers, and a respirator, to decontaminate the immediate area of the spill with a suitable disinfectant and to wipe down nearby surfaces onto which some material may have splashed, or areas where the ventilation system in the laboratory may have carried aerosols. Once the preliminary clean up is completed, it may be sufficiently clean to warrant others entering the room wearing protective outer garments, gloves, and felt protective masks to proceed with a more thorough clean up. All disposable materials that are used in the clean up should go into hazardous waste containers, and materials which are to be kept in autoclavable containers. It probably will be desirable to have tests made of the air to determine if there are still organisms present. All of these actions are not necessarily steps that should be taken in every case, but do represent a conservative approach which might be considered at the time of an incident bepending upon the circumstances.

For an organism which is highly infectious to humans or for a massive spill, it may be necessary to seal off the area thoroughly, including shutting off ventilation ducts to and from the area to ensure that the organisms remain isolated. A specially designed decontamination program may be necessary. Consultation with specialists at the CDC and elsewhere may be desirable for advice and aid on the care of the exposed individuals and correction of the situation. Two things should be kept in mind: equipment can be replaced, people cannot; if the situation is not deteriorating, it is best to leave it alone while a plan of correction is developed which will provide maximum safety for everyone involved.

Custodial personnel should not be asked to assist in cleaning up a spill involving materials posing biohazards. They are not qualified to perform the work safely. Decontamination is a technical problem, not a custodial one.

5.6.6. Generation of Aerosols

It is generally conceded that aerosols are the primary means by which infectious diseases are contracted or spread in the microbiological laboratory, although some cases are known to have occurred due to animal bites, needle sticks, and similar situations where direct contact can happen.

There are many opportunities for aerosols to be generated through normal laboratory procedures. Studies have been conducted of the average number of droplets created by many typical operations and some procedures are prolific generators of aerosols. Each droplet often contains several organisms. There are far more of these daily releases than there are accidents, where the potential for many thousands of infectious organisms being released exists.

Only about 20% of all known laboratory infections have been traced to specific incidents. The majority of the remainder probably came from these ordinary, routine activities. In one comprehensive study, it was found that over 70% of the infections occurring in the laboratory were to scientific personnel, and that 98% of all laboratory-acquired infections were in institutions doing research or diagnostic work.

Some of the laboratory operations which release a substantial number of droplets are almost trivial in nature, such as breaking bubbles on the surface of a culture as it is stirred, streaking a rough agar plate with a loop, a drop falling off the end of a pipette, inserting a hot loop into a culture, pulling a stopper or a cotton plug from a bottle or flask, taking a culture sample from a vaccine bottle, opening and closing a petri dish in some applications, or opening a lyophilized culture, among many others. Most of these only take a few seconds and are often repeated many times daily. Other, more complicated procedures might be considered to be more likely to release organisms into the air, such as grinding tissue with a mortar and pestle, conducting an autopsy on a small animal, harvesting infected tissue from animals or eggs, intranasal inoculation of small animals, opening a blender too quickly, etc. Some incidents have occurred by failing to take into account the possibility that accidents can happen, such as a tube breaking in a centrifuge when not using a safety centrifuge cup. The possibility of aerosol production should always be considered while working with infectious organisms.

5.6.7. Infectious Waste

Any item that has been in contact with infectious organisms, or with materials such as blood, serums, excreta, tissue, etc., that may be infected must be considered infectious unless it has been treated. There are individuals who have excellent laboratory technique and take every precaution to avoid infection themselves and do not think of the possibility that someone who handles the materials which they discard may become infected from these materials. Some of the items used to carry out the procedures discussed in the previous paragraph are potentially infectious but are sometimes discarded as ordinary trash, in which case custodial workers can come into contact with them. Infected tissue has been found in clothing sent to a local laundry. Animal carcasses from animals that had rabies have been sent for disposal with no warnings about careful handling.

Waste from areas where the potential exists for coming into contact with infectious diseases should be treated as if it were hazardous and prepared so that it can be handled safely. It should be bagged or put in a container which is not likely to break or rupture, then incinerated, steam sterilized, or perhaps chemically treated before disposal. Every organization which generates infectious waste should establish procedures to make sure that the waste is treated to render it harmless and that all personnel follow these procedures. One method suitable for a small generator of biological waste to safely collect waste is to double-bag it in heavy-duty plastic bags and freeze it until enough has been collected to run an incinerator economically or to justify the cost of a pickup by a biological waste disposal firm. There are now a number of commercial firms available for disposing of dangerous biological waste just as there are firms for disposing of hazardous chemical wastes. Almost every state has now adopted regulations governing the handling and disposal of infectious waste. The regulations are not equally stringent in every state so that laboratories generating infectious waste need to become familiar with the regulations that could affect their practices. There are many alternative technologies to steam sterilization and incineration for processing infectious waste. These are discussed earlier in this book. Laboratories working with large animals have the most serious problems in dealing with routine infectious waste as existing autoclaving and incineration techniques are not as suitable for handling large animal carcasses and bedding as they are for processing blood, small tissue samples, "sharps," and other wastes characteristic of most biological facilities.

5.6.8. Laboratory Facilities—Design and Equipment

A microbiological laboratory designed for the level of activity expected to be housed within it is a major factor in protection of the employees. It must be designed properly to allow for safe working conditions. It will share many of the same basic features of a good chemistry laboratory. The ventilation should be a 100% fresh air system in most cases, with perhaps more stringent temperature and humidity controls than other types of laboratories. The temperature should be controlled over a relatively narrow range around 72°F (22°C) and the humidity should be maintained between 45 and 60%. There should be more than one means of egress in an emergency, although this is typically not a code requirement in most cases. The interior layout should be conducive to free movement of personnel. Aisles should be sufficiently wide so that stools, chairs, or equipment placed temporarily in the aisles will not block them. The floors, walls, and surfaces of the equipment should be of easily decontaminated material. The junctures of the floors with the walls and the equipment should be designed to be as seamless as possible to avoid cracks in which organic materials could collect and microorganisms thrive. There should be self-contained areas, either near the entrance to the laboratory or adjacent to it, where paperwork and records can be kept and processed, and where social conversation, studying, eating, etc., can be done safely outside the work area. Adequate utilities should be provided. Proper experimental equipment, which will allow the laboratory operations to be done safely, should be provided and properly maintained. An eyewash fountain and deluge shower should be provided, and emergency equipment likely to be needed in the laboratory should be kept in a readily accessible place. A permanent shower and transition area between "clean" and "dirty" areas may be essential at some levels of risk. The laboratory should be at a negative pressure with respect to the corridor servicing it and the airflow within the laboratory should be away from the "social" area toward the work area.

Equipment is considered the primary barrier for protection of the employees. Items such as biosafety cabinets, safety centrifuges, enclosed containers, impervious work surfaces, autoclaves, foot-operated sinks, and other equipment specifically designed to prevent direct contact with infectious organisms or with aerosols must be available. Personal protective equipment can also be considered as a primary barrier if engineering controls are not sufficient. These items can include at minimum a lab coat or wrap-around gown, possibly gloves, masks, or respirators, goggles, and head and foot covers.

5.6.9. Biosafety Levels

This section has been adapted with only minor changes, primarily syntax, from *Biosafety in Microbiological and Biomedical Laboratories* (BMBL) published by the U.S. Department of Health and Human Services. The concept of protection in the following sections is based on a combination of three factors: a properly designed facility, appropriate equipment, and appropriate procedures. All three factors are the basis for the minimum level of needed protection not only for the workers but for others as well. One other factor must be added, and that is the attitude of everyone that safety is a very high priority that cannot be sacrificed to attain a scientific goal.

Microbiological laboratories are divided by the BMBL into four different classifications or levels, with level 1 intended for work with the lowest risk and level 4 designated for work with the highest risk. An ordinary laboratory would approximate a level 2 facility. Each level features a combination of design, standard and special laboratory practices and procedures, and standard and special equipment needed to allow the work appropriate for each level to be done safely.

The following four sections describe each of these biosafety levels, after which several lists of organisms will be provided which are appropriate for each level.

5.6.9.1. Biosafety Level 1

The work at this level normally involves well-characterized agents not known to cause disease in healthy adult humans and of minimal risk to lab personnel and the environment. However, individuals who, because of poor health or who may be immunodeficient or immunosuppressed for any reason, may be at a higher risk should inform the person responsible for the laboratory operation, to assist him in reducing the possibility of their acquiring an infection. The laboratory need not be separated from the remainder of the building. Work is generally conducted on open bench tops, using standard microbiological practices. No special containment equipment is required. The facilities and equipment are appropriate for undergraduate and secondary instruction and training in microbiological techniques. Laboratory personnel have been trained in the specific procedures used in the facility and are supervised by a person with general training in microbiology or a related science.

5.6.9.1.1. Standard Microbiological Practices

- Access to the laboratory may be limited or restricted, at the discretion of the laboratory director, when experiments are in progress.
- Persons must wash their hands with a disinfectant soap or detergent after they handle viable materials and animals, after removing gloves, and before leaving the laboratory.
- Work surfaces are to be decontaminated after each day in which operations are performed and after any spill of viable material.
- All contaminated liquid or solid wastes must be decontaminated before disposal by an approved decontamination method, such as autoclaving. Materials to be decontaminated outside the laboratory are to be placed in a durable, leakproof container and closed for transport from the laboratory. Materials to be decontaminated off-site from the laboratory must be packaged in accordance with applicable local, state, and federal regulations before removal from the facility.
- Mechanical or automatic pipetting devices must be used; mouth pipetting is prohibited.
- Eating, drinking, smoking, handling contact lenses, and applying cosmetics are not permitted in work areas. Persons who wear contact lenses should also wear goggles or face masks while in the active work area. Food can only be stored in cabinets or refrigerators specifically designated for that purpose and not used for storage of any laboratory materials. These must be located outside the work area.
- All procedures must be performed carefully to minimize the creation of aerosols.
- There must be an effective insect and rodent control program.

5.6.9.1.2. Special Practices

- None

5.6.9.1.3. Safety Equipment (Primary Barriers)

- No special containment devices or equipment such as a biological safety cabinet are usually required for manipulations of agents assigned to biosafety level 1.
- It is recommended that laboratory coats, gowns, or uniforms be worn to prevent contamination or soiling of street clothes.
- Gloves should be worn if the skin of the hands is broken or if a rash exists.
- Protective eyewear (chemical splash goggles and/or a face mask) should be worn for anticipated splashes of microorganisms or other hazardous materials to the face.

5.6.9.1.4. Laboratory Facilities (Secondary Barriers)

- Each laboratory must contain a sink for handwashing.
- The laboratory is designed so that it is easily cleaned. Seamless or poured floor coverings are recommended. Epoxy paint is recommended for the walls. Rugs in laboratories are not appropriate and should not be used because proper decontamination of them after a spill is extremely difficult.
- Laboratory furniture should be sturdy and incorporate as few seams and cracks as possible. Spaces between benches, cabinets, and equipment should be easily cleaned.
- Bench tops must be impervious to water and resistant to acids, alkalies, organic solvents, and moderate heat.
- Laboratory windows that can be opened must be equipped with fly screens.

5.6.9.2. Biosafety Level 2

This class of laboratory is suitable for work involving agents of moderate potential hazard to personnel and the environment. In addition to the characteristics of a biosafety level 1 facility: (1) laboratory personnel have specific training in handling pathogenic agents and are directed by competent scientists, (2) access to the laboratory is limited whenever work is being conducted, (3) extreme precautions are taken with contaminated sharp items, (4) some procedures in which infectious aerosols or splashes may be created or conducted will be performed in biological safety cabinets or other physical containment equipment. The laboratory facilities, equipment, and practices are suitable for clinical, diagnostic, and teaching programs involving the typical broad spectrum of indigenous moderate-risk agents present in the community. If the potential for aerosol production or splashes is low, work with these agents can be done safely on the open work bench.

5.6.9.2.1. Standard Microbiological Practices

- Access to the laboratory is to be limited or restricted, at the discretion of the laboratory director, when experiments are in progress.
- Persons must wash their hands with a disinfectant soap or detergent after they handle viable materials and animals, after removing gloves, and before leaving the laboratory.

- Eating, drinking, smoking, handling contact lenses, and applying cosmetics are not permitted in the laboratory. Persons who wear contact lenses should also wear goggles or face masks while in the active work area. Food is to be stored outside the work area in cabinets or refrigerators specifically designated for that purpose only.
- Mechanical or automatic pipetting devices must be used; mouth pipetting is prohibited.
- All procedures must be performed carefully to minimize the creation of aerosols.
- Work surfaces are to be decontaminated at least once each day in which operations are performed and after any spill of viable material.
- All contaminated liquid or solid wastes must be decontaminated before disposal by an approved decontamination method, such as autoclaving. Materials to be decontaminated outside the laboratory are to be placed in a durable, leakproof container and closed for transport from the laboratory. Materials to be decontaminated off-site from the laboratory must be packaged in accordance with applicable local, state, and federal regulations before removal from the facility.
- There must be an effective insect and rodent control program.

5.6.9.2.2. Special Practices

- The laboratory director limits or restricts access to the laboratory when work with infectious agents is in progress. In general, persons who are at increased risk of acquiring infection or for whom infection might be unusually hazardous are not allowed in the laboratory or animal rooms. For example, persons who are immuno-compromised or immunosuppressed may be at risk of acquiring infections. The laboratory director has the final responsibility for assessing each circumstance and determining who may enter or work in the laboratory.
- The laboratory director establishes policies and procedures whereby only persons who have been advised of the biohazards and meet any specific entry requirements (e.g., special training or immunization) enter the laboratory or animal rooms.
- When the infectious agents is in use in the laboratory requires special provisions for entry (e.g., immunization). A hazard warning sign incorporating the universal biohazard symbol is to be posted on the door to the laboratory work area. The warning sign identifies the infectious agent, lists the name of the laboratory director or other responsible persons, and identifies the special requirements for entering the laboratory.
- Laboratory personnel are to receive appropriate immunizations or tests for the agents handled or potentially present in the laboratory (e.g., HBV vaccine or TB skin testing).
- When appropriate, considering the agents handled, baseline serum samples for laboratory and other at-risk personnel are to collected and stored. Additional serum samples may be taken periodically, depending on the agents handled or the function of the facility.
- A biosafety manual is to be prepared and adopted. Personnel are to be advised of special hazards and are required to read the manual and to follow instructions on practices and procedures.
- Laboratory personnel are to receive appropriate training on the potential hazards associated with the work involved, the necessary precautions to prevent exposure, and the exposure evaluation procedures. Personnel receive annual updates or additional training as necessary for procedural or policy changes.
- Laboratory personnel must always exercise a high degree of precaution with any contaminated sharp items, including needles and syringes, capillary tubes, and scalpels. Needles and syringes or other sharp instruments should be restricted in the laboratory,

for use only when there is no alternative, such as parenteral injection, phlebotomy, or aspiration of fluids from laboratory animals and diaphragm bottles. Plasticware should be substitutes for glassware whenever possible.

- Only needle-locking syringes or disposable syringe-needle units (i.e., needle is integral to the syringe) are to be used for the injection or aspiration of infectious fluids. Needles must not be bent, sheared, broken, recapped, removed from disposable syringes, or otherwise manipulated by hand before disposal; rather, they must be carefully placed in conveniently located, puncture-resistant containers used for sharps disposal. Nondisposable sharps must be placed in a hard-walled container for transport to a processing area for decontamination, preferably by autoclaving.

- Syringes which resheathe the needle, needleless systems, and other safe devices should be used when appropriate.

- Broken glassware must not be handled directly by hand, but must be removed by mechanical means such as a brush and dustpan, tongs, or forceps. Containers of contaminated needles, sharp equipment, and broken glass are to be decontaminated before disposal, according to any local, state, or federal regulations.

- Cultures, tissues, or specimens of body fluids are to be placed in a container that prevents leakage during collection, handling, processing, storage transport, transport, or shipping.

- Laboratory equipment and work surfaces should be decontaminated with an appropriate disinfectant on a routine basis, after work with infectious materials is finished, and especially after any overt spills, splashes, or other contamination by infectious materials. Contaminated equipment must be decontaminated before it is sent for repair or maintenance or packaged for transport in accordance with applicable local, state, or federal regulations before removal from the facility.

- Spills and accidents which result in overt exposures to infectious materials are to be immediately reported to the laboratory director. Medical evaluation, surveillance, and treatment are to be provided as appropriate and written records maintained.

- Animals not involved in the research being performed are not permitted in the laboratory.

5.6.9.2.3. Safety Equipment (Primary Barriers)

- Properly maintained biological safety cabinets, preferably class II, or other appropriate personal protective or physical containment devices are to be used whenever:

  - Procedures with a potential for creating infectious aerosols or splashes are conducted. These may include centrifuging, grinding, blending, vigorous shaking or mixing, sonic disruption, opening pressurized containers, inoculating animals intranasally, and harvesting infected tissues from animals or eggs.
  - High concentrations or large volumes of infectious agents are used. Such materials may be centrifuged in the open laboratory if sealed heads or centrifuge safety cups are used and if these rotors or safety cups are opened only in a biological safety cabinet.

- Protective laboratory coats, gowns, smocks, or uniforms are to be worn while in the laboratory. This protective clothing is removed and left in the laboratory before leaving the laboratory for nonlaboratory areas (e.g., cafeteria, library, administrative offices). All protective clothing is either disposed of in the laboratory or laundered by the institution; it should never be taken home by personnel.

- Gloves are to be worn when handling infected animals and when hands may contact infectious materials, contaminated surfaces or equipment. Wearing two pairs of gloves may be appropriate; if a spill or spatter occurs, the hand will be protected after the contaminated glove is removed. Gloves are disposed of when contaminated, removed when work with infectious materials is completed, and are not to be worn outside the laboratory. Disposable gloves are not to be washed or reused.

5.6.9.2.4. Laboratory Facilities (Secondary Barriers)

- Each laboratory is to contain a sink for handwashing.
- The laboratory is designed so that it is easily cleaned. Seamless or poured floor coverings are recommended. Epoxy paint is recommended for the walls. Rugs in laboratories are not appropriate and should not be used because proper decontamination of them after a spill is extremely difficult.
- Bench tops must be impervious to water and resistant to acids, alkalies, organic solvents, and moderate heat.
- Laboratory furniture should be sturdy and incorporate as few seams and cracks as possible. Spaces between benches, cabinets, and equipment should be easily cleaned.
- Laboratory windows that can be opened shall be equipped with fly screens.
- A method for decontamination of infectious or regulated laboratory waste is available (e.g., autoclave, chemical disinfection, incinerator, or other approved decontamination system).
- An eyewash facility (preferably one combined with a deluge shower) is to be readily available.

5.6.9.3. Biosafety Level 3

This level is applicable to clinical, diagnostic, teaching, and research facilities in which work is done with indigenous or exotic agents which may cause serious or potential lethal disease as a result of exposure by inhalation. All laboratory personnel are to be specifically trained in handling pathogenic and potentially lethal agents and are to be supervised by competent scientists who are experienced in working with these agents. All procedures involving manipulation of infectious agents are conducted within biological safety cabinets or other physical containment devices or by personnel wearing appropriate personal protective clothing or devices. The laboratory must be equipped with specific design features.

Many existing facilities may not have all the facility safeguards recommended for biosafety level 3 (e.g., access zone, sealed penetrations, and directional airflow, etc.). In these circumstances, acceptable safety may be achieved for routine or repetitive operations (e.g., diagnostic procedures involving the propagation of an agent for identification, typing, and susceptibility testing) in biosafety level 2 facilities. However, the recommended Standard Microbiological Practices, Special Practices, and Safety Equipment for Biosafety Level 3 must be rigorously followed. The decision to implement this modification of biosafety level 3 recommendations should be made only by the laboratory director.

5.6.9.3.1. Standard Microbiological Practices

- Access to the laboratory is to be limited or restricted, at the discretion of the laboratory director, when experiments are in progress.

- Persons must wash their hands with a disinfectant soap or detergent after they handle viable materials and animals, after removing gloves, and before leaving the laboratory.
- Eating, drinking, smoking, handling contact lenses, and applying cosmetics are not permitted in the laboratory. Persons who wear contact lenses should also wear goggles or face shields while in the active work area. Food is to be stored outside the work area in cabinets or refrigerators specifically designated for that purpose only.
- Mechanical or automatic pipetting devices must be used; mouth pipetting is prohibited.
- All procedures must be performed carefully to minimize the creation of aerosols.
- Work surfaces are to be decontaminated at least once each day in which operations are performed and after any spill of viable material.
- All contaminated liquid or solid wastes must be decontaminated before disposal by an approved decontamination method, such as autoclaving. Materials to be decontaminated outside the laboratory are to be placed in a durable, leakproof container and closed for transport from the laboratory. Materials to be decontaminated off-site from the laboratory must be packaged in accordance with applicable local, state, and federal regulations before removal from the facility.
- There must be an effective insect and rodent control program.

5.6.9.3.2. Special Practices

- Laboratory doors are to be kept closed when experiments are in progress.
- The laboratory director controls access to the laboratory and restricts access to persons whose presence is required for program or support purposes. For example, persons who are immunocompromised or immunosuppressed may be at risk of acquiring infections. Persons who are at increased risk of acquiring infection or for whom infection may be unusually hazardous are not to be allowed in the laboratory or animal rooms. The laboratory director has the final responsibility for assessing each circumstance and determining who may enter or work in the laboratory.
- The laboratory director establishes policies and procedures whereby only persons who have been advised of the biohazards, who meet any specific entry requirements (e.g., special training or immunization), and who comply with all entry and exit procedures enter the laboratory or animal rooms.
- When the infectious materials or infected animals are present in the laboratory or containment module, a hazard warning sign incorporating the universal biohazard symbol is to be posted on the door to the laboratory and animal room access doors. The warning sign identifies the infectious agent, lists the name and telephone number of the laboratory director or other responsible persons, and indicates any special requirements for entering the laboratory, such as the need for immunizations, respirators, or other personal protective measures.
- Laboratory personnel are to receive appropriate immunizations or tests for the agents handled or potentially present in the laboratory (e.g., HBV vaccine or TB skin testing).
- Baseline serum samples are to be collected and stored for all laboratory and other at-risk personnel. Additional serum samples may be taken periodically, depending on the agents handled or the function of the facility.
- A biosafety manual is to be prepared and adopted. Personnel are to be advised of special hazards and are required to read the manual and to follow instructions on practices and procedures.

- Laboratory personnel are to receive appropriate training on the potential hazards associated with the work involved, the necessary precautions to prevent exposure, and the exposure evaluation procedures. Personnel are to receive annual updates or additional training as necessary for procedural or policy changes.
- The laboratory director is responsible for ensuring that before working with organisms at biosafety level 3, all personnel must demonstrate proficiency in standard microbiological practices and techniques and in the practices and operations specific to the laboratory facility. This might include prior experience in handling human pathogens or cell cultures, or a specific training program provided by the laboratory director or other competent scientist proficient in safe microbiological practices and techniques.
- Laboratory personnel must always exercise a high degree of precaution with any contaminated sharp items, including needles and syringes, capillary tubes, and scalpels. Needles and syringes or other sharp instruments should be restricted in the laboratory for use only when there is no alternative, such as parenteral injection, phlebotomy, or aspiration of fluids from laboratory animals and diaphragm bottles. Plasticware should be substitutes for glassware whenever possible.

  - Only needle-locking syringes or disposable syringe-needle units (i.e., needle is integral to the syringe) are to be used for the injection or aspiration of infectious fluids. Used disposable needles must not be bent, sheared, broken, recapped, removed from disposable syringes, or otherwise manipulated by hand before disposal; rather, they must be carefully placed in conveniently located puncture-resistant containers used for sharps disposal. Nondisposable sharps must be placed in a hard-walled container for transport to a processing area for decontamination, preferably by autoclaving.
  - Syringes which resheathe the needle, needleless systems, and other safe devices should be used when appropriate.
  - Broken glassware must not be handled directly by hand, but must be removed by mechanical means such as a brush and dustpan, tongs, or forceps. Containers of contaminated needles, sharp equipment, and broken glass are to be decontaminated before disposal, according to any local, state, or federal regulations.

- All manipulations involving infectious materials are to be conducted in biological safety cabinets or other physical containment devices within the containment module. No work in open vessels is to be conducted on the open bench.
- Laboratory equipment and work surfaces should be decontaminated with an appropriate disinfectant on a routine basis, after work with infectious materials, and especially after overt spills, splashes, or other contamination with infectious materials. Contaminated equipment must be decontaminated before it is sent for repair or maintenance or packaged for transport in accordance with applicable local, state, or federal regulations, before removal from the facility. Plastic-backed paper toweling used on nonperforated work surfaces within biological safety cabinets facilitates clean up.
- Cultures, tissues, or specimens of body fluids are to be placed in a container that prevents leakage during collection, handling, processing, storage, transport, or shipping.
- All potentially contaminated waste materials (e.g., gloves, lab coats, etc.) from laboratories or animal rooms are to be decontaminated before disposal or reuse.
- Spills of infectious materials are to be decontaminated, contained, and cleaned up by appropriate professional staff or others properly trained and equipped to work with concentrated infectious materials.

- Spills and accidents which result in overt or potential exposures to infectious materials are to be immediately reported to the laboratory director. Medical evaluation, surveillance, and treatment are to be provided as appropriate and written records maintained.
- Animals and plants not related to the work being conducted are not permitted in the laboratory.

5.6.9.3.3. Safety Equipment (Primary Barriers)

- Properly maintained biological safety cabinets (class II or III) are to be used for all manipulations of infectious materials.
- For work done outside biological safety cabinets, appropriate combinations of personal protective equipment are to be used (e.g., special protective clothing, masks, gloves, face protection, or respirators), in combination with physical containment devices (e.g., centrifuge safety cups, sealed centrifuge rotors, or containment caging for animals).
- This equipment must be used for manipulation of cultures and of those clinical or environmental materials which may be a source of infectious aerosols, the aerosol challenge of infected animals, harvesting of tissues or fluids from infected animals and embryonated eggs, and necropsy of infected animals.
- Face protection (goggles and mask or face shield) is to be worn for manipulations of infectious materials outside of a biological safety cabinet.
- Respiratory protection is worn when aerosols cannot be safely contained (i.e., outside of a biological safety cabinet), and in rooms containing infected animals.
- Protective laboratory clothing such as solid-front or wraparound gowns, scrub suits, or coveralls must be worn in the laboratory, but is not to be worn outside.
- Gloves must be worn when handling infected animals and when hands may contact infectious materials and contaminated surfaces or equipment. Disposable gloves should be discarded when contaminated and never washed for reuse.

5.6.9.3.4. Laboratory Facilities (Secondary Barriers)

- The laboratory is to be separated from areas which are open to unrestricted traffic flow within the building. Passage through two sets of self-closing doors is the basic requirement for entry into the laboratory from access corridors or other contiguous areas is required. A clothes change room (shower optional) may be included in the passage way.
- Each laboratory must contain a sink for handwashing. The sink is to be foot-, elbow-, or automatically operated.
- The interior surfaces of walls, floors, and ceilings are to be water resistant so that they can be easily cleaned. Penetrations in these surfaces are to be sealed or are to be capable of being sealed to facilitate decontamination.
- Bench tops must be impervious to water and resistant to acids, alkalies, organic solvents, and moderate heat.
- Laboratory furniture is to be sturdy, and spaces between benches, cabinets, and equipment must be easily cleaned.
- Windows in the laboratory are to be closed and sealed.
- A method of decontaminating all laboratory wastes must be available, preferably within the laboratory (i.e., autoclave, chemical disinfection, incineration, or other approved decontamination method).

- A ducted exhaust air ventilation system is to be provided. The system is to be designed so as to create directional airflow that draws air from "clean" areas into the laboratory toward "contaminated" areas. The exhaust air is not to be recirculated to any other area of the building, and is to be discharged to the outside with filtration and other treatment optional. The outside exhaust must be dispersed away from occupied areas and air intakes. Laboratory personnel must verify that the direction of the airflow (into the laboratory) is proper (a static pressure gauge with an alarm between the laboratory and the adjacent spaces can ensure this).
- The high-efficiency particulate air (HEPA)-filtered exhaust air from class II or class III biological safety cabinets is to be exhausted directly to the outside or through the building exhaust system. If the HEPA-filtered exhaust air from class II or III biological safety cabinets is to be discharged through the building exhaust air system, it is to be connected to this system in a manner (e.g., thimble unit connection) that avoids any interference with the air balance of the cabinets or building exhaust system. Exhaust air from class II biological safety cabinets may be recirculated within the laboratory if the cabinet is tested and certified at least every 12 months.
- Continuous flow centrifuges or other equipment that may produce aerosols are to be contained in devices that exhaust air through HEPA filters before discharge into the laboratory.
- Vacuum lines are protected with liquid disinfectant traps and HEPA filters, or their equivalent, which are routinely maintained and replaced as needed.
- An eyewash facility (preferably one combined with a deluge shower) is to be readily available.

5.6.9.4. Biological Safety Level 4

Biosafety level 4 is required for work with dangerous and exotic agents which pose a high individual risk of aerosol-transmitted laboratory infections and life-threatening disease. Agents that have a close or identical antigenic relationship to biosafety level 4 agents are to be handled at this level until sufficient data are available to either confirm continued work at this level or to work with them at a lower level. Members of the laboratory staff are to receive specific and thorough training in handling extremely hazardous infectious agents. They are to understand the primary and secondary containment functions of the standard and special practices, the containment equipment, and the laboratory design characteristics. They are to be supervised by competent scientists who are trained and experienced in working with these agents. Access to the laboratory is strictly controlled by the laboratory director. The facility is to be in either a separate building or in a controlled area within the building that is completely isolated from all other areas of the building. A specific facility operations manual is to be prepared or adopted.

Within the work areas of the facility, all activities are confined to Class III biosafety cabinets, or to class II biological safety cabinets used with one-piece positive pressure personnel suits ventilated by a life support system. The facility is designed to prevent discharge of microorganisms into the environment.

5.6.9.4.1. Standard Microbiological Practices

- Access to the laboratory is limited or restricted at the discretion of the laboratory director when experiments are in progress.

- Persons must wash their hands after handling infectious materials and animals; they must take a decontaminating shower when they leave the laboratory.
- Eating, drinking, smoking, handling contact lenses, and applying cosmetics are not permitted in the laboratory. Persons who wear contact lenses should also wear goggles or face shields while in the active work area. Food is to be stored outside the work area in cabinets or refrigerators specifically designated for that purpose only.
- Mechanical or automatic pipetting devices must be used; mouth pipetting is prohibited.
- All procedures must be performed carefully to minimize the creation of aerosols.
- Work surfaces are to be decontaminated at least once each day in which operations are performed and after any spill of viable material.
- Work surfaces are to be decontaminated at least once a day and after any spill of viable material.
- An insect and rodent control program is to be in effect.

5.6.9.4.2. Special Practices

- Only persons whose presence in the facility or individual laboratory rooms is required for program or support purposes are authorized to enter. Persons who are immunocompromised or immunosuppressed may be at risk of acquiring infections. Therefore, persons who may be at increased risk of acquiring infection or for whom infection may be unusually hazardous, such as children or pregnant women, are not allowed in the laboratory or animal rooms. The supervisor has the responsibility for assessing each circumstance and determining who may enter or work in the laboratory. Access to the facility is limited by means of secure, locked doors. Accessibility is managed by the laboratory director, biohazards control officer, or other person responsible for the physical security of the facility. All persons entering the facility are to be informed in advance of the potential biohazards and instructed as to the appropriate safeguards for ensuring their safety. Authorized persons must comply with the instructions and all other applicable entry and exit procedures. A logbook, to be signed by all personnel, indicates the date and time of each entry and exit. Practical and effective emergency protocols for emergencies are to be established.
- When infectious material or infected animals are present in the laboratory or animal rooms, hazard warning signs incorporating the universal biohazard symbol are to be prominently posted on all access doors. The sign is to identify the agent, list the name of the laboratory director or other responsible persons, and indicate any special requirements for entering the area, such as the need for immunizations or respirators.
- The laboratory director is responsible for ensuring that, before working with organisms at biosafety level 4, all personnel must demonstrate proficiency in standard microbiological practices and techniques, and in the special practices and operations specific to the laboratory facility. This might include prior experience in handling human pathogens or cell cultures, or a specific training program provided by the laboratory director or other competent scientist proficient in these unique safe microbiological practices and techniques.
- Laboratory personnel are to receive available immunizations for the agents handled or potentially present in the laboratory.
- Baseline serum samples are to be collected and stored for all laboratory and other at-risk personnel. Additional serum samples may be collected periodically, depending on the agents handled or the function of the laboratory. The decision to establish a serologic surveillance program takes into account the availability of methods for the

assessment of antibody to the agents of concern. The program must make provision for the testing of serum samples at each collection interval and the communication of results to the participants.

- A biosafety manual is to be prepared or adopted. Personnel are to be advised of special hazards and are required to read the manual and to follow instructions on practices and procedures.
- Laboratory personnel are to receive appropriate training on the potential hazards associated with the work involved, the necessary precautions to prevent exposure, and the exposure evaluation procedures. Personnel are to receive annual updates or additional training as necessary for procedural or policy changes.
- Personnel are to enter and leave the facility only through the clothing change and shower rooms, and are to shower each time they leave the facility. Personnel are to use the airlocks to enter or leave the laboratory only in an emergency.
- Personal clothing is to be removed in the outer clothing change room and kept there. Complete laboratory clothing, including underclothes, pants, shirts or jumpsuits, shoes, and gloves, is to be provided and used by all personnel entering the facility. When leaving the laboratory and before proceeding into the shower area, personnel are to remove their laboratory clothing in the inner change room and store it in a locker or hamper in the inner change room.
- Supplies and materials needed in the facility are to be brought in by way of the double-doored autoclave, fumigation chamber, or airlock which is appropriately decontaminated between each use. After the outer doors are secured, personnel within the facility are to retrieve the materials by opening the interior doors of the autoclave, fumigation chamber, or airlock. The interior doors are to be secured after the materials are brought into the facility.
- Laboratory personnel must always exercise a high degree of precaution with any contaminated sharp items, including needles and syringes, capillary tubes, and scalpels. Needles and syringes or other sharp instruments should be restricted in the laboratory, for use only when there is no alternative, such as parenteral injection, phlebotomy, or aspiration of fluids from laboratory animals and diaphragm bottles. Plasticware should be substitutes for glassware whenever possible.

    - Only needle-locking syringes or disposable syringe-needle units (i.e., needle is integral to the syringe) are to be used for the injection or aspiration of infectious fluids. Used disposable needles must not be bent, sheared, broken, recapped, removed from disposable syringes, or otherwise manipulated by hand before disposal; rather, they must be carefully placed in conveniently located puncture-resistant containers used for sharps disposal. Nondisposable sharps must be placed in a hard-walled container for transport to a processing area for decontamination, preferably by autoclaving.
    - Syringes which resheathe the needle, needleless systems, and other safe devices should be used when appropriate.
    - Broken glassware must not be handled directly by hand, but must be removed by mechanical means such as a brush and dustpan, tongs, or forceps. Containers of contaminated needles, sharp equipment, and broken glass are to be decontaminated before disposal, according to any local, state, or federal regulations.

- Biological materials to be removed from the class III cabinet or from the biosafety level 4 laboratory in a viable or intact state are to be transferred to a nonbreakable, sealed primary container and then enclosed in a nonbreakable, sealed secondary container, which is removed from the facility through a disinfectant dunk tank, fumigation chamber, or an airlock designed for this purpose.

- No materials, except for biological materials that are to remain in a viable or intact state, are to be removed from the maximum containment laboratory unless they have been autoclaved or decontaminated before they leave the facility. Equipment or materials which may be damaged by high temperatures or steam may be decontaminated by gaseous or vapor methods in an airlock or chamber designed for the purpose.
- Laboratory equipment and work surfaces is to be decontaminated routinely after work with infectious materials is finished, and especially after overt spills, splashes, or other contamination with infectious materials. Contaminated equipment must be decontaminated before it is sent for repair or maintenance.
- Spills of infectious materials are to be decontaminated, contained, and cleaned up by appropriate professional staff or others properly trained and equipped to work with concentrated infectious materials.
- A system is to be established for proper reporting of laboratory accidents, exposures, and employee absenteeism, and for medical surveillance of potential laboratory-associated illnesses. Written records are to be prepared and maintained of the absentee records, exposures, and incidents. An essential adjunct to such a reporting-maintenance system is to have available a facility for the quarantine, isolation, and medical care of personnel with potential or known laboratory-associated illnesses.
- Materials, such as plants, animals, and clothing, not related to the experiment being conducted, are not to be permitted in the facility.

5.6.9.4.3. Safety Equipment (Primary Barriers)

- All procedures within the facility with agents assigned to biosafety level 4 are to be conducted in class III biological safety cabinets or in Class II biological safety cabinets used in conjunction with one-piece positive pressure personnel suits ventilated by a life support system. Work with viral agents that require biosafety level 4 secondary containment capabilities can be conducted in class II biological safety cabinets within the facility without the one-piece positive pressure personnel suit if: (1) the facility has been decontaminated, (2) no work is being conducted in the facility with other agents assigned to biosafety level 4, (3) all personnel are immunized against the specific agent being manipulated and demonstrate protective antibody levels, and (4) all other standard and special practices are followed.
- All personnel entering the facility must don complete laboratory clothing, including undergarments, pants, shirts or jumpsuits, shoes, and gloves. All personnel protective equipment is to be removed in the change room before showering and leaving the laboratory.

5.6.9.4.4. Laboratory Facility (Secondary Barriers)

- The biosafety level 4 facility consists of either a separate building or a clearly demarcated and isolated zone within a building. Outer and inner change rooms separated by a shower are to be provided for personnel entering or leaving the facility. A double-doored autoclave, fumigation chamber, or ventilated airlock is to be provided for passage of those materials, supplies, or equipment which are not to be brought into the facility through the change room.
- Walls, floors, and ceilings of the facility are be constructed to form a sealed internal shell which facilitates fumigation and is animal- and insect-proof. The internal surfaces of the shell are to be resistant to liquids and chemicals to facilitate cleaning

and decontamination. All penetrations in these structures and surfaces are to be sealed. Any drains in the floor are to contain traps filled with a chemical disinfectant of demonstrated efficacy against the target agent. The drains are to be connected directly to the liquid waste decontamination system. Sewer vents and other ventilation lines are to contain HEPA filters.

- Internal facility appurtenances, such as light fixtures and utility pipes, are to be installed in such a way as to minimize the horizontal surface area on which dust can settle.
- Bench tops are to have seamless surfaces which are impervious to water and resistant to acids, alkalies, organic solvents, and moderate heat.
- Laboratory furniture is to be of simple and sturdy construction and spaces between benches, cabinets, and equipment are to be easily cleaned.
- A foot-, elbow-, or automatically operated handwashing sink is to be provided near the door of each laboratory room in the facility.
- Any installed central vacuum system is not to serve any areas outside the facility. In-line HEPA filters are to be placed as near as practicable to each use point or service cock. Filters are to be installed to permit in-place decontamination and replacement. Other liquid and gas services to the facility are to be protected by backflow prevention devices.
- If water fountains are provided, they are to be foot operated and are to be located in the facility corridors outside the laboratory. The water service to the fountains is not to be connected to the backflow protected system supplying water to the laboratory areas.
- Access doors to the laboratory are to be self-closing and lockable.
- Any windows are to be sealed and breakage resistant.
- A double-doored autoclave is to be provided for decontaminating materials passing out of the facility. The autoclave door that opens to the area external to the facility is to be sealed to the outer wall and is automatically controlled so that the outside door cannot be opened until after the "sterilization" cycle has been completed.
- A pass-through dump tank, fumigation chamber, or an equivalent decontamination method is to be provided so that materials and equipment that cannot be decontaminated in the autoclave can be safely removed from the facility.
- Liquid effluents from laboratory sinks, biological safety cabinets, floor drains (if used), and autoclave chambers are to be decontaminated by heat treatment before being discharged to the sanitary sewer. Effluents from showers and toilets may be discharged to the sanitary sewer without treatment. The process used for decontaminating liquid wastes must be validated physically and biologically by use of a constant recording temperature sensor in conjunction with an indicator microorganism having a defined heat susceptibility profile.
- A dedicated nonrecirculating ventilation system is to be provided. The supply and exhaust components of the system are to be balanced to assure directional airflow from the area of at least hazard to the areas of greatest potential hazard. The differential pressure/directional airflow between adjacent areas is to be monitored and alarmed to indicate malfunction of the system. The airflow in the supply and exhaust components is to be monitored and the components interlocked to assure inward (or zero) airflow is maintained.
- The general room exhaust air from the facility in which the work is conducted in a class III cabinet system is to be treated by passage through HEPA filters prior to discharge to the outside. The air is to be discharged away from occupied spaces and air intakes. The HEPA filters are be located as near as practicable to the source in

order to minimize the length of potentially contaminated ductwork. The HEPA filter housings are to be designed to allow for *in situ* decontamination of the filter prior to removal, or removal of the filter in a sealed gas-tight primary container for subsequent decontamination and/or destruction by incineration. The design of the HEPA filter housing should facilitate validation of the filter installation. The use of pre-certified HEPA filters can be an advantage. The service life of the exhaust HEPA filters can be extended through adequate filtration of the supply air.

• A specially designed suit area may be provided in the facility to provide personnel protection equivalent to that provided by class III cabinets. Personnel who enter this area are to wear a one-piece positive-pressure suit that is ventilated by a life support system. The life support system is to include alarms and emergency backup breathing air tanks. Entry into this area is to be through an airlock fitted with airtight doors. A chemical shower is to be provided to decontaminate the surface of the suit before the worker leaves the area. The exhaust air from the suit area is to be filtered by two sets of HEPA filters installed in series. A duplicate filtration unit, exhaust fan, and an automatically starting emergency power source are to be provided. The air pressure within the suit area is to be lower than that of any adjacent area. Emergency lighting and communication systems are to be provided. All penetrations into the internal shell of the suit area are to be sealed. A double-doored autoclave is to be provided for decontaminating waste materials to be removed from the suit area.

• The treated exhaust air from class II biological safety cabinets located in a facility in which workers wear a positive-pressure suit may be discharged into the animal room environment or to the outside through the facility air exhaust system. The biological safety cabinets are to be tested and certified at 12-month intervals. The exhaust air from class III biological safety cabinets is to be passaged through two sets of HEPA filters in series prior to discharge to the outside. If the treated exhaust air is discharged to the outside through the facility exhaust air system, it is to be connected to this system in such a manner as to avoid any interference with the air balance of the cabinets or the facility exhaust air system.

5.6.10. Recommended Biosafety Levels

The subsections of this part will contain, for a large number of potentially infectious agents, the biosafety level appropriate for typical laboratory-scale operations involving these agents. However, selection of an appropriate safety level for work with a specific agent or animal study depends upon a large number of factors. Some of the most important are (1) the virulence, pathogenicity, biological stability, route of spread, and communicability of the agent; (2) the nature or function of the laboratory, the procedures and manipulations involving the agent, and the endemicity of the agent; and (3) the availability of effective vaccines or therapeutic measures. The following sections will not present the rationale for the recommendations. For additional information, the reader should consult the original document, Reference 15, from which the lists in this section were taken.

The risk assessments and biosafety levels recommended presuppose a population of healthy, immunocompetent individuals. There are a number of parameters which influence this, including age, heredity, race, sex, pregnancy, predisposing diseases, surgery, and prior exposure to immunosuppressing agents.

Recommendations for the use of vaccines and toxoids are included where effective and safe versions are available. Appropriate precautions should be taken in the administration of live attenuated virus vaccines to individuals with altered immunocompetence. However, these specific recommendations should in no way preclude the routine use of such products as diphtheria-tetanus toxoids, poliovirus vaccine, or influenza vaccine.

The basic biosafety level assigned to an agent is based on the activities typically associated with the growth and manipulation of quantities and concentrations of infectious agents required to accomplish identification or typing. If activities with clinical materials pose a lower risk to personnel than those activities associated with the manipulation of cultures, a lower biosafety level is recommended. On the other hand, if the activities involve large volumes or highly concentrated preparations ("production quantities") or manipulations which are likely to produce aerosols or which are otherwise intrinsically hazardous, additional personnel precautions and increased levels of primary containment may be indicated. It may be possible to adapt biosafety levels up or down to compensate for the appropriate level of safety.

As noted in many places in the preceding material, it is the responsibility of the laboratory director to make these decisions. Risk assessment is ultimately a subjective process, but it is recommended that decisions should be biased toward more safety rather than less.

5.6.10.1. Agent Summaries

All of the recommendations in the following exclude working with production quantities or in situations where substantial amounts of aerosols may be generated. In such cases, a higher level of protection is recommended.

Parasitic Agents

- Cestode Parasites of Humans
  Echinoccus granulosus   Level 2
  Taenia solium (cysticercus cellulosae)   Level 2
  Hymenolepsis nana   Level 2
- Nematode Parasites of Humans
  Strongyloides spp.   Level 2
  Ascaris spp.   Level 2
- Protozoal Parasites of Humans
  Toxoplasma spp.   Level 2
  Plasmodium spp. (including *P. cynomologi*)   Level 2
  Trypanosoma spp.   Level 2
  Leishmania spp.   Level 2
  Sarcocystis spp.   Level 2
  Coccidia spp.   Level 2
  Entamoeba spp.   Level 2
  Giardia spp.   Level 2
  Cryptosporidia spp.   Level 2
  Naegleria fowleri   Level 2
  Toxoplasma cruzi   Level 2
- Trematode Parasites of Humans
  Fasciola spp. (metacercaria)   Level 2
  Schistosoma spp.   Level 2

Fungal Agents

- *Blastomyces dermatitidis* — Level 2
- *Coccidioides immitis*
  Clinical specimens and animal tissue — Level 2
  Animal studies when route of challenge is parenteral — Level 2
  Sporulating mold form cultures, samples likely to contain — Level 3
      infectious arthoconidia
- *Cryptococcus neoformans* — Level 2
- *Histoplasma capsulatum*
  Clinical specimens, animal tissues and animal tissues when — Level 2
      route of challenge is parenteral
  Processing of mold cultures, soil, etc., when likely to contain — Level 3
      infectious conidia
- *Sporothrix schenckii* — Level 2
- Pathogenic Members of the Genera *Epidermophyton, Microsporum,* — Level 2
      and *Trichophyton*
- Miscellaneous Molds, *Cladosporium (Xylohypha) trichoides,* — Level 2
      Cladosporium bantium, Penicillium marnefil, Exophiala
      (Wangiella) dermatitidis, Fonsecaea pedrosoi, and *Dactylaria*
      gallopava (Ochroconis gallopavum)

Bacterial Agents

- *Bacillus anthracis* — Level 2
  A licensed vaccine is available. Not normally recommended
      for ordinary use except for workers having frequent contact
      with clinical specimens or diagnostic specimens
- *Bordetella pertussis* — Level 2
- *Brucella (B. abortus, B. canis, B. melitensis, B. suis)*
  Activities with clinical materials of human or animal origin — Level 2
  Manipulation of cultures of pathogenic *Brucella* spp. — Level 3
- *Ampylobacter (C. jejuni/C. coli, C. fetus* ssp. *fetus* — Level 2
- *Chlamydia psittaci, C. trachomatis*
  Diagnostic examination of tissues or cultures, necropsies, — Level 2
      contact with clinical materials
  Activities with concentrations of infectious materials or — Level 3
      high potential for aerosol production
- *Clostridium botulinum*
  All activities; a botulism toxoid is available from the CDC — Level 2
      as an experimental new drug (ND)
- *Clostridium tetani* — Level 2
  Administration of an adult diphtheria-tetanus toxoid at 10-year — Level 2
      intervals is highly recommended
- *Corynebacterium diphtheria* — Level 2
  Administration of an adult diphtheria-tetanus toxoid at
      10-year intervals may be desirable
- *Francisella tularensis*
  Activities with clinical materials of human or animal origin — Level 2
  Manipulations of cultures and for experimental animal studies — Level 3

An investigational live attenuated virus is available, and is
    recommended for those working with the agent or having
    potential contact in the laboratory or with infected animals

- *Leptospira interrogans*—all serovars
  - All activities — Level 2
- *Legionella pneumophila*; other legionella-like agents
  - All activities — Level 2
- *Mycobacterium leprae*
  - All activities (special care with syringes) — Level 2
- *Mycobacterium* spp. other than *M. tuberculosis, M. bovis*
    or *M. leprae*
  - Activities involving clinical materials and cultures — Level 2
- *Mycobacterium tuberculosis, M. bovis*
  - Working with acid-fast smears or culturing sputa, other — Level 2
    clinical specimens provided aerosol generating manipulations
    are done in class I or II biosafety cabinet; a few other
    restricted exceptions
  - Propagation and manipulation of cultures and studies using — Level 3
    nonhuman primates
- *Neisseria gonorrhoeae*
  - All activities, with following exception: — Level 2
  - Activities generating aerosols or droplets — Level 3
° *Neisseria meningitidis*
  - All activities with following exception: — Level 2
  - Activities generating aerosol or droplets — Level 3
  - Use of a licensed vaccine should be considered for work with
    high concentrations of infectious materials
- *Pseudomonas pseudomallei*
  - All activities with following exception: — Level 2
  - Activities with high potential for aerosol or droplet production — Level 3
- *Salmonella*—all serotypes except *typhi*
  - Clinical materials, cultures, and potentially contaminating agents — Level 2
- *Salmonella typhi*
  - All activities with following exception: — Level 2
  - Activities with high potential for aerosol or droplet production — Level 3
  - A reasonably effective licensed vaccine is available and should
    be considered
- *Shigella* spp.
  - All activities — Level 2
- *Treponema pallidum*
  - All activities — Level 2
  - Periodic serological monitoring should be considered
    for personnel
- Vibrionic enteritis *(Vibrio cholerae, V. parahaemolyticus)* — Level 2
  - Short-term vaccine available but not recommended for routine
    use by laboratory personnel
- *Yersinia pestis*
  - All activities with following exception: — Level 2
  - Activities with a high potential for aerosol or droplet — Level 3
    production and for work with antibiotic-resistant strains
  - Licensed inactivated vaccines available and use recommended

Rickettsial Agents

- *Coxiella burnetii*
  Serological examinations, staining of impression smears   Level 2
  Inoculation, incubation, etc., of eggs or tissue cultures, necropsies,   Level 3
      manipulation of tissue cultures
  New Q fever vaccine (IND) available from Fort Detrick, use
      should be limited to those at high risk, no demonstrated
      sensitivity to Q fever antigen
- *Rickettsia prowazekii, Rickettsia typhi (R. mooseri), Rickettsia*
  tsutsugamushi, Rickettsia canada, and spotted fever group agents
  Rickettsia rickettsii, Rickettsia conorii, Rickettsia akari,
  Rickettsia australis, and *Rickettsia siberica*
  Nonpropagative laboratory procedures including serological and   Level 2
      fluorescent antibody procedures, staining of impression smears
  All other manipulations of known or potentially infected   Level 3
      materials
  Access to medical program and antibiotic therapy very important

Viral Agents

- Hepatitis A virus, Hepatitis E virus
  All activities   Level 2
  A licensed inactivated vaccine for hepatitis A is available in
      Europe, and as an investigational vaccine in the U.S.;
      it is recommended for laboratory personnel
- Hepatitis B virus, Hepatitis C virus (formerly known as
  nonA-nonB virus), Hepatitis D virus
  All activities with following exception: Level 2
  Activities with high potential for aerosol or droplet production   Level 3
  Licensed recombinant vaccines against hepatitis B are
      available and are highly recommended; if the requirements
      for the OSHA bloodborne pathogen standard are met, it is
      required that laboratory personnel be offered this vaccine
      and other measures.
- Herpesvirus simiae (B-virus)
  All activities involving the use or manipulation of tissues,   Level 2
      body fluids, and primary tissue culture materials from
      macaques
  Activities involving the use or manipulation of any material   Level 3
      known to contain Herpesvirus simiae.
- Human Herpesviruses
  All activities   Level 2
- Influenza
  All activities   Level 2
- Lymphocytic chloriomeningitis (LCM) virus
  All activities utilizing possibly infectious body fluids or   Level 2
      tissues and for tissue culture passage of mouse-brain
      passaged strains; manipulation of possibly infectious
      passage and clinical materials should be done in a
      biosafety cabinet
  Activities with high potential for aerosol or droplet production   Level 3

- Poliovirus
  All activities   Level 2
  All laboratory personnel working with the agent must have
      documented polio vaccinations or demonstrated evidence of
      immunity to all three types of poliovirus
- Poxviruses
  All activities   Level 2
  Persons working in or entering facilities where activities
      with vaccinia, monkeypox, or cowpox should have documented
      evidence of vaccination within the preceding 3 years
- Rabies virus
  All activities   Level 2
  Preexposure immunization for personnel working in facilities
      involved with diagnostic activities or research with rabies-
      infected materials; level 3 precautions should be used
      for activities with a high potential for droplet or
      aerosol production
- Retroviruses, including human and simian immunodeficiency
  viruses (HIV and SIV)
  Follow "Universal Precautions" and other provisions of the OSHA
      bloodborne pathogen standard; most activities   Level 2
  Preparation of concentrated HIV or SIV   Level 3
- Transmissible spongiform encephalopathies (Creutzfeldt-Jakob
  and kura agents)
  All activities   Level 2
- Vesicular stomatitis virus (VSV)
  Activities utilizing laboratory-adapted strains of   Level 2
      demonstrated low virulence
  Activities involving the use or manipulation of infected   Level 3
      tissues and virulent isolates from infected livestock

Arboviruses Assigned to Biosafety Level 2. The classification of two of the following, marked with an asterisk, depend upon personnel being immunized. When performing some operations with highly infectious materials, some adaptation of biosafety level 3 conditions may be needed:

Acado	Aruac	Barur	Bouboui
Acara	Arumowot	Batai	Bujaru
Aguacate	Aura	Batu	Bunyamwera
Alfuy	Avalon	Batama	Bunyip
Almpiwar	Abras	Bauline	Burg el Arab
Amapari	Abu Hammad	Bebaru	Bushbush
Ananindeua	Aabahoyo	Belmont	Bussuquara
Anhanga	Bagaza	Benevides	Buttonwillow
Anhembi	Bahig	Benfica	Bwamba
Anopheles A	Bakau	Bertioga	Cacao
Anopheles B	Baku	Bimiti	Cache Valley
Apeu	Bandia	Birao	Caimito
Apoi	Bangoran	Bluetongue	California
Aride	Bangui	Boracela	Encephalomyelitis
Arkonam	Banzi	Botambi	Calovo
Aroa	Barmah Forest	Boteke	Candiru

Cape Wrath
Capim
Caraparu
Carey Island
Catu
Chaco
Chagres
Chandipura
Changuinola
Charleville
Chenuda
Chilibre
Chobar Gorge
Clo Mor
Colorado Tick Fever
Corriparta
Cotia
Cowbone Ridge
Csiro Village
Culaba
D'aguilar
Dakar Bat
Dengue-1
Dengue-2
Dengue-3
Dengue-4
Dera Ghazi Khan
Eastern Equine
  Encephalomyelitis[a]
Edge Hill
Entebbe Bat
Epizootic
  Hemorrhagic
  Disease
Erve
Eubenangee
Eyach
Flanders
Fort Morgan
Frijoles
Gamboa
Gan Gan
Gomoka
Gossas
Grand Arbaud
Great Island
Guajara
Guama
Guaratuba
Guaroa
Gumbo Limbo
Hart Park
Hazara

Huacho
Hughes
Icoaraci
Ieri
Ilesha
Ilheus
Ingwavuma
Inkoo
Ippy
Irituia
Isfahan
Itaporanga
Itaqui
Jamestown Canyon
Japanaut
Jerry Slough
Johnston Atoll
Joinjakaka
Juan Diaz
Jugra
Jurona
Jutiapa
Kadam
Kaeng Khoi
Kaikalur
Kaisodi
Kamese
Kammavan Pettai
Kannaman Galam
Kao Shuan
Karimabad
Karshi
Kasba
Kemerovo
Kern Canyon
Ketapang
Keterah
Keuraliba
Keystone
Kismayo
Klamath
Kokabera
Kolongo
Koongol
Kowanyama
Kunjin
Kununurra
Kwatta
La Crosse
La Joya
Lagos Bat
Landjia
Langat

Lanjan
Las Maloyas
Latino
Le Dantec
Lebombo
Lednice
Lipovnik
Lokern
Lone Star
Lukuni
M'Poko
Madrid
Maguari
Mahogany
  Hammock
Main Drain
Malakal
Manawa
Manzanilla
Mapputta
Maprik
Marco
Marituba
Marrakai
Matariya
Matruh
Matucare
Melao
Mermet
Minatitlan
Minnal
Mirim
Mitchell River
Modoc
Moju
Mono Lake
Mont. Myotis
  leukemia
Moriche
Mosqueiro
Mossuril
Mount Elgon Bat
Murutucu
Mykines
Navarro
Nepuyo
Ngaingan
Nique
Nkolbisson
Nola
Ntaya
Nugget
Nyamanini

Nyando
O'nyong-nyong
Okhotskiy
Okola
Olifantsvlei
Oriboca
Ossa
Pacora
Pacui
Pahayokee
Palyam
Parana
Pata
Pathum Thani
Patois
Phom-Penh Bat
Pichinde
Pixuna
Pongola
Ponteves
Precarious
Point
Pretoria
Prospect Hill
Puchong
Punta Salinas
Punta Toro
Qalyub
Quaranfil
Restan
Rio Bravo
Rio Grande
Ross River
Royal Farm
Sabo
Saboya
Saint Floris
Sakhalin
Salehabad
San Angelo
Sandfly f. (Naples)
Sandfly F. (Sicilian)
Sandjimba
Sathuperi
Sawgrass
Sebokele
Seletar
Sembalam
Serra do Navio
Shamonda
Shark River
Shuni
Silverwater

Simbu
Simian Hemorrhagic
  Fever
Sindbis
Sixgun City
Snowshoe Hare
Sokuluk
Soldado
Sororoca
Stratford
Sunday Canyon
Tacaiuma
Tacaribe
Taggert
Tahyna
Tamiami
Tanga
Tanjong Rabok

Tataguine
Tembe
Tembusu
Tensaw
Tete
Tettnang
Thimiri
Thottapalayam
Tibrogargan
Timbo
Timboteua
Tindholmur
Toscana
Toure
Tribec
Triniti
Trivittatus
Trubanaman

Tsuruse
Turlock
Tyuleniy
Uganda S
Umatilla
Umbre
Una
Upola
Urucuri
Usutu
Uukuniemi
Vellore
Venkatapuram
Vesicular Stomatitis-
  Indiana
Vesicular Stomatitis-
  New Jersey
Wad Medani

Wallal
Wanowrie
Warrego
Western Equine
  Encephalomyelitis[a]
Whataroa
Witwatersrand
Wongal
Wongorr
Wyeomyia
Yaquina Head
Yata
Yogue
Zaliv Terpeniya
Zegla
Zika
Zingilamo
Zirqa

Vaccine Strains of Biosafety Level 3/4 Viruses which May Be Handled at Biosafety Level 2

Virus

Vaccine Strain

Virus	Vaccine Strain
Chikungunya	131/25
Junin	Candid #1
Rift Valley fever	Mp-12
Venezuelan equine encephalomyelitis	Tc-83
Yellow fever	17-D

Arboviruses and Certain Other Viruses Assigned to Biosafety Level 3 (on the basis of insufficient experience)

Adfelaide River
Agua Preta
Alenquer
Almeirim
Altamira
Andasibe
Antequeuera
Araguari
Aransas Bay
Arbia
Arboledas
BabankiBatken
Belem
Berrimah
Bimbo

Bobaya
Bobia
Bozo
Buenventura
Cabassue[a,b]
Cacipacore
Calchaqui
Cananela
Caninde
Chim
Coastal Plain
Conneccticut
Corfu
Dabakala
Douglas

Ensenada
Estero Real
Fomede
Forecariah
Fort Sherman
Gabek Forest
Gadgets Gully
Garba
Gordil
Gray Lodge
Gurupi
Iaco
Ibaraki
Ife
Ingangapi

Inini
Issyk-Kul
Itatuba
Itimirim
Itupiranga
Jacareacanga
Jamanxi
Jari
Kedougou
Khassan
Kindia
Kyzlagach
Lake Clarendon
Llano Seco
Macaua

Mapuera	Omo	Razdan	Tai
Mboke	Oriximina	Resistencia	Tamdy
Meaban	Ouango	Rochambeau	Telok Forest
Mojui Dos	Oubangui	Salanga	Termeil
Compos	Oubi	San Juan	Thiafora
Monte Dourado	Ourem	Santa Rosa	Tilligerry
Munguba	Palestina	Santarem	Tinaroo
Naranjal	Para	Saraca	Tlacotalpan
Nariva	Paramushir	Saumarez Reef	Tonate[a,b]
Nasoule	Paroo River	Sedlec	Ttinga
Ndelle	Perinet	Sena Madureira	Xiburema
New Minto	Petevo	Sepik	Yacaaba
Ngari	Picola	Shokwe	Yaounde
Ngoupe	Playas	Slovaakia	Yoka
Nodamura	Pueblo Viejo	Somone	Yug
Northway	Purus	Spipur	Bogkanovac
Odrenisrou	Radi		

[a] It is recommended that work with this agent should be done only in biosafety level 3 facilities that provide for HEPA filtration of all exhaust air prior to discharge from the laboratory.

[b] A vaccine is available and is recommended for all persons working with this agent.

Arboviruses and Certain Other Viruses Assigned to Biosafety Level 3

Aino	Junin[c,d]	Negishi	ThogotoTocio
Akabane	Kairi	Oropouche[b]	Turuna
Bhanja	Kimberley	Orungo	Venezuelan
Chikungunya[c,d]	Koutango	Peaton	Equine
Cocal	Louping III[a,b]	Piry	Encephalitis[c,d]
Dhori	Mayaro	Powassan	Vesicular
Dugbe	Middelburg	Piumala	Stomatitis
Everglades[c,d]	Mobala	Rift Valley[a,b,c,d]	(Alagoas)
Flexal	Mopeia[e]	Sagiyama	Wesselsbron[a,c]
Germiston[c]	Mucambo[c,d]	SalVieja	West Nile
Getah	Murray Valley	San Perlita	Yellow Fever[c,d]
Hantaan	Encephalomy-	Semliki Forest	Zinga[b]
Israel Turkey	elitis	Seoul	
Meningitis	Nairobi Sheep	Spondweni	
Japanese	Disease	St. Louis	
Encephalomy-	Ndumu	Encephalomy-	
elitis		elitis	

[a] The importation, possession, or use of this agent is restricted by USDA regulation or administrative policy.

[b] Zinga virus is now recognized as being identical to Rift Valley Fever virus.

[c] The Subcommittee on Arbovirus Laboratory Safety recommends that work with this agent should only be done in a biosafety level 3 facility that provides for HEPA filtration of all exhaust air from the laboratory.

[d] A vaccine is available for this agent and is recommended for all persons working with this agent.

[e] This agent is presently being registered in the *Catalogue of Arboviruses*.

Arboviruses, Arenaviruses, and Filoviruses Assigned to Biosafety Level 4

Congo-Crimean hemorraghic fever
Tick-borne encephalitis virus complex
   (Absettarov, Hanzalova, Hypr.,
   Kumlinge, Kyasanur Forest disease,
   Omsk hemorrhagic fever,
   Russian Spring-Summer encephalitis)

Marburg
Ebota
Lassa
Junin
Machupo
Guanarito

Restricted Animal Pathogens. Nonindigenous pathogens of domestic livestock and poultry may require special laboratory design, operation, and containment features not generally addressed in CDC guidelines. The importation, possession, or use of the following agents is prohibited or restricted by law or by U.S. Department of Agriculture regulations or administrative policies.

African Horse Sickness Virus
African Swine Fever Virus
Akabane Virus
Besnoitia besnoiti
Borna disease virus
Bovine spongiform encephalopathy
Bovine infectious petechial fever agent
Brucellosis melitensis
Camelpox virus
Cochliomyia hominivorax
  (screwworm)
Ephemeral fever virus
Foot and mouth disease virus
Fowl plague virus (lethal avian influenza)
Hog cholera virus
Histoplasma (Zymonema)
 farciminosum
Louping III virus
Lumpy skin disease virus
 Viral hemorrhagic disease of rabbits
Vesicular exanthema virus

Mycoplasma agalactiae
Mycoplasma mycoides
Nairobi Sheep Disease Virus
  (Ganjam Virus)
Newcastle disease virus
  (velogenic strains)
Peste des petits ruminants
Pseudomonas ruminantium
  (heartwater)
Rift Valley fever virus
Riderpest virus
Sheep and goat pox
Swine vesicular disease virus
Teschen disease virus
Theileria annulata
Theileria lawerencia
Theileria bovis
Theileria hirci
Trypanosoma evansi
Trypanosoma vivax
Wesselsbron disease virus

The importation, possession, use or interstate shipment of animal pathogens other than those listed above may also be subject to regulations of the U.S. Department of Agriculture. Additional information may be obtained by writing to:

Chief Staff Veterinarian
Organisms and Vectors
Veterinary Services, Import-Export Products Staff
Room 756, Federal Building
6505 Belcrest Road
Hyattsville, Maryland 20782
Telephone: (301) 436-7830
        (301) 436-8499
Fax: (301) 436-8226

5.6.10.2.   Importation and Interstate Shipment of Human Pathogens and Related Materials

The importation or subsequent receipt of etiologic agents and vectors of human disease is subject to the Public Health Service Foreign Quarantine Regulations (42 CFR, Section 71.156). Permits authorizing the importation or receipt of regulated materials and specifying conditions under which the agent or vector is shipped, handled, and used are issued by the CDC.

The interstate shipment of indigenous etiologic agents, diagnostic specimens, and biological products is subject to applicable packaging, labeling, and shipping requirements of the Interstate Shipment of Etiologic Agents (42 CFR Part 72).

Additional information on the importation and interstate shipment of etiologic agents of human disease and other related materials may be obtained by writing to:

Centers for Disease Control
Attention: Biosafety Branch
Office of Health and Safety, Mail Stop F-05
1600 Clifton Road, N.E.
Atlanta, Georgia 30333
Telephone: (404)-329-3883
Fax: (404) 639-2294

REFERENCES

1. **Favero, M.S.,** Biological hazards in the laboratory, in *Proceedings of Institute on Critical Issues in Health Laboratory Practice*, Richardson, J.W., Schoenfeld, E., Tullis, J.W., and Wagner, W.W., Eds., The DuPont Company, Wilmington, DE, 1986, 1.
2. **Pike, R.M.,** Laboratory-associated infections: incidence, fatalities, cases and prevention, *Annu. Rev. Microbiol.*, 33, 41, 1979.
3. **Pike, R.M.,** Past and present hazards of working with infectious agents, *Arch. Path. Lab. Med.*, 102, 333, 1978.
4. **Pike, R.M.,** Laboratory-associated infections: summary and analysis of 3,921 cases, *Health Lab. Sci.*, 13, 105, 1976.
5. **Litsky, B.Y.,** Microbiology of sterilization, *AORN J.*, 26, 340, 1977.
6. Laboratory Safety at the Centers for Disease Control, U.S. Department of Health, Education and Welfare Publ. No. CDC 75-8118, September, 1974.
7. National Cancer Institute Safety Standards for Research Involving Oncogenic Viruses, U.S. Department of Health, Education and Welfare Publ. No. (NIH)75-790, October, 1974.
8. National Institutes of Health Biohazards Safety Guide, U.S. Department of Health, Education and Welfare, Public Health Service, National Institutes of Health, U.S. Government Printing Office, Stock No. 1740-00383, 1974.
9. **Hellman, A., Oxman, M.N., and Pollack, R., Eds.,** *Biohazards in Biological Research*, Cold Spring Harbor Laboratory, Cold Spring Harbor, NY, 1974.
10. **Steere, N.V., Ed.,** *CRC Handbook of Laboratory Safety*, 2nd ed., CRC Press, Inc., Boca Raton, FL, 1971.
11. **Bodily, J.L.,** General administration of the laboratory, in *Diagnostic Procedures for Bacterial, Mycotic and Parasitic Infections*, Bodily, H.L., Updyke, EL., and Mason, J.O., Eds., American Public Health Association, New York, 1970, 11.

12.   **Darlow, H.M.,** Safety in the microbiological laboratory, in *Methods in Microbiology,* Norris, J.R., and Robbins, D.W., Eds., Academic Press, Inc., New York, 1969, 169.

13.   **Collins, C.H., Hartley, E.G., and Pilsworth, R.,** *The Prevention of Laboratory Acquired Infection,* Public Health Laboratory Service, Monograph Series No. 6, 1974.

14.   **Chatigny, M.A.,** Protection against infection in the microbiological laboratory: devices and procedures, in *Advances in Applied Microbiology,* 3, Umbreit, W.W., Ed., Academic Press, New York, 1961, 131.

15.   **Richmond, J.Y. and McKinney, R.W., Eds.,** Biosafety in Microbiological and Biomedical Laboratories, 3rd ed., U.S. Department of Health and Human Services, Centers for Disease Control, and National Institutes of Health, HHS Publication No. (CDC) 93-8395, Washington, D.C., 1993.

5.7. RECOMBINANT DNA LABORATORIES

The basic guidelines now in effect for recombinant DNA research appeared in the *Federal Register* on May 7, 1986 (51 CFR 16958). The following material represents selected portions taken directly from the Guidelines. There will be a few additional comments and occasional bridging comments where intermediate material is deleted. Only those portions which apply to research organization personnel are included, although the deletions are not extensive. These are evolving regulations, and any organization working in recombinant DNA research or planning to do so should conform to the latest revisions of the Guidelines. Changes have been published in the *Federal Register* on several occasions since the original Guidelines were published. These will be found in the references. However, the basic Guidelines have not changed to a major degree since their original publication in 1986. The changes have addressed primarily large-scale uses of the technique, deliberate releases, gene therapy in humans, application to work outside the United States, and detailed changes in a fairly small number of sections for clarity. For most scientists working with the technique, or who wish to do so, additional guidance should be sought from the facilities's Institutional Biosafety Committee or a local committee charged with the responsibility for reviewing an organization's recombinant DNA research.

5.7.1. Definition of Recombinant DNA Molecules

The definition of recombinant DNA molecules has been changed slightly from that found in the original Guidelines.

In the context of these Guidelines, recombinant DNA molecules are defined as either (i) molecules which are constructed outside living cells by joining natural or synthetic DNA segments to DNA molecules that can replicate in a living cell, or (ii) DNA molecules that result from the replication of those described in (i) above.

Synthetic DNA segments likely to yield a potentially harmful polynucleotide or polypeptide (e.g., a toxin or a pharmacologically active agent) shall be considered as equivalent to their natural DNA counterpart. If the synthetic DNA segment is not expressed *in vivo* as a biologically active polynucleotide or polypeptide product, it is exempt from the Guidelines.

Genomic DNA of plants and bacteria that has acquired a transposable element, even if the latter was donated from a recombinant vector no longer present, is not subject to these Guidelines unless the transposon itself contains recombinant DNA.

5.7.2. General Applicability

The guidelines are applicable to all recombinant DNA research within the United States or its territories which is conducted at or sponsored by an institution that receives any support for recombinant DNA research from the National Institutes of Health (NIH). This includes research performed by NIH directly.

An individual receiving support for research involving recombinant DNA must be associated with or sponsored by an institution that can and does assume the responsibilities assigned in these guidelines. The Guidelines are also applicable to projects done abroad if (1) they are supported by NIH funds, or (2) they involve deliberate release into the environment or testing in humans of materials containing recombinant DNA developed with NIH funds, and if the institution that developed those materials sponsors or participates in those projects. Participation includes research collaboration or contractual agreements, but not mere provision of research materials.

If the host country, however, has established rules for the conduct of recombinant DNA projects, then the project must be in compliance with those rules. If the host country does not have such rules, the proposed project must be reviewed by an NIH-approved institutional biosafety committee (IBC) and accepted in writing by an appropriate national governmental authority of the host country.

The safety practices to be employed abroad must be reasonably consistent with the NIH Guidelines.

5.7.3. Containment

Effective biological safety programs have been operative in a variety of laboratories for many years. Considerable information, therefore, exists for the design of physical containment facilities and the selection of laboratory procedures applicable to organisms carrying recombinant DNA. Existing programs rely upon mechanisms that, for convenience, can be divided into two categories (i) a set of standard practices that are generally used in microbiological laboratories;" [see Section 5.5 and subsections thereof], and (ii) special procedures, equipment, and laboratory installations that provide physical barriers which are applied in varying degrees according to the estimated biohazard. [again, see Section 5.5 and subsections thereof].

5.7.4. Guidelines in Covered Experiments

This part discusses experiments involving recombinant DNA. The experiments have been divided into four classes:

- III-A. Experiments which require specific RAC[*] review and NIH and IBC[**] approval before initiation of the experiment.
- III-B. Experiments which require IBC approval before initiation of the experiment.

---

[*]   The Recombinant DNA Advisory Committee (RAC) advises the Secretary, the Assistant Secretary of Health, and the Director, NIH, concerning recombinant DNA research.

[**] Institutional Biosafety Committee reviews, approves, and oversees projects involving recombinant DNA at the institutional level.

- III-C. Experiments which require IBC notification at the time of initiation of the experiment.
- III-D. Experiments which are exempt from the procedures of the Guidelines.

IF AN EXPERIMENT FALLS INTO BOTH CLASSES III-A AND ONE OF THE OTHER CLASSES, THE RULES PERTAINING TO CLASS III-A MUST BE FOL-LOWED. If an experiment falls into Class III-D and into either of Class III-B or III-C as well, it can be considered exempt from the requirements of the Guidelines.

Changes in containment levels from those specified in the Guidelines cannot be made without the express approval of the Director, NIH.

5.7.4.1.   III-A. Experiments that Require RAC Review, NIH Review, and NIH and IBC Approval Before Initiation

Experiments in this category cannot be initiated without submission of relevant information on the proposed experiment to NIH, the publication of the proposal in the *Federal Register* for 30 days of comment, review by the RAC, and specific approval by NIH. The containment conditions for such experiments will be recommended by RAC and set by NIH at the time of approval. Such experiments also require the approval of the IBC before initiation. Specific experiments already approved in this section and the appropriate containment conditions are listed in Appendices D and F. [All appendices referred to are those in the Guidelines.] If an experiment is similar to those in Appendices D and F, the NIH Office of Recombinant DNA Activities (ORDA) may determine appropriate containment conditions according to case precedents.

III-A-1. Deliberate formation of recombinant DNAs containing genes for the biosynthesis of toxic molecules lethal for vertebrates at an LD_{50} of less than 100 nanograms per kilogram body weight (e.g., microbial toxins such as the botulinum toxins, tetanus toxin, diphtheria toxin, *Shigella dysenteriae* neurotoxin). Specific approval has been given for the cloning in *Escherichia. coli* K-12 of DNAs containing genes coding for the biosynthesis of toxic molecules which are lethal to vertebrates at 100 ng to 100 µg/kg body weight. Containment levels for these experiments are given in Appendix F.

III-A-2. Deliberate release into the environment of any organism containing recombinant DNA except those listed below. The term "deliberate release" is defined as a planned introduction of recombinant DNA-containing microorganisms, plants, or animals into the environment.

III-A-2-a. Introductions conducted under conditions considered to be accepted scientific practices in which there is adequate evidence of biological and/or physical control of the recombinant DNA-containing organisms. The nature of such evidence is described in Appendices L, M, N, and O.

III-A-2-b. Deletion derivatives and single base changes not otherwise covered by the Guidelines.

III-A-2-c. For extrachromosomal elements and microorganisms (including viruses), rearrangements, and amplifications within a single genome. Rearrangements involving the introduction of DNA from different strains of the same species would not be covered by this exemption.

III-A-3. Deliberate transfer of a drug resistance trait to microorganisms that are not known to acquire it naturally, if such acquisition could compromise the use of the drug to control disease agents in human or veterinary medicine or agriculture.

III-A-4. Deliberate transfer of recombinant DNA or DNA or RNA derived from recombinant DNA into human subjects. The requirement for RAC review should not be considered to preempt any other required review of experiments with human subjects. Institutional Review Board (IRB) review of the proposal should be completed before submission to NIH. A brochure, intended as educational material and entirely separate from the Guidelines, was included in the material issued on March 1, 1990.

5.7.4.2. III-B. Experiments that Require IBC Approval Before Initiation

Investigators performing experiments in this category must submit to their IBC, prior to initiation of the experiments, a registration document that contains a description of: (i) the source(s) of DNA; (ii) the nature of the inserted DNA sequences; (iii) the hosts and vectors to be used; (iv) whether a deliberate attempt will be made to obtain an expression of the foreign gene, and if so, what protein will be produced; and (v) the containment conditions specified in these Guidelines. This registration document must be dated and signed by the investigator and filed only with the local IBC. The IBC shall review all such proposals prior to initiation of the experiments. Requests for lowering of containment for experiments in this category will be considered by NIH.

5.7.4.2.1.   III-B-1. Experiments Using Human or Animal Pathogens (Class 2, Class 3, Class 4, or Class 5 Agents as Host-Vector Systems)

III-B-1-a. Experiments involving the introduction of recombinant DNA into Class 2 agents can be carried out in a BL2 containment.
III-B-1-b. Experiments involving the introduction of recombinant DNA into Class 3 agents can be carried out in a BL3 containment.
III-B-1-c. Experiments involving the introduction of recombinant DNA into Class 4 agents can be carried out in a BL4 containment.
III-B-1-d. Experiments involving the introduction of recombinant DNA into Class 5 agents will be set on a case-by-case basis following ORDA review. A U.S. Department of Agriculture (USDA) permit is required for work with Class 5 agents.

5.7.4.2.2.   III-B-2. Experiments in Which DNA from Human or Animal Pathogens (Class 2, Class 3, Class 4, or Class 5 Agents) is Cloned in Nonpathogenic Prokaryotic or Lower Eukaryotic Host-Vector Systems

III-B-2-a. Recombinant DNA experiments in which DNA from Class 2 or Class 3 agents is transferred into nonpathogenic prokaryotes or lower eukaryotes may be performed under BL2 containment. Recombinant DNA experiments in which DNA from Class 4 agents is transferred into nonpathogenic prokaryotes or lower eukaryotes can be performed at BL2 containment after demonstration that only a totally and irreversibly defective fraction of the agent's genome is present in a given recombinant. In the absence of such a demonstration, BL4 must be used. Specific lowering of containment BL1 for particular experiments can be approved by the IBC. Many experiments in this category will be exempt from the Guidelines.

Experiments involving the formation of recombinant DNAs for certain genes coding for molecules toxic for vertebrates require RAC review and NIH approval or must be carried out under conditions as described in Appendix F.

III-B-2-b. Containment conditions for experiments in which DNA from class 5 agents is transferred into nonpathogenic prokaryotes or lower eukaryotes will be determined by ORDA following a case-by-case review. A USDA permit is required for work with Class 5 agents.

5.7.4.2.3. III-B-3. Experiments Involving the Use of Infectious Animal or Plant DNA or RNA Viruses or Defective Animal or Plant DNA or RNA Viruses in the Presence of Helper Virus in Tissue Culture Systems

Caution: Special care should be used in the evaluation of containment levels for experiments which are likely to either enhance the pathogenicity (e.g., the insertion of a host oncogene) or to extend the host range (e.g., introduction of novel control elements) of viral vectors under conditions which permit a productive infection. In such cases, serious consideration should be given to raising the physical containment by at least one level.

Note: Recombinant DNA molecules or RNA molecules derived therefrom, which contain less than two thirds of the genome of any eukaryotic virus (all virus from a single family being considered identical), may be considered defective and can be used in the absence of a helper virus under the conditions specified in Section III-C.

III-B-3-a. Experiments involving the use of infectious Class 2 animal viruses or defective Class 2 animal viruses in the presence of a helper virus can be performed at BL2 containment.

III-B-3-b. Experiments involving the use of infectious Class 3 animal viruses or defective Class 3 animal viruses in the presence of a helper virus can be carried out at BL3 containment.

III-B-3-c. Experiments involving the use of infectious Class 4 animal viruses or defective Class 4 animal viruses in the presence of a helper virus may be carried out under BL4 containment.

III-B-3-d. Experiments involving the use of infectious Class 5 animal viruses or defective Class 5 animal viruses in the presence of a helper virus will be determined on a case-by-case basis following ORDA review. A USDA permit is required for work with Class 5 pathogens.

III-B-3-e. Experiments involving the use of infectious animal or plant viruses in the presence of helper viruses not covered by [the four preceding sections] may be carried out under BL1 containment.

5.7.4.2.4. III-B-4. Recombinant DNA Experiments Involving Whole Animals or Plants

III-B-4-a. Recombinant DNA, or RNA molecules derived therefrom, from any source except for greater than two thirds of a eukaryotic viral genome may be transferred to any nonhuman vertebrate organism and propagated under conditions of physical containment comparable to BL1 and appropriate to the organism under study. It is important that the investigator demonstrate that the fraction of the viral genome being utilized does not lead to productive infection. A USDA permit is required for work with Class 5 agents.

III-B-4-b. For all experiments not covered by II-B-4-a, the appropriate containment will be determined by the IBC.

5.7.4.2.5. Experiments Involving More Than 10 Liters of Culture

The appropriate containment will be decided by the IBC. Where appropriate, Appendix K, *Physical Containment for Large-Scale Uses of Organisms Containing Recombinant DNA Molecules,* of the Guidelines should be used.

5.7.4.3. III-C. Experiments that Require IBC Notice Simultaneously with Initiation of Experiments

Experiments not included in III-A to C and subsections of those sections are included in this section. All such experiments can be carried out at BL1 containment...a registration document [as previously described] must be dated and signed by the investigator and filed with the local IBC at the time of initiation of the experiment. The IBC shall review all such proposals, but IBC review prior to initiation of the experiment is not required.

Caution (experiments involving formation of recombinant DNA molecules containing no more than two thirds of the genome of any eukaryotic virus): Recombinant DNA molecules containing no more than two thirds of the genome of any eukaryotic virus (all viruses from a single family being considered identical) may be propagated and maintained in cells in tissue culture using BL1 containment. For such experiments, it must be shown that the cells lack helper virus for the specific families of defective viruses being used. If helper virus is present, procedures specified under Section III-B-3 should be used. The DNA may contain fragments of the genome of viruses from more than one family but each fragment must be less than two thirds of a genome.

5.7.4.4. III-D. Exempt Experiments

The following recombinant DNA molecules are exempt from these guidelines and no registration with the IBC is necessary.

III-D-1. Those that are not in organisms or viruses.

III-D-2. Those that consist entirely of DNA segments from a single nonchromosomal or viral DNA source, though one or more of the segments may be a synthetic equivalent.

III-D-3. Those that consist entirely of DNA from a prokaryotic host, including its indigenous plasmids or viruses when propagated only in that host (or a closely related strain of the same species) or when transferred to another host by well-established physiological means. (Also, those that consist entirely of DNA from an eukaryotic host including its chloroplasts, mitochondria, or plasmids (but excluding viruses) when propagated only in that host (or a closely related strain of the same species).

III-D-4. Certain specified recombinant DNA molecules that consist entirely of DNA segments from different species that exchange DNA by known physiological processes though one or more of the segments may be a synthetic equivalent. A list of such exchangers will be prepared and periodically revised by the Director, NIH, with advice of the RAC, after appropriate notice and opportunity for public comment. Certain classes are exempt as of publication of these revised Guidelines [May 7, 1986]. This list is in Appendix A. An updated list may be obtained from the Office of Recombinant DNA Activities, National Institutes of Health, Building 31, Room 3B10, Bethesda, Maryland 20892.

III-D-5. Other classes of recombinant DNA molecules. If the Director, NIH, with advice of the RAC, after appropriate notice and opportunity for public comment, finds that they do not present a significant risk to health or the environment. Certain classes are exempt as of publication of these revised Guidelines [May 7, 1986]. This list is in Appendix C. An updated list may be obtained from the Office of Recombinant DNA Activities, National Institutes of Health, Building 31, Room 4B11, Bethesda, Maryland 20892-0031, Telephone No. 301-436-7601, Fax No. 301-436-8724.

5.7.5. Roles and Responsibilities

5.7.5.1. IV-A. Policy

Safety in activities involving recombinant DNA depends on the individual conducting them. The Guidelines cannot anticipate every possible situation. Motivation and good judgment are the key essentials to protection of health and the environment.

The Guidelines are intended to help the Institution, Institutional Biosafety Committee (IBC), Biological Safety Officer (BSO), and Principal Investigator (PI) determine the safeguards that should be implemented. These Guidelines will never be complete or final, since all conceivable experiments involving recombinant DNA cannot be foreseen. Therefore, *it is the responsibility of the Institution and those associated with it to adhere to the intent of the Guidelines as well as to their specifics.*

Each Institution (and the IBC acting on its behalf) is responsible for ensuring that recombinant DNA activities comply with the Guidelines. General recognition of institutional authority and responsibility properly establishes accountability for safe conduct of the research at the local level.

5.7.5.2. IV-B. Responsibility of the Institution

5.7.5.2.1. IV-B-1. General Information

Each Institution conducting or sponsoring recombinant DNA research covered by these Guidelines is responsible for ensuring that the research is carried out in full conformity with the provisions of the Guidelines. In order to fulfill this responsibility, the Institution shall:

IV-B-1-a. Establish and implement policies that provide for the safe conduct of recombinant DNA research and that ensure compliance with the Guidelines. The Institution, as part of its general responsibilities for implementing the Guidelines, may establish additional procedures as deemed necessary to govern the Institution and its components in the discharge of its responsibilities under the Guidelines.

IV-B-1-b. Establish an IBC that meets the requirements set forth in Section IV-B-2 and carries out the functions detailed in Section IV-B-3.

IV-B-1-c. If the Institution is engaged in recombinant DNA research at the BL3 or BL4 containment level, appoint a BSO, who shall be a member of the IBC and carry out the duties specified in Section IV-B-4.

IV-B-1-d. Require that investigators responsible for research covered by these Guidelines comply with the provisions of Section IV-B-5 and assist investigators to do so.

IV-B-1-e. Ensure appropriate training for the IBC chairperson and members, the BSO, PIs, and laboratory staff regarding the Guidelines, their implementation, and laboratory safety. Responsibility for training IBC members may be carried out through the IBC chairperson. Responsibility for training laboratory staff may be carried out through the PI. The Institution is responsible for seeing that the PI has sufficient training, but may delegate this responsibility to the IBC.

IV-B-1-f. Determine the necessity in connection with each project for health surveillance of recombinant DNA research personnel, and conduct, if found appropriate, a health surveillance program for the project. [If it appears that a medical surveillance program might be appropriate, a copy of the Laboratory Safety Monograph should be obtained from ORDA for guidance.]

IV-B-1-g. Report within 30 days to ORDA any significant problems with and violations of the Guidelines and significant research-related accidents and illnesses, unless the institution determines that the PI or IBC has done so.

5.7.5.2.2. IV-B-2. Membership and Procedures of the IBC

The Institution shall establish an IBC whose responsibilities need not be restricted to recombinant DNA. The committee shall meet the following requirements:

IV-B-2-a. The IBC shall comprise no fewer than five members so selected that they collectively have experience and expertise in recombinant DNA technology and the capability to assess the safety of recombinant DNA research experiments and any potential risk to public health or the environment. At least two members shall not be affiliated with the Institution (apart from their membership on the IBC) and shall represent the interest of the surrounding community with respect to health and protection of the environment. Members meet this requirement if, for example, they are officials of state or local public health or environmental protection agencies, members of other local governmental bodies, or persons active in medical, occupational health, or environmental concerns in the community. The BSO, mandatory when research is being conducted at the BL3 and BL4 levels, shall be a member.

IV-B-2-b. In order to ensure the competence necessary to review recombinant DNA activities, it is recommended that: (i) the IBC include persons with expertise in recombinant DNA technology, biological safety, and physical containment; (ii) the IBC include or have available as consultants, persons knowledgeable in Institutional commitments and policies, applicable law, standards of professional conduct and practice, and community attitudes, and at least one member be from the laboratory technical staff.

IV-B-2-c. The Institution shall identify the committee by name in a report to ORDA and shall include relevant background information on each member in such form and at such times as ORDA may require.

IV-B-2-d. No member of an IBC may be involved (except to provide information requested by the IBC) in the review or approval of a project in which he or she has been or expects to be engaged or has a direct financial interest.

IV-B-2-e. The Institution, who is ultimately for the effectiveness of the IBC, may establish procedures that the IBC will follow in its initial and continuing review of applications, proposals, and activities.

IV-B-2-f. Institutions are encouraged to open IBC meetings to the public whenever possible, consistent with protection of privacy and proprietary interests.

IV-B-2-g. Upon request, the Institution shall make available to the public all members of IBC meetings and any documents submitted to or received from funding agencies which the latter are required to make available to the public. If comments are made by members of the public on IBC actions, the Institution shall forward to NIH both the comments and the IBCs response.

5.7.5.2.3. IV-B-3. Functions of the IBC

On behalf of the Institution, the IBC is responsible for:

IV-B-3-a. Reviewing for compliance with the NIH Guidelines recombinant DNA research as specified in Part III conducted at or sponsored by the Institution, and approving those research projects that it finds in conformity with the Guidelines. This review shall include:

IV-B-3-a-(1). An independent assessment of the containment levels required by these Guidelines for the proposed research, and

IV-B-3-a-(2). An assessment of the facilities, procedures, and practices, and of the training and expertise of recombinant DNA personnel.

IV-B-3-b. Notifying the PI of the results of their review.

IV-B-3-c. Lowering containment levels for certain experiments as specified in Sections III-B-2.

IV-B-3-d. Setting containment levels as specified in Section III-B-4-b and III-B-5.

IV-B-3-e. Reviewing periodically recombinant DNA research being conducted at the Institution to ensure that the requirements of the Guidelines are being fulfilled.

IV-B-3-f. Adopting emergency plans covering accidental spills and personnel contamination resulting from such research. [Information on this topic is also available in the Laboratory Safety Monograph.]

IV-B-3-g. Reporting within 30 days to the appropriate Institutional official and to ORDA any significant problems with or violations of the Guidelines and any significant research-related accidents or illnesses unless the IBC determines that the PI has done so.

IV-B-3-h. The IBC may not authorize initiation of any experiments not explicitly covered by the Guidelines until NIH (with the advice of the RAC when required) establishes the containment requirement.

IV-B-3-i. Performing such other functions as may be delegated to the IBC under Section IV-B-1.

5.7.5.2.4. IV-B-4. Biological Safety Officer

The Institution shall appoint a BSO if it engages in recombinant DNA research at the BL3 or BL4 containment level. The officer shall be a member of the IBC and his duties shall include (but need not be limited to):

IV-B-4-a. Ensuring through periodic inspections that laboratory standards are rigorously followed;

IV-B-4-b. Reporting to the IBC and the Institution all significant problems with and violations of the Guidelines and all significant research-related accidents and illnesses of which the BSO becomes aware unless the BSO determines that the PI has done so;

IV-B-4-c. Developing emergency plans for dealing with accidental spills and personnel contamination and investigating recombinant DNA research laboratory accidents;

IV-B-4-d. Providing advice on laboratory security;

IV-B-4-e. Providing technical advice to the PI and the IBC on research safety procedures.

Note: See the LSM for additional information on the duties of the BSO.

5.7.5.2.5. IV-B-5. Principal Investigator (PI)

On behalf of the Institution, the PI is responsible for complying fully with the Guidelines in conducting any recombinant DNA research.

IV-B-5.PI—General. As part of this general responsibility, the PI shall:

IV-B-5-a-(1). Initiate or modify no recombinant DNA research requiring approval by the IBC prior to initiation (see Sections III-A and III-B) until that research or the proposed modification thereof has been approved by the IBC and has met all other requirements of the Guidelines;

IV-B-5-a-(2). Determine whether experiments are covered by Section III-C and follow the appropriate procedures;

IV-B-5-a-(3). Report within 30 days to the IBC and NIH(ORDA) all significant problems with and violations of the Guidelines and all research-related accidents and illnesses;

IV-B-5-a-(4). Report to the IBC and NIH(ORDA) new information bearing on the Guidelines;

IV-B-5-a-(5). Be adequately trained in good microbiological techniques;

IV-B-5-a-(6). Adhere to IBC-approved emergency plans for dealing with accidental spills and personnel contamination; and

IV-B-5-a-(7). Comply with shipping requirements for recombinant DNA molecules. (See Appendix H for shipping requirements and the LSM for technical recommendations.)

IV-B-5-b. Submissions by the PI to NIH. The PI shall:

IV-B-5-b-(1). Submit information to NIH(ORDA) in order to have new host-vector systems certified;

IV-B-5-b-(2). Petition NIH with notice to the IBC for exemptions to these Guidelines;

IV-B-5-b-(3). Petition NIH with concurrence of the IBC for approval to conduct experiments specified in Section III-A of the Guidelines;

IV-B-5-b-(4). Petition NIH for determination of containment for experiments requiring case-by-case review;

IV-B-5-b-(5). Petition NIH for determination of containment for experiments not covered by the Guidelines.

IV-B-5-c. Submissions by the PI to the IBC. The PI shall:

IV-B-5-c-(1). Make the initial determination of the required levels of physical and biological containment in accordance with the Guidelines;

IV-B-5-c-(2). Select appropriate microbiological practices and laboratory techniques to be used in the research;

IV-B-5-c-(3). Submit the initial research protocol if covered under guidelines Sections III-A to C (and also subsequent changes, e.g., changes in the source of DNA or host-vector system) to the IBC for review and approval or disapproval; and

IV-B-5-c-(4). Remain in communication with the IBC throughout the conduct of the project.

IV-B-5-d. PI Responsibilities Prior to Initiating Research. The PI is responsible for:

IV-B-5-d-(1). Making available to the laboratory staff copies of the protocols that describe the potential biohazards and the precautions to be taken;

IV-B-5-d-(2). Instructing and training staff in the practices and techniques required to ensure safety and in the procedures for dealing with accidents and

IV-B-5-d-(3). Informing the staff of the reasons and provisions for any precautionary medical practices advised or requested, such as vaccinations or serum collection.

IV-B-5-e. PI Responsibilities During the Conduct of the Research. The PI is responsible for:

IV-B-5-e-(1). Supervising the safety performance of the staff to ensure that the required safety practices and techniques are employed;

IV-B-5-e-(2). Investigating and reporting in writing to ORDA, the BSO (where applicable), and the IBC any significant problems pertaining to the operation and implementation of containment practices and procedures;

IV-B-5-e-(3). Correcting work errors and conditions that may result in the release of recombinant DNA materials;

IV-B-5-e-(4). Ensuring the integrity of the physical containment (e.g., biological safety cabinets) and the biological containment (e.g., purity and genotypic and phenotypic characteristics).

The sections of the Guidelines dealing with the responsibilities of the NIH, RAC, and ORDA are omitted here. The areas that directly affect the research facility are generally incorporated in the instructions to the Institutions, the IBC, PI, and BSO.

5.7.5.3. IV-D. Compliance

As a condition for NIH funding of recombinant DNA research, Institutions must ensure that such research conducted at or sponsored by the Institution, irrespective of the source of funding, shall comply with these Guidelines. The policies on noncompliance are as follows.

IV-D-1. All NIH-funded projects involving recombinant DNA techniques must comply with the NIH Guidelines. Noncompliance may result in (i) suspension, limitation, or termination of financial assistance for such projects and of NIH funds for other recombinant DNA research at the Institution, or (ii) a requirement for prior NIH approval of any or all recombinant DNA projects at the Institution.

IV-D-2. All non-NIH funded projects involving recombinant DNA techniques conducted at or sponsored by an Institution that receives NIH funds for projects involving such techniques must comply with the NIH Guidelines.

Noncompliance may result in (i) a requirement for prior approval of any or all recombinant DNA projects at the Institution.

IV-D-3. Information concerning noncompliance with the Guidelines may be brought forward by any person. It should be delivered to both NIH(ORDA) and the relevant Institution. The Institution, generally through the IBC, shall take appropriate action. The Institution shall forward a complete report of the incident to ORDA, recommending any further action.

IV-D-4. In cases where NIH proposes to suspend, limit, or terminate, financial assistance because of noncompliance with the Guidelines, applicable DHHS and Public Health Service procedures shall govern.

IV-D-5. Voluntary Compliance. Any individual, corporation, or Institution that is not otherwise covered by the Guidelines is encouraged to conduct recombinant DNA research activities in accordance with the Guidelines through the procedures set forth in Part VI of the Guidelines. The most significant components of Part VI of the Guidelines are the sections dealing with protection of proprietary data, in addressing the concerns of Institutions or individuals as to whether their interests will be protected. These sections are given immediately below. Section V, which consists of references and some notes to Sections I–IV will be combined with the references to the other sections and moved to the end of the recombinant DNA material.

5.7.5.4. Selected Portions of Section VI. Voluntary Compliance

VI-C. Certification of Host-Vector Systems

A host-vector system may be proposed for certification by the Director, NIH, in accordance with the procedures set forth in Appendix I-II-A.

In order to ensure protection for proprietary data, any public notice regarding a host-vector system which is designated by the Institution as proprietary under Section-VI-E-1 will be issued only after consultation with the Institution as to the content of the notice.

VI-D. Requests for Exemptions and Approvals

Requests for exemptions or other approvals required by the Guidelines should be requested by following the procedures set forth in the appropriate sections in Parts I–IV of the Guidelines.

In order to ensure protection for proprietary data, any public notice regarding a request for an exemption or other approval which is designated by the Institution as proprietary under Section VI-E-1 will be issued only after consultation with the Institution as to the content of the notice.

VI-E. Protection of Proprietary Data

In general, the Freedom of Information Act requires federal agencies to make their records available to the public upon request. However, this requirement does not apply to, among other things, "trade secrets and commercial and financial information obtained from a person and privileged or confidential." 18 U.S.C. 1905, in turn makes it a crime for an officer or employee of the United States or any federal department or agency to publish, divulge, disclose, or make known "in any manner or to any extent not authorized by law any information coming to him in the course of his employment or official duties or by reason of any examination or investigation made by, or return, report or record made to filed with, such department or agency or officer or employee thereof, which information concerns or relates to the trade secrets, (or) processes...of any person, firm, partnership, corporation or association." This provision applies to all employees of the federal government, including special government employees. Members of the Recombinant DNA Advisory Committee are "special Government employees."

VI-E-1. In submitting to NIH for purposes of complying voluntarily with the Guidelines, an Institution may designate those items of information which the Institution believes constitute trade secrets, privileged, confidential commercial, or financial information.

VI-E-2. If NIH receives a request under the Freedom of Information Act for information so designated, NIH will promptly contact the Institution to secure its views as to whether the information (or some portion) should be released.

VI-E-3. If the NIH decides to release this information (or some portion) in response to a Freedom of Information Act request or otherwise, the Institution will be advised; and the actual release will not be made until the expiration of 15 days after the Institution is so advised except to the extent that earlier release in the judgment of the Director, NIH is necessary to protect against an imminent hazard to the public or the environment.

VI-E-4. Presubmission Review

VI-E-4-a. Any Institution not otherwise covered by the Guidelines, which is considering submission of data or information voluntarily to NIH, may request presubmission review of the records involved to determine whether, if the records are submitted, NIH will or will not make part or all of the records available upon request under the Freedom of Information Act.

VI-E-4-b. A request for presubmission review should be submitted to ORDA along with the records involved. These records must be clearly marked as being the property of the Institution on loan to NIH solely for the purpose of making a determination under the Freedom of Information Act. The ORDA will then seek a determination from the HHS Freedom of Information Officer, the responsible official under HHS regulations (45 CFR Part 5), as to whether the records involved (or some portion) are or are not available to members of the public under the Freedom of Information Act. Pending such a determination the records will be kept separate from ORDA files, will be considered records of the Institution and not ORDA, and will not be received as part of ORDA files. No copies will be made of the records.

VI-E-4-c. The ORDA will inform the Institution of the HHS Freedom of Information Officer's determination and follow the Institution's instruction as to whether some or all of the records involved are to be returned to the Institution or to become a part of ORDA files. If the Institution instructs ORDA to return the records, no copies or summaries of the records will be made or retained by the HHS, NIH, or ORDA.

VI-E-4-d. The HHS Freedom of Information Officer's determination will represent that official's judgment at the time of the determination as to whether the records involved (or some portion) would be exempt under the Freedom of Information Act if at the time of the determination the records were in ORDA files when a request was received for them under the Act.

5.7.6. Appendices

5.7.6.1. Appendix A. Exemptions Under Section III-D-4

Under Section II-D-4 of these Guidelines are recombinant DNA molecules that are (1) composed entirely of DNA segments from one or more of the organisms within a sublist and (2) to be propagated in any of the organisms within a sublist. (Classification of Bergey's Manual of Determinative Bacteriology, 8th ed., Buchanan R.E. and Gibbons, N.E. Eds., Williams and Wilkins Company, Baltimore, 1974.)

Although these experiments are exempt, it is recommended that they be performed at the appropriate biosafety level for the host or recombinant organism.

Sublist A

1. Genus *Escherichia*
2. Genus *Shigella*
3. Genus *Salmonella* (including *Arizona*)
4. Genus *Enterobacter*
5. Genus *Citrobacter* (including *Levinea*)
6. Genus *Klebsiella* (including oxytoca)
7. Genus *Erwinia*
8. *Pseudomonas aeruginosa, Pseudomonas putida* and *Pseudomonas fluorescens*

9. *Pseudomonas Mendocina*
10. *Serratia marcescens*
11. *Yersinia enterocolitica*

Sublist B

1. *Bacillus subtilis*
2. *Bacillus licheniformis*
3. *Bacillus pumilus*
4. *Bacillus globigii*
5. *Bacillus niger*
6. *Bacillus nato*
7. *Bacillus amyloliquefaciens*
8. *Bacillus aterrimus*

Sublist C

1. *Streptomyces aureofaciens*
2. *Streptomyces rimosus*
3. *Streptomyces coelicolor*

Sublist D

1. *Streptomyces griseus*
2. *Streptomyces cyaneus*
3. *Streptomyces venezuelae*

Sublist E

1. One-way transfer of *Streptococcus mutans* or *Streptococcus lactis* DNA into *Streptococcus sanguis*

Sublist F

1. *Streptococcus sanguis*
2. *Streptococcus pneumoniae*
3. *Streptococcus faecalis*
4. *Streptococcus pyogenes*
5. *Streptococcus mutans*

5.7.6.2. Appendix B. Classification of Microorganisms on the Basis of Hazard

Appendix B-I. Classification of Etiologic Agents

The original reference for this classification was the publication *Classification of Etiological Agents on the Basis of Hazard,* 4th edition, July 1974, U.S. Department of Health, Education and Welfare, Public Health Service, Centers for Disease Control, Office of Biosafety, Atlanta, GA 30333. For the purposes of these Guidelines, this list has been revised by the NIH.

Appendix B-I-A. Class 1 Agents

All bacterial, parasitic, fungal, viral, rickettsial, and chlamydial agents not included in higher classes.

Appendix B-I-B. Class 2 Agents

Appendix B-I-B-1. Bacterial Agents

Acinetobacter calcoaceticus
Actinobacillus—all species
Aeromonas hydrophila
Arizona hinshawii—all serotypes
Bacillus anthracis
Bordetella—all species
Borrelia recurrentis, B. vincenti
Campylobacter fetus
Campylobacter jejuni
Chlamydia psittaci
Chlamydia trachomatis
Clostridium botulinum, Cl. chauvoei, Cl. haemolyticum, Cl. histolyticum, Cl. novyi, Cl. septicum, Cl. tetani
Corynebacterium diphtheriae, C. equi, C. haemolyticum, C. pseudotuberculosis, C. pyogenes, C. renale
Edwardsiella tarda
Erysipelothrix insidiosa
Escherichia coli—all enteropathogenic, enterotoxigenic, enteroinvasive, and strains bearing K1 antigen
Haemophilus ducreyi, H. influenzae
Klebsiella—all species except *oxytoca*
Legionella pneumophila
Leptospira interrogens—all serotypes
Listeria—all species
Moraxella—all species
Mycobacteria all species except those listed in Class 3
Mycoplasma—all species except *Mycoplasma mycoides* and *Mycoplasma agalactiae,* which are in Class 5
Neisseria gonorrhoeae, N. meningitidis
Pasteurella—all species except those listed in Class 3
Salmonella—all species and all serotypes
Shigella—all species and all serotypes
Sphaerophorus necrophorus
Staphylococcus aureus
Streptobacillus moniliformis
Streptococcus pneumoniae
Streptococcus pyogenes
Treponema carateum, T. pallidum, and *T. pertenue*
Vibrio cholerae
Vibrio parahemolyticus
Yersinia enterocolitica

Appendix B-I-B-2. Fungal Agents

Actinomycetes (including *Nocardia* sp., *Actinomyces* sp., and *Arachnia propionica)*
Blastomyces dermatitidis
Cryptococcus neoformans
Paracoccidioides braziliensis

Appendix B-I-B-3. Parasitic Agents

Endamoeba histolytica
Leishmania sp.
Naegleria gruberi
Schistosoma mansoni
Toxoplasma gondii
Toxocara canis
Trichinella spiralis
Trypanosoma cruzi

Appendix B-I-B-4. Viral, Rickettsial, and Chlamydial Agents

Adenoviruses—human—all types
Cache Valley virus
Corona viruses
Coxsackie A and B viruses
Cytomegaloviruses
Echoviruses—all types
Encephalomyocarditis virus (EMC)
Flanders virus
Hart Park virus
Hepatitus—associated antigen material
Herpes virus—except Herpesvirus simiae (monkey B virus) which is in Class 4
Influenza viruses—all types except A/PR8/34 which is in Class I
Langat virus
Lymphogranuloma venereum agent
Measles virus
Mumps virus
Parainfluenza virus—all types except parainfluenza virus 3, SF4 strain, which is in Class 1
Polioviruses—all types, wild and attenuated
Poxviruses—all types except Alastrim, smallpox, and whitepox, which are Class 5, and monkeypox which, depending on experiments, is in Class 3 or Class 4.
Rabies virus—all strains except rabies street virus, which should be classified in Class 3
Reoviruses—all types
Respiratory syncytical virus
Rhinoviruses—all types
Rubella virus
Simian viruses—all types except herpesvirus simiae (monkey B virus) and Marburg virus, which are in Class 4
Sindbis virus
Tensaw virus
Turlock virus
Vaccinia virus

Varicella virus
Vesicular stomata virus
Vole rickettsia
Yellow fever virus, 17D vaccine strain

Appendix B-I-C. Class 3 Agents

Appendix B-I-C-1. Bacterial Agents
Bartonella—all species
Brucella—all species
Francisella tularensis
Mycobacterium avium, M. bovis, M. tuberculosis
Pasteurella multocide type B ("buffalo" and other foreign virulent strains)

Appendix B-I-C-2. Fungal Agents
Coccidioides immitis
Histoplasma capsulatum
Histoplasma capsulatum var. *duboisii*

Appendix B-I-C-3. Parasitic Agents
None

Appendix B-I-C-4. Viral, Rickettsial, and Chlamydial Agents
Monkeypox virus, when used *in vitro*
Arboviruses—all strains except those used in Class 2 and 4 (arboviruses indigenous to the
    United States are in Class 3 except those listed in Class 2. West Nile and Semliki Forest
    viruses may be classified up or down, depending on the condition of use and geograph-
    ical location of the laboratory.)
Dengue virus, when used for transmission or animal inoculation experiments
Lymphocytic choriomeningitis (LCM)
Rickettsia—all species except *Vole rickettsia,* when used for transmission or animal in-
    oculation experiments
Yellow fever virus—wild when used *in vitro*

Appendix B-I-D. Class 4 Agents

Appendix B-I-D-1. Bacterial Agents
None

Appendix B-I-D-2. Fungal Agents
None

Appendix B-I-D-3. Parasitic Agents
None

Appendix B-I-D-4. Viral, Rickettsial, and Chlamydial Agents

Ebola fever virus

Monkeypox, when used for transmission or animal inoculation experiments

Hemorrhagic fever agents, including Crimean hemorrhagic fever, (Congo), Junin, and machupo viruses, and others as yet unidentified.

Herpesvirus simiae (monkey B virus)

Lassa virus

Marburg virus

Tick-borne encephalitis virus complex, including Russian spring-summer encephalitis, Kyasanur Forest disease, Omsk hemorrhagic fever, and Central European encephalitis viruses

Venezuelan equine encephalitis virus, epidemic strains, when used for transmission or animal inoculation experiments

Yellow fever virus—wild, when used for transmission or animal inoculation experiments

Appendix B-II. Classification of Oncogenic Viruses on the Basis of Potential Hazard

Appendix B-II-A. Low-Risk Oncogenic Viruses

Rous sarcoma	Rat leukemia
SV-40	Hamster leukemia
CELO	Bovine leukemia
Ad7-SV40	Dog sarcoma
Polyoma	Mason-Pfizer monkey virus
Bovine papilloma	Marek's
Rat mammary tumor	Guinea pig herpes
Avian leukosis	Lucke (frog)
Murine leukemia	Adenovirus
Murine sarcoma	Shope fibroma
Mouse mammary tumor	Shope papilloma

Appendix B-II-B. Moderate-Risk Oncogenic Viruses

Ad2-SV40	EBV	HV ateles
FeLV	SSV-1	Yaba
HV Saimiri	GaLV	FeSV

Appendix B-III. Class 5 Agents

Appendix B-III-A. Animal Disease Organisms which are Forbidden Entry into the U.S. by Law

Foot and mouth disease virus

Appendix B-III-B. Animal Disease Organisms which are Forbidden Entry into the U.S. by USDA Policy

African horse sickness virus	Borna disease virus
African swine fever virus	Bovine infectious petechial fever
Besnoitia besnoiti	Camelpox virus

Ephemeral fever virus
Fowl plague virus
Goatpox virus
Hog cholera virus
Louping ill virus
Nairobi sheep disease virus
Newcastle disease virus (Asiatic strains)
Mycoplasma mycoides (contagious
  bovine pleuropneumonia)
Mycoplasma agalactiae (contagious
  agalactia of sheep)
Rickettsia ruminatium (heartwater)
Rift valley fever virus

Rhinderpest virus
Sheep pox virus
Swine vesicular disease virus
Trypanosoma vivax (Nagana)
Trypanosoma evansi
Theileria parva (East Coast fever)
Theileria annulata
Theileria lawrencei
Theileria bovis
Theileria hirci
Vesicular exanthema virus
Wesselsbron disease virus
Zyonema

Appendix B-III-C. Organisms which May Not Be Studied in the U.S. Except at Specified Facilities

Smallpox   Alastrim   Whitepox

5.7.6.3. Appendix C. Exemptions Under Section III-D-5

The following classes of experiments are exempt under Section III-D-5 of the Guidelines.

Appendix C-I. Recombinant DNAs in Tissue Culture

Recombinant DNA molecules containing less than one half of any eukaryotic genome (all viruses from a single family being considered identical) that are propagated and maintained in cells in tissue culture are exempt from these Guidelines with the exceptions listed below.

Exceptions: Experiments described in Section III-A which require specific RAC review and NIH approval before initiation of the experiment.

Experiments involving DNA from Class 3, 4, or 5 organisms or cells known to be infected with these agents.

Experiments involving the deliberate introduction of genes coding for the biosynthesis of molecules toxic for vertebrates (see Appendix F).

Appendix C-II. Experiments Involving *E. coli* K-12 Host-Vector Systems

Experiments which use *E. coli* K-12 host-vector systems, with the exception of those experiments listed below, are exempt from these Guidelines provided that (i) the *E. coli* host shall not contain conjugation proficient plasmids or generalized transducing phages; and (ii) lambda or lambdoid or Ff bacteriophages or nonconjugative plasmids shall be used as vectors. However, experiments involving the insertion into *E. coli* K-12 of DNA from prokaryotes that exchange genetic information with *E. coli* may be performed with any *E. coli* K-12 vector (e.g., conjugative plasmid). When a nonconjugative vector is used, the *E. coli* K-12 host may contain conjugative-proficient plasmids either autonomous or integrated, or generalized transducing phages.

For these exempt laboratory experiments, BL1 physical containment conditions are recommended.

For large-scale (LS) fermentation experiments BL1-LS physical containment conditions are recommended. However, following review by the IBC of appropriate data for a particular host-vector system, some latitude in the application of BL1-LS requirements as outlined in Appendix K-II-A through K-II-F is permitted.

Exceptions: Experiments described in Section III-A which require specific RAC review and NIH approval before initiation of the experiment.

Experiments involving DNA from Class 3, 4, or 5 organisms or from cells known to be infected with these agents may be conducted under containment conditions specified in Section III-B-2 with prior IBC review and approval.

Large-scale experiments (e.g., more than 10 liters of culture) require prior IBC review and approval (see Section III-B-5).

Experiments involving the deliberate cloning of genes coding for the biosynthesis of molecules toxic for vertebrates (see Appendix F).

Appendix C-III. Experiments Involving *Saccharomyces* Host-Vector Systems

Experiments which use *Saccharomyces cerevisiae* host-vector systems, with the exception of experiments listed below, are exempt from these Guidelines.

Experiments which use *Saccharomyces uvarum* host-vector systems, with the exception of experiments listed below, are exempt from these Guidelines.

For these exempt laboratory experiments, BL1 physical containment conditions are recommended.

For large-scale (LS) fermentation experiments BL1-LS physical containment conditions are recommended. However, following review by the IBC of appropriate data for a particular host-vector system, some latitude in the application of BL1-LS requirements as outlined in Appendix K-II-A through K-II-F is permitted.

Exceptions: Same as for Appendix C-II.

Appendix C-IV. Experiments Involving *Bacillus subtilis* or *Bacillus lichenformis* Host-Vector Systems

Any asporogenic *Bacillus subtilis* strain which does not revert to a spore former with a frequency greater than 10^{-7} can be used for cloning DNA with the exception of those experiments listed below.

For these exempt laboratory experiments, BL1 physical containment conditions are recommended.

For large-scale (LS) fermentation experiments BL1-LS physical containment conditions are recommended. However, following review by the IBC of appropriate data for a particular host-vector system, some latitude in the application of BL1-LS requirements as outlined in Appendix K-II-A through K-II-F is permitted.

Exceptions: Same as for Appendix C-II.

Appendix C-V. Extrachromosomal Elements of Gram-Positive Organisms

Recombinant DNA molecules derived entirely from extrachromosomal elements of the organisms listed below (including shuttle vectors constructed from vectors described in Appendix C), propagated and maintained in organisms listed below are exempt from these Guidelines:

Bacillus subtilis	*Bacillus natto*	
Bacillus pumilus	*Bacillus niger*	
Bacillus licheniformis	*Bacillus aterrimus*	
Bacillus thuringiensis	*Bacillus amylosacchariticus*	
Bacillus cereus	*Bacillus anthracis*	
Bacillus amyloliquefaciens	*Bacillus globigii*	
Bacillus brevis	*Bacillus megaterium*	
Bacillus stearothermophilis		

Staphylococcus aureus	*Staphylococcus carnosus*
Staphylococcus epidermidis	*Clostridium acetobutylicum*

Pediococcus damnosus	*Listeria grayi*	*Lactobacillus casei*
Pediococcus pentosaceus	*Listeria murrayi*	
Pediococcus acidilactici	*Listeria monocytogenes*	

Streptococcus pyogenes	*Streptococcus avium*	*Streptococcus equisimilis*
Streptococcus agalactiae	*Streptococcus faecalis*	*Streptococcus thermophylus*
Streptococcus sanguis	*Streptococcus anginosus*	*Streptococcus milleri*
Streptococcus salivarious	*Streptococcus sobrinus*	*Streptococcus durans*
Streptococcus cremoris	*Streptococcus lactis*	*Streptococcus mitior*
Streptococcus pneumoniae	*Streptococcus mutans*	*Streptococcus ferus*

Exceptions: Experiments described in Section III-A which require specific RAC review and NIH approval before initiation of the experiment.

Large-scale experiments (e.g., more than 10 liters of culture) require prior IBC review and approval (see Section III-B-5).

Experiments involving the deliberate cloning of genes coding for the biosynthesis of molecules toxic for vertebrates (see Appendix F).

5.7.6.4. Appendix D. Actions Taken Under the Guidelines

These are usually specific to a given Institution for specific procedures, and are deleted for that reason.

5.7.6.5. Appendix E. Certified Host-Vector Systems

While many experiments using *E. coli* K-12, *Saccharomyces cerevisiae,* and *Bacillus subtilis* are currently exempt from the Guidelines under Section III-D-5, some derivatives of these host vector-systems were previously classified as HV1 or HV2. A listing of those systems follows.

Appendix E-I. *Bacillus subtilis*

HV1. The following plasmids are accepted as the vector components of certified *B. subtilis* HV1 systems: pUB110, pC194, pS194, pSA2100, pE194, pT127, pUB112, pC221, pC223, and pAB124. *B. subtilis* strains RUB 331 and BGSC 1S53 have been certified as the host component of HV1 systems based on these plasmids.

HV2. The asporogenic mutant derivative of *Bacillus subtilis*, ASB 298 with the following plasmids as the vector component: pUB110, pC194, pS194, pSA2100, pE194, pT127, pUB112, pC221, pC223, and pAB124.

Appendix E-II. *Saccharomyces cerevisiae*

HV2. The following sterile strains of *Saccharomyces cerevisiae*, all of which have the ste-VC9 mutation, SHY1, SHY2, SHY3, and SHY4. The following plasmids are certified for use: YIp1, YEp2, YEp4, YIp5, YEp6, YRp7, YEp20, YEp21, YEp24, YIp25, YIp26, YIp27, YIp28, YIp29, YIp30, YIp31, YIp32, and YIp33.

Appendix E-III. *Escherichia coli*

EK2 Plasmid Systems. The *E. coli* K-12 strain chi-1776. The following plasmids are certified for use: pSC101, pMB9, pBR313, pBR322, pDH24, pBR325, pBR327, pGL101, and pHB1. The following *E. coli/S. cerevisiae* hybrid plasmids are certified as EK2 vectors when used in *E. coli* chi-1776 or in the sterile yeast strains, SHY1, SHY2, SHY3, and SHY4: YIp1, YEp2, YEp4, YIp5, YEp6, YRp7, YEp20, YEp21, YEp24, YIp25, YIp26, YIp27, YIp28, YIp29, YIp30, YIp31, YIp32, and YIp33.

EK2 Bacteriophage Systems. The following are certified EK2 systems based on bacteriophage lambda:

Vector	Host
λgtWES.λB'	DP50*sup*F
λgtWES.λB*	DP50*sup* F
λgtZJvir.λB'	*E. coli* K-12
λgtALO.λB	DP50*sup*F
Charon 3A	DP*sup*50 or DP50 F
Charon 4A	DP*sup*50 or DP50 F
Charon 3A	DP*sup*50 or DP50 F
Charon 16A	DP*sup*50 or DP50 F
Charon 21A	DP50 F
Charon 23A	DP*sup*50 or DP50 F
Charon 24A	DP*sup*50 or DP50 F

E. coli K-12 strains chi-2447 and chi-2281 are certified for use with lambda vectors that are certified for use with strain DP*sup*50 or DP50 F provided that the *sup* strain is not used as a propagation host.

Appendix E-IV. *Neurospora crassa*

HV1. The following specified strains of *Neurospora crassa* which have been modified to prevent aerial dispersion:

Inl (inositolless) strains 37102, 37401 46316, 64001, and 89601
Csp-1 strain UCLA37 and csp-2 strains FS 590, UCLA101 (these are conidial separation mutants)
Eas strain UCLA191 (an "easily wettable" mutant)

Appendix E-V. *Streptomyces*

HV1. The following *Streptomyces* species: *Streptomyces coelicolor, S. lividans, S. parvulus*, and *S. grieseus*. The following are accepted as vector components of certified *Streptomyces* HV1 systems: *Streptomyces* plasmids SCP2, SLP1.2, pJ101, actinophage phi C31, and their derivatives.

Appendix E-VI. *Pseudomonas putida*

HV1. *Pseudomonas putida* strains KT2440 with plasmid vectors pKT262, pKT263, and pKT264.

5.7.6.6. Appendix F. Containment Conditions for Cloning of Genes Coding for the Biosynthesis of Molecules Toxic for Vertebrates

(see Section III-A) ... No specific restrictions shall apply to the cloning of genes if the protein specified by the gene has an LD_{50} of 100 micrograms or more per kilogram of body weight. Experiments involving genes coding for toxic molecules with an LD_{50} of 100 micrograms or less per kilogram shall be registered with ORDA prior to initiating the experiments. A list of toxic molecules classified as to LD_{50} is available from ORDA. Testing procedures for determining toxicity of toxic molecules not on the list are available from ORDA. The results of such tests shall forwarded to ORDA which will consult with the RAC Working Group on Toxins to inclusion of the molecules on the list (see Section IV-C-1-b-(2)-(e)).

Appendix F-II. Containment conditions for Cloning of Toxic Molecule Genes in *E. coli* K-12

Appendix F-II-A. Cloning of genes coding for molecules toxic for vertebrates that have an LD_{50} in the range of 100 nanograms to 1000 nanograms per kilogram of body weight (e.g., *Clostridium perfringens* epsilon toxin) may proceed under BL2 + EK2 or BL3 + EK1 containment conditions.

Appendix F-II-B. Cloning of genes for the biosynthesis of molecules toxic for vertebrates with an LD_{50} in the range of 1 microgram to 100 micrograms per kilogram body weight may proceed under BL1 + EK1 containment conditions (e.g., *Staphylococcus aureus* alpha toxin, *Staphylococcus aureus* beta toxin, ricin, *Pseudomonas aeruginosa* exotoxin A, *Bordatella pertussis* toxin, the lethal factor of *Bacillus anthracis*, the oxygen-labile hemolysins such as streptolysin O, and certain neurotoxins present in snake venoms and other venoms).

Appendix F-II-C. Some enterotoxins are substantially more toxic when administered enterally than parenterally. The following enterotoxins shall be subject to BL1 + EK1 containment conditions: cholera toxin, the heat labile toxins of *E. coli* and of *Yersinia enterocolitica*.

Appendix-F-III. Containment Conditions for Cloning of Toxic Molecule Genes in Organisms Other than *E. coli* K-12

Requests involving the cloning of genes coding for molecules toxic for vertebrates in host-vector systems other than *E. coli* will be evaluated by ORDA which will consult with the Working Group on Toxins (see Section IV-C-1-b-(3)-(f)).

Appendix F-IV-A. Specific Approvals

This section is deleted because the material is generally specific to a given user application.

5.7.6.7. Physical Containment

This material has been presented for the most part in Section 5.5 , in subsection 5.5.11, and its subsections.

In most cases the material for recombinant DNA facilities is virtually identical, often requiring only the substitution of the words containing recombinant DNA material for containing viable infectious material.

As stated in Appendix G-II, the purpose of physical containment is to confine organisms containing recombinant DNA molecules and thus to reduce the potential for exposure of the laboratory worker, persons outside of the laboratory, and the environment to organisms containing recombinant DNA molecules. ... The selection of alternative methods of primary containment is dependent, however, on the level of biological containment provided by the host-vector system used in the experiment.

5.7.6.8. Appendix H. Shipment

Recombinant DNA molecules contained in an organism or viral genome shall be shipped under the applicable requirements of the U.S. Postal Service (39 CFR, Part III); U.S. Public Health Service (42 CFR, Part 72); the U.S. Department of Agriculture (9 CFR, subchapters D and E; 7 CFR, Part 340; and/or the U.S. Department of Transportation (49 CFR Parts 171-179).

For purposes of the NIH Guidelines: Host organisms or viruses will be shipped as etiologic agents regardless of whether or not they contain recombinant DNA, if they are regulated as human pathogens by the U.S. Public Health Service (42 CFR Part 72), or as animal pathogens under the Animal and Plant Health Inspection Service APHIS), U.S. Department of Agriculture (Titles 9 and 7 CFR, respectively).

Additionally, host organisms and viruses will be shipped as etiologic agents if they contain recombinant DNA when:

1. the recombinant DNA includes the complete genome of a host organism or virus regulated as a human or animal pathogen or a plant pest: or:
2. the recombinant DNA codes for a toxin or other factor directly involved in eliciting human, animal, or plant disease or inhibiting plant growth and is carried on an expression vector or within the host chromosome and/or when the host organism contains a conjugation proficient plasmid or a generalizing transducing phage, or
3. the recombinant DNA comes from a host organism or virus regulated as a human or animal pathogen or as a plant pest and has not been adequately characterized to demonstrate that it does not code for a factor involved in eliciting human, animal or plant disease.

5.7.6.9. Appendix I. Biological Containment

Appendix I-I. Levels of Biological Containment

In consideration of biological containment, the vector (plasmid, organelle, or virus) for the recombinant DNA and the host (bacterial, plant, or animal cell) in which the vector is propagated in the laboratory will be propagated together. Any combination of vector and host which is to provide biological containment must be chosen or constructed so that the following types of "escape" are minimized: (i) survival of the vector in its host outside in the laboratory and (ii) transmission of the vector from the propagation host to other nonlaboratory hosts.

The following levels of biological containment (HV, or host-vector, systems) for prokaryotes will be established; specific criteria will depend on the organisms to be used.

Appendix I-I-A. HVI.
A host-vector system which provides a moderate level of containment.

Specific Systems
Appendix I-I-A-1. EK1. The host is always *E. coli* K-12 or a derivative thereof, and the vectors include nonconjugative plasmids (e.g., pSC101, ColE1), or derivatives thereof and variants of bacteriophages, such as lambda. The K-12 hosts shall not contain conjugation-proficient plasmids, whether autonomous or integrated, or generalized transducing phages.

Appendix I-I-A-2. Other HV1. Hosts and vectors shall be, at minimum, comparable in containment to K-12 with a nonconjugative plasmid or bacteriophage vector. The data to be considered and a mechanism for approval of such HV1 systems are described below (Appendix I-II).

Appendix I-I-B. HV2. These are host-vector systems shown to provide a high level of biological containment as demonstrated by data from suitable tests performed in the laboratory. Escape of the recombinant DNA either via survival of the organisms or via transmission of recombinant DNA to other organisms should be less than 10^{-8} under specified conditions.

Specific Systems
Appendix I-I-B-1. For EK2 host-vector systems in which the vector is a plasmid, no more than 10^8 host cells should be able to perpetuate a cloned DNA fragment under the specified nonpermissive laboratory conditions designed to represent the natural environment, either by survival of the original host or as a consequence of transmission of the cloned DNA fragment.

Appendix I-I-B-2. For EK2 host-vector systems in which the vector is a phage, no more than 10^8 phage particles should be able to perpetuate a cloned DNA fragment under the specified nonpermissive laboratory conditions designed to represent the natural environment either; (i) as a prophage (in the inserted or plasmid form) in the laboratory host used for phage propagation or (ii) by surviving in natural environments and transferring a cloned DNA fragment to other hosts (or their resident prophages).

Appendix I-II. Certification of Host-Vector Systems

Appendix I-II-A. Responsibility

HV1 systems other than *E. coli* and HV2 host-vector systems may not be designated as such until they have been certified by the Director, NIH.

Application for certification of a host-vector system is made by written application to ORDA, NIH, Building 31, Room 4B11, Bethesda, Maryland 20892-0031, telephone: 301-496-9838.

Host-vector systems that are proposed for certification will be reviewed by the RAC (see Section IV-C-I-b-(i)-(e)). This will first involve review of the data on construction, properties, and testing of the proposed host-vector system by a Working Group composed of one or more members of the RAC and other persons chosen because of their expertise in evaluating such data.

When new host-vector systems are certified, notice of the certification will be sent by ORDA to the applicant and to all IBCs and will be published in the *Recombinant DNA Technical Bulletin.* Copies of a list of all currently certified host-vector systems may be obtained from ORDA at any time.

The Director, NIH, may at any time rescind the certification of any host-vector systems (see Section IV-C-I-b-(3)-(d)). If certification of a host-vector system is rescinded, NIH will instruct investigators to transfer cloned DNA into a different system, or use the clones at a higher physical containment level unless NIH determines that the already constructed clones incorporate adequate biological containment.

Certification of a given system does not extend to modifications of either the host or vector component of that system. Such modified systems must be independently certified by the Director, NIH. If modifications are minor, it may only be necessary for the investigator to submit data showing that the modifications have either improved or not impaired the major phenotypic traits on which the containment of the system depends. Substantial modifications of a certified system require the submission of complete testing data.

Appendix I-II-B. Data to be Submitted for Certification

Appendix I-II-B-1. HV1 Systems Other than *E. coli* K-12. The following types of data shall be submitted, modified as appropriate for the particular system under consideration: (i) a description of the organism and vector; the strain's natural habitat and growth requirements; its physiological properties; particularly those related to its reproduction and survival and the mechanisms by which it exchanges genetic information; the range of organisms with which this organism exchanges genetic information and what sort of information is exchanged; and any relevant information on its pathogenicity or toxicity; (ii) A description of the history of the particular strains and vectors to be used, including data on any mutations which render this organism less able to survive or transmit genetic information; (iii) a general description of the range of experiments contemplated, with emphasis on the need for developing such an HV1 system.

Appendix I-II-B-2. HV2 Systems. Investigators planning to request HV2 certifications for host-vector systems can obtain instructions from ORDA concerning data to be submitted. In general, the following types of data are required: (i) description of construction steps with indication of source, properties, and manner of introduction of genetic traits; (ii) quantitative data on the stability of genetic traits that contribute to the containment of the system; (iii) data on the survival of the system of the host-vector system under nonpermissive laboratory conditions designed to represent the relevant natural environment; (iv) data on transmissibility of the vector and/or a cloned DNA fragment under both permissive and

nonpermissive conditions; (v) data on all other properties of the system which affect containment and utility, including information on yields of phage or plasmid molecules, ease of DNA isolation, and ease of transfection or transformation; (vi) in some cases, the investigator may be asked to submit data on survival and vector transmissibility from experiments in which the host-vector is fed to laboratory animals and human subjects. Such *in vivo* data may be required to confirm the validity of predicting *in vivo* survival on the basis of *in vitro* experiments.

Data must be submitted in writing to ORDA. Approximately 10 to 12 weeks are normally required for review and circulation of the data prior to the meeting at which such data can be considered by the RAC. Investigators are encouraged to publish their data on the construction, properties, and testing of proposed HV2 systems prior to consideration of the system by the RAC and its subcommittee. More specific instructions concerning the type of data to be submitted to NIH for proposed EK2 systems involving either plasmids or bacteriophages in *E. coli* K-12 are available from ORDA.

Appendix J. Federal Interagency Advisory Committee on Recombinant DNA Research
[Omitted]

5.7.6.10.   Appendix K. Physical Containment for Large-Scale Uses of Organisms Containing Recombinant DNA Molecules

This part of the NIH Guidelines specifies physical containment guidelines for large-scale (greater than 10 liters of culture) research or production involving viable organisms containing recombinant DNA molecules. It shall apply to large-scale research or production activities as specified in Section III-B-5 of the Guidelines.

All provisions of the NIH Guidelines shall apply to large-scale research or production activities with the following modifications:

- Appendix K shall replace Appendix G when quantities in excess of 10 liters of culture are involved in research or production.
- The Institution shall appoint a BSO if it engages in large-scale research or production activities involving viable organisms containing recombinant DNA molecules.
- The Institution shall establish and maintain a health and surveillance program for personnel engaged in large-scale research or production activities involving viable organisms containing recombinant DNA molecules which require BL3 containment at the laboratory scale. The program shall include preassignment and periodic physical and medical examinations; collection, maintenance and analysis of serum specimens for monitoring serologic changes that may result from the employee's work experience and provisions for the investigation of any serious, unusual or extended illnesses of employees to determine possible occupational origin.

Since most laboratory-scale research will not involve this size operation, the remainder of this appendix is omitted. Investigators with programs which involve this level of use or are considering large-scale applications should refer to the published guidelines.

5.7.6.11. Appendix L. Release Into the Environment of Certain Plants

Appendix L-I. General Information

Appendix L specifies conditions under which certain plants, as specified below, may be approved for release into the environment. Experiments in this category cannot be initiated without submission of relevant information on the proposed experiment to NIH, review by the RAC Plant Working Group, and specific approval by NIH. Such experiments also require the approval of the IBC before initiation. Information on specific experiments which have been approved will be available in ORDA and will be listed in Appendix L-III when the Guidelines are republished. Experiments which do not meet the specifications of Appendix L-II fall under Section III-A and require RAC review and NIH and IBC approval before initiation.

Appendix L-II. Criteria Allowing Review by the RAC Plant Working Group without the Requirement for Full RAC Review

Approval may be granted by ORDA in consultation with the RAC Plant Working Group without the requirement for full RAC review (IBC review is also necessary) for growing plants containing recombinant DNA under the following conditions

Appendix L-II-A. The plant species is a cultivated crop of a genus that has no species known to be a noxious weed.

Appendix L-II-B. The introduced DNA consists of well-characterized genes containing no sequences harmful to humans, animals, or plants.

Appendix L-II-C. The vector consists of DNA: (i) from exempt host-vector systems (Appendix C); (ii) from plants of the same or closely related species; (iii) from nonpathogenic prokaryotes or nonpathogenic lower eukaryotic plants; (iv) from plant pathogens only if sequences causing disease have been deleted; or (v) chimeric vectors constructed from sequences defined in (i) to (iv) above. The DNA may be introduced by any suitable method.

Appendix L-II-D. Plants are grown in controlled access fields under specified conditions appropriate for the plant under study and the geographical location. Such conditions should include provisions for using good cultural and pest control practices, for physical isolation from plants of the same species outside of the experimental plot in accordance with pollination characteristics of the species, and for further preventing plants from becoming established in the environment. Review by the IBC should include an appraisal by scientists knowledgeable of the crop, its production practices, and the local geographical conditions. Procedures for accessing alterations in and the spread of organisms containing recombinant DNA must be developed. The results of the outlined tests must be submitted for review by the IBC. Copies must also be submitted to the Plant Working Group of the RAC.

Appendix L-III. Specific Approvals

As of publication of the revised Guidelines, no specific proposals have been approved. An updated list may be obtained from the Office of Recombinant DNA Activities, National Institutes of Health, Building 31, Room 3B10, Bethesda, Maryland 20892.

REFERENCES

1.   National Institutes of Health guidelines for research involving recombinant DNA molecules, *Fed. Reg.*, 51, May 7, 1986, 16958.

2.   Classification of Etiologic Agents on the Basis of Hazard, 4th Edition, U.S. Department of Health Education and Welfare, Center for Disease Control, Office of Biosafety, Atlanta, GA, July, 1974.

3.   Laboratory Safety at the Center for Disease Control, Publ. No. CDC 75-8118, U.S. Department of Health, Education and Welfare, 1974.

4.   National Cancer Institute Safety Standards for Research Involving Oncogenic Viruses, Publ. No. (NIH) 75-790, U.S. Department of Health, Education and Welfare, 1974.

5.   National Institutes of Health Biohazards Safety Guide, Stock No. 1740-00383, Public Health Service, National Institutes of Health, U.S. Department of Health, Education and Welfare, Washington, D.C., 1974.

6.   **Hellman, A., Oxman, M.N., and Pollack, R., Eds.,** *Biohazards in Biological Research,* Cold Spring Harbor Laboratory, New York, 1974.

7.   **Steere, N.V., Ed.,** *CRC Handbook of Laboratory Safety,* 2nd ed., CRC Press, Boca Raton, FL, 1971.

8.   **Bodily, J.L.,** General administration of the laboratory, in *Diagnostic Procedures for Bacterial, Mycotic and Parasitic Infections*, Bodily, H.L., Updyke, E.L., and Mason, J.O., Eds., American Public Health Association, New York, 1970, 11.

9.   **Darlow, H.M.,** Safety in the microbiological laboratory, in *Methods in Microbiology,* Norris, J.R., and Robbins, D.W., Eds., Academic Press, New York, 1969, 169.

10. **Collins, C.H., Hartley, E.G., and Pilsworth, R.,** The Prevention of Laboratory Acquired Infection,, Monogr. Ser. No. 6, Public Health Laboratory Service, 1974.

11. **Chatigny, M.A.,** Protection against infection in the microbiological laboratory: devices and procedures, in *Advances in Applied Microbiology*, Vol. 3, Umbreit, W.W., Ed., Academic Press, New York, 1961, 131.

12. Design Criteria for Viral Oncology Research Facilities, Publ. No. (NIH) 75-891, U.S. Department of Health, Education and Welfare, Washington, D.C., 1975.

13. **Kuehne, R.W.,** Biological containment facility for studying infectious disease, *Appl. Microbiol.*, 26, 239, 1973.

14. **Runkle, R.S. and Phillips, G.B.,** *Microbial Containment Control Facilities*, Van Nostrand Reinhold, New York, 1969.

15. **Chatigny, M.A. and Clinger, D.I.,** Contamination control in aerobiology, in *An Introduction to Experimental Aerobiology,* Dimmick, R.L. and Akers, A.B., Eds., John Wiley & Sons, New York, 1969, 194.

16. **Matthews, R.E.F., Ed.,** Third report of the international committee on taxonomy of viruses: classification and nomenclature of viruses, *Intervirology*, 12, 129, 1979.

17. **Buchanan, R.E. and Gibbons, N.E., Eds.,** *Bergey's Manual of Determinative Bacteriology,* 8th ed., Williams and Wilkins Co., Baltimore, MD, 1974.

18. **Richmond, J.Y. and McKinney, Eds.,** *Biosafety in Microbiological and Biomedical Laboratories,* 3rd ed., Publ.. No. (CDC) 84-8395, U.S. Department of Health, and Human Services, National Institutes of Health, Centers for Disease Control, Washington, D.C., 1993.

19. Laboratory Safety Monograph—A Supplement to the NIH Guidelines for Recombinant DNA Research, Office of Recombinant DNA Activities, National Institutes of Health, Bethesda, MD.

20. **Hershfield, V., Boyer, H.W., Yanofsky, C., Lovett, M.A., and Helinski, D.R.,** Plasmid Col E1 as a molecular vehicle for cloning and amplification of DNA, *Proc. Natl. Acad. Sci., USA,* 71, 3455, 1974.

21. **Wensink, P.E., Finnegan, D.J., Donelson, J.E., and Hogness, D.S.,** A system for mapping DNA sequences in the chromosomes of *Drosophila melanogaster, Cell*, 3, 315, 1974.

22. **Tanaka, T., and Weisblum, B.,** Construction of a colicin E1-R factor composite plasmid in vitro: means for amplification of deoxyribonucleic acid, *J. Bacteriol.,* 121, 354, 1975.

23. **Armstrong, K.A., Hershfield, V., and Helsinki, D.R.,** Gene cloning and containment properties of plasmid Col E1 and its derivatives, *Science,* 196, 172, 1977.

24. **Bolivar, F., Rodriguez, R.L., Batlach, M.C., and Boyer, H.W.,** Construction and characterization of new cloning vehicles. I. Ampicillin-resistant derivative of pMB9, *Gene,* 2, 75, 1977.

25. **Cohen, S.N., Chang, A.C.W., Boyer, H., and Helling, R.,** Construction of biologically functional plasmids in vitro, *Proc. Natl. Acad. Sci. USA,* 70, 3240, 1973.

26. **Bolivar, F., Rodriguez, R.L., Greene, R.J., Batlach, M.C., Reyneker, H.L., Boyer, W., Crosa, J.H., and Falkow, S.,** Construction and characterization of new cloning vehicles. II. A multi-purpose cloning system, *Gene,* 2, 95, 1977.

27. **Thomas, M., Cameron, I.R., and Davis, R.W.,** Viable molecular hybrids of bacteriophage lambda and eukaryotic DNA, *Proc. Natl. Acad. Sci. USA,* 71, 4579, 1974.

28. **Murray, N.E., and Murray, K.,** Manipulation of restriction targets in phage lambda to form receptor chromosomes for DNA fragments, *Nature,* 251, 476, 1974.

29. **Rambach, A., and Tiolais, P.,** Bacteriophage having ecor1 endonuclease sites only in the non-essential region of the genome, *Proc. Natl. Acad. Sci. USA,* 71, 3927, 1974.

30. **Blattner, F.R., Williams, G.G., Bleche, A.E., Denniston-Thompson, K., Faber, H.E., Furlong, L.A., Gunwald, D.J., Kiefer, D.O., Moore, D.D., Shumm, J.W., Sheldon, E.L., and Smithies, O.,** Charon phages: safer derivatives of bacteriophage lambda for DNA cloning, *Science,* 196, 163, 1977.

31. **Donoghue, D.J., and Sharp, P.A.,** An improved lambda vector: construction of model recombinants coding for kanamycin resistance, *Gene,* 1, 209, 1977.

32. **Leder, P., Tiemeier, D., and Enquist, L.,** EK2 derivatives of bacteriophage lambda useful in the cloning of DNA from higher organisms: the gt WES System, *Science,* 196, 175, 1977.

33. **Skalka, A.,** Current Status of Coliphage EK2 Vectors, *Gene,* 3, 29, 1978.

34. **Szybalski, W., Skalka, A., Gottesman, S., Campbell, A., and Botstein, D.,** Standardized laboratory tests for EK2 certification, *Gene,* 3, 36, 1978.

35. Recombinant DNA Research, Action Under Guidelines; Notices, *Fed. Reg.,* Vol. 52, No. 163, 1987.

36. Recombinant DNA Research, Action Under Guidelines; Notices, *Fed. Reg.,* Vol. 53, No. 134, July 29, 1988.

37. Recombinant DNA Research, Action Under Guidelines; Notices, *Fed. Reg.,* Vol. 53, No. 207, October 26, 1988.

38. Recombinant DNA Research, Action Under Guidelines; Notices, *Fed. Reg.,* Vol. 54, No. 47, March 13, 1989.

39. Recombinant DNA Research, Action Under Guidelines; Notices, *Fed. Reg.,* Vol. 55, No. 177, September 12, 1990.

5.8. RESEARCH ANIMAL CARE AND HANDLING*

5.8.1. Introduction

Animal use in research, teaching, and testing has provided advances in health care and preventive medicine for both animals and humans. Experimental results are greatly dependent upon the humane care and treatment of animals used in research. There is a

---

*   This section was written by David M. Moore, D.V.M.

large body of laws, regulations, and guidelines governing the use of animals in research to assure humane animal care and use. These regulations should not be feared as inhibitory to scientific freedom. Rather, as Aristotle said, "Shall we not like the archer who has a mark to aim at, be more likely to hit upon that which is right?" Compliance with these laws and guidelines assures healthy, high-quality animal models for use in research, assuring consistency from laboratory to laboratory throughout the nation, thus enhancing experimental reliability. The use of high-quality, healthy animals and experimental methodologies which seek to minimize or eliminate pain or discomfort should be incorporated because not only is it the law, but because it makes scientific sense and is the most humane thing to do. The following sections briefly describe the laws and regulations governing animal care and use, and programs for health maintenance of animals and research personnel who come into contact with those animals.

5.8.2. Laws and Regulations Relating to Animal Care and Use

Two major types of regulatory activities impacting the use of animals at a research facility involve voluntary and involuntary regulations. Involuntary regulations are statutory in nature and are uncompromising, and include federal and state laws which dictate minimum standards for the acquisitions of animals, provision of veterinary and husbandry care, use, and disposition of laboratory animals, and personal and institutional compliance is mandatory. Voluntary regulations are those which a research facility imposes on itself above and beyond the minimum standards set forth by the government. Knowledge of and compliance with applicable institutional, state, and federal policies, regulations, and laws will assure humane care of animals, improve scientific reliability, and deflect criticism from the small segment of society which questions whether animals used in research are humanely treated.

5.8.2.1. Animal Welfare Act

The major federal law affecting and regulating use of animals in research is the Federal Animal Welfare Act (P.L. 89-544, and its amendments—P.L. 91-579, P.L. 94-279, and P.L. 99-198). Full text copies of the Act are published in the Code of Federal Regulations (CFR), Title 9—Animals and Animal Products, Subchapter A—Animal Welfare, Parts 1, 2, and 3, and copies of these laws and regulations can be obtained from the Director, Regulatory Enforcement and Animal Care, USDA, APHIS, Room 207, Federal Building, 6505 Belcrest Road, Hyattsville, MD 20782, or the U.S. Department of Agriculture, Animal and Plant Health Inspection Service, Veterinary Services, Import-Export Products Staff, Room 756, Federal Building, 6505 Belcrest Road, Hyattsville, MD 20782.

The Act is administered and enforced by the United States Department of Agriculture Animal, Plant Health Inspection Service (USDA/APHIS). All research institutions using animals defined by the Act (dog, cat, nonhuman primate, guinea pig, hamster, or rabbit) must complete and submit VS Form 18-11 "Applications for Registration of a Research Facility" to the USDA-APHIS-VS Veterinarian-in-Charge for the state in which the facility is located (contact the national office listed above for the local address). The form asks for the location of animal holding facilities, and the species and numbers of animals to be used. Failure to register as a research facility using covered animals may result in fines or sanctions prohibiting future use of animals at that institution. Currently, rodent species, birds, farm animals, and exotic species are not considered "Animals" under the definition contained in the Act. These may be included in subsequent amendments, so it would be advisable to contact the veterinarian-in-charge for your state to determine if you must register. The Act addresses and sets minimum standards for the care and use of animals in research in the areas of facilities, construction, caging, and operations of the facility; animal health and husbandry standards covering feeding, watering, sanitation, employee qualifications, separation of species, record keeping, and provision of adequate veterinary care; and transportation standards of animals to and from the facility. Please refer to a copy of the Act for detailed specifications. The size of a facility and its research program may determine whether a full-time veterinarian is required on the staff, or if the animal health care needs can be met with a part-time or consulting veterinarian with laboratory animal training or experience. The institution must file a "Program of Veterinary Care" with the USDA/APHIS-VS Veterinarian-in-Charge for that state, detailing programs of disease control and prevention, euthanasia methods, and use of appropriate anesthetic, analgesic, or tranquilizing drugs when necessary as determined by the institutional veterinarian.

Each registered research facility must submit VA 18-23 "Annual Report of Research Facility" with USDA/APHIS each year, listing the numbers of animals by species and by category of potential pain or discomfort used during that year, with the signatures of the attending institutional veterinarian and a designated senior administrative official at the institution. Federal penalties may be invoked against the signatories for falsification of information in the body of the report.

The Improved Standards for Laboratory Animals Act (P.L. 99-198), the most recent amendment to the Animal Welfare Act, sets even more important requirements. Each research facility must have an institutional animal care committee of not fewer than three members to be appointed by the chief executive officer of the facility, to include a doctor of veterinary medicine (usually the institutional veterinarian), another facility employee, and a nonemployee, community member who has no family member affiliated with the facility. The committee must inspect all animal study areas and animal facilities twice annually, keeping all inspection reports on file for 3 years, and notifying the administrative representative of the facility of any deficiencies or deviations from the Act. Additionally, each research facility must establish a program for the training of scientists, animal technicians, and other personnel involved with animal care and treatment to include instruction on:

1. Humane practice of animal maintenance and experimentation
2. Research or testing methods that minimize or eliminate animal pain or distress
3. Utilization of the information service at the National Agricultural Library
4. Methods whereby deficiencies in animal care and treatment should be reported

This amendment also calls for establishing institutional standards for exercise of dogs, and provision for environmental enrichment for nonhuman primates.

5.8.2.2. The Good Laboratory Practices Act

The Good Laboratory Practices Act (22 December 1978 issue of the *Federal Register* 43 FR 59986-60025) regulates "nonclinical laboratory studies that support applications for research or marketing permits for products regulated by the Food and Drug Administration, including food and color additives, animal food additives, human and animal drugs, medical devices for humane use, biological products, and electronic products."

Standards addressed by GLP regulations include: (i) compliance with the "NIH Guide for the Care and Use of Laboratory Animals" (to be discussed later); (ii) establishment of standard operating procedures (SOPs) for animal husbandry and experimental treatment;

(iii) meticulous record keeping and documentation of activities; and the establishment of a functional quality assurance unit reporting to the highest administrative levels of the facility. Please consult this document for further details and applicability to your studies.

5.8.2.3. "The Guide for the Care and Use of Laboratory Animals"

The "Guide" (NIH Publications 85-23) was prepared by the Committee on Care and Use of Laboratory Animals of the Institute of Laboratory Animal Resources, National Research Council. The Guide presents recommendations and basic guidelines for: (i) appropriate cage and enclosure sizes for a variety of commonly used laboratory species; (ii) social environmental enrichment; (iii) appropriate environmental temperature and humidity ranges; (iv) ventilation of animal facilities, levels of illumination, noise; (v) separation of species, sanitation of caging and facilities; (vi) provision for quarantine, and provision of adequate veterinary care.

Whereas the standards in the Guide are more stringent than those in the Animal Welfare Act, they are simply recommendations and do not by themselves carry legal penalties. However, other government agencies use the Guide as a measure for animal care, and will terminate funding support for an Institution for noncompliance with the Guide. Single copies of the Guide can be obtained from the Animal Resources Program, Division of Research Resources, National Institutes of Health, Bethesda, Maryland 20205.

5.8.2.4. Public Health Service Policies

Institutions receiving federal grant support for animal research activities from the Public Health Service (including the National Institutes of Health) must file an Animal Welfare Assurance statement with the Office for Protection from Research Risks (OPRR), National Institutes of Health, 9000 Rockville Pike, Building 31, Room 4B09, Bethesda, Maryland 20892. The components of the Assurance are listed in the publications "Public Health Service Policy on Humane Care and Use of Laboratory Animals" which can be obtained from OPRR. The intent of this policy is to "require Institutions to establish and maintain proper measures to ensure the appropriate care and use of all animal involved in research, research training, and biological lasting activities."

The PHS "requires that institutions, in their Assurance Statement, use the Guide for the Care and Use of Laboratory Animals as the basis for developing and implementing an institutional program for activities involving animals." In contrast to the Animal Welfare Act, PHS mandates an institutional animal care and use committee with a minimum of five members (the membership credentials mirror those specified in the Act). Institutions with approved assurance statements must file an annual report with OPRR detailing any changes in the animal program, and listing the dates of the twice-yearly facility inspections by the committee. Failure to comply with the provision of this policy will result in nonfunding or withdrawal of funding for ongoing activities.

As with the Animal Welfare Act, investigators submitting proposals to PHS agencies must submit a research protocol detailing animal use for review of and approval by the institutional animal care and use committee. This committee must approve the project before funding is released, and may require modification of the project if it is not in compliance with Animal Welfare Act standards or PHS policies. The institutional animal care and use committee may halt or terminate an ongoing project for noncompliance with federal laws and policies.

5.8.2.5. Voluntary Regulations

Institutions may develop their own internal policies regarding animal care and use, providing that they are equal to or more stringent than those contained in the Guide or the Animal Welfare Act. Internal policies might include SOPs for animal care and use, mechanisms for selection of and purchase from commercial animal vendors, quarantine policies, and human health monitoring programs.

5.8.3. Personnel

"The Guide for the Care and Use of Laboratory Animals" promotes Institutional personnel policies requiring the use of technicians qualified to provide proper, humane animal care and husbandry, and recommends that these individuals apply for and receive certification from the American Association for Laboratory Animal Science (AALAS, 70 Timbercreek Drive, Suite 5, Cordova, TN 38018). There are three levels of certification based upon educational background and training, and on-the-job experience dealing with laboratory animals: Assistant Technician, Technician, and Laboratory Animal Technologist. Most facilities require facility supervisors or managers to have the AALAS Technologist certification. In-house training of technicians using AALAS course materials will satisfy the training requirement set forth in the Improved Standards for Laboratory Animals Act. Training of the scientific staff in humane animal care would be best accomplished by the lab animal veterinarian or an AALAS technologist. The qualifications for full-time or consulting veterinarians to the research facility should include either specialty board certification by the American College of Laboratory Animal Medicine (ACLAM) or indications of postdoctoral training or experience with laboratory animals. The role of the veterinarian in assuring the provision of "adequate, veterinary care" (as referred to in the laws and policies section of this chapter) is described in a report by ACLAM on "Adequate Veterinary Care" issued in October 1966.

5.8.4. Animal Holding Facilities

Animal holding facilities should be designed and constructed following the recommendations of "The Guide for the Care and Use of Laboratory Animals," which also assures compliance with the Animal Welfare Act. Consult the references at the end of this section for further information on animal facility design and management.

Facilities should be designed and operated for the comfort of the animals and the convenience of the investigator. Another critical factor in facility design operations involves the prevention of transmission of latent diseases from animal to animal, or animals to humans. The first step involves purchase of animals from "clean" sources who have a documented animal health quality assurance program. Newly arrived animals should be held in a quarantine area in the facility to prevent potential contamination of existing research animal populations. The quarantine facility should be located in an area adjacent to the main colony, but with separate access to prevent cross-contamination of the colony as would be the case with common traffic flows.

The second step for maintaining clean animals involves control of the microenvironment through the use of appropriate housing units (i.e., micro-isolator caging), laminar airflow housing racks, or mass air displacement "clean" rooms. The items suggested above may be cost prohibitive for some facilities, and adequate care could involve simply following sanitation and hygiene recommendations in the "Guide." Air pressure in the room can be changed to protect either the animals or humans working in the facility. Making the room air pressure slightly positive with respect to the hallways will minimize the chances of entry of airborne disease agents into an animal room. However, if the research involves animals infected with animal or human pathogens, or they are treated with toxic or carcinogenic agents, then the room air pressure should be made slightly negative as compared to the adjacent hallways, to prevent contamination of animals in other nearby rooms or humans who use that hallway. Hessler and Moreland[5] discuss the use of HEPA filtration in rooms using nonvolatile carcinogens.

5.8.5. Animal Care and Handling

The Improved Standards for Laboratory Animals Act requires that experimental procedures "ensure that animal pain and distress are minimized." A stressed animal is not a good experimental model, since its biochemical and physiological attributes are altered during stress. Minimizing animal stress can be easily accomplished by:

1. Purchasing animals free of latent and overt clinical diseases, and providing adequate veterinary care to maintain their health.
2. Familiarizing animals with experimental devices or rooms prior to the start of the experiment.
3. Limiting restraint to that which is necessary to accomplish the experimental goals, preconditioning the animals to the restraint apparatus, or using other nonrestraint alternatives.
4. Controlling or eliminating environmental stressors (i.e., inappropriate temperature, humidity, light, noise, aggressive cagemates, cage size).
5. Providing environmental enrichment programs and/or exercise for animals, especially dogs, cats, and nonhuman primates.
6. Selecting and using appropriate anesthetics, tranquilizers, sedatives, and analgesic drugs for procedures in which pain or distress are likely.
7. Allowing only skilled, trained individuals to perform surgery
8. Providing training for technicians and investigators in humane animal care and use techniques.

5.8.6. Human Health Monitoring

It is imperative that a human health monitoring program be established for those individuals having limited or full-time contact with research animals. Caretakers and investigators can be exposed to hazardous aerosols, bites, scratches, bodily wastes and discharges, and fomites contaminated with zoonotic agents. A preemployment physical should be conducted to obtain baseline physical and historical data, and a serum sample should be drawn and frozen for future reference. Additional examinations should be scheduled periodically depending on the nature and risks in the work environment.

Training programs should be established to acquaint personnel with biologic (zoonotic), chemical, and physical hazards within the animal facility. Appropriate hygiene should be stressed, and protective clothing and equipment (gloves, protective outer garments, masks, respirators, face shields or eye protectors) should be made available, and their use made mandatory where appropriate in SOPs. Technicians should be aware of clinical signs of disease, notifying the facility veterinarian for confirmation, treatment, isolation of the animal(s), or euthanasia of the affected animal.

REFERENCES

1. **Clark, J.D.,** Regulation of animal use: voluntary and involuntary, *J. Vet. Med. Educ.*, 6(2), 86, 1979.
2. **McPherson, C.W.,** Legislation regulations pertaining to laboratory animals - United States, in *Handbook of Laboratory Animal Science*, Vol. I, Melby, E.C. and Altman, N.H., Eds., CRC Press, Cleveland, 1974, 3.
3. **McPherson, C.W.,** Laws, regulations, and policies affecting the use of laboratory animals, in *Laboratory Animal Medicine,* Fox, J.G. et al., Eds., Academic Press, Orlando, FL, 1984, 19.
4. **Poiley, S.M.,** Housing requirements - general considerations, in *Handbook of Laboratory Animal Science*, Vol. I., Melby, E.C. and Altman, N.H., Eds., CRC Press, Cleveland, 1974, 21.
5. **Hessler, J.R., and Moreland, A.F.,** Design and management of animal facilities in *Laboratory Animal Medicine,* Fox, J.G. et al., Eds., Academic Press, Orlando, FL, 1984, 505.
6. **Simmonds, R.C.,** The design of laboratory animal homes, in *Aeromedical Review*, Vol 2, USAF School of Aerospace Medicine, Brooks Air Force Base, San Antonio, TX, 1973.
7. **Runkle, R.S.,** Laboratory animal housing - part II, *Am. Inst. Architect. J.*, 41, 77, 1964.
8. Comfortable Quarters for Laboratory Animals, Animal Welfare Institute, Washington, D.C., 1979.
9. **Richmond, J.Y. and McKinney, R.W.,** Biosafety in Microbiological and Biomedical Laboratories, 3rd ed., U.S. Department of Health and Human Services, HHS Publ. 93-8395, 1993.

CHAPTER 6

Personal Protective Equipment

6.1. INTRODUCTION

This chapter is intended to be a reasonably extensive but not exhaustive treatment of personal protective equipment. It is designed to provide an overview of the topic in the context of normal laboratory usage. However, the concept of laboratories will be extended to include field studies, where workers are often exposed to hazardous materials such as agricultural chemicals.

6.2. RESPIRATORY PROTECTION

Respiratory protective devices range all the way from the simple soft felt mask, which is often used to provide protection against nuisance levels of dusts and particulates, to the self-contained, positive-pressure, fully enclosing suit, which, if properly matched to the anticipated exposure, offers total body and respiratory protection from the toxic substance involved. As will be noted in a later section on gloves and protective clothing, there is no single material that will protect against all possible chemicals in the workplace.

If a laboratory is designed and operated properly, with adequate ventilation and efficient fume hoods or other types of safety cabinets, additional respiratory protection normally will not be needed for most procedures. However, in the event of an accident or an unusual operation which cannot be performed in a hood, laboratory persons working with toxic materials should be included in a respiratory protection program, managed by the institution or corporation.

The selection of the proper respiratory protective device will depend upon a number of factors, the most important of which relate to the properties of the chemical or material for which protection is needed. What is the permissible exposure limit (PEL) or threshold level value (TLV) of the material? Does it pose a skin absorption problem? Is it immediately dangerous to life and health (IDLH)? What type of material is it, i.e., acid, solvent, dust, radioactive, carcinogen, asbestos, ammonia, etc? Will air or oxygen need to be supplied? What levels of the material are expected to be present? Does the material provide an adequate warning of its presence by an odor or by irritation of the respiratory system or the eyes, at a level well below the legal limits? The units to be used must comply with the specifications of ANSI Z88.2, bear an appropriate NIOSH or Mine Safety Administration (MSA) approval number, and provide the degree of protection needed under the existing working conditions.

The Occupational Safety and Health Administration (OSHA) respirator standard, 29 CFR Part 1910.134(b)(10), requires that the individual who is to wear the respirator must be capable of wearing the respirator under working conditions, i.e., a person must be physically able to perform their work and safely use the equipment. A physician must determine if the respirator user's medical condition permits wearing a respirator. Normally, this will include a pulmonary function or spirometer test. An individual with poor respiratory function should not be asked or permitted to wear many types of respirators which require breathing through a protective filter or cartridge. The medical status of the respirator user should be checked on a regular basis. Several medical conditions would preclude the wearing of a respirator. Among these are emphysema, asthma, reduced pulmonary function (variety of causes other than the preceding two), severe hypertension, coronary artery disease, cerebral blood vessel disease, epilepsy, claustrophobia (brought on by the wearing of the unit), or other relevant conditions as determined by the examining physician. Since the worker is being asked to wear a respirator to protect his health by reducing or eliminating inhalation of noxious vapors, it would appear logical that the employee should participate in a comprehensive medical surveillance program to check on the status of the individual's health.

The respirator must be properly fitted to the user. It is not possible to obtain a proper seal of the respirator to the user's face if there is facial hair where the respirator comes into contact with the face. Facial hair also may interfere with the operation of the exhalation valve. There are skin conditions which make it very difficult to obtain a proper seal. Facial structure also will have a bearing on the quality of fit. For example, a small or petite face is often difficult to fit, as is a narrow face with a prominent nasal bridge. In recent years, significant advances have been made in the design of half-face cartridge respirators, which are the most common type used, to improve the seal to the face. Different size units are now available. There are units which provide a second seal to the skin to aid in keeping toxic fumes from entering. The means of holding respirator units more securely in place have been improved. Hypoallergenic materials can be used in the construction of respirators to prevent the skin from becoming irritated when wearing a unit for an extended period is required.

Each person wearing a respirator must be individually fitted to ensure that the respirator is providing the needed protection to the wearer. The worker should be trained in the proper use of the respirator to enable it to fulfill its function and to maintain it in good working condition. Respirators kept for common use in a facility are not acceptable. An individual should be issued a personal respirator, and should be held responsible for maintaining the unit in a clean, good operating condition.

The fitting program, at a minimum, must include a qualitative fit test administered by a knowledgeable person (usually someone from the safety or medical departments), using a particulate irritant, such as a nontoxic smoke generated by a smoke tube, or an organic solvent generating a distinctive odor, such as isoamyl acetate, to challenge the respirator. It would be preferable to perform a quantitative test, using known concentrations of an appropriate test material while the wearer performs simulated work movements. Even in qualitative tests, simulated work movements are helpful in detecting poor fits.

The training program should include, at a minimum: (a) how to care for the unit, including how to inspect it for proper functioning, as well as normal care; (b) how to put the respirator on and how to check to see that it is performing its function; (c) the function and limitations of the respirator; and (d) the health risks associated with either not using the protection or failure to use it properly. Refresher training should be given on an annual basis.

6.2.1. "Dust" Masks

The use of the term "dust" mask for the nonrigid soft wool felt mask is somewhat of a misnomer since, in modified form, they can be used for other applications such as protection against paint fumes, moderate levels of organics, acid fumes, mercury, etc., although the biggest use is for protection against dust. These units are the simplest form of the air-purifying respirator. These respirators normally should not be employed for hazardous dusts, but are helpful for exposures to inert or nuisance dust levels below 15 mg/M^3. More elaborate versions of these felt masks include such features as exhalation valves, molded bridge pieces, and small sections of metal over the bridge of the nose, which can be bent to help the mask remain in contact with the nose and cheeks. Some have chemical absorbent incorporated in the mask to absorb some fumes and gases. In most cases these inexpensive masks are meant to be worn at relatively low levels of air pollutants (although the better versions which provide good facial contact can theoretically provide a protection factor of 10) and disposed of after a limited period of wear. For particulates, the felt mask tends to become more effective in removing particulates as they are worn in a contaminated atmosphere, since the space for air to pass through the filter becomes more limited as loading of the filter increases. This increases the difficulty in breathing as the filter offers more resistance to the flow of air. The use of an exhalation valve eases the outflow of air, but does not decrease the effort during the intake of air.

Filters used to absorb chemical gases or vapors do not use a mechanical action to trap the material but use an absorbent material (or in some cases, a chemical reactant) to prevent the material from passing through. When the absorbent is saturated or the reactant exhausted, the filter will no longer be effective. The relatively small amount of absorbent material incorporated in a simple felt mask limits their lifetime. Since they are usually discarded after use, they are typically intended to be used for about 8 hours or less.

Unless the contaminant in the air has an effective warning property such as a distinctive odor or acts as an irritant, respirators which only purify the air should not be worn for protection. The sensations experienced by an individual due to lack of oxygen do not constitute a sufficient warning signal for this hazard. At oxygen concentrations of 10 to 16%, it is possible to continue to function for short intervals but with significantly impaired judgment as the oxygen supply to the brain is decreased. At levels below 6%, death occurs in only a few minutes. It is quite likely that more persons have died wearing air-purifying respirators because of the failure to recognize the lack of oxygen in the air than have died from the direct effects of toxic materials. As noted earlier, air-purifying respirators are not approved for use in atmospheres which are IDLH.

6.2.2. Half-Face Cartridge Respirators

The half-face cartridge respirator is the type most frequently used, especially in atmospheres in which there is little or no problem of irritation or absorption of material through the skin. The facepiece of most of these units is molded of a flexible plastic or silicone rubber, which provides a seal to the face when adjusted properly. As noted earlier, facial hair between the mask and the face will prevent the seal from being effective, and it is not permitted for a person with a beard or extended sideburns in the area of the seal to be fitted with a respirator. The facepieces of most of these units are provided with receptacles for two sets of cartridges and/or filters. The respirators are certified as complete units, i.e., the facepiece equipped with specific filters. Cartridges from one vendor cannot be used on another manufacturer's facepiece. The major advantage of this type of unit is that by interchanging cartridges and filters or by putting one or more filters and cartridges in series, a single facepiece can be adapted to provide protection against a large variety of contaminants. However, some cartridge respirators are now being built with nonremovable and noninterchangeable cartridges. These are disposable units since the protective devices cannot be replaced. However, the capacity of the cartridges is considerably larger than the felt mask type, and often can be worn for several shifts if the levels of contaminants challenging the filter are not excessive. The models of these disposable cartridge respirators currently available are shaped somewhat differently than the usual half-face respirator and some individuals prefer them for this reason. The price is usually competitive with the replacement costs of cartridges for dual-cartridge half-face respirators. Since combinations of filters and cartridges cannot be modified, a large variety of models must be maintained in stock to fit a variety of exposure conditions.

The normal protection factor provided by a half-face respirator, which is accepted by OSHA and the Environmental Protection Agency (EPA), is either ten times the PEL or the cartridge limit, whichever is lower. In order to maintain the usefulness of the cartridge respirators, they must be maintained and stored properly. Fit tests should be repeated periodically to ensure that they still provide the required protection. Replacement parts for the exhalation valves should be maintained.

A problem alluded to in an earlier part of this section is the effort required to breathe through the filters. Although check valves can be designed so as to require little effort during the exhalation cycle, breathing air must pass through the cartridges on the intake cycle. Power-assisted breathing units are now available from a number of vendors which provide the flexibility of movement provided by the independent respirator, but remove much of the additional breathing effort incurred by wearing a respirator. In this system air is fed into the facepiece by a small battery-operated pump. The intake air is passed through cartridge filters located on the pump instead of on the facepiece. The pump is usually attached to the belt, and including the weight of the battery, is still light enough not to represent any significant problem. The batteries are usually designed to provide a nominal 8 hours of continuous operation, if they are maintained properly. Most of the batteries are nickel-cadmium which can lose capacity if they are not routinely taken through a complete discharge-charge cycle.

Since a power-assisted air-purifying (PAAP) respirator is a positive-pressure system, i.e., the air within the facepiece is at a higher pressure than the outside air, this type of unit intrinsically provides more protection than the ordinary half-face respirator. If provision is made for an "escape" mode of operation, i.e., the wearer can continue to breathe through the filters should the pump fail and escape from the contaminated atmosphere, the ANSI Z88.2 standard (American National Standard Institute) would permit the use of this type of respirator in an IDLH atmosphere. Some of the early models of this type of unit had some problems with the seals on the pumps, but these problems have been corrected and they represent a desirable alternative if the wearer is to remain in a contaminated atmosphere for extended periods. There have been a few problems with the pumps overheating while being used at elevated temperatures.

Since a PAAP type of unit will permit entry into an IDLH atmosphere, the user should be aware of a number of essential safety practices associated with this use. At least one standby person with the proper equipment for entry and rescue must be present outside the affected area. Communications must be continuously maintained between the worker within the area and the standby person. Persons within the area must be equipped with devices such as harnesses and safety lines to facilitate rescue operations should they be necessary.

In the best circumstances, speaking is difficult while wearing a respirator. Throat microphones may be connected to amplifiers or radios so that the wearer can communicate with others without having to use the hands to activate a microphone.

If the possibility of chemical splashing exists, the eyes can be protected by wearing chemical splash goggles and/or face masks in addition to the half-face respirator. Goggles which provide ports or other means to allow air into the space behind the lens will not protect the eyes from vapors and gases in the air since the air behind the lens is room air. Some goggles and respirators are physically incompatible, and if both are needed, they must be selected taking compatibility into account.

The use of any type of respirator in almost any type of interior laboratory should be the exception rather than the rule, since these spaces should be engineered to be normally safe as far as atmospheric contaminants are concerned, if workers follow safe laboratory practices. However, field workers often must depend upon the proper selection of personal protective equipment to protect them from the hazards of exposure to contaminants. In many cases, the half-face respirator is the minimum acceptable respiratory protection. It is often worn under adverse personal comfort conditions as well. The weather may be very hot and the respirator may be very uncomfortable, so that the user may wish to forego wearing a unit and "take his chances." There is also the tendency for some workers to dismiss the need for the units and to make fun of the persons asking for a respirator or wearing one. Management sometimes shares this attitude. It is extremely important that responsible respiratory protection programs be made available and the use of protective equipment required. This is especially true in academic institutions in which future managers of commercial farm operations and agricultural research operations are being trained. They should be taught by formal instruction and example to follow good safety practices, including some that might appear inconvenient at the time, such as wearing the right respiratory protection in contaminated environments. The contaminated environments may not be immediately obvious. It is clear that when spraying operations are being conducted, respiratory protection is likely to be needed, but it is also likely to be needed when entering a treated field a day or two later, depending upon the rapidity of the biodegradation of the material used. Often, in agricultural research, experimental chemicals are used and the data to determine exposure problems may be incomplete or unavailable. An even more serious problem in encouraging field workers to wear personal protective equipment is the need to provide overall body protection for dispersal of pesticides and herbicides. This will be discussed further in a later section.

6.2.3. Full-Face Respirators

Full-face air-purifying respirators are similar in many respects to half-face respirators, with the obvious difference that the mask covers the upper part of the face, protecting the eyes. This has advantages and disadvantages. It is often easier to obtain a fit to the user than with a half-face unit. As a result both the current ANSI Z88.2 standard and OSHA allow a higher protection factor (note that the present OSHA selection standards are based on the 1969 ANSI Z88.2 version instead of the current one), generally by a factor of 5 or less depending upon the contaminant. OSHA has announced that in the very near future from the time of this writing, it intents to accept current standards for personal protection equipment (PPE).

A major difficulty with wearing a full-face respirator exists for persons who require prescription lenses for seeing. The temple piece extending back over the ears will interfere with the seal at those points. Some units are built to accommodate eyeglasses. In other cases, the wearer may decide to temporarily remove the temple pieces from the glasses and tape them to the bridge of the nose. This solution is acceptable for occasional, sporadic wearing but not for extended periods or use in which the facepiece is taken off and on frequently. One solution which is not acceptable is the wearing of contact lenses. These are not allowed under the OSHA standards.

Another problem with a full-face respirator is fogging due to the warm, moist air exhaled in breathing. Full-face units are designed so that the incoming air flows across the lens of the unit. This feature, plus antifogging coatings on the lens, will normally prevent fogging at normal temperatures. However, at temperatures below freezing, fogging becomes an increasingly serious problem as the temperature decreases. Some full-face respirators include nose cups so that the warm air from the nose is directed through the exhalation valve and does not come into contact with the lens. These units should be able to go down to about -32°C (-25°F) and still allow adequate vision through the lens. At very low temperatures, the warm, moisture-laden air passing out through the exhalation valve may be a problem. The valve may stick open or closed because of ice, or the moisture may freeze and block the free flow of air through the valve.

The cost of full-face respirators is substantially more than the half-face units, typically by a factor of 4 or 5.

A variation of the full-face respirator is the PAAP unit discussed in the previous section. In one version, the only difference is that it supplies a full-face mask instead of a half-face mask. However, a useful variation is for the PAAP unit to supply air into the top of a hood which has a cape extending down to the shoulders. Some of these have a transparent section extending all the way around the upper part of the hood and provide an unusual degree of flexibility in vision and comfort, with a minimal loss in protection.

For other features of full-face respirators other than those covered here, the information on half-face units in Section 6.2.2 will apply.

6.2.4. Air-Supplied Respirators

An air-supplied respirator is intended to provide a source of breathing air to the user independent of the air in the surrounding space so that they can be used in oxygen-deficient atmospheres. However, depending upon the design they may or may not be approved for IDLH atmospheres.

There are two basic designs, one in which the supply of air is from a source outside the contaminated area, while in the other the wearer of the respirator unit carries his air supply with him. There are subdivisions within these two major types. A major subdivision, common to both, is whether the units operate as "demand" or "pressure demand" units. For the former, the demand valve permits the flow of air only during inhalation, and a negative pressure exists at that time within the facepiece, which may allow leakage from the contaminated atmosphere. The pressure-demand type maintains a positive pressure within the facepiece at all times and is unlikely to allow leakage of outside air into the respirator. Pressure demand units are much more desirable and should be used in most applications. Demand units are not approved for IDLH atmospheres.

Air-supplied units are supplied with air through a hose from a source outside the contamination area, either from cylinders or air compressors, and must be of high purity. Cylinders may be used to supply oxygen instead of compressed air. Compressed air may contain low concentrations of oil. Since contact of high-pressure oxygen with oil may result in a fire or explosion, it is not permissible to use oxygen with air-supplied units that have previously used compressed air. A compressor used to supply breathing air must be equipped with a high-temperature alarm or carbon-monoxide alarm, or both, if the compressor is lubricated with oil. The air provided by the compressor must be passed through an absorbent and filter to ensure that the air supplied is pure. A hose up to a maximum length of 300 feet long is permissible.

A major difficulty with the units supplied by air through a hose is the hose itself. The person wearing the unit is constrained to move only as permitted by the hose, the hose is subject to damage or kinking, and one must retrace one's steps when leaving the area. The major advantage is that the external source effectively provides an infinite supply of air, if an ample number of cylinders are available or if the supply is from a compressor. An acceptable provision for escape from a contaminated area, should the pump fail or the hose fail or become constricted, would be an auxiliary tank of air to be carried by the user and connected to the respirator. This would also permit the wearer to disconnect from the supply hose and leave the contaminated area by perhaps a shorter and safer route in an emergency.

There are two different types of units in which the user carries his own air supply (in addition to the demand and pressure-demand versions). These self-contained breathing apparatus (SCBA) units have a basic limitation in that they provide only that amount of air that the user can carry with him. There are considerable differences in the useful life among different types of units within this class. Most units incorporate a full-facepiece, but other styles are available. Again, pressure-demand versions are the most desirable for most applications and, if equipped with escape provisions, are acceptable for IDLH atmospheres.

The basic type of SCBA unit is a tank of breathing air carried on the user's back and fed into the facepiece through appropriate regulators and valves. If the tank holds ordinary air, a nominal "30 minute" tank may last only 15 minutes for a person engaged in strenuous activity. These units also suffer from being bulky and relatively heavy. Because of their limited life, often there is little time for productive labor while wearing one.

There are several commercial units which use pure compressed oxygen instead of compressed air. After reduction in pressure from that in the oxygen cylinder, the oxygen is used in a system in which the air exhaled by the wearer is passed through a chemical pack which removes the carbon dioxide from the exhaled air and returns the purified air to the system to supplement the oxygen from the portable canister. The system is often much smaller and lighter than the basic system incorporating an air tank and, by appropriate sizing of the oxygen tank and the air purifying chemical, can be designed to last a fixed amount of time. A typical system will last 1 hour, although there are systems which are designed to last considerably longer. This is usually long enough to allow a significant amount of productive work. Some persons do not like to use this type where a fire is involved because of the pure oxygen. The system is designed so that the "scrubber" chemical will outlast the oxygen supply so there is no danger that the purification process will cease too early. A disadvantage of the scrubber units is that the purification action releases heat so that the air becomes warm and moist. This is a problem for some users.

Pressure-demand units which are self-contained and which use full-facepieces are approved for IDLH atmospheres.

Exposure to contaminants can damage respirator components, even after they have been "cleaned" and put in storage, due to permeation of chemicals into the materials of which they are made. An examination of the units should be made each time they are worn and a careful check made on a definite schedule. This is important for all types of units, but especially for those which are intended to be used in unusually hazardous applications. Records should be maintained of all maintenance.

6.3. EYE PROTECTION

The primary concerns for laboratory workers are impact and chemical splash protection. There are numerous commercial products available which meet the standards for eye protection in OSHA 29 CFR Part 1910.133. This standard is based on ANSI standard Z87.1-1968. Later revisions of this standard have been issued, and as other revisions are adopted, eye protection programs should incorporate more protective portions of these standards. Any protective devices purchased after July 5, 1994, must comply with the ANSI Z87.1-1989 standard or be demonstrated to be equally effective. Devices purchased earlier must comply with the 1968 version of the same standard.

Eye protection should be worn at all times while working with chemicals in the laboratory. A large percentage of the issues of *Chemical and Engineering News* contain letters reporting unexpected explosions in the laboratory, many of which could cause injury to the eye for persons not protected by goggles. Visitors should be provided with temporary protective goggles or glasses if they are allowed in any area in which the occupational use of eye protection is required.

6.3.1. Contact Lenses

Contact lens provide no physical or chemical protection for the eye. However, by making it possible to trap contaminants under the lens, they are likely to substantially reduce (or eliminate) the effectiveness of flushing with water from an eyewash station from removing the contaminant. In addition, they may, by capillary action, increase the amount of chemical trapped on the surface of the eye which otherwise might normally be removed by tearing. There are some eye conditions (other than solely for correction of vision) which are improved or controlled by wearing contact lens so that these individuals need to wear contact lenses. Except for these individuals, it is recommended that contact lens not be worn in the laboratory. Where it is necessary for medical reasons to wear contact lenses, the wearer should wear a chemical splash goggle as a minimum at all times while in the laboratory, and under conditions of risk, supplement this practice by wearing a face mask.

6.3.2. Chemical Splash Goggles

There are dozens of brands of chemical splash goggles available, almost all of which meet the basic standards in ANSI Z87.1 for this type of eye protection. However, there are wide variations in the degree of acceptance of these goggles by the users. Chemical splash goggles should fit snugly and comfortably around the eyes. The goggles should "breathe," i.e., the wearer should not overheat under them and perspire. They should not fog. They should provide good peripheral vision. Preferably, they should be compatible with the wearing of respirators. It would be desirable (if not essential) if prescription glasses could be worn under the goggles, although some styles of glasses are too big for most chemical splash goggles. The goggles should be easy to clean. Goggles made of hypoallergenic materials are available for wearers irritated by specific materials.

Just because a pair of chemical splash goggles is provided with "ports" to allow air into the space behind the lens of the goggle and an antifogging coating does not necessarily mean that all models with these features will be comparably effective. Ports on the side of the goggles appear to be less successful in many cases in eliminating fogging than openings around the edge of the lens, where motion of the head moves air directly across the lens.

The latter configuration also appears to work well in removing heat. Not all antifogging coatings appear equally effective, nor in some cases does product control quality appear to be uniform, even within a single brand. In at least one case, involving over 1000 pairs of goggles, the coating worked very well on about half of the goggles, while for the remainder, the lens fogged up within 15 minutes of the users' donning them.

Because it is so important to ensure that personnel in laboratories wear eye protection under circumstances in which it is needed, it is desirable to test goggles under actual use conditions. Not only must chemical splash goggles meet the required physical specifications, but they must be sufficiently comfortable to be accepted by the users as well. A pair of goggles pushed up on the forehead or lying on the work bench does not afford eye protection. Price is not necessarily a valid guarantee of quality. A mid-price unit may perform as well as or better than a higher price unit. Products change over time, and newer products are continually coming on the market. Prior to selecting a specific chemical splash goggle, it is recommended that it be tested under actual use conditions in comparison with a selection of other units which meet not only the OSHA and ANSI standards, but the criteria mentioned in the introductory paragraph to this section. A vendors' claim that a given product will perform as well as a tested unit needs to be verified. Making sure of the quality and efficacy of a goggle is especially important if buying large quantities to be used in instructional laboratories in school or in a process involving a large number of employees.

If the probability of a vigorous reaction appears to be substantial, or the material involved in the work in progress is very corrosive to tissue, a face mask should be used to supplement the splash goggles and provide additional protection to the face and throat. In the event that there is a risk of a minor explosion, an explosion shield should be placed between the worker and the reaction vessel. A wraparound shield will provide protection to the sides as well as directly in front of the shield.

The most commonly used lens material in safety goggles and safety glasses is polycarbonate. Typically the material used in the goggles is 0.060 inches (1.52 mm) thick. It is also used in face masks and, in somewhat greater thicknesses, in explosion shields. This material is lightweight and tough and resists impacts and scratches. Models coated with silicone are resistant to a number of chemicals.

The close fit to the face provided by chemical splash goggles and the strap around the head provides good stability against lateral impacts which might knock ordinary safety spectacles off.

6.3.3. Safety Spectacles

Safety spectacles (which resemble ordinary prescription glasses) that meet ANSI Z87.1 for impact protection offer very limited protection against chemicals. They do not fit tightly against the face and would not prevent chemicals from running down from the forehead into the eyes. They would provide some protection against flying glass in the event of a reaction vessel exploding. Sideshields are used to protect the eyes from flying objects from the side. However, as noted in the previous section, chemical splash goggles are form fitting and are held on tightly, so that they resist lateral impacts better than most safety spectacles.

Because they do not fit snugly against the face, safety spectacles do not have any more problems with heat and fogging than would ordinary glasses. In laboratory facilities which do not use chemicals, but do offer opportunities for mechanical injuries, safety glasses are acceptable. Many companies offer safety spectacles as prescription glasses and in attractive choices of frames. Some individuals who would resist wearing safety spectacles because they think that they must be unattractive can be provided with a choice of glasses which should prove satisfactory to almost any taste.

6.4. MATERIALS FOR PROTECTIVE APPAREL

As Keith[7] stated, "How many times have we seen the phrase 'use appropriate protective materials'?" This is true in some other areas as well for specific types of protective gear. However, in the field of chemically protective materials, where personnel must depend on protective gear, it is essential that they select effective protective clothing and articles, such as gloves, which will provide protection to the wearer against contact with the hazardous substances which they use in their laboratories. An examination of the catalog description of many of the items of protective apparel shows that most of the advertisements at best provide only qualitative descriptions of the efficacy of the materials used in the products.* However, some firms will provide technical support data on request.

6.4.1. Recommended Information Sources

Two recent publications (one of which is a two-component computerized book and expert selection system) provide a substantial amount of detailed information, not only by types of materials but by brand names, since not all versions of a given material have identical properties. These two publications are briefly described below. Both are recommended.

The first component of the computerized system is a computerized reference book compiled by Kristan Forsberg[8]: *Chemical Permeation and Degradation Database and Selection Guide for Resistant Protective Materials.* According to the vendor, the database contains over 4200 permeation tests on more than 540 compounds and mixtures. Included are more than 6000 breakthrough times or permeation rates and a total of over 20,000 pieces of associated data, which includes information on the test material, manufacturer, model number, thickness, comments, a safety guide number, and references. The stress is on gloves. The program is an outgrowth of "Guidelines for the Selection of Chemical Protective Clothing," which appeared in *Performance of Protective Clothing* edited by R.L. Barker and G.C. Coletta, American Society for Testing and Materials, 1986.

The second component of the computerized system is an "expert" system called GlovEs, which, given a set of initial conditions or parameters, screens the information in the data base and makes recommendations based on the needs as defined by the user of the system. The system allows considerable flexibility in seeking information and quickly provides the data in a useful form. The program runs on a standard IBM or compatible PC.

The second source of data is the third edition of *Guidelines for the Selection of Chemical Protective Clothing.* The work was sponsored by the U.S. EPA and the U.S. Coast Guard. The publication is organized in two volumes, the first of which contains an

---

*   The advertisements of NORTH HAND PROTECTION (Siebe North, Inc.) for their Silver Shield™ glove provide detailed breakthrough times and permeation rates for 49 chemicals for their gloves and give comparison data for several other materials. They claim, "It resists permeation and breakthrough by more toxic/hazardous chemicals than any other type of glove on the market today." Tests seem to support this statement; see Reference 10.

overview of the general topic of chemical protective clothing (CPC) and tables which can be used to select and use CPC properly. Twelve major clothing materials are evaluated in the context of about 500 different chemicals and permeation data, including 25 multiple-component organic solutions. Of particular interest are Appendices I and J, which define equipment needed to provide different levels of protection, and procedures for using protective gear in decontamination efforts. The latter information is of value to organizations who may have to respond occasionally to emergencies involving chemical spills but, fortunately, not sufficiently frequently so that they find it easy to maintain their skills.

The second volume is a technical support base for Volume I. It provides the data on which the recommendations in Volume I are based and some of the theoretical material used in arriving at the recommendations.

6.4.2. Overview of Chemical Protective Clothing[*]

The purpose of CPC is to prevent chemicals from reaching the skin. The chemicals can do this in two ways: permeate the material of which the clothing is made, or enter through penetrations in the clothing. The two sources described in the previous section concentrate on the problem of permeation and breakthrough, although the second of the two, in the introductory portions, discusses the penetration issue.

Much of the data in the two resources is based on manufacturers' subjective evaluations, i.e., the material provided "excellent," "good," "fair," or "poor" protection, and much of these subjective evaluations were based on visible degradation of the product. However, as noted in the *Guidelines,* "...it has been found that chemicals can permeate a material without there being any visible sign of problems." In addition, in the reported evaluations, the temperatures at which the tests were performed often are not given and the permeation rate has been found to depend significantly on the temperature. The thicknesses of the materials tested also are not given in many cases. More scientific means of comparative evaluations need to be adopted, and the techniques are available.

In some cases, the chemical to which the barrier material is exposed will simply diffuse through the barrier. In others, the chemical will react with the barrier material and degrade the performance of the barrier, for example, by changing the chemical properties of the material or by leaching out some of the components in the material. As far as penetration is concerned, it is desirable for the material be a poor absorber of the challenging chemical and the rate of permeation to be slow.

Chemicals, once they have begun to permeate a material will continue to do so, even after the challenge has been removed from the surface, because of the chemical absorbed within the material. If the amount of material absorbed is large, and the rate of permeation is relatively high, the chemical may eventually penetrate through the barrier material after the protective clothing has been removed and stored. The contaminant may cause exposure the next time it is used if it has not had an opportunity to diffuse away. If the clothing is carefully folded and placed in a container, the possibility of contaminant being trapped in the clothing is substantial. Another possibility is that the breakthrough will result in pinhole leaks and the item, which still afforded protection when removed, will no longer do so the next time it is used. If the amount absorbed is small and the permeation rate is

---

[*]   Many of the concepts in this section are developed more fully in Reference 9.

slow, the diffusion back through the entering surface may result in a negligible amount penetrating to the interior surface.

Some articles of protective clothing, such as gloves, have no openings which could come into contact with chemicals. However, other items such as coats, jackets, and trousers usually have one or more edges which are intended to be opened and closed, and all of these garments and others have seams where sections of the materials are joined together.

Protective clothing is normally fabricated of sections of materials which are often welded together chemically or by heat. The resulting garments should have no means for chemicals to penetrate due to the fabrication process, but the seams do have a tendency to split under strain. Some have sewn seams for added strength. The sewn seams, unless sealed afterwards, will leave pinholes through which chemicals can penetrate.

The normal openings such as the fronts of coats, or overalls, are fastened by either zippers, pressure-locking lips, or buttons. All of these should be supplemented by inner and outer flaps if they are intended to eliminate the possibility of chemicals penetrating through these openings. For total encapsulating clothing, boots, hoods, and gloves may be integral parts of the suits with a minimal number of openings for which secure seals have to be provided. Where separate items are used, arrangements for seals at the neck, feet, and hands must be provided. Many of these totally encapsulating suits will be used at a slight positive pressure to keep vapors which might occur from penetrating through any small gaps in the protective clothing.

The visor is a key component of any chemical splash suit. Not only must the visor material withstand the effects of the chemicals and maintain good visibility, but the seal to the fabric portion of the suit also should be checked frequently. Normally visors are made of materials such as polycarbonates, acrylics, fluorinated ethylene propylene (FEP), and clear polyvinyl chloride. The first two of these are subject to "crazing" upon contact with some chemicals. Covering these two types of material with a thin layer of FEP is sometimes done to protect them. A good seal between the visor and the suit is essential. The seals tend to be checked frequently.

A major problem with a totally enclosing chemical protective suit is elimination of body heat. If the air supply is carried with the user (see the section on respiratory protection), a common practice is to wear a cooling vest which has pockets for ice cubes and cooling liquid. If the source of air is an external supply, a vortex tube cooling unit is often incorporated in the system. In order to protect the tubes, valves, etc., associated with the air supply system from the effects of chemicals, the air supply components are often worn inside the protective suit. High body temperature is an especially troublesome problem in extended field uses of protective clothing. Workers should take a break for at least 10 minutes of each hour during normal summer use, open the clothes in a safe location, and replenish the salt in the body with cool liquids containing a modest level of sodium. Heat exhaustion or heat stroke could result from the overheating due to wearing air-tight protective clothing, unless care is taken.

The second reference source described earlier provides a wealth of information on the properties of protective materials for individual chemicals. Reference 10 reports on several later studies of many of the same materials challenged by many of the same chemicals. They generally support the findings in reference 9. Table 6.1 is adapted from this source for the resistance of materials for classes of chemicals.

The nomenclature used in the table is the following. RR, R, rr, and r represent various positive degrees of resistance while NN, N, nn, and n represent various degrees of poor resistance. Double characters indicate that the rating is based on test data, and single characters, on qualitative data. Upper case letters indicate a large body of consistent data, while lower case letters indicate either a small quantity of data or inconsistent information. Asterisks (**) mean that the material varied considerably in its resistance to chemicals within a class and data for specific chemicals should be used if available, or an alternative selected.

The column headings in Table 6.1 stand for the following materials:

Butyl—Butyl rubber
CPE—Chlorinated polyethylene
Viton/Neoprene—Layered material, first material on surface
Natural Rubber—same
Neoprene—same
Nitrile + PVC—Nitrile rubber + polyvinyl chloride
Nitrile—Nitrile rubber
PE—Polyethylene
PVA—Polyvinyl alcohol
PVC—Polyvinyl chloride
Viton—same
Butyl/Neoprene—Layered material, first material on surface

In addition to the chemical properties of materials, a number of physical properties of materials are also of importance in selecting an appropriate material. Table 6.2 is also adapted from Reference 9. The qualities listed in the table may also be affected by factors such as thickness, formulation, and whether there is a fabric backing to the material.

Much of the information used in the preparation of both resource books came from the use of the tested materials in the fabrication of gloves, and there are some differences in how gloves are made and how items of protective clothing are made. However, the basic information should be similar, if not identical.

The severity of the application will govern the choice of protective clothing in most cases. If the clothing is to be exposed to severe abuse and the environment is unusually hazardous, the choice should be the most durable and protective material available. In other cases, less protective and/or less durable units, or even disposable items, might prove entirely acceptable. Experience should assist in selecting brands. There have been instances where well-known brands have experienced runs of high levels of problems with seam failures in their moderately priced lines, while other brands of comparable price have not demonstrated these difficulties. Such problems should be documented and, unless the difficulties are resolved, used in the selection process. User confidence in the protection offered by the protective items by the users is too important to allow competitive cost to be the only, or even the primary factor, in selecting CPC items.

An important area not touched upon as yet as far as protective clothes are concerned is an alternative to the asbestos gloves used in laboratories to handle hot objects. Asbestos gloves observed in the laboratory often are in poor condition, i.e., the material of which they are made has become very friable. Any asbestos items in poor condition, especially those employed as asbestos gloves are, should be discarded as hazardous material through the organization's hazardous waste program. It would be desirable to eliminate asbestos gloves entirely. Alternatives that have been found acceptable for many high-temperature laboratory applications are gloves made of materials such as Kevlar™, Nomex™, Zetex™, and fiberglass or combinations of these materials. Zetex™, for example, is specifically advertised as a replacement for asbestos for high-temperature applications.

Table 6.1.   Resistant Properties of Selected Materials by Chemical Class

Chemical	Butyl	CPE	Viton/ Neoprene	Natural Rubber	Neoprene	Nitrile + PVC	Nitrile	PE	PVA	PVC	Viton	Butyl/ Neoprene
Acids, carboxylic, aliphatic												
Unsubstituted	R	r	r	**	rr	**	rr	NN	**	**	**	r
Polybasic					rr	rr	rr	rr	n	rr		r
Aldehydes												
Aliphatic and alicyclic	RR	NN	r	**	NN	nn	NN	**	NN	**	**	r
Aromatic and heterocyclic	rr		n	nn	nn	n	nn	NN	rr	N		r
Amides	rr			**	nn		nn	nn			nn	
Amines, aliphatic, alicyclic												
Primary	**	**	n	NN	**		rr		NN	**	**	
Secondary	**		n	NN	nn		**		**	NN	nn	n
Tertiary	**	**		**	**	**	**		**	**	rr	
Polyamine	**			NN	**	nn				NN	rr	
Cyanides					r							
Esters, carboxylic												
Formate			n									
Acetates	**	**	n	NN	nn	nn	NN	NN	**	NN	n	**
Higher monobasic	nn	nn	**	NN	nn		nn	NN	rr	NN		**
Polybasic			r	r	r		**			rr		r
Aromatic phthalate	rr		r	**	**		**			nn	rr	r
Ethers												
Aliphatic	**	rr	**	NN	**	**	**		**	**		**
Halogen compounds												
Aliphatic, unsubstituted	nn	nn	r	NN	NN	NN	NN	NN	**	NN	**	n
Aliphatic, substituted	**			NN	rr		nn		**	NN	rr	
Aromatic, unsubstituted	nn	nn	r	NN		n	nn	NN		N	rr	n
Polynuclear				NN						n	rr	
Vinyl halides				NN	nn					n	rr	
Heterocyclic compounds												
Epoxy compounds	**			**	nn		nn	NN	**	nn	NN	n
Furan derivatives	nn		nn									n

Table 6.1.   Resistant Properties of Selected Materials by Chemical Class (continued)

Chemical	Butyl	CPE	Viton/ Neoprene	Natural Rubber	Neoprene	Nitrile + PVC	Nitrile	PE	PVA	PVC	Viton	Butyl/ Neoprene
Hydrazines	**	nn	n	**	**	**	**		nn	**	**	n
Hydrocarbons												
Aliphatic and alicyclic	N	r	r	NN	**	**	**	**	**	NN	RR	n
Aromatic	**	rr	r	NN	NN	NN	**	NN	**	NN	RR	r
Hydroxyl compounds												
Aliphatic and alicyclic	RR	rr	rr	nn	**	nn	**	**	**	**	rr	**
Primary	rr	rr	r	**	**	**	rr		rr	**	rr	r
Secondary	r			**	rr	rr	rr			**		
Tertiary	r		**	rr	rr	rr	rr	**		**	rr	**
Polyois	r		r	rr	rr	rr		**		**	rr	r
Aromatic	**	**	rr	**	**	**	**	**	nn	**	rr	r
Inorganic acids	**			RR	RR	RR	RR	**		**	rr	**
Inorganic bases	r	r		nn	RR	**	RR	**	n	**	**	**
Inorganic gases	**	r	n	n	r			**		**		
Inorganic salts	r		n	**	r	r	r		rr	R		
Isocyanates			n	NN	n	N						**
Ketones, aliphatic	**	NN		NN	NN	Z	**	NN	**	NN	NN	
Nitriles, aliphatic	rr			NN	**			NN	rr	NN	rr	
Nitro compounds												
Unsubstituted	rr	r		NN	**		nn		**	**	**	
Organophosphorus compounds			r									r
Peroxides				r								
Sulfur compounds												
Thiois			**									n

Note: See text for explanation of abbreviations.

Table 6.2.   Physical Characteristics of Chemical-Resistant Materials

Material	Abrasion Resistance	Cut Resistance	Flexibility	Heat Resistance	Ozone Resistance	Puncture Resistance	Tear Resistance	Relative Cost
Butyl rubber	F	G	G	E	E	G	G	High
Chlorinated polyethylene (CPE)	E	G	G	G	E	G	G	Low
Natural rubber	E	E	E	F	P	E	E	Medium
Nitrile-butadiene rubber (NBR)	E	E	E	G	F	E	G	Medium
Neoprene	E	E	G	G	E	G	G	Medium
Nitrile rubber (nitrile)	E	E	E	G	F	E	G	Medium
Nitrile rubber + polyvinyl chloride (nitrile + PVC)	G	G	G	F	E	G	G	Medium
Polyethylene	F	F	G	F	F	P	F	Low
Polyurethane	E	G	E	G	G	G	G	High
Polyvinyl alcohol (PVA)	F	F	P	G	E	F	G	Very high
Polyvinyl chloride (PVC)	G	P	F	P	E	G	G	Low
Styrene-butadiene rubber (SBR)	E	G	G	G	F	F	F	Low
Viton	G	G	G	G	E	G	G	Very high

Note: E = excellent, G = good, F = fair, and P = poor.

6.5. HEARING PROTECTION

The noise levels in most laboratories are usually not excessive, but there are laboratory facilities in which noise can reach levels for which hearing protection should be provided or the employees required to be involved in a hearing conservation program. It would be preferable, of course, if the noise levels could be lowered rather than to depend upon personal protective devices.

OSHA has adopted a comprehensive hearing conservation program in 29 CFR Part 1910.95. Under this standard, any employee exposed to an 8-hour time weighted average of 85 dB, as measured on a properly calibrated sound level instrument on the slow response (A) scale, must be placed in an employer-run hearing conservation program. Among other requirements, an annual audiometric test is required. Loss of hearing as one grows older is normal, and there are diseases resulting in hearing losses which are not occupationally related. A properly administered audiometric test should be able to distinguish these two causes from a loss of hearing due to external factors. One of the difficulties in determining if the loss of hearing is occupationally related is that the individual's lifestyle during nonworking hours can affect his hearing as well. Prolonged listening to loud music can cause problems, and so can frequent shooting of firearms for recreation. Worker's compensation claims are sometimes disallowed because the evidence is not clear that the hearing loss is occupationally related. Sound level measurements and sound dosimetry measurements in the workplace, as well as periodic audiometric tests, are all needed. At this time, the record of enforcement of the OSHA standard is not impressive, and the number of worker's compensation cases processed has been small. Hearing is vital, especially to a laboratory employee, since loss of hearing may make it impossible to perform many laboratory operations.

According to Table 6.3, at average noise levels over an 8-hour day in excess of 90 dB, an unprotected worker must have their working hours reduced. The table provides that the work interval should be cut in half if the sound level goes up by 5 dB, implying that the sound level goes up by two when the measured level increases by 5 dB. Actually, the sound level increases by a factor of two for a measured increase of 3 dB, so that the sound level at 105 dB is 32 times that at 90 dB instead of 8, as could be inferred from the table.

There are many different types of hearing protection on the market. The simplest type is earplugs which are placed within the ear. Typically, these are soft foam which conform to the ear canal and are effective in reducing noise levels. Many can be washed and reused, if desired. Some persons do not like to use this type of hearing protection because they do not like to keep the plugs in their ears and do not like to be continually taking them in and out as they go from noisy areas to quieter ones. There are many inexpensive earmuffs on the market which attenuate sounds by 20 to 30 dB. The attenuation of the hearing protection devices varies with frequency. The ear muffs can be taken on and off and kept available for reuse. Over a relatively short interval, they will be more economical than ear plugs, and often have a higher degree of acceptance by the users.

6.6. FIRST AID KITS

A first aid kit for a laboratory or for most areas should be intended to provide immediate treatment for most minor injuries or burns, not serve as a substitute for the family medical cabinet. It also need not be a large unit. Most injuries within a laboratory are to individuals. If a major accident were to occur, it would be necessary to call in emergency

Table 6.3.   Permissible Noise Exposure

Duration (Hours/Day)	Sound Level (dBA Slow Response)
8	90
6	92
4	95
3	97
2	100
1.5	102
1	105
0.5	110
0.25 or less	115

medical personnel rather than to try local treatment. The contents of a first kit should include the supplies listed in Table 6.4.

Note that there are no tourniquets, aspirin (or other medicines to be taken internally), iodine, or merthiolate. Possible additions to the above list could be Ipecac, used to induce vomiting, and activated charcoal, to help absorb poisons internally, but if these are supplied, persons should receive specific training in how to use them properly. As noted in Chapter 1, it would be highly desirable if a number of persons in a laboratory facility would receive formal training in first aid and CPR. A specific person should be designated to be responsible for maintaining the supplies in the first aid kit.

REFERENCES

1.   General Industry Standards, 29 CFR Section 1910.134, Respiratory Protection, OSHA, Washington, D.C.
2.   Practices for Respiratory Protection, ANSI Z88.2, American National Standards Institute, New York, 1980.
3.   Practices for Occupational and Educational Eye and Face Protection, ANSI Z87.1, American National Standards Institute, New York, 1989.
4.   Recommendations for Prescription Ophthalmic Lenses, ANSI Z80.1, American National Standards Institute, New York, 1989.
5.   **Keith, L.H.,** Technology blends computers, books, for comparing protective materials, *Occup. Health Saf.,* 56(11), 74, 1987.
6.   **Keith, L.H., Johnston, M.T., Hudak, C.E., and Conoley,** *Gloves,* Lewis Publishers, Boca Raton, FL, 1992.
7.   **Forsberg, K.,** *Chemical Permeation and Degradation Database and Selection Guide for Resistant Protective Materials,* Instant Reference Sources, Inc., Austin, TX, 1987.
8.   **Forsberg, K.** *Chemical Protective Clothing Permeation and Degradation Database,* Keith, L.H., Ed., Lewis Publishers, Boca Raton, FL, 1992.
9.   **Schwope, A.D., Costas, P.P., Jackson, J.O., Stull, J.O., and Weitzman, D.J., Eds.,** *Guidelines for the Selection of Chemical Protective Clothing,* 3rd ed., Arthur D. Little, Inc., U.S. Environmental Protection Agency and U.S. Coast Guard, American Conference of Governmental Industrial Hygienists, Cincinnati, OH, 1987.
10.   **McBriarty, J.P. and Henry, N.W.,** *Performance of Protective Clothing,* Vol. 4, American Society for Testing and Materials, Philadelphia, PA, 1992.

Table 6.4.   Laboratory First Aid Kit

Adhesive bandages, various sizes	Antiseptic wipes
Sterile pads, various sizes	Cold packs
Sterile sponges	Burn cream
Bulk gauze	Antiseptic cream
Eye pads	Absorbent cotton
Adhesive tape	Scissors
First aid booklet	Tweezers

Appendix: Laboratory Checklist

A. General

1. Housekeeping satisfactory
2. Aisles not cluttered, paths of egress maintained free of obstructions
3. Hazard warning signs at entrance(s)
4. Laboratory Authority List at entrances
5. Work area separated from study/social areas
6. Laboratory Hygiene Plan written and available, including
    Written Standard Operating Procedures available
    Material Safety Data Sheets available
    Emergency response and evacuation plan
7. Food not stored in laboratory refrigerators
8. Equipment in good condition, preventive maintenance plan in place

B. Fire Safety

1. Flammables stored in flammable material storage cabinets
2. Class ABC fire extinguisher in laboratory, near door or on path of egress
3. Fire blankets available
4. Flammable material stocks maintained at minimal levels, within limits
5. Refrigerators used for flammables are flammable material storage units or explosion-proof
6. Two well-separated exits, doors swing outward for hazard class labs
7. Flammables not stored along path of egress

C. Chemical Handling

1. Chemicals stored according to compatibility
2. Ethers identified by date of receipt and latest date for disposal
3. All containers properly labeled, including secondary containers
4. Quantities of chemicals not excessive
5. Chemicals stored at safe levels, in cabinets or on stable shelving; no chemicals on floor
6. Chemical waste properly labeled and segregated prior to disposal, removed frequently

7. Gas cylinders strapped firmly in place, cylinders not in use capped, oxidizing and reducing gases properly segregated
8. Perchloric acid quantities maintained at minimum levels; perchloric acid hoods available for hot perchloric acid applications
9. Apparatus marked with warning signs or protected by barriers if susceptible to damage
10. All work generating toxic and hazardous fumes done in hoods
11. Work capable of causing an explosion behind protective barriers, vacuum vessels taped, etc.

D. Ventilation

1. Ventilation 100% fresh air, 6 to 12 air changes per hour
2. Laboratory at negative pressure with respect to corridors
3. Hoods located in low traffic draft-free zones
4. Hoods capable of maintaining 100 fpm face velocity with sash fully open
5. Low-velocity warning alarm on hoods
6. Fume generating apparatus with hoods at least 20 cm from sash opening
7. Local exhaust units used where hoods not suitable
8. Hoods not used for storage of surplus materials

E. Electrical

1. All electrical circuits three-wire
2. No circuits overloaded with extension cords or multiple connection
3. No extension cords used unsafely, cords protected or in raceways
4. Apparatus equipped with three-prong plugs or double-insulated
5. Motors are nonsparking
6. Heating apparatus equipped with redundant temperature controls
7. Adequate lighting, lights in hoods protected from vapors
8. Circuits, equipment provided with ground-fault interrupters as needed
9. Electrical equipment properly covered
10. Breaker panel accessible
11. Deluge shower located so water will not splash on electrical equipment or circuits

F. Safety Devices

1. Eyewash station available, checked at least semiannually
2. Deluge shower available, checked at least semiannually
3. First aid kit available and fully maintained
4. Personal protective equipment—goggles, face masks, gloves, aprons, respirators, explosion shields, escape breathing masks, available and used as needed
5. Evacuation route marked
6. No smoking, other safety signs posted and observed

6. Chemical waste properly labeled and segregated prior to disposal, removed frequently
7. Gas cylinders strapped firmly in place, cylinders not in use capped, oxidizing and reducing gases properly segregated
8. Perchloric acid quantities maintained at minimum levels; perchloric acid hoods available for hot perchloric acid applications
9. Apparatus marked with warning signs or protected by barriers if susceptible to damage
10. All work generating toxic and hazardous fumes done in hoods
11. Work capable of causing an explosion behind protective barriers, vacuum vessels taped, etc.

D. Ventilation

1. Ventilation 100% fresh air, 6 to 12 air changes per hour
2. Laboratory at negative pressure with respect to corridors
3. Hoods located in low traffic draft-free zones
4. Hoods capable of maintaining 100 fpm face velocity with sash fully open
5. Low-velocity warning alarm on hoods
6. Fume generating apparatus with hoods at least 20 cm from sash opening
7. Local exhaust units used where hoods not suitable
8. Hoods not used for storage of surplus materials

E. Electrical

1. All electrical circuits three-wire
2. No circuits overloaded with extension cords or multiple connection
3. No extension cords used unsafely, cords protected or in raceways
4. Apparatus equipped with three-prong plugs or double-insulated
5. Motors are nonsparking
6. Heating apparatus equipped with redundant temperature controls
7. Adequate lighting, lights in hoods protected from vapors
8. Circuits, equipment provided with ground-fault interrupters as needed
9. Electrical equipment properly covered
10. Breaker panel accessible
11. Deluge shower located so water will not splash on electrical equipment or circuits

F. Safety Devices

1. Eyewash station available, checked at least semiannually
2. Deluge shower available, checked at least semiannually
3. First aid kit available and fully maintained
4. Personal protective equipment—goggles, face masks, gloves, aprons, respirators, explosion shields, escape breathing masks, available and used as needed
5. Evacuation route marked
6. No smoking, other safety signs posted and observed

INDEX